AF717365

Plant Spatial Biology

CABI BIOTECHNOLOGY SERIES

Biotechnology, in particular the use of transgenic organisms, has a wide range of applications including agriculture, forestry, food and health. There is evidence that it could make a major impact in producing plants and animals that are able to resist stresses and diseases, thereby increasing food security. There is also potential to produce pharmaceuticals in plants through biotechnology, and provide foods that are nutritionally enhanced. Genetically modified organisms can also be used in cleaning up pollution and contamination. However, the application of biotechnology has raised concerns about biosafety, and it is vital to ensure that genetically modified organisms do not pose new risks to the environment or health. To understand the full potential of biotechnology and the issues that relate to it, scientists need access to information that not only provides an overview of and background to the field, but also keeps them up to date with the latest research findings.

This series, which extends the scope of CABI's successful "Biotechnology in Agriculture" series, addresses all topics relating to biotechnology including transgenic organisms, molecular analysis techniques, molecular pharming, in vitro culture, public opinion, economics, development and biosafety. Aimed at researchers, upper-level students and policy makers, titles in the series provide international coverage of topics related to biotechnology, including both a synthesis of facts and discussions of future research perspectives and possible solutions.

Titles Available

1. Animal Nutrition with Transgenic Plants
   Edited by G. Flachowsky
2. Plant-derived Pharmaceuticals: Principles and Applications for Developing Countries
   Edited by K.L. Hefferon
3. Transgenic Insects: Techniques and Applications
   Edited by M.Q. Benedict
4. Bt Resistance: Characterization and Strategies for GM Crops Producing Bacillus thuringiensis Toxins
   Edited by Mario Soberón, Yulin Gao and Alejandra Bravo
5. Plant Gene Silencing: Mechanisms and Applications
   Edited by Tamas Dalmay
6. Ethical Tensions from New Technology: The Case of Agricultural Biotechnology
   Edited by Harvey James
7. GM Food Systems and their Economic Impact
   Tatjana Brankov and Koviljko Lovre
8. Endophyte Biotechnology: Potential for Agriculture and Pharmacology
   Edited by Alexander Schouten
9. Forest Genomics and Biotechnology
   Richard Meilan and Matias Kirst
10. Transgenic Insects: Techniques and Applications, 2nd Edition
    Edited by M.Q. Benedict and Maxwell J. Scott
11. Plant Omics: Advances in Big Data Biology
    Edited by Hajime Ohyanagi, Eiji Yamamoto, Ai Kitazumi and Kentaro Yano

12. Next-generation Sequencing and Agriculture
Edited by Phillip Bayer and Dave Edwards

13. Aquaculture and Fisheries Biotechnology: Genetic Approaches
Rex A Dunham

14. Algal Biotechnology
Edited by Qiang Wang

15. Genetic Engineering of Vegetable Crops
Edited by Pritam Kalia

16. Plant Breeding Technology: Future Trends and Challenges
Edited by Zeba Khan

17. Stress-Resilient Crops: Coordinated Omics-CRISPR-Nanotechnology Strategies
Edited by Jen-Tsung Chen

18. High Throughput Plant Metabolomics
Edited by Jen-Tsung Chen

19. Plant Biotechnology and Sustainable Agriculture: Bridging the Gap for Global Food Security
Edited by Tariq Aftab

20. Biotechnology of Orchids and Other Ornamentals
Edited by Pablo Bolaños-Villegas, Jian-Zhi Huang

21. Deep Learning in Plant Science and Agricultural Biotechnology
Edited by Jen-Tsung Chen

22. AI-Enhanced Plant-omics
Edited by Jen-Tsung Chen

23. Plant Spatial Biology
Edited by Jen-Tsung Chen

Plant Spatial Biology

Edited by:

Jen-Tsung Chen
National University of Kaohsiung, Taiwan

CABI

CABI is a trading name of CAB International

CABI
Nosworthy Way
Wallingford
Oxfordshire OX10 8DE
UK

Tel: +44 (0)1491 832111
E-mail: info@cabi.org
Website: www.cabi.org

CABI
200 Portland Street
Boston
MA 02114
USA

T: +1 (617)682-9015
E-mail: cabi-nao@cabi.org

© CAB International 2026. All rights, including for text and data mining, AI training, and similar technologies, are reserved. No part of this publication may be reproduced in any form or by any means, electronically, mechanically, by photocopying, recording or otherwise, without the prior permission of the copyright owners.

The views expressed in this publication are those of the author(s) and do not necessarily represent those of, and should not be attributed to, CAB International (CABI). Any images, figures and tables not otherwise attributed are the author(s)' own. References to internet websites (URLs) were accurate at the time of writing.

CAB International and, where different, the copyright owner shall not be liable for technical or other errors or omissions contained herein. The information is supplied without obligation and on the understanding that any person who acts upon it, or otherwise changes their position in reliance thereon, does so entirely at their own risk. Information supplied is neither intended nor implied to be a substitute for professional advice. The reader/user accepts all risks and responsibility for losses, damages, costs and other consequences resulting directly or indirectly from using this information.

CABI's Terms and Conditions, including its full disclaimer, may be found at https://www.cabidigitallibrary.org/terms-and-conditions.

A catalogue record for this book is available from the British Library, London, UK.

ISBN-13: 9781836994558 (hardback)
9781836994565 (ePDF)
9781836994572 (ePub)

DOI: 10.1079/9781836994572.0000

Commissioning Editor: Ward Cooper
Editorial Assistant: Emma McCann
Production Editor: Helen Elliott

Typeset by Exeter Premedia Services Pvt Ltd, Chennai, India
Printed and bound in the UK by CPI Group (UK) Ltd, Croydon, CR0 4YY

Contents

Editor's Biography

Dr Jen-Tsung Chen is a professor of cell biology at the National University of Kaohsiung in Taiwan. He teaches cell biology, genomics, proteomics, plant physiology, and plant biotechnology. Dr Chen's research interests include bioactive compounds, chromatography techniques, plant molecular biology, plant biotechnology, bioinformatics, and systems pharmacology. He is an active editor of academic books and international journals to advance the exploration of multidisciplinary knowledge involving plant physiology, plant biotechnology, nanotechnology, materials science, ethnopharmacology, and systems biology. He serves as an associate editor, editorial board member, and guest editor in reputed journals. Dr Chen published books in collaboration with international publishers, and he is handling book projects on diverse topics such as AI technologies, drug discovery, drug development, herbal medicine, medicinal biotechnology, nanotechnology, bioengineering, plant functional genomics, plant speed breeding, epigenetics, functional RNAs, and CRISPR-based genome editing. Dr Chen is a productive author in academic publications and was recognized as one of the World's Top 2% Scientists 2023, 2024, and 2025 by Elsevier and Stanford University. In 2025, Dr Chen received the Springer Nature Editor of Distinction Award.

Contributors

David Appiah Abban, Council for Scientific and Industrial Research-Crops Research Institute (CSIR-CRI), Fumesua, P.O. Box 3785, Kumasi, Ghana; Email: davidappiahabban@gmail.com

Shabbar Abbas, Department of Biosciences, COMSATS University Islamabad, Islamabad 45550, Pakistan; Email: shabbar.abbas@comsats.edu.pk

Banzeer Ahsan Abbasi, Department of Botany, Rawalpindi Women University, 6th Road, Satellite Town, Rawalpindi 46300, Pakistan; Email: benazirahsanabbasi786@gmail.com

Joseph Adomako, Council for Scientific and Industrial Research-Crops Research Institute (CSIR-CRI), Fumesua, P.O. Box 3785, Kumasi, Ghana; Department of Crop and Soil Science Education, Akenten Appiah-Menka University of Skill Training and Entrepreneurial Development (AAMUSTED), P.O. Box 40, Mampong, Ashanti, Ghana; Email: joeadomako@gmail.com

Kennedy Agyeman, Council for Scientific and Industrial Research-Crops Research Institute (CSIR-CRI), Fumesua, P.O. Box 3785, Kumasi, Ghana; Email: agyemanken@yahoo.com

Anees Ali, Department of Biotechnology, Faculty of Sciences, Necmettin Erbakan Üniversitesi Türkiye, Türkiye; Email: aneesalisindher@gmail.com

Priscilla Amponsah, State Key Laboratory of Agricultural and Forestry Biosecurity, College of Plant Protection, Nanjing Agricultural University, Nanjing, 211800, PR China; Email: amponsahpriscillaa@gmail.com

Arumugam Vijaya Anand, Department of Human Genetics and Molecular Biology, Bharathiar University, Coimbatore 641 046, Tamil Nadu, India; Email: avahgmb@buc.edu.in

Muhammad Anas, Department of Soil Science & Plant Nutrition, University of Selcuk Campus, Konya 42079, Türkiye; Email: muhammed.anas@selcuk.edu.tr

Esther Afoley Annang, Council for Scientific and Industrial Research-Crops Research Institute (CSIR-CRI), Fumesua, P.O. Box 3785, Kumasi, Ghana; Email: esther.annang1@gmail.com

Augustine Antwi-Boasiako, Council for Scientific and Industrial Research-Crops Research Institute (CSIR-CRI), Fumesua, P.O. Box 3785, Kumasi, Ghana; Key Laboratory of Biology and Genetics Improvement of Soybean, Ministry of Agriculture, Zhongshan Biological Breeding Laboratory (ZSBBL), National Innovation Platform for Soybean Breeding and Industry-Education

Integration, State Key Laboratory of Crop Genetics & Germplasm Enhancement and Utilization, College of Agriculture, Nanjing Agricultural University, Nanjing 210095, China; Email: bbbantwi@yahoo.com

Mariadhas Valan Arasu, Department of Botany and Microbiology, College of Science, King Saud University, P.O. Box 2455, Riyadh 11451, Saudi Arabia; Email: mvalanarasu@gmail.com

Ernest Baafi, Council for Scientific and Industrial Research-Crops Research Institute (CSIR-CRI), Fumesua, P.O. Box 3785, Kumasi, Ghana; Email: e.baafi@gmail.com

Balasubramanian Balamuralikrishnan, Department of Food Science and Biotechnology, College of Life Science, Sejong University, Seoul 05006, South Korea; Email: bala.m.k@sejong.ac.kr

Madhumita Barooah, Department of Agricultural Biotechnology, Assam Agricultural University, Jorhat, India; DBT-NECAB, Assam Agricultural University, Jorhat, India; Email: madhumita.barooah@aau.ac.in

Sana Basharat, School of Breeding and Multiplication (Sanya Institute of Breeding and Multiplication), School of Tropical Agriculture and Forestry, Hainan University, Sanya 572025, China; Email: sanabasharat143@gmail.com

Khadija Benamar, Phytopathology Unit, Department of Plant Protection, Ecole National of Agriculture Meknes, Km10, Rte Haj Kaddour, BP S/40, Meknes 50001, Morocco; Email: khadija.benamar2@usmba.ac.ma

Paramashivan Bhavani, Department of Animal Science, Manonmaniam Sundaranar University, Tirunelveli 627 012, Tamil Nadu, India; Email: pmbbavi18@gmail.com

Augustine Boakye Boateng, Council for Scientific and Industrial Research-Crops Research Institute (CSIR-CRI), Fumesua, P.O. Box 3785, Kumasi, Ghana; Email: aboakyeboateng@gmail.com

Isaac Osei Bonsu, Council for Scientific and Industrial Research-Crops Research Institute (CSIR-CRI), Fumesua, P.O. Box 3785, Kumasi, Ghana; Email: ob.isaac@gmail.com

Ishani Borthakur, Department of Agricultural Biotechnology, Assam Agricultural University, Jorhat, India; Email: ishaaneeb@gmail.com

Nihar Ranjan Chakraborty, Department of Genetics and Plant Breeding, Institute of Agriculture, Visva-Bharati University, Sriniketan, West Bengal, India; Email: nrchakraborty@gmail.com

Ajithan Chandrasekaran, Department of Horticulture, Chungnam National University, Daejeon, 34134, Korea; ajithancbiotech@gmail.com

Urjashi Chatterjee, Department of Biotechnology, St. Xavier's College (Autonomous), 30, Mother Teresa Sarani, Kolkata - 700016, West Bengal, India; Email: urjashic0315@gmail.com

Debdeep Chattopadhyay, Department of Biotechnology, St. Xavier's College (Autonomous), 30, Mother Teresa Sarani, Kolkata - 700016, West Bengal, India; Email: chattopadhyay.debdeep777@gmail.com

Daouda Coulibaly, Laboratory of Agro-physio-genetics and plant biotechnology, Department of Agricultural Technical Science (STA), Rural Polytechnic Institute for Training and Applied Research (IPR/IFRA) of Katibougou, Koulikoro - Mali; Email: cdaouda40@yahoo.fr

Yaw Danso, Council for Scientific and Industrial Research-Crops Research Institute (CSIR-CRI), Fumesua, P.O. Box 3785, Kumasi, Ghana; Email: ydanso219@gmail.com

Swayanshu Satyaprajna Das, Odisha University of Agriculture and Technology, Bhubaneswar, Odisha, India; Email: swayanshudas123@gmail.com

Prahlad Deb, Department of Horticulture and Postharvest Technology, Institute of Agriculture, Visva-Bharati University, Sriniketan, West Bengal, India; Email: prahlad.deb@visva-bharati.ac.in

Sandip Debnath, Department of Genetics and Plant Breeding, Institute of Agriculture, Visva-Bharati University, Sriniketan, West Bengal, India; Email: sandip.debnath@visva-bharati.ac.in

Ilham Dehbi, Phytopathology Unit, Department of Plant Protection, Ecole National of Agriculture Meknes, Km10, Rte Haj Kaddour, BP S/40, Meknes 50001, Morocco; Laboratory of Biotechnology and Bioresources Valorization. Moulay Ismail University of Meknes, Zitoune, BP 11201, Meknes, Morocco; Email: il.dehbi@edu.umi.ac.ma

Richard Dormatey, Council for Scientific and Industrial Research-Crops Research Institute (CSIR-CRI), Fumesua, P.O. Box 3785, Kumasi, Ghana; Email: rmddormatey@gmail.com

Sahul Ameed Nihaal Fathima, Department of Human Genetics and Molecular Biology, Bharathiar University, Coimbatore 641 046, Tamil Nadu, India; Email: nihaal2822@gmail.com

Kofi Frimpong-Anin, Council for Scientific and Industrial Research-Crops Research Institute (CSIR-CRI), Fumesua, P.O. Box 3785, Kumasi, Ghana; Email: nanakofy@yahoo.com

Bhoirob Gogoi, DBT-NECAB, Assam Agricultural University, Jorhat, India; Email: bhoirobgogoi1998@gmail.com

Rahul Gogoi, Department of Agricultural Biotechnology, Assam Agricultural University, Jorhat, India; Email: rahulgogoi.biotech@gmail.com

Kalibulla Syed Ibrahim, Department of Biotechnology, PSG College of Arts & Science, Coimbatore 641 014, Tamil Nadu, India; Email: syedibrahim.k@gmail.com

Madiha Imtiaz, Department of Botany, Bacha Khan University, Charsadda, 24420; Email: madiha92bkuc@gmail.com

Javed Iqbal, Department of Botany, Bacha Khan University, Charsadda, 24420, Khyber Pakhtunkhwa, Pakistan; Email: javed89qau@gmail.com

Mouna Janati, Laboratory of Biotechnology and Bioresources Valorization, Moulay Ismail University of Meknes, Zitoune, BP 11201, Meknes, Morocco; Email: m.janati@edu.umi.ac.ma

Subramaniam Kalidass, Department of Animal Science, Manonmaniam Sundaranar University, Tirunelveli 627 012, Tamil Nadu, India; Email: kalidass@msuniv.ac.in

Karuppasamy Kaviarasan, Department of Animal Science, Manonmaniam Sundaranar University, Tirunelveli 627 012, Tamil Nadu, India; Email: drkkavi.4711@gmail.com

Amjid Khan, Department of Plant Sciences, Faculty of Biological Sciences, Quaid-i-Azam University, Islamabad- 45320, Pakistan; UNESCO UNISA Africa Chair in Nanoscience and Nanotechnology, College of Graduate Studies, University of South Africa, Pretoria, South Africa; Email: amjid-khan@bs.qau.edu.pk

Rashid Abbas Khan, Department of Biology, Faculty of Science, Prince of Songkhla University, Hat Yai, Songkhla, 90110, Thailand; Email: rashidabbaskhan98@gmai.com

Aritabha Kole, Department of Genetics and Plant Breeding, Institute of Agriculture, Visva-Bharati University, Sriniketan, West Bengal, India; Email: aritabha@gmail.com

Ravi Kiran Reddy Kondi, SPESS, LSU AgCenter, Louisiana State University, LA, USA; Email: Rkondi1@lsu.edu

Rachid Lahlali, Phytopathology Unit, Department of Plant Protection, Ecole National of Agriculture Meknes, Km10, Rte Haj Kaddour, BP S/40, Meknes 50001, Morocco; Email: rlahlali@enameknes.ac.ma

Mohanraj Lavanya, PG & Research Department of Botany, PSG College of Arts & Science, Coimbatore 641 014, Tamil Nadu, India; Email: lavanyasrimohan288@gmail.com

Pingwu Liu, School of Breeding and Multiplication (Sanya Institute of Breeding and Multiplication), School of Tropical Agriculture and Forestry, Hainan University, Sanya 572025, China; Email: hnulpw@hainanu.edu.cn

Shilpa Malakar, Department of Microbiology, Kalinga University, Raipur, Chhattisgarh – 492 001 (C.G), India; Email: shilpamalakardhn123@gmail.com

A.K. Mall, ICAR – Indian Sugarcane Research Institute Lucknow - 226 002, U.P., India; Email: ashutoshkumarmall@gmail.com

Mani Manoj, Department of Human Genetics and Molecular Biology, Bharathiar University, Coimbatore 641 046, Tamil Nadu, India; Email: manojinthravathi@gmail.com

Arumugam Maruthupandian, Department of Botany, School of Life Sciences, Periyar University, Salem 636 011, Tamil Nadu, India; Email: maruthubot@periyaruniversity.ac.in

Kasi Rao Mediga, Department of Plant Pathology, School of Agriculture, Mohan Babu University, Tirupati, Andhra Pradesh-517102, India; Email: mediga.kasirao@mbu.asia

Varucha Misra, ICAR – Indian Sugarcane Research Institute Lucknow - 226 002, U.P., India; Email: misra.varucha@gmail.com

Gyana Ranjan Mohanty, Department of Genetics and Plant Breeding, Institute of Agriculture, Visva-Bharati University, Sriniketan, West Bengal, India; Email: mgyanaranjan09@gmail.com

Saroj Kumar Mohanty, Odisha University of Agriculture and Technology, Bhubaneswar, Odisha, India; Email: sarojkumarmohanty070269@gmail.com

Soumya Ranjan Mohanty, College of Horticulture, OUAT, Chiplima, Odisha, India; Email: soumyaran999@gmail.com

Sabavat Raju Naik, International Crops Research Institute for Semi-Arid Tropics, Patancheru, Hyderabad, Telangana - 502 324, India; Department of Genetics and Plant Breeding, Vasantrao Naik Marathwada Krishi Vidyapeeth, Parbhani, Maharashtra- 431 402, India; Email: rajuplantbreeder@gmail.com

Joshua Obeng, Council for Scientific and Industrial Research - Oil Palm Research Institute, Box 74, Kade, Ghana; College of Agriculture, Tennessee State University, 3500 John A. Merritt Blvd, Nashville, TN, USA; Email: josh_139@yahoo.com

Godfried Ohene-Mensah, Council for Scientific and Industrial Research-Crops Research Institute (CSIR-CRI), Fumesua, P.O. Box 3785, Kumasi, Ghana; Email: godthegee@yahoo.co.uk

Harry Okyere, Council for Scientific and Industrial Research-Crops Research Institute (CSIR-CRI), Fumesua, P.O. Box 3785, Kumasi, Ghana; Email: toro192001@yahoo.com

Daniel Oppong-Sekyere, Department of Ecological Agriculture, School of Agriculture, Bolgatanga Technical University, Post Office Box 767, Bolgatanga, Upper East Region, Ghana; Email: doppong-sekyere@bolgatu.edu.gh

Soumyajit Paikray, Department of Genetics and Plant Breeding, College of Agriculture, OUAT, Bhubaneswar, Odisha; Email: soumyaran999@gmail.com

Sudin Kumar Pandey, DBT-NECAB, Assam Agricultural University, Jorhat, India; Email: sudink816@gmail.com

Karuppiah Ponmurugan, Department of Botany and Microbiology, College of Science, King Saud University, PO Box 2455, Riyadh 11451, Saudi Arabia; Email: pkaruppiah@ksu.edu.sa

Raju Prabakaran, PG & Research Department of Botany, PSG College of Arts & Science, Coimbatore 641 014, Tamil Nadu, India; Email: raju.prabakaran@gmail.com

Jeyabal Philomenathan Antony Prabhu, Department of Human Genetics and Molecular Biology, Bharathiar University, Coimbatore 641 046, Tamil Nadu, India; Email: jantony.prabhu@gmail.com

Vijayan Praveena, PG & Research Department of Botany, PSG College of Arts & Science, Coimbatore 641 014, Tamil Nadu, India; Email: praveenavijayan24@gmail.com

Tauqeer Ahmed Qadri, Department of Biosciences, COMSATS University Islamabad, Islamabad 45550, Pakistan; Email: Mrtauqeer0027@gmail.com

Asirvatham Alwin Robert, Department of Endocrinology and Diabetes, Prince Sultan Military Medical City, Ministry of Defense, Riyadh 11159, Kingdom of Saudi Arabia; Email: aalwinrobert@gmail.com

Aryadeep Roychoudhury, Discipline of Life Sciences, School of Sciences, Indira Gandhi National Open University, Maidan Garhi, New Delhi – 110068, India; Email: aryadeep.rc@gmail.com

Aqsa Shabbir, Department of Horticulture, Faculty of Agriculture, Ordu University, Ordu 52200, Türkiye; Email: aqsashabbir228@gmail.com

Khomdram Niren Singh, DBT-NECAB, Assam Agricultural University, Jorhat, India; Email: nirenkhomdram@gmail.com

Sudenaz Soylu, Department of Field Crops, Faculty of Agriculture, Selcuk University, 42079 Konya, Türkiye; Email: sudenaz.5151s@gmail.com

Gunna Suresh Babu Suruthi, Department of Human Genetics and Molecular Biology, Bharathiar University, Coimbatore 641 046, Tamil Nadu, India; Email: sruthigunna11@gmail.com

Ashwini Talakayala, International Crops Research Institute for Semi-Arid Tropics, Patancheru, Hyderabad, Telangana - 502 324, India; Email: Ashwini.Talakayala@icrisat.org

Nadia Tehseen, Department of Botany, Lahore College for Women University (LCWU), Lahore–54000, Pakistan; Email: nadiatehseen909@gmail.com

Thiruvengadam Thirumurugan, Department of Genetics & Plant Breeding, Anbil Dharmalingam Agricultural College & Research Institute, Tamil Nadu Agricultural University, Trichy - 620 027, Tamil Nadu, India; Email: thirumurugant@tnau.ac.in

Sathyalingam Sathya Trisha, Department of Human Genetics and Molecular Biology, Bharathiar University, Coimbatore 641 046, Tamil Nadu, India; Email: sstrisha2002@gmail.com

Abdul Wahab, Department of Biosciences, COMSATS University Islamabad, Islamabad 45550, Pakistan; Email: abdulwahabuaf@gmail.com

Muhammad Waseem, School of Breeding and Multiplication (Sanya Institute of Breeding and Multiplication), School of Tropical Agriculture and Forestry, Hainan University, Sanya 572025, China; Email: m.waseem.botanist@gmail.com

Farishta Zarshan, Department of Botany, Bacha Khan University, Charsadda, 24420, Khyber Pakhtunkhwa, Pakistan; Email: farishta24bkuc@gmail.com

Hajar Zennouhi, Phytopathology Unit, Department of Plant Protection, Ecole National of Agriculture Meknes, Km10, Rte Haj Kaddour, BP S/40, Meknes 50001, Morocco; Email: zennouhihajar53@gmail.com

Preface

Systems biology and multiple omics have been widely recognized as cutting-edge approaches in biological research. In recent years, they have been applied in a variety of research fields for plant sciences, reaching spatial and single-cell resolution. Plants consist of a huge number of cells that can be grouped into specific cell types that possess complex controlling machinery and regulatory networks. Thus, in plant biology, approaches that can reveal biological activities spatially at a single-cell resolution are becoming critical and it is unavoidable they will be conducted intensively in the future, including firstly transcriptomics, then metabolomics, and eventually integrated multiple omics.

In nature or the field, plants have to face a range of stressors that are classified into biotic (pathogenic) and abiotic (environmental) factors, which can cause huge losses in agricultural/food production. This largely damages global food security and can be worse in the scenario of climate change and increasing environmental pollution worldwide. Fortunately, modern crop breeding approaches have upgraded rapidly, and one of the promising technologies is based on spatial and single-cell technology or spatial multi-omics, which leads to precision crop improvement by molecular profiling.

Spatial multi-omics has been conducted for years and has achieved some successes in the field of biomedical science. By contrast, for plants it is an emerging approach and has the potential to be applied to plant biology and advance crop improvement through its high-throughput characteristics. In plant biology, for decades, functional genomics has led to significant findings and indeed is facing a gap to further expand knowledge. A cell-level resolution via spatial and single-cell methods can aid a breakthrough in molecular profiling and consequently provide highly reliable data on integrative omics, and could be a promising window to open the way for multiple areas of plant biology. For organizing spatial omics in plants, one of the essential tools is bioinformatics, particularly artificial intelligence (AI), including machine learning, deep learning, and generative models, for modeling and simulating cellular activities at high-dimensional and temporal levels.

Adding the technology of spatial multi-omics can advance conventional functional genomics and crop improvement, and thus would enable a greater understanding of how plants, environmental factors, and pathogenic microbes interact globally while maintaining its reliability based on high-resolution techniques. Indeed, plant functional genomics, including genomics, transcriptomics, proteomics, and interactomics, can be integrated through the application of spatial multi-omics technology, but at first, a comprehensive summary and discussion of the relationship between these technologies is needed to offer an efficient overview of updated knowledge for the readers in plant biology and crop breeding.

This book explores current findings from spatial multi-omics and digital biological tools that are expected to advance plant physiology and crop breeding for the development of future stress-resilient and high-yield crops to produce more food and biomass. It covers approaches to combat many types of critical stressors, such as temperature, salt, drought, pollutants, pests, fungi, bacteria, viruses, etc. This book is an ideal reference to provide updated information on spatial multi-omics using plant high-throughput technology, advanced functional genomics technologies, and AI models for graduate students, teachers, researchers, and experts in the fields of plant sciences, plant breeding, as well as agronomy and agriculture.

The editor, Dr Jen-Tsung Chen, appreciates all contributors' innovative chapters and the publisher's guidance and assistance.

1 Introduction to Plant Multiple Omics: Technical Advancement and Current Applications

Sabavat Raju Naik[1,2], Kasi Rao Mediga[3], Ravi Kiran Reddy Kondi[4], Ajithan Chandrasekaran[5], Ashwini Talakayala[1], Sana Basharat[6], Pingwu Liu[6] and Muhammad Waseem[6*]

[1]International Crops Research Institute for Semi-Arid Tropics, Patancheru, Hyderabad, India; [2]Department of Genetics and Plant Breeding, Vasantrao Naik Marathwada Krishi Vidyapeeth, Parbhani, India; [3]Department of Plant Pathology, School of Agriculture, Mohan Babu University, Tirupati, India; [4]SPESS, LSU AgCenter, Louisiana State University, Louisiana, USA; [5]Department of Horticulture, Chungnam National University, Daejeon, Korea; [6]School of Breeding and Multiplication (Sanya Institute of Breeding and Multiplication), School of Tropical Agriculture and Forestry, Hainan University, Sanya, China

1.1 Overview of Multi-omics in Plant Biology

Multi-omics represents a transformative approach in modern plant biology, encompassing both comprehensive and integrative analyses of multi-omics, that is, genomics, transcriptomics, proteomics, metabolomics, and epigenomics, to understand biological systems (Hasin *et al.*, 2017). In plant biology, this omics technology enables the exploration of the complex regulatory networks that control growth, development, and stress adaptation. Multi-omics integration connects genetic variation with transcriptional regulation, protein function, and metabolite accumulation, offering a complete view of plant cellular responses (Weckwerth, 2011; Rajasundaram and Selbig, 2016). In plants, its scope extends from basic functional genomics to advanced programs in plants. It correlates molecular data with physiological traits and multi-omics by deciphering genotype–phenotype connectivity, identifying candidate genes for breeding, and revealing biomarkers that are linked to yield, stress tolerance, and nutrient efficiency (Rhaman *et al.*, 2024). Therefore, integrative studies facilitate precision breeding and synthetic biology applications, accelerating the development of climate-resilient and resource-efficient crop varieties.

Multi-omics contributes unique but complementary insights into plant systems. Genomics relies on the complete DNA sequence and structural variations within the plant genome, with high-throughput sequencing technologies enabling genome assembly and comparative analyses to identify genes

*Corresponding author: m.waseem.botanist@gmail.com; drwaseem@hainanu.edu.cn

© CAB International 2026. *Plant Spatial Biology* (ed. Chen)
DOI: 10.1079/9781836994572.0001

that control key agronomic traits (Michael and VanBuren, 2020). Transcriptomics reveals gene expression profiles through RNA-seq and microarray, leading to the exploration of transcriptional responses to developmental and environmental cues (Wang *et al.*, 2009). Proteomics provides protein abundance, modifications, and interactions, with quantitative mass spectrometry (MS) allowing large-scale profiling of plant proteomes to reveal stress signaling and metabolic regulation (Bantscheff *et al.*, 2012). Metabolomics analyzes cellular end products, such as sugars, amino acids, and secondary metabolites, using liquid chromatography (LC)–MS/MS and gas chromatography (GC)–MS to uncover pathways linked to growth, stress, and defense (Kumar *et al.*, 2017), while epigenomics explores heritable gene expression changes, such as DNA methylation and histone modification, that regulate plant adaptability (Springer and Schmitz, 2017).

1.2 Technological Advancements

1.2.1 Next-generation sequencing

In plant biology, next-generation sequencing (NGS) is a revolutionary process that enables high-throughput sequencing of entire multi-omics with exceptional accuracy (Goodwin *et al.*, 2016). NGS platforms such as Illumina, Pacific Biosciences (PacBio), and Oxford Nanopore facilitate the discovery of genetic variations, that is, single-nucleotide polymorphisms (SNPs), insertions/deletions, and structural variants across diverse plant species (Michael and VanBuren, 2020). In transcriptomics, RNA sequencing (RNA-seq) is a widely used NGS platform that provides deep insight into gene expression under various developmental stages and environmental stresses (Wang *et al.*, 2009). This technology enables the identification of key candidate genes related to stress tolerance, yield traits, and metabolic regulation. Moreover, NGS has enhanced our understanding of small RNAs, epigenetic modifications, and plant microbiome interactions, thereby contributing to various omics studies in plants (Weckwerth, 2011).

1.2.2 Quantitative mass spectrometry

In plant proteomic and metabolomic studies, quantitative MS serves as a central analytical technology that precisely enables the identification and quantification of biomolecules (Bantscheff *et al.*, 2012). The techniques used include:

1. **Label-free quantification:** This method measures peptide ion intensities directly from spectra without isotopic labeling. It is cost-effective and suitable for large comparative proteomic analyses across various plant samples under different stress conditions (Zhang *et al.*, 2013).
2. **Isobaric tagging methods (iTRAQ and TMT):** These methods use stable isotope labels by multiplexing to analyze multiple samples simultaneously. Isobaric tags for relative and absolute quantification (iTRAQ) and tandem mass tags (TMT) provide enhanced accuracy and throughput, allowing the quantification of proteins across various plant genotypes within treatments (Rauniyar and Yates, 2014). High-throughput-MS-based techniques have become indispensable for studying protein expression patterns and post-translational modifications involved in stress response and metabolic regulation.
3. **High-resolution metabolomics (LC–MS/MS and GC–MS):** Metabolomic analysis of small molecular metabolites provides direct insights into the biochemical components of plants under specific physiological or stress conditions. Advanced analytical platforms, such as liquid chromatography – tandem mass spectrometry (LC–MS/MS) and gas chromatography – mass spectrometry (GC–MS), have been employed for qualitative and quantitative metabolite profiling (Dettmer *et al.*, 2007). LC–MS/MS is used to detect non-volatile and thermally unstable metabolites, such as flavonoids, alkaloids, and phytohormones, whereas GC–MS is used for volatile compounds, such as amino acids, sugars, and fatty acids (Sawada *et al.*, 2009). Metabolomic tools facilitate various metabolic fingerprinting, functional gene annotation, and elucidation of biochemical

pathways associated with stress responses, growth, and yield in plants (Kumar *et al.*, 2017), and are applicable for advancing systems biology and crop improvement through metabolic engineering.

1.2.3 Single-cell profiling technologies

Single-cell profiling technologies are mainly applicable in plant omics research, offering unprecedented resolution of cellular heterogeneity, developmental processes, and environmental responses (Rhaman *et al.*, 2024). An average signal across various mixtures of cell types can mask certain tissue-specific responses, whereas single-cell approaches enable the measurement of gene expression, chromatin accessibility, proteomes, and metabolites at the level of individual cells (Vandereyken *et al.*, 2023). The application of single-cell RNA sequencing (scRNA-seq) in plants has enabled the dissection of root and shoot cell populations, reconstruction of developmental cells, and identification of cell types for hormones and nutrient availability (Efroni and Birnbaum, 2016). Moreover, single-cell omics technologies have been utilized to identify plant–environment interactions by gaining new insights into how specific cell tissues, types, or individual cells respond to various environmental stimuli (Tenorio Berrío and Dubois, 2024). In plants, single-cell transcriptomics is applicable for profiling gene expression at the resolution of individual cells, allowing for the identification of distinct cell types and transcriptional states within complex plant tissues (Shaw *et al.*, 2021). The core principle of single-cell profiling involves isolating single cells or nuclei, capturing their mRNA, and forming complementary DNA (cDNA) for high-throughput sequencing. In plants, the steps involved are enzymatic digestion or protoplasting, which is used to release cells by single-nucleus RNA sequencing (snRNA-seq) (Ryu *et al.*, 2019). The integration of single-cell data with spatial and multi-omic approaches offers a more comprehensive understanding of cell-type-specific regulatory and signaling mechanisms in plants (Rich-Griffin *et al.*, 2020; Nobori, 2025) (Fig. 1.1).

1.2.4 Computational tools for multi-omics data analysis

In plant biology, recent advancements in multi-omics technology have produced vast and complex datasets that demand computational tools efficient in integration as well as interpretation (Yang *et al.*, 2021). Bioinformatics platforms and systems biology approaches enable

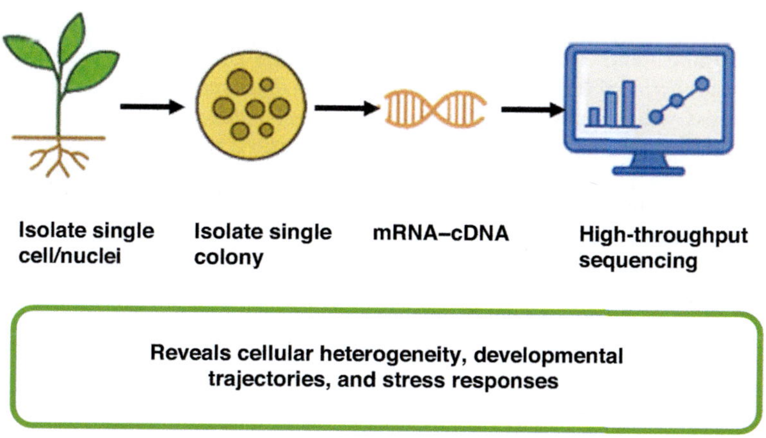

Fig. 1.1. Workflow for single-cell sequencing in plants. The process begins with a plant model from which individual cells are isolated. The genetic material (DNA/RNA) is then sequenced, and the resulting data are analyzed computationally.

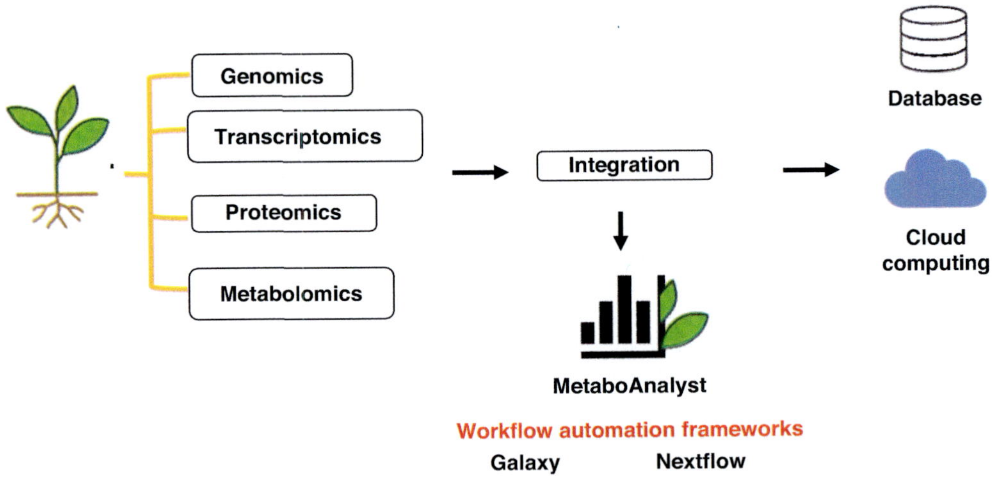

Fig. 1.2. Integrative plant omics workflow. Diagrammatic illustration of the process of generating and integrating omics data from a plant. Data from genomics, transcriptomics, proteomics, and metabolomics are collected and integrated. The integrated data are stored in a database via cloud computing, facilitated by workflow automation frameworks to deliver new biological insights.

researchers to combine genomics, transcriptomics, proteomics, and metabolomics data to uncover plant regulatory networks and predict complex traits (Abdullah-Zawawi *et al.*, 2022). Some of the most important databases and tools, such as PLAZA, MetaboAnalyst, Cytoscape, and OmicsFusion, enable data visualization, pathway mapping, and cross-omics integration, which facilitate the functional relationship or discovery of genes, proteins, and other metabolites (Maghsoudi *et al.*, 2022) (Fig. 1.2).

Computational tools play a vital role in managing, analyzing, and visualizing these datasets, enabling functional insights into plant growth and stress adaptation mechanisms. Van Bel *et al.* (2022) presented an integrative online environment for comparative and evolutionary genomics in plants, allowing genome alignment, gene family identification, and functional annotation across various species; PLAZA integrates genomic and transcriptomic data to uncover conserved gene networks. For instance, in *Arabidopsis* and maize, orthologous stress-responsive transcription factors have been identified, aiding cross-species gene function prediction and crop improvement (Van Bel *et al.*, 2012). Chong *et al.* (2019) used another web-based tool, MetaboAnalyst, for metabolomic data analysis and integration with transcriptomic or proteomic data to reveal key biochemical pathways. In rice and tomatoes, it has been used to link metabolite profiles with gene expression under drought and salinity stress, highlighting stress-specific metabolic reprogramming (Chowdhury *et al.*, 2023). Cytoscape enables the visualization and analysis of biological interaction networks by integrating multi-omics data (Shannon *et al.*, 2003). In soybean and *Arabidopsis*, it has been used to build gene–metabolite co-expression networks and identify hub genes linked to stress tolerance and yield regulation (Shen and Chen, 2022). Han *et al.* (2023) described the use of OmicsFusion, which integrates genomics, transcriptomics, proteomics, and metabolomics data to identify cross-omic correlations. In *Arabidopsis* and maize, it has been used to link transcripts and metabolites, revealing key regulators of biomass and stress responses for predictive breeding purposes (Table 1.1).

Table 1.1. Summary of key computational tools used in plant multi-omics research, highlighting their primary functions, example applications, and references.

Tool	Primary function	Example in plant multi-omics	Reference
PLAZA	Comparative genomics and functional annotation	Identified conserved drought-responsive genes in maize and *Arabidopsis*	Van Bel *et al.*, 2022
MetaboAnalyst	Metabolite profiling and pathway integration	Linked metabolite changes to drought-induced transcriptional shifts in rice	Chong *et al.*, 2019
Cytoscape	Network visualization and systems biology	Built co-expression and metabolite networks in soybean	Shannon *et al.*, 2003
OmicsFusion	Cross-omic data integration	Combined transcriptomic and metabolomic data to predict regulatory hubs in maize	Han *et al.*, 2023

1.3 Applications of Multi-omics

A multi-omic or panomic approach can play a significant role in crop research and improvement (Weckwerth *et al.*, 2020). It helps identify genes, proteins, epigenomes, and transcripts involved in abiotic stress (Roychowdhury *et al.*, 2023). This can be useful for identifying biological elements and their roles in phenotypic trait expression (Patel *et al.*, 2024). It offers the possibility of identifying genetic variations in plant genotypes (Fan *et al.*, 2025). The complex regulatory networks in plant tissues can be explored by combining metabolomics, transcriptomics, and proteomics (Wang *et al.*, 2021; Zhang *et al.*, 2021). Plant physiological processes and quality control under severe stress conditions can be identified using multi-omic studies (Raza *et al.*, 2021; Ashraf *et al.*, 2022; Muthuramalingam *et al.*, 2022; Fan *et al.*, 2025). It is possible to gain knowledge on polygenic traits and key genes/variants or proteins responsible for tolerance (Roychowdhury *et al.*, 2023). A multi-omics approach identifies pathways through which the trait is expressed in crop plants (Derbyshire *et al.*, 2022). Multi-omic technologies offer the opportunity to study the whole plant system and complex interactions between genomics, transcriptomics, proteomics, and metabolomics (Vennapusa *et al.*, 2023). For example, Cai *et al.* (2021) observed species formation and stress adaptation in a study on wild barley using a multi-omic approach, and Sheoran *et al.* (2022) identified drought tolerance genes in maize using multi-omics.

The selection of plants under stress can be quick and easy with a multi-omic study based on gene expression and/or the composition of proteins or metabolites (Gayacharan *et al.*, 2023). The chances of success are higher with multi-omics because it integrates omics data and enables the identification of candidate genes responsible for stress in crop varieties (Roychowdhury *et al.*, 2023). The time and cost of breeding can be reduced with multi-omics, because it precisely identifies genes in crop plants (Fan *et al.*, 2025). Single-cell multi-omics and spatial transcriptomics can be used to identify, characterize, and understand cell type, function, and cellular interactions, respectively (Jovic *et al.*, 2022; Fan *et al.*, 2025). Such studies have been widely conducted in *Arabidopsis* (Denyer *et al.*, 2019), rice (Zhang *et al.*, 2021), tomato (Tian *et al.*, 2020), and maize (Nelms and Walbot, 2019).

Several important quantitative trait loci (QTLs) for yield, yield-related traits, and abiotic and biotic stresses in major cereals such as rice, wheat, and maize have been identified using multi-omics (Kulkarni *et al.*, 2020). A combined approach of QTL-seq, QTL-mapping and RNA-seq has identified a major QTL for seed weight in groundnut (Wang *et al.*, 2022). In a multi-omics study (genome-wide association study (GWAS), transcriptome-wide association study (TWAS), and genomic selection) on Brassica, QTLs and gene networks involved in seed oil content were identified (Tang *et al.*, 2021). Similarly, in soybean, combined transcriptomics, metabolomics, and proteomics have been used to identify genes associated with seed oil content (Jong *et al.*, 2024).

In Brassica, Zhou *et al.* (2021) identified salt stress-related genes using a multi-omics approach. Flavonoid synthesis in tomatoes has been observed using metabolomics, transcriptomics, and lipidomics (Meng *et al.*, 2020). In rice, the *OsSEH1* gene and phenylpropanoid and flavonoid pathways have been identified using transcriptomic and metabolomic analyses, respectively (Gu *et al.*, 2022). In date palms, salt avoidance by osmotic regulation has been observed through ion analysis, proteomics, metabolomics, and hormone analysis (Mueller *et al.*, 2024). The nitrogen uptake efficiency-related gene *OsGATA8* has been observed in a GWAS study, which influences tillering in rice (Wu *et al.*, 2024). A similar type of study was conducted by Ichihashi *et al.* (2020) using a multi-omic approach to investigate the importance of soil nitrogen in Brassica. Nitrogen-use efficiency genes have been identified using integrated multi-omics in maize (Xing *et al.*, 2024). In rice, multi-omics has been used to identify crucial genes and mRNA for blast disease (Zhao *et al.*, 2023). Ye *et al.* (2017) identified the genes *NlEG1* and *NlSEF1* that confer defense against brown planthoppers in rice using proteomic and transcriptomic studies.

1.3.1 Stress response and environmental adaptation: Understanding plant responses to abiotic stresses, such as drought and salinity

Plants usually undergo climatic changes, that is, abiotic stress (mainly drought and salinity) and biotic stress (pathogens, pests, and parasitic weeds). Traditional studies often fail to capture the complexity of plant responses, adaptation, and survival under stress conditions. Thus, unraveling these complex processes requires omics approaches (genomics, epigenomics, transcriptomics, proteomics, metabolomics, lipidomics, phenomics, and microbiomics).

Plant multi-omics provides a comprehensive framework for unraveling the complex molecular and physiological mechanisms underlying abiotic stress tolerance (Debnath *et al.*, 2011; Ali *et al.*, 2022). Abiotic stress causes changes in plant structure, function, chemistry, and genetics. Stress factors such as drought and excessive heat can cause such changes, as can cold conditions. The ability of plants to use sensors to detect stress and activate pathways signaling Ca^{2+}, reactive oxygen species, and phytohormones such as abscisic acid, jasmonic acid, salicylic acid, and ethylene, as well as various types of MAPKs, has been well elaborated (Golldack *et al.*, 2014; Zhu, 2016). These activate the transcription factors DREB, NAC, WRKY, MYB, and bZIP, which in turn target and regulate protective genes, such as LEA proteins and antioxidants (Hirt and Shinozaki, 2004; Fujita *et al.*, 2006). Plant responses to stress include stomatal closure, osmolyte production, and ion concentration. The actions of the key ion transporters SOS1, NHX, and HKT, as well as the activation of reactive oxygen species (ROS)-scavenging pathways, have been described (Chinnusamy *et al.*, 2007; Gupta *et al.*, 2020). Structural changes are also responses to stress and include the development of deeper roots, waxy leaves, and the adjustment of flowering time. Markers of stress, such as epigenetic mechanisms, DNA methylation and hydroxymethylation, histone modifications, and small RNAs, can also help plants 'memorize' stressful situations (Fujita *et al.*, 2006; Zhu, 2016). These adaptations help plants not only to survive but also to flourish in difficult environments. Using genomics and transcriptomics, stress-responsive genes, transcription factors, and regulatory networks are identified, while proteomics and metabolomics reveal dynamic changes in proteins, metabolites, osmolytes, and antioxidants that help plants mitigate stresses such as drought, salinity, and heat. Phenomics connects molecular responses to visible traits by screening tolerant genotypes (Derbyshire *et al.*, 2022). The integration of these datasets using systems biology and machine learning helps build models of stress adaptation, which ultimately accelerates breeding and genetic engineering strategies for developing climate-resilient crops (Joshi *et al.*, 2024; Varadharajan *et al.*, 2025). Plant growth and adaptation are shaped by gene–environment–metabolite interactions, which provide system biology frameworks that go beyond single-layer approaches. Some abiotic stresses, with examples of multi-omics approaches, are described in Table 1.2.

Various omics tools provide comprehensive insights into biotic stress tolerance by linking molecular and physiological responses to stress. Plants undergo biotic stress from pathogens (fungi, bacteria, viruses, and nematodes), insects, weeds, and parasitic plants, which negatively affect plant growth, development, and yield (Varadharajan *et al.*, 2025). These stresses trigger complex defense responses, including the production of antimicrobial compounds such as phytoalexins, and signaling via plant hormones such as salicylic acid, jasmonic acid, and ethylene (Appu *et al.*, 2021). Therefore, plants have innate immunity systems; however, biotic stress can still lead to major agricultural losses, making host–pathogen interactions, defense mechanisms, and the role of beneficial microbes essential for developing resistant and sustainable crop varieties (Ijaz *et al.*, 2024a). The highly organized defense machinery in plants is directed towards resistance against biotic stress factors, such as pathogens (fungi, bacteria, and viruses), herbivores, and parasitic weeds. The initial interaction involves the recognition of pathogen associated molecular patterns (PAMPs) by pattern recognition receptors (PRRs), followed by PAMP-triggered immunity (PTI). In contrast, pathogens secrete a group of molecules known as effectors that directly or indirectly suppress PTI. Plant R proteins can neutralize effectors, resulting in effector-triggered immunity (ETI) (Jones and Dangl, 2006; Boller and Felix, 2009; Dodds and Rathjen, 2010). However, both PTI and ETI pathways converge to a downstream set of signal transduction components, including Ca^{2+} influx, ROS burst, MAPK cascades, salicylic acid, jasmonic acid, and ethylene (Pieterse *et al.*, 2012; Meng and Zhang, 2013). Induced defense responses commonly comprise physical barriers, such as callose deposition, lignification, and cuticle reinforcement (Underwood, 2012). Biochemical reactions are also responsible for the induction of pathogenesis-related proteins, chitinases, glucanases, and the production of antimicrobial compounds such as phytoalexins and phenolics (van Loon *et al.*, 2006; Bednarek and Osbourn, 2009). Localized programmed cell death or hypersensitive response (HR) also restricts pathogen spread (Coll *et al.*, 2011). Genomics, transcriptomics, and epigenomics are important for identifying resistance genes, elucidating regulatory networks, and deciphering stress memory. In contrast, proteomics and metabolomics have focused on defense proteins, phytoalexins, and hormone-mediated signaling. Phenomics is used to track the development of disease, and microbiomics focuses on the positive microorganisms that strengthen immunity. Together with the techniques of systems biology and genome editing, they open the door to the rapid production of disease-resistant crops (Varadharajan *et al.*, 2025). Some details of biotic stress applications with examples are described in Table 1.2.

1.4 Current Challenges

Our understanding of plant biology has considerably improved by integrating multi-omics technologies; however, it has not been explored effectively due to numerous challenges. The main challenge is the complex data integration from different omics technologies, which requires advanced computational and bioinformatics resources to handle and interpret large datasets (Hina *et al.*, 2024; Ijaz *et al.*, 2024b). The significant financial investment and technological requirements of omics technologies may limit access to these tools, especially in resource-limited settings, thereby hindering their widespread adoption in plant research (Sirangelo, 2019). Another challenge is the cellular structure of plants, which complicates the adoption of single-cell omics technologies; the existence of stiff walls in plant cells hinders the isolation of single cells for study (Rhaman *et al.*, 2024). In addition, although multi-omics approaches have successfully identified novel signaling molecules in plants and metabolic pathways, a thorough understanding of the role of many known and unknown metabolites is needed to promote crop development and address food security concerns (Luo *et al.*, 2022; Nasar *et al.*, 2024). The integration of omics data with systems biology for the prediction of complex traits and calibration of agricultural genes is in its nascent stages and

Table 1.2. The application of omics techniques to investigate abiotic and biotic stresses in plants.

S. No.	Omics	Application	Relevance to abiotic stress	Examples or case studies	References
Abiotic stress					
1	Genomics	Genome sequencing, GWAS, QTL mapping	Identifies genes, allelic variants, and regulatory elements for stress tolerance and yield	Drought: DREB, NAC, AREB transcription factors regulating ABA-dependent responses. Salinity: HKT1;5, SOS1, NHX transporters for Na^+/K^+ homeostasis. Heat: HSP gene families; allelic variation in ARGOS8 (maize) for yield stability. Cold: CBF/DREB1 genes at QTL hotspots in rice and wheat. Nutrient deficiency: PHT1 (phosphate transporters), NRT1/2 (nitrate transporters).	Ali *et al.*, 2022; Joshi *et al.*, 2023
2	Epigenomics	DNA methylation, histone modifications, and non-coding RNAs	Regulates stress-responsive genes; stress memory and transgenerational adaptation	Drought/heat: Stress-induced DNA hypomethylation in promoters of stress genes. Salinity: siRNAs and miRNAs regulate ion transporter expression. Cold: Histone acetylation (H3K9ac) enhances CBF gene expression in Arabidopsis. Nutrient stress: DNA methylation changes alter phosphate starvation response genes (PHR1 network).	Springer and Schmitz, 2017; Zhang *et al.*, 2018
3	Transcriptomics	RNA-seq, single-cell transcriptomics	Reveals gene expression reprogramming, identifies stress-inducible pathways and transcripts	Drought: Abscisic acid (ABA) biosynthesis (NCED genes) upregulated; suppression of photosynthetic genes. Salinity: Upregulation of ion transporters (SOS, HKT, NHX). Heat: Heat shock factors (HSFs) and heat shock proteins (HSPs) were strongly induced. Cold: ICE1-CBF-COR pathway genes are activated. Nutrient deficiency: Upregulation of phosphate transporters (PHT), nitrogen assimilation genes (GS, GOGAT).	Mochida and Shinozaki, 2011; Shaar-Moshe *et al.*, 2015

4	Proteomics	Protein abundance and post-translational modifications	Identifies stress proteins (HSPs, kinases, antioxidants) for protection and signaling	Drought: Accumulation of dehydrins, LEA proteins, and aquaporins. Salinity: Ion-binding proteins, vacuolar ATPases, ROS detox enzymes. Heat: HSP70, HSP90, ubiquitin-proteasome proteins for protein quality control. Cold: Antifreeze proteins, enzymes stabilizing membranes. Nutrient deficiency: High-affinity transporters (NRT2, PHT), enzymes for scavenging/redistribution.	Kosová *et al.*, 2015; Jha *et al.*, 2025
5	Metabolomics and lipidomics	Profiling of metabolites and lipids	Detects osmolytes, hormones, and flavonoids for osmotic adjustment, antioxidant defense, and membrane stability	Drought/salinity: Proline, glycine betaine, trehalose, raffinose for osmotic adjustment. Heat: Accumulation of polyamines, flavonoids, ROS-scavenging metabolites. Cold: Sugars (raffinose, sucrose), fatty acid desaturation to maintain membrane fluidity. Nutrient deficiency: Organic acids (malate, citrate) are secreted under P deficiency; amino acids (glutamine, asparagine) are altered under N deficiency.	Fernie and Schauer, 2009; Obata and Fernie, 2012
6	Phenomics	High-throughput imaging, drones, sensors	Links molecular responses with field-level physiological and agronomic traits	Drought: Reduced canopy temperature, increased root depth detected by imaging. Salinity: Early leaf chlorosis and reduced biomass. Heat: Altered flowering time, reduced pollen viability. Cold: Reduced chlorophyll fluorescence (Fv/Fm). Nutrient deficiency: visible deficiency symptoms (chlorosis, stunted growth) and altered root system architecture.	Furbank and Tester, 2011; Araus *et al.*, 2018

Continued

Table 1.2. Continued

S. No.	Omics	Application	Relevance to abiotic stress	Examples or case studies	References
7	Microbiomics	Plant–microbe interactions (rhizosphere and endophytes)	Enhances nutrient uptake, growth, and stress resilience	–	Compant *et al.*, 2019; Vives-Peris *et al.*, 2020
Biotic stress					
1	Genomics	Genome sequencing, GWAS, QTL mapping	Identifies resistance (*R*) genes, allelic variants	*NBS-LRR* genes in rice conferring blast resistance	Dodds and Rathjen, 2010; Varshney *et al.*, 2021
2	Epigenomics	DNA methylation, histone modifications, and small RNAs	Regulates defense genes and stress memory	DNA methylation regulates pathogen-responsive genes in *Arabidopsis*	Dodds and Rathjen, 2010; Kourelis and van der Hoorn, 2018
3	Transcriptomics	RNA-seq, single-cell transcriptomics	Reveals gene expression changes and defense signaling pathways	Upregulation of WRKY and MYB TFs during pathogen attack in tomato	Dodds and Rathjen, 2010; Varshney *et al.*, 2021
4	Proteomics	Protein abundance, modifications	Identifies defense proteins (HSPs, PR proteins) and kinases	PR proteins and kinases accumulation in barley infected with powdery mildew	Kourelis and van der Hoorn, 2018; Jha *et al.*, 2025
5	Metabolomics	Profiling of metabolites and hormones	Reveals antimicrobial compounds and signaling molecules	Accumulation of phytoalexins, salicylic acid, and jasmonic acid in *Arabidopsis*	Dodds and Rathjen, 2010; Obata and Fernie, 2012
6	Phenomics	High-throughput imaging, drones, sensors	Monitors disease progression and host response	Imaging of fungal spread and leaf necrosis in wheat	Furbank and Tester, 2011; Araus *et al.*, 2018
7	Microbiomics	Plant–microbe interactions	Identifies beneficial microbes that enhance immunity	*Pseudomonas* and *Trichoderma* spp. promoting disease resistance in tomato	Compant *et al.*, 2019; Vives-Peris *et al.*, 2020

requires advancement through the development of robust models and methodologies (Yang *et al.*, 2021). Lastly, the ethereal characteristics and intellectual property of omics data used in agriculture serve as further hurdles that need to be carefully addressed to ensure equitable gains from technology (Nasar *et al.*, 2024). Overcoming these issues will be vital for the sustained development and application of multi-omics in plant sciences, and sustainable agriculture and greater food security (Carper *et al.*, 2022).

1.4.1 Data integration across omics platforms

The integration of data from multiple omics platforms in plant research is confronted with various challenges and limitations, primarily owing to the variability and complexity of the data itself. The volume and heterogeneity of data from high-throughput technologies, such as genomics, transcriptomics, proteomics, and metabolomics, are some of the largest challenges. Such enormous volumes of data require advanced bioinformatics capabilities and computational models for appropriate analysis and integration, which are resource intensive and technically demanding (James *et al.*, 2021; Chao *et al.*, 2024). In addition, the non-standardization of analysis and data collection protocols at different omics platforms renders integration even more difficult, risking inconsistent and incorrect interpretation of the data (Papadaki *et al.*, 2024). Another constraint is the incomplete annotation of plant genomes, which makes it difficult to map and integrate omics data accurately, particularly in plants with large and complex genomes (Jamil *et al.*, 2020). In addition, multi-omics data integration requires interdisciplinary collaboration by combining bioinformatics, systems biology, and plant science knowledge, which can be difficult to organize. To overcome these challenges, advanced analytical frameworks that incorporate artificial intelligence and machine learning for the easier interpretation and handling of complex datasets are being formulated (Cembrowska-Lech *et al.*, 2023). In addition, the establishment of elaborate databases and platforms that can enable seamless omics data sharing and integration is pivotal for surmounting the standardization of data issues (Papadaki *et al.*, 2024). Scientific guidelines and protocols for systematic multi-omics integration have also been presented to ensure the correct representation of data and to help researchers, particularly those untrained in the field (Jamil *et al.*, 2020). By overcoming these hurdles, multi-omics data can be utilized to substantially advance the understanding of plant biology and improve crop breeding, stress management, and sustainable cultivation (Chao *et al.*, 2024).

1.4.2 Standardization of data formats and protocols

The standardization of data formats and protocols has been significantly influenced by advancements in multi-omics, which require advanced and compatible systems. Different data formats are generated by different NGS technologies. This scenario calls for the establishment of standard procedures and data formats to ensure successful data integration and reuse, as evidenced by the importance of data standardization in the omics field (Kumuthini *et al.*, 2023). In plant phenotyping, the lack of standard protocols for storage and data collection has hindered interoperability, as variations in the methodologies of phenotyping and environmental parameters make it challenging to integrate the data (Krajewski *et al.*, 2015). Software such as OmicsSuite has emerged to solve these hurdles by providing an integrated toolkit for multi-omics analysis and visualization that supports different data formats and allows for seamless data transfer and analysis of pipelines (Miao *et al.*, 2023). The FAIR data principles, which emphasize the findability, accessibility, interoperability, and reusability of data, have been extended to plant research, and projects such as MIAPPE (Minimum Information About a Plant Phenotyping Experiment) and Breeding API facilitate the curation of metadata and data lifecycle management (Pommier *et al.*, 2023). These projects are crucial for the advancement of plant science, as they allow for the integration of complex datasets in the context of systems biology, and hence allow for revolutionary insights into plant biology (Misra *et al.*, 2019). In addition, the combination of multi-omics data with artificial intelligence provides a robust approach for plant characterization that combats the difficulties that are universal to plant biology while boosting precision horticulture and predictive breeding (Cembrowska-Lech *et al.*, 2023). In short, the progress of plant multi-omics is propelling the establishment of normalized data formats and protocols that are important for the fruitful combination and exploitation of multi-dimensional data in the plant science discipline (Haidar *et al.*, 2024).

1.4.3 Interpretation of high-dimensional datasets

Breakthroughs in NGS and high-throughput technologies have significantly facilitated the consolidation of high-dimensional multi-omics datasets in plants. This has rendered multi-omics data less expensive and more obtainable, thus promoting knowledge of plant systems at the genomic, transcriptomic, proteomic, and metabolomic levels (Singh *et al.*, 2020). Consolidation of these datasets is needed to elucidate complicated biological activities, such as plant development, growth, and environmental reactions, by unveiling the interrelationships between biomolecules and their functions (Mahmood *et al.*, 2023). Multi-omics techniques elucidate molecular markers, functional genes, and medicinal plant biosynthetic routes that single-omics research cannot explain (Yang *et al.*, 2023). Consolidation facilitates complete models that predict complicated traits, augmenting crop genetics and breeding under stress conditions (Yang *et al.*, 2021). However, the complexity of omics datasets hinders consolidation and interpretation, requiring novel computational platforms and bioinformatics resources (Abdullah-Zawawi *et al.*, 2022). These technologies are significant for disease classification, predicting biomarkers, and providing molecular insights into plant mechanisms (Subramanian *et al.*, 2020). Additionally, the integration of omics data with systems biology fosters comprehensive knowledge of plant phenotypic responses at the cellular and molecular levels, boosting stress tolerance and tenacity (Singh *et al.*, 2020). Despite the persistent problems with valid outcomes and the need for more experimentation to corroborate multi-omics knowledge, the integration of high-dimensional plant omics data continues to surge, opening new avenues for plant research and crop improvement.

1.5 Future Directions

Multi-omics heterogeneous data require more advanced computational models to perform tasks such as data alignment, normalization, and understanding biological functions, networks, regulatory mechanisms, trait expression, and the interplay between omics. Future research should focus on matching pan-genomic data with phenotypic expression to enable variants for stress tolerance, along with genomic prediction. Plants characterized using multi-omics can be a source for future plant breeding programs, with information on processes at both the cellular and molecular level. The research focus must be maintained on single-cell multi-omics and spatial transcriptomics to identify variance and its translation under abiotic stress. Further research on cellular-level variation in phenotypic traits can be an advanced study in the field of plant sciences. Artificial intelligence and machine learning can also help in understanding multi-omics data. Future research must also focus on microbiome omics.

1.6 Conclusion

The incorporation of various omics technologies into plant studies enhances our understanding of the intricacies of plant biology and has potential practical benefits for addressing international agricultural issues. The swift breeding of new crop cultivars with desired traits, alongside customizing the underlying biosynthetic pathways for specific production processes, are examples of how multiple omics approaches are applied to plants, which fosters a fresh surge in sustainable and cutting-edge agriculture. These advances not only tackle the challenges of food security and environmental preservation but also unlock new frontiers in plant biology research. The integration of multi-omics approaches in plant biology has significantly advanced our understanding of the complex molecular mechanisms underlying plant development, adaptation, and productivity. Technological innovations in sequencing, mass spectrometry, and computational analysis have enabled the generation and interpretation of multi-layered datasets with increasing resolution and throughput. These tools have proven instrumental in decoding gene regulatory networks, metabolic dynamics, and stress responses, offering powerful insights into crop improvement and precision breeding. However, challenges such as the integration and standardization of heterogeneous data and the need for robust analytical frameworks remain critical barriers. Continued progress in bioinformatics, particularly the application of machine learning and systems biology models, is essential for harnessing the full potential of multi-omics data in functional genomics and agricultural innovationChao *et al.* (2024).

Through spatial multi-omics, researchers can observe cell–cell interactions, gain deeper insights into cellular functions, identify rare cell populations, and characterize complex metabolites, thereby laying a strong foundation for the advancement of plant developmental biology.

References

Abdullah-Zawawi, M.R., Govender, N., Harun, S., Muhammad, N.A.N., Zainal, Z. *et al.* (2022) Multi-omics approaches and resources for systems-level gene function prediction in the plant kingdom. *Plants* 11(19), 2614. DOI: 10.3390/plants11192614.

Ali, A., Altaf, M.T., Nadeem, M.A., Karaköy, T., Shah, A.N. *et al.* (2022) Recent advancement in OMICS approaches to enhance abiotic stress tolerance in legumes. *Frontiers in Plant Science* 13, 952759. DOI: 10.3389/fpls.2022.952759.

Appu, M., Ramalingam, P., Sathiyanarayanan, A. and Huang, J. (2021) An overview of plant defense-related enzymes responses to biotic stresses. *Plant Gene* 27, 100302. DOI: 10.1016/j.plgene.2021.100302.

Araus, J.L., Kefauver, S.C., Zaman-Allah, M., Olsen, M.S. and Cairns, J.E. (2018) Translating high-throughput phenotyping into genetic gain. *Trends in Plant Science* 23(5), 451–466. DOI: 10.1016/j.tplants.2018.02.001.

Ashraf, U., Mahmood, S., Shahid, N., Imran, M., Siddique, M. *et al.* (2022) Multi-omics approaches for strategic improvements of crops under changing climatic conditions. In: *Principles and Practices of OMICS and Genome Editing for Crop Improvement*. Springer International Publishing, Cham, pp. 57–92. DOI: 10.1007/978-3-030-96925-7_3.

Bantscheff, M., Lemee, S., Savitski, M.M. and Kuster, B. (2012) Quantitative mass spectrometry in proteomics: Critical review update from 2007 to the present. *Analytical and Bioanalytical Chemistry* 404, 939–965.

Bednarek, P. and Osbourn, A. (2009) Plant-microbe interactions: Chemical diversity in plant defense. *Science (New York, N.Y.)* 324(5928), 746–748. DOI: 10.1126/science.1171661.

Boller, T. and Felix, G. (2009) A renaissance of elicitors: Perception of microbe-associated molecular patterns and danger signals by pattern-recognition receptors. *Annual Review of Plant Biology* 60, 379–406. DOI: 10.1146/annurev.arplant.57.032905.105346.

Cai, S., Shen, Q., Huang, Y., Han, Z., Wu, D. *et al.* (2021) Multi-omics analysis reveals the mechanism underlying the edaphic adaptation in wild barley at evolution slope (tabigha). *Advanced Science (Weinheim, Baden-Wurttemberg, Germany)* 8(20), e2101374. DOI: 10.1002/advs.202101374.

Carper, D.L., Appidi, M.R., Mudbhari, S., Shrestha, H.K., Hettich, R.L. *et al.* (2022) The promises, challenges, and opportunities of omics for studying the plant holobiont. *Microorganisms* 10(10), 10. DOI: 10.3390/microorganisms10102013.

Cembrowska-Lech, D., Krzemińska, A., Miller, T., Nowakowska, A., Adamski, C. *et al.* (2023) An integrated multi-omics and artificial intelligence framework for advance plant phenotyping in horticulture. *Biology* 12(10), 1298. DOI: 10.3390/biology12101298.

Chao, H., Zhang, S., Hu, Y., Ni, Q., Xin, S. *et al.* (2024) Integrating omics databases for enhanced crop breeding. *Journal of Integrative Bioinformatics* 20(4), 20230012. DOI: 10.1515/jib-2023-0012.

Chinnusamy, V., Zhu, J. and Zhu, J.K. (2007) Cold stress regulation of gene expression in plants. *Trends in Plant Science* 12(10), 444–451. DOI: 10.1016/j.tplants.2007.07.002.

Chong, J., Wishart, D.S. and Xia, J. (2019) Using metaboanalyst 4.0 for comprehensive and integrative metabolomics data analysis. *Current Protocols in Bioinformatics* 68(1), e86. DOI: 10.1002/cpbi.86.

Chowdhury, R.H., Eti, F.S., Ahmed, R., Gupta, S.D., Jhan, P.K. *et al.* (2023) Drought-responsive genes in tomato: Meta-analysis of gene expression using machine learning. *Scientific Reports* 13(1), 19374. DOI: 10.1038/s41598-023-45942-2.

Coll, N.S., Epple, P.D.J.L. and Dangl, J.L. (2011) Programmed cell death in the plant immune system. *Cell Death & Differentiation* 18(8), 1247–1256. DOI: 10.1038/cdd.2011.37.

Compant, S., Samad, A., Faist, H. and Sessitsch, A. (2019) A review on the plant microbiome: Ecology, functions, and emerging trends in microbial application. *Journal of Advanced Research* 19, 29–37. DOI: 10.1016/j.jare.2019.03.004.

Debnath, M., Pandey, M. and Bisen, P.S. (2011) An omics approach to understand the plant abiotic stress. *Omics* 15(11), 739–762. DOI: 10.1089/omi.2010.0146.

Denyer, T., Ma, X., Klesen, S., Scacchi, E., Nieselt, K. *et al.* (2019) Spatiotemporal developmental trajectories in the *Arabidopsis* root revealed using high-throughput single-cell RNA sequencing. *Developmental Cell* 48(6), 840–852. DOI: 10.1016/j.devcel.2019.02.022.

Derbyshire, M.C., Batley, J. and Edwards, D. (2022) Use of multiple 'omics techniques to accelerate the breeding of abiotic stress tolerant crops. *Current Plant Biology* 32, 100262. DOI: 10.1016/j.cpb.2022.100262.

Dettmer, K., Aronov, P.A. and Hammock, B.D. (2007) Mass spectrometry-based metabolomics. *Mass Spectrometry Reviews* 26(1), 51–78. DOI: 10.1002/mas.20108.

Dodds, P.N. and Rathjen, J.P. (2010) Plant immunity: Towards an integrated view of plant-pathogen interactions. *Nature Reviews. Genetics* 11(8), 539–548. DOI: 10.1038/nrg2812.

Efroni, I. and Birnbaum, K.D. (2016) The potential of single-cell profiling in plants. *Genome Biology* 17(1), 65. DOI: 10.1186/s13059-016-0931-2.

Fan, B.L., Chen, L.H., Chen, L.L. and Guo, H. (2025) Integrative multi-omics approaches for identifying and characterizing biological elements in crop traits: Current progress and future prospects. *International Journal of Molecular Sciences* 26(4), 1466. DOI: 10.3390/ijms26041466.

Fernie, A.R. and Schauer, N. (2009) Metabolomics-assisted breeding: A viable option for crop improvement. *Trends in Genetics* 25(1), 39–48. DOI: 10.1016/j.tig.2008.10.010.

Fujita, M., Fujita, Y., Noutoshi, Y., Takahashi, F., Narusaka, Y. *et al.* (2006) Crosstalk between abiotic and biotic stress responses: A current view from the points of convergence in the stress signaling networks. *Current Opinion in Plant Biology* 9(4), 436–442. DOI: 10.1016/j.pbi.2006.05.014.

Furbank, R.T. and Tester, M. (2011) Phenomics–technologies to relieve the phenotyping bottleneck. *Trends in Plant Science* 16(12), 635–644.

Gayacharan, Parida, S.K., Mondal, N., Yadav, R., Vishwakarma, H. *et al.* (2023) Mining legume germplasm for genetic gains: An Indian perspective. *Frontiers in Genetics* 14, 996828. DOI: 10.3389/fgene.2023.996828.

Golldack, D., Li, C., Mohan, H. and Probst, N. (2014) Tolerance to drought and salt stress in plants: Unraveling the signaling networks. *Frontiers in Plant Science* 5, 151.

Goodwin, S., McPherson, J.D. and McCombie, W.R. (2016) Coming of age: Ten years of next-generation sequencing technologies. *Nature Reviews Genetics* 17(6), 333–351.

Gu, S., Zhuang, J., Zhang, Z., Chen, W., Xu, H. *et al.* (2022) Multi-omics approach reveals the contribution of *OsSEH1* to rice cold tolerance. *Frontiers in Plant Science* 13, 1110724. DOI: 10.3389/fpls.2022.1110724.

Gupta, A., Rico-Medina, A. and Caño-Delgado, A.I. (2020) The physiology of plant responses to drought. *Science (New York, N.Y.)* 368(6488), 266–269. DOI: 10.1126/science.aaz7614.

Haidar, S., Hooker, J., Lackey, S., Elian, M., Puchacz, N. *et al.* (2024) Harnessing multi-omics strategies and bioinformatics innovations for advancing soybean improvement: A comprehensive review. *Plants (Basel, Switzerland)* 13(19), 2714. DOI: 10.3390/plants13192714.

Han, L., Zhong, W., Qian, J., Jin, M., Tian, P. *et al.* (2023) A multi-omics integrative network map of maize. *Nature Genetics* 55(1), 144–153. DOI: 10.1038/s41588-022-01262-1.

Hasin, Y., Seldin, M. and Lusis, A. (2017) Multi-omics approaches to disease. *Genome Biology* 18(1), 83.

Hina, A., Abbasi, A., Arshad, M., Imtiaz, S., Shahid, S. *et al.* (2024) Using machine learning approaches for multi-omics data analysis: A review. *Biotechnology Advances* 91–121. DOI: 10.1002/9781394209156.

Hirt, H. and Shinozaki, K. (2004) *Plant Responses to Abiotic Stress*. Springer, Berlin.

Ichihashi, Y., Date, Y., Shino, A., Shimizu, T., Shibata, A. *et al.* (2020) Multi-omics analysis on an agroecosystem reveals the significant role of organic nitrogen to increase agricultural crop yield. *Proceedings of the National Academy of Sciences* 117(25), 14552–14560. DOI: 10.1073/pnas.1917259117.

Ijaz, S., Iqbal, J., Abbasi, B.A., Yaseen, T., Rehman, S. *et al.* (2024a) Role of OMICS-based technologies in plant sciences. *OMICs-based Techniques for Global Food Security* 45–66. DOI: 10.1002/9781394209156.

Ijaz, U., Zhao, C., Shabala, S. and Zhou, M. (2024b) Molecular basis of plant-pathogen interactions in the agricultural context. *Biology* 13(6), 421. DOI: 10.3390/biology13060421.

James, D., Rennya, P.R., Mallavarapu, M.D., Panigrahi, R.C. and Patel, H.K. (2021) Integrating pan-omics data in a systems approach for crop improvement: Opportunities and challenges. *Omics Technologies for Sustainable Agriculture and Global Food Security* 215–246.

Jamil, I.N., Remali, J., Azizan, K.A., Nor Muhammad, N.A., Arita, M. *et al.* (2020) Systematic multi-omics integration (MOI) approach in plant systems biology. *Frontiers in Plant Science* 11, 944. DOI: 10.3389/fpls.2020.00944.

Jha, U.C., Naik, Y.D., Priya, M., Nayyar, H., Sofi, P.A. *et al.* (2025) Chickpea (*Cicer arietinum* L.) battling against heat stress: Plant breeding and genomics advances. *Plant Molecular Biology* 115(4), 101. DOI: 10.1007/s11103-025-01628-z.

Jones, J.D.G. and Dangl, J.L. (2006) The plant immune system. *Nature* 444(7117), 323–329. DOI: 10.1038/nature05286.

Jong, C., Yu, Z., Zhang, Y., Choe, K., Uh, S. *et al.* (2024) Multi-omics analysis of a chromosome segment substitution line reveals a new regulation network for soybean seed storage profile. *International Journal of Molecular Sciences* 25(11), 5614. DOI: 10.3390/ijms25115614.

Joshi, B., Singh, S., Tiwari, G.J., Kumar, H., Boopathi, N.M. *et al.* (2023) Genome-wide association study of fiber yield-related traits uncovers the novel genomic regions and candidate genes in Indian upland cotton (*Gossypium hirsutum* L.). *Frontiers in Plant Science* 14, 1252746.

Joshi, S., Patil, S., Shaikh, A., Jamla, M. and Kumar, V. (2024) Modern omics toolbox for producing combined and multifactorial abiotic stress tolerant plants. *Plant Stress* 11, 100301. DOI: 10.1016/j.stress.2023.100301.

Jovic, D., Liang, X., Zeng, H., Lin, L., Xu, F. *et al.* (2022) Single-cell RNA sequencing technologies and applications: A brief overview. *Clinical and Translational Medicine* 12(3), e694. DOI: 10.1002/ctm2.694.

Kosová, K., Vítámvás, P., Urban, M.O., Klíma, M., Roy, A. *et al.* (2015) Biological networks underlying abiotic stress tolerance in temperate crops—A proteomic perspective. *International Journal of Molecular Sciences* 16(9), 20913–20942. DOI: 10.3390/ijms160920913.

Kourelis, J. and van der Hoorn, R.A.L. (2018) Defended to the nines: 25 years of resistance gene cloning identifies nine mechanisms for R protein function. *The Plant Cell* 30(2), 285–299. DOI: 10.1105/tpc.17.00579.

Krajewski, P., Chen, D., Ćwiek, H., van Dijk, A.D.J., Fiorani, F. *et al.* (2015) Towards recommendations for metadata and data handling in plant phenotyping. *Journal of Experimental Botany* 66(18), 5417–5427. DOI: 10.1093/jxb/erv271.

Kulkarni, S.R., Balachandran, S.M., Ulaganathan, K., Balakrishnan, D., Praveen, M. *et al.* (2020) Molecular mapping of QTLs for yield related traits in recombinant inbred line (RIL) population derived from the popular rice hybrid KRH-2 and their validation through SNP genotyping. *Scientific Reports* 10(1), 13695. DOI: 10.1038/s41598-020-70637-3.

Kumar, R., Bohra, A., Pandey, A.K., Pandey, M.K. and Kumar, A. (2017) Metabolomics for plant improvement: Status and prospects. *Frontiers in Plant Science* 8, 1302. DOI: 10.3389/fpls.2017.01302.

Kumuthini, J., Zass, L., Chaouch, M., Gill, Z., Ras, V. *et al.* (2023) Data standardization in the omics field. In: *Genomic Data Sharing*. Academic Press, pp. 137–155.

Luo, F., Yu, Z., Zhou, Q. and Huang, A. (2022) Multi-omics-based discovery of plant signaling molecules. *Metabolites* 12(1), 76. DOI: 10.3390/metabo12010076.

Maghsoudi, Z., Nguyen, H., Tavakkoli, A. and Nguyen, T. (2022) A comprehensive survey of the approaches for pathway analysis using multi-omics data integration. *Briefings in Bioinformatics* 23(6), bbac435. DOI: 10.1093/bib/bbac435.

Mahmood, A., Imran, M., Jamshaid, M.U., Riaz, U., Arif, M. *et al.* (2023) Integrated OMIC approaches for bioenergy crops. In: *Biotechnology and Omics Approaches for Bioenergy Crops*. Springer Nature, Singapore, pp. 77–84. DOI: 10.1007/978-981-99-4954-0_4.

Meng, X. and Zhang, S. (2013) MAPK cascades in plant disease resistance signaling. *Annual Review of Phytopathology* 51(1), 245–266. DOI: 10.1146/annurev-phyto-082712-102314.

Meng, X., Zhao, X., Ding, X., Li, Y., Cao, G. *et al.* (2020) Integrated functional omics analysis of flavonoid-related metabolism in *AtMYB12* transcript factor overexpressed tomato. *Journal of Agricultural and Food Chemistry* 68(24), 6776–6787. DOI: 10.1021/acs.jafc.0c01894.

Miao, B.B., Dong, W., Gu, Y.X., Han, Z.F., Luo, X. *et al.* (2023) OmicsSuite: A customized and pipelined suite for analysis and visualization of multi-omics big data. *Horticulture Research* 10(11), uhad195. DOI: 10.1093/hr/uhad195.

Michael, T.P. and VanBuren, R. (2020) Building near-complete plant genomes. *Current Opinion in Plant Biology* 54, 26–33. DOI: 10.1016/j.pbi.2019.12.009.

Misra, B.B., Langefeld, C., Olivier, M. and Cox, L.A. (2019) Integrated omics: Tools, advances and future approaches. *Journal of Molecular Endocrinology* 62(1), R21–R45. DOI: 10.1530/JME-18-0055.

Mochida, K. and Shinozaki, K. (2011) Advances in omics and bioinformatics tools for systems analyses of plant functions. *Plant and Cell Physiology* 52(12), 2017–2038. DOI: 10.1093/pcp/pcr153.

Mueller, H.M., Franzisky, B.L., Messerer, M., Du, B., Lux, T. *et al.* (2024) Integrative multi-omics analyses of date palm (*Phoenix dactylifera*) roots and leaves reveal how the halophyte land plant copes with sea water. *The Plant Genome* 17(1), e20372. DOI: 10.1002/tpg2.20372.

Muthuramalingam, P., Jeyasri, R., Rakkammal, K., Satish, L., Shamili, S. *et al.* (2022) Multi-omics and integrative approach towards understanding salinity tolerance in rice: A review. *Biology* 11(7), 1022. DOI: 10.3390/biology11071022.

Nasar, S., Muhammad, M., Siddiqi, E.H., Majeed, M., Ameen, N. *et al.* (2024) OMICS-based knowledge for achieving food and nutritional security. In: *OMICs-based Techniques for Global Food Security*. pp. 67–90.

Nelms, B. and Walbot, V. (2019) Defining the developmental program leading to meiosis in maize. *Science (New York, N.Y)* 364(6435), 52–56. DOI: 10.1126/science.aav6428.

Nobori, T. (2025) Exploring the untapped potential of single-cell and spatial omics in plant biology. *The New Phytologist* 247(3), 1098–1116. DOI: 10.1111/nph.70220.

Obata, T. and Fernie, A.R. (2012) The use of metabolomics to dissect plant responses to abiotic stresses. *Cellular and Molecular Life Sciences* 69(19), 3225–3243. DOI: 10.1007/s00018-012-1091-5.

Papadaki, E., Kakkos, I., Vlamos, P., Petropoulou, O., Miloulis, S.T. *et al.* (2024) Recent web platforms for multi-omics integration unlocking biological complexity. *Applied Sciences* 15(1), 329. DOI: 10.3390/app15010329.

Patel, R., Menon, J., Kumar, S., Nóbrega, M.B.M., Patel, D.A. *et al.* (2024) Modern day breeding approaches for improvement of castor. *Heliyon* 10(5), e27048. DOI: 10.1016/j.heliyon.2024.e27048.

Pieterse, C.M.J., Van der Does, D., Zamioudis, C., Leon-Reyes, A. and Van Wees, S.C.M. (2012) Hormonal modulation of plant immunity. *Annual Review of Cell and Developmental Biology* 28(1), 489–521. DOI: 10.1146/annurev-cellbio-092910-154055.

Pommier, C., Coppens, F., Ćwiek-Kupczyńska, H., Faria, D., Beier, S. *et al.* (2023) Plant science data integration, from building community standards to defining a consistent data lifecycle. *Towards Responsible Plant Data Linkage: Data Challenges for Agricultural Research and Development* 149–160.

Rajasundaram, D. and Selbig, J. (2016) More effort - more results: Recent advances in integrative "omics" data analysis. *Current Opinion in Plant Biology* 30, 57–61. DOI: 10.1016/j.pbi.2015.12.010.

Rauniyar, N. and Yates, J.R., III (2014) Isobaric labeling-based relative quantification in shotgun proteomics. *Journal of Proteome Research* 13(12), 5293–5309. DOI: 10.1021/pr500880b.

Raza, A., Tabassum, J., Kudapa, H. and Varshney, R.K. (2021) Can omics deliver temperature resilient ready-to-grow crops? *Critical Reviews in Biotechnology* 41(8), 1209–1232. DOI: 10.1080/07388551.2021.1898332.

Rhaman, M.S., Ali, M., Ye, W. and Li, B. (2024) Opportunities and challenges in advancing plant research with single-cell omics. *Genomics, Proteomics & Bioinformatics* 22(2), qzae026. DOI: 10.1093/gpbjnl/qzae026.

Rich-Griffin, C., Stechemesser, A., Finch, J., Lucas, E., Ott, S. *et al.* (2020) Single-cell transcriptomics: A high-resolution avenue for plant functional genomics. *Trends in Plant Science* 25(2), 186–197. DOI: 10.1016/j.tplants.2019.10.008.

Roychowdhury, R., Das, S.P., Gupta, A., Parihar, P., Chandrasekhar, K. *et al.* (2023) Multi-omics pipeline and omics-integration approach to decipher plant's abiotic stress tolerance responses. *Genes* 14(6), 1281.

Ryu, K.H., Huang, L., Kang, H.M. and Schiefelbein, J. (2019) Single-cell RNA sequencing resolves molecular relationships among individual plant cells. *Plant Physiology* 179(4), 1444–1456. DOI: 10.1104/pp.18.01482.

Sawada, Y., Akiyama, K., Sakata, A., Kuwahara, A., Otsuki, H. *et al*. (2009) Widely targeted metabolomics based on large-scale MS/MS data for elucidating metabolite accumulation patterns in plants. *Plant and Cell Physiology* 50(1), 37–47. DOI: 10.1093/pcp/pcn183.

Shaar-Moshe, L., Hübner, S. and Peleg, Z. (2015) Identification of conserved drought-adaptive genes using a cross-species meta-analysis approach. *BMC Plant Biology* 15(1), 111. DOI: 10.1186/s12870-015-0493-6.

Shannon, P., Markiel, A., Ozier, O., Baliga, N.S., Wang, J.T. *et al*. (2003) Cytoscape: A software environment for integrated models of biomolecular interaction networks. *Genome Research* 13(11), 2498–2504. DOI: 10.1101/gr.1239303.

Shaw, R., Tian, X. and Xu, J. (2021) Single-cell transcriptome analysis in plants: Advances and challenges. *Molecular Plant* 14(1), 115–126. DOI: 10.1016/j.molp.2020.10.012.

Shen, Z. and Chen, M. (2022) Deciphering novel transcriptional regulators of soybean hypocotyl elongation based on gene co-expression network analysis. *Frontiers in Plant Science* 13, 837130.

Sheoran, S., Kaur, Y., Kumar, S., Shukla, S., Rakshit, S. *et al*. (2022) Recent advances for drought stress tolerance in maize (*Zea mays* L.): Present status and future prospects. *Frontiers in Plant Science* 13, 872566.

Singh, A.K., Kumar, K., Singh, L.B., Ahuja, A., Choudhury, S. *et al*. (2020) Omics technology: Role and future in providing biotic and abiotic stress tolerance to plants. In: *Microbes and Signaling Biomolecules against Plant Stress: Strategies of Plant-Microbe Relationships for Better Survival*. Springer Singapore, Singapore, pp. 151–168. DOI: 10.1007/978-981-15-7094-0_8.

Sirangelo, T.M. (2019) Multi-omics approaches in the study of plants. *International Journal of Advanced Research in Botany* 5, 1–7.

Springer, N.M. and Schmitz, R.J. (2017) Exploiting induced and natural epigenetic variation for crop improvement. *Nature Reviews. Genetics* 18(9), 563–575. DOI: 10.1038/nrg.2017.45.

Subramanian, I., Verma, S., Kumar, S., Jere, A. and Anamika, K. (2020) Multi-omics data integration, interpretation, and its application. *Bioinformatics and Biology Insights* 14, 1177932219899051. DOI: 10.1177/1177932219899051.

Tang, S., Zhao, H., Lu, S., Yu, L., Zhang, G. *et al*. (2021) Genome- and transcriptome-wide association studies provide insights into the genetic basis of natural variation of seed oil content in *Brassica napus*. *Molecular Plant* 14(3), 470–487. DOI: 10.1016/j.molp.2020.12.003.

Tenorio Berrío, R. and Dubois, M. (2024) Single-cell transcriptomics reveals heterogeneity in plant responses to the environment: A focus on biotic and abiotic interactions. *Journal of Experimental Botany* 75(17), 5188–5203. DOI: 10.1093/jxb/erae107.

Tian, C., Du, Q., Xu, M., Du, F. and Jiao, Y. (2020) Single-nucleus RNA-seq resolves spatiotemporal developmental trajectories in the tomato shoot apex. *BioRxiv 2020-09*.

Underwood, W. (2012) The plant cell wall: A dynamic barrier against pathogen invasion. *Frontiers in Plant Science* 3, 85. DOI: 10.3389/fpls.2012.00085.

Van Bel, M., Proost, S., Wischnitzki, E., Movahedi, S., Scheerlinck, C. *et al*. (2012) Dissecting plant genomes with the PLAZA comparative genomics platform. *Plant Physiology* 158(2), 590–600. DOI: 10.1104/pp.111.189514.

Van Bel, M., Silvestri, F., Weitz, E.M., Kreft, L., Botzki, A. *et al*. (2022) PLAZA 5.0: Extending the scope and power of comparative and functional genomics in plants. *Nucleic Acids Research* 50(D1), D1468–D1474. DOI: 10.1093/nar/gkab1024.

van Loon, L.C., Rep, M. and Pieterse, C.M.J. (2006) Significance of inducible defense-related proteins in infected plants. *Annual Review of Phytopathology* 44(1), 135–162. DOI: 10.1146/annurev.phyto.44.070505.143425.

Vandereyken, K., Sifrim, A., Thienpont, B. and Voet, T. (2023) Methods and applications for single-cell and spatial multi-omics. *Nature Reviews. Genetics* 24(8), 494–515. DOI: 10.1038/s41576-023-00580-2.

Varadharajan, V., Rajendran, R., Muthuramalingam, P., Runthala, A., Madhesh, V. *et al*. (2025) Multi-omics approaches against abiotic and biotic stress-A review. *Plants (Basel, Switzerland)* 14(6), 865. DOI: 10.3390/plants14060865.

Varshney, R.K., Bohra, A., Yu, J., Graner, A., Zhang, Q. *et al*. (2021) Designing future crops: Genomics-assisted breeding comes of age. *Trends in Plant Science* 26(6), 631–649. DOI: 10.1016/j.tplants.2021.03.010.

Vennapusa, A.R., Nimmakayala, P., Zaman-Allah, M.A. and Ratnakumar, P. (2023) Editorial: Physiological, molecular and genetic perspectives of environmental stress response in plants. *Frontiers in Plant Science* 14, 1213762. DOI: 10.3389/fpls.2023.1213762.

Vives-Peris, V., de Ollas, C., Gómez-Cadenas, A. and Pérez-Clemente, R.M. (2020) Root exudates: From plant to rhizosphere and beyond. *Plant Cell Reports* 39(1), 3–17. DOI: 10.1007/s00299-019-02447-5.

Wang, Y., Huan, Q., Li, K. and Qian, W. (2021) Single-cell transcriptome atlas of the leaf and root of rice seedlings. *Journal of Genetics and Genomics = Yi Chuan Xue Bao* 48(10), 881–898. DOI: 10.1016/j.jgg.2021.06.001.

Wang, Z., Gerstein, M. and Snyder, M. (2009) RNA-seq: A revolutionary tool for transcriptomics. *Nature Reviews Genetics* 10(1), 57–63. DOI: 10.1038/nrg2484.

Wang, Z., Yan, L., Chen, Y., Wang, X., Huai, D. *et al.* (2022) Detection of a major QTL and development of KASP markers for seed weight by combining QTL-seq, QTL-mapping and RNA-seq in peanut. *Theoretical and Applied Genetics* 135(5), 1779–1795. DOI: 10.1007/s00122-022-04069-0.

Weckwerth, W. (2011) Green systems biology — from single genomes, proteomes and metabolomes to ecosystems research and biotechnology. *Journal of Proteomics* 75(1), 284–305. DOI: 10.1016/j.jprot.2011.07.010.

Weckwerth, W., Ghatak, A., Bellaire, A., Chaturvedi, P. and Varshney, R.K. (2020) PANOMICS meets germplasm. *Plant Biotechnology Journal* 18(7), 1507–1525. DOI: 10.1111/pbi.13372.

Wu, W., Dong, X., Chen, G., Lin, Z., Chi, W. *et al.* (2024) The elite haplotype OsGATA8 coordinates nitrogen uptake and productive tiller formation in rice. *Nature Genetics* 56(7), 1516–1526. DOI: 10.1038/s41588-024-01795-7.

Xing, J., Zhang, J., Wang, Y., Wei, X., Yin, Z. *et al.* (2024) Mining genic resources regulating nitrogen-use efficiency based on integrative biological analyses and their breeding applications in maize and other crops. *The Plant Journal* 117(4), 1148–1164. DOI: 10.1111/tpj.16550.

Yang, L., Yang, Y., Huang, L., Cui, X. and Liu, Y. (2023) From single- to multi-omics: Future research trends in medicinal plants. *Briefings in Bioinformatics* 24(1). DOI: 10.1093/bib/bbac485.

Yang, Y., Saand, M.A., Huang, L., Abdelaal, W.B., Zhang, J. *et al.* (2021) Applications of multi-omics technologies for crop improvement. *Frontiers in Plant Science* 12, 563953. DOI: 10.3389/fpls.2021.563953.

Ye, W., Yu, H., Jian, Y., Zeng, J., Ji, R. *et al.* (2017) A salivary EF-hand calcium-binding protein of the brown planthopper nilaparvata lugens functions as an effector for defense responses in rice. *Scientific Reports* 7(1), 40498.

Zhang, H., Lang, Z. and Zhu, J.K. (2018) Dynamics and function of DNA methylation in plants. *Nature Reviews Molecular Cell Biology* 19(8), 489–506.

Zhang, T.Q., Chen, Y., Liu, Y., Lin, W.H. and Wang, J.W. (2021) Single-cell transcriptome atlas and chromatin accessibility landscape reveal differentiation trajectories in the rice root. *Nature Communications* 12(1), 2053. DOI: 10.1038/s41467-021-22352-4.

Zhang, Y., Fonslow, B.R., Shan, B., Baek, M.-C. and Yates, J.R. (2013) Protein analysis by shotgun/bottom-up proteomics. *Chemical Reviews* 113(4), 2343–2394. DOI: 10.1021/cr3003533.

Zhao, E., Dong, L., Zhao, H., Zhang, H., Zhang, T. *et al.* (2023) A relationship prediction method for *Magnaporthe oryzae*-rice multi-omics data based on WGCNA and graph autoencoder. *Journal of Fungi (Basel, Switzerland)* 9(10), 1007. DOI: 10.3390/jof9101007.

Zhou, T., Yue, C.P., Liu, Y., Zhang, T.Y., Huang, J.Y. *et al.* (2021) Multiomics reveal pivotal roles of sodium translocation and compartmentation in regulating salinity resistance in allotetraploid rapeseed. *Journal of Experimental Botany* 72(15), 5687–5708. DOI: 10.1093/jxb/erab215.

Zhu, J.K. (2016) Abiotic stress signaling and responses in plants. *Cell* 167(2), 313–324. DOI: 10.1016/j.cell.2016.08.029.

2 Spatial Omics in Plants: Current Achievements, Challenges, and Future Directions

Mani Manoj[1], Subramaniam Kalidass[2], Karuppiah Ponmurugan[3], Paramashivan Bhavani[2], Karuppasamy Kaviarasan[2], Asirvatham Alwin Robert[4]*, Jeyabal Philomenathan Antony Prabhu[1] and Arumugam Vijaya Anand[1]*

[1]Department of Human Genetics and Molecular Biology, Bharathiar University, Coimbatore, India; [2]Department of Animal Science, Manonmaniam Sundaranar University, Tirunelveli, India; [3]Department of Botany and Microbiology, College of Science, King Saud University, Riyadh, Saudi Arabia; [4]Department of Endocrinology and Diabetes, Prince Sultan Military Medical City, Ministry of Defense, Riyadh, Kingdom of Saudi Arabia

2.1 Introduction to Spatial Omics in Plant Systems

Spatial omics is a novel and multidisciplinary field that combines molecular profiling technologies with spatial data, thus allowing visualization and analysis of biomolecular data in its native spatial environment. The technique combines advanced techniques such as imaging, sequencing, mass spectrometry, and bioinformatics to generate spatially resolved molecular maps over a range of biological scales from subcellular organization to whole organisms (Alexandrov *et al.*, 2023). Spatial transcriptomics and spatial proteomics are major technologies of spatial omics and are particularly useful in interrogating complex biological systems like the tumor microenvironment, allowing insights beyond the resolution of standard bulk and single-cell omics technologies (Xie *et al.*, 2023). A combination of spatial data with molecular data allows comprehensive tissue microenvironmental and cellular network analysis, hence allowing correlation of spatial architecture with transcriptomic and proteomic expression. Programs such as Aquila allow management, visualization, and analysis of spatial omics data using advanced computational tools (Zheng *et al.*, 2023). A combination of spatial omics with imaging-based artificial intelligence (AI) also contributes to the extraction of morphological features complementary to molecular data, thus allowing tissue characterization and prediction of gene expression based on morphological indicators (Chelebian *et al.*, 2024). Pipelines like Smoother have been engineered to overcome challenges posed by spatial data analysis by incorporating spatial dependencies in imputation and cell-type deconvolution protocols, thus boosting analytical accuracy (Su *et al.*, 2023a).

In plant biology, spatial omics – more precisely, spatial transcriptomics – has been a

*Corresponding author: avahgmb@buc.edu.in; aalwinrobert@gmail.com

© CAB International 2026. *Plant Spatial Biology* (ed. Chen)
DOI: 10.1079/9781836994572.0002

pioneering strategy, providing unprecedented views of the spatial organization and functional dynamics of plant tissues. This technology allows for *in situ* probing of gene expression and metabolic events at the single-cell level, thus overcoming the limitations of conventional transcriptomics that mask cellular heterogeneity through bulk averaging (Lv *et al.*, 2024). Spatial transcriptomics has played a central role in the discovery of heterogeneous cell types, reconstruction of developmental pathways, and unscrambling of intercellular communication networks, thus advancing our understanding of plant development, differentiation, and environmental stress responses (Yin *et al.*, 2023). Integration of spatial transcriptomics with other omics strategies (e.g. metabolomics) has also made it possible to conduct a full-fledged analysis of plant tissues, disclosing specialized metabolic pathways and complex gene regulatory networks, especially for medicinal plants (Wang *et al.*, 2024). Despite data integration bottlenecks, high-throughput processing, and technical standardization, spatial omics has tremendous potential for streamlining the sustainable production of bioactive compounds and cracking cell-specific dynamics of biosynthesis (Zhang *et al.*, 2024).

The shift from traditional omics approaches to spatially resolved methods is a paradigm shift across the whole field of biological research. Traditional omics approaches, such as genomics and transcriptomics, were primarily focused on bulk analyses, averaging molecular signals over cell populations at the expense of losing cellular and spatial resolution. While single-cell technologies allowed the resolution of cellular heterogeneity, they did not preserve the spatial organization of cells within tissues (Williams *et al.*, 2022). Spatial omics addresses this deficiency by allowing *in situ* probing of biomolecules like RNA and proteins while preserving anatomical and histological contexts (Kulasinghe *et al.*, 2025). This has allowed high-resolution spatial mapping of gene expression and protein localization, providing valuable information regarding tissue structure and cellular interactions across areas of oncology, neuroscience, and developmental biology (Zhuang, 2021).

Despite these improvements, challenges remain in key areas. These include the need for standardized procedures, scalable computation platforms, and greater accessibility of spatial omics technologies to a broader range of research environments (Bressan *et al.*, 2023). In the plant biology discipline, spatial omics has proven especially useful in revealing how plants respond to multifactorial environmental stresses – drought and herbivory – through enabling spatially resolved analysis of metabolic activity and gene expression (Hu *et al.*, 2024a). When combined with single-cell approaches, spatial omics can distinguish between heterogeneity manifest in plant cell populations and reveal cell-specific adaptive responses, which are of critical concern in understanding the mechanisms of plant resilience (Shafi *et al.*, 2024). Spatial omics also facilitates the discovery of new genes and regulatory networks that can potentially be exploited for crop improvement and hence aid in the enhancement of yield, quality, and resilience to environmental perturbations. Synergy of spatial omics with genomics, proteomics, and metabolomics is a synergetic data integration that provides an integrated strategy for understanding the molecular basis of plant development and stress adaptation (Sahoo *et al.*, 2020).

2.2 Spatial Omics Technologies: Tools Driving Innovation

2.2.1 Spatial transcriptomics

Spatial transcriptomics is currently an emerging tool in plant biology, enabling spatially resolved descriptions of tissue architecture and gene expression (Fig. 2.1). Several new platforms, such as Slide-seq, Visium, Stereo-seq, and MERFISH, have advanced the field by enabling high-resolution, spatially contextualized transcriptomic profiling. Slide-seq is a prime example of such an emerging spatial transcriptomics tool, enabling high-resolution mapping of gene expression without sacrificing tissue architecture integrity. Significantly, the HiFi-Slide variants enable submicron resolution, thus enabling detailed analysis of RNA transcripts in fresh-frozen plant tissues (Xu *et al.*, 2023). In contrast to traditional transcriptome approaches that average gene expression across cell populations, Slide-seq provides cellular-level precision,

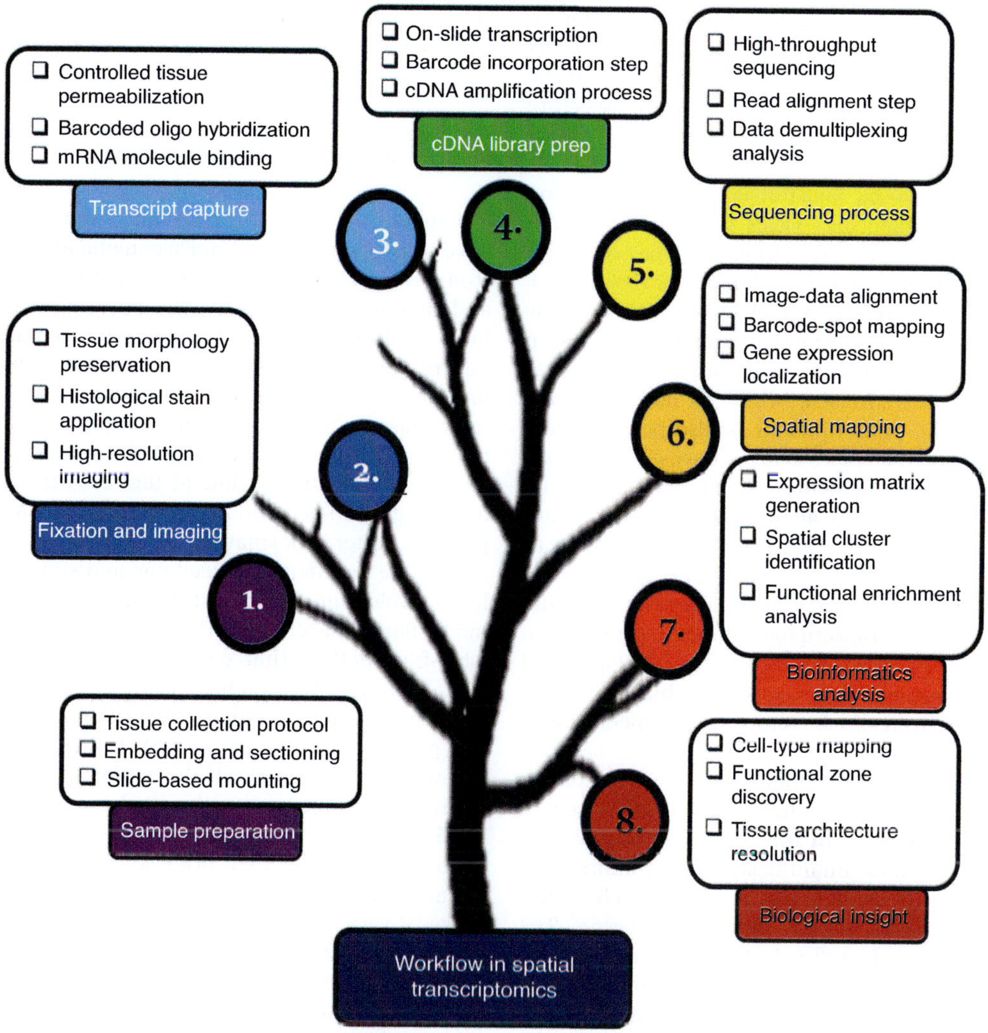

Fig. 2.1. Workflow in spatial transcriptomics.

thereby uncovering gene expression variability, which is crucial for understanding plant growth and physiological processes (Lv *et al.*, 2024). Recent advances in computational methods, specifically the application of spectral graph theory, have enhanced the detection of spatially variable genes by converting coordinate systems to tissue morphology and thus enhancing detection and interpretation of spatial patterns (Nicol *et al.*, 2024).

By connecting gene expression and histological context through spatial transcriptomics, the Visium Spatial Gene Expression technology significantly improves the integration of genomic and transcriptomic investigations. The Visium platform combines high-resolution imaging and RNA sequencing, allowing scientists to generate comprehensive spatial gene expression maps within intact plant tissues. The analysis software, such as VistoSeg, enables the extraction and visualization of image-derived features, which can be combined with transcriptomic information through programming interfaces like R and Python (Tippani *et al.*, 2023). Additionally, pipelines such as ENACT improve cell-type classification accuracy, thus enriching our understanding of cellular microenvironments and tissue-specific expression patterns (Kamel *et al.*, 2025). These integrations enable more informative inferences about the structure and function of plant tissues (Radkevich *et al.*, 2024).

Stereo-seq (SpaTial Enhanced REsolution Omics-sequencing) yields high spatial resolution and extensive transcriptome coverage, overcoming some of the limitations of single-cell RNA sequencing (scRNA-seq), which typically lacks spatial information. Stereo-seq has been successfully used in model organisms such as *Arabidopsis thaliana*, where it has uncovered intricate transcriptional landscapes and cell-type-specific expression patterns (Xia *et al.*, 2022). Development of the Stereo-seq Analysis Workflow (SAW) allows scalable and efficient analysis of large-scale spatial data, an important component of integrative multi-omics studies (Gong *et al.*, 2024). By integrating spatial data with transcriptomic and genomic data, Stereo-seq opens up our view of developmental programs and stress responses, thus facilitating novel strategies for crop improvement (Bawa *et al.*, 2024).

MERFISH (multiplexed error-robust fluorescence *in situ* hybridization) takes spatial transcriptomics to the next level by allowing simultaneous quantification of hundreds to thousands of RNA species at single-cell resolution without losing spatial information. The technology is well suited to detecting intricate gene expression profiles in intact plant tissues (Emanuel and He, 2021). For correcting noise and biases, a Bayesian statistical framework has been suggested in transcript detection, thus improving the quantitative accuracy of expression profiling (Köster *et al.*, 2019). Moreover, recent breakthroughs in three-dimensional (3D) imaging have expanded MERFISH applications to thicker plant tissues to examine spatial gene expression in 3D tissue architecture (Fang *et al.*, 2024). These spatial transcriptomics tools, such as Slide-seq, Visium, Stereo-seq, and MERFISH, collectively are transforming plant molecular biology by making accurate, spatially-resolved measurements of gene expression possible. Their connection to computation and other omics technologies offers an integrated understanding of plant tissue architecture, developmental biology, and environmental interfaces that provides new frontiers for basic research and applied agriculture innovation.

2.2.2 Spatial proteomics

Advances in spatial proteomics methods have made significant improvements in the study of protein localization, functionality, and dynamics in plant tissues by providing high-resolution, context-dependent molecular information. Compared to conventional bulk proteomic methods, novel methods such as matrix-assisted laser desorption/ionization mass spectrometry (MALDI-MS) imaging, imaging cytometry by time of flight (CyTOF), and imaging co-detection by indexing (CODEX) provide better information on protein and metabolite distribution at cellular and subcellular levels, and significantly augment plant molecular research. MALDI-MS imaging is a label-free method that enables direct spatial visualization of proteins, peptides, lipids, and metabolites in plant tissues. One of the greatest advantages of MALDI-MS over conventional proteomics is the milder requirements for sample preparation and the ability to discriminate among multiple molecules simultaneously without prior knowledge of the analytes (Susniak *et al.*, 2020). This feature makes it very valuable for untargeted exploratory studies. MALDI-MS produces qualitative and semi-quantitative data, which makes it easier to identify and locate metabolites and putative biomarkers in various plant tissues (Milosavic, 2015). Versatility in terms of diverse plant organs and environments, such as situations with biotic and abiotic stress, augments its use in unveiling plant responses and metabolic pathways *in situ* (Mehta and Silva, 2015). Through the combination of spatial and molecular data, MALDI-MS imaging augments the proteomics workflow and improves the holistic understanding of plant physiology and pathology at a systems level (Peukert *et al.*, 2012).

Imaging CyTOF combines time-of-flight mass spectrometry and metal-tagged antibodies to facilitate high-dimensional proteomic analysis with spatial resolution at submicron resolutions. In contrast to conventional methods involving homogenized samples and average bulk measurements, imaging CyTOF preserves tissue architecture, which allows for the precise localization of proteins in their native microenvironments (Gnann *et al.*, 2021). This is needed for the study of numerous processes such as vesicular trafficking, signal transduction, and intercellular communication in plant systems (Zhang *et al.*, 2023b). Imaging CyTOF also facilitates the multiplexed detection of protein targets and post-translational modifications in a single tissue section, which improves the study of complex regulatory networks and protein–protein interactions (Sang *et al.*, 2025). The ability to map multiple proteins and their functional states simultaneously in intact plant tissues greatly improves our understanding of dynamic proteomic landscapes, which allows for the identification of regulatory hubs and mechanisms involved in stress responses (Wang *et al.*, 2023).

CODEX imaging is a very multiplexed spatial proteomics platform that offers subcellular level resolution and single-cell level sensitivity. In contrast to bulk proteomics methods that average protein signals across cell populations, CODEX imaging enables detailed examination of protein expression and localization in single cells of plants (Lundberg and Borner, 2019). The capability of high-resolution imaging is necessary to record dynamic biological processes such as photosynthesis, cell differentiation, and environmental response. This innovative approach uses repeated cycles of fluorescence detection and DNA-conjugated antibodies to image more than 40 proteins per sample simultaneously, while maintaining spatial context (Borner, 2020). In plant systems, this capability enables the study of protein heterogeneity, intracellular compartmentalization, and spatial signaling events that are vital for growth and responsiveness (Shahzad *et al.*, 2016). CODEX imaging thus enables researchers to study the cellular architecture of plant tissues at unparalleled resolution, thereby fueling basic and applied research in plant science. MALDI-MS imaging, imaging CyTOF, and CODEX imaging are potent tools for spatial proteomics in plant science. These methods complement the limitations of traditional bulk analyses by offering high-resolution, context-aware, and multiplexed detection of proteins and metabolites.

2.2.3 Spatial metabolomics

Spatial metabolomics, a discipline aimed at elucidating the localization and dynamics of metabolites in biological tissues, has been significantly propelled by advances in mass spectrometry imaging (MSI) technologies. Recent advances, such as DESI-MSI, LAESI, SIMS, and nano-DESI, have facilitated high-resolution, *in situ* metabolic profiling, which is essential to unravel the biochemical complexity of plant systems. These methods enable scientists to investigate metabolic pathways, identify low-abundance bioactive compounds, and visualize metabolite distribution under physiological and stress conditions with ever-growing accuracy. Desorption electrospray ionization MSI (DESI-MSI) has proven to be a versatile ambient ionization method for high-resolution, label-free imaging of plant metabolites directly from tissue sections. It supports spatial mapping of multiple metabolite classes with minimal sample manipulation, yielding real-time information on metabolic distribution and biosynthesis. Importantly, applications in *Cyclocarya paliurus* and *Panax quinquefolius* have proven their capacity to disclose tissue-specific localization patterns of flavonoids and ginsenosides, attesting to their importance in phytochemical and pharmacological research (Xi *et al.*, 2024). The pairing of DESI-MSI with optimized sample preparation protocols has further amplified the visualization of secondary metabolites, especially in traditional Chinese medicinal plants, to support quality control and compound standardization (Xu *et al.*, 2022; Zemaitis *et al.*, 2023; Sembiyaa *et al.*, 2025).

Both laser ablation electrospray ionization (LAESI) and secondary ion mass spectrometry (SIMS) offer distinctive advantages in the application of spatial metabolomics, but with distinct advantages. LAESI is especially advantageous for *in situ* metabolic profiling of intact, fresh, living plant tissues with a resolution of around 100 μm without the necessity of complex sample preparation (Etalo *et al.*, 2015). This feature makes it extremely useful for intact plant organ analysis and for tracking metabolite dynamics under natural physiological conditions. Its sensitivity, however, might be found to be poor in detecting metabolites that occur in very low concentrations (Etalo *et al.*, 2018). SIMS, in contrast, has excellent sensitivity and the ability to detect isotopic and elemental distributions at parts-per-billion (ppb) level (Audinot *et al.*, 2022). SIMS is especially advantageous for subcellular imaging and nano-scale mapping of both metabolic and elemental compositions. However, despite its advantages, SIMS typically demands stricter sample preparation and often sacrifices spatial resolution in softer biological matrices than in the case of LAESI (Bartels and Svatoš, 2015). Therefore, the choice between the application of LAESI or SIMS typically depends on the particular requirements for sensitivity or resolution and the corresponding biological context.

Nanospray desorption electrospray ionization (nano-DESI) is one of the latest advances in ambient mass spectrometry imaging that offers higher spatial resolution and sensitivity to enable *in situ* metabolomic analysis of plant tissues. The technique allows soft and local desorption of analytes from the plant surface with minimal disturbance of the sample. Nano-DESI has been shown to have superior ability in the analysis of chemically labile or low-abundance analytes, with studies showing successful analysis of less than 10 ng of plant material (Cha *et al.*, 2022). Its high sensitivity also allows it to be used to investigate metabolite heterogeneity in different tissue regions. The compatibility of nano-DESI with real-time, label-free imaging also allows researchers to define the spatial arrangement and temporal dynamics in major metabolites under different environmental conditions (Wang *et al.*, 2025). This method is the most beneficial for the analysis of dynamic metabolic profiles and identification of bioactive molecules with major roles in plant stress responses and adaptation mechanisms (Marinello *et al.*, 2022).

2.2.4 Emerging spatial epigenomics tools

Spatial epigenomics has, in recent years, emerged as a revolutionary method for deciphering the intricate regulatory maps of plant genomes. By combining high-resolution spatial mapping and epigenetic profiling, these new-generation technologies yield unprecedented insights into how chromatin structure and modification influence gene regulation in various cellular contexts. Two of the most promising tools in this domain – spatial assay for transposase-accessible chromatin with sequencing (spATAC-seq) and spatial CUT&Tag – are transforming our understanding of plant developmental biology, environmental responses, and the regulation of specialized metabolism. The spATAC-seq tool enables the genome-wide profiling of chromatin accessibility in specific tissue and cell-type contexts, with fine-scale resolution of regulatory activity. In plants, this tool has served to map epigenetic landscapes that underlie development and adaptive response. By identifying open chromatin regions *in situ*, spATAC-seq reveals active regulatory elements, such as enhancers and promoters, that are differentially used across plant tissues (Fan *et al.*, 2021). This ability to spatially resolve chromatin accessibility enables the uncovering of epigenetic mechanisms that orchestrate gene expression in response to environmental cues, critical to deciphering plant stress responses as well as phenotypic plasticity. Furthermore, spATAC-seq enables the discovery of cell-type-specific regulatory sequences linked to specialized metabolism, suggesting potential targets for epigenome-based crop improvement (Conneely *et al.*, 2022). Its application in agricultural biotechnology promises novel routes to enhancing plant resilience, productivity, and metabolic engineering.

Spatial CUT&Tag allows cell-type-specific and high-resolution profiling of transcription factor (TF) binding sites and histone modifications. The technique allows *in situ* examination of epigenomic marks like TF occupancy and H3K27me3 and H3K4me3 without sonication and large amounts of input material, thus making it highly adaptable for plant tissues of complex architectures and with limited sample availability (Ouyang and Li, 2023). Spatial CUT&Tag has revolutionized the capability to track chromatin changes that regulate gene repression and activation during developmental transitions and in response to environmental stresses. By facilitating the production of cell-resolved epigenomic data, the technique allows the resolution of the regulatory networks underlying specialized metabolic processes, tissue differentiation, and stress signaling. Moreover, its high efficiency and low background noise make it well-equipped for the precise identification of cis-regulatory elements, thus obtaining crucial insights into the 3D structure of the plant genome (Tao *et al.*, 2022).

2.2.5 Multi-omics-ready platforms and their adaptation for plant tissues

The integration of multi-omics technologies has revolutionized plant biology by allowing one to gain a systemic understanding of molecular networks regulating plant growth, development, and stress response. Modern multi-omics platforms provide the analytical depth required to investigate the genome, transcriptome, proteome, and metabolome simultaneously, thus unmasking intricate regulatory networks and biosynthetic paths, especially in medicinally and agriculturally important plants. Multi-omics approaches have proven to be highly effective in investigating the biosynthesis of secondary metabolites in medicinal plants. As Pandita *et al.* (2021) demonstrated, the integration of genomics, transcriptomics, proteomics, and metabolomics discovery has identified new regulatory genes and biosynthetic enzymes that are central in the production of therapeutically valuable phytocompounds. These findings not only improve our understanding of plant metabolism but also enable metabolic engineering and synthetic biology approaches towards enhancing the production of such compounds (Pandita *et al.*, 2021).

The PlantæViz platform represents a significant advancement in terms of ease of access and usefulness of plant omics information. As a comprehensive web-based information source, PlantæViz supports integration, visualization, and analysis of multi-omics data in a wide variety of plant species. Its functions include gene co-expression networks and transcriptome and metabolome integration, supporting researchers in making hypotheses and characterizing complex biological processes with better accuracy (Santiago *et al.*, 2024). Democratization of omics information supports better cross-species comparison and functional annotation in model and non-model species. Single-cell multi-omics has further improved understanding of plant tissues by providing the ability to map cellular heterogeneity and intercellular communication at high resolution. The single-cell approach is of particular importance in horticultural and developmental biology, where the ability to find cell-specific expression patterns and metabolic functions is critical. Zhang *et al.* highlight the revolutionary potential of single-cell approaches in horticulture through functional activity mapping at the cellular level (Zhang *et al.*, 2023a).

In addition, Yu *et al.* (2023b) highlight the importance of spatial and single-cell multi-omics in the investigation of tissue-specific gene expression and metabolite distribution, which offer deep insights into differentiation mechanisms and developmental gradients. Such approaches enable a more precise comprehension of plant ontogeny and regulatory platforms that control tissue specialization (Yu *et al.*, 2023b). Multi-omics approaches have also been instrumental in uncovering how plants adapt to abiotic and biotic stressors. MS and liquid chromatography–mass spectrometry (LC–MS) techniques play a crucial role in monitoring dynamic alterations in the proteome and metabolome during stress. According to Gupta *et al.*, these technologies help to unveil important physiological and biochemical pathways involved in stress adaptation, thus facilitating the determination of stress biomarkers and the creation of resilient crop lines (Gupta *et al.*, 2023).

2.3 Advanced Applications in Plant Science

The progress in multi-omics-ready platforms specific to plant tissues has been significantly boosted by the integration of various omics technologies, thus offering unprecedented insights into the complex biological processes of plants. The multi-omics approaches consist of proteomics, genomics, transcriptomics, and metabolomics, which offer a holistic understanding of cellular and molecular networks. The platforms described by Pandita *et al.* have enabled researchers to perform holistic analyses of medicinal plants, revealing novel genetic pathways and biosynthetic pathways for therapeutic secondary metabolites (Pandita *et al.*, 2021). Among the major tools in this discipline is the PlantæViz platform, which enables web-based integration, visualization, and analysis of multi-omics data from a diverse range of plant species. The platform offers improved functionalities, such as gene co-expression network mapping and metabolomics–transcriptomics integration, both of which are important for hypothesis generation and aiding in-depth biological interpretation (Santiago *et al.*, 2024).

In progressing the discipline of plant biology, single-cell multi-omics technologies have been a revolutionary discovery. The technologies enable high-resolution examination of cellular heterogeneity and intercellular communication within the context of plant tissues, helping researchers to investigate regulatory systems at the level of a single cell. Such precision has had a deep impact on horticultural science and plant developmental research (Zhang *et al.*, 2023a). In addition, spatial multi-omics technologies have made tissue-specific examinations possible, uncovering how gene expression and metabolic activities vary in individual anatomical areas of plants, resulting in a comprehensive understanding of plant developmental processes (Yu *et al.*, 2023b). These multi-omics technologies have also played an important role in identifying how plants respond to biotic and abiotic stress factors (Gupta *et al.*, 2023).

Mapping organ-specific plant development, especially in root tips, meristems, and leaf gradients, has come a long way with the use of multi-scale and multi-dimensional analytical techniques. Quantitative root biology has seen new developments recently, enabling 3D reconstruction and analysis of root morphology at single-cell resolution. This has facilitated accurate mapping of root zonation, chromatin structure, and developmental asymmetry in plants such as *A. thaliana* and *Medicago sativa*, which can assist in furthering our knowledge of lateral root initiation and vascular patterning (Pasternak and Pérez-Pérez, 2021). Plant root development undergoes distinct phases, such as root apical meristem initiation, exploratory foraging growth, and eventual senescence. These are controlled by a mix of autonomous genetic programs, non-autonomous signaling pathways, hormonal signals, and periodic oscillatory pathways. This control is at the root of the plasticity of root architecture, enabling plants to respond to environmental heterogeneity and maximize resource capture (Siqueira *et al.*, 2022).

One of the most striking features of plant organogenesis is that cell fate is more specified by positional information than by lineage. This is based on the spatial arrangement of signals such as phytohormones, intercellular communication, and transcription factor activity. These converge to create spatial domains of gene expression that regulate differentiation and tissue patterning (Pallakies and Simon, 2010). In addition, recent research has revealed that environmental conditions control transcriptional states of root meristem cells, guiding developmental pathways. These environmentally induced states reveal another complexity and plasticity in root development, which demonstrates that external cues can reprogram cell identities and behaviors (Oliva *et al.*, 2022). Exploration of spatial plant–pathogen interactions has also been aided by a suite of new technologies. Subcellular proteomics now enables dissection of host–pathogen interactions at the level of the extracellular matrix, playing an essential role during initial phases of pathogen recognition and immune activation (Figueiredo *et al.*, 2024). To support this, protoplast-based systems enable transient gene expression and functional analysis, enabling researchers to capture rapid signaling events and downstream transcriptional reprogramming caused by pathogen-associated molecular patterns (Xing *et al.*, 2017).

In addition, the integration of high-resolution digital imaging with genome-wide association mapping (GWA) has made it possible to quantify complex phenotypes, for instance, lesion color and shape, which are genetically unique traits that combine to quantify resistance, as seen in the case of *A. thaliana* and *Botrytis cinerea* (Fordyce *et al.*, 2018). The sustainability of illnesses and evolutionary pressures is also influenced by spatial pathogen dissemination dynamics; small-scale dispersal increases genetic diversity and the frequency of host–pathogen contact, which in turn influences resistance deployment tactics (Thrall and Burdon, 1999). The deployment of image-based phenotyping, along with computational methods such as the level set method, is a valuable tool for observing disease progression in real time. The method enhances throughput as well as accuracy in disease symptom quantification, as seen in the case of *Peyronellaea pinodes* infections in pea cultivars (Permanes *et al.*, 2024).

The emergence of spatial transcriptomics has opened up new avenues for research into plant responses to abiotic stresses such as drought, salinity, and heat. Unlike conventional transcriptomics, this method preserves the spatial structure of gene expression in tissues, allowing researchers to study localized responses and tissue-specific gene regulation under stress (Hu *et al.*, 2024b). When paired with other omics platforms, spatial transcriptomics is able to reveal the coordinated molecular change networks governing plant adaptation. This integrated approach has revealed dynamic regulatory networks that modulate the phenotypes of plants to changing environments, as well as influencing the underlying gene regulation (Soni *et al.*, 2015). Transcriptomic advancements have made it possible to identify stress-inducible genes as well as to dissect pathways conferring tolerance. These findings are of importance in the breeding of climate-resilient cultivars through targeted breeding (Govind *et al.*, 2022). Additionally, machine-learning-assisted meta-analyses have combined large-scale transcriptomic datasets to recognize a conserved set of stress-response genes, which has paved the way for engineering crops with enhanced tolerance to multiple stress factors (Sanchez-Munoz *et al.*, 2024). The combination of spatial and single-cell transcriptomics further broadens our ability to discover new regulatory elements and transcriptional circuits governing the complex traits of plants (Hu *et al.*, 2024a).

The understanding of developmental zonation and cellular tissue specification has been central to the manipulation of vascular differentiation and plant growth. The development of xylem and phloem tissue is controlled by a highly interconnected network of transcription factors, including HD-ZIP III, KANADI, MYB, and NAC, that control asymmetric cell divisions and tissue identity determination (Jung and Park, 2007). In *A. thaliana*, embryogenesis has been studied following early vascular patterning to oriented divisions and cell fate choices in embryonic stages and illustrated how tissue organization emerges from intensely regulated morphogenetic signals (De Rybel *et al.*, 2016). Determination of tissue identity during embryogenesis is strongly controlled by positional information that is mediated by auxin gradients, intercellular communication, and mechanical stress from the cell wall. These determine the spatial expression of transcription factors like WOX, which play important roles in the definition of embryonic domains and primary tissue lineages (Armenta-Medina *et al.*, 2021). Moreover, sophisticated live-imaging techniques have illustrated how shoot apical meristems (SAMs) actively integrate hormonal and mechanical signals through feedback mechanisms to maintain stem cell niches and organize meristems (Sijacic and Liu, 2010).

The complex regulation of hormone transport and signaling at both the tissue and cellular levels has been dramatically altered by technological advancements. Improved MS and biosensor platforms have enabled real-time measurement and localization of phytohormones, including auxin, within specific plant structures (Novák *et al.*, 2017). Such experimental approaches have described how hormone gradients influence developmental processes, including root branching and shoot architecture (Swarup and Bennett, 2014). The use of mathematical modeling has been instrumental in explaining the spatial dynamics involved with auxin distribution, as well as its interaction with transport proteins, including PINs and regulators involved in the ARF-Aux/IAA signaling cascade (Allen and Ptashnyk, 2020). Additionally, the identification of multiple hormone transporters with tissue-specific expression patterns highlights the biochemical flexibility plants employ to attain complex and highly tuned developmental regulation (Zhang *et al.*, 2023c). Such knowledge is proving central to programs designed to improve crop productivity, adaptability, and overall plant health under a variety of environmental conditions.

Dissection of crop traits according to spatial expression quantitative trait loci (eQTL) and quantitative trait loci (QTL) mapping constitutes a strong foundation for the explanation of the genetic basis of complex traits. This combined approach combines genetic variation and transcriptome data in an effort to detect regulatory loci influencing gene expression and phenotype in different spatial and developmental contexts. A notable model proposed by Sang *et al.* (2019) laid out a dissection approach especially tailored for composite traits, detecting the pleiotropic effects of QTLs on individual component traits. This model was successfully applied to dissect Populus tree and seedling traits, facilitating a more complex understanding of trait architecture (Sang *et al.*, 2019). Similarly, Holtan and Hake's experiment in tomato detected QTLs responsible for the control of leaf dissection patterns, demonstrating the genetic origin of morphological variation in leaf development (Holtan and Hake, 2003). Progress in high-throughput sequencing technologies and genome-wide association studies (GWAS) has enhanced the efficiency of QTL discovery and cloning significantly. These developments are critical in marker-assisted selection (MAS) and QTL pyramiding application, both crucial strategies in modern crop improvement (Zhao *et al.*, 2014). For fine-mapping regulatory variants, the TreeMap algorithm is a breakthrough. This approach facilitates the detection of causal eQTLs by examining gene expression in multiple tissues and has been successfully applied to brain and colon tissue datasets (Liu *et al.*, 2021). Such methods play a critical role in the explanation of tissue-specific regulation of genes and spatial dynamics of gene expression. In a move to overcome the limitations associated with traditional 1D-trait eQTL mapping, Sun *et al.* (2008) proposed a 2D-trait eQTL mapping approach. This approach clarifies co-expression patterns between pairs of genes and hence identifies interactions and regulatory variants that are involved in complex biological networks. This approach enhances the resolution of regulatory network evaluation and provides more insight into the multigenic control mechanisms that regulate multigenic traits (Sun *et al.*, 2008). In general, the integration of spatial transcriptomics with eQTL and QTL mapping enables researchers to analyze trait development at the cellular level, which is crucial for functional genomics, trait prediction, and precision breeding. Such approaches are indispensable for the development of climate-resilient and high-yielding crops under the umbrella of sustainable agriculture.

2.4 Spatial Multi-omics Integration and Systems-level Insights

The concurrent profiling of transcriptome, proteome, metabolome, and epigenome in plant biology is a groundbreaking approach to understanding plant biology and solving the challenges of world agriculture. Multi-omics technologies have proven to be invaluable in understanding the complex molecular networks that regulate plant responses to environmental stimuli. In the study of cellular heterogeneity and adaptive responses to biotic and abiotic stress, single-cell multiomics systems provide a fresh resolution by enabling the parallel analysis of many omic layers produced from a single cell. Because it makes it possible to identify important molecular regulators and signaling pathways involved in stress adaptation mechanisms, this type of study is especially important for crop development through increased resistance (Srivastava and Singh, 2022). One of the intriguing advances in this direction is the development of a universal protocol that is capable of effectively isolating metabolites, DNA, long RNAs, small RNAs, and proteins from one plant sample. With the protocol, comprehensive multi-omics analysis is facilitated by using limited biological samples, thereby enabling the integration of heterogeneous datasets for comprehensive systems biology applications (Valledor *et al.*, 2014). Integration of transcriptomic, proteomic, and metabolomic data enhances understanding of plant systems, particularly in plants with large, poorly annotated genomes or plants that accumulate a wide variety of secondary metabolites. Integration of these data solves existing plant molecular biology challenges and opens the gateway to the deciphering of complex regulatory networks (Jamil *et al.*, 2020). The overall workflow encompassing multi-omics data acquisition, spatial mapping, and integrative analysis in plants is illustrated in Fig. 2.2.

Recent technologies, such as scMeT-seq significantly enhance the utility of multi-omics by enabling simultaneous profiling of metabolomics and transcriptomics at single-cell

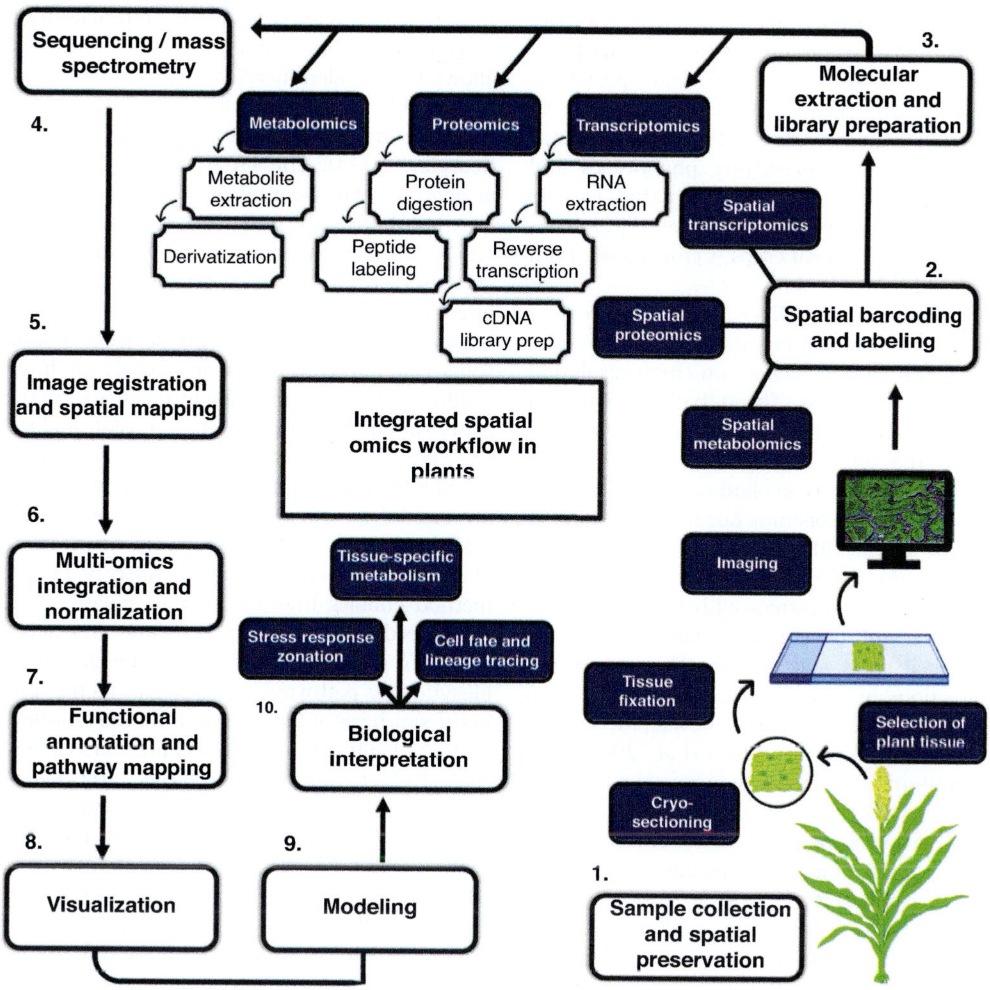

Fig. 2.2. Integrated spatial omics workflow in plants.

resolution. Such a combined analytical platform illuminates the dynamic correlations between metabolites and gene expression, thereby gaining insight into the functional states of specific cells and further enhancing the interpretative potential of metabolomics in a cellular context (Mao *et al.*, 2025). Such advancements in methodology are significant in the investigation of plant responses to abiotic stresses, including drought, salinity, and temperature stress, thereby contributing to the clarification of stress-responsive pathways and guiding breeding programs for climate-resilient crops (Roychowdhury *et al.*, 2023).

In further enhancing the usability of multi-omics approaches, advanced computational packages such as MOFA+ (multi-omics factor analysis plus) have been invented. MOFA + enables the integration of spatial multi-omics data through simultaneous modeling of heterogeneous molecular layers, including genomics, transcriptomics, and metabolomics, at single-cell resolution. This statistical technique employs variational inference along with the incorporation of flexible sparsity constraints, thereby enabling the recovery of low-dimensional representations from complex datasets. Employing such a technique, scientists are able to reveal complex molecular interplay and regulatory mechanisms that influence plant responses to abiotic stresses and specialized metabolic processes (Argelaguet *et al.*, 2020; Sravanalakshmi *et al.*, 2025). Such findings are significant in predicting phenotypes and the identification of biomarkers that can be utilized in breeding programs to develop climate-resilient horticultural crops (Cembrowska-Lech *et al.*, 2023).

In addition to MOFA+, Seurat V5 provides other capabilities to integrate spatial multi-omics data, with special focus on high-resolution plant tissue analysis. Through the integration of transcriptomically and proteomically derived data at the single-cell level, Seurat V5 allows spatial heterogeneity in plant organs to be investigated, thereby deepening our understanding of cell-type-specific functions and developmental processes (Lv *et al.*, 2024). The advancements in spatial transcriptomics with Seurat V5 have allowed scientists to differentiate between different cell types and decipher their molecular interactions in tissues, an essential foundation for understanding plant development and environmental responses (Yin *et al.*, 2023). The integration also allows demarcation of gene regulatory networks, especially in the context of plant–pathogen interactions, thereby deepening our understanding of immunity and disease resistance in plants (Nobori *et al.*, 2023). The integration of AI into these analytical systems further enhances data interpretation through the facilitation of high-throughput phenotyping and predictive modeling, which underpins sustainable agriculture and crop improvement programs (Cembrowska-Lech *et al.*, 2023). Similarly, the totalVI (Total Variational Inference, https://scvi-tools.org) approach has been an essential advancement in plant systems biology through its facilitation of the simultaneous investigation of transcriptomic and proteomic data at the single-cell level. This approach allows correlations of gene expression profiles with protein abundance directly, giving insights into cellular phenotypes and functional conditions that would not be possible through analyses confined to a single layer. Through the integration of spatial transcriptomics, totalVI positions cell types in their native tissue environments, thereby understanding developmental pathways and stress responses (Gayoso *et al.*, 2021). The comprehensive view that totalVI offers by way of genomic, proteomic, and metabolomic data allows for the identification of genetic markers of desirable agronomic traits as well as guiding the planning of agricultural practices (Ali and Georgiev, 2023). The LIGER (linked inference of genomic experimental relationships) method greatly enhances the integration of spatial multi-omics data in plant biology. The method easily overcomes previous methodological limitations by enabling cell type detection and cell–cell interactions in plant tissues through the use of spatial transcriptomic data (Yin *et al.*, 2023). By integrating different omics dimensions, including transcriptomics and proteomics, LIGER enables the inference of gene regulatory networks and the identification of molecular interactions that are crucial for plant growth and development (Clark *et al.*, 2023). In addition, the integration of AI and multi-omics analysis under the LIGER method enables high-throughput phenotyping and speeds up precision breeding programs. This is of particular value in the elucidation of biosynthetic pathways in medicinal plants and in the synthesis of large datasets for the formulation of sustainable agricultural practices (James *et al.*, 2021; Cembrowska-Lech *et al.*, 2023).

Despite these technological advances, integration of multi-omics data at multiple spatial scales and heterogeneous plant tissues remains an ongoing challenging issue. The biggest challenge lies in integrating disparate omics layers into a shared molecular framework that is representative of cellular processes like transcription, translation, and metabolism. The inherent heterogeneity of the cell types and developmental stages found in the plant tissues requires a resolution at the single-cell level to capture the complexity of molecular programming (Kalia and Sharma, 2019). Due to these integration challenges, novel computational tools like CAST and SIMO have been conceptualized. CAST allows spatial-to-spatial data alignment at the level of single cells, while SIMO integrates spatial transcriptomics with other epigenetic modalities such as chromatin accessibility and DNA methylation (Tang *et al.*, 2024). Such tools enable the detection of spatial heterogeneity and interpretation of regulatory mechanisms across different omics layers, thereby providing improved insights into plant tissue organization and the regulation of genes.

However, ongoing barriers remain in the form of technical challenges related to sequencing accuracy and the need for more sophisticated computational infrastructures (Yang *et al.*, 2025). Therefore, further progress in spatial multi-omics technologies and computational approaches is imperative to unraveling plant development and enhancing crop resilience in environmental stress conditions (Kalia and Sharma, 2019). Integrated omics approaches have become pivotal for understanding the molecular basis of plant stress responses and developmental biology, as inferred through a range of case studies on various plant species. These approaches integrate genomic, transcriptomic, proteomic, and metabolomic data to determine a multifaceted understanding of plant physiological and molecular functions. For example, in bioenergy crops such as switch grass, maize, and sugarcane, integrated omics has been utilized to discover genetic components and metabolic pathways that enhance biomass yield and stress tolerance, thereby helping to establish sustainable bioenergy systems (Mahmood *et al.*, 2023). In stress biology, integrated omics help bridge gene expression to protein abundance under different conditions of stress, gaining valuable information on the molecular bases of stress tolerance and enabling the development of stress-resistant crops (Waseem and Aslam, 2024). The combination of multi-omics technologies and high-speed breeding technologies enables stress-tolerant cultivars to be developed through allowing multiple breeding cycles per year, thereby enhancing resistance to drought, pathogens, and other stress-inducing agents (Azeem *et al.*, 2025).

In particular, metabolomics excels in identifying metabolite profiles linked to stress tolerance, which helps to clarify the physiological and biochemical mechanisms by which plants adapt to less-than-ideal conditions (Singh *et al.*, 2021a). Multi-omics integrated studies have also revealed responsible genes like *DREB1A* and *HSP70* that are core members of stress response pathways and perform crucial genetic marker functions in crop improvement (Chavhan *et al.*, 2024). All these multi-omics resources are the foundation of current plant systems biology, enabling overall understanding of plant–environment interactions and facilitating the creation of the next generation of crops possessing the ability to address the challenges of climate change and global food insecurity.

2.5 Artificial Intelligence and Machine Learning in Spatial Omics

Deep learning (DL) models have significantly enhanced spatial feature detection in plants, particularly in plant disease detection and plant species identification. Convolutional neural networks such as ResNet34, ResNet50, and VGG16 have been effectively used in plant disease recognition systems. DL models leverage transfer learning to enhance classification performance through enhanced accuracy and precision measures, hence enabling credible and independent plant disease monitoring systems (Gupta *et al.*, 2024). Similarly, the Xception architecture and DenseNet have performed well in leaf image classification. The models surmount varied lighting and image tilt issues by calculating the correlation between global and low-level features, resulting in enhanced accuracy in a variety of leaf recognition tasks (Gladston and Sucithra, 2023). A feature fusion pyramid focal network (PFN) supplemented with attention modules performed better in herbal plant identification than conventional DL models, especially on Chinese herbal medicine plant sets (Jiang *et al.*, 2024). An investigation of plant image identification technology identifies DL as being superior to manually created feature extraction and suggests neural network architecture and data pre-processing to improve plant image analysis (Chen *et al.*, 2023). In addition, explainable AI (XAI) integrated into deep convolutional neural networks attained near-perfect accuracy and AUC (area under the curve) scores in plant disease diagnosis, reflecting the capability of interpretable DL in early disease diagnosis and classification tasks (Natarajan *et al.*, 2024).

Unsupervised learning methods have become increasingly important in plant system deconvolution and cell type annotation, especially in spatial transcriptomics. TACIT, an unsupervised method, allows cell type annotation with pre-specified gene signatures without labeled training sets. The process is extremely scalable and accurate with multi-omic data use cases, outperforming existing annotation strategies (Huynh *et al.*, 2024). Additionally, STdeconvolve enables reference-free deconvolution of spatial transcriptomics information, retrieving gene manifestation profiles according to cell types and relative geographical positions without the need for external annotations (Miller *et al.*, 2022). Marker gene application in an unsupervised clustering framework, as illustrated by ACAM, improves annotation accuracy by identifying biologically relevant clusters, which are annotated with gene manifestation profiles (Chen and Zhang, 2022). Furthermore, the use of methods such as principal feature analysis and mutual information increases cell type-specific feature identification, with interpretable knowledge of phenotypic heterogeneity and tissue structure (Caliskan *et al.*, 2023). Unsupervised methods are essential in the analysis of high-dimensional, large datasets where supervised learning is in distress due to the unavailability of adequate annotated examples.

In predictive modeling, the discovery of gene regulatory hotspots in plant genomes has been greatly aided by the application of DL techniques and multi-layer neural networks. Predmoter, a neural network architecture-based algorithm, predicts chromatin accessibility and histone modification maps from DNA sequence data. The model is beneficial in annotating cis-regulatory elements (CREs) across different organisms, such as *A. thaliana* and *Oryza sativa* (Kindel *et al.*, 2024). Similarly, PlantBind uses an attention-based, multi-label neural network for predicting transcription factor binding sites (TFBSs) for a collection of 315 transcription factors from DNA sequence data and shape profiles, enhancing the modeling of transcriptional regulatory networks (Yan *et al.*, 2022). Other DL tools are capable of predicting gene expression profiles directly from non-coding sequences, with high predictive accuracy in the majority of plant species, including *Arabidopsis*, *Solanum lycopersicum*, *Sorghum bicolor*, and *Zea mays* (Peleke *et al.*, 2024). Chromatin modifications, such as H3K4me3 and H3K36me3, are predictive indicators of gene expression patterns in several plant species. Machine learning (ML) algorithms that use epigenetic data to take tissue-specific transcription into account confirm these findings (Wu *et al.*, 2019). Additionally, transformer-based models trained on long-length genomic sequence fragments are central to regulatory element identification in crops like maize and sorghum, enhancing scalable genome annotation and functional interpretation of non-coding regions (Tross *et al.*, 2024).

The combination of AI with image-omics alignment and spatial domain classification has created revolutionary innovation in plant phenotyping and disease surveillance. AI techniques, particularly ML and deep learning, are well-suited to multi-omics data analysis, enabling end-to-end analysis beyond traditional reductionist approaches (Cembrowska-Lech *et al.*, 2023). In phenotyping, AI-assisted imaging technology, from 2D to 3D reconstruction, enables measurement of intricate traits and detection of minute phenotypic variation critical to precision breeding (Maraveas, 2024). In phytopathology, AI enables disease diagnosis with image-based models, achieving over 95% accuracy. Such models enable early disease diagnosis and intervention, for example, disease outbreak predictive modeling (González-Rodríguez *et al.*, 2024). Deployed platforms such as AICropCAM employ edge computing and DL for image classification, segmentation, and object detection tasks, enabling low-cost, scalable provision for high-throughput phenotyping and measurement of stress responses (Chamara *et al.*, 2023).

2.6 Spatial Epigenomics and Chromatin Architecture

The nascent field of spATAC-seq in plant biology aims to determine chromatin accessibility at high spatial resolution, thereby yielding critical insights into the 3D genome structure and regulatory architectures. Despite spATAC-seq being optimized for application in animal systems, its application in plant biology remains in an early stage of development. Conventional ATAC-seq has been successfully applied in several plant species to identify open chromatin regions and transcription factor binding sites, unveiling conserved regulatory features as well as cell type-specific variation in chromatin accessibility (Maher *et al.*, 2018). However, one major limitation with conventional ATAC-seq is its low spatial resolution, which is crucial to understanding the spatial structure and regulatory dynamics of plant genomes in different tissues and developmental contexts. Recent improvements in technologies, including spatial chromatin accessibility sequencing (SCA-seq), have been shown to be able to capture chromatin accessibility and the 3D genome structure simultaneously. These novel strategies employ techniques such as methylation labeling and nanopore sequencing to map chromatin accessibility and spatial interactions at single-molecule resolution, thus providing an integrated view of genome regulation and complexity of higher-order chromatin structure (Tang *et al.*, 2022; Xie *et al.*, 2024). While these strategies have been applied extensively in animal models to date, their application in plant systems could significantly illuminate our understanding of chromatin dynamics, particularly in the context of developmental and environmental stimuli. Application of spatially resolved epigenomics in plant research could facilitate the discovery of novel regulatory modules and the dissection of the function of chromatin topology in gene expression, developmental processes, and stress responses (Fan *et al.*, 2021).

Among these, *in situ* chromatin conformation capture techniques, i.e. Hi-C and Micro-C, have particularly advanced plant genomics by allowing for the investigation of long-range chromatin contacts and genome spatial organization. The Hi-C method uses chromatin cross-linking, enzymatic digestion, and proximity ligation followed by high-throughput sequencing to produce genome-wide maps of chromatin interaction frequencies. This has been adapted for use in many plant species, such as important crops like barley, wheat, and rice, to not only investigate chromatin structure but also enhance genome assemblies through the identification of structural components and mis-assemblies (Padmarasu *et al.*, 2019). Micro-C is an improvement over Hi-C, using micrococcal nuclease digestion along with longer cross-linkers, which can determine chromatin interactions at single-nucleosome resolution, thereby allowing for the investigation of fine-scale chromatin features like enhancer-promoter loops and local chromatin boundaries, as in *Arabidopsis* research (Sun *et al.*, 2024). Additionally, improvements like low-input Hi-C protocols have enabled the investigation of chromatin conformation in limited or specialized plant tissues. For example, by isolating nuclei based on specified endopolyploidy levels, scientists were able to investigate chromatin interactions in a tissue-type-specific context (Wang and Liu, 2020). Separately, these processes have demonstrated that chromatin spatial organization is crucial for transcription control, with transcription factors and RNA polymerases being key players in the establishment and upkeep of localized chromatin areas. Thanks to these improvements, the understanding of 3D genome architecture in plants is being established at a higher level, shedding light on its functional significance for processes like development, stress response, and evolutionary diversification (Šimková *et al.*, 2024).

Epigenetic heterogeneity between different tissue layers and organs in plants is mainly induced by differences in DNA methylation patterns, which play crucial roles in the regulation of gene expression and transposon silencing maintenance. DNA methylation, being a heritable and stable epigenetic modification, occurs in three distinct sequence contexts – CG, CHG, and CHH – each controlled by distinct enzymatic pathways and methyltransferases that impart specificity and heterogeneity to methylation in the plant genome (Ashapkin *et al.*, 2016). Increasing evidence suggests that methylation patterns are not only organ specific but also show considerable heterogeneity between different cell types of the same tissue. For example, cell-type-specific methylomes of the *Arabidopsis* root meristem have detected unique methylation signatures, especially in the CHH context, showing the complexity of intra-organ epigenetic landscapes (Kawakatsu *et al.*, 2016). Such epigenetic heterogeneity is also controlled by the CLASSY family of chromatin remodeler proteins, which are differentially expressed during the different stages of plant development and play crucial roles in the establishment of tissue-specific DNA methylation patterns (Zhou *et al.*, 2022).

The dynamic character inherent in DNA methylation enables reversible changes regulated by developmental phases, especially in decisive phases of development such as gametogenesis and embryogenesis. Spatial and temporal changes in the localization of 5-methyl-deoxy-cytidine (5mdC) take place during these phases to coincide with cell proliferation and differentiation events (Testillano *et al.*, 2013). Such plasticity constitutes a basic mechanism for the provision of phenotypic diversity and flexibility in plants, enabling rapid response to environmental stress by reversible and inheritable epimutations. Such epigenetic mechanisms are thus essential for developmental processes as well as the formation of long-term evolutionary patterns. These epigenomic landscapes, which show the close relationships by which chromatin-based control of gene expression reacts to intrinsic and extrinsic inputs, have become much easier to navigate with the development of high-resolution sequencing methods (Köhler and Springer, 2017).

Concurrently, the merging of 3D genome structure with spatial gene expression landscapes is proving to be a strong platform to unravel how chromatin topology affects transcriptional regulation and cell identity in plants. Advances in 3D genomics have recently presented strong evidence that the spatial organization of chromatin directly affects gene expression. For example, research in groundnuts has revealed that the conversion of genomic regions from inactive to active chromatin states in mutant lines correlates robustly with shifts in patterns of gene expression, especially in critical biological pathways like flavonoid biosynthesis and circadian rhythm regulation (Zhang *et al.*, 2021). The 3D genome is structured by the intricate interaction between genetic sequence factors and epigenetic marks, and the movement of heterochromatin markings significantly affects transcriptional fates and chromatin connections (He *et al.*, 2024).

The use of chromosome conformation capture technologies, for example, Hi-C, together with high-resolution imaging instrumentation, facilitated the detection of chromatin domains and their correlation with particular transcriptional processes (Pei *et al.*, 2021). Additionally, developments in ligation-free technologies like split-pool recognition of interactions by tag extension (SPRITE) and genome architecture mapping (GAM) made it possible to detect multi-way chromatin interactions and, consequently, characterize the spatial organization of the genome at higher resolutions in relation to functional gene networks (Tavallaee and Orouji, 2025). These advances have revealed the criticality of 3D chromatin structure in the plant developmental processes and environmental tolerance. All these cumulatively provide new horizons for the manipulation of genome organization to yield better plant traits and durability, further cementing the importance of spatial epigenomic analysis in driving plant biology (Zhang *et al.*, 2021).

2.7 4D Spatial Omics and Time-resolved Mapping

The usage of temporal–spatial omics in plant biology has been an innovative approach to uncover the intricate molecular mechanisms that regulate development, immunity, and biosynthesis of metabolites. Through the tracking of molecular changes in both space and time, scientists are now able to interpret intricate biological processes with unprecedented resolution. Recent advances in single-cell and spatial transcriptomics have greatly enhanced our ability to profile the expression of genes in single cells, detecting heterogeneity that bulk methods are unable to distinguish. Of particular mention is time-resolved single-cell and spatial transcriptomic analysis in *A. thaliana*, which has identified spatially varying patterns of immune gene expression in response to infection by pathogens, underlining the importance of spatial context in plant immune responses (Nobori *et al.*, 2023). In addition, the combination of laser-capture microdissection and RNA sequencing has facilitated tissue- and time-specific spatial profiling of barley embryos, shedding light on the regulation of germination processes at the transcriptional and regulatory network level (Liew *et al.*, 2020). Similarly, spatial transcriptome profiling in model plant species has discovered differences in gene expression among distinct tissue domains, offering fundamental insights into developmental as well as evolutionary dynamics (Giacomello *et al.*, 2017).

Alongside transcriptomics, plant metabolomics has also developed with the use of high-resolution mass spectrometry and spatial imaging technology, enabling non-invasive localization and characterization of secondary metabolites. These newer methods have enabled the identification of new bioactive compounds possessing pharmacological activity, thus bridging basic disciplines of plant biology with applications in phytochemistry (Jaiswal and Williams, 2019). Furthermore, complementing resolved analysis spatially, live-tissue photoactivatable barcoding and labeling tools enabled the possibility of monitoring cellular and subcellular processes with caution in real-time. Advances such as photochemical programming and photoactivatable dyes enabled the creation of distinctive fluorescent barcodes in specific cells, enabling the study of spatiotemporal pathways in complex tissues. For instance, the cucurbit[7]uril (CB[7]) system enables spatial regulation of fluorophore localization, thus allowing multiplex optical barcoding at the cellular scale (Som *et al.*, 2021). Similarly, the SPACECAT method utilizes photocaged fluorophores to label and distinguish viable cells with high precision, enabling their subsequent analysis using tools such as scRNA-seq (Genshaft *et al.*, 2021).

In the attempt to expand the toolkit for spatial omics further, photocontrolled oligonucleotide-based barcoding techniques have been designed to encode spatial information through sequencing, and thus enable high-accuracy mapping of mRNA localization (Venkatramani *et al.*, 2024). Photoconvertible fluorescent proteins, such as Eos, which photoactivate through a green-to-red fluorescence switch, have added a new dimension of accuracy to subcellular imaging, enabling scientists to investigate cellular processes *in vivo* with high resolution (Mathur *et al.*, 2012).

Concurrently, the role of the circadian clock in modulating the regulation of plant response to environmental cues has gained more attention. As an inherent timekeeping mechanism, the circadian rhythm enables plants to anticipate and prepare for daily fluctuations and, therefore, become more stress resistant. It has broad regulatory control over physiological processes such as hormone signaling, carbon metabolism, and photosynthesis, primarily through interlinked transcriptional and translational feedback networks (Venkat and Muneer, 2022). Circadian regulation of secondary metabolite biosynthesis, including phenolic compounds and terpenoids, responsible for stress protection against abiotic and biotic stress, has recently emerged (Pérez-Llorca and Müller, 2024). Coordination of the circadian clock with the cellular redox state introduces additional complexity, influencing metabolic activity as well as stress response, although the mechanistic underpinning remains to be fully elucidated (Guadagno *et al.*, 2018). Transcriptomics in *Ulmus pumila* under salt stress has demonstrated how circadian regulation controls core pathways involved in photosynthesis and hormone signaling, thereby rendering the plant more resistant to stress (Chen *et al.*, 2022). Such evidence points towards the potential for integrating circadian information into prediction models to engineer crops with enhanced environmental stress tolerance, particularly in the context of climate change (Pérez-Llorca and Müller, 2024). In combination, these concurrent lines of investigation – temporal–spatial omics, state-of-the-art imaging and labeling technologies, and circadian biology – are revolutionizing our understanding of plant systems. They offer a multi-faceted platform for the dissection of plant complexity in responding to internal and external cues and provide a window of opportunity for novel approaches in crop improvement and sustainable agriculture.

Visualization of developmental progression at single-cell resolution in plants has been greatly improved through the combination of scRNA-seq with high-end visualization techniques. In *A. thaliana*, scRNA-seq has enabled high-resolution mapping of gene expression across heterogeneous tissues and developmental stages, thus allowing the detection of discrete cellular subpopulations, including rare cell types such as quiescent center cells (Ryu *et al.*, 2019). This technique has been crucial in recreating developmental pathways, such as root epidermal cells, from meristematic to differentiated stages, allowing for a comprehensive understanding of cell destiny decisions and the impact of genetic alterations. To better navigate and interpret these rich datasets, new tools like Cella have been created. By leveraging 3D modeling techniques like Blender, Cella, which converts huge scRNA-seq datasets into intuitive 3D heatmaps, enables the spatial representation of gene expression in anatomical structures, for example, the *Arabidopsis* root meristem (Su *et al.*, 2023b). Such approaches not only improve interpretability but also represent an accessible portal to evaluate thousands of gene expression profiles *in situ*. Beyond use in *Arabidopsis*, such technologies represent a scalable platform for root architecture and gene activity profiling across a range of plant species (Shahan *et al.*, 2021). Overall, the combination of single-cell transcriptomics and high-end visualization technologies represents a powerful platform for deconstructing plant developmental biology with unprecedented depth and accuracy.

2.8 Spatial Metabolite–Microbiome 'Interactomics'

Understanding the spatial and temporal dynamics of root exudates and how they interact with soil and microbes has been critical in demarcating exudate gradients in root zones and the rhizosphere. Root exudates, which are dominated by sugars, organic acids, and amino acids, create localized microdomains along the root axis that have an important role in soil carbon storage, mobilization of nutrients, and microbial activity (Garcia Arredondo *et al.*, 2024). These chemicals create radial gradients whose magnitude and spatial arrangement are controlled by parameters like soil depth, diffusion properties, and sorption capacity, whose impact is more significant in deeper soils where physical constraints are higher (Proctor and He, 2021). Technological advancements have facilitated the *in situ* mapping of chemical gradients using correlative imaging pipelines, providing comprehensive visualizations of elements such as carbon and nitrogen surrounding roots (Lippold *et al.*, 2023). The role of exudates goes beyond chemical mediation; they are also important determinants of microbial community structure and activity, acting as energy sources and signaling molecules that control rhizosphere interactions (Yetgïn, 2023). The composition of exudates can exert control on enzyme activities, with some components of exudates selectively inducing the expression of microbial enzymes critical in nutrient cycling (Zhang *et al.*, 2019). These observations point to the complexity of root exudation and its critical role in shaping rhizosphere ecology.

In parallel, host–microbe interface analysis through spatial metabolomics has also become a highly effective tool for interrogating biochemical interactions at the resolution of micrometers within plant tissues. Such innovations include the metaFISH in which fluorescence *in situ* hybridization (FISH) microscopy is combined with atmospheric-pressure MALDI-MSI (AP-MALDI-MSI) at high resolution in an attempt to attain accurate spatial co-localization of host and symbiont metabolites on a single section of tissue (Geier *et al.*, 2020). MSI, particularly with modalities such as time-of-flight secondary ion mass spectrometry (ToF-SIMS) and MALDI-MSI, has become an essential tool in plant science for mapping the spatial organization of metabolic compounds and unraveling localized chemical interactions involving different parts of a plant, like the rhizosphere and the phyllosphere (Yin *et al.*, 2024). Despite some limitations involving spatial resolution and metabolite annotation confidence, innovations such as microfluidic integration and ML-based image analysis are greatly increasing the ability of MSI to unravel complex plant–microbe metabolic networks, thereby creating new insights into processes pivotal to plant sustainability and biomass yield (Parker *et al.*, 2023).

The combined use of spatial metagenomics to investigate plant microbiome environments has improved microbial-biogeography knowledge of microhabitats of different plants, e.g. the rhizosphere, phyllosphere, and endosphere. New technologies such as spatial metatranscriptomics and spatial host–microbiome sequencing (SHM-seq) have enabled researchers to acquire spatial data on microbial communities and host transcriptomes simultaneously, providing unparalleled information on the interactions and functionalities of plant-associated microbiomes (Mu *et al.*, 2024). For instance, the analysis of spatial metatranscriptomics in *A. thaliana* leaves identified bacterial and fungal hotspots, which revealed a multifaceted pattern of inter- and intrakingdom microbial interactions (Saarenpää *et al.*, 2024). In another experiment, the use of long-read metagenomic sequencing of the rice phyllosphere revealed uncharacterized bacterial diversity of taxa, including novel phages and plasmids, demonstrating the effectiveness of metagenomics in revealing new aspects of microbial communities (Masuda *et al.*, 2024). Spatially resolved studies in the rhizosphere of *Brachypodium distachyon* have further elucidated the heterogeneity of root exudation gradients forming microbial niches, which supported the role of root metabolic patterns in defining microbial distribution (Acharya *et al.*, 2023). Such data indicate the importance of spatial resolution in revealing microbial assemblages and their functional effects, especially in applications in crop improvement and disease resistance (Rosselli and Squartini, 2015).

The interplay between microbial colonization and metabolite flux is intrinsic to plant–microbiome interactions with far-reaching consequences for plant productivity and well-being. Metabolite chemo-attractants and modulators in the form of root exudates selectively induce colonization by beneficial microbial taxa, which in turn enhances the stress tolerance of the plant towards abiotic and biotic stressors (Ali and Glick, 2024). Pre-symbiotic metabolic responses in *Eucalyptus grandis* are specific towards diverse microbial lifestyles and hence modulate subsequent colonization success and symbiotic outcome (Wong *et al.*, 2019). This symbiotic interaction, mediated by mutual exchange of signal molecules, orchestrates complex communication networks governing microbiome composition and plant fitness. Metabolomics-guided research further reveals the potential of plant-associated microbes to induce stress tolerance and metabolic plasticity (Olanrewaju *et al.*, 2024). Model experiments demonstrate that resource availability, particularly the availability of carbon sources, has far-reaching effects on microbial community structure, where the availability of metabolically flexible strains dictates dominance by either competitive or cooperative interactions (Schäfer *et al.*, 2023). Furthermore, microbial colonization can modulate secondary metabolism in the plant, which in turn affects the involved biosynthetic pathways towards defense as well as growth regulation (Singh *et al.*, 2021b). These findings together emphasize the central role of metabolite flux in the organization of the rhizobiome and designing the functional dynamics between microbes and plants as prerequisites for sustainable agricultural practices.

2.9 Synthetic Biology and Spatial Control in Plants

Construction and design of synthetic gene circuits from spatial mapping is a major advance in the science of synthetic biology, in the context of plant systems. An increasing number of studies highlight the intentional reprogramming of plant morphology and development through the design of synthetic genetic circuits, allowing for precise, tissue-specific expression of genes. Such advances are of particular importance in root systems, where spatially patterned growth is involved in plant–environment interaction. In one of the first studies of its kind, Brophy *et al.* (2022) constructed a library of synthetic transcriptional regulators with Boolean logic capabilities that are essential to the spatially controlled activation or repression of genes. By using these circuits in *A. thaliana*, the authors demonstrated their ability to reprogram root architecture by adjusting lateral root density, thus highlighting the promise of spatially targeted genetic engineering in controlling plant development. Boolean logic gates, i.e. AND, OR, and NOT, make gene expression contingent upon different biological inputs, allowing for high precision in reprogramming development. Such control not only reflects predictive knowledge of genetic regulation in multicellular plant systems but also provides a foundation for the breeding of crops with improved traits (Brophy *et al.*, 2022).

The use of spatial gene circuits is not only for regulation in development but also for environmental stress adaptation. Alamos and Shih have emphasized that the ability to control transcriptional patterns over a spectrum of tissues allows the specific optimization of plant resistance to abiotic stress, such as drought and nutrient starvation (Alamos and Shih, 2022). The evolution of recombinase-mediated gene circuits has paved the way for long-term and irreversible modification of gene expression. These systems use site-specific recombinases to trigger rearrangements of DNA, thus allowing targeted, programmable transcriptional modification independent of transient signals (Lloyd *et al.*, 2022). Together, these advances constitute a paradigm shift in plant biotechnology, from descriptive to predictive and programmable design. A unifying feature enabling these advances is the use of tissue-specific promoters, now indispensable to the realization of spatial precision in gene expression. In contrast to constitutive promoters, which activate gene expression in a promiscuous manner, tissue-specific promoters limit activity to a specific cellular environment, thus suppressing off-target effects and improving the precision of genetic modification. This approach has been highly effective in CRISPR (clustered regularly interspaced short palindromic repeats)-Cas9-mediated gene editing, where spatial specificity significantly enhances the heritability and efficiency of genetic modification (Rahman *et al.*, 2022).

For instance, recent research in soybean genomics has employed transcriptome-wide analysis to identify 288 genes with tissue-specific expression patterns, which in turn identified ten candidate promoters enriched for tissue-specific cis-regulatory elements (Yu *et al.*, 2023a). These regulatory elements are at the heart of the assembly of highly specific genetic circuits with negligible pleiotropic effects. Cell-type-specific promoters, such as those for WOX5, CYCD6;1, and EN7, in *Arabidopsis* have been effectively employed to facilitate cell-type-specific gene knockouts, thus allowing researchers to investigate cell-specific gene functions without the systemic perturbations usually associated with conventional mutagenesis (Li *et al.*, 2023). In woody trees, tissue-specific promoters targeted to the xylem have allowed specific manipulation of lignification and hydraulic traits, thus allowing for greater potential for timber improvement and carbon sequestration (Hertzberg *et al.*, 2007). Centralized databases like the tissue-specific gene promoter (TGP) database provide curated information on developmentally regulated and tissue-preferential promoters, thus allowing for streamlined design of personalized genetic constructs (Smirnova and Kochetov, 2020). Overall, these advances allow for multilayered control of gene function, thus underpinning the modular nature of plant synthetic biology.

Spatial resolution is critical to the engineering of plant secondary metabolism, particularly in defense and stress response pathways. For example, induction of the phenylpropanoid pathway upon pathogen attack leads to the biosynthesis of antimicrobial phenolics, structural compounds like lignin, and signaling molecules like salicylic acid. Optimization of the pathway allows for enhanced disease resistance without enhancing overall plant fitness. In accordance with Yadav *et al.* (2020), deliberate metabolic reprogramming may yield the creation of cultivars that possess defense reactions suited to particular environmental circumstances (Yadav *et al.*, 2020). Additionally, primary metabolism – usually regarded as distinct from defense processes – has increasingly been seen as complementary to stress responses through the provision of energy and secondary metabolite precursors (Bolton, 2009). In addition, the spatial organization of metabolic fluxes is critical to the provision of functional specificity. Glucosinolate biosynthesis – forming a class of nitrogen- and sulfur-containing defense compounds – in *Arabidopsis* is an example of extensive spatial compartmentalization. Indole and aliphatic glucosinolate biosynthesis take place in distinct cell types within vascular tissues, and this highlights the significance of cell-specific localization for their bioactivity (Nintemann *et al.*, 2018).

Advanced synthetic biology tools, including CRISPR-based transcription activators and repressors, enable the control of cell-type-specific metabolic pathways. For instance, it is now feasible to apply inducible systems to transiently elevate salicylic acid or jasmonate levels in localized tissues, thus facilitating plants to induce rapid and localized defense responses against biotic stressors. In addition, crosstalk between signaling pathways, particularly involving salicylate, jasmonate, and ethylene, tightly controls these defense networks, and hence the need for systems-level modeling and control (Peng and Huang, 2005). With advances in genome editing and promoter engineering, it is now feasible to engineer plants to demonstrate enhanced resistance, enhanced nutrient utilization, and enhanced tolerance to stress in an eco-safe manner (Fan, 2024). The design of biosensors with the ability to measure *in vivo* localized concentrations of hormones and metabolites has also further increased the understanding and manipulation of plant signaling networks. Genetically encoded biosensors based on ratiometric and fluorescence technologies provide high spatial and temporal resolution for monitoring dynamic physiological change. For instance, Andres *et al.* (2024) have designed gibberellin biosensors that employ DELLA proteins fused to luciferase enzymes. These ratiometric sensors provide real-time and accurate quantification of gibberellin activity in *Arabidopsis* protoplasts, thus facilitating real-time information on hormonal signaling pathways (Andres *et al.*, 2024).

Complementary systems that leverage fluorescent protein fusions with hormone-binding domains enable the live-cell imaging of hormone gradients. Upon hormone binding to these sensors, they undergo conformational changes that regulate fluorescence resonance energy transfer (FRET) signals, enabling high-resolution, non-invasive measurement of hormone dynamics (Balcerowicz *et al.*, 2021). Furthermore, modern electrochemical and spectroscopic biosensors enable portable and inexpensive means for field detection, in line with the emerging concept of the 'Internet of Plants,' which involves plants as nodes in a sensor network sensing environmental parameters (Naqvi *et al.*, 2023). These biosensing platforms not only enable elementary research but also serve as feedback units in synthetic regulatory circuits. Biosensors, when incorporated into synthetic pathways, enable gene expression changes in response to changes in metabolite or hormone levels, enabling closed-loop systems that emulate natural homeostatic responses (Martin-Arevalillo and Vernoux, 2019). Such real-time, adaptive systems are essential in precision agriculture, which allows plants to adapt autonomously to environmental fluctuations.

2.10 Digital Twins and Computational Modeling in Plant Spatial Biology

The combination of 3D reconstruction software and omics data is transforming the practice of studying plant tissues, providing an end-to-end perspective of plant development, structural organization, and functional dynamics. These combined methodologies have made it possible to reconstruct entire plant organelles and tissues from 2D electron and light microscopy images, thereby allowing accurate lineage tracing and accurate cellular architectural analysis in a variety of plant species, such as *Corydalis bracteata* and *Calla palustris* (Rudskiy and Khodorova, 2012). Such reconstruction plays a key role in the process of understanding spatial cellular differentiation and organogenesis during the developmental phases. Specifically, research into *Ginkgo biloba* has demonstrated how 3D reconstructions, when integrated with multi-omics profiling, can reveal the complex mechanisms of embryogenesis. These integrative approaches have made it possible to identify gene networks that govern vascular development and secretory cavity formation, providing valuable evolutionary insights into ancient plant lineages. High-resolution imaging instruments such as focused ion beam scanning electron microscopy (FIB-SEM) have been integral to this process through the ability to circumvent the resolution limitations of standard light microscopy and image intracellular structures at nanometer scales (Zechmann and Zellnig, 2022).

To efficiently analyze complicated datasets, computational tools such as large-volume fully automatic cell reconstruction (LVACR) have been developed to facilitate automatic segmentation and analysis of plant cell connections and morphology. This has been a significant step forward in the analysis of tissue organization and cellular dynamics *in situ* (Hu *et al.*, 2024b). Further, dual-optimization algorithms for 3D tessellation, which incorporate geometric and topological constraints, enable highly accurate reconstructions of plant tissues. Such models facilitate *in silico* simulations, enabling advanced quantitative analyses of cell growth and division patterns (Cerutti *et al.*, 2015). Combined with these physical reconstructions, virtual simulation of spatial gene expression and developmental gradients has emerged as the new frontier in systems plant biology. Spatial transcriptomics, especially as implemented in wheat inflorescences, has made it possible to profile gene expression at single-cell resolution, unveiling dynamic and highly structured spatiotemporal gene expression landscapes that control floral development (Long *et al.*, 2024). Data obtained from such analysis serve as the basis for computational modeling tools such as Virtual Plant Tissue (VPTissue) and VirtualLeaf, which combine biophysical, genetic, and mechanical modules to simulate plant tissue development under varying environmental and genetic conditions (De Vos *et al.*, 2017). These platforms enable researchers to investigate tissue growth by altering geometry, developmental parameters, and regulatory interactions, which is critical to the dissection of the multi-level regulation of morphogenesis (Merks *et al.*, 2011).

System-level phenomenon modeling is particularly crucial in the explanation of hormonal gradients, such as auxin distribution in *Arabidopsis* roots, where both spatial and temporal accuracy are important to the patterning of the roots. Computational models enable the prediction of hormonal interactions and feedback, thereby facilitating a more advanced understanding of the interactions between environmental and developmental signals (Moore *et al.*, 2015). The comprehensive simulations are not only used as analytical tools but also used as predictive tools for the design and testing of hypotheses in experimental plant biology. Based on this, the integration of digital phenotyping with augmented reality and virtual reality-based visualization interfaces is transforming the way we address the interpretation and interaction with plant-related data. Digital phenotyping involves high-throughput imaging tools, ranging from visible light cameras to hyperspectral sensors, to obtain morphological and physiological characteristics of plants in real environmental conditions, thereby enabling objective measurement of phenotypes (Ruckelshausen and Busemeyer, 2015). Analysis of these data involves the application of ML and computer vision algorithms in the identification of patterns, tracking of growth, and assessment of stress responses, thereby bridging the gap between genotype and phenotype (Scharr *et al.*, 2016).

Concurrently, software of an integrative nature like PlantæViz enables scientists to visualize and integrate multi-omics data, supporting sophisticated queries across transcriptomics, metabolomics, and protein interactions. These are crucial for hypothesis testing and systems-level inference of plant traits (Santiago *et al.*, 2024). Immersive software like VR-Omics takes the boundary further by allowing users to navigate and manipulate spatially resolved transcriptomic data in 2D and 3D virtual spaces. This software democratizes access to complex data structures, facilitating intuitive discovery and hypothesis generation (Bienroth *et al.*, 2023).

Sophisticated visualization platforms, such as NBFvis, play a pivotal role in spatial omics analysis by using neighborhood-based feature extraction and dimensionality reduction methods to detect spatial patterns of gene expression and chromatin structure. These advanced tools allow scientists to decipher context-dependent regulation of genes in the heterogeneous tissue microenvironments (Xu and Sankaran, 2021). In combination, the synergistic integration of these advanced technologies provides a robust digital framework for plant science that facilitates improved phenotyping, spatial omics integration, and mechanistic understanding of plant development. Possibly one of the most revolutionary applications of these digital technologies is their application in digital twins, computational models of physical agricultural systems that facilitate real-time simulation, monitoring, and management. In precision agriculture, digital twins combine heterogeneous datasets, such as soil chemistry, plant physiology, and climatic conditions, to dynamically optimize resource allocation for irrigation, nutrient supply, and pest management (Peladarinos *et al.*, 2023). Model-based management allows for adaptive decision making, ultimately enhancing productivity and sustainability.

In sophisticated agricultural contexts, digital twins enable personalized crop management, allowing for monitoring and compensation of micro-variations in orchards or even between single plants. In mandarin orchards, for instance, such digital twins have been used to customize fertilization and disease management treatments throughout intra-orchard zones, mirroring the personalized treatment strategies typical of precision medicine (Kim and Heo, 2024). The incorporation of ontologically arranged mathematical models also provides these digital twins with an added layer of biological understanding to simulate the growth phase and predict yield outcomes under varying environmental conditions (Skobelev *et al.*, 2021). The incorporation of remote sensing with AI-based analytics in such digital twin platforms also substantially raises their predictive strength regarding crop performance and stress response. Such integration enables real-time decision making and deployment of in-season management plans sensitive to changing climatic and agronomic conditions, as noted in some staple crops (Landivar-Bowles *et al.*, 2023). With the integration of computational modeling, spatial omics, and real-time phenotypic data, digital twins are transforming the discipline of plant research and precision breeding.

2.11 Technical and Computational Challenges

The discipline of plant spatial omics is fast developing but still struggles with major technical and computational issues associated with the inherent complexity and heterogeneity of plant tissue. One of the significant technical issues is obtaining high spatial resolution without sacrificing depth and breadth of transcriptomic coverage. Most existing spatial transcriptomics methods have to compromise on detailed spatial information over broad transcript detection at the expense of limited transcript coverage, and therefore, their utility for resolving intricate regulatory circuits and cell-type heterogeneity in plants is impaired (Gurazada *et al.*, 2021). Further, the integration of imaging modalities (e.g. electron/light microscopy) with omics platforms (e.g. RNA-seq, mass spectrometry) requires sophisticated instrumentation that can correlate molecular and structure-based information at the cellular and subcellular level. Such integrative strategies are especially challenging in plants, where cell wall stiffness, tissue heterogeneity, and autofluorescence make it difficult to prepare samples and image protocols (Alexandrov *et al.*, 2023). On the computational side, spatial omics produces high-dimensional, nonuniform data that are difficult to manipulate and analyze with standard tools. The lack of standard pipelines for data normalization, registration, and integration hampers reproducibility and cross-study comparison. This lacuna requires the use of sophisticated AI and ML strategies, which are increasingly employed to carry out activities like pattern recognition, noise removal, and predictive modeling in spatial omics (Dezem *et al.*, 2024).

Furthermore, the research community is marred by a lack of adequate standardization of experimental protocols and data analysis pipelines, thus inhibiting scaling and reproducibility of results across different plant species and research environments (Bressan *et al.*, 2023). In plant biology, sample preparation is a major bottleneck; for example, protoplasting and cell sorting often result in loss of the original tissue context and may generate artifacts, thus impairing data accuracy and spatial resolution. These technical, biological, and computational hurdles are summarized in Fig. 2.3, which highlights the multifaceted challenges currently limiting spatial omics research in plant systems. Notwithstanding these limitations, the embracing of multi-omics tools and the development of strong computational platforms and automated image setup analysis platforms is the passkey to future progress. Advances in *in situ* sequencing, spatial proteomics, and ML segmentation are poised to overcome existing constraints, thus enhancing the resolution and interpretative abilities of plant spatial omics. Such advances are expected to enable in-depth exploration of developmental processes, stress responses, and metabolic zoning in plant systems (Zhang *et al.*, 2024).

2.12 Future Directions and Vision

The future development in plant spatial omics is a groundbreaking advance in plant biology, driven by the intersection of spatial transcriptomics and multi-omics platforms. Recent breakthroughs in spatial transcriptomics have already yielded high-resolution views of cell differentiation, developmental processes, and stress response processes by capturing gene expression within the inherent spatial organization of plant tissues (Zhang *et al.*, 2024). Despite current limitations – technical issues with tissue sectioning, sample preparation, and signal integrity – spatial transcriptomics will advance rapidly in the future. Future breakthroughs will facilitate the accurate recognition of cell types, mapping of cellular lineage trajectories, and identification of intercellular signal networks (Yin *et al.*, 2023). Such features will enrich our understanding of the dynamic regulation mechanisms underlying plant growth, development, and environmental responsiveness.

Stepping back to look at the bigger picture, the emerging discipline of spatial omics – consisting of spatially resolved transcriptome, proteome, epigenome, and metabolome analyses – is potentially able to deliver an integrated, systems-level view of plant biology. The integrating paradigm allows researchers to expose intricate molecular interactions as they play out *in situ*, thereby supporting more realistic models of plant physiology and responses (Bressan *et al.*, 2023).

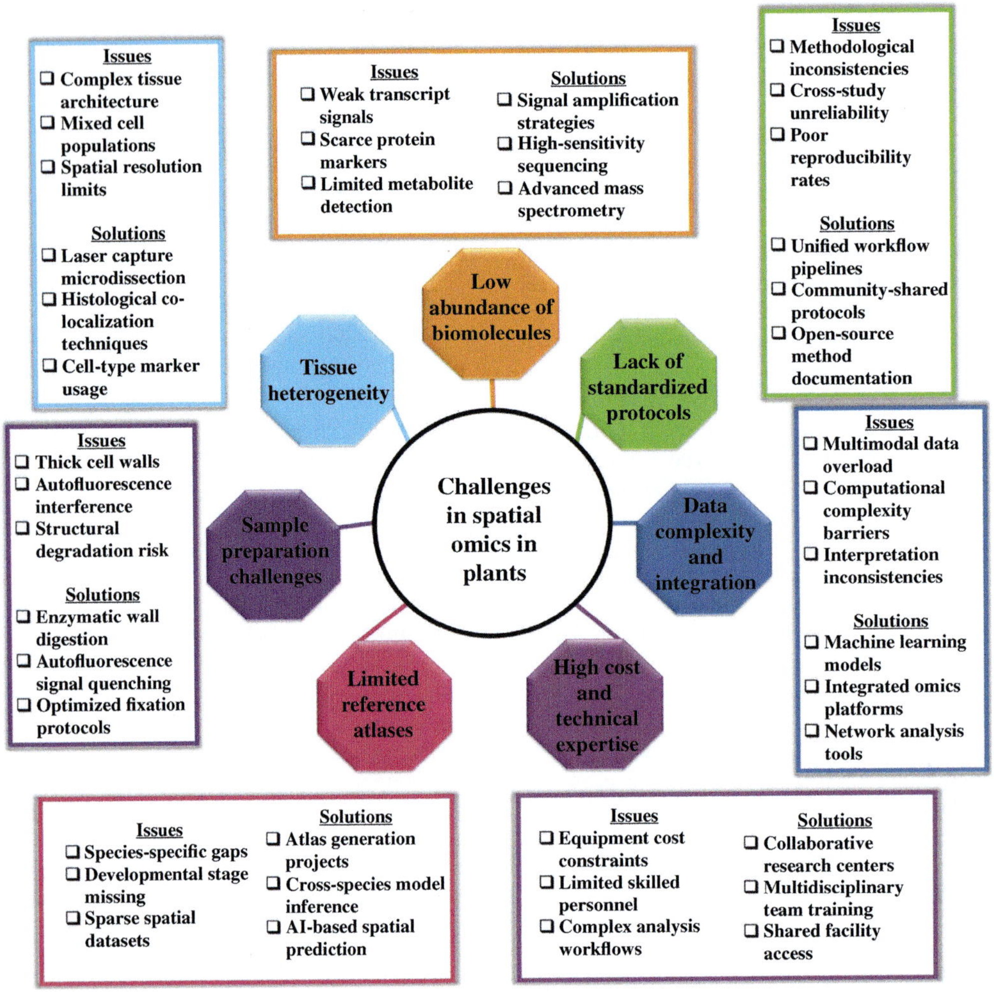

Fig. 2.3. Challenges in spatial omics in plants.

Future developments are moving towards the integration of different multi-omics modalities, which will enable the concurrent measurement of several molecular properties in an integrative spatial context. Such integration will potentially support more advanced explanations of gene regulation, metabolic flux, and stress signaling pathways in a more advanced way (Yang *et al.*, 2023). Further technological advances will enable 3D spatial omics as well as spatial–temporal resolution, and this will allow researchers to track the dynamics of molecular processes in whole organs or whole organisms over time.

The multidisciplinary character of spatial omics, which brings together molecular biology, cutting-edge imaging techniques, high-throughput sequencing, and computational analysis, presents challenges to be overcome unlike any other, but also unprecedented possibilities. With instrumentation becoming increasingly available and bioinformatics infrastructure improving, spatial omics is poised to transform plant biology by revealing the complex spatial organization and molecular activities of plant tissues (Alexandrov *et al.*, 2023).

2.13 Conclusion

Spatial omics is transforming plant systems biology by enabling researchers to explore cellular and molecular topographies at spatial resolution. This is a new methodology that enables researchers to dissect the complex networks of gene expression, metabolite dynamics, and protein localization patterns that control plant development and their environmental reactions. The combination of transcriptomics, proteomics, and metabolomics within a spatially defined framework has created new avenues for examining tissue-specific functions and intercellular communication in plant systems. Despite these advances, there are still certain challenges. High technical complexity, high cost, low spatial resolution, and the absence of robust computation platforms remain barriers to wider use. In addition, plant-specific problems, such as cell-wall hardness and cellular heterogeneity, require plant-specific solutions. Overcoming these challenges by constructing better imaging methods, data integration, and the establishment of reference atlases will be critical.

In future developments, affordable as well as scalable spatial omics platforms, integrated with artificial intelligence-based data analysis platforms, will most likely transform the field of plant science. Use of such technologies in enhancing crop quality, enhancing disease resistance, and enhancing stress tolerance is most likely to transform the landscape of plant research and agriculture. Spatial omics thus appears to be a frontrunner among next-generation plant biotechnology, offering a solid platform for the conversion of molecular data into useful biological information.

References

Acharya, S.M., Yee, M.O., Diamond, S., Andeer, P.F., Baig, N.F. *et al.* (2023) Fine scale sampling reveals spatial heterogeneity of rhizosphere microbiome in young *Brachypodium* plants. *Ecology*. DOI: 10.1101/2023.01.20.524947.

Alamos, S. and Shih, P.M. (2022) Synthetic gene circuits take root. *Science (New York, N.Y)* 377(6607), 711–712. DOI: 10.1126/science.add6805.

Alexandrov, T., Saez-Rodriguez, J. and Saka, S.K. (2023) Enablers and challenges of spatial omics, a melting pot of technologies. *Molecular Systems Biology* 19(11), e10571. DOI: 10.15252/msb.202110571.

Ali, K. and Georgiev, M.I. (2023) Editorial: Omics and its integration: A systems biology approach to understanding plant physiology. *Frontiers in Plant Science* 14, 1324901. DOI: 10.3389/fpls.2023.1324901.

Ali, S. and Glick, B.R. (2024) Root exudate metabolites alter food crops microbiomes, impacting plant biocontrol and growth. *Crops* 4(1), 43–54. DOI: 10.3390/crops4010004.

Allen, H.R. and Ptashnyk, M. (2020) Mathematical modelling of auxin transport in plant tissues: Flux meets signalling and growth. *Bulletin of Mathematical Biology* 82(2), 17. DOI: 10.1007/s11538-019-00685-y.

Andres, J., Schmunk, L.J., Grau-Enguix, F., Braguy, J., Samodelov, S.L. *et al.* (2024) Ratiometric gibberellin biosensors for the analysis of signaling dynamics and metabolism in plant protoplasts. *The Plant Journal* 118(1), 927–939. DOI: 10.1111/tpj.16725.

Argelaguet, R., Arnol, D., Bredikhin, D., Deloro, Y., Velten, B. *et al.* (2020) MOFA+: A statistical framework for comprehensive integration of multi-modal single-cell data. *Genome Biology* 21(1), 111. DOI: 10.1186/s13059-020-02015-1.

Armenta-Medina, A., Gillmor, C.S., Gao, P., Mora-Macias, J., Kochian, L.V. *et al.* (2021) Developmental and genomic architecture of plant embryogenesis: From model plant to crops. *Plant Communications* 2(1), 100136. DOI: 10.1016/j.xplc.2020.100136.

Ashapkin, V.V., Kutueva, L.I. and Vanyushin, B.F. (2016) Epigenetic variability in plants: Heritability, adaptability, evolutionary significance. *Russian Journal of Plant Physiology* 63(2), 181–192. DOI: 10.1134/S1021443716020059.

Audinot, J.-N., Ost, A.D., Stoffels, C., Philipp, P., De Castro, O. *et al.* (2022) SIMS imaging performed on focused ion beam - based platforms. *Microscopy and Microanalysis* 28(S1), 944–946. DOI: 10.1017/S1431927622004135.

Azeem, A., Tayyab, F., Mustafa, G., Khan, M.S. and Joyia, F.A. (2025) Integration of multi-omics approaches and speed breeding for plant stress tolerance: Challenges and opportunities. *Plant Stress Tolerance* 134–150.

Balcerowicz, M., Shetty, K.N. and Jones, A.M. (2021) Fluorescent biosensors illuminating plant hormone research. *Plant Physiology* 187(2), 590–602. DOI: 10.1093/plphys/kiab278.

Bartels, B. and Svatoš, A. (2015) Spatially resolved *in vivo* plant metabolomics by laser ablation-based mass spectrometry imaging (MSI) techniques: LDI-MSI and LAESI. *Frontiers in Plant Science* 6, 471. DOI: 10.3389/fpls.2015.00471.

Bawa, G., Liu, Z., Yu, X., Tran, L.S.P. and Sun, X. (2024) Introducing single cell stereo-sequencing technology to transform the plant transcriptome landscape. *Trends in Plant Science* 29(2), 249–265. DOI: 10.1016/j.tplants.2023.10.002.

Bienroth, D., Charitakis, N., Jaeger-Honz, S., Garkov, D., Elliott, D.A. *et al.* (2023) Spatially resolved transcriptomics mining in 3D and virtual reality environments with VR-omics. *bioRxiv* 2023–03.

Bolton, M.D. (2009) Primary metabolism and plant defense--fuel for the fire. *Molecular Plant-Microbe Interactions* 22(5), 487–497. DOI: 10.1094/MPMI-22-5-0487.

Borner, G.H.H. (2020) Spatial proteomics: A gateway to understanding cell biology. *Proteomics* 20(23), e1900328. DOI: 10.1002/pmic.201900328.

Bressan, D., Battistoni, G. and Hannon, G.J. (2023) The dawn of spatial omics. *Science (New York, N.Y.)* 381(6657), eabq4964. DOI: 10.1126/science.abq4964.

Brophy, J.A.N., Magallon, K.J., Duan, L., Zhong, V., Ramachandran, P. *et al.* (2022) Synthetic genetic circuits as a means of reprogramming plant roots. *Science (New York, N.Y.)* 377(6607), 747–751. DOI: 10.1126/science.abo4326.

Caliskan, A., Caliskan, D., Rasbach, L., Yu, W., Dandekar, T. *et al.* (2023) Optimized cell type signatures revealed from single-cell data by combining principal feature analysis, mutual information, and machine learning. *Computational and Structural Biotechnology Journal* 21, 3293–3314. DOI: 10.1016/j.csbj.2023.06.002.

Cembrowska-Lech, D., Krzemińska, A., Miller, T., Nowakowska, A., Adamski, C. *et al.* (2023) An integrated multi-omics and artificial intelligence framework for advance plant phenotyping in horticulture. *Biology* 12(10), 1298. DOI: 10.3390/biology12101298.

Cerutti, G., Ribes, S., Godin, C., Galvan-Ampudia, C. and Vernoux, T. (2015) 3-d tessellation of plant tissue - A dual optimization approach to cell-level meristem reconstruction from microscopy images. In: *2015 International Conference on 3D Vision (3DV)*, IEEE, Lyon, France, pp. 443–451. DOI: 10.1109/3DV.2015.57.

Cha, S., Jun, G., Park, Y., An, S.J. and Lee, D. (2022) Toward depth-resolved analysis of plant metabolites by nanospray desorption electrospray ionization mass spectrometry. *Molecules (Basel, Switzerland)* 27(21), 7582. DOI: 10.3390/molecules27217582.

Chamara, N., Bai, G.F. and Ge, Y. (2023) AICropCAM-an edge imaging platform with a stack of deep learning models for plant phenotyping. Authorea Preprints. DOI: 10.22541/essoar.169699683.34657894/v1.

Chavhan, R.L., Hinge, V.R., Wankhade, D.J., Deshmukh, A.S., Mahajan, N. *et al.* (2024) Bioinformatics for molecular breeding and enhanced crop performance: Applications and perspectives. *Bioinformatics for Plant Research and Crop Breeding* 21–74.

Chelebian, E., Avenel, C. and Wählby, C. (2024) What makes for good morphology representations for spatial omics. *arXiv Preprint arXiv:2407.20660*. DOI: 10.48550/arXiv.2407.20660.

Chen, Y. and Zhang, S. (2022) Automatic cell type annotation using marker genes for single-cell RNA sequencing data. *Biomolecules* 12(10), 1539. DOI: 10.3390/biom12101539.

Chen, P., Liu, P., Zhang, Q., Zhao, L., Hao, X. *et al.* (2022) Dynamic physiological and transcriptome changes reveal a potential relationship between the circadian clock and salt stress response in *Ulmus pumila*. *Molecular Genetics and Genomics* 297(2), 303–317. DOI: 10.1007/s00438-021-01838-2.

Chen, Y., Huang, Y., Zhang, Z., Wang, Z., Liu, B. *et al.* (2023) Plant image recognition with deep learning: A review. *Computers and Electronics in Agriculture* 212, 108072. DOI: 10.1016/j.compag.2023.108072.

Clark, N.M., Hurgobin, B., Kelley, D.R., Lewsey, M.G. and Walley, J.W. (2023) A practical guide to inferring multi-omics networks in plant systems. In: *Plant Gene Regulatory Networks: Methods and Protocols*. Springer US, New York, NY, pp. 233–257. DOI: 10.1007/978-1-0716-3354-0_15.

Conneely, L.J., Berkowitz, O. and Lewsey, M.G. (2022) Emerging trends in genomic and epigenomic regulation of plant specialised metabolism. *Phytochemistry* 203, 113427.

De Rybel, B., Mähönen, A.P., Helariutta, Y. and Weijers, D. (2016) Plant vascular development: From early specification to differentiation. *Nature Reviews Molecular Cell Biology* 17(1), 30–40.

De Vos, D., Dzhurakhalov, A., Stijven, S., Klosiewicz, P., Beemster, G.T.S. *et al.* (2017) Virtual plant tissue: Building blocks for next-generation plant growth simulation. *Frontiers in Plant Science* 8, 686. DOI: 10.3389/fpls.2017.00686.

Dezem, F.S., Morosini, N.S., Arjumand, W., DuBose, H. and Plummer, J. (2024) Spatially resolved single-cell omics: Methods, challenges, and future perspectives. *Annual Review of Biomedical Data Science* 7.

Emanuel, G. and He, J. (2021) MERFISH enables mapping of cellular diversity with spatial context at subcellular resolution. *Microscopy Today* 29(4), 30–33. DOI: 10.1017/S1551929521000894.

Etalo, D.W., De Vos, R.C.H., Joosten, M.H.A.J. and Hall, R.D. (2015) Spatially resolved plant metabolomics: Some potentials and limitations of laser-ablation electrospray ionization mass spectrometry metabolite imaging. *Plant Physiology* 169(3), 1424–1435. DOI: 10.1104/pp.15.01176.

Etalo, D.W., Díez-Simón, C., Vos, R.C. and Hall, R.D. (2018) Laser ablation electrospray ionization-mass spectrometry imaging (LAESI-MS) for spatially resolved plant metabolomics. *Plant Metabolomics: Methods and Protocols* 253–267.

Fan, K. (2024) Metabolic engineering in plants: Advancing crop productivity and sustainability through precision pathway manipulation. *Chemistry*. DOI: 10.26434/chemrxiv-2024-qkkg6.

Fan, R., Deng, Y., Bartosovic, M., Ma, S., Zhang, D. *et al.* (2021) Spatial-ATAC-seq: Spatially resolved chromatin accessibility profiling of tissues at genome scale and cellular level. *In Review*. DOI: 10.21203/rs.3.rs-593462/v1.

Fang, R., Halpern, A., Rahman, M.M., Huang, Z., Lei, Z. *et al.* (2024) Three-dimensional single-cell transcriptome imaging of thick tissues. *eLife* 12, RP90029. DOI: 10.7554/eLife.90029.

Figueiredo, J., Santos, R.B. and Figueiredo, A. (2024) Exploring plant-pathogen interactions through subcellular proteomics: Insights and challenges. In: *Plant Pathogen Interaction*. Springer Nature Singapore, Singapore, pp. 287–310. DOI: 10.1007/978-981-99-4890-1_11.

Fordyce, R.F., Soltis, N.E., Caseys, C., Gwinner, R., Corwin, J.A. *et al.* (2018) Combining digital imaging and genome wide association mapping to dissect uncharacterized traits in plant/pathogen interactions. *Plant Biology*. DOI: 10.1101/296939.

Garcia Arredondo, M., Kew, W., Chu, R., Jones, M.E., Boiteau, R.M. *et al.* (2024) Differential exudation creates biogeochemically distinct microenvironments during rhizosphere evolution. *Environmental Science & Technology* 58(42), 18713–18722. DOI: 10.1021/acs.est.4c04108.

Gayoso, A., Steier, Z., Lopez, R., Regier, J., Nazor, K.L. *et al.* (2021) Joint probabilistic modeling of single-cell multi-omic data with totalVI. *Nature Methods* 18(3), 272–282. DOI: 10.1038/s41592-020-01050-x.

Geier, B., Sogin, E.M., Michellod, D., Janda, M., Kompauer, M. *et al.* (2020) Spatial metabolomics of *in situ* host-microbe interactions at the micrometre scale. *Nature Microbiology* 5(3), 498–510. DOI: 10.1038/s41564-019-0664-6.

Genshaft, A.S., Ziegler, C.G.K., Tzouanas, C.N., Mead, B.E., Jaeger, A.M. *et al.* (2021) Live cell tagging tracking and isolation for spatial transcriptomics using photoactivatable cell dyes. *Nature Communications* 12(1), 4995. DOI: 10.1038/s41467-021-25279-y.

Giacomello, S., Salmén, F., Terebieniec, B.K., Vickovic, S., Navarro, J.F. *et al.* (2017) Spatially resolved transcriptome profiling in model plant species. *Nature Plants* 3(6), 17061. DOI: 10.1038/nplants.2017.61.

Gladston, A. and Sucithra, B. (2023) Hand-crafted feature extraction and deep learning models for leaf image recognition. In: *Handbook of Research on Computer Vision and Image Processing in the Deep Learning Era*. IGI Global, pp. 161–178. DOI: 10.4018/978-1-7998-8892-5.ch010.

Gnann, C., Cesnik, A.J. and Lundberg, E. (2021) Illuminating non-genetic cellular heterogeneity with imaging-based spatial proteomics. *Trends in Cancer* 7(4), 278–282. DOI: 10.1016/j.trecan.2020.12.006.

Gong, C., Li, S., Wang, L., Zhao, F., Fang, S. *et al.* (2024) SAW: An efficient and accurate data analysis workflow for stereo-seq spatial transcriptomics. *GigaByte (Hong Kong, China)* 2024, gigabyte111. DOI: 10.46471/gigabyte.111.

González-Rodríguez, V.E., Izquierdo-Bueno, I., Cantoral, J.M., Carbú, M. and Garrido, C. (2024) Artificial intelligence: A promising tool for application in phytopathology. *Horticulturae* 10(3), 197. DOI: 10.3390/horticulturae10030197.

Govind, G., Kulkarni, J., Shinde, H., Dudhate, A., Srivastava, A. *et al.* (2022) Plant abiotic stress tolerance on the transcriptomics atlas. In: *Advancements in Developing Abiotic Stress-Resilient Plants*. CRC Press, pp. 193–236.

Guadagno, C.R., Ewers, B.E. and Weinig, C. (2018) Circadian rhythms and redox state in plants: Till stress do us part. *Frontiers in Plant Science* 9, 247. DOI: 10.3389/fpls.2018.00247.

Gupta, S., Kaur, R., Sharma, T., Bhardwaj, A., Sharma, S. *et al.* (2023) Multi-omics approaches for understanding stressor-induced physiological changes in plants: An updated overview. *Physiological and Molecular Plant Pathology* 126, 102047. DOI: 10.1016/j.pmpp.2023.102047.

Gupta, S., Gilotra, S., Rathi, S., Choudhury, T., Kotecha, K. *et al.* (2024) Plant disease recognition using different CNN models. In: *2024 14th International Conference on Cloud Computing, Data Science & Engineering (Confluence)*, IEEE, Noida, India, pp. 787–792. DOI: 10.1109/Confluence60223.2024.10463383.

Gurazada, S.G.R., Cox, K.L., Jr, Czymmek, K.J. and Meyers, B.C. (2021) Space: The final frontier – achieving single-cell, spatially resolved transcriptomics in plants. *Emerging Topics in Life Sciences* 5(2), 179–188. DOI: 10.1042/ETLS20200274.

He, X., Dias Lopes, C., Pereyra-Bistrain, L.I., Huang, Y., An, J. *et al.* (2024) Genetic-epigenetic interplay in the determination of plant 3D genome organization. *Nucleic Acids Research* 52(17), 10220–10234. DOI: 10.1093/nar/gkae690.

Hertzberg, L., Izraeli, S. and Domany, E. (2007) STOP: Searching for transcription factor motifs using gene expression. *Bioinformatics (Oxford, England)* 23(14), 1737–1743. DOI: 10.1093/bioinformatics/btm249.

Holtan, H.E.E. and Hake, S. (2003) Quantitative trait locus analysis of leaf dissection in tomato using *Lycopersicon pennellii* segmental introgression lines. *Genetics* 165(3), 1541–1550. DOI: 10.1093/genetics/165.3.1541.

Hu, Y., Dash, L., May, G., Sardesai, N. and Deschamps, S. (2024a) Harnessing single-cell and spatial transcriptomics for crop improvement. *Plants* 13(24), 3476. DOI: 10.3390/plants13243476.

Hu, Z., Liu, J., Shen, S., Wu, W., Yuan, J. *et al.* (2024b) Large-volume fully automated cell reconstruction generates a cell atlas of plant tissues. *The Plant Cell* 36(12), 4840–4861. DOI: 10.1093/plcell/koae250.

Huynh, K.L.A., Tyc, K.M., Matuck, B.F., Easter, Q.T., Pratapa, A. *et al.* (2024) Spatial deconvolution of cell types and cell states at scale utilizing TACIT. *Research Square* rs–3. DOI: 10.21203/rs.3.rs-4536158/v1.

Jaiswal, Y.S. and Williams, L.L. (2019) Spatio-temporal imaging and high-resolution mass spectrometry analysis: New avenues in herbal drug discovery and plant metabolomics. In: *Herbal Medicine in India: Indigenous Knowledge, Practice, Innovation and Its Value*. Springer Singapore, Singapore, pp. 249–261. DOI: 10.1007/978-981-13-7248-3_18.

James, D., Rennya, P.R., Mallavarapu, M.D., Panigrahi, R.C. and Patel, H.K. (2021) Integrating pan-omics data in a systems approach for crop improvement: Opportunities and challenges. *Omics Technologies for Sustainable Agriculture and Global Food Security* 215–246.

Jamil, I.N., Remali, J., Azizan, K.A., Nor Muhammad, N.A., Arita, M. *et al.* (2020) Systematic multi-omics integration (MOI) approach in plant systems biology. *Frontiers in Plant Science* 11, 944.

Jiang, L., Sun, J., Zhang, X. and Wang, S. (2024) Deep learning based herbal plant recognition. In: *2024 6th International Conference on Electronic Engineering and Informatics (EEI)*, IEEE, Chongqing, China, pp. 1477–1481. DOI: 10.1109/EEI63073.2024.10696459.

Jung, J.H. and Park, C.M. (2007) Vascular development in plants: Specification of xylem and phloem tissues. *Journal of Plant Biology* 50(3), 301–305. DOI: 10.1007/BF03030658.

Kalia, A. and Sharma, S.P. (2019) Single-cell omics in crop plants: Opportunities and challenges. Single-Cell Omics 341–355.

Kamel, M., Song, Y., Solbas, A., Villordo, S., Sarangi, A. *et al.* (2025) ENACT: End-to-end analysis of visium high definition (HD) data. *Bioinformatics (Oxford, England)* 41(3), btaf094. DOI: 10.1093/bioinformatics/btaf094.

Kawakatsu, T., Stuart, T., Valdes, M., Breakfield, N., Schmitz, R.J. *et al.* (2016) Unique cell-type-specific patterns of DNA methylation in the root meristem. *Nature Plants* 2(5), 16058. DOI: 10.1038/nplants.2016.58.

Kim, S. and Heo, S. (2024) An agricultural digital twin for mandarins demonstrates the potential for individualized agriculture. *Nature Communications* 15(1), 1561. DOI: 10.1038/s41467-024-45725-x.

Kindel, F., Triesch, S., Schlüter, U., Randarevitch, L.A., Reichel-Deland, V. *et al.* (2024) Predmoter-cross-species prediction of plant promoter and enhancer regions. *Bioinformatics Advances* 4(1), vbae074. DOI: 10.1093/bioadv/vbae074.

Köhler, C. and Springer, N. (2017) Plant epigenomics-deciphering the mechanisms of epigenetic inheritance and plasticity in plants. *Genome Biology* 18(1), 132. DOI: 10.1186/s13059-017-1260-9.

Köster, J., Brown, M. and Liu, X.S. (2019) A Bayesian model for single cell transcript expression analysis on MERFISH data. *Bioinformatics* 35(6), 995–1001. DOI: 10.1093/bioinformatics/bty718.

Kulasinghe, A., Berrell, N., Donovan, M.L. and Nilges, B.S. (2025) Spatial-omics methods and applications. *Gene Expression Analysis: Methods and Protocols* 101–146.

Landivar-Bowles, J., Pal, P., Bhandari, M. and Landivar-Scott, J.L. (2023) *Digital Twins Models for Crop Phenotyping, Management and Yield Forecasting*. Authorea Preprints.

Li, M., Niu, X., Li, S., Fu, S., Li, Q. *et al.* (2023) CRISPR/cas9 based cell-type specific gene knock-out in *Arabidopsis* roots. *Plants* 12(12), 2365. DOI: 10.3390/plants12122365.

Liew, L.C., Wang, Y., Peirats-Llobet, M., Berkowitz, O., Whelan, J. *et al.* (2020) Laser-capture microdissection RNA-sequencing for spatial and temporal tissue-specific gene expression analysis in plants. *Journal of Visualized Experiments* 162, e61517. DOI: 10.3791/61517.

Lippold, E., Schlüter, S., Mueller, C.W., Höschen, C., Harrington, G. *et al.* (2023) Correlative imaging of the rhizosphere–a multimethod workflow for targeted mapping of chemical gradients. *Environmental Science & Technology* 57(3), 1538–1549. DOI: 10.1021/acs.est.2c07340.

Liu, L., Chandrashekar, P., Zeng, B., Sanderford, M.D., Kumar, S. *et al.* (2021) TreeMap: A structured approach to fine mapping of eQTL variants. *Bioinformatics* 37(8), 1125–1134. DOI: 10.1093/bioinformatics/btaa927.

Lloyd, J.P.B., Ly, F., Gong, P., Pflüger, J., Swain, T. *et al.* (2022) Synthetic memory circuits for programmable cell reconfiguration in plants. *Synthetic Biology*. DOI: 10.1101/2022.02.11.480167.

Long, K.A., Lister, A., Jones, M.R.W., Adamski, N.M., Ellis, R.E. *et al.* (2024) Spatial transcriptomics reveals expression gradients in developing wheat inflorescences at cellular resolution. *Plant Biology*. DOI: 10.1101/2024.12.19.629411.

Lundberg, E. and Borner, G.H. (2019) Spatial proteomics: A powerful discovery tool for cell biology. *Nature Reviews Molecular Cell Biology* 20(5), 285–302.

Lv, Z., Jiang, S., Kong, S., Zhang, X., Yue, J. *et al.* (2024) Advances in single-cell transcriptome sequencing and spatial transcriptome sequencing in plants. *Plants (Basel, Switzerland)* 13(12), 1679. DOI: 10.3390/plants13121679.

Maher, K.A., Bajic, M., Kajala, K., Reynoso, M., Pauluzzi, G. *et al.* (2018) Profiling of accessible chromatin regions across multiple plant species and cell types reveals common gene regulatory principles and new control modules. *The Plant Cell* 30(1), 15–36.

Mahmood, A., Imran, M., Jamshaid, M.U., Riaz, U., Arif, M. *et al.* (2023) Integrated OMIC approaches for bioenergy crops. In: *Biotechnology and Omics Approaches for Bioenergy Crops*. Springer Nature Singapore, Singapore, pp. 77–84. DOI: 10.1007/978-981-99-4954-0_4.

Mao, X., Xia, D., Xu, M., Gao, Y., Tong, L. *et al.* (2025) Single-cell simultaneous metabolome and transcriptome profiling revealing metabolite-gene correlation network. *Advanced Science (Weinheim, Baden-Wurttemberg, Germany)* 12(4), e2411276. DOI: 10.1002/advs.202411276.

Maraveas, C. (2024) Image analysis artificial intelligence technologies for plant phenotyping: Current state of the art. *AgriEngineering* 6(3), 3375–3407. DOI: 10.3390/agriengineering6030193.

Marinello, F., Scaramuzzo, F.A., Dinarelli, S., Passeri, D. and Rossi, M. (2022) Nanoscale characterization methods in plant disease management. In: *Nanotechnology-Based Sustainable Alternatives for the Management of Plant Diseases*. Elsevier, pp. 149–177. DOI: 10.1016/B978-0-12-823394-8.00002-0.

Martin-Arevalillo, R. and Vernoux, T. (2019) Shining light on plant hormones with genetically encoded biosensors. *Biological Chemistry* 400(4), 477–486.

Masuda, S., Gan, P., Kiguchi, Y., Anda, M., Sasaki, K. *et al.* (2024) Uncovering microbiomes of the rice phyllosphere using long-read metagenomic sequencing. *Communications Biology* 7(1), 357. DOI: 10.1038/s42003-024-05998-w.

Mathur, J., Griffiths, S., Barton, K. and Schattat, M.H. (2012) Green-to-red photoconvertible mEosFP-aided live imaging in plants. In: *Methods in Enzymology*, Vol. 504. Academic Press, pp. 163–181. DOI: 10.1016/B978-0-12-391857-4.00008-2.

Mehta, A. and Silva, L.P. (2015) MALDI-TOF MS profiling approach: How much can we get from it? *Frontiers in Plant Science* 6, 184. DOI: 10.3389/fpls.2015.00184.

Merks, R.M.H., Guravage, M., Inzé, D. and Beemster, G.T.S. (2011) VirtualLeaf: An open-source framework for cell-based modeling of plant tissue growth and development. *Plant Physiology* 155(2), 656–666. DOI: 10.1104/pp.110.167619.

Miller, B.F., Huang, F., Atta, L., Sahoo, A. and Fan, J. (2022) Reference-free cell type deconvolution of multi-cellular pixel-resolution spatially resolved transcriptomics data. *Nature Communications* 13(1), 2339.

Milosavic, N. (2015) MALDI MS imaging-molecular mapping of biological samples by MALDI mass spectrometry. *Journal of Microbial & Biochemical Technology* 07(3), e122. DOI: 10.4172/1948-5948.1000e122.

Moore, S., Zhang, X., Liu, J. and Lindsey, K. (2015) Modelling plant hormone gradients. *eLS*. DOI: 10.1002/047001590X.

Mu, P., Li, W., Tran, L.S.P. and Li, X. (2024) SmT/SHM-seq: Simultaneously capturing spatial transcriptome and microbiome information in plants. *Trends in Plant Science* 29(11), 1277–1278. DOI: 10.1016/j.tplants.2024.09.010.

Naqvi, S.M.Z.A., Zhang, Y., Tahir, M.N., Ullah, Z., Ahmed, S. *et al.* (2023) Advanced strategies of the *in-vivo* plant hormone detection. *TrAC Trends in Analytical Chemistry* 166, 117186. DOI: 10.1016/j.trac.2023.117186.

Natarajan, S., Chakrabarti, P. and Margala, M. (2024) Robust diagnosis and meta visualizations of plant diseases through deep neural architecture with explainable AI. *Scientific Reports* 14(1), 13695. DOI: 10.1038/s41598-024-64601-8.

Nicol, P.B., Ma, R., Xu, R.J., Moffitt, J.R. and Irizarry, R.A. (2024) Identifying spatially variable genes by projecting to morphologically relevant curves. *Bioinformatics*. DOI: 10.1101/2024.11.21.624653.

Nintemann, S.J., Hunziker, P., Andersen, T.G., Schulz, A., Burow, M. *et al.* (2018) Localization of the glucosinolate biosynthetic enzymes reveals distinct spatial patterns for the biosynthesis of indole and aliphatic glucosinolates. *Physiologia Plantarum* 163(2), 138–154. DOI: 10.1111/ppl.12672.

Nobori, T., Monell, A., Lee, T.A., Zhou, J., Nery, J. *et al.* (2023) Time-resolved single-cell and spatial gene regulatory atlas of plants under pathogen attack. *Plant Biology*. DOI: 10.1101/2023.04.10.536170.

Novák, O., Napier, R. and Ljung, K. (2017) Zooming in on plant hormone analysis: Tissue- and cell-specific approaches. *Annual Review of Plant Biology* 68(1), 323–348. DOI: 10.1146/annurev-arplant-042916-040812.

Olanrewaju, O.S., Glick, B.R. and Babalola, O.O. (2024) Metabolomics-guided utilization of beneficial microbes for climate-resilient crops. *Current Opinion in Chemical Biology* 79, 102427. DOI: 10.1016/j.cbpa.2024.102427.

Oliva, M., Stuart, T., Tang, D., Pflueger, J., Poppe, D. *et al.* (2022) An environmentally responsive transcriptional state modulates cell identities during root development. *Plant Biology*. DOI: 10.1101/2022.03.04.483008.

Ouyang, W. and Li, X. (2023) CUT&tag for mapping *in vivo* protein-DNA interactions in plants. In: *Plant Gene Regulatory Networks: Methods and Protocols*. Springer US, New York, NY, pp. 109–117. DOI: 10.1007/978-1-0716-3354-0_8.

Padmarasu, S., Himmelbach, A., Mascher, M. and Stein, N. (2019) *In situ* hi-C for plants: An improved method to detect long-range chromatin interactions. *Plant Long Non-Coding RNAs: Methods and Protocols* 441–472.

Pallakies, H. and Simon, R. (2010) Positional information in plant development. In: *Encyclopedia of Life Sciences*. Wiley, Hoboken, NJ, USA. DOI: 10.1002/047001590X.a0021239.

Pandita, D., Pandita, A., Wani, S.H., Abdelmohsen, S.A.M., Alyousef, H.A. *et al.* (2021) Crosstalk of multi-omics platforms with plants of therapeutic importance. *Cells* 10(6), 1296. DOI: 10.3390/cells10061296.

Parker, G.D., Hanley, L. and Yu, X.-Y. (2023) Mass spectral imaging to map plant–microbe interactions. *Microorganisms* 11(8), 2045. DOI: 10.3390/microorganisms11082045.

Pasternak, T. and Pérez-Pérez, J.M. (2021) Methods of *in situ* quantitative root biology. *Plants* 10(11), 2399. DOI: 10.3390/plants10112399.

Pei, L., Li, G., Lindsey, K., Zhang, X. and Wang, M. (2021) Plant 3D genomics: The exploration and application of chromatin organization. *New Phytologist* 230(5), 1772–1786. DOI: 10.1111/nph.17262.

Peladarinos, N., Piromalis, D., Cheimaras, V., Tserepas, E., Munteanu, R.A. *et al.* (2023) Enhancing smart agriculture by implementing digital twins: A comprehensive review. *New Phytologist* 23(16), 7128. DOI: 10.3390/s23167128.

Peleke, F.F., Zumkeller, S.M., Gültas, M., Schmitt, A. and Szymański, J. (2024) Deep learning the cis-regulatory code for gene expression in selected model plants. *Nature Communications* 15(1), 3488. DOI: 10.1038/s41467-024-47744-0.

Peng, J.Y. and Huang, Y.P. (2005) The signalling pathways of plant defence response and their interaction. Zhi wu Sheng li yu fen zi Sheng wu xue xue bao=. *Journal of Plant Physiology and Molecular Biology* 31(4), 347–353.

Pérez-Llorca, M. and Müller, M. (2024) Unlocking nature's rhythms: Insights into secondary metabolite modulation by the Circadian clock. *International Journal of Molecular Sciences* 25(13), 7308. DOI: 10.3390/ijms25137308.

Permanes, S.R.E., Mammeri, Y. and Leclerc, M. (2024) Spatio-temporal modeling of host-pathogen interactions using level set method. *Plant Biology*. DOI: 10.1101/2024.08.09.607313.

Peukert, M., Matros, A., Lattanzio, G., Kaspar, S., Abadía, J. *et al.* (2012) Spatially resolved analysis of small molecules by matrix-assisted laser desorption/ionization mass spectrometric imaging (MALDI-MSI). *New Phytologist* 193(3), 806–815.

Proctor, C. and He, Y. (2021) Modelling root exudate accumulation gradients to estimate net exudation rates by peatland soil depth. *Plants* 10(1), 106.

Radkevich, E., D'Souza, D., Mattiuz, R., Merand, R., Chen, R. *et al.* (2024) Informing biologically relevant signal from spatial transcriptomic data. *bioRxiv* 2024–09.

Rahman, F., Mishra, A., Gupta, A. and Sharma, R. (2022) Spatiotemporal regulation of CRISPR/cas9 enables efficient, precise, and heritable edits in plant genomes. *Frontiers in Genome Editing* 4, 870108.

Rosselli, R. and Squartini, A. (2015) Metagenomics of plant–microbe interactions. In: *Advances in the Understanding of Biological Sciences Using Next Generation Sequencing (NGS) Approaches*. pp. 135–153.

Roychowdhury, R., Das, S.P., Gupta, A., Parihar, P., Chandrasekhar, K. *et al.* (2023) Multi-omics pipeline and omics-integration approach to decipher plant's abiotic stress tolerance responses. *Genes* 14(6), 1281. DOI: 10.3390/genes14061281.

Ruckelshausen, A. and Busemeyer, L. (2015) Toward digital and image-based phenotyping. In: *Phenomics in Crop Plants: Trends, Options and Limitations*. pp. 41–60.

Rudskiy, I.V. and Khodorova, N.V. (2012) Reconstruction of the 3D structure and developmental history of plant cells and tissues. *arXiv Preprint arXiv:1205.0225*. DOI: 10.48550/arXiv.1205.0225.

Ryu, K.H., Huang, L., Kang, H.M. and Schiefelbein, J. (2019) Single-cell RNA sequencing resolves molecular relationships among individual plant cells. *Plant Physiology* 179(4), 1444–1456. DOI: 10.1104/pp.18.01482.

Saarenpää, S., Shalev, O., Ashkenazy, H., Carlos, V., Lundberg, D.S. *et al.* (2024) Spatial metatranscriptomics resolves host-bacteria-fungi interactomes. *Nature Biotechnology* 42(9), 1384–1393. DOI: 10.1038/s41587-023-01979-2.

Sahoo, J.P., Behera, L., Sharma, S.S., Praveena, J., Nayak, S.K. *et al.* (2020) Omics studies and systems biology perspective towards abiotic stress response in plants. *American Journal of Plant Sciences* 11(12), 2172–2194. DOI: 10.4236/ajps.2020.1112152.

Sanchez-Munoz, R., Depaepe, T., Samalova, M., Hejatko, J., Zaplana, I. *et al.* (2024) The molecular core of transcriptome responses to abiotic stress in plants: A machine learning-driven meta-analysis. *Plant Biology*. DOI: 10.1101/2024.01.24.576978.

Sang, M., Shi, H., Wei, K., Ye, M., Jiang, L. *et al.* (2019) A dissection model for mapping complex traits. *The Plant Journal* 97(6), 1168–1182. DOI: 10.1111/tpj.14185.

Sang, T., Zhang, Z., Liu, G. and Wang, P. (2025) Navigating the landscape of plant proteomics. *Journal of Integrative Plant Biology* 67(3), 740–761. DOI: 10.1111/jipb.13841.

Santiago, A., Orduna, L., Fernandez, J.D., Vidal, A., de Martin-Agirre, I. *et al.* (2024) The plantae visualization platform: A comprehensive web-based tool for the integration, visualization, and analysis of omic data across plant and related species. *Bioinformatics*. DOI: 10.1101/2024.12.19.629382.

Schäfer, M., Pacheco, A.R., Künzler, R., Bortfeld-Miller, M., Field, C.M. *et al.* (2023) Metabolic interaction models recapitulate leaf microbiota ecology. *Science (New York, N.Y.)* 381(6653), eadf5121. DOI: 10.1126/science.adf5121.

Scharr, H., Dee, H., French, A.P. and Tsaftaris, S.A. (2016) Special issue on computer vision and image analysis in plant phenotyping. *Machine Vision and Applications* 27(5), 607–609. DOI: 10.1007/s00138-016-0787-1.

Sembiyaa, A., Dharshini, J., Rashmika, B., Manoj, M., Velayuthaprabhu, S. *et al.* (2025) Integrated molecular strategies for improving traditional chinese medicinal plants. In: *Biotechnology, Multiple Omics, and Precision Breeding in Medicinal Plants*. CRC Press, pp. 231–247.

Shafi, I., Gautam, M. and Kariyat, R. (2024) Integrating ecophysiology and omics to unlock crop response to drought and herbivory stress. *Frontiers in Plant Science* 15, 1500773. DOI: 10.3389/fpls.2024.1500773.

Shahan, R., Nolan, T.M. and Benfey, P.N. (2021) Single-cell analysis of cell identity in the *Arabidopsis* root apical meristem: Insights and opportunities. *Journal of Experimental Botany* 72(19), 6679–6686. DOI: 10.1093/jxb/erab228.

Shahzad, M.A., Khan, A., Khalid, M. and Gul, A. (2016) Plant proteomics: An overview. *Plant Omics: Trends and Applications* 295–325. DOI: 10.1007/978-3-319-31703-8_12.

Sijacic, P. and Liu, Z. (2010) Novel insights from live-imaging in shoot meristem development. *Journal of Integrative Plant Biology* 52(4), 393–399. DOI: 10.1111/j.1744-7909.2010.00941.x.

Šimková, H., Câmara, A.S. and Mascher, M. (2024) Hi-C techniques: From genome assemblies to transcription regulation. *Journal of Experimental Botany* 75(17), 5357–5365. DOI: 10.1093/jxb/erae085.

Singh, A., Chaubey, R., Srivastava, S., Kushwaha, S. and Pandey, R. (2021a) Beneficial root microbiota: Transmogrifiers of secondary metabolism in plants. *Emerging Trends in Plant Pathology* 343–365.

Singh, N., Mansoori, A., Dey, D., Kumar, R. and Kumar, A. (2021b) Potential of metabolomics in plant abiotic stress management. *Omics Technologies for Sustainable Agriculture and Global Food Security* 193–214.

Siqueira, J.A., Otoni, W.C. and Araújo, W.L. (2022) The hidden half comes into the spotlight: Peeking inside the black box of root developmental phases. *Plant Communications* 3(1), 100246. DOI: 10.1016/j.xplc.2021.100246.

Skobelev, P., Mayorov, I., Simonova, E., Goryanin, O., Zhilyaev, A. *et al.* (2021) Development of digital twin of plant for adaptive calculation of development stage duration and forecasting crop yield in a cyber-physical system for managing precision farming. In: *Cyber-Physical Systems: Digital Technologies and Applications*. Springer International Publishing, Cham, pp. 83–96. DOI: 10.1007/978-3-030-67892-0_8.

Smirnova, O.G. and Kochetov, A.V. (2020) Choice of the promoter for tissue and developmental stage-specific gene expression. In: *Biolistic DNA Delivery in Plants: Methods and Protocols*. pp. 69–106. DOI: 10.1007/978-1-0716-0356-7_4.

Som, A., Pahwa, M., Bawari, S., Saha, N.D., Sasmal, R. *et al.* (2021) Multiplexed optical barcoding of cells via photochemical programming of bioorthogonal host–guest recognition. *Chemical Science* 12(15), 5484–5494. DOI: 10.1039/D0SC06860H.

Soni, P., Nutan, K.K., Soda, N., Nongpiur, R.C., Roy, S. *et al.* (2015) Towards understanding abiotic stress signalling in plants: Convergence of genomic, transcriptomic, proteomic, and metabolomic approaches. In: *Elucidation of Abiotic Stress Signalling in Plants: Functional Genomics Perspectives*. pp. 3–40. DOI: 10.1007/978-1-4939-2211-6_1.

Sravanalakshmi, M., Manoj, M., Kumari, P., Srinidhi, M., Vijayarani, K. *et al.* (2025) Statistical tools and data analysis for organizing plant high-throughput phenomics. In: *Plant High-Throughput Phenotyping and Functional Phenomics*. CRC Press, pp. 260–281. DOI: 10.1201/9781003505631-15.

Srivastava, U. and Singh, S. (2022) Approaches of single-cell analysis in crop improvement. In: *Genomics of Cereal Crops*. Springer US, New York, NY, pp. 271–294. DOI: 10.1007/978-1-0716-2533-0_14.

Su, C., Lyu, M., Mähönen, A.P., Helariutta, Y., Rybel, B. *et al.* (2023a) Cella: 3D data visualization for plant single-cell transcriptomics in blender. *Physiologia Plantarum* 175(6), e14068. DOI: 10.1111/ppl.14068.

Su, J., Reynier, J.-B., Fu, X., Zhong, G., Jiang, J. *et al.* (2023b) Smoother: A unified and modular framework for incorporating structural dependency in spatial omics data. *Genome Biology* 24(1), 291. DOI: 10.1186/s13059-023-03138-x.

Sun, L., Zhou, J., Xu, X., Liu, Y., Ma, N. *et al.* (2024) Apping nucleosome-resolution chromatin organization and enhancer-promoter loops in plants using micro-C-XL. *Nature Communications* 15(1), 35. DOI: 10.1038/s41467-023-44347-z.

Sun, W., Yuan, S. and Li, K.C. (2008) Trait-trait dynamic interaction: 2D-trait eQTL mapping for genetic variation study. *BMC Genomics* 9, 1–13. DOI: 10.1186/1471-2164-9-242.

Susniak, K., Krysa, M., Gieroba, B., Komaniecka, I. and Sroka-Bartnicka, A. (2020) Recent developments of MALDI MSI application in plant tissues analysis. *Acta Biochimica Polonica* 67(3), 277–281. DOI: 10.18388/abp.2020_5394.

Swarup, R. and Bennett, M. (2014) Auxin transport: Providing plants with a new sense of direction. *The Biochemist* 36(2), 12–15. DOI: 10.1042/BIO03602012.

Tang, C., Xie, Y., Ruan, F., Li, Y., Luo, M. *et al.* (2022) Spatial chromatin accessibility sequencing resolves next-generation genome architecture. *In Review*. DOI: 10.21203/rs.3.rs-2314753/v1.

Tang, Z., Luo, S., Zeng, H., Huang, J., Sui, X. *et al.* (2024) Search and match across spatial omics samples at single-cell resolution. *Nature Methods* 21(10), 1818–1829. DOI: 10.1038/s41592-024-02410-7.

Tao, X., Gao, M., Wang, S. and Guan, X. (2022) Profiling of H3K4me3 modification in plants using cleavage under targets and tagmentation. *Journal of Visualized Experiments* 182(182), e62534. DOI: 10.3791/62534-v.

Tavallaee, G. and Orouji, E. (2025) Mapping the 3D genome architecture. *Computational and Structural Biotechnology Journal* 27, 89–101. DOI: 10.1016/j.csbj.2024.12.018.

Testillano, P.S., Solís, M.T. and Risueño, M.C. (2013) The 5-methyl-deoxy-cytidine (5mdC) localization to reveal *in situ* the dynamics of DNA methylation chromatin pattern in a variety of plant organ and tissue cells during development. *Physiologia Plantarum* 149(1), 104–113. DOI: 10.1111/ppl.12015.

Thrall, P.H. and Burdon, J.J. (1999) The spatial scale of pathogen dispersal: Consequences for disease dynamics and persistence. *Evolutionary Ecology Research* 1(6), 681–701.

Tippani, M., Divecha, H.R., Catallini, J.L., II, Kwon, S.H., Weber, L.M. *et al.* (2023) *VistoSeg*: Processing utilities for high-resolution images for spatially resolved transcriptomics data. *Biological Imaging* 3, e23. DOI: 10.1017/S2633903X23000235.

Tross, M.C., Duggan, G., Shrestha, N. and Schnable, J.C. (2024) Models trained to predict differential expression across plant organs identify distal and proximal regulatory regions. *Plant Biology*. DOI: 10.1101/2024.06.04.597477.

Valledor, L., Escandón, M., Meijón, M., Nukarinen, E., Cañal, M.J. *et al.* (2014) A universal protocol for the combined isolation of metabolites, DNA, long RNA s, small RNA s, and proteins from plants and microorganisms. *The Plant Journal* 79(1), 173–180. DOI: 10.1111/tpj.12546.

Venkat, A. and Muneer, S. (2022) Role of Circadian rhythms in major plant metabolic and signaling pathways. *Frontiers in Plant Science* 13, 836244. DOI: 10.3389/fpls.2022.836244.

Venkatramani, A., Ciftci, D., Cohen, L., Li, C., Pham, K. *et al.* (2024) Multiplexed optical barcoding and sequencing for spatial omics. *Genomics*. DOI: 10.1101/2024.06.04.597495.

Wang, N. and Liu, C. (2020) Study of cell-type-specific chromatin organization: *In situ* hi-C library preparation for low-input plant materials. *Plant Epigenetics and Epigenomics: Methods and Protocols* 115–127.

Wang, X., Han, J., Li, Z., Li, B., Wan, Y. *et al.* (2023) Editorial: Insight into plant spatial omics: Mass spectrometry imaging. *Frontiers in Plant Science* 14, 1273010. DOI: 10.3389/fpls.2023.1273010.

Wang, J.B., Li, X.L., Fan, R.X., Lü, P. and Yin, R.C. (2025) Recent progress in mass spectrometry imaging using nanospray desorption electrospray ionization. *Se Pu = Chinese Journal of Chromatography* 43(1), 43–49. DOI: 10.3724/SP.J.1123.2024.07013.

Wang, M., Zhang, S., Li, R. and Zhao, Q. (2024) Unraveling the specialized metabolic pathways in medicinal plant genomes: A review. *Frontiers in Plant Science* 15, 1459533. DOI: 10.3389/fpls.2024.1459533.

Waseem, M. and Aslam, M.M. (2024) Integrated approaches of "omics" for the improvement of stress tolerance in plants. In: *Improving Stress Resilience in Plants*. pp. 321–329.

Williams, C.G., Lee, H.J., Asatsuma, T., Vento-Tormo, R. and Haque, A. (2022) An introduction to spatial transcriptomics for biomedical research. *Genome Medicine* 14(1), 68.

Wong, J.W.H., Lutz, A., Natera, S., Wang, M., Ng, V. *et al.* (2019) The influence of contrasting microbial lifestyles on the pre-symbiotic metabolite responses of *Eucalyptus grandis* roots. *Frontiers in Ecology and Evolution* 7, 10. DOI: 10.3389/fevo.2019.00010.

Wu, Z., Tang, J., Zhuo, J., Tian, Y., Zhao, F. *et al.* (2019) Chromatin signature and transcription factor binding provide a predictive basis for understanding plant gene expression. *Plant and Cell Physiology* 60(7), 1471–1486. DOI: 10.1093/pcp/pcz051.

Xi, H., Xu, W., He, F., Liu, Z., Wang, Y. *et al.* (2024) Spatial metabolome of biosynthesis and metabolism in *Cyclocarya paliurus* leaves. *Food Chemistry* 443, 138519. DOI: 10.1016/j.foodchem.2024.138519.

Xia, K., Sun, H.-X., Li, J., Li, J., Zhao, Y. *et al.* (2022) The single-cell stereo-seq reveals region-specific cell subtypes and transcriptome profiling in *Arabidopsis* leaves. *Developmental Cell* 57(10), 1299–1310. DOI: 10.1016/j.devcel.2022.04.011.

Xie, F., Xi, N., Han, Z., Luo, W., Shen, J. *et al.* (2023) Progress in research on tumor microenvironment-based spatial omics technologies. *Oncology Research* 31(6), 877–885. DOI: 10.32604/or.2023.029494.

Xie, Y., Ruan, F., Li, Y., Luo, M., Zhang, C. *et al.* (2024) Spatial chromatin accessibility sequencing resolves high-order spatial interactions of epigenomic markers. *eLife* 12. DOI: 10.7554/eLife.87868.4.

Xing, T., Li, X.Q., Laroche, A., Tian, L., Tubei, K. *et al.* (2017) Protoplasts in the analysis of early plant-pathogen interactions: Current applications and perspectives. *European Journal of Plant Pathology* 149(4), 1001–1010. DOI: 10.1007/s10658-017-1230-9.

Xu, T. and Sankaran, K. (2021) Interactive visualization of spatial omics neighborhoods. *arXiv Preprint arXiv:2112.00902*. DOI: 10.48550/arXiv.2112.00902.

Xu, B., Chen, L., Lv, F., Pan, Y., Fu, X. *et al.* (2022) Visualization of metabolites identified in the spatial metabolome of traditional chinese medicine using DESI-MSI. *Journal of Visualized Experiments* 190(190), e64912. DOI: 10.3791/64912.

Xu, T., Zhu, E., Zhang, C., Calandrelli, R., Lin, P. *et al.* (2023) High-resolution characterization of human brain cortex with high-fidelity spatial transcriptomic slides (hifi-slides). *Bioengineering*. DOI: 10.1101/2023.06.12.544625.

Yadav, V., Wang, Z., Wei, C., Amo, A., Ahmed, B. *et al.* (2020) Phenylpropanoid pathway engineering: An emerging approach towards plant defense. *Pathogens (Basel, Switzerland)* 9(4), 312. DOI: 10.3390/pathogens9040312.

Yan, W., Li, Z., Pian, C. and Wu, Y. (2022) PlantBind: An attention-based multi-label neural network for predicting plant transcription factor binding sites. *Briefings in Bioinformatics* 23(6), bbac425. DOI: 10.1093/bib/bbac425.

Yang, L., Yang, Y., Huang, L., Cui, X. and Liu, Y. (2023) From single- to multi-omics: Future research trends in medicinal plants. *Briefings in Bioinformatics* 24(1), bbac485. DOI: 10.1093/bib/bbac485.

Yang, P., Jin, K., Yao, Y., Jin, L., Shao, X. *et al.* (2025) Spatial integration of multi-omics single-cell data with SIMO. *Nature Communications* 16(1), 1265. DOI: 10.1038/s41467-025-56523-4.

Yetgin, A. (2023) The dynamic interplay of root exudates and rhizosphere microbiome. *Soil Studies* 12(2), 111–120. DOI: 10.21657/soilst.1408089.

Yin, R., Xia, K. and Xu, X. (2023) Spatial transcriptomics drives a new era in plant research. *The Plant Journal* 116(6), 1571–1581. DOI: 10.1111/tpj.16437.

Yin, Z., Huang, W., Li, K., Fernie, A.R. and Yan, S. (2024) Advances in mass spectrometry imaging for plant metabolomics—expanding the analytical toolbox. *The Plant Journal* 119(5), 2168–2180. DOI: 10.1111/tpj.16924.

Yu, L., Zhang, H., Guan, R., Li, Y., Guo, Y. *et al.* (2023a) Genome-wide tissue-specific genes identification for novel tissue-specific promoters discovery in soybean. *Genes* 14(6), 1150. DOI: 10.3390/genes14061150.

Yu, X., Liu, Z. and Sun, X. (2023b) Single-cell and spatial multi-omics in the plant sciences: Technical advances, applications, and perspectives. *Plant Communications* 4(3), 100508. DOI: 10.1016/j.xplc.2022.100508.

Zechmann, B. and Zellnig, G. (2022) 3D reconstruction of plant leaf cells using TEM and FIB-SEM. *Microscopy and Microanalysis* 28(S1), 1098–1099. DOI: 10.1017/S1431927622004640.

Zemaitis, K., Lin, V., Ahkami, A., Winkler, T., Anderton, C. *et al.* (2023) Expanded molecular imaging of phytocompounds by mass spectrometry using novel on-tissue chemical derivatization. *ChemRxiv*. DOI: 10.26434/chemrxiv-2023-v0xj4.

Zhang, C., Zhou, C., Tian, C. and Guo, Y. (2024) Spatial transcriptomics techniques and its applications in plant research. *Chinese Journal of Biotechnology* 40(4), 971–987.

Zhang, J., Ahmad, M. and Gao, H. (2023a) Application of single-cell multi-omics approaches in horticulture research. *Molecular Horticulture* 3(1), 18. DOI: 10.1186/s43897-023-00067-y.

Zhang, Y., Berman, A. and Shani, E. (2023b) Plant hormone transport and localization: Signaling molecules on the move. *Annual Review of Plant Biology* 74(1), 453–479. DOI: 10.1146/annurev-arplant-070722-015329.

Zhang, L., Liang, X., Takáč, T., Komis, G., Li, X. *et al.* (2023c) Spatial proteomics of vesicular trafficking: Coupling mass spectrometry and imaging approaches in membrane biology. *Plant Biotechnology Journal* 21(2), 250–269. DOI: 10.1111/pbi.13929.

Zhang, X., Dippold, M.A., Kuzyakov, Y. and Razavi, B.S. (2019) Spatial pattern of enzyme activities depends on root exudate composition. *Soil Biology and Biochemistry* 133, 83–93. DOI: 10.1016/j.soilbio.2019.02.010.

Zhang, X., Pandey, M.K., Wang, J., Zhao, K., Ma, X. *et al.* (2021) Chromatin spatial organization of wild type and mutant peanuts reveals high-resolution genomic architecture and interaction alterations. *Genome Biology* 22(1), 315. DOI: 10.1186/s13059-021-02520-x.

Zhao, Z., Gu, H., Sheng, X., Yu, H., Wang, J. *et al*. (2014) Advances and applications in crop quantitative trait loci. *Journal of Nuclear Agricultural Sciences* 28, 1615–1624.

Zheng, Y., Chen, Y., Ding, X., Wong, K.H. and Cheung, E. (2023) Aquila: A spatial omics database and analysis platform. *Nucleic Acids Research* 51(D1), D827–D834. DOI: 10.1093/nar/gkac874.

Zhou, M., Coruh, C., Xu, G., Martins, L.M., Bourbousse, C. *et al*. (2022) The CLASSY family controls tissue-specific DNA methylation patterns in *Arabidopsis*. *Nature Communications* 13(1), 244. DOI: 10.1038/s41467-021-27690-x.

Zhuang, X. (2021) Spatially resolved single-cell genomics and transcriptomics by imaging. *Nature Methods* 18(1), 18–22. DOI: 10.1038/s41592-020-01037-8.

3 Plant High-throughput Transcriptomics: Fundamentals and Advances

Aryadeep Roychoudhury[1]*, Debdeep Chattopadhyay[2] and Urjashi Chatterjee[2]
[1]Discipline of Life Sciences, School of Sciences, Indira Gandhi National Open University, Maidan Garhi, New Delhi, India; [2]Department of Biotechnology, St. Xavier's College (Autonomous), Kolkata, India

3.1 Introduction

Transcriptomics refers to the study of the entire repertoire of RNA transcripts from a particular cell, tissue, or organism under a given condition. In plants, transcriptomics is a genome-wide snapshot of gene expression and isoform usage, which includes mRNA and non-coding RNAs; such snapshots are critical to link phenotype to genotype, map regulatory networks, and dissect responses to biotic and abiotic stresses. Modern high-throughput transcriptomics has advantages over older low-throughput methods as digital readouts quantify transcripts based on sequencing counts, providing consistent and reproducible measures throughout a very large dynamic range (Mortazavi *et al.*, 2008).

Historically, transcript profiling relied on hybridization microarrays (in the 1990s), which came with constraints stemming from the prior knowledge about transcripts, low dynamic range, and cross-hybridization. Next-generation sequencing (NGS) for high-throughput sequencing fundamentally transformed the field by introducing RNA-seq as an unbiased sequencing-based method that interrogated known and novel transcripts, splice isoforms, and allele-specific expression at greatly improved sensitivity compared to arrays (Wang *et al.*, 2009). Practical implementation of RNA-seq at scale was contingent on NGS platforms that reduced cost per base and increased throughput (Margulies *et al.*, 2005), and these technological achievements facilitated whole-transcriptome studies in both non-model and crop plants, at sequencing depths and involving number of samples previously unimaginable.

Generalized plant RNA-seq analyses generally consist of experimental design (replication, randomization), RNA extraction, and quality control (QC), library construction (polyA selection or rRNA depletion, strandedness), sequencing, read cleaning, alignment, or quasi-mapping, transcript quantification, normalization, and downstream analyses of differential expression, isoform analysis, and functional enrichment. Best practices and challenges (batch effects, library biases, quality issues, and multiple testing) are now well described (Conesa *et al.*, 2016).

NGS addressed array discovery and dynamic range challenges but presented an array of new conveniences engendering issues

*Corresponding author: aryadeep.rc@gmail.com

© CAB International 2026. *Plant Spatial Biology* (ed. Chen)
DOI: 10.1079/9781836994572.0003

such as library and sequencing biases (e.g. GC content, transcript length), the requirement for increasingly deeper sequencing to identify low abundance transcripts and isoforms, the need for proper quantitation of multi-mapping reads (gene families and paralogs very common in plant genomes), and the proverbial issues of computational loads elongated by potentially massive datasets. Accurately resolving isoforms, specifically for allele-specific expression in polyploid crops, is especially problematic. Many of these challenges have been alleviated through methodological advancements (e.g. strand-specific libraries, long-read sequencing, normalization improvements, and more sophisticated aligners and quantifiers), but overall experimental design requires developing trade-offs between costs, depth, and number of samples. Single-cell RNA-seq (scRNA-seq) uniquely adapted deep sequencing to single cells, where heterogeneity of cell types, developmental trajectories, and rare cell populations masked by bulk profiles were uncovered (Tang *et al.*, 2009). Spatial transcriptomics can overcome the tradition of losing tissue architecture to provide a measure of gene expression, while allowing a cell-resolved map of expression across a set of tissue slices (Ståhl *et al.*, 2016). Adapting these technologies for plant applications required addressing cell wall dissociation or protoplasting; overcoming biases associated with the protoplasting; and addressing trade-offs with resolution. Single-cell and spatial studies on plant systems are, however, rapidly developing.

High-throughput transcriptomics has revolutionized our understanding of how plants respond to pathogens and environmental stresses; for instance, comparative transcriptome profiling uncovers defense-related gene expression pathways, candidate resistance genes, and linkages between biotic and abiotic response networks. In studies of abiotic stress, transcriptomics can uncover shifts in hormone signaling pathways, shifts in transcription factor families, and metabolic reprogramming that underlie plant tolerance to stress; transcriptomic information can then also inform discovery of plant-derived biomarkers, breeding target selection, and timing of transgenic manipulation. The combination of bulk, single-cell, spatial, long-read, and multi-omics approaches holds promise to advance mechanistic, spatially resolved models of plant responses to environmental stressors and pathogens. With continued methodology refinement and development of better computational tools to analyze polyploid genomes, we anticipate that the utility of transcriptomics will accelerate basic discovery and agricultural innovation.

3.2 History of Plant Transcriptomics

Plant transcriptomics originated from early molecular biology discoveries involving messenger RNA and gene expression; however, the field progressed rapidly when expressed sequence tags (ESTs) enabled large-scale snapshot sequencing of expressed genes in the early 1990s, e.g. the EST concept and 'pilot' projects (Fig. 3.1) (Adams *et al.*, 1991). Then came microarray technology, originally developed for the parallel profiling of thousands of cDNAs on glass slides in the mid-1990s, that allowed plant scientists (and others) to ultimately compare expression

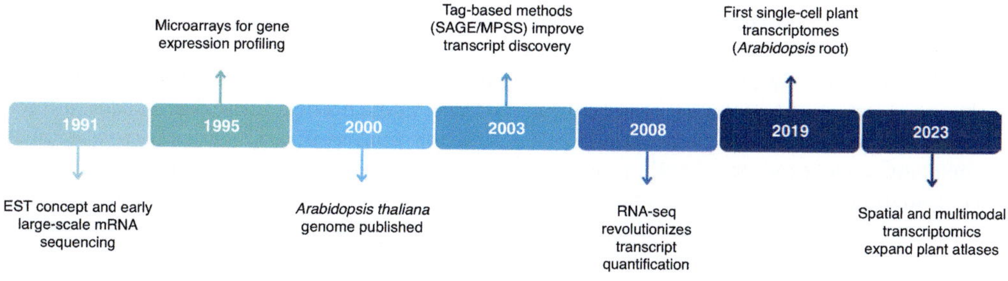

Fig. 3.1. Timeline showing major milestones in plant transcriptomics history, from EST discovery to the current spatial and multimodal era.

patterns across tissues, developmental stages, and stresses at a genome-scale by the emergence of EST catalogs and increasing sequence availability (Schena *et al.*, 1995). From the perspectives of genome-referenced gene modeling, the resulting public release of complete plant genomes, most importantly, the reference sequence of *Arabidopsis thaliana* in 2000 (The Arabidopsis Genome Initiative, 2000), matured transcriptome studies away from EST-anchored catalogs to genome-anchored transcriptional landscapes. In the 2000s, tag-based methods (SAGE/MPSS) and tiling arrays improved transcript discovery and isoform mapping; however, the major methodological innovation came with high-throughput sequencing applied to RNA (RNA-seq), which afforded digital, nucleotide-level quantification and discovery of new transcripts and splice variants (Mortazavi *et al.*, 2008). RNA-seq quickly became the new standard for bulk plant transcriptomics research. It created opportunities to conduct large-scale investigations into gene regulation in response to abiotic and biotic stress, development, and evolutionary processes across species. In the past five years, single-cell and single-nucleus transcriptomics have introduced an entirely new dimension to plant biology. Since 2019, a number of studies have been published on specialized high-resolution atlases of distinct *Arabidopsis* root cell types and their trajectories of development (Denyer *et al.*, 2019; Shulse *et al.*, 2019), which collectively illustrated that through either droplet-based or plate-based single-cell methods, it is now possible to resolve cell identity, lineage, and cell-type-specific responses to the environment in plants, despite challenges posed by cell walls and artifacts of protoplasting. At the same time, innovations in spatial transcriptomics, multimodal single-cell assay for transposase-accessible chromatin with sequencing (ATAC) +RNA assays, isolation protocols that improve plant nucleus yield, and cross-species atlasing approaches have vastly expanded our toolkit, the technologies available for mapping gene expression with positional and chromatin context, whilst also allowing for system-level views of developmental programs and stress responses that were unfathomable previously. New reviews and databases that summarize these datasets underscore notions of reproducibility, standardized pipelines, and integration across labs and species, suggesting a time in the not-so-distant future that comparative, cell-resolved transcriptomics across crops will be able to inform breeding designs, synthetic biology innovations, and precision agriculture applications (Jin *et al.*, 2023).

3.3 Problems Faced in RNA-seq for Plant Transcriptomics and the Improvements with NGS Advances

RNA-seq has been a central technique in plant transcriptomics since its introduction in the late 2000s, providing an unparalleled method to profile gene expression at the level of the genome. However, RNA-seq workflows, especially in the early days, were not without challenges. One of the most basic problems was the introduction of bias during library preparation. One of the main sources of either directional or random bias came from using random hexamer priming and/or PCR amplification, which could cause certain transcript regions to be over-represented and quantification of expression to be skewed, especially for highly expressed transcripts (Mortazavi *et al.*, 2008). Added complications arose from the fact that plant RNA extractions are often difficult due to the presence of polysaccharides, polyphenols, and secondary metabolites that co-precipitate with the nucleic acids and inhibit enzyme reactions and cDNA synthesis efficiency (Zhao *et al.*, 2014). The genomes of many plant species are large, with a large number of repetitive sequences with the potential for extensive segmental duplications and large gene families, which makes short-read mapping many times impossible and makes isoform discrimination difficult. Another significant hurdle in the early days of plant transcriptomics has been the lack of high-quality reference genomes. While well-annotated reference sequences were available for model organisms like *Arabidopsis thaliana*, a considerable number of crops and non-model plants had either incomplete or fragmented reference sequences. As a result, detecting new transcripts, isoforms, or alternative splicing events was difficult, which was particularly important to consider in plants, where splicing is known to regulate and thus be important for environmental adaptation (Leschner *et al.*, 2012). Methods for *de novo* transcriptome assembly were utilized to account for a lack of references, but many *de novo* transcriptome assembly methods create incomplete and redundant contigs, making it difficult to reliably assign functional annotation or quantify expression. Additionally, early Illumina platforms had an output read length limitation of (~35–50 bp), which contributed to errors in assembly and the inability to span full transcriptions of genes, ultimately hindering accurate gene model reconstruction.

NGS technology has already resolved many of the bottlenecks above, causing improvements in the accuracy of mapping reads, providing more accurate strand-specific RNA-seq protocols to preserve transcript orientation, and leading to the identification of antisense transcription and overlapping genes (Table 3.1). A novel advancement with third-generation sequencing platforms, such as PacBio Iso-Seq, and Oxford Nanopore sequencing, produces reads that span full-length transcripts, alleviating the assembly step and allowing resolution of complex alternative splicing events in plants (Sheng *et al.*, 2017). While long-read platforms have higher initial error rates, there are base-calling algorithms that continue to improve the accuracy of long-read platforms, as well as hybrid sequencing platforms that generate short-read Illumina data alongside long-read data, which allow for both high accuracy and high completeness of transcriptome projects (Sandoval-Velasco *et al.*, 2020). Multiplexing methods have decreased per-sample sequencing costs in many studies, and therefore have allowed for many more effective project experimental designs that sample both larger amounts of biological variability and greater power for statistical analysis. These advancements have significantly expanded the possibilities for plant transcriptomics. Researchers can now profile complete transcriptomes of polyploid crops, identify low-abundance or condition-specific transcripts, and produce cell-type-specific expression profiles using single-cell and single-nucleus RNA-seq (with some technical difficulties related to plant cell walls). Meanwhile, the incorporation of spatial transcriptomics and multimodal assays with NGS platforms is adding position and epigenomic context to transcriptional data and linking gene expression to development or environmental responses. While early RNA-seq research in plants faced technical biases, biochemical limitations, and incomplete genomic databases, the steady evolution of NGS technologies has made plant transcriptomics a multi-dimensional, high-resolution domain, able to deal with complex biological questions with unprecedented sophistication and depth.

3.4 RNA-seq

Before high-throughput sequencing, gene expression in plants and animals was mainly studied using DNA microarrays and SAGE. Since 2008, NGS has enabled transcriptome-wide analyses, allowing researchers to profile gene expression across tissues and conditions. RNA

Table 3.1. Comparative table summarizing the problems in early plant RNA-seq vs solutions provided by modern NGS technologies.

Problems in early RNA-seq for plants	NGS improvements/solutions
PCR and priming biases leading to uneven transcript coverage	Optimized library preparation and reduced PCR cycles; use of random priming corrections
RNA extraction issues due to polysaccharides and secondary metabolites	Improved RNA purification protocols and kits designed for plants
Short reads hindered isoform resolution and transcript assembly	Long-read sequencing (PacBio Iso-Seq, Nanopore) for full-length transcripts
Incomplete or poor-quality reference genomes in non-model plants	Hybrid sequencing combining short and long reads for accuracy
Low detection sensitivity for low-abundance transcripts	Deep sequencing and improved normalization protocols
High per-sample cost limiting replication and experimental scale	Multiplexing and cost reduction enabling large-scale studies

sequencing (RNA-seq) is one of the most widely used approaches. Based on second-generation sequencing, RNA-seq captures all mRNAs in a tissue or cell, enabling genome-wide assessment of gene expression, transcript structure, and regulatory networks. It also detects novel transcripts and sequence variants, including single-nucleotide polymorphisms (SNPs) and alternative splice isoforms. Compared with microarrays, RNA-seq offers a wider dynamic range, lower background noise, higher sensitivity, and improved reproducibility (Wang *et al.*, 2024a). By providing both quantitative and qualitative insights into coding and non-coding transcripts, RNA-seq has become a fundamental tool in modern molecular biology, enabling researchers to map gene regulatory networks, uncover novel genes, and investigate the molecular mechanisms underlying cellular function and adaptation (Tyagi *et al.*, 2022).

3.4.1 Methodology of standard RNA-seq

The RNA-seq workflow encompasses sections such as selection of sequencing platform, RNA extraction, library construction, sequencing, and data analysis. Illumina short-read sequencing is widely used due to its high throughput, accuracy, and cost efficiency. RNA must be carefully extracted to preserve its integrity and prevent degradation. Total RNA is isolated from biological samples using TRIzol or commercial kits, and mRNA is enriched, fragmented, and reverse-transcribed into cDNA, which is followed by second-strand synthesis, A-tailing, adapter ligation, and limited PCR amplification. In library construction, RNA is converted into sequencing libraries via oligo-dT enrichment, end-specific selection, or the use of unique molecular identifiers (UMIs) for duplicate detection. The fragment sizes are adjusted based on the sequencing platform. Long-read technologies like Oxford Nanopore MinION and PacBio Iso-Seq enable full-length transcript sequencing, reducing assembly errors and false splice isoforms. In short-read RNA-seq, cDNA is fragmented, ligated to adapters, amplified if required, and sequenced using single or paired-end methods. Libraries are then quantified and assessed for quality using a bioanalyzer and spectrophotometry before NGS. Raw reads are filtered for adapters, ambiguous bases, and low-quality sequences, and finally, the high-quality reads are used for transcriptome assembly via *de novo* methods such as Trinity, Oases, Rnnotator, or SOAPdenovo-Trans with Cogent applied for non-redundant transcript clustering and reconstruction. The assembly type can be reference-based or *de novo*, and it depends on the availability of genomic data. Single k-mer methods are preferred for accuracy, with Trinity widely used for high-quality outputs while Cufflinks and Scripture are used for reference-guided assembly. Data analysis includes transcript assembly, quality control, read alignment, and quantification of gene expression. Functional annotation of genes is performed by comparison to public databases such as Swiss-Prot, Pfam, GO, KEGG, etc. The genes are categorized by biological processes, cellular components, and molecular functions, and functional interpretation may include GO and KEGG enrichment or gene co-expression networks (WGCNA) (Tyagi *et al.*, 2022; Wang *et al.*, 2024a).

3.4.2 Applications of RNA-seq

Transcriptomics, particularly through RNA-seq, has transformed our understanding of gene expression by offering an in-depth, quantitative view of cellular transcription under diverse environmental conditions. Unlike earlier methods such as microarrays, which rely on hybridization, RNA-seq provides higher sensitivity and resolution that ensures the detection of low-abundance transcripts and subtle changes in gene expression that microarrays often fail to capture. This increased sensitivity allows researchers to measure the absolute number of RNA molecules present in a cell population and directly compare expression levels between different samples or experimental conditions. It facilitates the discovery of previously unannotated genes, assembles transcripts *de novo*, and identifies novel coding and non-coding RNA sequences without relying solely on existing genome annotations. It is highly effective in detecting sequence-level variations, such as polymorphisms within coding regions and potential gene fusions. Furthermore, it can capture alternative splicing events. Sequencing depth, defined as the ratio of total sequenced bases to the genome size, ensures sufficient coverage to detect these diverse transcripts, including those spanning splice junctions. Finally, RNA-seq enables the identification and characterization of non-coding RNAs (ncRNAs), which play important regulatory roles in cellular processes, metabolism, and stress responses (Tyagi *et al.*, 2022).

3.5 Transcriptomics of Single Cell

Single-cell transcriptomics has emerged as a powerful tool in plant research for detailed study and analysis of gene expression, cellular heterogeneity, development, and differentiation. This technology allows the study of plant responses to external stimuli like nutrient or environmental stress. Integrating single-cell transcriptomics with other omics approaches such as spatial transcriptomics, ATAC sequencing, and multi-omics techniques will enhance spatial resolution, remove problems of gene redundancy, and provide deeper insights into gene regulation and chromatin accessibility. However, challenges arising due to thick cell walls, tissue specificity, genotype, and environmental conditions make the isolation of single cells difficult. Recent advances in nuclei and protoplast isolation methods have improved cell yield and cost-effectiveness, enabling broader application of single-cell transcriptomics in plant biology and contributing to the construction of detailed cell atlases and understanding of plant developmental and stress-response mechanisms (Chen *et al.*, 2023).

3.6 Single-Cell RNA Transcriptomics

Single-cell RNA sequencing (scRNA-seq) has transformed plant transcriptomics by overcoming the limitations of bulk RNA-seq, which averages expression profiles across many cells, masking low-abundance transcripts and ignoring cellular heterogeneity. By profiling gene expression at single-cell resolution, scRNA-seq enables the identification of rare cell types, developmental trajectories, and adaptive responses to environmental stress. Its application in plants is challenging due to the rigid cell wall that requires enzymatic digestion to obtain single-cell suspensions, which often induce stress responses, causing dissociation biases and limiting access to certain tissues (Sun *et al.*, 2024). In plants, scRNA-seq allows systematic analysis of stem cells at the cellular and molecular level, revealing differentiation pathways and cell-type interactions. Early applications focused on *Arabidopsis* root tips, where studies showed that signaling pathways, such as brassinosteroids (BRs), influence cell division and phenotypic variation without affecting proliferation. Single-cell profiling has also uncovered developmental trajectories in rice root tissues, mechanisms of cell fate determination, and identification of marker genes for epidermal and ground tissue lineages. Overall, scRNA-seq provides unprecedented resolution to study heterogeneity between cells, differentiation, and lineage-specific gene regulation, thus offering valuable insights for plant development, functional genomics, and agricultural applications (Tyagi *et al.*, 2022). Single-nucleus RNA sequencing (snRNA-seq) offers a partial solution, avoiding protoplasting while still capturing transcriptomic information. Initially, it was focused on model species such as *A. thaliana* and rice, but these methods are now increasingly being applied to crops, including soybean, *Medicago truncatula*, and groundnut. Technological advances in scRNA-seq have progressed from low-throughput to high-throughput platforms. Earlier methods, such as STRT-seq, Smart-seq, Smart-seq2, and CEL-seq, improved transcript coverage and sensitivity but typically analyzed only tens to hundreds of cells. More recent developments include MARS-seq, Drop-seq, inDrop CytoSeq, and the commercial 10 × Genomics Chromium platform. These platforms use microfluidics, droplet encapsulation, and barcoding to profile tens of thousands of cells with higher reproducibility and rare cell detection capability. Certain cost-effective approaches like Seq-Well, SPLiT-seq, and Microwell-seq have expanded accessibility, and Smart-seq3 has enhanced sensitivity, isoform detection, and allele-specific resolution. Overall, scRNA-seq has evolved into a high-resolution, high-throughput tool that enables detailed exploration of plant cellular diversity, facilitating novel insights into development, stress adaptation, and evolutionary processes (Sun *et al.*, 2024).

3.6.1 Methodology

scRNA-seq technology can be divided into two components: the experimental components involved in generating scRNA-seq data and the computational components related to the analysis of scRNA-seq data. In plants, scRNA-seq has been performed using protoplasts or nuclei from roots, leaves, shoot apical meristems, stems, and inflorescences, with the majority of studies involving protoplasts. Protoplast-based scRNA-seq manages to capture both nuclear and cytoplasmic RNAs, providing a complete transcriptome view, but the enzymatic cell wall digestion can induce stress responses that alter gene expression. In contrast, snRNA-seq selectively captures nuclear transcripts, enabling profiling of cell types sensitive to enzymatic digestion. It allows the use of frozen samples via fluorescence-activated cell sorting (FACS), though it omits cytoplasmic RNA and associated post-transcriptional information. Both sample types can be processed on microfluidics-based platforms such as 10x Genomics, but require high-quality and debris-free suspensions. Protoplast isolation is most optimized for *A. thaliana*, but still certain tissues like xylem remain resistant to digestion. Finally, the choice between protoplast and nuclei preparations depends on the research objectives, species, and trade-offs between transcriptome coverage, cell-type representation, and technical feasibility. In plant research, the 10x Genomics system is the most widely used platform for scRNA-seq library construction and profiling between 5000 and 20,000 cells from various tissues. Other methods also show promise, such as the Nextera XT DNA Library Preparation Kit, which is used to profile over 12,000 *Arabidopsis* root cells, and the BD Rhapsody system, which captures a wide range of single cells and can measure both gene expression and protein abundance. The latter has been applied to profile over 37,000 cells from rice inflorescences, generating a detailed developmental atlas. Libraries from these platforms are typically sequenced using Illumina systems such as HiSeq, NextSeq, or NovaSeq (Islam *et al.*, 2024).

3.6.2 Some scRNA databases

The Plant Cell Marker DataBase (PCMDB) was developed using a standardized annotation process. It includes visualization tools, such as eFP Browser, bar graphs, and UMAP/t-SNE plots, to explore where and how these genes are expressed. PCMDB also offers analysis tools for identifying cell types and can suggest possible markers for plant species not yet included by finding matches to related species. With its organized design and multiple search options, PCMDB is a valuable resource for annotating plant cell types and predicting their functions (Jin *et al.*, 2022). The Plant Single-Cell Transcriptome Database (PlantscRNAdb) is a comprehensive resource compiling 26,326 marker genes from 128 cell types across four plant species: *A. thaliana*, *Oryza sativa*, *Zea mays*, and *Solanum lycopersicum*. The database allows users to explore gene expression patterns at both single-cell-type and genome-wide scales, supporting the identification of reference and marker genes for further functional studies. It addresses the current lack of multi-species plant scRNA-seq resources and provides a valuable tool for annotating cell types and understanding plant cellular functions at single-cell resolution (Chen *et al.*, 2021). Cellular heterogeneity during growth and differentiation in developing organs can be characterized at unprecedented resolution through single-cell transcriptomics. A method was developed in which genome-wide gene expression patterns of individual cells in a 3D flower meristem are reconstructed by combining snRNA-seq with microscopy-based 3D spatial reconstruction. By this approach, gene expression differences among meristematic domains giving rise to different tissues and organs can be determined. As a proof of principle, the method was applied to trace the initiation of vascular identity within the floral meristem, and the power of spatially reconstructed single-cell transcriptome atlases to enhance understanding of plant morphogenesis (Neumann *et al.*, 2022). The Plant Single

Cell Transcriptome Hub (PsctH) is a curated database of experimentally validated plant cell markers. It compiles 98 markers from 51 cell types across nine tissues in five plant species, all supported by RNA *in situ* hybridization or green fluorescent protein (GFP) reporter evidence. Roots contribute the largest share of entries, with lateral root primordia being the most represented cell type. PsctH enables keyword searches, provides detailed evidence, and offers downloadable data for plant single-cell research (Xu *et al.*, 2022).

3.6.3 Applications of scRNA transcriptomics

In 2019, the Plant Cell Atlas (PCA) initiative was launched to integrate scRNA-seq with spatial mapping of nucleic acids, proteins, and metabolites across plant cell types. However, plant-specific structural differences, such as variable cell wall thickness and composition, introduce unique technical challenges for scRNA-seq, making research design and sample preparation critical (Bawa *et al.*, 2022). scRNA-seq on roots and aerial tissues at defined times of day, capturing nearly 70,000 cells, was performed. This enabled the identification of shared markers between organs, detection of temporal transcriptome shifts, and discovery of non-annotated and dicistronic transcripts. A novel dividing cell marker providing a temporally resolved atlas of *Arabidopsis* tissues at single-cell resolution was also developed (Apelt *et al.*, 2022). It was used to study BR signaling in *briTRIPLE* mutants and found that BR influences cell division orientation and anisotropy without altering proliferation. Other studies profiled *rhd6* and *gl2* mutants to reveal partial identity shifts between hair and non-hair epidermal cells. Pseudotime analyses have enabled reconstruction of continuous developmental trajectories producing root cell atlases that detail gene expression cascades from stem cell niches through differentiation (Bawa *et al.*, 2022). In the maize plant, scRNA-seq has mapped ear and bundle sheath cells, revealing transcriptional differentiation, gene regulatory networks, and loci associated with yield traits, highlighting its potential for crop improvement (Tyagi *et al.*, 2022). Transcriptome analysis in *Arabidopsis* has been widely applied to understand gametophyte development, stress responses, nutrient signaling, and pathogenic defense responses (Imadi *et al.*, 2015). scRNA-seq studies in *Arabidopsis* have provided detailed insights into cellular diversity, developmental trajectories, and stress responses across multiple tissues. In roots, early work profiled over 12,000 cells, identifying markers and revealing dynamic gene regulation, followed by comprehensive root atlases and reconstruction of differentiation pathways. Significant heterogeneity has been observed even within the same cell type, and gene expression changes during lateral root initiation have been mapped, identifying key regulators and developmental trajectories. Stress-response analyses have revealed cell-type-specific transcriptional regulatory networks under heat and osmotic stress, while functional studies have linked regulatory programs to xylem formation, root hair density, and developmental stage-specific gene expression. In leaves, stomatal lineage development has been characterized, a detailed leaf vasculature atlas created, cell fate regulation mechanisms elucidated, and immune and susceptible states during *Pseudomonas syringae* infection mapped. Other tissues have also been explored, with findings showing ploidy-dependent transcriptome scaling, detailed profiling of the vegetative shoot apex, and the creation of an organism-wide atlas including lncRNA signatures. Spatial approaches have demonstrated the utility of snRNA-seq for profiling floral meristems and roots, revealing additional cell subtypes not captured by protoplast-based methods (Sun *et al.*, 2024).

scRNA-seq can provide powerful insights into plant hormone functions and transcriptional regulation. For instance, scRNA-seq revealed that ethylene (ET) and reactive oxygen act as wound signals promoting new root regeneration in *Arabidopsis*. By constructing gene expression networks, key transcription factors involved in development and differentiation can be identified. Examples include SHORT-ROOT (SHR) in maize roots, which regulates cortical tissue expansion and exhibits hypermobility in cortical cells, and LATERAL ORGAN BOUNDARIES DOMAIN (LBD) in tomato, which controls the initiation of roots. scRNA-seq has identified critical transcription factors in *Arabidopsis*, groundnut, tomato, and maize, highlighting its broad applicability in deciphering plant developmental programs. It is increasingly applied to uncover gene pathways underlying metabolite biosynthesis. In *Nicotiana attenuata*, cell clusters associated with benzylacetone production were identified through co-expression analysis and functional validation via gene silencing. In *Catharanthus roseus*, scRNA-seq enabled mapping the spatial organization of monoterpenoid indole alkaloid (MIA) metabolism. Similarly, in tea leaves, a novel catechin ester glucosyltransferase was discovered, uncovering a metabolic pathway for catechin ester synthesis. These studies demonstrate that it can potentially link cell-type-specific transcriptional activity to specialized metabolic functions, enabling a detailed understanding of plant secondary metabolism (Zheng *et al.*, 2023). It allows the classification of both common and rare cell types based on marker genes, like the transcriptional heterogeneity observed in the quiescent center of *Arabidopsis* roots and phloem parenchyma in leaves. It enables validation of known markers, discovery of novel markers, and identification of conserved and divergent transcriptomic features. It also facilitates identification of differentially expressed genes (DEGs) and gene regulatory networks (GRNs) at single-cell resolution. Integrating GRN analysis with trajectory inference uncovers cell-specific regulatory mechanisms underlying development, growth, and stress responses (Wang *et al.*, 2024a).

3.7 Plant Spatial Transcriptomics

The scRNA-seq technique has provided valuable insights into plant biology; however, its application in plant tissues faces several challenges. First, protoplasting is technically difficult for certain organs and can introduce artifacts that alter gene expression profiles. Second, some cell types are resistant to protoplasting, which may skew the representation of cell populations. Finally, tissue dissociation results in the loss of spatial context, which is critical for identifying cell types without known marker genes and for analyzing cell–cell and cell–environment interactions. These limitations underscore the need for newer methods that preserve transcript localization while still providing high-resolution gene expression profiling (Sang and Kong, 2024). Spatial transcriptomics was originally developed for mammalian systems but is increasingly being adapted for plants despite challenges posed by cell walls and tissue complexity. Research focuses on optimizing methods for plants and applying them to study gene expression *in situ*. Spatial transcriptomics integrates histological imaging with RNA sequencing to quantify and map gene expression simultaneously. This technique was adapted for plants to address the challenges posed by cell walls, vacuoles, chloroplasts, and secondary metabolites in plants (Giacomello, 2021).

Spatial transcriptomic technologies can be categorized into three main types based on how spatial information is obtained: laser capture microscopy (LCM)-based methods, imaging-based approaches, and *in situ* capture sequencing-based methods. LCM-based methods, such as LCM-seq and Geo-seq, provide anatomical or cellular resolution by isolating specific tissue regions but are limited by cumbersome processing and low throughput. Imaging-based approaches, including FISSEQ, MERFISH, osmFISH, BaristaSeq, and STARmap, achieve single-cell or subcellular resolution through *in situ* hybridization and high-resolution microscopy, though their throughput and sensitivity are constrained by tissue size. *In situ* capture sequencing methods, such as Visium, Slide-seq, HDST, DBiT-seq, Seq-Scope, and Stereo-seq, encode spatial information via barcoded mRNA capture, allowing high-throughput sequencing of larger tissue areas with near single-cell or subcellular resolution. Among these, only Visium and Stereo-seq have been successfully applied to plant research, enabling differentiation of morphologically similar cell types and the mapping of tissue complexity. Advancements such as Stereo-seq extend the field of view to centimeter-scale tissue slices, permitting large-scale spatial mapping of organs and developmental stages while preserving single-cell resolution. These high-resolution and high-throughput methods allow researchers to investigate cell heterogeneity, subcellular organization, receptor–ligand interactions, and spatial gene expression patterns across both small and large plant tissues, providing a critical tool for understanding plant development, physiology, and responses to environmental stimuli (Yin *et al.*, 2023).

3.7.1 Methods in plant spatial transcriptomics

Spatial transcriptomics in plants is limited by cell structures such as rigid cell walls, large vacuoles, and variable lignification as seen in any transcription procedure involving plants. Early studies preserved spatial information by dissecting tissues in order, e.g. separating leaves from outer to inner layers or sorting root and aerial tissues and sequencing them individually. While this maintains general spatial context, resolution is still low due to the pooling of multiple cells. Single-molecule fluorescence *in situ* hybridization (smFISH) is a classical technique for spatial transcriptomics. It employs libraries of fluorescently labeled 20-base oligonucleotides to visualize individual RNA molecules within intact tissues, providing precise localization and quantification of transcripts. smFISH is more sensitive than scRNA-seq and preserves the spatial context of cells, making it a better method for experimental validation. However, its application in plants is constrained by tissue autofluorescence, limitations in probe design, and the inability to profile the entire transcriptome. Combinatorial labeling strategies have been developed to increase multiplexing; yet, plant applications remain limited.

To overcome these limitations, untargeted spatial transcriptomic approaches have emerged, enabling transcriptome-wide mapping without prior probe selection (Sang and Kong, 2024). High-resolution and near single-cell data have been achieved using high-throughput methods involving targeted tissue collection, embedding followed by cryosectioning, fixation, staining, and finally imaging, permeabilization, and sequencing. Embedding typically uses optimal cutting temperature (OCT) compounds to support tissues during cryosectioning, which must be carefully optimized for thickness and temperature (−15 to −20 °C) to avoid breakage. Fixation protocols vary, but mainly formaldehyde or methanol is used, and staining dyes such as toluidine blue or hematoxylin–eosin are utilized. Imaging captures tissue organization followed by permeabilization, which is critical for efficient mRNA release, and some protocols include pre-permeabilization to partially denature proteins. Sequencing then proceeds through reverse transcription, tissue removal, and probe release using nuclease-free water to protect cDNA–mRNA hybrids. Although plant-specific studies are limited, methods from mammalian spatial transcriptomics and manufacturer protocols provide a framework that can be adapted and optimized for plant tissues (Chen *et al.*, 2023). In the method described by Giacomello (2021), cryosections are placed on arrays coated with reverse transcription oligo(dT) primers carrying unique spatial barcodes. Following fixation, staining, and permeabilization, mRNA is locally captured and reverse transcribed to cDNA. The tissue is then removed, followed by mRNA–cDNA hybrid release and gene analysis of expression data linked to tissue morphology.

To optimize the method for plants, modifications included milder fixation to prevent irreversible binding, Toluidine blue staining for morphology visualization, tissue-specific enzymatic permeabilization, additional steps to account for secondary metabolites, and an enzyme cocktail to degrade tissue without affecting captured transcripts can be followed. The protocol was validated across diverse plant tissues, including the *Arabidopsis* inflorescence meristem, *Populus* leaf buds, and *Picea* female cones, etc. The method demonstrated high biological and technical reproducibility. Localization of known genes and hierarchical clustering confirmed expected spatial patterns, such as flowering-related genes in floral tissues. The approach also enabled quantification of differences in gene expression and functional pathway analysis across different tissue regions and replicates. Overall, the method provides scalable, high-throughput, and spatially resolved transcriptomics in plant tissues, offering clear advantages over existing techniques like FACS or laser capture microdissection, which require transgenic lines or have limited scalability (Lieben, 2017). Current strategies include NGS-based methods like 10 × Visium, which combine barcoded RNA capture with sequencing for transcriptome-wide spatial mapping. Each read carries a spatial barcode, allowing transcript localization. Initial arrays had 100 µm spot resolution, whereas current platforms, such as 10x Genomics Visium, provide 55 μm spot diameters and 5000 spots per capture area. Probe sequences incorporate spatial barcodes, UMIs, and poly(dT) regions to enable transcript capture and mapping (Sang and Kong, 2024). Some imaging-based methods, such as STARmap, *in situ* hybridization, *in situ* sequencing, and MERFISH, visualize gene expression in tissue but are limited by low mRNA abundance. Other techniques like INTACT (isolation of nuclei tagged in specific cell types), FACS, and LCM offer cell-specific data but often compromise spatial context or require transgenic lines. While spatial transcriptomics preserves tissue context and reveals cellular heterogeneity, its use in plant research remains limited due to technical challenges such as resolution, sensitivity, and RNA quality (Chen *et al.*, 2023).

3.7.2 Applications of spatial transcriptomics

Using spatial transcriptomics, a study demonstrated that high-resolution dissection of *A. thaliana* leaves allows precise mapping of tissue-specific and site-specific gene expression, thus revealing patterns that bulk RNA-seq cannot capture. Using spatial transcriptomics through the GaST-seq (grid-assisted, spatial transcriptome sequencing) method, the study established a robust, low-cost workflow to generate high-resolution maps of gene expression from very small plant tissue sections. Applying GaST-seq to *A. thaliana* leaves challenged with bacterial peptide flg22 revealed spatially localized immune and defense responses, including known *FLARE* genes and potential novel candidates, highlighting regulatory elements with short-range signaling potential. Clustering of these differentially expressed genes uncovered spatially defined regulatory patterns and associated biological processes, highlighting local vs adjacent tissue responses. Further analysis identified key transcription factors, particularly WRKY family members, that likely control spatially resolved immune responses (Giolai *et al.*, 2019). This technology is particularly powerful in plant research as it allows the analysis of tissue-specific gene expression patterns across diverse species, ranging from herbaceous plants to trees. It can identify genes associated with fundamental biological processes such as translation, photosynthesis, cell division, differentiation, and morphogenesis. Additionally, the technique can detect homologs of known developmental regulators involved in leaf and floral development, providing new findings on spatially coordinated organogenesis (Giacomello, 2021). It enables mapping of cell types and developmental lineages in plants that can reveal subtle differences in cell fate. It can trace shoot progenitor cells and monitor key regulators like WUS and WOX5 during regeneration. The technology also helps to dissect the Kranz structure in C4 plants, mapping bundle sheath and mesophyll cells to uncover molecular regulators of tissue differentiation. Comparative studies across C3 and C4 species provided insights into the evolution and functional adaptation of specialized structures (Yin *et al.*, 2023). It is especially valuable for studying plant tissues where single-cell approaches face challenges. The technology also facilitates developmental and pathway studies, such as mapping stamen filament elongation or pollen wall formation, and performing spatial network enrichment analysis to uncover active biological pathways in specific tissue regions (Giacomello, 2021). Spatial transcriptomics allows mapping of plant–microbe interactions at the cellular level, such as distinguishing infected and uninfected cells in soybean–rhizobia nodules to study nitrogen fixation and signaling. It also identifies cell types susceptible to pathogens and tracks immune signal propagation, providing insights into symbiosis and disease resistance. It can also identify the specific cell types and molecular regulators involved in the biosynthesis of valuable plant metabolites such as vinblastine in *C. roseus*. The technology can also be applied to tea, rubber, and salidroside-producing plants, and, when combined with spatial metabolomics, it can provide detailed findings on enzyme activity, metabolite localization, and the organization of metabolic pathways (Yin *et al.*, 2023).

Spatial transcriptomics has differentiated upper and lower epidermal cell subtypes in *Arabidopsis* leaves and uncovered fine-scale expression patterns among groundnut fruit cells of the same type. In orchids, it has traced the developmental differentiation of phloem cells into both vegetative structures and reproductive structures, highlighting tissue-specific developmental trajectories. It has also revealed distinct expression domains in MADS-box genes across different floral regions. Spatial modeling of C4 and CAM photosynthesis in amaranth has provided insight into the cellular organization of metabolic pathways. Similarly, temporal and spatial profiling in germinating barley seeds identified dynamic expression of auxin transport genes in specific embryonic regions, pointing towards their role in axis formation during early development (Wang *et al.*, 2024a). Finally, it supports exploratory and comparative research, allowing untargeted analysis without preselecting gene targets. This makes it suitable for species with complex or partially annotated genomes and enables comprehensive studies of spatial gene expression changes associated with growth, differentiation, and organ development (Giacomello, 2021).

3.8 Spatial Enhanced Resolution Omics Sequencing

For higher resolution and single-cell-level analyses, spatial enhanced resolution omics sequencing (Stereo-seq) integrates scRNA-seq with high-resolution spatial transcriptomics. Stereo-seq uses DNA nanoball (DNB) chips with spots ~ 220 nm in diameter and center distances of 500–715 nm. Each DNB contains a coordinate identity (CID) barcode, molecular identifiers (MIDs) and polyT oligos, enabling spatially resolved transcriptomic profiling at subcellular resolution. Tissue sections are placed on the chips, mRNA captured, and *in situ* sequencing performed. High-resolution images combined with MIDs allow the extraction of single-cell transcriptomes and identification of cell subtypes.

Stereo-seq has been applied in *Arabidopsis* leaves to resolve upper and lower epidermal cell-specific transcriptomic differences and to map gradients from main veins to leaf edges to find out spatial developmental trajectories of vascular and guard cells. It has also been used to study onion bulb enlargement, generating spatiotemporal transcriptomic maps that identified sponge cells as key contributors and highlighted regulatory pathways governing growth. Stereo-seq surpasses other techniques in detecting resolution (down to 500 nm), spot density, and capture area, offering useful insights into heterogeneity in plant cell types. There are limitations of this technique that include the efficiency of recognition of single-cell boundaries and resolution improvements for small or densely packed cells, such as shoot apical meristems (Sang and Kong, 2024).

3.9 Data Integration of scRNA-seq with Spatial Transcriptomics

Spatial transcriptomics provides information on the spatial organization of gene expression, thus complementing scRNA-seq. In plants, its application is limited by lower resolution, incomplete coverage, and other technical challenges such as cell fragility, RNA diffusion, and difficulty in distinguishing cell types. Integrating scRNA-seq with spatial data enables high-resolution mapping of cellular gene expression while leveraging the strength of both approaches.

Computational tools for integration are essential, particularly for handling large datasets and correcting batch effects between technologies. Methods include anchor-based annotation transfer, canonical correlation analysis, machine learning approaches such as Bayesian models, neural networks, and deep generative models that impute gene expression at the pseudo-cell level. These approaches allow spatial transcriptomics data to be combined with scRNA-seq or snRNA-seq, thereby improving gene expression mapping, tissue composition analysis, and cell-type assignment within histological sections. Pipelines and platforms streamline such an analysis and visualization process for providing standardized formats and interactive tools for end-to-end spatial data interpretation.

Recent studies in plants have demonstrated the power of these integrative approaches. Researchers have successfully captured transitional cell states during nodule development in soybean and mapped spatiotemporal gene expression dynamics during shoot regeneration in tomato callus. Despite these advances, challenges remain, such as the mismatch between cross-sectional tissue data and complete single-cell profiles, limited RNA recovery from water-rich tissues, batch effect overcorrection, and difficulties in integrating datasets without common anchors. Addressing these challenges requires robust computational models that identify shared features while preserving sample-specific biological variation (Sang and Kong, 2024).

3.10 Application of Transcriptomics in Plant Research

Studies on male and female gametophytes revealed stage-specific and cell-type-specific expression patterns such as pollen and sperm-specific mRNAs, early factors responsible for pollen tube growth, and genes controlling first mitotic divisions and seed dormancy. Transcriptomics has elucidated responses to environmental stimuli such as carbon dioxide, light, hydrogen peroxide, and oxidative stress, highlighting crosstalk between redox signaling, photoreceptor pathways, and stress-responsive genes. Analysis under nutrient deprivation conditions, specifically including sulfur, phosphorus, magnesium, potassium, and sucrose, led to the discovery of regulatory networks for nutrient sensing, hormone signaling, autophagy, and metabolic adaptation. Transcriptome profiling of roots and seedlings exposed to abiotic stress or growth regulators revealed the induction of genes involved in detoxification, cell rescue, transport, and hormone-mediated signaling. Additionally, pathogen and pest interactions showed transcriptional reprogramming of defense pathways mediated by SA, JA, and ET that revealed differential responses to phloem-feeding insects vs pathogens. Overall, transcriptomic applications in *Arabidopsis* provided insights into defense mechanisms, environmental adaptation, nutrient management, and reproductive development, establishing a framework for functional genomics and stress biology (Imadi *et al.*, 2015). Transcriptomic and metabolomic approaches have significantly advanced our understanding of sulfur uptake and assimilation in plants, which are crucial for the biosynthesis of cysteine and methionine. By applying high-throughput technologies to *Arabidopsis* under sulfur-deficiency stress, researchers have been able to capture a systems-level view of the plant response. Transcriptomics has allowed the identification of gene expression changes across the sulfur assimilation network and when combined with metabolomic profiling, these studies have uncovered temporal dynamics of sulfur stress responses, distinguishing an initial short-term state which is marked by enhanced lateral root development to scavenge for sulfur to a later long-term state, where the plant reallocates resources toward vital seed production when nutrient recovery is insufficient. This integrated data-driven approach provides a comprehensive framework for understanding the state transitions in sulfur metabolism and offers potential strategies for manipulating the bioproduction of sulfur-containing compounds in plants (Hoefgen and Nikiforova, 2008). Transcriptome sequencing of medicinal plants (e.g. *Vitis amurensis*, *Aloe vera*, *Trillium govanianum*, *Polygonum minus*, and *Callerya speciosa*) identified genes and transcription factors involved in stress responses, secondary metabolite biosynthesis, and hormone signaling. Transcriptomic analyses of other medicinal plants (e.g. *Entada phaseoloides*, *Lantana camara*, *Tetrastigma hemsleyanum*, *Saussurea lappa*, *Arisaema heterophyllum*, *Ginkgo biloba*, *Artemisia argyi*, and *Panax ginseng*) have revealed genes and pathways involved in the biosynthesis of flavonoids, saponins, terpenes, and other bioactive compounds. Simple sequence repeats (SSRs) have been widely used for assessing genetic diversity, mapping, and breeding. Studies on *Asparagus*, *Salvadora oleoides*, *Populus alba*, *Gastrodia elata*, *Tinospora cordifolia*, *Menispermum*, and *Paeonia lactiflora* identified numerous SSRs and polymorphic markers providing information for future conservation, breeding, and functional genomics research (Guo *et al.*, 2021; Tyagi *et al.*, 2022).

3.11 Application of Plant Transcriptomics in the Study of Diseases

3.11.1 Introduction to plant pathogen interactions

Based on a genome-wide analysis approach, plant transcriptomics provides the opportunity to better understand host–pathogen interactions by gaining an inclusive understanding of how plants recognize, transduce, and respond to infection. Through a comparative transcriptome approach, transcriptomes of healthy tissue vs infected tissue can be compared in order to identify pathways and genes concerning pathogen recognition (like pattern recognition receptors), signal transduction, and defensive responses. These data will allow researchers to further explore the molecular basis of disease resistance and susceptibility (Kawahara *et al.*, 2012). Further, by using time-resolved transcriptome studies, we can track the timing of the activation of the pattern-triggered immunity (PTI) and effector-triggered immunity (ETI) pathways using specific windows of time (time from first pathogen contact, mitogen-activated protein kinase (MAPK) cascade activation, hormonal cross-talk, etc.). As time-resolved transcriptome studies allow for the simultaneous tracking of both transient defense responses and unique early transcriptional changes that coincide with any late transcriptional changes, there may be interesting regulatory changes occurring early that could have been missed in static expression studies; for example, an early burst in transcription of signaling and MAPK pathway components, and pathogenesis related (*PR*) genes as primary markers of PTI. In rice infected by *Magnaporthe oryzae*, the early whole transcriptome showed increased expression of signaling and secondary metabolism genes associated with defense, mainly during the incompatible interaction. Some of the genes encoding WRKY transcription factors were significantly upregulated, for example, *OsWRKY47*, which increased resistance to rice blast fungus (Wei *et al.*, 2013). Additionally, *OsWRKY67* was activated when rice plants were infected by leaf or panicle blast, and its overexpression conferred resistance to both rice blast and bacterial blight by activating an SA-responsive *PR* gene (Liu *et al.*, 2018).

3.11.2 Identifying defense-related genes and pathways

RNA-seq has demonstrated that defense-related genes are dynamic concerning regulation, including of PR proteins, receptor-like kinases, transcription factors (WRKY, NAC, MYB families), and genes involved in signaling of phytohormones like SA, JA, and ET, for example, transcriptomic profiling in *Oryza sativa* infected with *Magnaporthe oryzae* showed the rapid induction of WRKY transcription factors and JA/ET signaling components involved with resistance (Wei *et al.*, 2013). Transcriptomics, coupled with co-expression and epigenomic profiling, can also be leveraged to determine transcriptional hubs or transcriptional regulators. One can learn about key transcription factors marked by histone modifications (for example, H3K9ac) during the activation of plant defenses, or find co-expression modules that include hormones and signaling that can be used as candidates for functional validation. In rice and its responses to *M. oryzae*, transcriptomic profiling revealed a plethora of WRKY transcription factors that were differentially regulated and acted as central player to defense. *OsWRKY47* overexpression led to resistance, while the integrated analyses of transcriptomic and epigenetic studies uncovered several of the WRKYs, namely seven WRKYs in total, including OsWRKY45 and OsWRKY76, associated with changes in H3K9ac, supporting active regulation during defense (Wei *et al.*, 2013; Peng *et al.*, 2023). OsWRKY76 has also been demonstrated to act like a transcriptional repressor, its overexpression leading to reduced *PR* gene induction and phytoalexin production during blast infection, thus increasing susceptibility while providing additional insights into negative regulatory roles in defense networks (Yokotani *et al.*, 2013).

3.11.3 Characterizing resistance mechanisms in host plants

Comparative transcriptomics between resistant and susceptible cultivars can help identify *resistance* (*R*) genes and R-related expression patterns, for example, in wheat, transcriptomics done on the host wheat post infection with the pathogen *Puccinia triticina* identified genes related to nucleotide binding site leucine-rich repeat (*NBS-LRR*) and downstream signaling, as it transitioned to a hypersensitive response prior to reactive oxygen species (ROS) accumulation (Yadav *et al.*, 2016).

Comparative transcriptomics among resistant and susceptible genotypes reveals specific expression dynamics that are dependent on the genotype, including *R* gene expressions, signaling pathways, and metabolite-associated gene sets. Comparative transcriptomics directs pinpointing the genetic basis for durable or hypersensitive resistance and potential novel sources of resistance within wild forms or local landraces. Using a pan-genome-wide association study (Pan-GWAS) and transcriptome analysis, a new broad-spectrum quantitative trait locus (QTL) (qPBR1) was identified in rice, associated with six candidate differentially expressed genes (DEGs) for resistance to leaf and panicle blast. The Pan-GWAS transcriptome association identified genetic loci and expression patterns directly associated with durable blast resistance; ultimately, this research was suitable for breeding purposes (Wang *et al.*, 2024b).

3.11.4 Understanding pathogen virulence strategies

Dual RNA-seq approaches allow the detection of both host and pathogen transcripts, and demonstrate how pathogens modulate their own gene expression to promote virulence and inhibit plant immunity at the same time. In the *Arabidopsis thaliana – Botrytis cinerea* interaction, the pathogen transcriptomes identified the key virulence-promoting transcripts conjugated to either secreted effectors or enzymes involved with degrading the cell wall (Wei *et al.*, 2024). Dual RNA-seq allows the simultaneous tracking of transcriptional dynamics in both the host and the pathogen, providing insights into how pathogens deploy effectors and manipulate hosts in real-time. Dual RNA-seq enables identification of co-regulation patterns, structural similarities between effectors, and additional information about the impact of genetic variation on the pathogen and host transcriptomes. In the *Arabidopsis thaliana – Botrytis cinerea* system, dual RNA-seq demonstrated a dynamic reprogramming of transcription in both organisms. The authors identified 591 candidate effector proteins (CEPs) from *B. cinerea*, with many showing temporal co-regulation. Functional studies in *Nicotiana benthamiana* demonstrated that 11 of these cell-death-inducing proteins were essential for virulence (Wei *et al.*, 2024). In addition, a GWAS of dual transcriptomic data from 96 *B. cinerea* isolates revealed two genetic loci (trans-eQTL hotspots) in *B. cinerea* that influenced gene expression patterns in both *B. cinerea* and *Arabidopsis*, revealing more layers of host manipulation by the pathogen (Soltis *et al.*, 2020).

3.11.5 Contribution to breeding and crop improvement

The inherent advantage of transcriptomics is to basically provide molecular markers, which can be SNPs or SSRs in disease resistance loci, speeding the process of marker-assisted selection (MAS) and genomic selection. In *Solanum lycopersicum*, RNA-seq led to the identification of candidate genes and expression markers associated with resistance against *Phytophthora infestans* that facilitated breeding programs aimed to increase late blight resistance in potato cultivars (Naik *et al.*, 2023). Markers based on expression (eQTLs, expression levels of *R* genes, and regulatory hubs) obtained from transcriptomics can be integrated into MAS or genomic selection systems. When transcriptome-derived markers can be prescribed alongside GWAS or QTL mapping, the selection of cultivars for new levels of disease resistance can be accelerated enormously. In rice, the facilitation by Pan-GWAS and transcriptomics of the discovery of a new QTL, qPBR1, that conferred broad resistance, along with DEGs identified with qPBR1, provided breeders with candidate markers for breeding disease-resilient new varieties (Wang *et al.*, 2024b). There is still further work in relation to tomato late-blight, but the mechanistic similarities lead us to believe that any transcriptomic markers could be identified in the future.

3.11.6 Integration of multi-omics

By integrating transcriptomic data with proteomic and metabolomic data, we are able to further increase our understanding of disease processes (Table 3.2). By employing integrated analysis, the transcriptional activation of defense genes can be viewed relative to changes in protein accumulation and metabolite changes. This integrated (systems) approach is useful for identifying important nodal components of disease resistance/ network signaling (Naik *et al.*, 2023). Multi-omics approaches (transcriptome, proteome, and phospho-proteome) can provide a more comprehensive understanding of plant defense mechanisms, across multiple regulatory levels, by identifying signaling hubs (i.e. kinases, ROS-associated proteins) that integrate multiple signals, and help reveal post-transcriptional regulation pathways (Fig. 3.2). A systems-based approach takes into account changes in phosphorylation events, gene products, and transcriptional regulation within a single framework. In the durable blast-resistant rice landrace Mowanggu, integration of transcriptomics, proteomics, and phospho-proteomics following infection by *M. oryzae* uncovered the coordinated upregulation of genes and proteins involved in protein phosphorylation, MAPK signaling, ROS metabolism, phenylpropanoid pathways, and amino acid biosynthesis. Phospho-proteome analysis also indicated the identification of several modified proteins which were linked to ROS, and metabolic-associated signaling, providing insights into durable resistance mechanisms in plants (Peng *et al.*, 2023).

3.12 Application of Plant Transcriptomics in the Study of Plant Stress

3.12.1 Abiotic stress

In sweet potato (*Ipomoea batatas*) exposed to drought, DEGs were identified, revealing the central roles of abscisic acid (ABA), ET, and JA signaling pathways, with proteins such as ABI-like protein and Ca^{2+}-ATPase contributing to drought tolerance. Similarly, transcriptome analysis of *Cicer arietinum* and wheat under drought highlighted transcription factors (AP2-EREBP, bHLH, MYB, NAC, WRKY) and regulatory genes mediating stress adaptation (Wang *et al.*, 2020). Transcriptomic studies have highlighted the pivotal role of histidine kinases such as AHK1/ATHK1, which enhance drought tolerance through ABA-dependent pathways during both vegetative and reproductive stages, while cytokinin receptors, including AHK2, AHK3, and CRE1, appear to act as negative regulators of drought responses. The SNF-1-like kinase family (SnRKs) and OST1 kinase orchestrate ABA-mediated stomatal closure, while phosphatases such as ABI1 and ABI2 serve as key negative regulators. It was seen that calcium signaling and secondary messengers, including phospholipase D, phosphatidic acid, IP_3, and diacylglycerol, further integrate environmental cues to control stomatal dynamics.

Calcium-dependent protein kinases, calmodulins, and calcineurin B-like proteins modulate downstream gene expression, forming a sophisticated regulatory network that enhances plant resilience to water deficit (Javaid *et al.*, 2022). Transcriptomic analysis played a pivotal role in elucidating the molecular mechanisms underlying drought stress and recovery in tea plants by providing a comprehensive high-resolution view of gene expression dynamics. Through RNA-seq, over 378 million high-quality reads were generated and *de novo* assembled, which enabled the identification and annotation of a large number of DEGs. This extensive dataset allowed precise mapping of stress-responsive pathways, including phytohormone signaling, regulation of transcription factors, activity of protein kinase/phosphatase, and osmolyte metabolism. It managed to reveal temporal regulation of key biosynthetic and signaling genes under drought stress and rehydration, such as the SnRK2-mediated ABA responses, NPR1-independent SA pathways, JA-Ile biosynthesis, and cytokinin metabolism (Liu *et al.*, 2016).

Under temperature stress, *Manihot esculenta* showed upregulation of ROS metabolism and cold response genes, enhancing cold tolerance, while *Eucalyptus globulus*, rice, grape (*Vitis vinifera*), and *Dendranthema morifolium* transcriptomics revealed cold-induced activation of transcription factors (bHLH, CBF/ DREB, WRKY) and other metabolic pathways. High-temperature stress in *Abies koreana* and

Table 3.2. Application of transcriptomics in plant disease research.

Type of plant	Type of disease under study	Application of transcriptomics	References
Rice (*Oryza sativa*)	Rice blast (*Magnaporthe oryzae*)	Time-resolved transcriptome revealed early induction of signaling and WRKY TFs (OsWRKY47, OsWRKY67), mediating resistance through PTI/ETI and SA-responsive *PR* genes	Wei *et al*., 2013; Liu *et al*., 2018
Rice (*O. sativa*)	Rice blast (*M. oryzae*)	Integrated transcriptomics and epigenomics identified WRKY TFs (OsWRKY45, OsWRKY76) with histone H3K9ac regulation; WRKY76 acted as a negative regulator of defense	Wei *et al*., 2013; Yokotani *et al*., 2013; Peng *et al*., 2023
Wheat (*Triticum aestivum*)	Leaf rust (*Puccinia triticina*)	Comparative transcriptomics among resistant vs susceptible cultivars revealed *NBS-LRR* genes and hypersensitive response with ROS accumulation	Yadav *et al*., 2016
Rice (*Oryza sativa*)	Leaf and panicle blast (*M. oryzae*)	Pan-GWAS + transcriptomics identified novel QTL (qPBR1) with six candidate DEGs conferring broad-spectrum resistance	Wang *et al*., 2024b
Arabidopsis (*Arabidopsis thaliana*)	Gray mold (*Botrytis cinerea*)	Dual RNA-seq tracked host–pathogen dynamics, identifying 591 candidate effector proteins (CEPs) and host defense responses	Wei *et al*., 2024
Arabidopsis (*A. thaliana*) and *B. cinerea* isolates	Gray mold	Dual RNA-seq + GWAS identified pathogen trans-eQTL hotspots influencing expression in both host and pathogen	Soltis *et al*., 2020
Tomato (*Solanum lycopersicum*)	Late blight (*Phytophthora infestans*)	RNA-seq identified resistance-associated candidate genes and eQTL markers aiding marker-assisted selection	Naik *et al*., 2023
Rice (*O. sativa*)	Broad blast resistance	Pan-GWAS + transcriptomics discovered qPBR1 and expression markers supporting genomic selection	Wang *et al*., 2024b
Rice landrace (Mowanggu)	Blast (*M. oryzae*)	Multi-omics (transcriptome, proteome, phospho-proteome) revealed MAPK signaling, ROS metabolism, and phenylpropanoid pathways in durable resistance	Peng *et al*., 2023

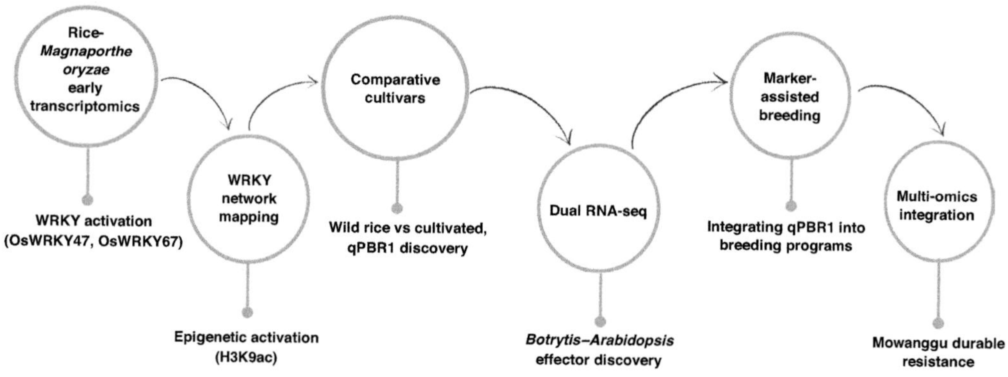

Fig. 3.2. Interconnected case study in plant transcriptomics research.

Camellia sinensis demonstrated transcriptional regulation of light responses, stress signaling, and various post-translational modifications (Wang *et al.*, 2020). Transcriptomic analyses revealed that the genes responsible for proline biosynthesis are upregulated to maintain osmotic balance, whereas genes encoding proline dehydrogenase (ProDH) and MYB60 are downregulated to conserve cellular resources. Stress-responsive proteins like TAS14 (a LEA/dehydrin protein) help in osmotic adjustment, while MYB60 regulates stomatal behavior to balance root growth and transpiration. ROS function as signaling molecules that transmit stress cues from roots to shoots via calcium channels and NADPH oxidases and activate transcription factors, including SERF1 and DREB2A. Antioxidant systems, comprising glutathione-S-transferases, peroxidases, and thioredoxins, are transcriptionally upregulated to mitigate the oxidative damage that takes place. Furthermore, sulfur assimilation pathways via enzymes like ATP sulfurylase and APS reductase function to supply cysteine for glutathione biosynthesis, showing a tightly coordinated network that enables plants to tolerate high salinity (Javaid *et al.*, 2022).

Brassica juncea exposed to cadmium and apple (*Malus domestica*) under salt and osmotic stress showed differential regulation of metal transporters, antioxidant enzymes, and transcription factors (MYB, WRKY, NAC). Transcriptome profiling in maize, eggplant, rice (*O. sativa*), and *Salix* showed genotype and organ-specific DEGs controlling hormone signaling, carbohydrate metabolism, ROS scavenging, and secondary metabolite pathways (Wang *et al.*, 2020). Exposure to heavy metals imposes toxicity by disturbing metabolic processes and generating oxidative stress. Transcriptome studies have elucidated the regulatory networks that govern metal uptake, transport, and its detoxification. Iron homeostasis, which is controlled by FER/FIT transcription factors, ensures sufficient Fe uptake while avoiding toxicity by transcriptionally activating transporters such as IRT1 and IRT2. Detoxification mechanisms that involve transporters like HMA3 and MTP, along with enzymes including peroxiredoxins and glutathione-S-transferases that neutralize ROS and sequester metals via phytochelatins, are also upregulated. Cadmium exposure, in particular, induces the activation of MAPK and also triggers NAC and ERF family transcription factors that coordinate these stress responses and help to reduce metal accumulation. These findings not only highlight plant survival strategies under heavy metal stress but also provide strategies required for phytoremediation and genetic enhancement of metal tolerance (Javaid *et al.*, 2022). Finally, waterlogging stress in grapes demonstrated the upregulation of genes involved in antioxidant systems, glycolysis, fermentation, etc., whereas chlorophyll synthesis genes were downregulated. Across all conditions, transcriptomics provided mechanistic insights into plant adaptation and resilience under environmental stress (Wang *et al.*, 2020).

3.12.2 Biotic stress

In woodland strawberry (*Fragaria vesca*), scRNA-seq of leaves during early *B. cinerea* infection revealed heterogeneous expression of disease-resistance genes across hydathode, epidermal, and mesophyll cells. Abiotic stresses such as heat, salt, or nutrient deprivation similarly induce cell-type-specific transcriptional changes. In *Arabidopsis* and rice roots, outer-layer cells showed stronger transcriptional responses to heat or nutrient stress, while root hair cells adapted to low phosphate through cytokinin-mediated gene regulation. Single-cell assay for transposase-accessible chromatin with sequencing (scATAC-seq) further revealed stress-induced chromatin remodeling, as seen in rice root tips under heat stress (Zheng *et al.*, 2023). Under biotic stress, pathways, particularly JA, SA, and ABA signaling, were observed in resistant and susceptible genotypes, indicating their roles in basal resistance and pathogen-specific responses. Transcriptomics allowed the identification of conserved gene modules responsive to multiple stresses, offering potential targets for enhancing broad-spectrum tolerance. Abiotic stresses, including drought, salinity, low temperature, and oxidative stress, were found to induce coordinated regulation of hormonal pathways (ABA, ET, SA, and auxin) and transcription factors (ERF, NAC, MYB, bZIP, WRKY, AP2/ERF) along with genes of calcium signaling and cell wall biosynthesis. Modulation of cell wall components and receptor-like kinases (RLKs), receptor-like proteins (RLPs), and calcium-dependent kinases exhibited stress-specific expression patterns, suggesting their involvement in the fine-tuning of defense signaling. A set of shared differentially expressed genes during both biotic and abiotic stresses was identified, highlighting a conserved regulatory network mediating stress responses and providing potential targets for the enhancement of broad-spectrum tolerance in tomato (Amoroso *et al.*, 2023). Transcriptomic profiling of *A. thaliana* under single and sequential stresses, including infection by the necrotrophic pathogen *B. cinerea*, herbivory by *Pieris rapae*, and drought, enabled a detailed understanding of plant stress responses. Analysis across multiple time points revealed that approximately one-third of DEGs were shared among the stresses. Transcriptomic data emphasized the central role of hormone-mediated pathways (JA, ABA, SA, and ET) in orchestrating adaptive responses (Coolen *et al.*, 2016). Plants use complex defense mechanisms to combat insects and pathogens mediated by signaling molecules like SA, JA, and ET, and these can be uncovered using transcriptome analysis. In *A. thaliana*, pathogen and herbivore attacks trigger differential hormone production and stress-related gene expression. Cotton responds to whitefly infestation via protein kinases, transcription factors, and phytohormone signaling, where WRKY40 and transport proteins act as key regulators. Tobacco overexpressing *A. thaliana* AtMYB12 accumulates flavonols, enhancing resistance to *Spodoptera litura* and *Helicoverpa armigera*, while flavonoid inhibition reduces resistance. Disease resistance studies showed that powdery mildew-resistant pumpkin upregulates genes like *bHLH87*, *WRKY21*, *ERF014*, *HSF*, *MLO3*, and *SGT1*, while wheat–*Thinopyrum* hybrids showed increased expression of kinases, proteases, and synthases. Wheat transcriptomes during powdery mildew and stripe rust reveal pathogen-specific defense gene expression patterns. In grape, VvTIFY9 induced by SA and methyl jasmonate (MeJA) enhanced resistance to *Erysiphe necator* by upregulating defense genes (Wang *et al.*, 2020).

ScRNA-seq enables high-resolution analysis of cell-to-cell heterogeneity in plants under environmental stresses. It reveals how cells that are genetically identical respond differently to abiotic stimuli such as heat, nutrient deprivation, or sucrose availability. scRNA-seq performed in *Arabidopsis* roots under sucrose or low phosphate conditions showed specific gene expression among cell types, accompanied by changes like increased root hair density and cytokinin signaling in vascular cells, reflecting adaptation mechanisms. Similarly, profiling rice roots under abiotic stress highlighted differences among major cell types in stress responses. In *Arabidopsis* leaf detachment experiments, gene expression patterns in wounded areas were seen after analysis of transcriptional networks that control adventitious rooting. scRNA-seq revealed transcriptional heterogeneity in microbes as well. In bacteria such as *Salmonella* and *Pseudomonas*, individual cells showed variable expression of stress-response genes. Similarly, yeast cells exhibited heterogeneous transcription factor activity under stress with distinct high and low response cell populations. scRNA-seq is continuously transforming our understanding of phenotypic variation and stress adaptation, and integration of scRNA-seq with spatial transcriptomics and spatial metabolomics, to link transcriptome and metabolome dynamics to precise tissue and cell locations, is being performed (Bawa *et al.*, 2022). The various applications of these transcriptomic techniques for studying plant stress responses are summarized in Table 3.3.

Plant stress responses are essential for survival under adverse conditions. Integrated molecular approaches such as genomics, transcriptomics, proteomics, and metabolomics allow the identification of stress-responsive genes, proteins, and metabolites. Transcriptomics successfully reveals changes in gene expression and signaling pathways, while proteomics and metabolomics provide insights into protein function and metabolic adjustments. Together, these tools collectively enable a systems-level understanding of stress adaptation that will help in guiding the development of resilient crops in the future (Satrio *et al.*, 2024).

3.13 Conclusion and Future Perspectives

Plant cells present unique challenges for single-cell transcriptome analysis due to their structural and functional characteristics. Unlike animal cells, which are typically 10–30 µm in diameter, plant cells can be much larger, sometimes reaching up to 100 µm, and are surrounded by a rigid cell wall, the composition and thickness of which varies according to species, tissue type, developmental stage, and environmental conditions. Protoplasting is required for the enzymatic removal of the cell wall to isolate individual plant cells. The efficiency and viability of protoplasts are influenced by various factors such as cell wall composition, anatomical location, and digestion protocols, which can lead to less representation of certain resistant cell types.

To address some of these limitations, snRNA-seq has emerged as an alternative that bypasses whole-cell isolation and reduces bias toward easily dissociable cells, and minimizes the stress responses, even though it manages to capture fewer genes per nucleus. There exists a large variation in plant cell sizes that complicates both isolation and transcriptional quantification. The oversized cells may not be efficiently captured in droplet-based systems, and normalization methods can introduce biases. Increasing the ploidy in larger cells also affects transcription, which needs computational correction to accurately interpret gene expression patterns. Methods that can accommodate size variability, such as microwell-based platforms or nucleus-based approaches, together with careful data normalization, can help to mitigate these challenges and allow more accurate single-cell transcriptomic profiling in plants (Shaw *et al.*, 2021).

Transcriptome sequencing has emerged as a powerful tool in plant research that has made comprehensive analysis of gene expression possible without the need for reference genomes. It facilitates the identification of functional genes, characterization of developmental pathways, and exploration of secondary metabolite biosynthesis, critical for understanding their medicinal potential and plant metabolism. Despite the vast diversity of plants, the majority of species remain underexplored at the genomic level, highlighting the importance of transcriptomic studies in bridging this knowledge gap. NGS has revolutionized this field by providing high-throughput, efficient, and high-coverage data. Single-cell transcriptomics further enhances resolution by capturing cell-to-cell heterogeneity that reveals dynamic gene expression changes that bulk RNA-seq may obscure. Integration of transcriptomic data with multi-omics approaches, including proteomics and metabolomics, allows a deeper understanding of gene function, regulation, and metabolic networks. As sequencing technologies advance, costs decline, and computational tools improve, transcriptome analysis is poised to modernize plant research, supporting gene discovery, functional genomics, and the rational design of medicinal and economically important plants (Tyagi *et al.*,

Table 3.3. Application of transcriptomic techniques for the study of plant stress responses.

Type of plant	Type of stress	Application of transcriptomics	References
Sweet potato (*Ipomoea batatas*)	Drought	Identification of DEGs revealing ABA, ET and JA signaling; proteins like ABI-like protein and Ca^{2+}-ATPase contributing to drought tolerance	Wang *et al*., 2020
Chickpea (*Cicer arietinum*), wheat (*Triticum aestivum*)	Drought	Transcription factors (AP2-EREBP, bHLH, MYB, NAC, WRKY) and regulatory genes mediating stress adaptation	Wang *et al*., 2020
Tea (*Camellia sinensis*)	Drought	High-resolution RNA-seq revealed temporal regulation of phytohormone signaling, transcription factors, kinase/phosphatase activity, osmolyte metabolism during stress and rehydration	Liu *et al*. (2018)
Manihot esculenta	Cold	Upregulation of ROS metabolism and cold response genes enhancing cold tolerance	Wang *et al*., 2020
Eucalyptus globulus, rice (*Oryza sativa*), grape (*Vitis vinifera*), *Dendranthema morifolium*	Cold	Activation of transcription factors (bHLH, CBF/DREB and WRKY) and metabolic pathways	Wang *et al*., 2020
Abies koreana, *Camellia sinensis*	Heat	Transcriptional regulation of light responses, stress signaling, post-translational modifications	Wang *et al*., 2020
Brassica juncea, apple (*Malus domestica*)	Salinity, heavy metals	Differential regulation of metal transporters, antioxidant enzymes, transcription factors (MYB, WRKY, NAC)	Javaid *et al*., 2022
Maize, eggplant, rice, *Salix*	Drought, salinity, heavy metals	Genotype- and organ-specific DEGs controlling hormone signaling, carbohydrate metabolism, ROS scavenging, and secondary metabolite pathways	Wang *et al*., 2020
Grapes	Waterlogging	Upregulation of genes involved in antioxidant systems, glycolysis, fermentation, downregulation of chlorophyll synthesis genes	Wang *et al*., 2020
Woodland strawberry (*Fragaria vesca*)	Biotic (*Botrytis cinerea* infection)	scRNA-seq revealed heterogeneous expression of disease-resistance genes across hydathode, epidermal, and mesophyll cells	Zheng *et al*., 2023
Arabidopsis, rice	Abiotic (heat, nutrient, salt stress)	Cell-type-specific transcriptional changes; root hair cells adapted to low phosphate via cytokinin-mediated regulation	Zheng *et al*., 2023

Continued

Table 3.3. Continued

Type of plant	Type of stress	Application of transcriptomics	References
Tomato	Biotic and abiotic	Identification of conserved DEGs across stresses; hormonal pathways (ABA, SA, JA, and ET) and transcription factors (ERF, NAC, MYB, bZIP, WRKY, and AP2/ERF)	Amoroso *et al.*, 2023
Arabidopsis thaliana	Biotic and abiotic	Stress responses to *Botrytis cinerea*, *Pieris rapae*; many DEGs shared among stresses; hormone-mediated pathways orchestrated adaptive responses	Coolen *et al.*, 2016
Cotton	Biotic (whitefly infestation)	Identification of protein kinases, transcription factors, phytohormone signaling; WRKY40 and transport proteins as regulators	Wang *et al.*, 2020
Tobacco	Biotic (herbivory)	Expression of AtMYB12 enhances flavonol accumulation and resistance to *Spodoptera litura* and *Helicoverpa armigera*	Wang *et al.*, 2020
Pumpkin	Biotic (powdery mildew)	Upregulation of bHLH87, WRKY21, ERF014, HSF, MLO3 and SGT1	Wang *et al.*, 2020
Wheat–*Thinopyrum* hybrids	Biotic (powdery mildew, stripe rust)	Upregulation of kinases, proteases, synthases; pathogen-specific defense gene expression	Wang *et al.*, 2020
Grape	Biotic (*Erysiphe necator*)	VvTIFY9 induced by SA and MeJA enhances resistance via upregulating defence genes	Wang *et al.*, 2020

2022). RNA-seq enables rapid identification of functional genes and analysis of developmental and stress-response pathways, while scRNA-seq provides transcriptional insights of individual cells. Spatial transcriptomics complements these single-cell approaches by preserving tissue context and enabling precise mapping of gene expression in space and time. Third-generation sequencing technologies offer longer reads and improved detection of transcript isoforms, but cost, throughput, and error rates remain as limiting factors. Integration of scRNA-seq, spatial transcriptomics, and multi-omics approaches, including proteomics, epigenomics, and metabolomics, allows for comprehensive characterization of cell types, states, and regulatory mechanisms. Together, these technologies are transforming plant research by enabling detailed exploration of gene expression, cellular heterogeneity, and associated regulatory networks (Wang *et al.*, 2024a, b).

Future directions in single-cell transcriptomics point towards expanding studies beyond *Arabidopsis* roots and to include diverse tissues and plant species, which will provide insights into evolutionary analysis and reveal traits relevant for agriculture and breeding. Comparative transcriptomics will further allow alignment of cell types across species that can potentially be used in the identification of conserved and species-specific cell types and also for the construction of cell-type phylogenies. Plant research is currently limited by dataset availability and computational tools. Integration of single-cell transcriptomics with chromatin accessibility assays, as well as emerging multi-omics technologies that combine genomic, transcriptomic, epigenomic, proteomic, and metabolomic profiling at the single-cell level, will, in the future, enable a more comprehensive understanding of how genotype drives the phenotype. These techniques, when combined with advanced imaging strategies, are expected to significantly advance plant biology, functional genomics, and breeding applications. Finally, plant transcriptomics is also a key component of systems biology and is widely applied to both model and crop plants. With the ongoing advances in systems biology technologies, transcriptomic analysis is expected to expand rapidly in the near future. It enables the discovery of molecular markers, functional characterization of genes across different plant varieties, and can also be used to compare wild relatives with domesticated crops, along with efforts to support crop improvement. The emerging field of multispecies transcriptomics will offer valuable insights into complex plant–microbe interactions. As plant molecular biology, computational biology, and systems biology continue to advance, transcriptomics is poised to play a central role in future plant science research (Imadi *et al.*, 2015).

References

Adams, M.D., Kelley, J.M., Gocayne, J.D., Dubnick, M., Polymeropoulos, M.H. *et al.* (1991) Complementary DNA sequencing: Expressed sequence tags and human genome project. *Science (New York, N.Y.)* 252(5013), 1651–1656. DOI: 10.1126/science.2047873.

Amoroso, C.G., D'Esposito, D., Aiese Cigliano, R. and Ercolano, M.R. (2023) Comparison of tomato transcriptomic profiles reveals overlapping patterns in abiotic and biotic stress responses. *International Journal of Molecular Sciences* 24(4), 4061. DOI: 10.3390/ijms24044061.

Apelt, F., Mavrothalassiti, E., Gupta, S., Machin, F., Olas, J.J. *et al.* (2022) Shoot and root single cell sequencing reveals tissue- and daytime-specific transcriptome profiles. *Plant Physiology* 188(2), 861–878. DOI: 10.1093/plphys/kiab537.

Bawa, G., Liu, Z., Yu, X., Qin, A. and Sun, X. (2022) Single-cell RNA sequencing for plant research: Insights and possible benefits. *International Journal of Molecular Sciences* 23(9), 4497. DOI: 10.3390/ijms23094497.

Chen, C., Ge, Y. and Lu, L. (2023) Opportunities and challenges in the application of single-cell and spatial transcriptomics in plants. *Frontiers in Plant Science* 14, 1185377. DOI: 10.3389/fpls.2023.1185377.

Chen, H., Yin, X., Guo, L., Yao, J., Ding, Y. *et al.* (2021) PlantscRNAdb: A database for plant single-cell RNA analysis. *Molecular Plant* 14(6), 855–857. DOI: 10.1016/j.molp.2021.05.002.

Conesa, A., Madrigal, P., Tarazona, S., Gomez-Cabrero, D., Cervera, A. *et al.* (2016) A survey of best practices for RNA-seq data analysis. *Genome Biology* 17(1), 13. DOI: 10.1186/s13059-016-0881-8.

Coolen, S., Proietti, S., Hickman, R., Davila Olivas, N.H., Huang, P. *et al.* (2016) Transcriptome dynamics of *Arabidopsis* during sequential biotic and abiotic stresses. *The Plant Journal* 86(3), 249–267. DOI: 10.1111/tpj.13167.

Denyer, T., Ma, X., Klesen, S., Scacchi, E., Nieselt, K. *et al.* (2019) Spatiotemporal developmental trajectories in the *Arabidopsis* root revealed using high-throughput single-cell RNA sequencing. *Developmental Cell* 48(6), 840–852. DOI: 10.1016/j.devcel.2019.02.022.

Giacomello, S. (2021) A new era for plant science: Spatial single-cell transcriptomics. *Current Opinion in Plant Biology* 60, 102041. DOI: 10.1016/j.pbi.2021.102041.

Giolai, M., Verweij, W., Lister, A., Heavens, D., Macaulay, I. *et al.* (2019) Spatially resolved transcriptomics reveals plant host responses to pathogens. *Plant Methods* 15(1), 114. DOI: 10.1186/s13007-019-0498-5.

Guo, J., Huang, Z., Sun, J., Cui, X. and Liu, Y. (2021) Research progress and future development trends in medicinal plant transcriptomics. *Frontiers in Plant Science* 12, 691838. DOI: 10.3389/fpls.2021.691838.

Hoefgen, R. and Nikiforova, V.J. (2008) Metabolomics integrated with transcriptomics: Assessing systems response to sulfur-deficiency stress. *Physiologia Plantarum* 132(2), 190–198. DOI: 10.1111/j.1399-3054.2007.01012.x.

Imadi, S.R., Kazi, A.G., Ahanger, M.A., Gucel, S. and Ahmad, P. (2015) Plant transcriptomics and responses to environmental stress: An overview. *Journal of Genetics* 94(3), 525–537. DOI: 10.1007/s12041-015-0545-6.

Islam, M.T., Liu, Y., Hassan, M.M., Abraham, P.E., Merlet, J. *et al.* (2024) Advances in the application of single-cell transcriptomics in plant systems and synthetic biology. *Biodesign Research* 6, 0029. DOI: 10.34133/bdr.0029.

Javaid, M.H., Khan, A.R., Salam, A., Neelam, A., Azhar, W. *et al.* (2022) Exploring the adaptive responses of plants to abiotic stresses using transcriptome data. *Agriculture* 12(2), 211. DOI: 10.3390/agriculture12020211.

Jin, J., Lu, P., Xu, Y., Tao, J., Li, Z. *et al.* (2022) PCMDB: A curated and comprehensive resource of plant cell markers. *Nucleic Acids Research* 50(D1), D1448–D1455. DOI: 10.1093/nar/gkab949.

Jin, J., Yu, S., Lu, P. and Cao, P. (2023) Deciphering plant cell-cell communications using single-cell omics data. *Computational and Structural Biotechnology Journal* 21, 3690–3695. DOI: 10.1016/j.csbj.2023.06.016.

Kawahara, Y., Oono, Y., Kanamori, H., Matsumoto, T., Itoh, T. *et al.* (2012) Simultaneous RNA-seq analysis of a mixed transcriptome of rice and blast fungus interaction. *PLoS ONE* 7(11), e49423. DOI: 10.1371/journal.pone.0049423.

Leschner, S., Deyneko, I.V., Lienenklaus, S., Wolf, K., Bloecker, H. *et al.* (2012) Identification of tumor-specific *Salmonella typhimurium* promoters and their regulatory logic. *Nucleic Acids Research* 40(7), 2984–2994. DOI: 10.1093/nar/gkr1041.

Lieben, L. (2017) Plant genetics: Spatial transcriptomics in plants. *Nature Reviews. Genetics* 18(7), 394–394. DOI: 10.1038/nrg.2017.41.

Liu, Q., Li, X., Yan, S., Yu, T., Yang, J. *et al.* (2018) OsWRKY67 positively regulates blast and bacteria blight resistance by direct activation of *PR* genes in rice. *BMC Plant Biology* 18(1), 257. DOI: 10.1186/s12870-018-1479-y.

Liu, S.-C., Jin, J.-Q., Ma, J.-Q., Yao, M.-Z., Ma, C.-L. *et al.* (2016) Transcriptomic analysis of tea plant responding to drought stress and recovery. *PLoS ONE* 11(1), e0147306. DOI: 10.1371/journal.pone.0147306.

Margulies, M., Egholm, M., Altman, W.E., Attiya, S., Bader, J.S. *et al.* (2005) Genome sequencing in microfabricated high-density picolitre reactors. *Nature* 437(7057), 376–380. DOI: 10.1038/nature03959.

Mortazavi, A., Williams, B.A., McCue, K., Schaeffer, L. and Wold, B. (2008) Mapping and quantifying mammalian transcriptomes by RNA-seq. *Nature Methods* 5(7), 621–628. DOI: 10.1038/nmeth.1226.

Naik, B., Kumar, V., Rizwanuddin, S., Chauhan, M., Choudhary, M. *et al.* (2023) Genomics, proteomics, and metabolomics approaches to improve abiotic stress tolerance in tomato plant. *International Journal of Molecular Sciences* 24(3), 3025. DOI: 10.3390/ijms24033025.

Neumann, M., Xu, X., Smaczniak, C., Schumacher, J., Yan, W. *et al.* (2022) A 3D gene expression atlas of the floral meristem based on spatial reconstruction of single nucleus RNA sequencing data. *Nature Communications* 13(1), 2838. DOI: 10.1038/s41467-022-30177-y.

Peng, W., Wang, Y., Zeng, X., Li, W., Song, N. *et al.* (2023) Integrative transcriptomic, proteomic, and phosphoproteomic analysis on the defense response to *Magnaporthe oryzae* reveals different expression patterns at the molecular level of durably resistant rice cultivar Mowanggu. *Frontiers in Plant Science* 14, 1212510. DOI: 10.3389/fpls.2023.1212510.

Sandoval-Velasco, M., Rodríguez, J.A., Perez Estrada, C., Zhang, G., Lieberman Aiden, E. *et al.* (2020) Hi-C chromosome conformation capture sequencing of avian genomes using the BGISEQ-500 platform. *GigaScience* 9(8), giaa087. DOI: 10.1093/gigascience/giaa087.

Sang, Q. and Kong, F. (2024) Applications for single-cell and spatial transcriptomics in plant research. *New Crops* 1, 100025. DOI: 10.1016/j.ncrops.2024.100025.

Satrio, R.D., Fendiyanto, M.H. and Miftahudin, M. (2024) Tools and techniques used at global scale through genomics, transcriptomics, proteomics, and metabolomics to investigate plant stress responses at the molecular level. In: Shahid, M. and Gaur, R. (eds) *Molecular Dynamics of Plant Stress and Its Management*. Springer Nature Singapore, pp. 555–607. DOI: 10.1007/978-981-97-1699-9_25.

Schena, M., Shalon, D., Davis, R.W. and Brown, P.O. (1995) Quantitative monitoring of gene expression patterns with a complementary DNA microarray. *Science (New York, N.Y.)* 270(5235), 467–470. DOI: 10.1126/science.270.5235.467.

Shaw, R., Tian, X. and Xu, J. (2021) Single-cell transcriptome analysis in plants: Advances and challenges. *Molecular Plant* 14(1), 115–126. DOI: 10.1016/j.molp.2020.10.012.

Sheng, K., Cao, W., Niu, Y., Deng, Q. and Zong, C. (2017) Effective detection of variation in single-cell transcriptomes using MATQ-seq. *Nature Methods* 14(3), 267–270. DOI: 10.1038/nmeth.4145.

Shulse, C.N., Cole, B.J., Ciobanu, D., Lin, J., Yoshinaga, Y. *et al.* (2019) High-throughput single-cell transcriptome profiling of plant cell types. *Cell Reports* 27(7), 2241–2247. DOI: 10.1016/j.celrep.2019.04.054.

Soltis, N.E., Caseys, C., Zhang, W., Corwin, J.A., Atwell, S. *et al.* (2020) Pathogen genetic control of transcriptome variation in the *Arabidopsis thaliana – Botrytis cinerea* Pathosystem. *Genetics* 215(1), 253–266. DOI: 10.1534/genetics.120.303070.

Ståhl, P.L., Salmén, F., Vickovic, S., Lundmark, A., Navarro, J.F. *et al.* (2016) Visualization and analysis of gene expression in tissue sections by spatial transcriptomics. *Science (New York, N.Y.)* 353(6294), 78–82. DOI: 10.1126/science.aaf2403.

Sun, Y., Sun, J., Lin, C., Zhang, J., Yan, H. *et al.* (2024) Single-cell transcriptomics applied in plants. *Cells* 13(18), 1561. DOI: 10.3390/cells13181561.

Tang, F., Barbacioru, C., Wang, Y., Nordman, E., Lee, C. *et al.* (2009) mRNA-seq whole-transcriptome analysis of a single cell. *Nature Methods* 6(5), 377–382. DOI: 10.1038/nmeth.1315.

The Arabidopsis Genome Initiative (2000) Analysis of the genome sequence of the flowering plant *Arabidopsis thaliana*. *Nature* 408(6814), 796–815. DOI: 10.1038/35048692.

Tyagi, P., Singh, D., Mathur, S., Singh, A. and Ranjan, R. (2022) Upcoming progress of transcriptomics studies on plants: An overview. *Frontiers in Plant Science* 13, 1030890. DOI: 10.3389/fpls.2022.1030890.

Wang, H., Xu, Y., Zhang, Z., Zhang, G., Tan, C. *et al.* (2024a) Development and application of transcriptomics technologies in plant science. *Crop Design* 3(2), 100057. DOI: 10.1016/j.cropd.2024.100057.

Wang, J., Hu, H., Jiang, X., Zhang, S., Yang, W. *et al.* (2024b) Pangenome-wide association study and transcriptome analysis reveal a novel QTL and candidate genes controlling both panicle and leaf blast resistance in rice. *Rice (New York, N.Y.)* 17(1), 27. DOI: 10.1186/s12284-024-00707-x.

Wang, X., Li, N., Li, W., Gao, X., Cha, M. *et al.* (2020) Advances in transcriptomics in the response to stress in plants. *Global Medical Genetics* 7(2), 30–34. DOI: 10.1055/s-0040-1714414.

Wang, Z., Gerstein, M. and Snyder, M. (2009) RNA-seq: A revolutionary tool for transcriptomics. *Nature Reviews. Genetics* 10(1), 57–63. DOI: 10.1038/nrg2484.

Wei, J., Zhou, Q., Zhang, J., Wu, M., Li, G. *et al.* (2024) Dual RNA-seq reveals distinct families of co-regulated and structurally conserved effectors in *Botrytis* cinerea infection of *Arabidopsis thaliana*. *BMC Biology* 22(1), 239. DOI: 10.1186/s12915-024-02043-4.

Wei, T., Ou, B., Li, J., Zhao, Y., Guo, D. *et al.* (2013) Transcriptional profiling of rice early response to *Magnaporthe oryzae* identified OsWRKYs as important regulators in rice blast resistance. *PLoS ONE* 8(3), e59720. DOI: 10.1371/journal.pone.0059720.

Xu, Z., Wang, Q., Zhu, X., Wang, G., Qin, Y. *et al.* (2022) Plant single cell transcriptome hub (PsctH): An integrated online tool to explore the plant single-cell transcriptome landscape. *Plant Biotechnology Journal* 20(1), 10–12. DOI: 10.1111/pbi.13725.

Yadav, I.S., Sharma, A., Kaur, S., Nahar, N., Bhardwaj, S.C. *et al.* (2016) Comparative temporal transcriptome profiling of wheat near isogenic line carrying Lr57 under compatible and incompatible interactions. *Frontiers in Plant Science* 7, 1943. DOI: 10.3389/fpls.2016.01943.

Yin, R., Xia, K. and Xu, X. (2023) Spatial transcriptomics drives a new era in plant research. *The Plant Journal* 116(6), 1571–1581. DOI: 10.1111/tpj.16437.

Yokotani, N., Sato, Y., Tanabe, S., Chujo, T., Shimizu, T. *et al.* (2013) WRKY76 is a rice transcriptional repressor playing opposite roles in blast disease resistance and cold stress tolerance. *Journal of Experimental Botany* 64(16), 5085–5097. DOI: 10.1093/jxb/ert298.

Zhao, X., Si, J., Miao, Y., Peng, Y., Wang, L. *et al.* (2014) Comparative proteomics of *Euphorbia kansui* Liou milky sap at two different developmental stages. *Plant Physiology and Biochemistry* 79, 60–65. DOI: 10.1016/j.plaphy.2014.03.008.

Zheng, D., Xu, J., Lu, Y., Chen, H., Chu, Q. *et al.* (2023) Recent progresses in plant single-cell transcriptomics. *Crop Design* 2(2), 100041. DOI: 10.1016/j.cropd.2023.100041.

4 Machine Learning-enhanced Plant Functional Genomics

Rashid Abbas Khan[1], Tauqeer Ahmed Qadri[2], Amjid Khan[3,4], Nadia Tehseen[5], Muhammad Anas[6]*, Aqsa Shabbir[7], Javed Iqbal[8], Banzeer Ahsan Abbasi[9], Sana Basharat[10], Pingwu Liu[10] and Muhammad Waseem[10]*

[1]Department of Biology, Prince of Songkhla University, Hat Yai, Songkhla, Thailand; [2]Department of Biosciences, COMSATS University Islamabad, Islamabad, Pakistan; [3]Department of Plant Sciences, Quaid-i-Azam University, Islamabad, Pakistan; [4]UNESCO UNISA Africa Chair in Nanoscience and Nanotechnology, College of Graduate Studies, University of South Africa, Pretoria, South Africa; [5]Department of Botany, Lahore College for Women University (LCWU), Lahore, Pakistan; [6]Department of Soil Science & Plant Nutrition, University of Selcuk, Konya, Türkiye; [7]Department of Horticulture, Ordu University, Ordu, Türkiye; [8]Department of Botany, Bacha Khan University, Charsadda, Khyber Pakhtunkhwa, Pakistan; [9]Department of Botany, Rawalpindi Women University, Satellite Town, Rawalpindi, Pakistan; [10]School of Breeding and Multiplication (Sanya Institute of Breeding and Multiplication), School of Tropical Agriculture and Forestry, Hainan University, Sanya, China

4.1 Introduction: Overview of Functional Genomics in Plants

Functional genomics is a new discipline in science that describes how genes operate in biology and how they are regulated, rather than merely naming genes. This field focuses on the study of functions, interactions, and regulatory systems that control genes within the operating environment of the entire genome (Du *et al.*, 2023). Functional genomics, in the botanical context, is specifically applicable to the effects of genes within the environment, responding to an environmental stimulus, and documenting the interactions among environmentally responsive genes in controlling growth, development, or stress tolerance. The ability of plants to adapt to various stresses can be elucidated through experimental evidence provided by functional genomics on how to respond and adapt to situations that may have dangerous effects or even lead to death caused by the invasion of other living organisms when faced with a lack of water and climate change consequences. By using the methodically researched molecular basis of the phenotypic characteristics, yield potential, resistance to pathogens, and nutritional value, plant researchers can translate this knowledge into strategies that increase the effectiveness of plants to succeed beyond the

*Corresponding authors: muhammed.anas@selcuk.edu.tr and m.waseem.botanist@gmail.com

© CAB International 2026. *Plant Spatial Biology* (ed. Chen)
DOI: 10.1079/9781836994572.0004

given circumstances. Consequently, the framework is the basic layer of functional genomics that has been driving the evolution of crops via inventions and advocating the evolutionary development of livestock (Liu *et al.*, 2023; Garg *et al.*, 2024).

Conventional approaches to functional genomics have involved technologies such as gene expression analysis, whereby mRNA patterns are compared to analyze the expression of a gene under different circumstances. In the process of determining the functions of genes, other techniques, including gene knockout experiments, in which genes of interest can be knocked out to determine their effects on the phenotype, have also been extensively applied. In the context of early plant genomics, the process of assigning functional roles to genes based on sequence similarity and expression data is referred to as functional annotation (Jakutis and Stainier, 2021; Tian *et al.*, 2022). However, the majority of these methods have disadvantages, as plant genomes are often complex, large, polyploid, or consist of many repetitive fragments, and it is difficult to analyze data to reach a definite conclusion (Sun *et al.*, 2022). In addition, old methods cannot be used to scale up large and complicated plant genomic projects, in which high-throughput approaches are necessary because the size of plant genomic projects is increasing (Shi *et al.*, 2023).

4.2 The Emergence of Machine Learning in Genomics

The introduction of machine learning (ML) has transformed plant genomics, replacing the newest methods that must deal with the massive dimensionality of the genomic age. ML processes enable programs to identify patterns and make predictions without necessarily being instructed, unlike traditional analysis. In particular, these algorithms have been used to study complex and high-volume data, identify biologically significant results, and achieve correct predictions in the presence of noise or missing values (van Dijk *et al.*, 2021).

In plant genomics, MLs are used to predict gene function, map regulatory networks, and detect associations responsible for complex traits (Erbe *et al.*, 2022). Many subtle non-linear relationships based on ML may be missed, as massive data generated from high-throughput sequencing, transcriptomics, and other omics-based new technologies allow the discovery of many non-linear relationships that had hitherto been undiscovered. Harnessing this potential is of particular relevance to plants, where the combinatorial nature of gene–environment interactions and growing genomic sequence data present an opportunity and a constraint to classical analytical responses (Kassem, 2025; Wang *et al.*, 2025a).

This chapter reviews the central role of ML in the development of plant functional genomics. It emphasizes the changing landscape of plant genetics over recent decades and highlights the possibilities of leveraging ML to identify new ways to solve existing challenges. Among others, supervised learning (i.e. gene function prediction), deep learning-based high-throughput phenotyping (i.e. deep learning-based functional enrichment), and multi-omics integrative gene network models are discussed. Selected themes are addressed with illustrations and commentaries on recent advances and ongoing developments, as applicable. The inventions detailed in this document should be seen as illustrative and non-limitative of the types of precision breeding and crop optimization applications of ML to the subject matter. Finally, the emergence of ML in plant functional genomics has marked the beginning of a new computational era for plant functional genomics studies and manipulation. This chapter demonstrates how ML can transform our understanding of what is genetically feasible in plants and how this knowledge can be applied to achieve sustainable agriculture, food security, and environmental resilience. It also offers a broad review of methodological improvements, opportunities, and limitations.

4.3 Foundations: Plant Genomics and Functional Genomics

4.3.1 Introduction to plant genomics

Plant genomics is the field concerned with the molecular structure of plant genomes and studies that focus on the structure and organization of genetic material in plant species from the plant kingdom, as well as the formulation of computational and statistical data analysis tools to answer plant species-related biological questions. It also considers the additional contribution of genetic variation to phenotypic diversity and the dual nature of genetic variation in trait expression and behavioral phenotypes. Plant genomes are generally complex because most plants have large genomes and exhibit a large degree of variation, both in absolute terms (large genome size) and in relative terms (polyploidy or multiple chromosome sets). For example, wheat (*Triticum aestivum*) and cotton (*Gossypium hirsutum*) are hexaploid, which means they have six sets of pairs of chromosomes (Renny-Byfield and Wendel, 2014; Blasio *et al.*, 2022). Polyploidy predominates in the plant kingdom, presenting a formidable challenge for genomic analysis by increasing genetic redundancy and genome size. Furthermore, some plants (e.g. maize: *Zea mays*) have large genomes comprising much repetitive DNA; these very large genomes are harder than average to sequence and interpret (Panchy *et al.*, 2016).

With the advent of sequencing technologies, plant genomes can now be sequenced using more cost-effective and accurate methods than previously possible. Plant genomes can now be sequenced quickly and cheaply, and the so-called next-generation sequencing (NGS) technologies, including Illumina and PacBio sequencing, have changed the paradigm from the original technologies (Sanger sequencing), which were slow and expensive (Rhoads and Au, 2015; Dmitriev *et al.*, 2022). The cost and time required to generate sequencing data have been significantly reduced, and next-generation sequencing (NGS) has made sequencing and assembling even the most difficult plant genomes a practical endeavor (Sun *et al.*, 2022). Additionally, long-read sequencing (LRS) has been used successfully to solve the problem of repetitive sequences in plant genomes and their resulting fragmentation of genome assemblies (Pucker *et al.*, 2022), notably PacBio and Oxford Nanopore.

However, plant genomics is a challenging area because the plant kingdom is very diverse, and plant genomes are very large. Because plant genomes are large, frequently pervaded by repetitive DNA elements, and punctuated by polyploidy in many taxa, they provide a significant hurdle to unambiguous assembly and correct annotation, and some grounds for questioning the functional interpretation of the genome.

4.3.2 Functional genomics approaches in plants

Functional genomics is a methodology used to determine the functions of genes and their networks involved in the regulation of biological functions *in vivo*. It is a core discipline of plant biology, particularly in deciphering the molecular events that frame the general cycle of growth, development, and adaptive reactions to environmental cues. Functional genomics complements gene discovery by annotating biological functions and organizing gene pathways within the systems context of plant physiology, instead of as monolithic gene units (Przybyla and Gilbert, 2022; Wang and Han, 2022). The various uses of functional genomics are discussed below.

4.3.2.1 Transcriptomics

Transcriptomics is a systematic study of RNA transcripts that aims to determine patterns of gene expression from a given tissue type, developmental stage, or state in response to an environmental stimulus. The introduction of high-throughput approaches, including RNA sequencing (RNA-seq), which produces high-resolution quantitative gene expression profiles, has been transformative in several of these fields. Transcriptomics has been demonstrated to be a powerful method not only for characterizing the molecular basis of physiological processes and plant responses to stresses, diseases, and developmental signals, but also for capturing the encoded repertoire of genes expressed in organisms (Lowe *et al.*, 2017; Tzec-Interián *et al.*, 2025).

4.3.2.2 Proteomics

Proteomics is the complete analysis of the plant proteome, which is a set of all proteins synthesized under defined conditions. Multimodal approaches of mass spectrometry, such as liquid chromatography–tandem mass spectrometry (LC–MS/MS), are widely used for detecting and quantifying proteins, and can provide important information about the structure and functional behavior of proteins and protein–protein interactions. This would enable researchers to study the translation of mRNAs to functioning proteins and to understand the importance of post-translational modifications (PTMs) for the function and/or stabilization of activities and regulatory effects (Mergner and Kuster, 2022; Neagu *et al.*, 2022).

4.3.2.3 Metabolomics

The purpose of metabolomics is to identify the physiological and biochemical states of plants through a systematic analysis of the small molecules or metabolites involved in plant metabolism. It is widely used for the detection and quantification of metabolites associated with central pathways from various analytical platforms (gas chromatography–mass spectrometry (GC–MS) and liquid chromatography–mass spectrometry (LC–MS)) of photosynthesis, stress adaptation, and secondary metabolite biosynthesis (Patel *et al.*, 2021; Putri *et al.*, 2022).

4.3.2.4 CRISPR-Cas-based functional studies

CRISPR (clustered regularly interspaced short palindromic repeats)-Cas9 genome engineering technologies have revolutionized functional genomics and enabled genome editing to modify plant genomes in a non-homogeneous and local manner on a global scale. By knocking out a particular gene (or knock-ins of foreign genes), scientists can determine the function of a particular gene under a clearly defined set of conditions. This new, highly efficient device accelerates functional genomic gene annotation and gene validation, as well as opening up new opportunities for high-throughput gene editing-based crop improvement and resistance engineering in plants (Guo *et al.*, 2023; Tsakirpaloglou *et al.*, 2023). While these common functional genomics tools have worked well for us, they too have numerical limitations in handling large datasets produced by high-throughput sequencing and other omics-based technologies. Typically, these data overshadow the size and depth of analysis that traditional data analysis models can currently handle, and defining the clues and predicting functions or interactions of the genes is not possible with much confidence (Nayak *et al.*, 2021). Moreover, there is an increasing need for the integration of data obtained at different omics levels (for instance, combining genomic, transcriptomic, proteomic, and metabolomic data), which remains one of the main challenges arising from the different measurement protocols, data formats, and data volumes.

4.3.3 Intersection of ML and functional genomics

ML has become a powerful tool for overcoming the limitations of traditional functional genomics methodologies. One fundamental strength of ML is its ability to process, analyze, and interpret high-dimensional data, which is a characteristic of modern plant functional genomics. In uncovering the underlying structure and correlation of vast amounts of data, ML has the potential to mine biologically relevant data that would not otherwise be visible via traditional statistical methods (Hesami *et al.*, 2022; Whalen *et al.*, 2022). The application of ML in plant genomics is widespread. It can be applied to predicting the function of a gene based on a combination of interrogation of sequence data and gene expression profiles (Wang *et al.*, 2025b). In addition, it can support the identification of interactions between genes and environmental conditions, which is a critical factor in the path towards elucidating plant responses to external stresses, such as drought, temperature shifts, and viral attacks. Interestingly, ML algorithms (including supervised learning), such as decision trees (DTs), random forests (RFs), and support vector machines (SVMs), can also be used to produce and/or teach decisions about the desired unknowns/predictions of gene functions or plant classification using phenotypic characteristics and genomic information (Danilevicz *et al.*, 2022; Rico-Chávez *et al.*, 2022).

Furthermore, unsupervised learning techniques, such as clustering and dimensional reduction, can be applied to identify the underlying dependencies among genes, metabolites, and proteins, which can be used to develop gene regulatory networks or metabolic pathways (Yan and Wang, 2022). Furthermore, integrating multiple omics data and applying the capabilities of ML to determine the details of plant biology could easily become a true force for crop improvement, stress tolerance, and prediction of crop productivity (Yang *et al.*, 2021).

The benefits of ML for dealing with high-dimensional data and characterizing latent structures have made it a major component of functional plant genomics. The integration of ML with standard methods in genomics can also dramatically improve researchers' ability to interpret plant biology, improve the prediction of gene function, identify biomarkers related to desirable traits, and improve designs for plant breeders (Mahood *et al.*, 2020; Yan and Wang, 2022).

4.4 Introduction to Machine Learning Techniques

4.4.1 Machine learning overview

ML has emerged as a significant contributor to plant genomics and can help researchers extract value from big data, which is typically high-dimensional and difficult to interpret using traditional methods. ML is a group of algorithms and statistical models that enable computers to learn patterns and make decisions from data without programming (Razzaq and Shah, 2025). Owing to its ability to learn and improve its estimates based on data, ML has become an important component of modern genomic research, particularly when studying plant genomes (Kell *et al.*, 2001). There are four main types of ML techniques, each with its own features and uses.

4.4.1.1 Supervised learning

In supervised learning, a model is taught using labeled data, where the label is present as part of each input. The basic idea is to train a mapping between inputs (e.g. genetic and environmental data) and outputs (e.g. gene function and trait prediction). Supervised learning has been widely applied to address plant genomic issues, such as gene functional inferences, trait classifications, and genotype–phenotype associations (Mahood *et al.*, 2020). ML algorithms, such as DTs, RFs, and SVMs, are commonly used to predict results in this scenario based on observed dependencies between the input (such as expression data) and the target label (such as resistance to disease) (Nitze *et al.*, 2012).

4.4.1.2 Unsupervised learning

Unsupervised learning, on the other hand, deals with unlabeled data and attempts to find some structure in the information or to figure out the clues hidden among the data without any specific categories. In the context of plant genomics, this type of learning is the ideal method for clustering similar genes, inferring novel gene regulatory networks, or analyzing plant phenotypic data for which no clear outcome is obvious (Yan and Wang, 2022). Popular unsupervised learning algorithms include unsupervised k-means clustering, hierarchical clustering, and principal component analysis (PCA). These algorithms are used to identify latent junctions and classes in genomic, transcriptomic, or phenotypic data, which could lead to new biological discoveries (Yan and Wang, 2022).

4.4.1.3 Semi-supervised learning

Semi-supervised learning is a combination of supervised and unsupervised learning, using some labeled data and a large amount of unlabeled data. In the genomics domain, this method is of particular importance because labeled data (e.g. gene function annotations) might be costly and time-consuming to obtain, whereas unlabeled data (e.g. genomic sequences and gene expression profiles) are readily available. Semi-supervised learning algorithms can benefit from every source of information to generate more accurate learning. Self-training and co-training are widely applied to improve the corresponding models for plant genomic purposes (Zeebaree *et al.*, 2021; Yan and Wang, 2022).

4.4.1.4 Reinforcement learning

Reinforcement learning (RL) is an emerging subfield of ML in plant genomics. In RL, some type of environment is interacted with, and the model is trained based on feedback (reward or penalty) after some type of interaction. This strategy is primarily recognized for use in decision-making routines, such as perspective optimization, resource organization, and environmental adaptation model routines. Another application that is gaining importance in plant genomics, although less common, is that RL will learn the best practices in plant breeding, agronomic crop use processes, and agricultural development through trial and error over time.

4.4.2 Key ML algorithms relevant to plant genomics

In plant genomics, various ML algorithms have been demonstrated to be useful for processing large-scale biological data. Communities of living things may incorporate complex signaling and interaction mechanisms that can only be detected by computationally powerful methods capable of identifying promyelinating patterns, internal structures, and relations among genomic, transcriptomic, and apparent phenotypes. Some of the excellent ML algorithms widely used in plant genomics are described here.

4.4.2.1 Decision trees (DTs)

DTs are a simple and effective classical classification and regression method. In plant genomics, DTs are used to sort plant traits based on genetic data, such as patterns of gene expression or genetic polymorphisms. Given the importance of interpretability in biological research owing to the need to understand the decision-making process leading to the prediction of the model, the interpretable nature of decision trees makes this method suitable for use in biological research.

4.4.2.2 Random forests (RFs)

RF is an ensemble learning tool that integrates the results of multiple decision trees to improve accuracy and reduce the likelihood of overfitting. Genome information has been extensively used in plant genomics to predict complex traits of plant breeding importance, such as yield potential, tolerance to abiotic stress, and resistance to pathogens, using RF. Owing to its capability to effectively and reliably select the features of high-dimensional data, it can be considered an important means of solving complicated problems in plant genome research (Yan and Wang, 2023).

4.4.2.3 Support vector machines (SVMs)

SVM is a powerful classification technique used to define a hyperplane that maximally discriminates between classes in high-dimensional data spaces. Currently, they are widely used for plant genomic data, which deals with the classification of a plant, the prediction of its susceptibility or tolerance/disgust to environmental stress, or environmental stress to a plant from its genomic data. With their powerful performance in high-dimensional analysis, they are indispensable tools for explaining gene expression patterns and for tasks such as the prediction of complex phenotypic traits (Shafik *et al.*, 2023; Dolatabadian *et al.*, 2025).

4.4.2.4 k-Nearest neighbors (k-NN)

The k-nearest neighbors (k-NN) algorithm is a popular yet simple classifier that provides the majority class label of an observation based on the most similar observations. The use of k-NN has already been proposed in plant genomics for clustering functions or to identify and distinguish plant species/cultivars using similarities based on genetics and phenotype. This aspect of the stochastic data model becomes especially prominent for complex and high-dimensional datasets that cannot be described by linear models connecting points (Goel and Nagpal, 2023; Gerakari *et al.*, 2025).

4.4.2.5 Neural networks (NNs)

Neural networks (NNs), especially deep-learning architectures, have become the heart of plant genomics, because they offer both ML models for modeling nonlinear relationships and a means to exploit the availability of large amounts of data (Farooq *et al.*, 2024). As such, both feedforward neural networks (FNNs) and convolutional neural networks (CNNs) have already proven useful tools for a number of applications, such as gene expression profiling analysis, prediction of phenotypic traits, and interpretation of NGS data. Furthermore, owing to their multilayer architecture, deep NNs have sufficient expressivity to record intricate genomic data patterns that surpass the detection power of standard ML tools (Farooq *et al.*, 2024; Escamilla *et al.*, 2025).

4.4.3 Machine learning tools and frameworks

Partly due to the availability of a wide range of open-source libraries and platforms, the development and application of ML in plant genomics has increased considerably. Several software packages offer a mature implementation of thousands of such ML algorithms for research that can easily be run on genomic data and have the potential to make the entire research process much more productive.

4.4.3.1 TensorFlow

In parallel, the development and application of ML in plant genomics has also accelerated exponentially, in part, due to a rich array of open-source libraries and platforms. Thousands of such ML algorithms have already been developed for research use, and these software packages implement them in a mature fashion suitable for wide use on genomic data, so that the entire scientific process can become more efficient.

4.4.3.2 Keras

Keras is a high-level NN application processing interface (API) that can be used with TensorFlow to build and train deep learning models. One of its most promising applications in plant genomics is its modularity and user-friendliness, enabling scientists to conduct efficient experiments over diverse NN architectures with different tasks, such as gene function prediction, phenotypic classification, and protein structure model validation (Pang *et al.*, 2020; Abdolrasol *et al.*, 2021).

4.5 Application of Deep Learning and Convolutional Neural Networks to Genomic Data Analysis

Plant genomics has also seen tremendous scalability in terms of the development and availability of various open-source libraries and platforms to support the development of ML algorithms and applications. These software packages provide a mature implementation of thousands of ML algorithms for research that are easy to use on genomic data and can make the entire research process much more effective. CNNs have attracted much attention for their ability to represent features of grid-like data, such as images, in plant genomics. CNNs can be used to analyze plant phenotypes using plant feature images (i.e. leaf shape, disease symptoms, and plant growth patterns). Using deep learning, large datasets with plant imagery can be high-throughput phenotypes to assess the phenotypic feature(s) related to the underlying genetic information (Pérez-Enciso and Zingaretti, 2019; Chicho and Sallow, 2021). Furthermore, deep learning methods, such as CNNs, are used in applications such as DNA and genomic sequence analysis. Considering genomic data as one-dimensional data, we can use CNNs to learn motifs, structures, and functional regions in DNA or RNA sequences or protein sequences. This is also very useful for gene function, regulatory regions, and even predicting protein structures from genomes only (Jiang and Li, 2020; Wang and Su, 2022), which is what makes deep learning valuable (Table 4.1).

4.6 Applications for Machine Learning in Plant Functional Genomics

4.6.1 Gene function prediction

Predicting the function of a gene based on its sequence is one of the most groundbreaking applications of ML in plant functional genomics. In the past, functional mapping had largely been a question of homology, with the expectation that genes with similar sequences had similar

Table 4.1. Overview of machine learning algorithms and their applications in plant functional genomics.

Machine learning algorithm	Type of learning	Key applications in plant genomics	Advantages	Reference
Decision trees (DT)	Supervised	Gene function prediction, classification of plant traits	Easy to interpret, fast to train, handles both numerical and categorical data	Wang *et al.*, 2020, 2025b
Random forests (RF)	Supervised	Trait prediction, disease resistance, genomic selection	Robust to overfitting, handles large datasets, and feature importance analysis	Geurts *et al.*, 2009; Mahood *et al.*, 2020
Support vector machines (SVM)	Supervised	Classification of plant varieties, disease susceptibility, and gene expression analysis	High accuracy, effective in high-dimensional spaces	Upadhyaya *et al.*, 2024; Zhang *et al.*, 2025
k-Nearest neighbors (k-NN)	Supervised	Phenotypic classification, clustering of gene expression profiles	Simple, effective for small datasets, no training required	Dixit and Nema, 2018; Gerakari *et al.*, 2025
Neural networks (NN)	Supervised/deep learning	Gene expression pattern prediction, phenotype–genotype association, protein structure prediction	Can model non-linear relationships, handle large datasets	Dixit and Nema, 2018; Escamilla *et al.*, 2025
Convolutional neural networks (CNN)	Deep learning	Image-based phenotyping, disease detection, protein structure prediction	Excellent for spatial data, high accuracy in image analysis	Cembrowska-Lech *et al.*, 2023; Farooq *et al.*, 2024
Reinforcement learning (RL)	Reinforcement	Crop optimization, precision breeding, and resource allocation in agriculture	Learns optimal strategies from real-time feedback	Qadri *et al.*, 2024; Ma *et al.*, 2025

functions. Although this approach has been a valuable tool, it often fails when attempting to characterize seasonally new genes, especially those added to complex plant genomes (Maghdid *et al.*, 2024).

ML, in particular supervised learning, such as DTs, RFs, and SVMs, is being increasingly used for training purposes for the prediction of gene functions based on derived features within sequence data (Abdolrasol *et al.*, 2021). Such features may include nucleotide coding sequence motifs, gene expression patterns, and evolutionary relatedness. Since observed data currently possess a variety and scale unprecedented in biology, ML can mine and describe underlying, non-obvious patterns that can be useful in correlating specific sequence features and gene activity. This is particularly useful in identifying the function of a gene where the function is unknown and/or the function of the gene cannot be easily predicted using more traditional methods.

4.6.1.1 Case studies

ML approaches have been employed in *Arabidopsis* to obtain a list of inferred functions of uncharacterized genes based on co-expression profiles across several tissues and environments combined with sequence data (Mahood *et al.*,

2020). For example, using these models in *Oryza sativa* (rice), they have been able to identify genes linked to the stress tolerance regulatory phenotype, the central metabolism/low-strength pathway, and, in particular, have made use of very large repositories of genomic sequence and phenotypic data. There are multiple examples of the widespread impact of ML and its utilization for significantly enhancing the accuracy of gene annotation, while also expediting the provision of functional insights and experimental validation (Ma *et al.*, 2013; Sharma *et al.*, 2024).

4.6.2 Gene regulatory networks and pathway modeling

The discovery of gene regulatory networks (GRNs) and metabolic pathways is one of the most successful efforts from a molecular perspective in plant development and plant–environment interactions. Transcription factors (TFs), genes, and other regulatory elements are deeply connected to the overall coordination of gene expression (called a transcription regulatory network, or GRN) (Zemlyanskaya *et al.*, 2021; Hu *et al.*, 2022). However, the experimental reconstruction of these networks is extremely challenging because of the large combinatorial space of possible interactions and the complexity of interactions within plant gene regulation. Unsupervised learning algorithms and ML applications provide effective tools for revealing the underlying relationships in GRNs. Late genes or novel regulators are predicted to have TF-target gene interactions based on transcriptomic data collected from different plant tissues or plant growth cycles. As such, method-derived networks can provide useful tools for understanding the underlying regulatory functional mechanisms of stress tolerance, growth, and metabolism (Zeebaree *et al.*, 2021; Yan and Wang, 2022).

4.6.2.1 Example

In the context of *Arabidopsis thaliana*, we have been able to expand transcription factor–gene regulatory relationships using RNA-sequencing data by adapting supervised learning frameworks (Huo *et al.*, 2024). Using an RF classifier, the researchers isolated potential regulatory relationships between genes encoding transcription factors and genes related to the defense response. This produced a DNA regulatory network that revealed novel, likely views of pathogen resistance molecular circuitry and allowed experimental tests for inferred interactions, in addition to others. This illustrative case provides a window into how ML can be used to construct robust and biologically meaningful gene regulatory networks that can support functional genomics studies as well as crop improvement programs.

4.6.3 Trait prediction and crop improvement

One of the main challenges in breeding plants for drought resistance, yield, and stress tolerance is the predicted quantification of complex traits. Such features are most likely polygenic, with multiple genome loci required to contribute to each property, which are further regulated through interactions with GbyE (genotype by environment). Despite significant improvements, conventional breeding approaches can only achieve moderate throughput because of the availability of specific genetic markers (Derbyshire *et al.*, 2022; Anilkumar *et al.*, 2023).

ML methods provide an alternative and promising approach that makes it possible to predict phenotypes directly from genomic profiles. In general, ML algorithms applied to large datasets of both genotypes and phenotypes can be used to discover functionally significant genetic variations linked to beneficial phenotypes. This is known as genomic selection, and its use helps breeders accurately predict the performance of untested genotypes and implement more accurate and timely selection programs (van Dijk *et al.*, 2021; Danilevicz *et al.*, 2022).

4.6.4 Integration of machine learning into precision breeding

ML models, libraries, and algorithms can incorporate genomic – e.g. single-nucleotide polymorphisms (SNPs) – and phenotypic information into predictive models, which can then be used for trait improvement. Finally, genomic sets have also been used to predict numeric traits, such as maize yield and drought tolerance, by identifying molecular markers related to these traits. Thus, hereditary indices can be effective for the indiscriminate simplification of the breeding process and can greatly reduce the time and economic resources required for breeding new cultivars with improved performance (Rasheed *et al.*, 2023; Chen *et al.*, 2024).

4.6.5 Phenotypic and genotypic data integration

Combining phenotypic and genotypic data is important to clarify the issues of genotype–phenotype mapping in plants. SNPs and other SNP-associated markers can complement genomic sequences and gene expression profiles, while morphological phenotypes, developmental pathways, and resistance to diseases can be information sources that complement genotypic data. This suite of data, when integrated, can be used to train automated predictive models of complex traits that function as inputs for crop improvement breeding (Ritchie *et al.*, 2015; Zhang *et al.*, 2024). ML techniques (such as regression approaches and artificial NNs) are widely used tools for modeling genotype–phenotype associations. Strategic models can be used to predict the consequences of genetic variation on plant traits under a wide range of environmental conditions and, ultimately, allow breeding cycles for variant types of plants, generating and producing food that is valued by human consumers.

4.6.5.1 Example

Similar to the rice study, even though it may be extended to a large number of traits, ML algorithms were used to relate genotypic information (SNPs) and phenotypic traits (for instance, grain size and disease resistance). These models have provided reliable phenotypic predictions based on genotype and demonstrated promise in the study of genetic architecture and in helping breeders identify the genotypes that are more favorable for improvement. Including large numbers of genotypic and phenotypic data may increase the efficiency and quality of crop improvement interventions by identifying underlying genetic drivers of complex traits using ML (Immanuel *et al.*, 2011; Begum *et al.*, 2015; Grinberg *et al.*, 2020).

4.6.6 High-throughput phenotyping

High-throughput phenotyping, which facilitates rapid data collection for large numbers of plants, has become a cornerstone of modern plant genomics and breeding packages. However, classical phenotyping methods are quite tedious and time-consuming; therefore, screening under a wide variety of environmental conditions is highly limited (Krishna *et al.*, 2023; Angidi *et al.*, 2025).

This area has been revolutionized by ML (especially image-based approaches). The extraction of more complex phenotypic traits can be performed automatically from plant images assisted by computer vision and deep learning technology, allowing highly scalable and efficient phenotypic analysis. Based on a computer vision approach and deep learning, ML can automatically measure phenotypic characteristics of images, including leaf area, plant height, flower count, and disease symptoms. This allows for the phenotyping of large numbers of individuals in a short time, generating large amounts of phenotypic data that can be matched to the genotype (Krishna *et al.*, 2023; Angidi *et al.*, 2025).

4.6.6.1 Example

In wheat breeding, drones can travel above the fields and provide images to confidently analyze characteristics such as plant height or resistance to diseases with the help of ML models. We found that applying CNNs to images to identify relevant phenotypic features greatly improved the phenotyping workflow. The latest technologies allow breeders to screen plant performance in the field at a very-high-throughput level, in contrast to the classical method of supplying seed selection of better genotypes for further breeding (Zheng *et al.*, 2021; Qadri *et al.*, 2024).

4.6.7 Predicting environmental stress responses

Drought, salinity, extreme temperatures, and other environmental stresses rank high in limiting agricultural crop production worldwide. Breeding resistant crop varieties that are tolerant to the adverse effects of climate change necessitates the best-informed knowledge of how plants cope with such challenges at the molecular level. The ability to capture and learn the interactions of genomic-level information with transcriptomics and the environmental context at the tissue or organ level paves the way for the development of robust approaches for the prediction of plant stress resilience (Zhang *et al.*, 2024). ML models combining climatic, genetic, and phenotypic data can also predict plant performance in different environmental scenarios, which can support breeders in deploying more appropriate crops to specific stresses and help fulfill future food demands (Harfouche *et al.*, 2019; Khan *et al.*, 2022).

4.6.7.1 *Example*

Various environmental variables (such as soil moisture and temperature) and transcriptomic data of maize were coupled with training to predict the behavior of maize affected by drought using ML algorithms. Both models used successfully identified genetic markers for drought tolerance and, therefore, were able to reduce the selection of genotypes with higher tolerance. Thus, predicting specific plant responses to abiotic stress through the advancement of ML is expected to improve the predictability of the climate resilience of crop species, which in turn will help accelerate the characterization of climatically tolerant varieties (Table 4.2) (Mahood *et al.*, 2020; Zhakypbek *et al.*, 2025).

4.7 Challenges and Future Prospects

Although ML is a promising, completely new technique for assisting plant functional genomics, two difficulties have so far restricted its extensive application. The major difficulties are associated with limitations of data quality and availability, interpretation of complex models, and, certainly, the inability to ensure that the model can be generalized beyond the species and environmental circumstances in which it has been tested. These obstacles must be removed to fulfill the research and development potential of ML in plants and agriculture.

One area where there are upcoming challenges is the availability and quality of data. Noisy, missing, and unbalanced genomic data streams lead to false ML hazard predictions. Although techniques have been developed to deal with clean-data issues, such as error correction, imputation, and synthetic data generation (i.e. SMOTE), high-quality data continue to be a challenge. In addition, the high complexity of modern ML models, particularly deep learning models, leads to their own faults. However, owing to their black-box property, they may be limited in their interpretability; thus, the potential for biological insights to be made by researchers may be diminished. Emerging techniques such as Shapley additive explanation (SHAP) and local interpretable model-agnostic explanation (LIME) have proven promising in providing a higher level of transparency; however, explanations that are rational in a biological way have been elusive.

Another important limitation is that ML models cannot be transferred to another species and/or setting. However, the cost of this is a large genotype and environmental difference between species; therefore, it is quite common that models trained on one species/environment do not translate well to other species/environments. Recently, to make these methods generally applicable and more robust, various techniques have been proposed based on transfer learning of ML models, generalization via training on augmented datasets, and multi-species training.

Several promising lines of research provide potential avenues for overcoming these challenges, many of which are likely to succeed. Computational enhancements, such as quantum computing, could allow these ML models to accommodate vast inputs and the high complexity of data, which could resemble a genetic structure. Reinforcement learning can also hold great promise in precision agriculture, as it encourages agile programs that can react to real-time environmental signals. Furthermore, the use of ML and the combination of ML with the latest bio-models, such as CRISPR-Cas9 and synthetic biology, will result in faster progress in precision breeding because

Table 4.2. Applications for machine learning in plant functional genomics.

Application area	ML methods used	Outcome/benefit	Example study	Reference
Gene function prediction	Random forests (RFs), Support vector machines (SVMs), deep neural networks	Accurate annotation of uncharacterized genes; discovery of novel gene functions	Predicted functions of unknown *Arabidopsis* genes using RF-based models	Mahood *et al.*, 2020
Gene regulatory network (GRN) modeling	Bayesian networks, neural networks, decision trees	Inference of transcription factor–gene interactions; modeling regulatory pathways	Constructed GRNs for rice stress response using ML-based inference	Mochida *et al.*, 2018; Yan and Wang, 2023
Trait prediction and crop improvement	Gradient boosting, genomic selection models, deep learning	Prediction of complex traits (yield, drought tolerance, stress resilience); accelerated breeding	ML models predicted maize yield under water-limited conditions	Harfouche *et al.*, 2019; Amin *et al.*, 2025
Phenotypic and genotypic data integration	k-Nearest neighbors, multi-omics integration models, deep learning	Correlation of phenotype with genotype; identification of markers for crop improvement	Integrated SNP and transcriptome data to predict wheat grain quality	Fritsche-Neto *et al.*, 2024; Gerakari *et al.*, 2025
High-throughput phenotyping	Convolutional neural networks (CNNs), image recognition models	Automated analysis of plant images (leaf area, disease resistance, growth patterns)	A CNN-based model classified rice blast disease with high accuracy	Jiang and Li, 2020; Wang and Su, 2022
Predicting environmental stress responses	SVMs, recurrent neural networks, ensemble methods	Prediction of plant responses to drought, salinity, pests, and climate change	ML models forecasted soybean stress tolerance under varying environments	Jianing *et al.*, 2022; Gou *et al.*, 2024

the prediction and building of editing strategies will be more precise and will pre-design novel design routes.

The potential of precision agriculture and tailored crop genomics is likely to be equally revolutionary. Combined with environmental, phenotypic, and genomic data, ML can be used to develop varieties that are optimized for a region, climate, or farm system. Therefore, it would be more productive, sustainable, and better able to cope with the changes we are seeing in the climate. Therefore, the issues and challenges related to societal acceptability, information privacy, and technology availability need to be addressed if the technology is to be appropriately and responsibly applied to ensure food security for the global population.

In conclusion, although the roadblocks to advancing data management, interpretability, and generalization are currently absolute, they appear to be alleviated, at least to a degree, by advances in computation together with the emerging integrative biotechnological applications. ML can help mold plant functional genomics and modern agriculture in easily assimilable and environmentally responsible ways through creativity, vision, prudent or even wise controlling regulations, and global cooperation.

4.8 Conclusion

ML is radically transforming plant functional genomics, qualifying the plant functional genomics field with progress that may have been achieved previously. In addition to the investigation of complex and large datasets, ML has increased the precision of gene function prediction, allowed the identification of complex gene regulatory networks, and sped up the identification of necessary traits for crop improvement. New methods, such as deep learning, transfer learning, and multi-omics data integration, have brought a paradigm shift in the study of plant genomes, providing additional insights into plant biology and promoting innovations in the breeding process, adaptation to stress, and sustainable agriculture. These advances not only increase the precision and capacity of functional genomics but also create the basis for developing varieties of crops that are fit for specific parts or circumstances.

To ensure that the full potential of ML is realized for plant genomics, continued communication between computational scientists and plant biologists is required. This type of interdisciplinarity is needed to develop better predictive systems, improve the quality of data, and further solve the issue of plant genome complexity. If researchers work together, ML can be used to its full potential to increase crop resilience, support sustainability, and establish food security worldwide. Moreover, the inclusion of emerging technologies and their ethical and responsible implementation will unlock a new future for plant science, in which ML will emerge as a transformative power in addressing the world's agricultural challenges.

References

Abdolrasol, M.G.M., Hussain, S.M.S., Ustun, T.S., Sarker, M.R., Hannan, M.A. *et al.* (2021) Artificial neural networks based optimization techniques: A review. *Electronics* 10(21), 2689. DOI: 10.3390/electronics10212689.

Amin, A., Zaman, W. and Park, S. (2025) Harnessing multi-omics and predictive modeling for climate-resilient crop breeding: From genomes to fields. *Genes* 16(7), 809. DOI: 10.3390/genes16070809.

Angidi, S., Madankar, K., Tehseen, M.M. and Bhatla, A. (2025) Advanced high-throughput phenotyping techniques for managing abiotic stress in agricultural crops—A comprehensive review. *Crops* 5(2), 8. DOI: 10.3390/crops5020008.

Anilkumar, C., Sah, R.P., Beena, R., Azharudheen, T.P.M., Kumar, A. *et al.* (2023) Conventional and contemporary approaches for drought tolerance rice breeding: Progress and prospects. *Plant Breeding* 142(4), 418–438. DOI: 10.1111/pbr.13119.

Begum, H., Spindel, J.E., Lalusin, A., Borromeo, T., Gregorio, G. *et al.* (2015) Genome-wide association mapping for yield and other agronomic traits in an elite breeding population of tropical rice (*Oryza sativa*). *PloS One* 10(3), e0119873.

Blasio, F., Prieto, P., Pradillo, M. and Naranjo, T. (2022) Genomic and meiotic changes accompanying polyploidization. *Plants* 11(1), 125.

Cembrowska-Lech, D., Krzemińska, A., Miller, T., Nowakowska, A., Adamski, C. *et al.* (2023) An integrated multi-omics and artificial intelligence framework for advance plant phenotyping in horticulture. *Biology* 12(10), 1298.

Chen, Q., Ying, Q., Lei, K., Zhang, J. and Liu, H. (2024) The integration of genetic markers in maize breeding programs. *Bioscience Methods* 15.

Chicho, B.T. and Bibo Sallow, A. (2021) A comprehensive survey of deep learning models based on Keras framework. *Journal of Soft Computing and Data Mining* 2(2), 49–62. DOI: 10.30880/jscdm.2021.02.02.005.

Danilevicz, M.F., Gill, M., Anderson, R., Batley, J., Bennamoun, M. *et al.* (2022) Plant genotype to phenotype prediction using machine learning. *Frontiers in Genetics* 13, 822173.

Derbyshire, M.C., Batley, J. and Edwards, D. (2022) Use of multiple 'omics techniques to accelerate the breeding of abiotic stress tolerant crops. *Current Plant Biology* 32, 100262. DOI: 10.1016/j.cpb.2022.100262.

Dixit, A. and Nema, S. (2018) Wheat leaf disease detection using machine learning method-A review. *International Journal of Computer Science and Mobile Computing* 7(5), 124–129.

Dmitriev, A.A., Pushkova, E.N. and Melnikova, N.V. (2022) Plant genome sequencing: Modern technologies and novel opportunities for breeding. *Molecular Biology* 56(4), 495–507. DOI: 10.1134/S0026893322040045.

Dolatabadian, A., Neik, T.X., Danilevicz, M.F., Upadhyaya, S.R., Batley, J. *et al.* (2025) Image-based crop disease detection using machine learning. *Plant Pathology* 74(1), 18–38. DOI: 10.1111/ppa.14006.

Du, H., Fang, C., Li, Y., Kong, F. and Liu, B. (2023) Understandings and future challenges in soybean functional genomics and molecular breeding. *Journal of Integrative Plant Biology* 65(2), 468–495. DOI: 10.1111/jipb.13433.

Erbe, R., Gore, J., Gemmill, K., Gaykalova, D.A. and Fertig, E.J. (2022) The use of machine learning to discover regulatory networks controlling biological systems. *Molecular Cell* 82(2), 260–273. DOI: 10.1016/j.molcel.2021.12.011.

Escamilla, D.M., Li, D., Negus, K.L., Kappelmann, K.L., Kusmec, A. *et al.* (2025) Genomic selection: Essence, applications, and prospects. *The Plant Genome* 18(2), e70053.

Farooq, M.A., Gao, S., Hassan, M.A., Huang, Z., Rasheed, A. *et al.* (2024) Artificial intelligence in plant breeding. *Trends in Genetics* 40(10), 891–908.

Fritsche-Neto, R., Al-Nowibet, K., Montesinos-López, A. and Crossa, J. (2024) A review of multimodal deep learning methods for genomic-enabled prediction in plant breeding. *GENETICS* 228, 4.

Garg, V., Bohra, A., Mascher, M., Spannagl, M., Xu, X. *et al.* (2024) Unlocking plant genetics with telomere-to-telomere genome assemblies. *Nature Genetics* 56(9), 1788–1799. DOI: 10.1038/s41588-024-01830-7.

Gerakari, M., Katsileros, A., Kleftogianni, K., Tani, E., Bebeli, P.J. *et al.* (2025) Breeding of solanaceous crops using AI: Machine learning and deep learning approaches—A critical review. *Agronomy* 15(3), 757. DOI: 10.3390/agronomy15030757.

Geurts, P., Irrthum, A. and Wehenkel, L. (2009) Supervised learning with decision tree-based methods in computational and systems biology. *Molecular bioSystems* 5(12), 1593–1605. DOI: 10.1039/b907946g.

Goel, L. and Nagpal, J. (2023) A systematic review of recent machine learning techniques for plant disease identification and classification. *IETE Technical Review* 40(3), 423–439. DOI: 10.1080/02564602.2022.2121772.

Gou, C., Zafar, S., Fatima, N., Hasnain, Z., Aslam, N. *et al.* (2024) Machine and deep learning: Artificial intelligence application in biotic and abiotic stress management in plants. *Frontiers in Bioscience-Landmark* 29(1). DOI: 10.31083/j.fbl2901020.

Grinberg, N.F., Orhobor, O.I. and King, R.D. (2020) An evaluation of machine-learning for predicting phenotype: Studies in yeast, rice, and wheat. *Machine Learning* 109(2), 251–277.

Guo, Y., Zhao, G., Gao, X., Zhang, L., Zhang, Y. *et al.* (2023) CRISPR/Cas9 gene editing technology: A precise and efficient tool for crop quality improvement. *Planta* 258(2), 36. DOI: 10.1007/s00425-023-04187-z.

Harfouche, A.L., Jacobson, D.A., Kainer, D., Romero, J.C., Harfouche, A.H. *et al.* (2019) Accelerating climate resilient plant breeding by applying next-generation artificial intelligence. *Trends in Biotechnology* 37(11), 1217–1235.

Hesami, M., Alizadeh, M., Jones, A.M.P. and Torkamaneh, D. (2022) Machine learning: Its challenges and opportunities in plant system biology. *Applied Microbiology and Biotechnology* 106(9), 3507–3530.

Hu, Y., Chen, X. and Shen, X. (2022) Regulatory network established by transcription factors transmits drought stress signals in plant. *Stress Biology* 2(1), 26.

Huo, Q., Song, R. and Ma, Z. (2024) Recent advances in exploring transcriptional regulatory landscape of crops. *Frontiers in Plant Science* 15, 1421503.

Immanuel, S.C., Pothiraj, N., Thiyagarajan, K., Bharathi, M. and Rabindran, R. (2011) Genetic parameters of variability, correlation and pathcoefficient studies for grain yield and other yield attributes among rice blast disease resistant genotypes of rice (*Oryza sativa* L.). *African Journal of Biotechnology* 10(17), 3322–3334. DOI: 10.5897/AJB10.2575.

Jakutis, G. and Stainier, D.Y. (2021) Genotype–phenotype relationships in the context of transcriptional adaptation and genetic robustness. *Annual Review of Genetics* 55(1), 71–91.

Jiang, Y. and Li, C. (2020) Convolutional neural networks for image-based high-throughput plant phenotyping: A review. *Plant Phenomics* 2020, 4152816.

Jianing, G., Yuhong, G., Yijun, G., Rasheed, A., Qian, Z. *et al*. (2022) Improvement of heat stress tolerance in soybean (*Glycine max* L.), by using conventional and molecular tools. *Frontiers in Plant Science* 13, 993189. DOI: 10.3389/fpls.2022.993189.

Kassem, M.A. (2025) Harnessing artificial intelligence and machine learning for identifying quantitative trait loci (QTL) associated with seed quality traits in crops. *Plants (Basel, Switzerland)* 14(11), 1727. DOI: 10.3390/plants14111727.

Kell, D.B., Darby, R.M. and Draper, J. (2001) Genomic computing. Explanatory analysis of plant expression profiling data using machine learning. *Plant Physiology* 126(3), 943–951. DOI: 10.1104/pp.126.3.943.

Khan, M.H.U., Wang, S., Wang, J., Ahmar, S., Saeed, S. *et al*. (2022) Applications of artificial intelligence in climate-resilient smart-crop breeding. *International Journal of Molecular Sciences* 23(19), 11156. DOI: 10.3390/ijms231911156.

Krishna, T.P.A., Veeramuthu, D., Maharajan, T. and Soosaimanickam, M. (2023) The era of plant breeding: Conventional breeding to genomics-assisted breeding for crop improvement. *Current Genomics* 24(1), 24–35. DOI: 10.2174/1389202924666230517115912.

Liu, Q., Cheng, L., Nian, H., Jin, J. and Lian, T. (2023) Linking plant functional genes to rhizosphere microbes: A review. *Plant Biotechnology Journal* 21(5), 902–917. DOI: 10.1111/pbi.13950.

Lowe, R., Shirley, N., Bleackley, M., Dolan, S. and Shafee, T. (2017) Transcriptomics technologies. *PLoS Computational Biology* 13(5), e1005457. DOI: 10.1371/journal.pcbi.1005457.

Ma, J., Cheng, Z. and Cao, Y. (2025) Artificial intelligence-assisted breeding for plant disease resistance. *International Journal of Molecular Sciences* 26(11), 5324. DOI: 10.3390/ijms26115324.

Ma, N.L., Rahmat, Z. and Lam, S.S. (2013) A review of the "omics" approach to biomarkers of oxidative stress in *Oryza sativa*. *International Journal of Molecular Sciences* 14(4), 7515–7541.

Maghdid, S., Askar, S., Xoshibi, F. and Hamad, S. (2024) Deep learning algorithms for IoT based crop yield optimization. *Indonesian Journal of Computer Science* 13(2). DOI: 10.33022/ijcs.v13i2.3846.

Mahood, E.H., Kruse, L.H. and Moghe, G.D. (2020) Machine learning: A powerful tool for gene function prediction in plants. *Applications in Plant Sciences* 8(7), e11376. DOI: 10.1002/aps3.11376.

Mergner, J. and Kuster, B. (2022) Plant proteome dynamics. *Annual Review of Plant Biology* 73(1), 67–92. DOI: 10.1146/annurev-arplant-102620-031308.

Mochida, K., Koda, S., Inoue, K. and Nishii, R. (2018) Statistical and machine learning approaches to predict gene regulatory networks from transcriptome datasets. *Frontiers in Plant Science* 9, 1770. DOI: 10.3389/fpls.2018.01770.

Nayak, S.N., Aravind, B., Malavalli, S.S., Sukanth, B.S., Poornima, R. *et al*. (2021) Omics technologies to enhance plant based functional foods: An overview. *Frontiers in Genetics* 12, 742095. DOI: 10.3389/fgene.2021.742095.

Neagu, A.-N., Jayathirtha, M., Baxter, E., Donnelly, M., Petre, B.A. *et al*. (2022) Applications of tandem mass spectrometry (MS/MS) in protein analysis for biomedical research. *Molecules* 27(8), 2411. DOI: 10.3390/molecules27082411.

Nitze, I., Schulthess, U. and Asche, H. (2012) Comparison of machine learning algorithms random forest, artificial neural network and support vector machine to maximum likelihood for supervised crop type classification. In: *Proceedings of the 4th GEOBIA*, Rio de Janeiro, p. 3540 (Vol. 79).

Panchy, N., Lehti-Shiu, M. and Shiu, S.-H. (2016) Evolution of gene duplication in plants. *Plant Physiology* 171(4), 2294–2316. DOI: 10.1104/pp.16.00523.

Pang, B., Nijkamp, E. and Wu, Y.N. (2020) Deep learning with tensorflow: A review. *Journal of Educational and Behavioral Statistics* 45(2), 227–248. DOI: 10.3102/1076998619872761.

Patel, M.K., Pandey, S., Kumar, M., Haque, M.I., Pal, S. *et al*. (2021) Plants metabolome study: Emerging tools and techniques. *Plants (Basel, Switzerland)* 10(11), 2409. DOI: 10.3390/plants10112409.

Pérez-Enciso, M. and Zingaretti, L.M. (2019) A guide for using deep learning for complex trait genomic prediction. *Genes* 10(7), 553. DOI: 10.3390/genes10070553.

Przybyla, L. and Gilbert, L.A. (2022) A new era in functional genomics screens. *Nature Reviews. Genetics* 23(2), 89–103. DOI: 10.1038/s41576-021-00409-w.

Pucker, B., Irisarri, I., de Vries, J. and Xu, B. (2022) Plant genome sequence assembly in the era of long reads: Progress, challenges and future directions. *Quantitative Plant Biology* 3, e5. DOI: 10.1017/qpb.2021.18.

Putri, S.P., Ikram, M.M.M., Sato, A., Dahlan, H.A., Rahmawati, D. *et al*. (2022) Application of gas chromatography-mass spectrometry-based metabolomics in food science and technology. *Journal of Bioscience and Bioengineering* 133(5), 425–435. DOI: 10.1016/j.jbiosc.2022.01.011.

Qadri, S.A.A., Huang, N.-F., Wani, T.M. and Bhat, S.A. (2024) Advances and challenges in computer vision for image-based plant disease detection: A comprehensive survey of machine and deep learning approaches. *IEEE Transactions on Automation Science and Engineering* 22, 2639–2670. DOI: 10.1109/TASE.2024.3382731.

Rasheed, A., Jie, H., Ali, B., He, P., Zhao, L. *et al.* (2023) Breeding drought-tolerant maize (*Zea mays*) using molecular breeding tools: Recent advancements and future prospective. *Agronomy* 13(6), 1459. DOI: 10.3390/agronomy13061459.

Razzaq, K. and Shah, M. (2025) Machine learning and deep learning paradigms: From techniques to practical applications and research frontiers. *Computers* 14(3), 93. DOI: 10.3390/computers14030093.

Renny-Byfield, S. and Wendel, J.F. (2014) Doubling down on genomes: Polyploidy and crop plants. *American Journal of Botany* 101(10), 1711–1725.

Rhoads, A. and Au, K.F. (2015) PacBio sequencing and its applications. *Genomics, Proteomics & Bioinformatics* 13(5), 278–289. DOI: 10.1016/j.gpb.2015.08.002.

Rico-Chávez, A.K., Franco, J.A., Fernandez-Jaramillo, A.A., Contreras-Medina, L.M., Guevara-González, R.G. *et al.* (2022) Machine learning for plant stress modeling: A perspective towards hormesis management. *Plants* 11(7), 970.

Ritchie, M.D., Holzinger, E.R., Li, R., Pendergrass, S.A. and Kim, D. (2015) Methods of integrating data to uncover genotype–phenotype interactions. *Nature Reviews Genetics* 16(2), 85–97.

Shafik, W., Tufail, A., Namoun, A., De Silva, L.C. and Apong, R.A.A.H.M. (2023) A systematic literature review on plant disease detection: Motivations, classification techniques, datasets, challenges, and future trends. *IEEE Access* 11, 59174–59203. DOI: 10.1109/ACCESS.2023.3284760.

Sharma, M., Kumar, C.J. and Bhattacharyya, D.K. (2024) Machine/deep learning techniques for disease and nutrient deficiency disorder diagnosis in rice crops: A systematic review. *Biosystems Engineering* 244, 77–92. DOI: 10.1016/j.biosystemseng.2024.05.014.

Shi, J., Tian, Z., Lai, J. and Huang, X. (2023) Plant pan-genomics and its applications. *Molecular Plant* 16(1), 168–186. DOI: 10.1016/j.molp.2022.12.009.

Sun, Y., Shang, L., Zhu, Q.-H., Fan, L. and Guo, L. (2022) Twenty years of plant genome sequencing: Achievements and challenges. *Trends in Plant Science* 27(4), 391–401.

Tian, Q., Li, B., Feng, Y., Zhao, W., Huang, J. *et al.* (2022) Application of CRISPR/Cas9 in rapeseed for gene function research and genetic improvement. *Agronomy* 12(4), 824. DOI: 10.3390/agronomy12040824.

Tsakirpaloglou, N., Septiningsih, E.M. and Thomson, M.J. (2023) Guidelines for performing CRISPR/Cas9 genome editing for gene validation and trait improvement in crops. *Plants (Basel, Switzerland)* 12(20), 3564. DOI: 10.3390/plants12203564.

Tzec-Interián, J.A., González-Padilla, D. and Góngora-Castillo, E.B. (2025) Bioinformatics perspectives on transcriptomics: A comprehensive review of bulk and single-cell RNA sequencing analyses. *Quantitative Biology* 13(2), e78. DOI: 10.1002/qub2.78.

Upadhyaya, S.R., Danilevicz, M.F., Dolatabadian, A., Neik, T.X., Zhang, F. *et al.* (2024) Genomics-based plant disease resistance prediction using machine learning. *Plant Pathology* 73(9), 2298–2309. DOI: 10.1111/ppa.13988.

van Dijk, A.D.J., Kootstra, G., Kruijer, W. and Ridder, D. (2021) Machine learning in plant science and plant breeding. *iScience* 24(1), 101890. DOI: 10.1016/j.isci.2020.101890.

Wang, C. and Han, B. (2022) Twenty years of rice genomics research: From sequencing and functional genomics to quantitative genomics. *Molecular Plant* 15(4), 593–619. DOI: 10.1016/j.molp.2022.03.009.

Wang, Y.-H. and Su, W.-H. (2022) Convolutional neural networks in computer vision for grain crop phenotyping: A review. *Agronomy* 12(11), 2659. DOI: 10.3390/agronomy12112659.

Wang, H., Cimen, E., Singh, N. and Buckler, E. (2020) Deep learning for plant genomics and crop improvement. *Current Opinion in Plant Biology* 54, 34–41. DOI: 10.1016/j.pbi.2019.12.010.

Wang, Z., Akber, M.A. and Aziz, A.A. (2025a) Application of remote sensing and geographic information systems for monitoring and managing chili crops: A systematic review. *Remote Sensing* 17(16), 2827.

Wang, Z., Yuan, H., Yan, J. and Liu, J. (2025b) Identification, characterization, and design of plant genome sequences using deep learning. *The Plant Journal* 121(1), e17190. DOI: 10.1111/tpj.17190.

Whalen, S., Schreiber, J., Noble, W.S. and Pollard, K.S. (2022) Navigating the pitfalls of applying machine learning in genomics. *Nature Reviews Genetics* 23(3), 169–181.

Yan, J. and Wang, X. (2022) Unsupervised and semi-supervised learning: The next frontier in machine learning for plant systems biology. *The Plant Journal* 111(6), 1527–1538. DOI: 10.1111/tpj.15905.

Yan, J. and Wang, X. (2023) Machine learning bridges omics sciences and plant breeding. *Trends in Plant Science* 28(2), 199–210.

Yang, Y., Saand, M.A., Huang, L., Abdelaal, W.B., Zhang, J. *et al.* (2021) Applications of multi-omics technologies for crop improvement. *Frontiers in Plant Science* 12, 563953. DOI: 10.3389/fpls.2021.563953.

Zeebaree, D.Q., Hasan, D.A., Abdulazeez, A.M., Ahmed, F.Y.H. and Hasan, R.T. (2021) Machine learning semi-supervised algorithms for gene selection: A review. In: *2021 IEEE 11th International Conference on System Engineering and Technology (ICSET)*, Shah Alam, Malaysia, p. 102058 (Vol. 63). DOI: 10.1109/ICSET53708.2021.9612526.

Zemlyanskaya, E.V., Dolgikh, V.A., Levitsky, V.G. and Mironova, V. (2021) Transcriptional regulation in plants: Using omics data to crack the *cis*-regulatory code. *Current Opinion in Plant Biology* 63, 102058. DOI: 10.1016/j.pbi.2021.10205834098218.

Zhakypbek, Y., Belkozhayev, A.M., Kerimkulova, A., Kossalbayev, B.D., Murat, T. *et al.* (2025) MicroRNAs in plant genetic regulation of drought tolerance and their function in enhancing stress adaptation. *Plants* 14(3), 410. DOI: 10.3390/plants14030410.

Zhang, D., Yang, F., Li, J., Liu, Z., Han, Y. *et al.* (2025) Progress and perspectives on genomic selection models for crop breeding. *Technology in Agronomy* 5(1). DOI: 10.48130/tia-0025-0002.

Zhang, X., Ibrahim, Z., Khaskheli, M.B., Raza, H., Zhou, F. *et al.* (2024) Integrative approaches to abiotic stress management in crops: Combining bioinformatics educational tools and artificial intelligence applications. *Sustainability* 16(17), 7651. DOI: 10.3390/su16177651.

Zheng, C., Abd-Elrahman, A. and Whitaker, V. (2021) Remote sensing and machine learning in crop phenotyping and management, with an emphasis on applications in strawberry farming. *Remote Sensing* 13(3), 531. DOI: 10.3390/rs13030531.

5 Artificial Intelligence Models for High-dimensional and Digital Biology in Plants

Rashid Abbas Khan[1], Nadia Tehseen[2], Aqsa Shabbir[3], Tauqeer Ahmed Qadri[4], Amjid Khan[5,6], Muhammad Anas[7]*, Javed Iqbal[8]*, Banzeer Ahsan Abbasi[9], Farishta Zarshan[8], Madiha Imtiaz[8], Sana Basharat[10], Pingwu Liu[10] and Muhammad Waseem[10]*

[1]Department of Biology, Prince of Songkhla University, Hat Yai, Thailand; [2]Department of Botany, Lahore College for Women University (LCWU), Lahore, Pakistan; [3]Department of Horticulture, Ordu University, Ordu, Türkiye; [4]Department of Biosciences, COMSATS University Islamabad, Islamabad, Pakistan; [5]Department of Plant Sciences, Quaid-i-Azam University, Islamabad, Pakistan; [6]UNESCO UNISA Africa Chair in Nanoscience and Nanotechnology, College of Graduate Studies, University of South Africa, Pretoria, South Africa; [7]Department of Soil Science & Plant Nutrition, Selçuk University, Konya, Türkiye; [8]Department of Botany, Bacha Khan University, Charsadda, Pakistan; [9]Department of Botany, Rawalpindi Women University, Satellite Town, Rawalpindi, Pakistan; [10]School of Breeding and Multiplication (Sanya Institute of Breeding and Multiplication), School of Tropical Agriculture and Forestry, Hainan University, Sanya, China

5.1 Introduction

Advancements in artificial intelligence (AI) are now transforming almost every area of science, including plant science (Khan, 2024). As biological data grow exponentially owing to next-generation sequencing, high-throughput phenotyping, imaging, and sensor technologies, AI has become a critical technology for interpreting complex patterns and extracting valuable insights (Anita Margret *et al.*, 2025). Compared to traditional statistics, AI algorithms can be learned through very large and multi-dimensional data, thus making them useful in predictive modeling, trait determination, and real-time decisions in plant biology. Whether it is identifying early signs of disease or predicting plant performance in a variable climate, AI is increasingly transforming plant research and sustainable agriculture (Cembrowska-Lech *et al.*, 2023).

The management and interpretation of high-dimensional biological data are among the most pressing problems in plant science. New technologies result in multi-omics datasets, including genomics, transcriptomics, proteomics, metabolomics, and phenomics, which are very large, heterogeneous, and in many cases cannot

*Corresponding authors: muhammed.anas@selcuk.edu.tr, javed89qau@gmail.com and m.waseem.botanist@gmail.com

© CAB International 2026. *Plant Spatial Biology* (ed. Chen)

DOI: 10.1079/9781836994572.0005

be integrated using traditional approaches to data analysis (Yetgin, 2025). Simultaneously, the emergence of digital biology has increased the complexity of data in digital twins, virtual simulations, and precision phenotyping. The combination of these data streams requires novel computational models that can capture nonlinear relationships at different biological scales, in which AI has demonstrated unprecedented potential (Möller and Pörtner, 2021).

Despite these developments, major knowledge gaps remain. The limitations of current AI models include data standardization problems, interpretability, and flexibility in dynamic biological processes (Papadimitroulas *et al.*, 2021). Additionally, comprehensive plant modeling, which combines digital biology and AI, remains an emerging area of innovation, with numerous potential applications yet to be explored. These limitations underscore the pressing need for comprehensive evaluations of the specific applications of AI in plant sciences. It is essential to ensure that the challenge of dimensionality is addressed and biological interpretability is achieved (Naskar *et al.*, 2025). This chapter aims to conduct a comprehensive examination of high-dimensional and digital AI models for plants, particularly their uses, shortcomings, and future potential. The innovation of this study lies in the integration of two advanced domains, high-dimensional data analytics and digital biology, to develop an AI model. By conducting a critical assessment of the concepts of machine learning, deep learning, and developing AI approaches, this chapter provides new perspectives on how digital and computational solutions can be used to accelerate crop improvement processes, increase stress resilience, and foster sustainable practices in the agricultural field.

5.2 High-dimensional Biology in Plants

5.2.1 Nature of high-dimensional data

The past 20 years have seen unparalleled growth in high-dimensional biological data in plant sciences, particularly owing to the advent of high-throughput and cost-effective methods.

The basis of this data explosion is in the areas of genomics, transcriptomics, proteomics, and metabolomics (Cembrowska-Lech *et al.*, 2023). High-throughput sequencing devices have allowed the generation of gigabases of genomic information all at once, allowing large-scale genome-wide association studies (GWAS) and genomic selection to improve crops (Cheng, 2024). High-resolution data on gene expression at various developmental stages and under stress have been obtained through transcriptome analysis, which involves the study of RNA sequences using RNA sequencing (Kamali and Singh, 2023). Proteomic analyses provide complementary data, such as protein abundance and localization, while metabolomics can determine the abundance, localization, and modifications of small-molecule intermediates and products that directly reflect plant physiology and biochemical activities. When combined, these datasets provide a multidimensional view of plant biology, encompassing the genome and phenotype (Wu *et al.*, 2023).

Simultaneously, phenomics and high-throughput imaging technologies have revolutionized the capture and analysis of phenotypic traits of plants. Advanced phenotyping tools are available to continuously and noninvasively measure phenotypic features at the organ, plant, and field levels, including unmanned aerial vehicles (UAVs), multispectral and hyperspectral imaging, and automated greenhouse systems. Such technologies are essential for filling the genotype–phenotype gap, providing a unique opportunity to associate molecular markers with apparent features such as growth rate, stress tolerance, and yield potential (Cembrowska-Lech *et al.*, 2023).

The combination of these layers into multi-omics approaches is what systems biology is about: the holistic vision of a plant as a living network rather than a compilation of independent units. Multi-omics also enables the interplay between different layers of molecular data, that is, mapping genomic variation into transcriptomic changes, proteomic changes, and metabolomic consequences of the stress response. Coupled data, although they provide a comprehensive picture of the plant, are intrinsically multidimensional and require advanced computational techniques for effective analysis (Fan *et al.*, 2025).

5.2.2 Computational challenges

Such high-dimensional data can, however, be very difficult to analyze and restrict the practical use of large amounts of information. Among the most important issues related to this area is the curse of dimensionality. As the number of features (e.g. genes, metabolites, or spectral bands) increases, the data space volume transcends exponentially, making normal analysis techniques less effective and prone to overfitting. This problem is especially severe in plant sciences, where the sample size is small compared to the number of variables being measured (Ali and Mohammed, 2024).

The other problem is the heterogeneity of the data and missing values. Multi-omics and imaging datasets are produced under various experimental conditions, platforms, or environments, and thus vary in quality and scale. Experimental errors or technological limitations leave blank data points, making it even more difficult to carry out analyses, which may lead to biased results in the model prediction. A fundamental challenge that does not apply across databases is standardization and normalization (Menges, 2022).

Finally, the complex interactions between plants and the environment complicate the matter. Plants exist in highly dynamic ecosystems in which environmental fluctuations (e.g. temperature, light, water, and soil nutrients) have profound molecular and phenotypic effects (Mansoor and Chung, 2024). The description of such interactions in high-dimensional spaces involves augmented algorithms that consider nonlinearities, feedback loops, and time dynamics. It is on these grounds that complexity needs AI solutions that can uniquely overcome these computational challenges, owing to their capacity to process large, high-volume, noisy, and disparate data in real time (Karaca and Baleanu, 2022).

5.3 Artificial Intelligence Models in Plant Biology

5.3.1 Traditional machine learning approaches

Machine learning (ML) is the first and most popular AI framework adopted for the analysis of high-dimensional datasets in plant sciences. ML primarily involves classification, regression, and clustering techniques to make predictions about categorical values (presence or absence of disease), real-valued variables (yield or biomass), and how to divide biological samples naturally (e.g. dividing genotypes into clusters based on omics profiles) (Hong *et al.*, 2025).

Support vector machines (SVMs) are among the most widely applied algorithms, especially suitable for high-dimensional omics data, and can create optimal decision boundaries within feature-rich spaces (Sahu and Pandey, 2023). Random forests (RFs) are particularly effective with noisy and heterogeneous data and can be used to rank feature importance and identify key genes, metabolites, or phenotypic traits. The k-nearest neighbors (k-NN) algorithm has the advantages of simplicity and interpretability, but a limitation of scalability in high-dimensional data (Imamoglu, 2024). Recently, gradient boosting machines (GBM), such as XGBoost and LightGBM, have been successfully used in crop yield prediction, stress classification, and quantitative trait prediction, as they are known to model non-linear relationships with high accuracy (Kumar *et al.*, 2024; Zhou *et al.*, 2025).

Examples of ML applications in plant biology include stress biology, where a classifier is trained on transcriptomic or metabolomic data to differentiate between tolerant and susceptible genotypes under drought, salinity, or heat stress (Singhal *et al.*, 2025). In yield prediction, ML models combine genomic, climate, and soil data to predict productivity in various environments. In trait mapping, ML-based feature selection techniques have been used to detect candidate loci and metabolic pathways underlying complex traits. Although ML has created the basis of computational plant biology, it is, in most cases, poor in scalability when applied to data that are highly unstructured and their dimensions are extremely high, such as images or time-series signals (Tamayo-Vera *et al.*, 2025).

5.3.2 Deep learning and neural architectures

The development of deep learning (DL) has opened the possibility of analyzing digitized, high-volume plant data directly and non-destructively. Compared to classic ML, which uses feature extraction, DL models do not require feature engineering and are thus well-suited to the high-dimensional data found in biology (Liu *et al.*, 2025).

Convolutional neural networks (CNNs) have emerged as the standard for image-based phenotyping and disease detection. With training on either spatial hierarchy in a leaf, canopy, or hyperspectral image, CNN models can detect changes missed by the naked eye, such as weak signs of stress, nutrient imbalances, or incipient disease (El Sakka *et al.*, 2024). Application areas include the automated mass scoring of plant phenotypes in breeding, real-time detection of pathogens in precision agriculture, and high-throughput sample sorting (Dang *et al.*, 2024).

Recurrent neural networks (RNNs) and their extensions, such as long short-term memory (LSTM) networks, can extract temporal dependencies and feedback loops in sequential or time-series biological data, such as growth progression or time-series omics data (Priyadarshini *et al.*, 2024). These models work especially well when forecasting crop yields, given the fluctuations in environmental conditions across the growing stages, as they learn through dynamic changes (Mienye *et al.*, 2024).

Autoencoders have also been proven to be effective for dimensionality removal and feature discovery during multi-omics integration. Autoencoders model high-dimensional inputs into low-dimensional latent representations to compress information and eliminate noise and redundancy in data. These latent features are also more often applied in downstream applications, such as trait prediction and genotype clustering, allowing scientists to reveal hidden patterns in complex biological systems (Zhang *et al.*, 2022).

DL frameworks offer computationally efficient, flexible, and scalable representations of plant biology in the digital world, outperforming traditional ML models because of their support for raw, high-dimensional data (Han and Wang, 2025).

5.3.3 Hybrid and emerging AI models

In addition to traditional ML and DL, hybrid and novel AI models have recently begun to transform high-dimensional plant science by merging the best features of different modalities (Katharria *et al.*, 2024). Ensemble learning methods, which combine the outputs of multiple models, typically produce better results than single models by lowering the variance and increasing the robustness. Ensemble approaches that incorporate RF, SVM, and CNNs can enhance the accuracy of yield prediction, disease classification, and genomic prediction in plant sciences (Ali *et al.*, 2024a; Taji *et al.*, 2024).

Transfer learning is an emerging method aimed at addressing the shortage of annotated data in plant biology. By modifying models based on large repositories of images (e.g. ImageNet) and retraining them on plant-specific tasks, researchers have obtained high levels of accuracy using small data repositories (Hassan *et al.*, 2021). Similarly, federated learning is appealing because it enables the training of multicentered AI and ensures data privacy, an important aspect of multicentered plant research and breeding programs (Mamba Kabala *et al.*, 2023).

One of these involves an increasingly important field, explainable AI (XAI), which addresses the long-standing challenge of interpretability. Although DL models can be considered black boxes, XAI provides researchers with the opportunity to trace the predictions back to biological inputs, such as identifying the most important genes, pathways, or image areas that contribute to model predictions (Table 5.1). This is particularly useful in high-dimensional datasets, where, in addition to accuracy, predictions should prove to be useful information in terms of biology (Porfírio *et al.*, 2025).

5.3.4 Digital biology and AI integration

5.3.4.1 Concept and scope of digital biology in plants

Digital biology represents a paradigm shift in plant science, where computational models and digital technologies are used to reconstruct, analyze, and predict biological systems. Its applications are broad, including the generation of digital twins of plants, virtual representations of real plants, and virtual crop models that are used to simulate genotype–environment interactions

Table 5.1. Overview of artificial intelligence models applied in plant biology.

AI approach	Key algorithms/ architectures	Applications in plant biology	Advantages	Limitations	References
Traditional ML	SVM, random forest, k-NN, gradient boosting	Trait prediction, stress classification, yield modeling	Handles medium-scale datasets, interpretable results	Limited scalability, feature engineering required	Ngugi *et al.*, 2024
Deep learning	CNNs, RNNs, LSTMs, autoencoders	Image-based phenotyping, disease detection, time-series growth analysis, dimensionality reduction	High accuracy, automatic feature extraction	Requires large datasets, is computationally intensive, and has limited interpretability	Amiri *et al.*, 2024
Hybrid models	ML + DL ensembles, transfer learning, federated learning	Multi-omics integration, cross-environment prediction, digital phenotyping	Improved robustness, adaptability to small datasets	Complexity, high computational demand	Danilevicz *et al.*, 2022; Sangjan *et al.*, 2025
Explainable AI	SHAP, LIME, attention mechanisms	Gene–trait association, biological interpretation of predictions	Enhances trust and interpretability	Still emerging, limited adoption in plant sciences	Mostafa *et al.*, 2023; Donmez *et al.*, 2025

(Datta, 2024). A digital twin bridges the gap between genomic, phenotypic, and environmental data to recreate plant growth and development and their interactions with stressors. These models can be used by researchers to simulate different conditions, such as drought or nutrient deficiency, without having to run large-scale physical field experiments (Choudhury *et al.*, 2024; Melesse, 2025).

The main structure of digital plant biology involves simulating growth, stress responses, and productivity. Modeling can forecast plant performance adaptations in different environments, which helps breeders and agronomists design stress-tolerant crops (Nguyen *et al.*, 2025). For example, digital wheat models of photosynthesis and canopy structure under heat stress have already shown the predictive validity of yield stability across a range of agro-climatic zones. Digital biology can enhance crop improvement by making it cost-effective and time-efficient through predictive experimentation in the virtual world (Kaur, 2023).

5.3.4.2 AI-powered digital platforms

The application of AI on digital platforms has revolutionized crop monitoring and management. Internet of Things (IoT) devices, drones, and intelligent sensors monitor soil moisture, canopy temperature, chlorophyll content, and disease incidence in real time (Kar and Chowdhury, 2024). By combining them with AI algorithms, these platforms can offer actionable insights, such as the optimal time to water the land or detect any possible stress before it becomes problematic. For example, AI-based UAV imaging has been used to delineate nitrogen-use efficiency within maize fields to provide site-specific fertilizer applications (Sow *et al.*, 2024).

Remote sensing and image recognition technologies also contribute to precision agriculture, as monitoring can be large-scale and high-resolution (Wang *et al.*, 2022). The use of AI models on hyperspectral and multispectral data can detect or identify nutrient deficiencies in plants and foresee biomass and abiotic stresses at the field or regional levels (Ali *et al.*, 2024b). Examples of this can be found in CNN-based analysis of drone imagery, which achieved more than 90% accuracy in detecting water stress in vineyards, demonstrating the usefulness of remote sensing in conjunction with AI (Almalki *et al.*, 2024).

The use of AI-based platforms to detect disease and stress in breeding programs and farm management is becoming popular. DL model-driven diagnosis tools using smartphones can detect several plant diseases in various light conditions or backgrounds in real time (Seng *et al.*, 2022). Similarly, cloud-based engines combine sensor and image data with AI models to deliver predictive notifications to farmers, who can intervene prior to the visible signs of stress. Together, such systems enable agricultural practices that are scalable, cost-effective, and sustainable (El-Tabakh *et al.*, 2024).

5.4 Multi-omics and Network Biology

The complexity of plant systems demands the combination of various molecular data, which is possible with AI-based multi-omics analysis (Yetgin, 2025). The integration of AI models enables the integration of genomics, transcriptomics, proteomics, and metabolomics data, extracting latent features that capture cross-talk between biological levels (Nguyen and Wang, 2020). For example, *Arabidopsis* and wheat multi-omics datasets have been used to identify regulatory modules related to flowering time and stress tolerance through DL-based autoencoders (Murmu *et al.*, 2024).

AI can also be used in systems biology to reconstruct gene regulatory, metabolic, and signaling networks. Probabilistic models/graph-based learning algorithms are used to infer causal interactions between genes and metabolites to gain a better understanding of the mechanisms underlying stress resilience. These modes further indicate regulatory hierarchies and control points that could be subject to genetic improvement (Manchel *et al.*, 2024).

Digital biology is another field of predictive modeling of complex plant traits that uses high-throughput data. When combined with environmental data, AI models can forecast the value of quantitative traits, including yield, biomass, and nutrient-use efficiency (Table 5.2) (Angidi *et al.*, 2025). For example, combined transcriptomic and environmental data were used in hybrid DL to effectively predict drought tolerance scores in maize, providing breeders with an efficient method for selecting more drought-tolerant genotypes (Kumar *et al.*, 2022).

Table 5.2. Digital biology and AI integration: tools, platforms, and applications.

Digital biology tool	AI integration	Application in plants	Impact on research/ agriculture	References
Digital twins	Simulation models, predictive AI	Growth, stress response, yield prediction	Virtual crop management, reduced experimental costs	Peladarinos *et al.*, 2023; Awais *et al.*, 2025
IoT and smart sensors	Edge AI, cloud-based ML	Real-time monitoring of soil, water, and nutrients	Precision agriculture, resource optimization	Avalekar *et al.*, 2024; Sharma, 2025
Drones and remote sensing	CNNs, image recognition	Disease detection, canopy structure, stress mapping	Scalable, field-wide monitoring	Umirzakova *et al.*, 2025
Multi-omics platforms	AI-driven data integration	Genome-to-phenome analysis, trait discovery	System-level insights, improved breeding strategies	Cembrowska-Lech *et al.*, 2023; Admas *et al.*, 2025
AI-based decision support systems	Predictive modeling, big data analytics	Crop management, climate adaptation strategies	Climate-smart, sustainable agriculture	Yousaf *et al.*, 2023; Chinchorkar, 2025

5.5 Case Studies and Practical Applications

5.5.1 AI in predicting plant stress tolerance

Abiotic stress prediction is one of the greatest prospects of AI in digital plant biology, with the most value added to crops under drought, salinity, and heat conditions (Angidi *et al.*, 2025). RF and GBM learning algorithms were used to categorize genotypes as either stress-tolerant or susceptible phenotypes using transcriptomic and metabolomic data (Wu *et al.*, 2025). For example, in wheat, multi-omics datasets established susceptibility and resilience to drought in AI models that defined fundamental transcription factors and metabolic processes in wheat correlated with water-use efficiency, resulting in the prediction of genotypes with a higher degree of resilience to drought and water-use efficiency (Le Roux *et al.*, 2024). Similarly, DL models based on hyperspectral imagery have been applied to determine salinity tolerance in rice, so that stress-resistant lines can be screened early in the breeding process. These strategies minimize the use of tedious field-testing criteria and shorten the process of choosing climate-resilient crops (Haque *et al.*, 2021).

5.5.2 Early disease detection via imaging and deep learning

The application of AI-based image analysis has shown that it is more accurate in detecting early plant diseases than scouting. By using CNNs trained on leaf images, early patterns of infection with fungi, bacteria, or viruses can be detected before they are visible to the naked eye. In maize, CNN-based approaches were found to be capable of exceeding 95% accuracy in northern leaf blight diagnosis under field conditions, whereas transfer learning methods allowed the successful detection of multiple soybean leaf diseases using smaller image datasets (Sibiya, 2021). UAV imagery combined with AI also increases scalability, as it allows the entire field to be monitored in real time and surveyed for diseases. These tools have become the focus of digitalized agrarian systems, where control at an early stage can minimize yield loss and chemical use (Ali *et al.*, 2024c).

5.5.3 Digital phenotyping for precision breeding

Digital phenotyping has advanced rapidly, and with AI-based algorithms it is now possible to evaluate traits in a way that is accurate, repeatable, and measurable (Bömer *et al.*, 2024). Computer-vision algorithms can analyze data collected using automated growth chambers, hyperspectral cameras, drones, and automated greenhouses to measure traits, including biomass, canopy architecture, water-use efficiency, and photosynthetic performance (Bian *et al.*, 2022; Jayasuriya *et al.*, 2024). For example, DL was used with time-series data collected by a UAV to monitor the dynamics of wheat development and robustly estimate grain yield. In rice, digital phenotyping platforms and ML have identified important spectral indices associated with drought tolerance, prioritizing promising genotypes for breeders. Through the combination of digital phenotyping with genomic data, AI is expected to facilitate precision breeding, in which the genotype is selected to best fit the target environment (Anand *et al.*, 2025).

5.5.4 Genome-wide prediction and trait selection in breeding programs

The application of AI has achieved breakthrough levels of genome-wide prediction (GWP) and trait selection, which are already quite advanced in traditional genomic selection models (Zhang and Li, 2024). CNNs and RNNs have been employed to identify non-linear relationships between genotype and phenotype using DL architectures and have outcompeted linear models in predicting complex traits such as yield, disease resistance, and flowering time. In maize, hybrid DL modeling of discrete genomic and continuous environmental data resulted in more accurate predictions of drought tolerance than conventional genomic selection approaches (Azrai *et al.*, 2024). In wheat, other genome-wide prediction models derived using

AI have also been able to identify marker-trait relationships with regard to heat tolerance, and in the process, have led to the accelerated identification of candidate genes for breeding (Mei *et al.*, 2022). AI allows breeders to make sound choices based on large amounts of phenotypic and environmental data and high-dimensional omics datasets to select the best lines to improve crops (Kassem, 2025).

5.6 Challenges and Future Prospects

Although significant advances have been made in the field, the implementation of AI in high-dimensional and digital plant biology has certain limitations that restrict its large-scale application. The lack of high-quality datasets that can be standardized is one of the most prominent drawbacks. Data derived from omics technology, phenotypic outcomes, and imaging statistics are regularly produced within various platforms and settings, frequently resulting in discrepancies in data format, resolution, and description. Missing values, the presence of noise, and imbalances in the dataset also negatively affected the model performance. Despite the presence of public repositories, much of the data are incomplete or poorly curated, which reduces their value and applicability to AI training. The postulate that AI predictions are reproducible across studies and species will not be realized without standardized data acquisition, preprocessing, and integration protocols.

Equally important are the computational pipelines used to curate and analyze high-dimensional plant datasets – such as whole-genome sequencing, hyperspectral imaging, and multi-omics – so that raw data can be converted into reliable, interpretable biological signals. Training DL models involves high-performance computing clusters, graphics processing unit (GPU)-based, or cloud-based infrastructures, which are not readily accessible everywhere, especially in developing areas where crop diversity is more prevalent. The digital divide is a limiting factor in the extensive usability, and the high energy demand of large AI models is also problematic when discussing sustainability in agricultural research.

The interpretability and reproducibility of AI models constitute another limitation. DL architectures have often been considered black boxes because of the inability to understand the biological reasoning behind their outputs, despite their predicted outcomes. This has become a barrier to plant science, where candidate genes, pathways, and stress response mechanisms must be understood. Moreover, models trained on particular datasets cannot be used in other settings or genotypes because they are sensitive to the training environment.

The use of AI raises ethical, privacy, and data-sharing issues, which complicate the integration of AI in plant biology. International collaborations and breeding schemes may require sensitive data on genomics, phenotypes, and the environment. Concerns over ownership, misuse, and unequal access to these data disincentivize such sharing practices, thus limiting the availability of diverse datasets needed to build robust AI training algorithms. The same problems will apply to the adoption of AI-based approaches in farming, as increased inequality in access may widen the divide between affluent and poor farmers. These limitations need to be addressed by establishing clearly understood data governance and access principles, access equity policies, and encouraging open-science programs.

Despite these challenges, they provide a foundation for future opportunities. The ability of AI to integrate with other emerging biotechnologies, such as CRISPR-Cas-based genome editing, holds great potential. Predictive models have the potential to sequence essential genetic loci and implement fine editing services to decrease stress sensitivity and enhance production and nutritional content. By connecting omics datasets with genome-editing technologies, the prototyping of crops with the ability to adapt to changing environments may be fast-tracked. Similarly, the design of second-generation digital twins, such as dynamic, data-driven, and virtual replicas of the plant, will allow modeling of growth, physiological reactions, and interaction with the environment to a novel degree of accuracy. In conjunction with real-time data in the field, such digital twins will enable researchers and breeders to test various scenarios, optimize strategies, and predict results without conducting expensive or time-consuming trials.

Innovations in computational structures also have transformative capabilities. The adoption of cloud computing has enabled the large-scale integration of multi-omics, phenomics, and environmental data across worldwide networks. Edge AI technology integrated into drones and field sensors can enable data processing at sites with reduced latency and the ability to provide site-specific recommendations in near real time. This two-pronged approach to cloud-edge integration provides both scalability and responsiveness for digital crop management.

Furthermore, AI is projected to be one of the core concepts for promoting sustainable and climate-smart agriculture. Predictive analytics can be applied to achieve optimal resource utilization and minimize the use of water, fertilizers, and pesticides, without compromising productivity. Plant–climate modeling using AI will help develop resilient crops and farming systems that can survive climatic fluctuations. Finally, worldwide networking and open-data projects are key to the sustainable prosperity of AI in plant biology. Developing diverse, standardized, and accessible datasets requires coordination of efforts between research institutions, governments, and industries. This type of collaborative structure will not only lead to an increase in the generalizability of the AI models but will also serve to level access to sophisticated tools, where they are needed the most, that is, in geographical areas affected by food security issues. By cultivating transparency, inclusiveness, and sustainability, the introduction of AI in plant biology can become not only a technological breakthrough but also a worldwide solution to agricultural issues in the 21st century.

5.7 Conclusion

The introduction of AI into plant science is a paradigm shift in the approach undertaken by scientists and breeders to resolve the complexity of high dimensionality and digital biology. In this chapter, we discuss the role of AI models, including traditional machine learning, more recent deep learning, and hybrid frameworks, in redefining data interpretation at the level of genomics, transcriptomics, proteomics, metabolomics, and phenomics. AI has addressed the shortcomings of traditional methods of analysis to identify specific traits, predict diseases, predict stress-resistant plants, and speed up genome-wide selection in crop breeding programs.

The use of digital biology combined with AI platforms takes these opportunities even further, as real-time monitoring, simulation, and predictive modeling capabilities are interconnected. Digital twinning, phenotyping using edge devices and IoT-connected sensors, and automated phenotyping systems mean that experimental systems that study plant growth and stress responses can now be recreated virtually, filling the gap between the lab and field. When combined with multi-omics integration and systems biology, AI is an inclusive method for decoding the complex interactions between plants and the environment that determine productivity and resilience.

Although the future of AI in plant biology appears bright, the reality of integrating AI into plant biology presents challenges. Issues of data standardization, computing and processing requirements, interpretability, and equitable access remain critical limitations. Filling these gaps will require international cooperation, open-data movements, and the construction of transparent, interpretable models that are biologically meaningful. However, the trend of AI-related research further proves its power to change.

In the future, AI will be a key pillar in spearheading food security and sustainable farming. With the combination of predictive analytics and genome editing, digital crop modeling, and climate-smart farming techniques, AI can accelerate the actualization of sturdy crops that are resilient to the conditions of a changing climate. With the generation of increasingly large and complex datasets at the molecular scale, the synergistic joint application of digital biology and AI will likely not only enhance our scientific understanding of plant systems but also provide trans-scale, real-world solutions to sustainably feed a growing population.

References

Ali, A.M. and Mohammed, M.A. (2024) A comprehensive review of artificial intelligence approaches in omics data processing: Evaluating progress and challenges. *International Journal of Mathematics, Statistics, and Computer Science* 2, 114–167. DOI: 10.59543/ijmscs.v2i.8703.

Admas, T., Jiao, S., Pan, R. and Zhang, W. (2025) Pan-omics insights into abiotic stress responses: Bridging functional genomics and precision crop breeding. *Functional & Integrative Genomics* 25(1), 128. DOI: 10.1007/s10142-025-01633-x.

Ali, A.H., Youssef, A., Abdelal, M. and Raja, M.A. (2024a) An ensemble of deep learning architectures for accurate plant disease classification. *Ecological Informatics* 81, 102618. DOI: 10.1016/j.ecoinf.2024.102618.

Ali, F., Razzaq, A., Tariq, W., Hameed, A., Rehman, A. *et al.* (2024b) Spectral intelligence: AI-driven hyperspectral imaging for agricultural and ecosystem applications. *Agronomy* 14(10), 2260. DOI: 10.3390/agronomy14102260.

Ali, Z.A., Deng, D., Shaikh, M.K., Hasan, R. and Khan, M.A. (2024c) AI-based UAV swarms for monitoring and disease identification of *brassica* plants using machine learning: A review. *Computer Systems Science and Engineering* 48(1), 1–34. DOI: 10.32604/csse.2023.041866.

Almalki, F.A., Salem, S.M., Fawzi, W.M., Alfeteis, N.S., Esaifan, S.A. *et al.* (2024) Coupling an autonomous UAV with a ML framework for sustainable environmental monitoring and remote sensing. *International Journal of Aerospace Engineering* 2024(1), 4285475. DOI: 10.1155/2024/4285475.

Amiri, Z., Heidari, A., Navimipour, N.J., Esmaeilpour, M. and Yazdani, Y. (2024) The deep learning applications in iot-based bio- and medical informatics: A systematic literature review. *Neural Computing and Applications* 36(11), 5757–5797. DOI: 10.1007/s00521-023-09366-3.

Anand, S., L., Visakh, R., Nalishma, R., Sah, R.P. and Beena, R. (2025) High throughput phenomics in elucidating drought stress responses in rice (*Oryza sativa* L.). *Journal of Plant Biochemistry and Biotechnology* 34(1), 119–132. DOI: 10.1007/s13562-024-00949-2.

Angidi, S., Madankar, K., Tehseen, M.M. and Bhatla, A. (2025) Advanced high-throughput phenotyping techniques for managing abiotic stress in agricultural crops—a comprehensive review. *Crops* 5(2), 8. DOI: 10.3390/crops5020008.

Anita Margret, A., Freeman, C., AS, M.M.D. and Ramesh, P. (2025) Next-gen biology. In: *AI's Transformative Impact on Life Sciences: AI Innovations in Biotechnology, Healthcare, and Agriculture*. OrangeBooks Publication.

Avalekar, U., Patil, D.J., Patil, D.S., Khot, P. and Prathapan, P. (2024) Optimizing agricultural efficiency: A fusion of Iot, AI, cloud computing, and wireless sensor network. In: *Prof.(Dr.) Kesava, Optimizing Agricultural Efficiency: A Fusion of Iot, Ai, Cloud Computing, and Wireless Sensor Network*. DOI: 10.2139/ssrn.4789232.

Awais, M., Wang, X., Hussain, S., Aziz, F. and Mahmood, M.Q. (2025) Advancing precision agriculture through digital twins and smart farming technologies: A review. *AgriEngineering* 7(5), 137. DOI: 10.3390/agriengineering7050137.

Azrai, M., Aqil, M., Andayani, N.N., Efendi, R., Suarni, O.O.O. *et al.* (2024) Optimizing ensembles machine learning, genetic algorithms, and multivariate modeling for enhanced prediction of maize yield and stress tolerance index. *Frontiers in Sustainable Food Systems* 8, 1334421. DOI: 10.3389/fsufs.2024.1334421.

Bian, L., Zhang, H., Ge, Y., Čepl, J., Stejskal, J. *et al.* (2022) Closing the gap between phenotyping and genotyping: Review of advanced, image-based phenotyping technologies in forestry. *Annals of Forest Science* 79(1), 22. DOI: 10.1186/s13595-022-01143-x.

Bömer, J., Esser, F., Marks, E., Rosu, R.A., Behnke, S. *et al.* (2024) A 3D printed plant model for accurate and reliable 3D plant phenotyping. *GigaScience* 13, giae035. DOI: 10.1093/gigascience/giae035.

Cembrowska-Lech, D., Krzemińska, A., Miller, T., Nowakowska, A., Adamski, C. *et al.* (2023) An integrated multi-omics and artificial intelligence framework for advance plant phenotyping in horticulture. *Biology* 12(10), 1298. DOI: 10.3390/biology12101298.

Cheng, F. (2024) Application of genome-wide association study in crop disease resistance breeding. *Field Crop* 7.

Chinchorkar, S. (2025) AI-enabled decision support systems for agrometeorological forecasting and climate change mitigation. *International Journal of Statistics and Applied Mathematics* 10(4), 13–20. DOI: 10.22271/maths.2025.v10.i4a.2015.

Choudhury, S.B., Kulat, R., Patil, R., Kumar, A., Sarangi, S. *et al.* (2024) On digital twin for a high yielding soybean variety towards optimal field recommendations. In: *2024 IEEE Global Humanitarian Technology Conference (GHTC)*, IEEE, pp. 251–311 (Vol. 90). DOI: 10.1109/GHTC62424.2024.10771522.

Dang, M., Wang, H., Li, Y., Nguyen, T.-H., Tightiz, L. *et al.* (2024) Computer vision for plant disease recognition: A comprehensive review. *The Botanical Review* 90(3), 251–311. DOI: 10.1007/s12229-024-09299-z.

Danilevicz, M.F., Gill, M., Anderson, R., Batley, J., Bennamoun, M. *et al.* (2022) Plant genotype to phenotype prediction using machine learning. *Frontiers in Genetics* 13, 822173. DOI: 10.3389/fgene.2022.822173.

Datta, S. (2024) Digital twin, Didymos, meets digital cousin, Didymium. From paradox to paradigm or a paradoxical paradigm? Preprints DOI: 10.20944/preprints202411.1638.v1.

Donmez, T.B., Kutlu, M., Mansour, M. and Yildiz, M.Z. (2025) Explainable AI in action: A comparative analysis of hypertension risk factors using SHAP and LIME. *Neural Computing and Applications* 37(5), 4053–4074. DOI: 10.1007/s00521-024-10724-y.

El Sakka, M., Mothe, J. and Ivanovici, M. (2024) Images and CNN applications in smart agriculture. *European Journal of Remote Sensing* 57(1), 2352386. DOI: 10.1080/22797254.2024.2352386.

El-Tabakh, M.A., Abd El-Samea, M.A., Roby, Y.A., Mohamed, Y.A., Harb, H.E. *et al.* (2024) Digital agriculture for enhancing yield, nutrition, and biological stress resistance. In: *Digital Agriculture: A Solution for Sustainable Food and Nutritional Security*. Springer, pp. 445–483. DOI: 10.1007/978-3-031-43548-5_15.

Fan, B.-L., Chen, L.-H., Chen, L.-L. and Guo, H. (2025) Integrative multi-omics approaches for identifying and characterizing biological elements in crop traits: Current progress and future prospects. *International Journal of Molecular Sciences* 26(4), 1466. DOI: 10.3390/ijms26041466.

Han, S. and Wang, H. (2025) Application of deep learning technology in monitoring plant attribute changes. *Sustainability* 17(17), 7602. DOI: 10.3390/su17177602.

Haque, M.A., Rafii, M.Y., Yusoff, M.M., Ali, N.S., Yusuff, O. *et al.* (2021) Advanced breeding strategies and future perspectives of salinity tolerance in rice. *Agronomy* 11(8), 1631. DOI: 10.3390/agronomy11081631.

Hassan, S.M., Maji, A.K., Jasiński, M., Leonowicz, Z. and Jasińska, E. (2021) Identification of plant-leaf diseases using CNN and transfer-learning approach. *Electronics* 10(12), 1388. DOI: 10.3390/electronics10121388.

Hong, K., Zhou, Y. and Han, H. (2025) The pipelines of deep learning-based plant image processing. *Quantitative Plant Biology* 6, e23. DOI: 10.1017/qpb.2025.10018.

Imamoglu, E. (2024) Artificial intelligence and/or machine learning algorithms in microalgae bioprocesses. *Bioengineering (Basel, Switzerland)* 11(11), 1143. DOI: 10.3390/bioengineering11111143.

Jayasuriya, N., Guo, Y., Hu, W. and Ghannoum, O. (2024) Image based crop monitoring technologies in protected horticulture: A review. *arXiv Preprint arXiv*.

Kamali, S. and Singh, A. (2023) Genomic and transcriptomic approaches to developing abiotic stress-resilient crops. *Agronomy* 13(12), 2903. DOI: 10.3390/agronomy13122903.

Kar, P. and Chowdhury, S. (2024) IoT and drone-based field monitoring and surveillance system. In: *Artificial Intelligence Techniques in Smart Agriculture*. Springer, pp. 253–266. DOI: 10.1007/978-981-97-5878-4_15.

Karaca, Y. and Baleanu, D. (2022) Evolutionary mathematical science, fractional modeling and artificial intelligence of nonlinear dynamics in complex systems. *Chaos Theory and Applications* 4, 111–118.

Kassem, M.A. (2025) Harnessing artificial intelligence and machine learning for identifying quantitative trait loci (QTL) associated with seed quality traits in crops. *Plants* 14(11), 1727. DOI: 10.3390/plants14111727.

Katharria, A., Pant, M., Velasquez, J.D., Snasel, V., Rajwar, K. *et al.* (2024) Information fusion in smart agriculture: Machine learning applications and future research directions. *arXiv Preprint arXiv:2405.17465*.

Kaur, M. (2023) Computation and validation of different agroclimatic indices for wheat. Doctoral dissertation, Punjab Agricultural University, Ludhiana, India.

Khan, A. (2024) Reshaping plant science: The power of ai and cutting-edge technologies *Available at SSRN 5250124*.

Kumar, A., Singh, V.K., Saran, B., Al-Ansari, N., Singh, V.P. *et al.* (2022) Development of novel hybrid models for prediction of drought- and stress-tolerance indices in teosinte introgressed maize lines using artificial intelligence techniques. *Sustainability* 14(4), 2287. DOI: 10.3390/su14042287.

Kumar, S., Sohail, M.A., Jadhav, S. and Gupta, R.K. (2024) Light gradient boosting machine for optimizing crop maintenance and yield prediction in agriculture. *ICTACT Journal on Soft Computing* 15(2), 3551–3555. DOI: 10.21917/ijsc.2024.0495.

Le Roux, M.S., Kunert, K.J., Cullis, C.A. and Botha, A.M. (2024) Unlocking wheat drought tolerance: The synergy of omics data and computational intelligence. *Food and Energy Security* 13(6), e70024. DOI: 10.1002/fes3.70024.

Liu, Z., Meyendorf, N., Blasch, E., Tsukada, K., Liao, M. *et al.* (2025) The role of data fusion in predictive maintenance using digital twins. In: *Handbook of Nondestructive Evaluation 4.0*. Springer, pp. 429–451.

Mamba Kabala, D., Hafiane, A., Bobelin, L. and Canals, R. (2023) Image-based crop disease detection with federated learning. *Scientific Reports* 13(1), 19220. DOI: 10.1038/s41598-023-46218-5.

Manchel, A., Gee, M. and Vadigepalli, R. (2024) From sampling to simulating: Single-cell multiomics in systems pathophysiological modeling. *iScience* 27(12), 111322. DOI: 10.1016/j.isci.2024.111322.

Mansoor, S. and Chung, Y.S. (2024) Functional phenotyping: Understanding the dynamic response of plants to drought stress. *Current Plant Biology* 38, 100331. DOI: 10.1016/j.cpb.2024.100331.

Mei, F., Chen, B., Du, L., Li, S., Zhu, D. *et al.* (2022) A gain-of-function allele of a DREB transcription factor gene ameliorates drought tolerance in wheat. *The Plant Cell* 34(11), 4472–4494. DOI: 10.1093/plcell/koac248.

Melesse, T.Y. (2025) Digital twin-based applications in crop monitoring. *Heliyon* 11(2), e42137. DOI: 10.1016/j.heliyon.2025.e42137.

Menges, P.R. (2022) Multi-omics analysis as a tool to investigate causes and consequences of impaired genome integrity. Doctoral dissertation, Technische Universität Kaiserslautern.

Mienye, I.D., Swart, T.G. and Obaido, G. (2024) Recurrent neural networks: A comprehensive review of architectures, variants, and applications. *Information* 15(9), 517. DOI: 10.3390/info15090517.

Möller, J. and Pörtner, R. (2021) Digital twins for tissue culture techniques—concepts, expectations, and state of the art. *Processes* 9(3), 447. DOI: 10.3390/pr9030447.

Mostafa, S., Mondal, D., Panjvani, K., Kochian, L. and Stavness, I. (2023) Explainable deep learning in plant phenotyping. *Frontiers in Artificial Intelligence* 6, 1203546. DOI: 10.3389/frai.2023.1203546.

Murmu, S., Sinha, D., Chaurasia, H., Sharma, S., Das, R. *et al.* (2024) A review of artificial intelligence-assisted omics techniques in plant defense: Current trends and future directions. *Frontiers in Plant Science* 15, 1292054. DOI: 10.3389/fpls.2024.1292054.

Naskar, S., Sharma, S., Kuotsu, K., Halder, S., Pal, G. *et al.* (2025) The biomedical applications of artificial intelligence: An overview of decades of research. *Journal of Drug Targeting* 33(5), 717–748. DOI: 10.1080/1061186X.2024.2448711.

Ngugi, H.N., Akinyelu, A.A. and Ezugwu, A.E. (2024) Machine learning and deep learning for crop disease diagnosis: Performance analysis and review. *Agronomy* 14(12), 3001. DOI: 10.3390/agronomy14123001.

Nguyen, N.D. and Wang, D. (2020) Multiview learning for understanding functional multiomics. *PLOS Computational Biology* 16(4), e1007677. DOI: 10.1371/journal.pcbi.1007677.

Nguyen, H.T., Khan, M.A.R., Nguyen, T.T., Pham, N.T., Nguyen, T.T.B. *et al.* (2025) Advancing crop resilience through high-throughput phenotyping for crop improvement in the face of climate change. *Plants (Basel, Switzerland)* 14(6), 907. DOI: 10.3390/plants14060907.

Papadimitroulas, P., Brocki, L., Christopher Chung, N., Marchadour, W., Vermet, F. *et al.* (2021) Artificial intelligence: Deep learning in oncological radiomics and challenges of interpretability and data harmonization. *Physica Medica* 83, 108–121. DOI: 10.1016/j.ejmp.2021.03.009.

Peladarinos, N., Piromalis, D., Cheimaras, V., Tserepas, E., Munteanu, R.A. *et al.* (2023) Enhancing smart agriculture by implementing digital twins: A comprehensive review. *Sensors (Basel, Switzerland)* 23(16), 7128. DOI: 10.3390/s23167128.

Porfírio, R.P., Madeira, R.N. and Santos, P.A. (2025) Exploring explainable AI techniques for plant disease classification in digital agriculture. *Procedia Computer Science* 265, 175–182. DOI: 10.1016/j.procs.2025.07.170.

Priyadarshini, K., Sikkandar, M.Y., AlDuraywish, A. and Alqahtani, T.M. (2024) Integrating relational and sequential information for enhanced detection of autoimmune disorders with relational neural networks and long short-term memory networks. *Biomedical Signal Processing and Control* 96, 106495. DOI: 10.1016/j.bspc.2024.106495.

Sahu, S.K. and Pandey, M. (2023) An optimal hybrid multiclass SVM for plant leaf disease detection using spatial fuzzy c-means model. *Expert Systems with Applications* 214, 118989. DOI: 10.1016/j.eswa.2022.118989.

Sangjan, W., Kick, D.R. and Washburn, J.D. (2025) Improving plant breeding through AI-supported data integration. *Theoretical and Applied Genetics* 138(6), 132.

Seng, K.P., Ang, L.M. and Ngharamike, E. (2022) Artificial intelligence internet of things: A new paradigm of distributed sensor networks. *International Journal of Distributed Sensor Networks* 18(3), 155014772110628. DOI: 10.1177/15501477211062835.

Sharma, S.K. (2025) Smart agriculture revolution: Cloud and iot-based solutions for sustainable crop management and precision farming. *AI-Based Advanced Optimization Techniques for Edge Computing* 253–288. DOI: 10.1002/9781394287062.ch10.

Sibiya, M. (2021) *Classification and Severity Prediction of Maize Leaf Diseases Using Deep Learning CNN Approaches*. University of South Africa, South Africa.

Singhal, R., Izquierdo, P., Ranaweera, T., Segura Abá, K., Brown, B.N.I. *et al.* (2025) Using supervised machine-learning approaches to understand abiotic stress tolerance and design resilient crops. *Philosophical Transactions of the Royal Society B* 380(1927), 20240252. DOI: 10.1098/rstb.2024.0252.

Sow, S., Ranjan, S., Seleiman, M.F., Alkharabsheh, H.M., Kumar, M. *et al.* (2024) Artificial intelligence for maximizing agricultural input use efficiency: Exploring nutrient, water and weed management strategies. *Phyton* 93(7), 1569–1598. DOI: 10.32604/phyton.2024.052241.

Taji, K., Sohail, A., Shahzad, T., Shoaib Khan, B., Adnan Khan, M. *et al.* (2024) An ensemble hybrid framework: A comparative analysis of metaheuristic algorithms for ensemble hybrid CNN features for plants disease classification. *IEEE Access* 12, 61886–61906. DOI: 10.1109/ACCESS.2024.3389648.

Tamayo-Vera, D., Mesbah, M., Zhang, Y. and Wang, X. (2025) Advanced machine learning for regional potato yield prediction: Analysis of essential drivers. *Npj Sustainable Agriculture* 3(1), 12. DOI: 10.1038/s44264-025-00052-6.

Umirzakova, S., Muksimova, S., Shavkatovich Buriboev, A., Primova, H. and Choi, A.J. (2025) A unified transformer model for simultaneous cotton boll detection, pest damage segmentation, and phenological stage classification from UAV imagery. *Drones* 9(8), 555. DOI: 10.3390/drones9080555.

Wang, D., Cao, W., Zhang, F., Li, Z., Xu, S. *et al.* (2022) A review of deep learning in multiscale agricultural sensing. *Remote Sensing* 14(3), 559. DOI: 10.3390/rs14030559.

Wu, M., Northen, T.R. and Ding, Y. (2023) Stressing the importance of plant specialized metabolites: Omics-based approaches for discovering specialized metabolism in plant stress responses. *Frontiers in Plant Science* 14, 1272363. DOI: 10.3389/fpls.2023.1272363.

Wu, X., Chen, H., Ding, R., Chen, G., Jia, H. *et al.* (2025) Integrating transcriptome, metabolome and microbiome to explore the molecular mechanism of phenotypic plasticity in *P. rotata* during low-altitude domestication. Plant, *Plant, Cell & Environment* 48(12), 8449–8468. DOI: 10.1111/pce.70138.

Yetgin, A. (2025) Revolutionizing multi-omics analysis with artificial intelligence and data processing. *Quantitative Biology* 13(3), e70002. DOI: 10.1002/qub2.70002.

Yousaf, A., Kayvanfar, V., Mazzoni, A. and Elomri, A. (2023) Artificial intelligence-based decision support systems in smart agriculture: Bibliometric analysis for operational insights and future directions. *Frontiers in Sustainable Food Systems* 6, 1053921. DOI: 10.3389/fsufs.2022.1053921.

Zhang, C., Chen, Y., Zeng, T., Zhang, C. and Chen, L. (2022) Deep latent space fusion for adaptive representation of heterogeneous multi-omics data. *Briefings in Bioinformatics* 23(2), bbab600. DOI: 10.1093/bib/bbab600.

Zhang, X. and Li, J. (2024) Genome-wide prediction and selection in plant and animal breeding: A systematic review of current techniques. *Computational Molecular Biology* 14. DOI: 10.5376/cmb.2024.14.0007.

Zhou, W., Ouyang, H., Yan, Z., Song, J., Li, Y. *et al.* (2025) Integrative machine learning approach for identifying genes associated with quantitative traits: A soybean yield case study. *Authorea Preprints*. DOI: 10.22541/au.175251398.84608512/v1.

6 Machine Learning-driven Plant Spatial and Multiple Omics

Tauqeer Ahmed Qadri[1], Amjid Khan[2,3], Rashid Abbas Khan[4], Anees Ali[5], Abdul Wahab[1], Sudenaz Soylu[6], Shabbar Abbas[1], Muhammad Anas[7]*, Javed Iqbal[8], Banzeer Ahsan Abbasi[9], Farishta Zarshan[8], Sana Basharat[10], Pingwu Liu[10] and Muhammad Waseem[10]*

[1]Department of Biosciences, COMSATS University Islamabad, Islamabad, Pakistan; [2]UNESCO UNISA Africa Chair in Nanoscience and Nanotechnology, College of Graduate Studies, University of South Africa, Pretoria, South Africa; [3]Department of Plant Sciences, Faculty of Biological Sciences, Quaid-i-Azam University, Islamabad-45320, Pakistan; [4]Department of Biology, Prince Sonkla University Thailand, Thailand; [5]Department of Biotechnology, Faculty of Sciences, Necmettin Erbakan Üniversitesi Türkiye, Türkiye; [6]Department of Field Crops, Selcuk University, Konya, Türkiye; [7]Department of Soil Science & Plant Nutrition, University of Selcuk, Konya, Türkiye; [8]Department of Botany, Bacha Khan University, Charsadda, Pakistan; [9]Department of Botany, Rawalpindi Women University, Satellite Town, Rawalpindi, Pakistan; [10]School of Breeding and Multiplication (Sanya Institute of Breeding and Multiplication), School of Tropical Agriculture and Forestry, Hainan University, Sanya, China

6.1 Introduction

Spatial and multi-omics methodologies represent significant advancements poised to initiate a paradigm shift within the domain of plant science (Yu *et al.*, 2023; Nobori, 2025). Spatial omics offers researchers a distinctive opportunity to examine molecular environments within native plant plasma. This approach can be characterized as the analysis of the spatial distribution of genes, proteins, and metabolites at the cellular and organ levels. Spatially resolved data have proven essential for elucidating cellular heterogeneity, tissue-specific regulation of signaling pathways, and processes critical for development, stress responses, and adaptation, in contrast to traditional bulk measurements. Concurrently, multi-omics techniques synthesize data from diverse omics disciplines, including genomics, transcriptomics, proteomics, metabolomics, ionomics, and phenomics, to provide a comprehensive understanding of plant systems (Cui *et al.*, 2023).

The recent emergence of high-throughput technologies has significantly enhanced the capacity to collect extensive data across various levels of biological measurement. Recent advancements in next-generation sequencing,

*Corresponding authors: muhammed.anas@selcuk.edu.tr and m.waseem.botanist@gmail.com

© CAB International 2026. *Plant Spatial Biology* (ed. Chen)
DOI: 10.1079/9781836994572.0006

mass spectrometry, imaging modalities, and high-resolution microscopy are generating multidimensional datasets that encompass interactions among factors, as well as specific responses. This technological progress has facilitated the transition from one-dimensional to multidimensional studies, thereby revealing molecular networks and regulatory pathways that govern plant phenotypes (Tresas *et al.*, 2025).

Despite significant advancements in the capacity to analyze omics data through modern computational methods, traditional analytical paradigms continue to face challenges in managing the large-scale molecular nature, complexity, and heterogeneity of omics data. Issues such as high dimensionality, the intrinsic nonlinear characteristics of biological systems, and insufficient information often limit the applicability of conventional statistical algorithms (Kasauli *et al.*, 2021). Furthermore, the necessity to process highly disparate omics strata, each with distinct data structures and units of measurement, in a standardized manner presents additional challenges in deriving meaningful data-driven insights from big data. Machine learning (ML) offers solutions to these limitations by providing algorithms capable of identifying patterns within the data, extracting significant features, and constructing predictive models, tasks that are challenging with large, multidimensional, complex datasets (Waqas *et al.*, 2025). Unlike classical statistical approaches, ML does not necessarily assume linear effects and can integrate heterogeneous datasets that contain biological relationships. In plant sciences, ML is increasingly utilized to integrate spatial and multi-omics data, facilitating biomarker discovery, elucidation of stress-responsive pathways, phenotype prediction, and modeling of genotype–environment interactions. This emerging synergy between ML and omics has the potential to accelerate advancements in crop breeding and biotechnology, contributing to more sustainable agricultural practices (Yu *et al.*, 2023).

The chapter delineates the intersection between ML and spatial and multi-omics methodologies within plant science. It commences with an established conceptual framework of spatial and multi-omics techniques, followed by an examination of the ML approaches that facilitate their integration and analysis. Case studies, accompanied by illustrative examples, demonstrate the application of ML-driven omics integration in areas such as plant stress biology, functional genomics, and precision agriculture (Panahi *et al.*, 2025). Addressing the challenges impeding the full realization of ML-based omics integration, the section on current obstacles – namely, data quality, computational limitations, and biological interpretation – highlights the factors constraining potential advancements. Finally, the chapter outlines prospective directions, particularly those focusing on explainable AI (XAI), digital twins, and real-time analytics, positing that such innovations possess the potential to transform the analysis and utilization of plant biological systems (Kreuzer *et al.*, 2024).

6.2 Fundamentals of Plant Spatial and Multiple Omics

Spatial omics encompasses a suite of methodologies designed to capture molecular data while preserving its inherent spatial characteristics at either the tissue-scale or single-cell resolution. Unlike traditional bulk sequencing techniques, which yield average profiles of heterogeneous samples, spatial omics methods retain positional information, thereby offering insights into intercellular heterogeneity in gene expression, protein abundance, modifications, and interactions (Wang *et al.*, 2023b). In the context of plant biology, spatial transcriptomics facilitates the identification of gene expression with high locational resolution within specific tissues, enabling the discernment and isolation of genetic programs active in roots, leaves, flowers, or regions exposed to abiotic stress. Concurrently, spatial proteomics provides a comprehensive understanding of the distribution of proteins and post-translational modifications across various cellular compartments, while spatial metabolomics allows for the mapping of metabolite locations, such as sugars, amino acids, and secondary compounds, within cells with high spatial coverage (Hu *et al.*, 2024).

Spatial omics has emerged as a pivotal focus in plant biology research, as numerous developmental transitions, stress responses, and metabolic processes occur within specific tissue or cell-type contexts. Spatial mapping techniques have been employed to trace the localization of secondary metabolites in trichomes, monitor nutrient movement across various root regions, and delineate hormone gradients during seed maturation. Such analyses facilitate the identification of developmental regulators, reveal tissue heterogeneity, and establish correlations between molecular and morphological phenotypes. Spatial omics posits that integrating molecular profiles within a biological context enables comprehensive methodologies for elucidating the dynamic architecture of plant systems (Liu *et al.*, 2024).

6.2.1 Multiple omics layers

Beyond spatial mapping, plant omics offers an extensive depiction of plant biology through a diverse array of analytical methodologies. Genomics provides insights into entire genome sequences and structural variations that contribute to phenotypic traits. Transcriptomics examines gene expression variability across different developmental stages or in response to environmental stressors (Baião *et al.*, 2025). Proteomics enables the assessment of protein levels, post-translational changes, and protein interactions, thereby emphasizing the functional manifestation of genetic information. Metabolomics identifies a broad spectrum of metabolites, delivering a quantitative analysis of biochemical activities and physiological conditions (Hajnajafi and Iqbal, 2025). Ionomics measures the concentrations of elements, including vital nutrients and harmful metals, thus indicating nutritional status and environmental exposure. Lastly, phenomics captures observable phenotypes, ranging from cellular characteristics to entire plant functions, often employing advanced imaging and sensor technologies. Together, these omics techniques can be synthesized to deliver a comprehensive understanding of plant biology (Donovan *et al.*, 2019).

The forthcoming advancement in systems-level plant biology lies in the integration of complementary omics platforms, encompassing the whole genome, transcriptome, and metabolome. This approach facilitates multifaceted data analysis and enhances the comprehension of complex biological phenomena through the dynamic interactions among genetic, molecular, biochemical, and phenotypic data (Suzuki, 2022). For instance, the coupling of genomics with metabolomics enables the identification of genetic loci associated with metabolite production, while the integration of transcriptomics with ionomics elucidates the regulatory mechanisms governing nutrient uptake and stress adaptation coordination. These synergistic methodological advancements not only expedite the identification of functional genes and trait prediction, thereby facilitating desired crop improvements, but also establish a direct link between molecular-level understanding and phenotypic outcomes (Bongaerts *et al.*, 2022).

6.2.2 Data complexity and challenges

Contemporary spatial and multi-omics methodologies offer unparalleled insights; however, they simultaneously generate complex and extensive datasets that pose significant analytical challenges. These datasets are characterized by exceptionally high dimensionality, with thousands of genes, proteins, or metabolites being measured concurrently within a large sample or spatial cohort (Luo *et al.*, 2024). Furthermore, they exhibit considerable heterogeneity, as different omics layers are defined by distinct data structures, metrics, characteristics, and technical variations. Inevitably, missing values will occur due to experimental errors, the non-detection of certain experimental attributes, and the potential saturation of genuine biological signals with noise.

Computational methodologies capable of handling heterogeneous and extensive datasets are essential for addressing biological challenges in contemporary plant research. Traditional statistical models often inadequately capture the nonlinear relationships and interdependencies present within and across omics layers. Consequently, advanced computational systems, particularly those utilizing ML, have become pivotal in the integration and interpretation of spatial and multi-omics data. These approaches facilitate feature extraction, dimensionality reduction, data fusion, and predictive modeling, thereby enhancing our understanding of complex plant systems (Jan *et al.*, 2025).

6.3 Machine Learning Approaches for Omics Integration

6.3.1 Supervised learning

Supervised learning represents a prevalent framework within plant-omics research. In this context, algorithms are trained on datasets where the response variables, which may include phenotypic or molecular measurements, are explicitly annotated. This annotation facilitates the approximation of response variables through measured variables (Reel *et al.*, 2021). Within the domain of plant sciences, supervised learning can be employed for: (i) classifying stress responses; (ii) diagnosing plant diseases; or (iii) predicting complex traits such as yield, biomass, or nutritional efficiency. For instance, labeled omics datasets containing stress and control samples enable models to differentiate between tolerant and susceptible genomes. Similarly, supervised methods facilitate the identification of molecular markers associated with pathogen resistance or abiotic stresses, including drought, salinity, and heavy metal exposure (Reel *et al.*, 2021).

In the realm of plant science, a diverse array of algorithms has proven to be highly effective. Random forests (RFs) stand out for their capacity to reduce the risk of overfitting, adeptly handle data with high dimensionality, and rank features for classification purposes. Support vector machines (SVMs) are esteemed for their expertise in binary classification and their ability to delineate optimal decision boundaries within intricate data environments (Yu *et al.*, 2017). Conversely, artificial neural networks (ANNs), with their distinctive layered architecture, are often utilized to model complex nonlinear interactions among omics layers, thereby facilitating predictions of traits influenced by a multitude of molecular and environmental factors (Han *et al.*, 2018). Together, these algorithms provide plant scientists with essential tools for both feature discovery and predictive modeling.

6.3.2 Unsupervised learning

The investigation of latent structures within high-dimensional omics datasets represents a domain where unsupervised learning has become indispensable, particularly in the absence of pre-existing labels. A prominent application of this approach is the employment of clustering techniques to categorize omics profiles into groups of genes, proteins, or metabolites that exhibit similar expression or abundance patterns. These clusters frequently correspond to co-regulated pathways or other biological functions, such as collections of stress-responsive genes or specific metabolic modules (Vahabi and Michailidis, 2022).

A significant application in the field is dimensionality reduction. The extensive feature data generated by omics necessitates mechanisms for condensing these data into a succinct collection of explanatory or indicative features, such as principal component analysis (PCA), to facilitate visual representation and simplified analysis (Jolliffe and Cadima, 2016). In addition to linear methods, nonlinear algorithms, including t-distributed Stochastic Neighbor Embedding (t-SNE) and uniform manifold approximation and projection (UMAP), offer more detailed visualizations capable of discerning subtle sample distribution contrasts, such as the differentiation between tolerant and susceptible genotypes under stress conditions (Marx, 2024). Autoencoders, a variant of neural networks, extend the concept of dimension reduction by learning compressed representations of high-dimensional data, with the added advantage of uncovering latent feature representations that may be challenging to infer using traditional statistical approaches.

6.3.3 Deep learning

Deep learning, a subset of machine learning, has emerged as a transformative approach in the domains of plant omics and phenotyping due to its ability to model highly complex and nonlinear patterns. Convolutional neural networks (CNNs) have been extensively utilized in various research areas involving plant imagery, such as phenotypic traits in microscopic images, scanned leaves, and aerial drone data. CNNs are capable of learning hierarchical feature representations, thereby reducing the necessity for manually specified features related to disease symptoms, nutritional deficiencies, or developmental changes (Alzubaidi *et al.*, 2021).

Beyond image data, deep learning is also applied to multi-omics integration. Graph neural networks (GNNs) interpret biological networks, such as gene regulatory or protein–protein interaction (PPI) networks, as topological relationships within omics-derived data. By modeling these networks, GNNs can identify key regulators, infer gene functions, and elucidate relationships that contribute to plant responses to environmental stressors. Deep learning methods offer the flexibility and scalability particularly suited for integrating diverse omics data into predictive models (Zhou *et al.*, 2020).

6.3.4 Transfer learning, multi-task learning

A significant challenge in plant research is the scarcity of detailed data concerning specific crops, traits, or experimental conditions. Transfer learning offers a solution to this issue by enabling the application of models trained on extensive datasets from one species or condition to another. For instance, models initially developed for well-studied species like *Arabidopsis thaliana* can be refined for use with crops such as wheat, rice, or maize, thus minimizing the need to compile a new, comprehensive dataset (Keceli *et al.*, 2022). In a similar vein, models crafted to estimate stress intensity under particular stress conditions can be adapted to assess stress intensity under varying stress circumstances.

Multi-task learning extends the potential of models to simultaneously predict multiple traits or outcomes, provided it is technically feasible. For example, a single, comprehensive model could be developed to forecast drought tolerance, biomass accumulation, and nutrient uptake through the use of integrated omics (Zhao *et al.*, 2024). This strategy promotes the transfer of knowledge across different tasks, often leading to more effective results and enhanced generalizability. This approach is particularly beneficial in crop improvement programs where multiple, sometimes interrelated, traits are of interest.

6.3.5 Explainable AI (XAI)

One of the most compelling critiques of advanced ML models, particularly deep learning, is their characterization as a 'black box,' rendering them essentially uninterpretable in physiological terms. In the field of plant sciences, where validation and mechanistic understanding are paramount, XAI has gained increasing significance. XAI endeavors to enhance the transparency of predictions by elucidating the most influential features or pathways through which the model arrives at decisions. For instance, feature attribution can assist in identifying critical genes, metabolites, or proteins that determine a plant's stress tolerance (Goche *et al.*, 2025).

XAI offers the advantage of rendering accurate predictions biologically meaningful. This capability assists researchers in formulating hypotheses that require experimental validation, thereby enhancing the credibility of ML models in the field of plant biology (Table 6.1). By harmonizing predictive accuracy with mechanistic explanations, XAI emerges as a crucial tool for integrating computational methods as valuable assets in the broader plant sciences. Furthermore, it accelerates the translation of omics knowledge into practical agricultural applications (Goche *et al.*, 2025).

6.4 Integration Strategies for Spatial and Multiple Omics

6.4.1 Data preprocessing and normalization

Effective integration of omics datasets necessitates a comprehensive preprocessing stage to ensure data consistency and reliability. Raw omics data may be affected by technical noise, sequencing errors, or variations introduced during sample preparation and measurement

Table 6.1. Machine learning methods used in plant spatial and multi-omics integration, occurrences and implications, advantages/disadvantages, and main references.

Approach/algorithm	Application in plant omics	Key advantages	Limitations/challenges	References
Random forest (RF)	Classification of tolerant vs susceptible genotypes under stress	Handles high-dimensional data, interpretable feature ranking	May overfit with noisy datasets	Zhang *et al.*, 2024
Support vector machine (SVM)	Disease diagnosis using transcriptomic markers	Effective in small to medium datasets, strong classification	Requires kernel selection, less scalable	Ozer *et al.*, 2020
Artificial neural networks (ANNs)	Prediction of plant yield from omics features	Models' nonlinear interactions, adaptable	Requires large datasets, risk of overfitting	Kassem, 2025
Gradient boosting machines (XGBoost, LightGBM)	Trait prediction from multi-omics profiles	High accuracy, fast computation	Sensitive to hyperparameter tuning	Rahaman *et al.*, 2025
Logistic regression	Marker-based classification of plant stress	Simple, interpretable	Poor performance with nonlinear data	Chang-Brahim *et al.*, 2024
k-Nearest neighbors (k-NN)	Grouping plant phenotypes based on omics	Simple, intuitive	Sensitive to noise, scalability issues	Cembrowska-Lech *et al.*, 2023
Principal component analysis (PCA)	Dimensionality reduction of transcriptomics	Reduces data complexity, easy visualization	Linear method misses nonlinear patterns	Du, 2019
Independent component analysis (ICA)	Separating mixed omics signals	Extracts independent signals	Less robust with noise	Sompairac *et al.*, 2019
t-SNE	Visualization of metabolomics data clustering	Captures local structure	Computationally expensive, not scalable	Garg and Dwivedi, 2024
UMAP	Visualizing multi-omics integration	Preserves both local and global structure	Sensitive to parameters	Castellano-Escuder *et al.*, 2025
Hierarchical clustering	Grouping co-expressed genes	Easy to implement, interpretable dendrograms	Struggles with very large datasets	Geeitha and Thangamani, 2018
k-Means clustering	Partitioning omics features	Fast, simple	Requires a pre-defined cluster number	Duan *et al.*, 2021
Autoencoders	Feature extraction from high-dimensional data	Nonlinear dimensionality reduction	Requires large datasets, black-box	Chaquet-Ulldemolins *et al.*, 2022

Convolutional neural networks (CNNs)	Image-based phenotyping, stress detection	Strong in pattern recognition	Requires large labeled datasets	Godinez *et al.*, 2017
Recurrent neural networks (RNNs, LSTMs)	Time-series omics data (gene expression dynamics)	Captures sequential dependencies	Training instability, vanishing gradients	Mienye *et al.*, 2024
Graph neural networks (GNNs)	Modeling gene/protein interaction networks	Captures relationships in networks	High computational cost	Al Musawi *et al.*, 2025
Transfer learning	Applying *Arabidopsis* models to crops	Reduces data needs in target species	May fail if biology differs strongly	Kaya *et al.*, 2019
Multi-task learning	Predicting multiple plant traits simultaneously	Improves generalization, shares information	Requires carefully balanced tasks	Chao *et al.*, 2025
Ensemble learning (bagging/boosting)	Combining multiple models for trait prediction	Improves accuracy and robustness	Computationally expensive	Rane *et al.*, 2024
Semi-supervised learning	Exploiting unlabeled omics data	Useful for small, labeled datasets	Sensitive to noise in pseudo-labels	Ran *et al.*, 2024
Explainable AI (XAI) – SHAP	Identifying key genes/metabolites driving predictions	Improves biological interpretability	Computational overhead	Praveen *et al.*, 2024
Explainable AI (XAI) – LIME	Local interpretability of ML decisions	Easy to apply, flexible	Local explanations may be unstable	Chinnaraju, 2025
Bayesian networks	Inferring gene regulatory networks	Captures probabilistic dependencies	Computationally heavy for large omics	Singh and Singh, 2024

processes. Normalization is conducted to adjust for variance due to sequencing depth, protein abundance, or metabolite intensities, thereby facilitating broad comparability across datasets. A common transformation for transcriptomics data involves converting TPM (transcripts per million) or RPKM/FPKM (reads/fragments per kilobase per million mapped reads), while protein and metabolomic data may undergo z-score normalization or quantile normalization (Picard *et al.*, 2021).

A significant challenge in multi-omics analysis is the presence of batch effects, which occur when various experiments are processed in distinct batches. These analyses are frequently affected by missing values, particularly in proteomics and metabolomics studies, due to the potential non-detection of small molecules. Approaches to address missing data include imputation-based methods such as k-nearest neighbor (k-NN) imputation, matrix factorization, or model-based inference. It is crucial to ensure that missing data do not introduce bias into subsequent analyses (Wu *et al.*, 2024).

6.4.2 Feature selection and engineering

Omics data often presents a challenge due to its high dimensionality, with thousands of features juxtaposed against a relatively small sample size. This complexity necessitates the application of feature selection methods to simplify the data and enhance its interpretability. The goal is to pinpoint biologically relevant features that play a pivotal role in the variation of plant traits or responses. Techniques such as variance filtering, mutual information ranking, and recursive feature elimination are instrumental in feature prioritization and noise reduction (Farheen and Kumar, 2024).

In addition to feature selection, feature engineering plays a pivotal role in the creation of composite variables or biologically significant descriptors from raw data. For instance, the ratios between metabolite concentrations may provide greater insight into pathway activity than individual metabolite levels. Furthermore, network-derived metrics, such as hub–gene connectivity or pathway enrichment scores, can enhance predictive models. By transforming raw measurements into biologically meaningful features, feature engineering is essential for linking data-driven algorithms to plant biology (Wu *et al.*, 2024).

6.4.3 Multi-omics integration techniques

The integration of diverse omics layers is essential for comprehensively characterizing the complexity inherent in plant biological systems (Jamil *et al.*, 2020). Typically, integration strategies are categorized into early, intermediate, and late approaches.

6.4.3.1 Early integration (feature level)

In the initial phase of integration, all features across the omics layers are consolidated into a single dataset prior to analysis. For instance, a unified profile comprising transcriptomic, proteomic, and metabolomic data (integrated into a single matrix) can be utilized by ML algorithms to train multi-task models through ensemble learning. This straightforward approach ensures that interactions between cross-omics are explicitly captured; however, dimensionality imbalances and noise may hinder this process (Zhan *et al.*, 2022).

6.4.3.2 Intermediate integration (latent representation)

Intermediate integration involves projecting omics data into shared latent spaces before data fusion. Techniques such as canonical correlation analysis (CCA), joint non-negative matrix factorization, and multi-omics autoencoders are employed to identify latent variables that represent common biological factors. This approach balances the datasets and mitigates the risk of one omics layer dominating the analysis. It is effectively utilized to capture complex regulatory dynamics, such as those between transcript profiles and metabolite accumulation (Jiang *et al.*, 2023).

6.4.3.3 Late integration (decision level)

Late combination refers to the integration of results derived from the analysis of distinct layers, which are organized independently. For instance, methodologies trained separately on transcriptomic, proteomic, and metabolomic data can be employed to construct an ensemble model, such as through majority voting or weighted averaging. While this approach preserves the integrity of individual data types, it may overlook cross-omics interactions. Nevertheless, decision-level integration demonstrates greater tolerance for information loss and biases specific to individual datasets.

6.4.3.4 Spatial data fusion

The integration of spatial omics introduces the concept of incorporating positional information within the context of plant tissue at a molecular level. Spatial transcriptomics, for instance, can be combined with proteomic or metabolomic maps to create a comprehensive molecular mapping of plant organs. This integration enables researchers to identify which molecules are expressed and their specific locations, thereby facilitating the establishment of connections between molecular profiles, cellular structure, and tissue function (Rampler *et al.*, 2020).

Instances of spatial fusion in plant biology are elucidated. One notable application involves the integration of spatial transcriptomics with metabolomics to examine the localized regulation of secondary metabolites, such as flavonoids in leaves or alkaloids in roots. Similarly, the combination of transcriptomics with spatial proteomics facilitates the elucidation of regulatory networks functioning in specific localized regions, including meristematic tissues or vascular bundles (Picard *et al.*, 2021). These methodologies are particularly advantageous in the investigation of developmental gradients, stress-responsive tissues, and specialized tissues such as trichomes or root hairs. This is attributed to the fact that the integration of spatial and multi-omics data can provide a systems-level perspective of the plant at cellular resolution. Such integration enhances the understanding of developmental biology, adaptive responses, and metabolic specialization, which can ultimately be applied in precision breeding and the design of stress-tolerant plants (Subramanian *et al.*, 2020).

6.5 Applications of ML in Plant Omics

6.5.1 Plant stress biology

ML has become integral to the analysis of molecular stress-related processes in plants. Plants encounter a variety of biotic and abiotic stresses, including drought, salinity, heavy metals, and pathogens, which lead to complex molecular and physiological responses. Traditional methodologies often struggle to effectively handle the transcriptomic, proteomic, and metabolomic data generated by such studies. In contrast, ML approaches are particularly adept at identifying hidden patterns, modeling stress conditions, and predicting tolerant genotypes (Panahi *et al.*, 2025).

RFs and SVM are examples of supervised ML algorithms that have been employed to classify drought-tolerant and drought-sensitive genotypes within a transcriptomic profile. Similarly, ML models can identify key biomarkers in proteomic and metabolomic data to predict plant responses to heavy metals, including detoxification pathways and antioxidant responses. CNNs have been utilized in pathogen studies, training algorithms to recognize disease symptoms on leaves. Additionally, the integration of omics data provides evidence of the molecular interrelation between these factors (Rico-Chávez *et al.*, 2022). ML technology facilitates the identification of stress-responsive genes and pathways, which are crucial for developing resilient crop varieties, due to its capacity for rapid classification and prediction.

6.5.2 Crop improvement and breeding

The application of ML in crop improvement represents a promising advancement within the field of plant science, primarily due to its capacity to expedite the enhancement of crop varieties. Breeding programs utilize the identification of molecular markers and trait-associated genes to guide selection processes. Agronomic data can potentially be predicted through ML by reverse-engineering genotype–phenotype associations, thereby offering a valuable application of ML to genotype–phenotype associations in crops (Yan and Wang, 2023).

ML models trained on genomics and phenomics data have been employed to predict grain yield in cereals, oil quality in oilseed crops, and sugar content in fruits. Ensemble-based methodologies, such as gradient boosting and deep learning frameworks, demonstrate superior performance compared to traditional methods, as they effectively capture nonlinear interactions between loci and traits (Zhang *et al.*, 2022). Furthermore, ML can be utilized to rank candidate genes, thereby enhancing marker-assisted or whole-genome selection and reducing the breeding cycle. The integration of multi-omics data with field-based performance records enhances predictive accuracy by facilitating the selection of unique genotypes tailored to specific environments (Mahmood *et al.*, 2022).

6.5.3 Functional genomics

Functional genomics significantly contributes to the field of genomics by facilitating the elucidation of gene regulatory networks and pathways, particularly those involved in plant development and regulation. High-throughput transcriptomic experiments typically yield data comprising thousands of differentially expressed genes, necessitating the use of advanced computational tools to discern the relationships among them (Bunnik and Le Roch, 2013). ML models that operate within a graph framework, such as GNNs and Bayesian networks, are among the most effective for uncovering regulatory interactions among genes, proteins, and metabolites.

These strategies facilitate the identification of key regulators or hub genes involved in biological processes such as flowering time, secondary metabolism, or stress adaptation. For instance, ML-based network analysis has been successfully applied to uncover transcription factors that drought-tolerant crops utilize to regulate drought tolerance or nutrient acquisition. By comparing multiple omics levels, it is possible to interpolate data, and such integrative approaches based on ML will enable functional genomics to achieve a comprehensive understanding of plant regulatory circuits, thereby enhancing the potential for targeted genetic manipulation (Yun *et al.*, 2024).

6.5.4 Spatial dynamics

The integration of ML and spatial omics offers researchers unparalleled insights into cellular and tissue-level heterogeneity. In plants, developmental processes and stress responses are often spatially regulated, with specific cells or tissue types fulfilling distinct roles. Spatial transcriptomic cell targeting, facilitated by ML models, enables the resolution of cell-type-specific spatial transcriptomic data, elucidation of tissue-specific transcriptomic expression patterns, and analysis of developmental gradients across plant organs (Schmitz, 2010).

For instance, UMAP and autoencoders can be employed in an unsupervised context to differentiate various cell types within roots or leaves, thereby illustrating their specialized functions, such as nutrient absorption in the root epidermis or photosynthesis in mesophyll cells. Additionally, ML facilitates the integration of spatial proteomics and metabolomics data with transcriptomic maps, offering a comprehensive view of the spatial heterogeneity of molecular activity across tissues. These insights enhance our understanding of plant development, organ formation, and tissue-specific adaptations to environmental changes.

6.5.5 Precision agriculture

ML plays a pivotal role in the integration of omics data with field-based phenotyping and environmental data to advance precision agriculture. Molecular-level data, when combined with high-throughput phenotyping, remote sensing, and environmental monitoring, facilitates informed decision-making for both researchers and farmers. For instance, ML models can be employed to assess crop performance under varying soil and climatic conditions by integrating genomic and field phenotyping data (Monteiro *et al.*, 2021).

CNNs applied to drone or satellite imagery have the potential to identify nutrient deficiencies, disease outbreaks, or water stress at their earliest stages. These observations can be integrated with omics data to provide molecular explanations for the phenotypes observed (Table 6.2). Site-specific management, including tailored irrigation, fertilization, and crop protection practices, can be implemented through decision-support systems based on ML. Ultimately, the integration of ML and omics in precision agriculture fosters sustainable agricultural practices by optimizing output per unit input, thereby reducing input costs and minimizing environmental impact.

6.6 Case Studies

Integrative metabolomics and transcriptomics offer a compelling framework for examining plant responses to abiotic stress at both cellular and molecular levels. While transcriptomics elucidates the activation and repression of genes involved in stress responses, a comprehensive understanding necessitates an exploration of the physiological manifestations of stress, including metabolic changes. The concurrent analysis of these datasets through ML techniques facilitates the elucidation of these complex interactions (Gluchowska *et al.*, 2025).

RF and SVM models have been investigated for their efficacy in classifying genotypes as either tolerant or sensitive to drought and saline environmental conditions. The integration of gene expression profiles with metabolite accumulation patterns in ML has yielded significant biomarkers, including transcription factors involved in osmolyte biosynthesis and metabolites associated with antioxidant defense (Geng *et al.*, 2025). This approach not only identifies stress-sensitive genes but also elucidates their association with biochemical mechanisms related to resilience. Such predictive models are instrumental in guiding the selection of stress-resistant cultivars and identifying potential targets for metabolic engineering or gene modification to enhance crop resistance.

Spatial transcriptomics is revolutionizing plant biology by enabling the localization of gene expression within intact tissues, thereby preserving positional landscapes that are not discernible through bulk RNA-sequencing technologies. The foundational computational tools required to effectively interpret and visualize spatial transcriptomic data are challenged by the vast number of cells and significant dimensionality. The pattern recognition capabilities of CNNs have been employed to analyze spatial gene expression maps, which have been successfully utilized (Zhang *et al.*, 2025).

CNNs have been shown to discern subtle variations in expression across different tissues, thereby enabling the identification of cell types and functional territories, such as vascular bundles, mesophyll, and meristematic cells. For instance, in root analysis, CNNs successfully identified spatial regions associated with nutrient uptake, hormonal regulation, and cellular differentiation, respectively (Wang *et al.*, 2025). CNN-based models offer the potential to elucidate the coordination of gene activity across tissues during developmental processes and stress responses by linking spatial transcriptomic data with phenotypic outcomes. This approach represents a significant advancement towards understanding cellular heterogeneity and provides valuable insights for the targeted manipulation of tissue-specific gene expression (Chen *et al.*, 2025).

6.7 Deep Learning Models of Estimating Plant Phenotypes on Multi-omics Data

Plant phenotypes, such as yield, growth rate, and stress tolerance, are influenced by the integration of various omics levels, which involve intricate and competing molecular interactions. Predicting these traits poses significant challenges due to the necessity for computational models capable of synthesizing and interpreting diverse datasets, including genomics, transcriptomics, proteomics, and metabolomics. Deep learning models, with their hierarchical architectures, are particularly well-suited for modeling the nonlinear and multidimensional relationships inherent in such data (Cembrowska-Lech *et al.*, 2023).

The application of autoencoders, recurrent neural networks (RNNs), and GNNs facilitates the integration of multi-omics data for phenotype prediction. For instance, transcriptomic and metabolomic data can be incorporated

Table 6.2. Machine learning in the spatial and multi-omics integration of plants.

Application area	Specific use case	ML approach/algorithm	Key outcomes	References
Plant stress biology	Drought tolerance classification	Random forest, SVM	Identified drought-tolerant vs sensitive genotypes using transcriptomic markers	Sprenger *et al.*, 2018
Plant stress biology	Salinity stress biomarker discovery	Gradient boosting, PCA	Selected key genes/ metabolites regulating salt adaptation	Sarfraz *et al.*, 2025
Plant stress biology	Heavy metal (Cd, Cr) response prediction	ANN, decision trees	Predicted plant detoxification pathways and antioxidant responses	Goncharuk and Zagoskina, 2023
Plant stress biology	Pathogen resistance detection	CNNs on leaf images	Automated early disease recognition and classification	Bhargava *et al.*, 2024
Plant stress biology	Integrated drought + heat stress modeling	Multi-task learning	Predicted combined stress responses from omics and phenomics	Esposito *et al.*, 2019
Crop improvement and breeding	Yield prediction in cereals	Ensemble models (RF, XGBoost)	Improved grain yield predictions from genomics and field phenotypes	Patil *et al.*, 2025
Crop improvement and breeding	Oil content in canola	SVM, ANN	Predicted oil quality traits using transcriptomic markers	Kassem, 2025
Crop improvement and breeding	Fruit sugar/acid profiling	Regression trees, partial least squares-discriminant analysis (PLS-DA)	Linked metabolomic signatures with sweetness and acidity	Ma *et al.*, 2025
Crop improvement and breeding	Genomic selection for disease resistance	Bayesian ML, RF	Identified loci linked to pathogen resistance	Chen *et al.*, 2024
Crop improvement and breeding	Marker-assisted breeding prioritization	Feature selection + RF	Ranked candidate genes for selection	Kumar *et al.*, 2024
Functional genomics	Gene regulatory network inference	Graph neural networks	Identified hub genes controlling stress adaptation	Bhati *et al.*, 2025

Functional genomics	Transcription factor-target prediction	Bayesian networks	Reconstructed gene regulatory interactions	Sauta *et al.*, 2020
Functional genomics	Pathway-level integration of omics	Autoencoders	Discovered hidden regulatory modules	Bansal *et al.*, 2023
Functional genomics	Epigenome–transcriptome integration	Multi-omics CCA	Linked DNA methylation with expression profiles	Santamarina-Ojeda *et al.*, 2023
Spatial dynamics	Cell type identification in roots	Unsupervised clustering (UMAP, k-means)	Distinguished root epidermis, cortex, and stele subpopulations	Wang *et al.*, 2024a
Spatial dynamics	Tissue-specific gene expression	Spatial transcriptomics + RF	Classified genes by localization patterns	Gulati *et al.*, 2025
Spatial dynamics	Developmental gradient mapping	PCA + CNN	Revealed gene expression waves in meristems	Ge *et al.*, 2025
Spatial dynamics	Secondary metabolite localization	Spatial metabolomics + SVM	Mapped alkaloid accumulation in root zones	Wang *et al.*, 2024b
Precision agriculture	Nutrient deficiency detection	CNN on drone images	Detected early nitrogen and phosphorus deficiency	Sudhakar and Swarna Priya, 2023
Precision agriculture	Water stress monitoring	Time-series RNN	Predicted water stress from field imaging and omics	Sudhakar and Swarna Priya, 2023
Precision agriculture	Disease outbreak forecasting	RF + environmental sensors	Linked omics susceptibility markers with climate data	Delfani *et al.*, 2024
Precision agriculture	Site-specific management	ML decision-support systems	Optimized fertilizer/irrigation scheduling	Ahmed *et al.*, 2023

into deep learning models to accurately predict drought tolerance or yield under stressful conditions. Similarly, CNNs, when combined with phenomics data from imaging-based platforms, have demonstrated efficacy in predicting biomass accumulation and disease severity (Wang *et al.*, 2023a). To identify desirable genotypes, researchers can leverage the predictive capabilities of these models to efficiently screen millions of genotypes, thereby prioritizing those worthy of further investigation in extended field trials. Furthermore, the implementation of XAI techniques ensures that the accuracy of these predictions is underpinned by biologically interpretable insights, identifying candidate genes, pathways, metabolites, and trait expressions. This advanced integration of omics data is paving the way for future breeding practices that are data-driven and precision-oriented, scalable to entire breeding populations (Wu and Xie, 2025).

6.8 Challenges and Limitations

6.8.1 Data quality and standardization

One of the primary challenges in applying ML to plant spatial and multi-omics data is the inconsistency in data generation and processing pipelines. Variations in the diversity of omics data formats, scaling, and technical biases among omics platforms often create communication barriers. Such inconsistencies hinder the direct comparison or integration of datasets across experiments, institutions, or species (Vasupalli *et al.*, 2024). For instance, transcriptomic data may be presented in a normalized TPM/FPKM state, whereas metabolomic data is represented as relative abundance, leading to mismatches during integration. In addition to sampling variations, differences in library preparation protocols and instrument calibration introduce variability that obscures the true biological variation. A lack of consistency in protocols and quality assurance mechanisms can result in downstream ML models confusing noise with signal, thereby producing unreliable predictions. Establishing community-wide standards for preprocessing, annotation, and data sharing is therefore essential to achieve consistency, particularly across studies (Hao *et al.*, 2025).

6.8.2 Computational constraints

Plant omics datasets are frequently characterized by high dimensionality, with sample sizes ranging from ten to several hundred. This skewness poses a challenge to machine learning, as models developed on small sample sizes are prone to overfitting. Additionally, the processing burden is exacerbated by the diverse omics layers, each possessing distinct characteristics and large data volumes (Luo *et al.*, 2024). For instance, the integration of spatial transcriptomics with proteomics can generate terabytes of data, necessitating high-performance computing infrastructure. Furthermore, advanced techniques such as deep neural networks and GNNs demand significant computational resources and specialized hardware, such as graphics processing units (GPUs). The scalability of algorithms to accommodate such extensive datasets remains a critical bottleneck, particularly in resource-constrained research environments. These challenges can be addressed through effective dimensionality reduction, cloud-based computation, and algorithmic optimization (Dale *et al.*, 2021).

6.8.3 Biological interpretability

While ML excels in predictive capabilities, it often falls short in data interpretation, particularly in the context of complex models such as deep learning. Although these models may yield accurate predictions, they frequently lack explanations regarding the underlying biological factors. For plant scientists, however, interpretability holds equal importance to accuracy, as the ultimate objective is to formulate testable biological hypotheses (Ahlquist *et al.*, 2023). For instance, a deep learning model might predict drought tolerance with high accuracy, yet without identifying the most influential genes, pathways, or metabolites, its utility in informing breeding or functional studies is limited. XAI methods, such as SHAP or LIME, offer potential solutions by assigning feature importance scores and providing easily interpretable visualizations. However, these methods are still evolving and may not fully capture biological complexity, and discrepancies often exist between computational predictions and experimental observations (Ahlquist *et al.*, 2023).

6.8.4 A combination of field data

The final significant limitation pertains to the reconciliation of controlled laboratory omics experiments with real-world field conditions. Omics datasets are frequently collected under highly controlled conditions to minimize variability, in contrast to natural plant cultivation, where environmental factors can fluctuate, soil may be heterogeneous, and interactions with other organisms occur. This discrepancy can reduce the practical applicability of laboratory-based ML models in agricultural settings (Holden *et al.*, 2015). For instance, a model trained with greenhouse transcriptomic data may not accurately predict field performance, considering that weather records and soil parameters are still under development. The seamless integration of molecular insights with field-scale applicability remains a work in progress, necessitating the multidisciplinary collaboration of plant biologists, agronomists, data scientists, and engineers due to unaccounted environmental interactions (Barbedo, 2022). Additionally, there is the challenge of integrating multi-omics data with field-based phenotyping platforms, remote sensing tools, and environmental monitoring. By interpolating unstructured field data with structured omics data, a comprehensive view of real-world conditions can be achieved. This requires the development of hybrid methods that combine omics data with high-throughput techniques (Wang *et al.*, 2024c).

6.9 Future Perspectives

6.9.1 AI-driven systems biology

The future trajectory of plant omics involves transitioning from isolated data analysis to an AI-enhanced systems biology framework. A promising direction is the development of digital twins for plants, which integrate multi-omics, spatial, and environmental data to simulate plant growth, development, and responses to mechanical, environmental, and other stressors in real-time. These models can predict plant behavior under specific conditions, such as drought or nutrient deficiency, constrained by molecular signatures known to produce physiological outcomes. Digital twins can be continuously refined with new experimental or field data, allowing hypotheses to be tested digitally before validation in the field. This approach reduces experimental costs and accelerates the discovery process, facilitating the advancement of precision breeding and the design of stress-resilient crops.

6.9.2 Edge computing and IoT in agriculture

The integration of ML with edge computing and the Internet of Things (IoT) is poised to revolutionize real-time plant monitoring and decision-making processes. The deployment of drones, ground-based sensors, and imaging platforms facilitates the collection of vast quantities of phenotypic and environmental data directly from the field. By processing these data locally at the 'edge' (i.e. on-site devices), rapid ML-driven analysis is enabled without dependence on centralized servers. For instance, IoT-enabled systems equipped with ML algorithms can identify nutrient deficiencies, water stress, or disease symptoms as they arise, thereby providing farmers with actionable insights in near real time. This approach reduces latency in decision making and facilitates dynamic interventions, such as adjusting irrigation schedules or applying targeted treatments. Furthermore, integrating these field-level observations with omics datasets will establish a closed feedback loop, linking molecular knowledge with practical agricultural applications.

6.9.3 Personalized crop management

A significant future direction in agricultural science is the advancement of personalized crop management, wherein omics-informed strategies are customized to specific genotypes and environmental conditions. Analogous to precision medicine, which utilizes patient-specific genomic data to inform treatment decisions, plant sciences are increasingly adopting genotype-specific omics signatures to enhance crop production. For instance, metabolomic or transcriptomic profiles of individual plant varieties could inform the selection of fertilizers, growth regulators, or stress-mitigation strategies most appropriate for that genotype. ML models, trained on integrated multi-omics and environmental data, will facilitate predictions of how different varieties respond to site-specific soil, climate, and management practices. This personalized approach holds the potential to maximize yield and quality while minimizing input costs and environmental impact, thereby aligning with the objectives of sustainable agriculture.

6.9.4 Ethical, data-sharing, and reproducibility concerns

While the integration of ML with spatial and multi-omics data holds transformative potential, it also presents significant ethical, data-sharing, and reproducibility challenges. Omics datasets are often large, sensitive, and generated under varying experimental standards, leading to concerns regarding accessibility and comparability. Open-source platforms and adherence to the FAIR principles (findable, accessible, interoperable, reusable) are critical to ensure that data can be reliably shared and reused across the scientific community. Furthermore, transparency in ML pipelines is essential to avoid bias, ensure reproducibility, and maintain trust in predictive models. Ethical considerations also extend to the ownership of data, particularly when integrating farmer-generated field data with omics information. Collaborative frameworks among researchers, policy makers, and industry stakeholders will be necessary to establish guidelines for data governance, protect privacy, and promote equitable access to technological benefits.

6.10 Conclusion

ML has emerged as a transformative force in the domains of plant spatial analysis and multiple omics, offering robust tools for the integration of complex datasets, the revelation of latent biological patterns, and the development of predictive models that enhance our comprehension of plant systems. By facilitating the identification of stress-responsive pathways, the mapping of tissue-specific gene expression, and the correlation of molecular profiles with phenotypic traits, ML is reshaping both fundamental research and applied crop improvement. To fully realize its potential, however, it is imperative to foster strong interdisciplinary collaboration among plant scientists, computer scientists, and statisticians to ensure robust data analysis, model interpretability, and biological validation. Looking forward, the integration of ML with advanced omics technologies, field-based phenotyping, and digital agriculture holds promise for accelerating next-generation breeding strategies, enhancing crop resilience, and supporting sustainable agricultural practices in the face of global challenges.

References

Ahlquist, K.D., Sugden, L.A. and Ramachandran, S. (2023) Enabling interpretable machine learning for biological data with reliability scores. *PLoS Computational Biology* 19(5), e1011175. DOI: 10.1371/journal.pcbi.1011175.

Ahmed, Z., Gui, D., Murtaza, G., Yunfei, L. and Ali, S. (2023) An overview of smart irrigation management for improving water productivity under climate change in drylands. *Agronomy* 13(8), 2113. DOI: 10.3390/agronomy13082113.

Al Musawi, A.F., Roy, S. and Ghosh, P. (2025) *A review of link prediction applications in network biology.* IEEE Access.

Alzubaidi, L., Zhang, J., Humaidi, A.J., Al-Dujaili, A., Duan, Y. *et al.* (2021) Review of deep learning: Concepts, CNN architectures, challenges, applications, future directions. *Journal of Big Data* 8(1), 53. DOI: 10.1186/s40537-021-00444-8.

Baião, A.R., Cai, Z., Poulos, R.C., Robinson, P.J., Reddel, R.R. *et al.* (2025) A technical review of multi-omics data integration methods: From classical statistical to deep generative approaches. *Brief Bioinform* 26(4). DOI: 10.1093/bib/bbaf355.

Bansal, H., Luthra, H. and Raghuram, S.R. (2023) A review on machine learning aided multi-omics data integration techniques for healthcare. *Data Analytics and Computational Intelligence: Novel Models, Algorithms and Applications* 211–239. DOI: 10.1007/978-3-031-38325-0_10.

Barbedo, J.G.A. (2022) Data fusion in agriculture: Resolving ambiguities and closing data gaps. *Sensors (Basel, Switzerland)* 22(6). DOI: 10.3390/s22062285.

Bhargava, A., Shukla, A., Goswami, O.P., Alsharif, M.H., Uthansakul, P. *et al.* (2024) Plant leaf disease detection, classification, and diagnosis using computer vision and artificial intelligence: A review. *IEEE Access* 12, 37443–37469. DOI: 10.1109/ACCESS.2024.3373001.

Bhati, U., Sharma, A., Gupta, S., Kumar, A., Pradhan, U.K. *et al.* (2025) Decoding stress specific transcriptional regulation by causality aware graph-transformer deep learning. *BioRxiv*. DOI: 10.1101/2025.03.11.642297.

Bongaerts, M., Bonte, R., Demirdas, S., Huidekoper, H.H., Langendonk, J. *et al.* (2022) Integration of metabolomics with genomics: Metabolic gene prioritization using metabolomics data and genomic variant (CADD) scores. *Molecular Genetics and Metabolism* 136(3), 199–218. DOI: 10.1016/j.ymgme.2022.05.002.

Bunnik, E.M. and Le Roch, K.G. (2013) An introduction to functional genomics and systems biology. *Advances in Wound Care (New Rochelle)* 2(9), 490–498. DOI: 10.1089/wound.2012.0379.

Castellano-Escuder, P., Zachman, D.K., Han, K. and Hirschey, M.D. (2025) GAUDI: Interpretable multi-omics integration with UMAP embeddings and density-based clustering. *Nature Communications* 16(1), 5771. DOI: 10.1038/s41467-025-60822-1.

Cembrowska-Lech, D., Krzemińska, A., Miller, T., Nowakowska, A., Adamski, C. *et al.* (2023) An integrated multi-omics and artificial intelligence framework for advance plant phenotyping in horticulture. *Biology* 12(10), 1298. DOI: 10.3390/biology12101298.

Chang-Brahim, I., Koppensteiner, L.J., Beltrame, L., Bodner, G., Saranti, A. *et al.* (2024) Reviewing the essential roles of remote phenotyping, GWAS and explainable AI in practical marker-assisted selection for drought-tolerant winter wheat breeding. *Frontiers in Plant Science* 15, 1319938. DOI: 10.3389/fpls.2024.1319938.

Chao, D., Wang, H., Wan, F., Yan, S., Fang, W. *et al.* (2025) MtCro: Multi-task deep learning framework improves multi-trait genomic prediction of crops. *Plant Methods* 21(1), 12. DOI: 10.1186/s13007-024-01321-0.

Chaquet-Ulldemolins, J., Gimeno-Blanes, F.-J., Moral-Rubio, S., Muñoz-Romero, S. and Rojo-Álvarez, J.-L. (2022) On the black-box challenge for fraud detection using machine learning (ii): Nonlinear analysis through interpretable autoencoders. *Applied Sciences* 12(8), 3856. DOI: 10.3390/app12083856.

Chen, C., Bhuiyan, S.A., Ross, E., Powell, O., Dinglasan, E. *et al.* (2024) Genomic prediction for sugarcane diseases including hybrid Bayesian-machine learning approaches. *Frontiers in Plant Science* 15, 1398903. DOI: 10.3389/fpls.2024.1398903.

Chen, X., Xu, H., Yu, S., Hu, W., Zhang, Z. *et al.* (2025) AI-driven transcriptome prediction in human pathology: From molecular insights to clinical applications. *Biology* 14(6), 651. DOI: 10.3390/biology14060651.

Chinnaraju, A. (2025) Explainable AI (XAI) for trustworthy and transparent decision-making: A theoretical framework for AI interpretability. *World Journal of Advanced Engineering Technology and Sciences* 14(3), 170–207. DOI: 10.30574/wjaets.2025.14.3.0106.

Cui, Z., Wei, H., Goding, C. and Cui, R. (2023) Stem cell heterogeneity, plasticity, and regulation. *Life Sciences* 334, 122240. DOI: 10.1016/j.lfs.2023.122240.

Dale, R., Oswald, S., Jalihal, A., LaPorte, M.F., Fletcher, D.M. *et al.* (2021) Overcoming the challenges to enhancing experimental plant biology with computational modeling. *Frontiers in Plant Science* 12, 687652. DOI: 10.3389/fpls.2021.687652.

Delfani, P., Thuraga, V., Banerjee, B. and Chawade, A. (2024) Integrative approaches in modern agriculture: Iot, ML and AI for disease forecasting amidst climate change. *Precision Agriculture* 25(5), 2589–2613. DOI: 10.1007/s11119-024-10164-7.

Donovan, B.M., Bastarache, L., Turi, K.N., Zutter, M.M. and Hartert, T.V. (2019) The current state of omics technologies in the clinical management of asthma and allergic diseases. *Annals of Allergy, Asthma & Immunology* 123(6), 550–557. DOI: 10.1016/j.anai.2019.08.460.

Du, T.Y. (2019) Dimensionality reduction techniques for visualizing morphometric data: Comparing principal component analysis to nonlinear methods. *Evolutionary Biology* 46(1), 106–121. DOI: 10.1007/s11692-018-9464-9.

Duan, R., Gao, L., Gao, Y., Hu, Y., Xu, H. *et al.* (2021) Evaluation and comparison of multi-omics data integration methods for cancer subtyping. *PLoS Computational Biology* 17(8), e1009224. DOI: 10.1371/journal.pcbi.1009224.

Esposito, S., Carputo, D., Cardi, T. and Tripodi, P. (2019) Applications and trends of machine learning in genomics and phenomics for next-generation breeding. *Plants (Basel, Switzerland)* 9(1), 34. DOI: 10.3390/plants9010034.

Farheen, O.O.O. and Kumar, R. (2024) Composition of feature selection for time-series prediction with deep learning. *Procedia Computer Science* 235, 1477–1488. DOI: 10.1016/j.procs.2024.04.139.

Garg, N. and Dwivedi, P. (2024) A novel approach for exploring data-driven nutritional insights using clustering and dimensionality reduction techniques. *SN Computer Science* 5(8), 1019. DOI: 10.1007/s42979-024-03397-w.

Ge, X., Yu, X., Liu, Z., Yuan, J., Qin, A. *et al.* (2025) Spatiotemporal transcriptome and metabolome landscapes of cotton somatic embryos. *Nature Communications* 16(1), 859. DOI: 10.1038/s41467-025-55870-6.

Geeitha, S. and Thangamani, M. (2018) Incorporating EBO-HSIC with SVM for gene selection associated with cervical cancer classification. *Journal of Medical Systems* 42(11), 225. DOI: 10.1007/s10916-018-1092-5.

Geng, S., Gao, W., Sun, F., Yang, N., Ma, T. *et al.* (2025) Integrated transcriptomic and metabolomic analyses provide new insights into alkaline stress tolerance in gossypium hirsutum. *Frontiers in Plant Science* 16, 1604606. DOI: 10.3389/fpls.2025.1604606.

Głuchowska, A., Zieniuk, B. and Pawełkowicz, M. (2025) Unlocking plant resilience: Metabolomic insights into abiotic stress tolerance in crops. *Metabolites* 15(6). DOI: 10.3390/metabo15060384.

Goche, T., Mavindidze, P. and Zenda, T. (2025) Advances in functional genomics for exploring abiotic stress tolerance mechanisms in cereals. *Plants* 14, 2459.

Godinez, W.J., Hossain, I., Lazic, S.E., Davies, J.W. and Zhang, X. (2017) A multi-scale convolutional neural network for phenotyping high-content cellular images. *Bioinformatics (Oxford, England)* 33(13), 2010–2019.

Goncharuk, E.A. and Zagoskina, N.V. (2023) Heavy metals, their phytotoxicity, and the role of phenolic antioxidants in plant stress responses with focus on cadmium: Review. *Molecules (Basel, Switzerland)* 28(9), 3921. DOI: 10.3390/molecules28093921.

Gulati, G.S., D'Silva, J.P., Liu, Y., Wang, L. and Newman, A.M. (2025) Profiling cell identity and tissue architecture with single-cell and spatial transcriptomics. *Nature Reviews. Molecular Cell Biology* 26(1), 11–31. DOI: 10.1038/s41580-024-00768-2.

Hajnajafi, K. and Iqbal, M.A. (2025) Mass-spectrometry based metabolomics: An overview of workflows, strategies, data analysis and applications. *Proteome Science* 23(1), 5. DOI: 10.1186/s12953-025-00241-8.

Han, S.H., Kim, K.W., Kim, S. and Youn, Y.C. (2018) Artificial neural network: Understanding the basic concepts without mathematics. *Dementia and Neurocognitive Disorders* 17(3), 83–89. DOI: 10.12779/dnd.2018.17.3.83.

Hao, Y., Zhang, Z., Luo, E., Yang, J. and Wang, S. (2025) Plant metabolomics: Applications and challenges in the era of multi-omics big data. *aBIOTECH* 6(1), 116–132. DOI: 10.1007/s42994-024-00194-0.

Holden, R.J., McDougald Scott, A.M., Hoonakker, P.L., Hundt, A.S. and Carayon, P. (2015) Data collection challenges in community settings: Insights from two field studies of patients with chronic disease. *Quality of Life Research* 24(5), 1043–1055. DOI: 10.1007/s11136-014-0780-y.

Hu, B., Zhu, J. and Zhao, F. (2024) The evolving landscape of spatial proteomics technologies in the AI age. *Fundamental Research*. DOI: 10.1016/j.fmre.2024.11.023.

Jamil, I.N., Remali, J., Azizan, K.A., Nor Muhammad, N.A., Arita, M. *et al.* (2020) Systematic multi-omics integration (MOI) approach in plant systems biology. *Frontiers in Plant Science* 11, 944. DOI: 10.3389/fpls.2020.00944.

Jan, R., Hussain, A., Assad, A., Khurshid, S. and Macha, M.A. (2025) Chapter 10 - challenges with multi-omics data integration. In: Macha, M.A., Bhat, A.A. and Masoodi, T.A. (eds) *Multi-Omics*

Technology in Human Health and Diseases. Academic Press, pp. 223–242. DOI: 10.1016/B978-0-443-13595-8.00010-6.

Jiang, M.Z., Aguet, F., Ardlie, K., Chen, J., Cornell, E. *et al*. (2023) Canonical correlation analysis for multi-omics: Application to cross-cohort analysis. *PLoS Genetics* 19(5), e1010517. DOI: 10.1371/journal.pgen.1010517.

Jolliffe, I.T. and Cadima, J. (2016) Principal component analysis: A review and recent developments. *Philosophical Transactions. Series A, Mathematical, Physical, and Engineering Sciences* 374(2065), 20150202. DOI: 10.1098/rsta.2015.0202.

Kasauli, R., Knauss, E., Horkoff, J., Liebel, G. and Oliveira Neto, F.G. (2021) Requirements engineering challenges and practices in large-scale agile system development. *Journal of Systems and Software* 172, 110851. DOI: 10.1016/j.jss.2020.110851.

Kassem, M.A. (2025) Harnessing artificial intelligence and machine learning for identifying quantitative trait loci (QTL) associated with seed quality traits in crops. *Plants* 14(11), 1727. DOI: 10.3390/plants14111727.

Kaya, A., Keceli, A.S., Catal, C., Yalic, H.Y., Temucin, H. *et al*. (2019) Analysis of transfer learning for deep neural network based plant classification models. *Computers and Electronics in Agriculture* 158, 20–29. DOI: 10.1016/j.compag.2019.01.041.

Keceli, A.S., Kaya, A., Catal, C. and Tekinerdogan, B. (2022) Deep learning-based multi-task prediction system for plant disease and species detection. *Ecological Informatics* 69, 101679. DOI: 10.1016/j.ecoinf.2022.101679.

Kreuzer, T., Papapetrou, P. and Zdravkovic, J. (2024) Artificial intelligence in digital twins—a systematic literature review. *Data & Knowledge Engineering* 151, 102304. DOI: 10.1016/j.datak.2024.102304.

Kumar, R., Das, S.P., Choudhury, B.U., Kumar, A., Prakash, N.R. *et al*. (2024) Advances in genomic tools for plant breeding: Harnessing DNA molecular markers, genomic selection, and genome editing. *Biological Research* 57(1), 80. DOI: 10.1186/s40659-024-00562-6.

Liu, L., Chen, A., Li, Y., Mulder, J., Heyn, H. *et al*. (2024) Spatiotemporal omics for biology and medicine. *Cell* 187(17), 4488–4519. DOI: 10.1016/j.cell.2024.07.040.

Luo, Y., Zhao, C. and Chen, F. (2024) Multiomics research: Principles and challenges in integrated analysis. *Biodesign Research* 6, 0059. DOI: 10.34133/bdr.0059.

Ma, J., Li, G., Chu, Y., Yue, H., Xu, Z. *et al*. (2025) Integrated analysis of the metabolome and transcriptome during apple ripening to highlight aroma determinants in ningqiu apples. *Plants* 14(8), 1165. DOI: 10.3390/plants14081165.

Mahmood, U., Li, X., Fan, Y., Chang, W., Niu, Y. *et al*. (2022) Multi-omics revolution to promote plant breeding efficiency. *Frontiers in Plant Science* 13, 1062952. DOI: 10.3389/fpls.2022.1062952.

Marx, V. (2024) Seeing data as t-SNE and UMAP do. *Nature Methods* 21(6), 930–933. DOI: 10.1038/s41592-024-02301-x.

Mienye, I.D., Swart, T.G. and Obaido, G. (2024) Recurrent neural networks: A comprehensive review of architectures, variants, and applications. *Information* 15(9), 517. DOI: 10.3390/info15090517.

Monteiro, A., Santos, S. and Gonçalves, P. (2021) Precision agriculture for crop and livestock farming—brief review. *Animals (Basel)* 11(8), 2345. DOI: 10.3390/ani11082345.

Nobori, T. (2025) Exploring the untapped potential of single-cell and spatial omics in plant biology. *The New Phytologist* 247(3), 1098–1116. DOI: 10.1111/nph.70220.

Ozer, M.E., Sarica, P.O. and Arga, K.Y. (2020) New machine learning applications to accelerate personalized medicine in breast cancer: Rise of the support vector machines. *Omics* 24(5), 241–246. DOI: 10.1089/omi.2020.0001.

Panahi, B., Hamid, R. and Mohammad Zadeh Jalaly, H. (2025) Deciphering plant transcriptomes: Leveraging machine learning for deeper insights. *Current Plant Biology* 41, 100432. DOI: 10.1016/j.cpb.2024.100432.

Patil, Y., Ramachandran, H., Sundararajan, S. and Srideviponmalar, P. (2025) Comparative analysis of machine learning models for crop yield prediction across multiple crop types. *SN Computer Science* 6(1), 64. DOI: 10.1007/s42979-024-03602-w.

Picard, M., Scott-Boyer, M.-P., Bodein, A., Périn, O. and Droit, A. (2021) Integration strategies of multi-omics data for machine learning analysis. *Computational and Structural Biotechnology Journal* 19, 3735–3746. DOI: 10.1016/j.csbj.2021.06.030.

Praveen, A., Bartelo, N. and Soni, V. (2024) Applications of artificial intelligence and machine learning in antimicrobial resistance study. In: *Antimicrobial Resistance: Factors to Findings: Omics and Systems Biology Approaches*. Springer, pp. 359–385. DOI: 10.1007/978-3-031-65986-7_11.

Rahaman, M.M., Islam, M.R., Bhuiyan, M.M.R., Noman, I.R., Aziz, M.M. *et al.* (2025) Harnessing big data in biotechnology: A machine learning approach to multi-omics. In: *2025 International Conference on Machine Learning and Autonomous Systems (ICMLAS)*, Prawet, Thailand. DOI: 10.1109/ICMLAS64557.2025.10967731.

Rampler, E., Abiead, Y.E., Schoeny, H., Rusz, M., Hildebrand, F. *et al.* (2020) Recurrent topics in mass spectrometry-based metabolomics and lipidomics—standardization, coverage, and throughput. *Analytical Chemistry* 93(1), 519–545. DOI: 10.1021/acs.analchem.0c04698.

Ran, L., Li, Y., Liang, G. and Zhang, Y. (2024) Pseudo labeling methods for semi-supervised semantic segmentation: A review and future perspectives. *IEEE Transactions on Circuits and Systems for Video Technology* 35(4), 3054–3080. DOI: 10.1109/TCSVT.2024.3508768.

Rane, N., Choudhary, S.P. and Rane, J. (2024) Ensemble deep learning and machine learning: Applications, opportunities, challenges, and future directions. *Studies in Medical and Health Sciences* 1(2), 18–41. DOI: 10.48185/smhs.v1i2.1225.

Reel, P.S., Reel, S., Pearson, E., Trucco, E. and Jefferson, E. (2021) Using machine learning approaches for multi-omics data analysis: A review. *Biotechnology Advances* 49, 107739. DOI: 10.1016/j.biotechadv.2021.107739.

Rico-Chávez, A.K., Franco, J.A., Fernandez-Jaramillo, A.A., Contreras-Medina, L.M., Guevara-González, R.G. *et al.* (2022) Machine learning for plant stress modeling: A perspective towards hormesis management. *Plants (Basel)* 11(7). DOI: 10.3390/plants11070970.

Santamarina-Ojeda, P., Tejedor, J.R., Pérez, R.F., López, V., Roberti, A. *et al.* (2023) Multi-omic integration of DNA methylation and gene expression data reveals molecular vulnerabilities in glioblastoma. *Molecular Oncology* 17(9), 1726–1743. DOI: 10.1002/1878-0261.13479.

Sarfraz, Z., Zarlashat, Y., Ambreen, A., Mujahid, M., Iqbal, M.S. *et al.* (2025) *Plant biochemistry in the era of omics: Integrated omics approaches to unravel the genetic basis of plant stress tolerance.* Plant Breeding. DOI: 10.1111/pbr.13277.

Sauta, E., Demartini, A., Vitali, F., Riva, A. and Bellazzi, R. (2020) A Bayesian data fusion based approach for learning genome-wide transcriptional regulatory networks. *BMC Bioinformatics* 21(1), 219. DOI: 10.1186/s12859-020-3510-1.

Schmitz, O.J. (2010) Spatial dynamics and ecosystem functioning. *PLoS Biology* 8(5), e1000378. DOI: 10.1371/journal.pbio.1000378.

Singh, V. and Singh, V. (2024) Inferring interaction networks from transcriptomic data: Methods and applications. In: *Transcriptome Data Analysis*. Springer, pp. 11–37. DOI: 10.1007/978-1-0716-3886-6_2.

Sompairac, N., Nazarov, P.V., Czerwinska, U., Cantini, L., Biton, A. *et al.* (2019) Independent component analysis for unraveling the complexity of cancer omics datasets. *International Journal of Molecular Sciences* 20(18), 4414. DOI: 10.3390/ijms20184414.

Sprenger, H., Erban, A., Seddig, S., Rudack, K., Thalhammer, A. *et al.* (2018) Metabolite and transcript markers for the prediction of potato drought tolerance. *Plant Biotechnology Journal* 16(4), 939–950. DOI: 10.1111/pbi.12840.

Subramanian, I., Verma, S., Kumar, S., Jere, A. and Anamika, K. (2020) Multi-omics data integration, interpretation, and its application. *Bioinformatics and Biology Insights* 14, 1177932219899051. DOI: 10.1177/1177932219899051.

Sudhakar, M. and Swarna Priya, R.M. (2023) Computer vision based machine learning and deep learning approaches for identification of nutrient deficiency in crops: A survey. *Nature Environment and Pollution Technology* 22(3), 1387–1399. DOI: 10.46488/NEPT.2023.v22i03.025.

Suzuki, T.K. (2022) Phenotypic systems biology for organisms: Concepts, methods and case studies. *Biophysics and Physicobiology* 19, 1–17. DOI: 10.2142/biophysico.bppb-v19.0011.

Tresas, T., Isaioglou, I., Roussis, A. and Haralampidis, K. (2025) A brief overview of the epigenetic regulatory mechanisms in plants. *International Journal of Molecular Sciences* 26(10), 10. DOI: 10.3390/ijms26104700.

Vahabi, N. and Michailidis, G. (2022) Unsupervised multi-omics data integration methods: A comprehensive review. *Frontiers in Genetics* 13, 854752. DOI: 10.3389/fgene.2022.854752.

Vasupalli, N., Bhat, J.A., Jain, P., Sri, T., Islam, M.A. *et al.* (2024) Omics big data for crop improvement: Opportunities and challenges. *The Crop Journal* 12(6), 1517–1532. DOI: 10.1016/j.cj.2024.10.007.

Wang, G., Ryu, K.H., Dinneny, A., Carlson, J., Goodstein, D. *et al.* (2024a) Diversification of gene expression across extremophytes and stress-sensitive species in the *Brassicaceae*. *BioRxiv*. DOI: 10.1101/2024.06.21.599952.

Wang, H., Zhang, Y., Zhang, Y., Zhao, X., Bai, Z. *et al.* (2025) SpaVGN: A hybrid deep learning framework for high-resolution spatial transcriptomics data reconstruction and spatial domain identification. *PLoS ONE* 20(8), e0329122. DOI: 10.1371/journal.pone.0329122.

Wang, J., Han, X., Zheng, Y., Zhao, Y., Wang, W. *et al.* (2024b) Spatial metabolomic profiling of pinelliae rhizoma from different leaf types using matrix-assisted laser desorption/ionization mass spectrometry imaging. *Molecules* 29(17), 4251. DOI: 10.3390/molecules29174251.

Wang, K., Abid, M.A., Rasheed, A., Crossa, J., Hearne, S. *et al.* (2023a) DNNGP, a deep neural network-based method for genomic prediction using multi-omics data in plants. *Molecular Plant* 16(1), 279–293. DOI: 10.1016/j.molp.2022.11.004.

Wang, X., Han, J., Li, Z., Li, B., Wan, Y. *et al.* (2023b) Editorial: Insight into plant spatial omics: Mass spectrometry imaging. *Frontiers in Plant Science* 14. DOI: 10.3389/fpls.2023.1273010.

Wang, Y., Hou, W., Sheng, N., Zhao, Z., Liu, J. *et al.* (2024c) Graph pooling in graph neural networks: Methods and their applications in omics studies. *Artificial Intelligence Review* 57(11), 294. DOI: 10.1007/s10462-024-10918-9.

Waqas, M., Naseem, A., Humphries, U.W., Hlaing, P.T., Dechpichai, P. *et al.* (2025) Applications of machine learning and deep learning in agriculture: A comprehensive review. *Green Technologies and Sustainability* 3(3), 100199. DOI: 10.1016/j.grets.2025.100199.

Wu, Y. and Xie, L. (2025) AI-driven multi-omics integration for multi-scale predictive modeling of genotype-environment-phenotype relationships. *Computational and Structural Biotechnology Journal* 27, 265–277. DOI: 10.1016/j.csbj.2024.12.030.

Wu, K., Bu, F., Wu, Y., Zhang, G., Wang, X. *et al.* (2024) Exploring noncoding variants in genetic diseases: From detection to functional insights. *Journal of Genetics and Genomics* 51(2), 111–132. DOI: 10.1016/j.jgg.2024.01.001.

Yan, J. and Wang, X. (2023) Machine learning bridges omics sciences and plant breeding. *Trends in Plant Science* 28(2), 199–210. DOI: 10.1016/j.tplants.2022.08.018.

Yu, P.-S., Yang, T.-C., Chen, S.-Y., Kuo, C.-M. and Tseng, H.-W. (2017) Comparison of random forests and support vector machine for real-time radar-derived rainfall forecasting. *Journal of Hydrology* 552, 92–104. DOI: 10.1016/j.jhydrol.2017.06.020.

Yu, X., Liu, Z. and Sun, X. (2023) Single-cell and spatial multi-omics in the plant sciences: Technical advances, applications, and perspectives. *Plant Communications* 4(3), 100508. DOI: 10.1016/j.xplc.2022.100508.

Yun, S., Noh, M., Yu, J., Kim, H.J., Hui, C.C. *et al.* (2024) Unlocking biological mechanisms with integrative functional genomics approaches. *Molecules and Cells* 47(8), 100092. DOI: 10.1016/j.mocell.2024.100092.

Zhan, C., Shen, S., Yang, C., Liu, Z., Fernie, A.R. *et al.* (2022) Plant metabolic gene clusters in the multi-omics era. *Trends in Plant Science* 27(10), 981–1001. DOI: 10.1016/j.tplants.2022.03.002.

Zhang, F., Shen, Z., Huang, S., Zhu, Y. and Yi, M. (2025) SpaInGNN: Enhanced clustering and integration of spatial transcriptomics based on refined graph neural networks. Methods (San Diego, Calif.) 233, 42–51. DOI: 10.1016/j.ymeth.2024.11.006.

Zhang, R., Zhang, C., Yu, C., Dong, J. and Hu, J. (2022) Integration of multi-omics technologies for crop improvement: Status and prospects. *Frontiers in Bioinformatics* 2, 1027457. DOI: 10.3389/fbinf.2022.1027457.

Zhang, X., Ibrahim, Z., Khaskheli, M.B., Raza, H., Zhou, F. *et al.* (2024) Integrative approaches to abiotic stress management in crops: Combining bioinformatics educational tools and artificial intelligence applications. *Sustainability* 16(17), 7651. DOI: 10.3390/su16177651.

Zhao, Z., Alzubaidi, L., Zhang, J., Duan, Y. and Gu, Y. (2024) A comparison review of transfer learning and self-supervised learning: Definitions, applications, advantages and limitations. *Expert Systems with Applications* 242, 122807. DOI: 10.1016/j.eswa.2023.122807.

Zhou, J., Cui, G., Hu, S., Zhang, Z., Yang, C. *et al.* (2020) Graph neural networks: A review of methods and applications. *AI Open* 1, 57–81. DOI: 10.1016/j.aiopen.2021.01.001.

7 Mass Spectrometry Imaging for Plant Spatial Metabolomics

Varucha Misra* and A.K. Mall
ICAR – Indian Sugarcane Research Institute, Lucknow, India

7.1 Introduction

Metabolomics plays a crucial role as a highly accurate approach for detecting and measuring low-molecular-weight compounds (metabolites) within living organisms. In plants, metabolite profiles provide important information about the plant's physiological condition, helping us understand changes at cellular and biochemical level. During the past decade, plant metabolomics has been widely utilized to investigate how metabolites change over time and how metabolic pathways drive growth, development, and stress responses. A particularly important area of research involves the remodeling of lipid metabolism, which is considered a key plant strategy for tolerating abiotic stresses. Alterations in lipid biosynthesis and accumulation can significantly impact cell membrane properties, including integrity, fluidity, permeability, and the functionality of embedded transport proteins. Changes in lipid biosynthesis and accumulation can significantly impact cell membrane properties, including membrane stability, flexibility, permeability, and the activity of membrane transport proteins. Under stresses such as salinity, many lipid classes and lipid molecules show stress-induced changes, which may help plants survive harsh environments (Darwish *et al.*, 2009; Mansour *et al.*, 2015). Thus, metabolomic and lipidomic analyses provided valuable data about which metabolites increase or decrease under stress, and such metabolites can act as biomarkers for predicting stress tolerance and explaining biochemical defense mechanisms. Traditional metabolomics workflows usually involve extracting target metabolites from plant tissues before analysis. While this method is effective for global metabolite profiling, it removes spatial information, implying that we lose knowledge about where metabolites are located within cells, tissues, or organs. This constraint may reduce our ability to correctly interpret metabolic responses because biological functions often depend on metabolite location. Although spatial information can be integrated with standard metabolomics, it often requires specialized techniques and skilled expertise, making the process time intensive and technically demanding (Holzlechner *et al.*, 2019; Alexandrov, 2020).

The development of spatial metabolomics, particularly through mass spectrometry imaging (MSI), has helped overcome this challenge by enabling *in situ* metabolite detection directly in tissues. MSI is a sophisticated, label-free analytical method that allows for the simultaneous mapping of both the abundance and distribution of endogenous biomolecules, such as lipids, primary and secondary metabolites, and proteins, directly from the surface

*Corresponding author: misra.varucha@gmail.com

© CAB International 2026. *Plant Spatial Biology* (ed. Chen)
DOI: 10.1079/9781836994572.0007

of thin plant tissue sections. This spatially resolved technique is invaluable for examining the molecular architecture of plant tissues and contributes significantly to our understanding of plant development, physiology, and plant–environment interactions under normal and stress conditions (Wang *et al.*, 2023). Mass spectrometry (MS), a core method in metabolomics, is widely recognized for its high sensitivity, precision, and capacity to identify a broad spectrum of molecular compounds. When MS is integrated with separation methods such as gas chromatography (GC–MS), liquid chromatography (LC–MS), capillary electrophoresis (CE–MS), or ion mobility spectrometry (IMS–MS), it can provide high-throughput, comprehensive metabolome profiling, even at low concentrations (Hong *et al.*, 2016; Maji *et al.*, 2023; Patidar and Kamble, 2025). This makes MS an essential tool for studying how plants respond to fluctuating environmental changes such as temperature shifts, drought, salinity, and pathogen infection (Hirayama and Shinozaki, 2010).

In addition to identifying and quantifying known metabolites, mass spectrometry also supports the discovery of novel compounds and helps characterize previously unknown biosynthetic pathways (Walker, 2021). High-resolution MS and tandem MS (MS/MS) generate accurate mass values and structural information, which improves confidence in metabolite annotation. When used in imaging mode, this technology becomes more powerful because MSI not only detects metabolite levels but also reveals their spatial distribution within plant tissues. This spatial information is important for understanding how metabolism is organized and coordinated across various cell types and developmental stages (Walker, 2021). Therefore, MSI has become an increasingly popular technique across multiple areas of plant science, including physiology, pathology, stress biology, and studies of plant interactions with microbes and insects (Wang *et al.*, 2023). By enabling the spatial localization of biomolecules *in situ*, MSI enhances the interpretability of metabolomic data and deepens our understanding of plant systems biology under both physiological and stressed conditions. This chapter provides an overview of MSI and its applications in plant development.

7.2 Fundamentals of Mass Spectrometry Imaging

MSI has become a leading tool in spatial metabolomics because it allows researchers to map the two- and three-dimensional localization of metabolites directly within tissue samples (Swales *et al.*, 2019; Alexandrov, 2020). By combining the high sensitivity and analytical precision of MS with spatial information, MSI generates mass spectral images that show the distribution of ions at specific positions within biological tissues (Holzlechner *et al.*, 2019; Alexandrov, 2020). One major advantage of MSI is its ability to detect multiple metabolites from individual cells in a short time, making it more efficient than many conventional metabolomic approaches. This capability has supported the growing use of MSI in both biological research and clinical investigations, as it enables the simultaneous detection and localization of many small molecules in different sample types (Fujii *et al.*, 2015; Föll *et al.*, 2019).

A standard MSI workflow includes four main steps: selecting and preparing the sample, desorbing and ionizing the analytes, analyzing ions using a mass spectrometer, and generating images and interpreting the data. Careful control of each step is necessary to obtain reliable and high-quality results. Among these steps, sample handling (selection, storage, and preparation) is especially important because problems at this step can greatly reduce data quality.

The basic principle of MSI involves placing a thin tissue section into an ion source, where molecules are desorbed and ionized. Mass spectra are then collected from across the tissue in a grid pattern using an appropriate ionization and detection method. Each location on the grid represents a defined spot on the tissue, and a mass spectrum is recorded that contains a molecular fingerprint based on the measured mass-to-charge (m/z) ratios. These spectra are aligned with an optical image of the same tissue section, obtained either before or after MSI analysis, often after staining to distinguish cell or tissue types. Each mass spectrum corresponds to one pixel in the ion image. By plotting the intensity of selected ions across all pixels, MSI produces a spatial distribution map that shows where specific compounds are located in the tissue. These molecular images can be directly compared with histological images. Three-dimensional (3D) MSI can also be achieved by collecting data from the serial tissue sections or by using depth profiling. The resulting data can then be computationally reconstructed to create a 3D volumetric ion map. MSI data acquisition generally follows two main strategies: the microprobe method and the microscope method (Soltwisch *et al.*, 2014). In the microprobe approach, which is the most common in commercial systems, mass spectra are collected point by point, and the full image is reconstructed from these sequential measurements. In contrast, the microscope method, which is still being developed, uses a spatially sensitive detector like Timepix. These detectors contain an array of individual pixels capable of recording ion patterns from a larger area in a single acquisition event (Ellis *et al.*, 2014).

Spatial resolution is a key feature of MSI and depends mainly on the ionization method and sample preparation. In 2D MSI, spatial resolution is often described as lateral resolution, which refers to the ability to distinguish closely spaced features. In practice, lateral resolution is commonly estimated from the pixel size or step size (50 μm × 50 μm). However, true spatial resolution can also include depth resolution, which becomes important in 3D imaging. The final spatial resolution of the MSI dataset is influenced by several linked factors: (i) the sampling area, or spot size, which depends on the size of the desorption source (e.g. laser beam, ion stream, or droplet size); (ii) the step size (raster interval), which defines the distance moved between measurements; and (iii) the pixel size and image sampling density, which determine how the data are converted into an image. Together, these factors control how accurately MSI can distinguish adjacent structures. Specialized software reconstructs intensity maps from the spatially indexed mass spectra, where each pixel holds a full mass spectrum. In the final MSI image, the x- and y-axes represent the spatial coordinates, and the color scale represents the intensity of the selected ion signal (Holzlechner *et al.*, 2019; Snel, 2019). In beam- or laser-based microprobe approaches, spatial detail can be further improved using oversampling, where the stage moves in steps smaller than the beam size. This approach can increase resolution beyond the nominal beam diameter. In microscope-based MSI, spatial resolution is primarily determined by the detector design and pixel density of the micro-detector array.

7.3 Recent Technologies Developed in Mass Spectrometry Imaging for Plants using Ionization

Ionization plays a central role in MSI because the mass spectrometer can only detect molecules after they are converted into ions. The efficiency and selectivity of ionization depend on the plant tissue type, the chemical properties of the metabolites, and the ionization method used. Matrix-assisted laser desorption/ionization (MALDI) remains the most widely applied ionization technique in MSI. In MALDI-MSI, a laser is used to desorb and ionize molecules from the tissue surface. However, newer MSI ionization technologies have expanded the range of analytes and sample types that can be studied. This includes desorption electrospray ionization (DESI), laser ablation electrospray ionization (LAESI), nanostructure-initiator mass spectrometry (NIMS), nanospray DESI (nano-DESI), infrared MALDI combined with electrospray ionization (IR-MALDESI), and secondary ion mass spectrometry (SIMS), which uses an ion beam for surface ionization (Swales *et al.*, 2019; Alexandrov, 2020).

7.3.1 Laser ablation electrospray ionization (LAESI)

LAESI MSI uses a mid-infrared (IR) laser to selectively excite the vibrational energy of water molecules present in the sample. This energy causes laser ablation, producing a plume of neutral molecules (analytes) from the sample surface. These analytes are then ionized when they interact with charged droplets generated by a nanoscale electrospray source. The resulting ions are transferred into the mass spectrometer, where they are separated and detected. To perform LAESI-MS analysis, a fresh or living sample is placed flat and stable on a sample stage and aligned with the focal point of the mid-IR laser. The laser beam is then used for ablation at defined points on the sample surface. The molecular distribution is visualized by mapping the intensity of total ion signals or selected ion signals onto a false color image. This image is generated by linking each signal intensity value with the spatial coordinates of each pixel across the imaged region (Etalo *et al.*, 2018).

7.3.2 Nanostructure-initiator mass spectrometry (NIMS)

NIMS is a matrix-free technique that provides high sensitivity and soft ionization, allowing rapid analysis that can be completed within seconds (Northen *et al.*, 2007). One major advantage of NIMS is that the surface can act like a reverse phase retention layer, which helps selectively retain certain target molecules. This improves sensitivity by reducing background interference and overlapping signals (Go *et al.*, 2007). Although NIMS was initially used mainly in positive ion mode, it has also been shown to generate negative ions, including negatively charged nucleotides (Amantonico *et al.*, 2009). During desorption and ionization, the porous silicon surface absorbs the laser energy, leading to a rapid increase in surface temperature and vaporization of the initiator. This process supports efficient ionization of analytes while minimizing fragmentation (Northen *et al.*, 2007). Even though NIMS silicon surfaces are not widely available commercially, they can be prepared using a simple three-step process: cleaning, etching, and coating with an initiator compound (Woo *et al.*, 2008). Proper cleaning before etching is critical to achieve high sensitivity and low background noise. The performance also depends on etching time, and around 30 min is considered optimal. NIMS can use different initiator compounds, which are chosen to remain non-ionizing, thereby reducing chemical noise in the spectra (Northen *et al.*, 2007; Woo *et al.*, 2008). This feature makes NIMS particularly useful for detecting low molecular weight metabolites. The initiator can also influence the ion patterns observed in mass spectra. Additionally, surfaces can be modified with salts like NaCl or $AgNO_3$ to enhance the detection of metabolites such as carbohydrates and steroids (Patti *et al.*, 2010). This flexibility, by changing initiators or surface coatings, allows NIMS to extract more detailed molecular information from complex biological samples. NIMS also enables direct analysis of tissue samples with little or no sample preparation. Compared to SIMS, NIMS provides a wider usable mass range and produces less fragmentation (Northen *et al.*, 2007). In MALDI imaging, even though matrix application methods have improved (Bouschen *et al.*, 2010), spatial resolution may still be restricted due to metabolite diffusion and matrix crystal formation. NIMS overcomes these limitations through its matrix-free design and improved sensitivity. Importantly, the laser beam in NIMS can be focused to a much smaller spot size (15–20 μm) compared to typical MALDI imaging (50–300 μm) (Cornett *et al.*, 2007). This smaller laser diameter can significantly enhance spatial resolution in imaging experiments.

7.3.3 Secondary ion mass spectrometry (SIMS)

SIMS was first developed in 1962 for the imaging of inorganic elements (Castaing and Slodzian, 1962). It was not until the 1970s that it was applied to biological samples (Galle, 1970; Morrison and Slodzian, 1975). In SIMS, a pulsed high-energy primary ion beam or gas cluster, usually accelerated at energies between 15 and 25 keV, is focused and directed into the surface of a sample. When it strikes the sample surface, this high-energy beam ejects secondary ions from there. By accumulating and analyzing these ions per pixel throughout the sample, a spatial map of metabolite distribution is created (Imai *et al.*, 2005; Saito *et al.*, 2008, 2012; Fletcher *et al.*, 2011a; Zhou *et al.*, 2011). Modern technological advancements in SIMS and its nanoscale counterpart, nano-SIMS, have greatly improved the technique to allow for ultra-high spatial resolution – nominally less than 100 nanometers – and 3D depth profiling capability. In addition to its strength, however, SIMS has some challenges when used for other biological MSIs. A key drawback is the requirement for analyzing high-vacuum conditions. This vacuum is also necessary to avoid collisions between the secondary ions and molecules of background gas, so both the sample and analytes have to be stable at vacuum conditions. Additionally, SIMS is a 'hard' ionization technique because of the high energy provided upon impact of primary ion beams. This tends to create extensive fragmentation of the target molecules, which can decrease detection capability for intact molecular ions and jeopardize chemical specificity. One of the significant analytical difficulties is being able to reliably match fragments.

7.3.4 Matrix-assisted laser desorption ionization (MALDI)

MALDI is one of the most commonly used ionization methods in MSI and is classified as a soft ionization technique. It can detect intact molecular ions with very little fragmentation (Caprioli *et al.*, 1997). MALDI requires a matrix, which is a compound that absorbs the laser energy that would otherwise be too strong to directly ionize the analytes. The matrix is typically a low-molecular-weight organic compound that helps molecules move from the solid to the gas phase and supports ion formation in the laser-induced plume (Dreisewerd, 2003). Because the matrix absorbs most of the energy, analyte fragmentation is usually minimal or negligible. MALDI instruments can detect compounds across a broad mass range, including large biomolecules above 100 kDa, making them suitable for analyzing many biological molecules such as proteins, peptides, and lipids. Various lasers can be used for MALDI, operating in either the ultraviolet (UV) or IR regions. Common laser systems include nitrogen lasers (337.1 nm), neodymium-doped yttrium aluminum garnet (Nd:YAG) lasers (355 nm), and tunable IR lasers, which can support efficient desorption (Park and Murray, 2012). A technique often compared with MALDI is laser desorption ionization (LDI), which does not involve a matrix. In LDI, analyte molecules directly absorb laser energy and are vaporized. However, this method generally requires more laser energy than MALDI, which can lead to substantial fragmentation during ionization. In addition, LDI usually produces fewer ions and is therefore less sensitive, with ion yields reported to be around 1% of MALDI (Hölscher *et al.*, 2009, 2014; Thiery-Lavenant *et al.*, 2013).

7.3.5 Desorption electrospray ionization (DESI)

DESI is an ambient ionization technique in which charged microdroplets are sprayed onto the sample surface using a spray capillary. When the droplets hit the surface, they desorb and ionize molecules from the sample (Ifa *et al.*, 2007). The generated ions enter the gas phase and are transported through an atmospheric interface into the MS for detection and analysis. DESI is a versatile technique and generally provides a lateral resolution of about 250 μm under standard operating conditions. However, with optimization settings, spatial resolution can be improved to 50 μm. A modified DESI approach involves transferring analytes from the sample surface onto another substrate, e.g. PTFE film, paper, or a thin-layer chromatography (TLC) plate, by pressing the sample against it. The transferred molecules can then be analyzed using standard DESI protocols (Müller *et al.*, 2011; Thunig *et al.*, 2011). A related method, nanospray DESI (nano-DESI), employs a self-aspirating nanospray capillary and typically achieves spatial resolutions ranging from 100 to 150 μm (Laskin *et al.*, 2012; Lanekoff *et al.*, 2013b). In nano-DESI, analytes are collected from the sample surface and directly delivered to a nanospray electrospray ionization source, where ionization occurs. Unlike conventional DESI, nano-DESI can partially separate the desorption and ionization steps, which can improve sample collection efficiency (Roach *et al.*, 2010). Both DESI and nano-DESI have the advantage of requiring minimal sample preparation and allowing analysis under ambient conditions without the use of high vacuum systems. Therefore, these techniques are highly suitable for studying the surface chemistry of biological samples, including plant tissues (Roach *et al.*, 2010).

7.3.6 Nanospray desorption electrospray ionization (nano-DESI)

Nano-DESI is an ambient MS technique that enables localized molecular analysis directly from surfaces (Roach *et al.*, 2010). The technique has developed rapidly and has proven useful for detecting a wide range of analytes. In particular, nano-DESI MSI has become a useful approach for mapping the spatial distribution of molecules in biological tissues, while requiring minimal or no sample preparation (Iqfath *et al.*, 2024). In nano-DESI, molecules are extracted from the sample into a small liquid bridge formed between the sample surface and the nano-DESI probe. The extracted analytes are then transported through the probe to the mass spectrometer by either self-aspiration or vacuum-assisted flow. Ionization occurs in a way that is similar to electrospray ionization (Roach *et al.*, 2010; Yin *et al.*, 2019). During MSI, the sample is moved at a constant speed beneath the probe, and mass spectra are collected continuously at fixed spatial intervals. This continuous sampling mode distinguishes nano-DESI from laser- or ion beam-based imaging methods, which typically generate ions in separate bursts at discrete points. Like other ambient ionization techniques, nano-DESI avoids complex sample preparation and is therefore suitable for laboratories that do not have matrix-coating facilities, as required for MALDI and related methods. Additionally, because nano-DESI does not require a matrix, it avoids matrix-related interference signals in the low m/z region, which improves its suitability for metabolite imaging (Lanekoff *et al.*, 2013a). Although MALDI and nano-DESI show similar performance for lipid imaging, nano-DESI has been reported to be more effective for detecting certain specific drugs and metabolites that may be difficult to observe using MALDI (Mesa *et al.*, 2022). Furthermore, nano-DESI can generate multiple charged ions, which supports the detection and identification of proteins (Yang *et al.*, 2022). The ability to use different solvents also enables semi-quantitative analysis. For example, internal standards can be added directly to the extraction solvent to reduce ion suppression effects, allowing relative quantification without major changes in the workflow (Bergman *et al.*, 2016).

Despite its strengths, non-DESI has some limitations (Table 7.1). The technique depends on the solubility of analytes in the extraction solvent, so molecules that are poorly soluble or strongly bound to biological components may be difficult to detect. In addition, compounds with low electrospray ionization efficiency may produce weak signals. Unlike MALDI-MSI, nano-DESI is not yet commercially available, and many laboratories must build custom non-DESI sources. However, several published protocols and tutorials are available to guide users in constructing nano-DESI systems and coupling them with commercial mass spectrometers (Yin *et al.*, 2019; Jiang *et al.*, 2024). Recent improvements have increased the sensitivity and selectivity of nano-DESI MSI. For instance, replacing the secondary self-aspirating probe with a nebulizer-assisted pneumatic probe can enhance signal intensity (Duncan *et al.*, 2017). High-resolution detection has also been achieved by coupling nano-DESI with Fourier transform ion cyclotron resonance (FT-ICR) MS, providing sub-ppm mass accuracy and enabling the resolution of isotopic fine structures (Vandergrift *et al.*, 2022). In addition, integrating nano-DESI with ion mobility spectrometry (IMS) and higher-order tandem MS (MS^2, MS^3, MS^4) improves separation and identification of isomeric and closely related isobaric compounds (Lanekoff *et al.*, 2013a; Lillja *et al.*, 2020; Unsihuay *et al.*, 2021; Lillja and Lanekoff, 2022). Online derivatization strategies have also been used to distinguish metabolite isomers, which helps clarify the biological roles of isomeric lipids and metabolites (Unsihuay *et al.*, 2021). Optimal nano-DESI performance depends on careful adjustment of several experimental parameters, including solvent composition, spray voltage, and precise positioning of the probe relative to both the sample and MS inlet. Spatial resolution is influenced by probe geometry, solvent flow rate, surface characteristics of the sample, and the stability of the probe-to-sample distance. To maintain consistent sampling across uneven surfaces, a three-point plane calibration method can be used to compensate for sample surface tilt, supporting spatial resolutions in the range of 20–200 μm (Lanekoff *et al.*, 2012). For higher resolution imaging, a shear force feedback probe can maintain constant probe to sample distance, enabling spatial resolutions below 10 μm (Nguyen *et al.*, 2017, 2018). With its ability to adapt solvent systems, high spatial resolution, and throughput, nano-DESI MSI stands out as a robust and flexible ambient imaging technique due to its solvent adaptability, good spatial resolution, and efficient sampling throughput. Iqfath *et al.* (2024) provide a detailed overview of nano-DESI MSI, including probe design, spatial

Table 7.1. Advantages and limitations of the recent technologies.

MSI technologies	Advantages and description	Limitations	References
Matrix-assisted laser desorption/ionization (MALDI)	Major mass spectrometry systems are presently employed for examining biological specimens.	The MALDI matrix often introduces substantial background signals below 1000 Da, which interfere with the analysis of low-mass compounds. Additionally, the dimensions of the matrix crystals restrict the spatial resolution attainable in tissue imaging	Stoeckli *et al.* (2001); Hanash (2003); Cornett *et al.* (2007)
Secondary ion MS (SIMS)	SIMS employs a high-energy primary ion beam to eject and produce secondary ions from a sample surface. Its main benefit lies in its exceptionally high spatial resolution – reaching down to 50 nanometers – making it a valuable tool for mass spectrometry-based tissue imaging.	It has not been widely adopted for analyzing biofluids and tissues due to its low sensitivity for molecules above 500 Da and the extensive fragmentation of analytes caused by the high-energy primary ion beam	McDonnell and Heeren (2007); Boxer *et al.* (2009)
Desorption electrospray ionization (DESI)	DESI is a matrix-free method for examining biological specimens that utilizes a stream of charged solvent droplets to release ions from a surface. Its key benefits include the ability to analyze samples without the need for a specialized surface and the capability to perform measurements under ambient conditions, allowing unrestricted access to the sample during data collection.	Achieving high spatial resolution is challenging due to the difficulty in precisely directing the charged solvent spray	Takáts *et al.* (2004); Cooks *et al.* (2006); Wiseman *et al.* (2008)
Desorption/ionization on silicon (DIOS)	A forerunner to NIMS, DIOS is a matrix-free laser-based desorption/ionization method. It employs a porous silicon surface to capture laser energy, facilitating the vaporization and ionization of analytes on the substrate with minimal fragmentation.	Limited ionization efficiency for certain analytes; lack of chemical versatility; surface degradation; lower sensitivity compared to enhanced techniques; surface uniformity and reproducibility issues	Wei *et al.* (1999); Shen *et al.* (2001); Thomas *et al.* (2001)

Continued

Table 7.1. Continued

MSI technologies	Advantages and description	Limitations	References
Nanostructure-initiator mass spectrometry (NIMS)	NIMS has largely superseded DIOS, as the use of initiators in NIMS enhances surface compatibility for imaging applications and offers greater consistency across experiments. Unlike MALDI matrices, NIMS initiators differ significantly; they typically do not absorb UV light, rarely undergo ionization themselves, and the analytes are not embedded in a co-crystallized form with the initiator.	Limited ionization of high mass molecules; requirement of high vacuum conditions; lower throughput compared to ambient techniques; initiator selection and compatibility	Shen *et al.* (2001); Lewis *et al.* (2003); Woo *et al.* (2008)

resolution, throughput, data acquisition modes, and data processing workflows. It also discusses solvent selection strategies and the influence of matrix effects on nano-DES results.

7.4 Application of MSI in Plant Metabolomics

MALDI-MSI has proven effective in visualizing a wide range of endogenous molecules across various plant organs, including roots, stems, leaves, flowers, fruits, and seeds. This technique provides valuable information about *in situ* molecular changes during plant growth, development, and environmental stress responses (Qin *et al.*, 2018). For instance, using on-tissue chemical derivatization with MALDI-MSI, Dueñas *et al.* (2019) detected approximately 600 metabolites in maize roots and leaves. Similarly, MSI of *Lychnophora salicifolia* leaves using laser desorption ionization (LDI) and MS/MS revealed that vicenin-2, a di-C-glycosyl flavonoid, accumulated mainly in the mesophyll and upper epidermal cells under intense sunlight, indicating its possible role in stress adaptation (Silva *et al.*, 2014).

Recent advances in MSI have greatly improved the detection and spatial mapping of plant secondary metabolites such as alkaloids, sesquiterpenes, and flavonoids. Liu *et al.* (2023c) combined ultra-performance LC/quadrupole time-of-flight MS (UPLC–QTOF-MS) with MALDI-TOF-MSI to investigate metabolite profiles, spatial localization, and biosynthetic pathways in *Dendrobium nobile* stems. They identified and semi-quantified 11 alkaloids, 10 sesquiterpenes, and 13 additional compounds, and their results clarified the biosynthesis and accumulation of dendrobine. Wu *et al.* (2022) used DESI-MSI to examine the spatiotemporal distribution of alkaloids in *Gelsemium elegans*. They detected 23 alkaloids in roots, stems, and leaves during the seedling stage, and 19 alkaloids during maturity, revealing changes linked to plant development and metabolite migration. Liu *et al.* (2023a) performed *in situ* imaging of flavonoid distribution in *Litchi chinensis* seed tissues using MALDI-MSI and detected 15 flavonoid-related ion signals in positive ion mode. In another study, Liu *et al.* (2023b) applied MALDI-TOF-MSI to differentiate *Pterocarpus santalinus* and *P. tinctorius* and identified 15 species-specific chemical markers. Liu *et al.* (2023d) further used MSI-based analysis to investigate the effects of stir-frying *Bombyx batryticatus* with wheat bran on protein-related metabolites. Their results showed that the processing reduced toxicity and enhanced certain bioactivities, emphasizing the importance of processing methods in traditional Chinese medicine research. In addition to chemical imaging, Jia *et al.* (2023) integrated optical imaging with machine learning to classify maturity stages and spatial patterns in *Elymus sibiricus* seeds. By using feature selection approaches and machine learning models with k-means clustering, they achieved accurate and low-cost seed maturity classification.

The use of MSI in plant–pathogen studies is also gaining momentum. Maia *et al.* (2022) investigated *Vitis vinifera* leaves infected with *Plasmopara viticola* and optimized sample preparation and matrix selection to enhance MALDI-MS performance. Their analysis revealed that sucrose accumulated near the leaf veins only in infected leaves and not in controls, suggesting that *P. viticola* may influence host sucrose metabolism during early infection. MALDI-MSI has also supported the *in-situ* visualization of both primary and secondary metabolites, helping to reveal spatially distinct biological processes in plant tissues (Lee *et al.*, 2012). For example, as mentioned, vicenin-2, a flavonoid, was reported to accumulate mainly in the upper epidermal layers of *Lychnophora salicifolia* leaves, supporting its proposed role in UV protection (Silva *et al.*, 2014). Liu *et al.* (2021) identified 17 metabolites and 125 lipids with different spatial distributions in two cottonseed cultivars exhibiting contrasting drought and salt tolerance. MSI revealed nine key lipids and four metabolites with clear differences in abundance and localization between cultivars. Spatial metabolomics has also been applied to identify specialized metabolites in *Euphorbia* spp. Ernst *et al.* (2019) reported that diterpenoids were mainly localized in young stems and roots, suggesting a defensive role against biotic stress. Kempthorne *et al.* (2024) used DESI-MS to detect 15 alkaloids at wounded regions of *Vincetoxicum rossicum* leaves, highlighting localized metabolite response to injury. Zhang *et al.* (2025) detected 3456 metabolites in root nodules and observed altered spatial patterns under drought and alkaline conditions. Transcriptomic profiling suggested that the isoflavone malonyltransferase gene *GmMaT2* had cortex-specific expression, consistent with the spatial distribution of related metabolites. Their study also identified *GmbZIP59* as a regulatory transcription factor that may enhance GmMaT2 expression, although its activity declined under stress. Ranner *et al.* (2025) identified three novel polyphenols in *Lotus japonica* roots using ultra-performance (UP) LC–electrospray ionization (ESI)–ion mobility (IM)-QTOF-MS during arbuscular mycorrhiza symbiosis. Oyarce *et al.* (2025) used MALDI-MSI to identify species- and tissue-specific metabolite distributions in the roots of *Avicennia marina* and *Phoenix dactylifera*.

Integrating spatial metabolomics with machine learning and cloud computing is expected to accelerate progress in plant metabolite research. Tools such as MSiReader (Bokhart *et al.*, 2018), when combined with machine learning algorithms, can improve data interpretation and reduce molecular ambiguity. Dimensionality reduction techniques such as t-SNE and UMAP support the visualization of complex MSI datasets and help reveal clusters and spatial patterns (Alexandrov, 2020; Ganesh *et al.*, 2021). Cloud platforms like METASPACE and OpenMSI support rapid metabolite annotation by organizing ion signals and assisting interpretation of ionization behavior, thereby expanding MSI applications (Alexandrov, 2020). Spatial metabolomics, therefore, provides a powerful framework to investigate plant responses to abiotic stress, offering high-resolution information on metabolites that may be lost during conventional extraction-based metabolomics. This approach also supports novel metabolite discovery across different organizational levels, ranging from subcellular regions to whole organs. In addition, new automated and high-throughput MSI technologies are emerging and are expected to improve scalability and precision in plant metabolomics research.

7.5 Conclusion

Plant metabolomics has advanced significantly in recent years due to the increasing use of MSI. MSI is a sensitive, label-free tool that enables the visualization and measurement of the spatial distribution of metabolites, lipids, proteins, and other biomolecules directly in plant tissues. MSI has improved our understanding of plant metabolic pathways by revealing where primary and secondary metabolites accumulate, how symbiotic interactions influence metabolism, and how metabolic patterns change during plant growth and development. Because MSI can link molecular information with tissue culture, it has become a valuable technique in modern plant research. Despite these advances, several challenges persist. Key limitations involve the need to improve spatial resolution, enhance sensitivity, strengthen quantitative reliability, and increase chemical specificity, particularly for distinguishing structurally similar compounds such as isomers. Additionally, optimizing sample preparation remains important because plant tissues are structurally complex and contain diverse metabolites. Future work should focus on developing MSI instruments with higher spatial and mass resolution to support single-cell and sub-cellular imaging. There is also a strong need for improved quantitative workflows and advanced ionization strategies to increase metabolite coverage and detection sensitivity. Moreover, the integration of MSI with other imaging approaches and omics technologies will provide a more complete understanding of plant biology. Advances in computational methods, including machine learning and artificial intelligence, will be essential for processing and interpreting the large and complex datasets generated by MSI. Finally, the development of standardized protocols for sample handling and data analysis will be critical to ensure consistency and comparability across MSI studies.

References

Alexandrov, T. (2020) Spatial metabolomics and imaging mass spectrometry in the age of artificial intelligence. *Annual Review of Biomedical Data Science* 3, 61–87. DOI: 10.1146/annurev-biodatasci-011420-031537.

Amantonico, A., Flamigni, L., Glaus, R. and Zenobi, R. (2009) Negative mode nanostructure-initiator mass spectrometry for detection of phosphorylated metabolites. *Metabolomics* 5(3), 346–353. DOI: 10.1007/s11306-009-0163-5.

Bergman, H.M., Lundin, E., Andersson, M. and Lanekoff, I. (2016) Quantitative mass spectrometry imaging of small-molecule neurotransmitters in rat brain tissue sections using nanospray desorption electrospray ionization. *The Analyst* 141(12), 3686–3695. DOI: 10.1039/c5an02620b.

Bokhart, M.T., Nazari, M., Garrard, K.P. and Muddiman, D.C. (2018) MSiReader v1.0: Evolving open-source mass spectrometry imaging software for targeted and untargeted analyses. *Journal of the American Society for Mass Spectrometry* 29(1), 8–18. DOI: 10.1007/s13361-017-1809-6.

Bouschen, W., Schulz, O., Eikel, D. and Spengler, B. (2010) Matrix vapor deposition/recrystallization and dedicated spray preparation for high-resolution scanning microprobe matrix-assisted laser desorption/ionization imaging mass spectrometry (SMALDI-MS) of tissue and single cells. *Rapid Communications in Mass Spectrometry* 24(3), 355–364. DOI: 10.1002/rcm.4401.

Boxer, S.G., Kraft, M.L. and Weber, P.K. (2009) Advances in imaging secondary ion mass spectrometry for biological samples. *Annual Review of Biophysics* 38, 53–74. DOI: 10.1146/annurev.biophys.050708.133634.

Caprioli, R.M., Farmer, T.B. and Gile, J. (1997) Molecular imaging of biological samples: Localization of peptides and proteins using MALDI-TOF MS. *Analytical Chemistry* 69(23), 4751–4760. DOI: 10.1021/ac970888i.

Castaing, R. and Slodzian, G. (1962) Microanalyse par emission ionique secondaire. *Journal of Microscopy* 1, 395–412. DOI: 10.1007/BF00465754.

Cooks, R.G., Ouyang, Z., Takats, Z. and Wiseman, J.M. (2006) Ambient mass spectrometry. *Science* 311(5767), 1566–1570. DOI: 10.1126/science.1119426.

Cornett, D.S., Reyzer, M.L., Chaurand, P. and Caprioli, R.M. (2007) MALDI imaging mass spectrometry: Molecular snapshots of biochemical systems. *Nature Methods* 4(10), 828–833. DOI: 10.1038/nmeth1094.

Darwish, E., Testerink, C., Khalil, M., El-Shihy, O. and Munnik, T. (2009) Phospholipid signaling responses in salt-stressed rice leaves. *Plant & Cell Physiology* 50(5), 986–997. DOI: 10.1093/pcp/pcp051.

Dreisewerd, K. (2003) The desorption process in MALDI. *Chemical Reviews* 103(2), 395–426. DOI: 10.1021/cr010375i.

Dueñas, M.E., Larson, E.A. and Lee, Y.J. (2019) Toward mass spectrometry imaging in the metabolomics scale: Increasing metabolic coverage through multiple on-tissue chemical modifications. *Frontiers in Plant Science* 10, 1–11. DOI: 10.3389/fpls.2019.00860.

Duncan, K.D., Bergman, H.M. and Lanekoff, I. (2017) A pneumatically assisted nanospray desorption electrospray ionization source for increased solvent versatility and enhanced metabolite detection from tissue. *The Analyst* 142(18), 3424–3431. DOI: 10.1039/c7an00901a.

Ellis, S.R., Soltwisch, J. and Heeren, R.M.A. (2014) Time-resolved imaging of the MALDI linear-TOF ion cloud: Direct visualization and exploitation of ion optical phenomena using a position- and time-sensitive detector. *Journal of the American Society for Mass Spectrometry* 25(5), 809–819. DOI: 10.1007/s13361-014-0839-6.

Ernst, M., Nothias, L.-F., van der Hooft, J.J.J., Silva, R.R., Saslis-Lagoudakis, C.H. *et al*. (2019) Assessing specialized metabolite diversity in the cosmopolitan plant genus *Euphorbia* L. *Frontiers in Plant Science* 10, 1–15. DOI: 10.3389/fpls.2019.00846.

Etalo, D.W., Diez-Simon, C., deVos, R.C.H. and Hall, R.D. (2018) Laser ablation electrospray ionization-mass spectrometry imaging (LAESI-MS) for spatially resolved plant metabolomics. In: António, C. (ed.) *Plant Metabolomics: Methods and Protocols, Methods in Molecular Biology*, Vol. 1778. Springer New York, New York, NY. DOI: 10.1007/978-1-4939-7819-9_18.

Fletcher, J.S., Lockyer, N.P. and Vickerman, J.C. (2011) Developments in molecular SIMS depth profiling and 3D imaging of biological systems using polyatomic primary ions. *Mass Spectrometry Reviews* 30(1), 142–174. DOI: 10.1002/mas.20275.

Föll, M.C., Moritz, L., Wollmann, T., Stillger, M.N., Vockert, N. *et al*. (2019) Accessible and reproducible mass spectrometry imaging data analysis in galaxy. *GigaScience* 8(12), 1–12. DOI: 10.1093/gigascience/giz143.

Fujii, T., Matsuda, S., Tejedor, M.L., Esaki, T., Sakane, I. *et al*. (2015) Direct metabolomics for plant cells by live single-cell mass spectrometry. *Nature Protocols* 10(9), 1445–1456. DOI: 10.1038/nprot.2015.084.

Galle, P. (1970) Sur une nouvelle methode d'analyse cellulaire utilisant le phenomene d'emission ionique secondaire. *Annales de Physique Biologie et Médecine* 42, 84–94.

Ganesh, S., Hu, T., Woods, E., Allam, M., Cai, S. *et al*. (2021) Spatially resolved 3D metabolomic profiling in tissues. *Science Advances* 7(5), 1–17. DOI: 10.1126/sciadv.abd0957.

Go, E.P., Uritboonthai, W., Apon, J.V., Trauger, S.A., Nordstrom, A. *et al*. (2007) Selective metabolite and peptide capture/mass detection using fluorous affinity tags. *Journal of Proteome Research* 6(4), 1492–1499. DOI: 10.1021/pr060608s.

Hanash, S. (2003) Disease proteomics. *Nature* 422(6928), 226–232. DOI: 10.1038/nature01514.

Hirayama, T. and Shinozaki, K. (2010) Research on plant abiotic stress responses in the post-genome era: Past, present and future. *The Plant Journal* 61(6), 1041–1052. DOI: 10.1111/j.1365-313X.2010.04124.x.

Hölscher, D., Shroff, R., Knop, K., Gottschaldt, M., Crecelius, A. *et al*. (2009) Matrix-free UV-laser desorption/ionization (LDI) mass spectrometric imaging at the single-cell level: Distribution of secondary metabolites of *Arabidopsis thaliana* and *Hypericum* species. *The Plant Journal* 60(5), 907–918. DOI: 10.1111/j.1365-313X.2009.04012.x.

Hölscher, D., Dhakshinamoorthy, S., Alexandrov, T., Becker, M., Bretschneider, T. *et al*. (2014) Phenalenone-type phytoalexins mediate resistance of banana plants (*Musa* spp.) to the burrowing nematode *Radopholus similis*. *Proceedings of the National Academy of Sciences* 111(1), 105–110. DOI: 10.1073/pnas.1314168110.

Holzlechner, M., Eugenin, E. and Prideaux, B. (2019) Mass spectrometry imaging to detect lipid biomarkers and disease signatures in cancer. *Cancer Reports* 2(6). DOI: 10.1002/cnr2.1229.

Hong, J., Yang, L., Zhang, D. and Shi, J. (2016) Plant metabolomics: An indispensable system biology tool for plant science. *International Journal of Molecular Sciences* 17(6), 767. DOI: 10.3390/ijms17060767.

Ifa, D.R., Wiseman, J.M., Song, Q. and Cooks, R.G. (2007) Development of capabilities for imaging mass spectrometry under ambient conditions with desorption electrospray ionization (DESI). *International Journal of Mass Spectrometry* 259(1–3), 8–15. DOI: 10.1016/j.ijms.2006.08.003.

Imai, T., Tanabe, K., Kato, T. and Fukushima, K. (2005) Localization of ferruginol, a diterpene phenol, in *Cryptomeria japonica* heartwood by time-of-flight secondary ion mass spectrometry. *Planta* 221(4), 549–556. DOI: 10.1007/s00425-004-1476-2.

Iqfath, M., Wali, S.N., Amer, S., Hernly, E. and Laskin, J. (2024) Nanospray desorption electrospray ionization mass spectrometry imaging (nano-DESI MSI): A tutorial review. *ACS Measurement Science Au* 4(5), 475–487. DOI: 10.1021/acsmeasuresciau.4c00028.

Jia, Z., Ou, C., Sun, S., Wang, J., Liu, J. *et al*. (2023) Integrating optical imaging techniques for a novel approach to evaluate Siberian wild rye seed maturity. *Frontiers in Plant Science* 14. DOI: 10.3389/fpls.2023.1170947.

Jiang, L.-X., Hilger, R.T. and Laskin, J. (2024) Hardware and software solutions for implementing nano-spray desorption electrospray ionization (nano-DESI) sources on commercial mass spectrometers. *Journal of Mass Spectrometry* 59(7), e5065. DOI: 10.1002/jms.5065.

Kempthorne, C.J., St Pierre, M., Le, A., Livingstone, S., McNulty, J. *et al.* (2024) Mass spectrometry-based metabolomics for the elucidation of alkaloid biosynthesis and function in invasive *Vincetoxicum rossicum* populations. *Phytochemistry* 221, 114051. DOI: 10.1016/j.phytochem.2024.114051.

Lanekoff, I., Heath, B.S., Liyu, A., Thomas, M., Carson, J.P. *et al.* (2012) Automated platform for high-resolution tissue imaging using nanospray desorption electrospray ionization mass spectrometry. *Analytical Chemistry* 84(19), 8351–8356. DOI: 10.1021/ac301909a.

Lanekoff, I., Burnum-Johnson, K., Thomas, M., Short, J., Carson, J.P. *et al.* (2013a) High-speed tandem mass spectrometric *in situ* imaging by nanospray desorption electrospray ionization mass spectrometry. *Analytical Chemistry* 85(20), 9596–9603. DOI: 10.1021/ac401760s.

Lanekoff, I., Geydebrekht, O., Pinchuk, G.E., Konopka, A.E. and Laskin, J. (2013b) Spatially resolved analysis of glycolipids and metabolites in living *Synechococcus* sp. PCC 7002 using nanospray desorption electrospray ionization. *The Analyst* 138(7), 1971–1978. DOI: 10.1039/c3an36716a.

Laskin, J., Heath, B.S., Roach, P.J., Cazares, L. and Semmes, O.J. (2012) Tissue imaging using nanospray desorption electrospray ionization mass spectrometry. *Analytical Chemistry* 84(1), 141–148. DOI: 10.1021/ac2021322.

Lee, Y.J., Perdian, D.C., Song, Z., Yeung, E.S. and Nikolau, B.J. (2012) Use of mass spectrometry for imaging metabolites in plants. *The Plant Journal* 70(1), 81–95. DOI: 10.1111/j.1365-313X.2012.04899.x.

Lewis, W.G., Shen, Z., Finn, M.G. and Siuzdak, G. (2003) Desorption/ionization on silicon (DIOS) mass spectrometry: Background and applications. *International Journal of Mass Spectrometry* 226(1), 107–116. DOI: 10.1016/S1387-3806(02)00973-9.

Lillja, J. and Lanekoff, I. (2022) Quantitative determination of sn-positional phospholipid isomers in ms^n using silver cationization. *Analytical and Bioanalytical Chemistry* 414(25), 7473–7482. DOI: 10.1007/s00216-022-04173-6.

Lillja, J., Duncan, K.D. and Lanekoff, I. (2020) Determination of monounsaturated fatty acid isomers in biological systems by modeling ms^3 product ion patterns. *Journal of the American Society for Mass Spectrometry* 31(12), 2479–2487. DOI: 10.1021/jasms.0c00194.

Liu, B., Wang, X., Li, K. and Cai, Z. (2021) Spatially resolved metabolomics and lipidomics reveal salinity and drought-tolerant mechanisms of cottonseeds. *Journal of Agricultural and Food Chemistry* 69(28), 8028–8037. DOI: 10.1021/acs.jafc.1c01598.

Liu, B., Chen, Q., Tang, L., Zhu, L., Zou, X. *et al.* (2023b) Screening of potential chemical marker with interspecific differences in *Pterocarpus* wood and a spatially-resolved approach to visualize the distribution of the characteristic markers. *Frontiers in Plant Science* 14, 1133848. DOI: 10.3389/fpls.2023.1133848.

Liu, P., Wang, J.-M., Guo, H.-C., Zhao, M.-W., Song, Y.-X. *et al.* (2023d) *In situ* detection and mass spectrometry imaging of protein-related metabolites in *Bombyx batryticatus* before and after frying with wheat bran. *Frontiers in Plant Science* 14. DOI: 10.3389/fpls.2023.1144556.

Liu, Q., Huang, Y., Linghu, C., Xiao, J. and Gu, R. (2023c) Metabolic profiling, *in situ* spatial distribution, and biosynthetic pathway of functional metabolites in *Dendrobium nobile* stem revealed by combining UPLC-QTOF-MS with MALDI-TOF-MSI. *Frontier in Plant Sciences* 13. DOI: 10.3389/fpls.2022.1125872.

Liu, Y., Nie, X., Wang, J., Zhao, Z., Wang, Z. *et al.* (2023a) Visualizing the distribution of flavonoids in litchi (*Litchi chinenis*) seeds through matrix-assisted laser desorption/ionization mass spectrometry imaging. *Frontiers in Plant Science* 14. DOI: 10.3389/fpls.2023.1144449.

Maia, M., McCann, A., Malherbe, C., Far, J., Cunha, J. *et al.* (2022) Grapevine leaf MALDI-MS imaging reveals the localisation of a putatively identified sucrose metabolite associated to *Plasmopara viticola* development. *Frontiers in Plant Science* 13. DOI: 10.3389/fpls.2022.1012636.

Maji, S.R., Roy, C. and Kumar, S. (2023) Gas chromatography (GC-MS): A comprehensive review of synergistic combinations and their applications in the past two decades. *Journal of Analytical Sciences and Applied Biotechnology* 5(2), 72–85. DOI: 10.48402/IMIST.PRSM/jasab-v5i2.40209.

Mansour, M.M.F., Salama, K.H.A. and Allam, H.Y.H. (2015) Role of the plasma membrane in saline conditions: Lipids and proteins. *The Botanical Review* 81(4), 416–451. DOI: 10.1007/s12229-015-9156-4.

McDonnell, L.A. and Heeren, R.M.A. (2007) Imaging mass spectrometry. *Mass Spectrometry Reviews* 26(4), 606–643. DOI: 10.1002/mas.20124.

Mesa Sanchez, D., Brown, H.M., Yin, R., Chen, B., Vavrek, M. *et al.* (2022) Mass spectrometry imaging of diclofenac and its metabolites in tissues using nanospray desorption electrospray ionization. *Analytica Chimica Acta* 1233, 340490. DOI: 10.1016/j.aca.2022.340490.

Morrison, G.H. and Slodzian, G. (1975) Ion microscopy. *Analytical Chemistry* 47(11), 932A–943A. DOI: 10.1021/ac60361a741.

Müller, T., Oradu, S., Ifa, D.R., Cooks, R.G. and Kräutler, B. (2011) Direct plant tissue analysis and imprint imaging by desorption electrospray ionization mass spectrometry. *Analytical Chemistry* 83(14), 5754–5761. DOI: 10.1021/ac201123t.

Nguyen, S.N., Liyu, A.V., Chu, R.K., Anderton, C.R. and Laskin, J. (2017) Constant-distance mode nanospray desorption electrospray ionization mass spectrometry imaging of biological samples with complex topography. *Analytical Chemistry* 89(2), 1131–1137. DOI: 10.1021/acs.analchem.6b03293.

Nguyen, S.N., Sontag, R.L., Carson, J.P., Corley, R.A., Ansong, C. *et al.* (2018) Towards high-resolution tissue imaging using nanospray desorption electrospray ionization mass spectrometry coupled to shear force microscopy. *Journal of the American Society for Mass Spectrometry* 29(2), 316–322. DOI: 10.1007/s13361-017-1750-8.

Northen, T.R., Yanes, O., Northen, M.T., Marrinucci, D., Uritboonthai, W. *et al.* (2007) Clathrate nanostructures for mass spectrometry. *Nature* 449(7165), 1033–1036. DOI: 10.1038/nature06195.

Oyarce, P., Xiao, T.T., Henkel, C., Frederiksen, S.F., Gonzalez-Kise, J.K. *et al.* (2025) Microscopy and spatial-metabolomics identify tissue-specific metabolic pathways uncovering salinity and drought tolerance mechanisms in *Avicennia marina* and *Phoenix dactylifera* roots. *Scientific Reports* 15(1), 1076. DOI: 10.1038/s41598-025-85416-1.

Park, S.G. and Murray, K.K. (2012) Infrared laser ablation sample transfer for MALDI imaging. *Analytical Chemistry* 84(7), 3240–3245. DOI: 10.1021/ac3006704.

Patidar, A. and Kamble, P. (2025) A comprehensive review on liquid chromatography-mass spectrometry (LC-MS): A hyphenated technique. *Asian Journal of Pharmaceutical Research and Development* 13(1), 95–103. DOI: 10.22270/ajprd.v13i1.1509.

Patti, G.J., Woo, H.-K., Yanes, O., Shriver, L., Thomas, D. *et al.* (2010) Detection of carbohydrates and steroids by cation-enhanced nanostructure-initiator mass spectrometry (NIMS) for biofluid analysis and tissue imaging. *Analytical Chemistry* 82, 121–128. DOI: 10.1021/ac9014353.

Qin, L., Zhang, Y., Liu, Y., He, H., Han, M. *et al.* (2018) Recent advances in matrix-assisted laser desorption/ionisation mass spectrometry imaging (MALDI-MSI) for *in situ* analysis of endogenous molecules in plants. *Phytochemical Analysis* 29(4), 351–364. DOI: 10.1002/pca.2759.

Ranner, J.L., Stabl, G., Piller, A., Paries, M., Sharma, S. *et al.* (2025) Untargeted metabolomics reveals novel metabolites in *Lotus japonicus* roots during arbuscular mycorrhiza symbiosis. *The New Phytologist* 246(3), 1256–1275. DOI: 10.1111/nph.70051.

Roach, P.J., Laskin, J. and Laskin, A. (2010) Nanospray desorption electrospray ionization: An ambient method for liquid-extraction surface sampling in mass spectrometry. *The Analyst* 135(9), 2233–2236. DOI: 10.1039/c0an00312c.

Saito, K., Mitsutani, T., Imai, T., Matsushita, Y. and Fukushima, K. (2008) Discriminating the indistinguishable sapwood from heartwood in discolored ancient wood by direct molecular mapping of specific extractives using time-of-flight secondary ion mass spectrometry. *Analytical Chemistry* 80(5), 1552–1557. DOI: 10.1021/ac7021162.

Shen, Z., Thomas, J.J., Averbuj, C., Broo, K.M., Engelhard, M. *et al.* (2001) Porous silicon as a versatile platform for laser desorption/ionization mass spectrometry. *Analytical Chemistry* 73(3), 612–619. DOI: 10.1021/ac000746f.

Silva, D.B., Turatti, I.C.C., Gouveia, D.R., Ernst, M., Teixeira, S.P. *et al.* (2014) Mass spectrometry of flavonoid vicenin-2, based sunlight barriers in *Lychnophora* species. *Scientific Reports* 4(1), 1–8. DOI: 10.1038/srep04309.

Snel, M.F. (2019) Ion mobility separation mass spectrometry imaging. In: *Comprehensive Analytical Chemistry*, 1st ed., Vol. 83. Elsevier B.V, Amsterdam, pp. 237–257. DOI: 10.1016/bs.coac.2018.09.001.

Soltwisch, J., Göritz, G., Jungmann, J.H., Kiss, A., Smith, D.F. *et al.* (2014) MALDI mass spectrometry imaging in microscope mode with infrared lasers: Bypassing the diffraction limits. *Analytical Chemistry* 86(1), 321–325. DOI: 10.1021/ac403421v.

Stoeckli, M., Chaurand, P., Hallahan, D.E. and Caprioli, R.M. (2001) Imaging mass spectrometry: A new technology for the analysis of protein expression in mammalian tissues. *Nature Medicine* 7(4), 493–496. DOI: 10.1038/86573.

Swales, J.G., Hamm, G., Clench, M.R. and Goodwin, R.J.A. (2019) Mass spectrometry imaging and its application in pharmaceutical research and development: A concise review. *International Journal of Mass Spectrometry* 437, 99–112. DOI: 10.1016/j.ijms.2018.02.007.

Takáts, S., Wiseman, J.M., Gologan, B. and Cooks, R.G. (2004) Mass spectrometry sampling under ambient conditions with desorption electrospray ionization. *Science* 306(5695), 471–473. DOI: 10.1126/science.1104404.

Thiery-Lavenant, G., Zavalin, A.I. and Caprioli, R.M. (2013) Targeted multiplex imaging mass spectrometry in transmission geometry for subcellular spatial resolution. *Journal of the American Society for Mass Spectrometry* 24(4), 609–614. DOI: 10.1007/s13361-012-0563-z.

Thomas, J.J., Shen, Z., Crowell, J.E., Finn, M.G. and Siuzdak, G. (2001) Desorption/ionization on silicon (DIOS): A diverse mass spectrometry platform for protein characterization. *Proceedings of the National Academy of Sciences* 98(9), 4932–4937. DOI: 10.1073/pnas.081069298.

Thunig, J., Hansen, S.H. and Janfelt, C. (2011) Analysis of secondary plant metabolites by indirect desorption electrospray ionization imaging mass spectrometry. *Analytical Chemistry* 83(9), 3256–3259. DOI: 10.1021/ac2004967.

Unsihuay, D., Su, P., Hu, H., Qiu, J., Kuang, S. *et al.* (2021) Imaging and analysis of isomeric unsaturated lipids through online photochemical derivatization of carbon–carbon double bonds. *Angewandte Chemie* 133(14), 7637–7641. DOI: 10.1002/ange.202016734.

Vandergrift, G.W., Kew, W., Lukowski, J.K., Bhattacharjee, A., Liyu, A.V. *et al.* (2022) Imaging and direct sampling capabilities of nanospray desorption electrospray ionization with absorption-mode 21 T fourier transform ion cyclotron resonance mass spectrometry. *Analytical Chemistry* 94(8), 3629–3636. DOI: 10.1021/acs.analchem.1c05216.

Walker, H.J. (2021) Metabolite imaging by mass spectrometry: A new discovery tool. *Advances in Botanical Research* 98, 271–296. DOI: 10.1016/bs.abr.2020.09.022.

Wang, X., Han, J., Li, Z., Li, B., Wan, Y. *et al.* (2023) Editorial: Insight into plant spatial omics: Mass spectrometry imaging. *Frontiers in Plant Science* 14, 1273010. DOI: 10.3389/fpls.2023.1273010.

Wei, J., Buriak, J.M. and Siuzdak, G. (1999) Desorption-ionization mass spectrometry on porous silicon. *Nature* 399(6733), 243–246. DOI: 10.1038/20400.

Wiseman, J.M., Ifa, D.R., Zhu, Y., Kissinger, C.B., Manicke, N.E. *et al.* (2008) Desorption electrospray ionization mass spectrometry: Imaging drugs and metabolites in tissues. *Proceedings of the National Academy of Sciences of the United States of America* 105(47), 18120–18125. DOI: 10.1073/pnas.0801066105.

Woo, H.-K., Northen, T.R., Yanes, O. and Siuzdak, G. (2008) Nanostructure-initiator mass spectrometry: A protocol for preparing and applying NIMS surfaces for high-sensitivity mass analysis. *Nature Protocols* 3(8), 1341–1349. DOI: 10.1038/nprot.2008.110.

Wu, Z.H., Wang, R.Z., Sun, Z.L., Su, Y. and Xiao, L.T. (2022) A mass spectrometry imaging approach on spatiotemporal distribution of multiple alkaloids in *Gelsemium elegans*. *Frontiers in Plant Science* 13. DOI: 10.3389/fpls.2022.1051756.

Yang, M., Hu, H., Su, P., Thomas, P.M., Camarillo, J.M. *et al.* (2022) Proteoform-Selective imaging of tissues using mass spectrometry. *Angewandte Chemie International Edition* 61, 29. DOI: 10.1002/anie.202200721.

Yin, R., Burnum-Johnson, K.E., Sun, X., Dey, S.K. and Laskin, J. (2019) High spatial resolution imaging of biological tissues using nanospray desorption electrospray ionization mass spectrometry. *Nature Protocols* 14(12), 3445–3470. DOI: 10.1038/s41596-019-0237-4.

Zhang, C., Zhao, Y., Yang, J., He, Y., Zhang, X. *et al.* (2025) Spatial metabolome mass spectrometry imaging reveals the flavonoid distribution change in soybean nodules under drought and alkaline stresses. *Plant Stress* 16, 100880. DOI: 10.1016/j.stress.2025.100880.

Zhou, C., Li, Q., Chiang, V.L., Lucia, L.A. and Griffis, D.P. (2011) Chemical and spatial differentiation of syringyl and guaiacyl lignins in poplar wood via time-of-flight secondary ion mass spectrometry. *Analytical Chemistry* 83(18), 7020–7026. DOI: 10.1021/ac200903y.

8 Plant Regulatory Networks Uncovered by Spatial and Multiple Omics Tools

Varucha Misra* and A.K. Mall
ICAR-Indian Sugarcane Research Institute, Lucknow, India

8.1 Introduction

Plants, being sessile, must quickly adjust to environmental changes by carefully controlling gene expression. Transcriptional regulation plays a central role in these adjustments, governing gene activity during development and in response to biotic and abiotic stresses. Transcription factors (TFs) regulate genes by binding to specific DNA motifs known as transcription factor binding sites (TFBSs). Through these interactions, TFs and their target genes form gene regulatory networks (GRNs). Knowledge of GRNs is crucial in genomics because it helps predict how plants respond to new or changing environmental conditions (Kulkarni and Vandepoele, 2020). GRNs are commonly represented as networks, logical models, or mathematical frameworks to describe changes in gene expression over time (Wang *et al.*, 2014). They help identify genes involved in specific biological processes and pathways (Krouk *et al.*, 2010). Constructing GRNs usually involves integrating experimental gene interaction data with computationally predicted interactions (Taylor-Teeples *et al.*, 2015; Ikeuchi *et al.*, 2018). Gene regulatory changes are important drivers of plant adaptation and evolution, contributing to phenotypic variation within and across species (Ichihashi *et al.*, 2014; Zheng *et al.*, 2023). Comparative analysis of GRNs is useful because it reveals how gene expression diverges and provides insight into evolutionary mechanisms (Wittkopp, 2007; Romero *et al.*, 2012; Necsulea and Kaessmann, 2014). Such analyses require large-scale gene expression datasets and robust computational tools to reconstruct and compare networks accurately. Gene regulation occurs at many molecular levels, but studies focus on transcription because mRNA data are widely available. GRNs represent highly dynamic systems involving multiple interacting components, making them difficult to model and a major challenge in systems biology (Ahmed and Kabir, 2022). By capturing regulatory interactions, GRNs provide a systems-level view of how gene expression is controlled.

Advances in computational power have enabled the reconstruction of GRNs involved in plant development (Chen *et al.*, 2018; Doroshkov *et al.*, 2019), responses to environmental stresses (Reynoso *et al.*, 2019; Van Moerkercke *et al.*, 2019), and the balance between growth and defense (Li *et al.*, 2020; Sanz-Carbonell *et al.*, 2020). However, the entire complexity of plant regulatory systems is still challenging to capture. Spatial omics has emerged as a powerful approach to address this challenge. Plant tissues contain multiple cell types with unique regulatory programs, and even identical cell types can show different gene expression

*Corresponding author: misra.varucha@gmail.com

© CAB International 2026. *Plant Spatial Biology* (ed. Chen)
DOI: 10.1079/9781836994572.0008

patterns depending on their spatial location. Spatial omics allows precise mapping of these spatial differences, revealing region-specific regulatory modules linked to key processes such as photosynthesis and environmental sensing (Robertson and Wilkins, 2023). These techniques also help identify TFs and regulatory interactions responsible for tissue-specific functions and spatial organization (Balaguer *et al.*, 2017). When combined with temporal data, spatial omics further improves the accuracy of GRN models, leading to better predictions of plant development under stress responses.

This chapter summarizes recent advances in spatial and multi-omics technologies for reconstructing plant GRNs, and emphasizes how integrated approaches enhance our understanding of plant development and environmental adaptation.

8.2 Tools for Understanding Plant Regulatory Networks Through Spatial and Omics Approaches

The tools used to study plant regulatory networks enable researchers to examine gene expression, protein abundance, and regulatory interactions within intact plant tissues. This makes it easier to understand plant development, physiological processes, and responses to environmental stress. These include spatial transcriptomics (ST), single-cell RNA sequencing (scRNA-seq), mass spectrometry imaging (MSI), and spatial multi-omics approaches. Recent developments in scRNA-seq and spatial transcriptomics have significantly transformed plant biology research. Instead of analyzing mixed signals from whole tissues, scientists can now study gene expression at the level of individual cell types and specific tissue locations. This progress has made it possible to identify cellular diversity and rebuild gene regulatory networks within their spatial environment in plant tissues and organs. Furthermore, combining different omics approaches such as transcriptomics, proteomics, metabolomics, and epigenomics provides a more complete picture of how regulatory networks function. For example, integrating spatial transcriptomics with spatial metabolomics (using MSI) allows researchers to link gene expression patterns with metabolite distribution. This approach reveals how genetic regulation and metabolism are coordinated during plant development and under stress conditions. Such multi-omics integration has led to the identification of new metabolic pathways and key regulatory genes that may not be detected using a single techniques approach (Roychowdhury *et al.*, 2023; Jain *et al.*, 2024).

8.2.1 Spatial transcriptomics

ST has emerged as an important complementary technique that allows genome-wide gene expression analysis while preserving the spatial position of transcripts in plant tissues, without the need for protoplast isolation. In the last few years, there have been substantial developments in ST that have overcome the main limitations inherent in single-cell transcriptomics. Unlike conventional bulk and single-cell approaches, which often lose spatial information during cell dissociation, ST retains the physical location of gene expression within tissues, providing a more realistic view of regulatory networks in plants. In traditional transcriptomic studies, gene expression data are often reassigned to tissues using computational clustering or cell sorting approaches, which lack true spatial resolution (Gurazada *et al.*, 2021). However, plant development comprises discrete, tightly regulated, and spatially restricted gene expression patterns across different tissues, leading to diverse phenotypic results. ST therefore provides a new perspective to study spatially dependent regulatory processes during plant growth and development.

Single-molecule fluorescence *in situ* hybridization (smFISH) laid the foundation for spatial RNA detection by enabling direct visualization of individual RNA molecules using fluorescently labeled probes (Femino *et al.*, 1998; Raj *et al.*, 2008). While smFISH provides high spatial precision and sensitivity, its application in plants is limited by the restricted number of distinct fluorophores, tissue autofluorescence, and the complexity of probe design for large plant genomes (Duncan *et al.*, 2016). To overcome these limitations, untargeted ST approaches, such as the approach reported by Ståhl *et al.* (2016), were adapted for plant systems. These technologies utilize barcoded oligo(dT) arrays to capture polyadenylated RNA from cryo-sectioned tissue, followed by spatially indexed reverse transcription and sequencing. Giacomello and coworkers (Giacomello *et al.*, 2017; Giacomello and Lundeberg, 2018) demonstrated the feasibility of this approach in multiple plant species, including *Arabidopsis thaliana*, *Populus tremula*, and *Picea abies*, highlighting its applicability across different tissues and genome sizes. One of the main benefits of ST is its capability to maintain spatial architecture during gene expression profiling, particularly in tissues that are recalcitrant to protoplasting. Plant tissue-specific challenges like thick cell walls, large vacuoles, and high contents of secondary metabolites can be addressed through optimized enzymatic digestion, cryo-sectioning methods, and chemical treatments. However, the spatial resolution of many ST platforms (~55 μm) is often larger than a single plant cell, resulting in mixed signals from multiple cell types within a single capture spot.

To improve the resolution, ST datasets are frequently integrated with scRNA-seq or single-nucleus RNA sequencing (snRNA-seq) using computational deconvolution approaches. This integration enables the estimation of cell type composition within spatial spots (Andersson *et al.*, 2020). Other ST methods include *in situ* sequencing (ISS), digital spatial profiling (DSP), and fluorescent *in situ* sequencing (FISSEQ). ISS uses padlock probes and rolling circle amplification to detect up to 99 genes within tissue sections (Ke *et al.*, 2013). DSP quantifies selected RNAs or proteins in defined tissue regions using affinity-labeled probes (Merritt *et al.*, 2020), while FISSEQ achieves transcriptome-wide mapping through random priming and sequencing-by-ligation (SOLiD) (Lee *et al.*, 2014). Though promising, these techniques face limitations in plants due to genome complexity, high cost, and restricted cross-species applicability. The combined use of scRNA-seq, snRNA-seq, and ST has the potential to transform plant biology by providing integrated spatial and cellular insights into development and environmental responses. Future development will depend on improvements in spatial resolution, probe designs, and advanced data integration strategies (Giacomello, 2021). These advancements enable researchers not only to identify where genes are expressed but also to understand how spatial organization and cell-to-cell interactions shape functional gene networks (Giacomello, 2021; Yu *et al.*, 2022; Chen *et al.*, 2023).

ST technologies are already generating valuable insights into plant growth and molecular regulation. For instance, Moreno-Villena *et al.* (2022) used the 10 × Genomics Visium platform to spatially map genes associated with C4 and crassulacean acid metabolism in *Portulaca*. Zhang *et al.* (2016) combined spatial and temporal transcriptomics to investigate the early developmental stages of tomato. In another study, Liu *et al.* (2022) utilized the 10 × Visium platform to analyze *Phalaenopsis* 'Big Chili'. Although early studies often relied on serial tissue sampling to approximate spatial information, this approach lacked the resolution provided by dedicated ST technologies (Guo *et al.*, 2021). Stereo-seq represents a next-generation ST platform that achieves submicron resolution (~500 nm) using DNA nanoball-patterned array chips combined with *in situ* RNA capture. Its large capture area of 13 × 13 cm ensures high sensitivity and uniform transcript detection (Xia *et al.*, 2022). Recently, Xia *et al.* (2022) successfully applied Stereo-seq to *Arabidopsis* leaf tissue, resolving closely-related cell types such as adaxial and abaxial epidermal layers and identifying spatial gradients in photosynthesis-related gene expression. By preserving tissue morphology, Stereo-seq overcomes limitations of earlier single-cell methods that struggled to differentiate structurally similar plant cells. The scalability of Stereo-seq enables organ-wide and spatiotemporal studies, with the largest chips supporting a capture area of up to 200 mm^2 (Chen *et al.*, 2022). This capability has facilitated the generation of spatiotemporal molecular atlases across diverse organisms, including *Arabidopsis*, *Mus musculus*, *Drosophila melanogaster*, and *Danio rerio* (Chen *et al.*, 2022; Liu *et al.*, 2022; Wang *et al.*, 2022; Xia *et al.*, 2022). Integration of Stereo-seq data with snRNA-seq improves cell type annotation and refines spatial localization of molecular signatures (Xia *et al.*, 2022). Advanced computational tools further strengthen spatial data interpretation. Methods such as SpatialDWLS efficiently filter irrelevant cell types during data integration (Dong and Yuan, 2021), while SPOTlight evaluates the accuracy of spatial cell-type assignments (Elosua-Bayes *et al.*, 2021). These approaches enhance GRN inference by ensuring that spatial analyses reflect biologically relevant cellular signals. Emerging platforms, including STARmap Plus, provide single-cell spatial resolution and allow investigation of how local microenvironments influence gene and intracellular GRNs (Shi *et al.*, 2023).

These technologies reveal spatially defined regulatory modules and illuminate interactions between intracellular signaling and cell-intrinsic regulatory networks. In addition, PHYTOMap (plant hybridization-based targeted observation of gene expression map), a low-cost and transgene-free method, enables multiplexed FISH in whole-mount plant tissues. Building on established plant *in situ* hybridization protocols (Brewer *et al.*, 2006; Hejátko *et al.*, 2006) and neuroscience-inspired spatial methods (Moses and Pachter, 2022), PHYTOMap achieves single-cell resolution without tissue sectioning. The method uses barcoded DNA probes, designed via intramolecular ligation (SNAIL) strategies, followed by *in situ* amplification and sequence by hybridization detection. Each imaging cycle detects four genes with repeated cycles enabling high-throughput spatial profiling across many targets (Gyllborg *et al.*, 2020). Overall, the rapid development of spatial omics technologies, from Stereo-seq to PHYTOMap, is reshaping how gene expression and regulation are studied in plants. By integrating spatial context with molecular depth, these tools overcome long-standing challenges in distinguishing similar cell types and uncover the spatial logic of gene regulatory networks. As analytical frameworks continue to improve, spatial multi-omics will play a central role in decoding the complex, context-dependent regulation underlying plant development, physiology, and stress adaptation (Fig. 8.1).

8.2.2 Single-cell RNA sequencing

Early applications of scRNA-seq were predominantly in root tips of *A. thaliana* because they have a relatively simple cell organization and can be obtained with well-defined marker genes (Birnbaum *et al.*, 2003; Brady *et al.*, 2007). In addition, standard protocols for isolating protoplasts using enzymatic digestion of the cell wall made these studies easier to perform (Birnbaum *et al.*, 2003). For instance, Shulse *et al.* (2019) showed that sucrose changes the composition of root tip cell populations by increasing root hair cells when sucrose is present, while favoring meristematic cells when sucrose is absent. Jean-Baptiste *et al.* (2019) reported that heat stress leads to higher expression of heat shock genes and lower expression of root hair cell-specific genes. Other research includes single-cell profiling of gametophytes in diploid and tetraploid *A. thaliana*, highlighting transcriptional differences linked to ploidy (Song *et al.*, 2020), and analyses of stomatal lineage cells in cotyledons (Liu *et al.*, 2020). Despite its usefulness, scRNA-seq is still difficult to apply widely across plant tissues and species because of rigid cell walls and the slow, stressful process of protoplast generation,

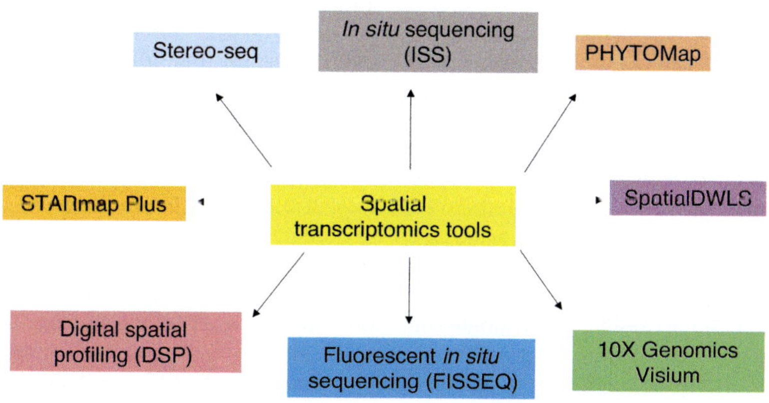

Fig. 8.1. Spatial transcriptomics tools for plant gene regulatory networks.

which can reduce RNA quality and introduce unwanted stress-related signals. To address this issue, snRNA-seq has emerged as a strong alternative because it avoids protoplasting. Conde *et al.* (2021) successfully applied snRNA-seq to *Populus* shoot apices and stems, preserving transcriptomic quality and reducing stress-induced experimental artifacts.

Single-cell transcriptomics is now increasingly used to investigate plant development, differentiation, and environmental responses. For instance, Lopez-Anido *et al.* (2021) mapped differentiation trajectories in leaf tissues, especially within the stomatal lineage, and identified distinct gene expression programs. Chen *et al.* (2022) tracked signaling events and gene expression dynamics to reconstruct developmental pathways. Wendrich *et al.* (2020) connected cytokinin signaling with phosphate stress responses, combining a single-cell atlas with target gene analysis to highlight tissue-specific regulatory modules. Similarly, Graeff *et al.* (2021) used single-cell data to map brassinosteroid signaling, identifying key transcriptional regulators and their tissue-specific roles.

Comparative single-cell profiling of less-well-characterized plant species against *A. thaliana* has improved gene function prediction and expedited building reference cell atlases. But extensive functional testing of target genes is still lacking, and the integration of target gene analysis within cell atlases will probably be an important area of research in the future. Beyond transcriptomics, multi-omics strategies are increasingly used to overcome the limitations of single datasets. Since scRNA-seq typically loses spatial information, spatial transcriptomics has become an important complementary approach. Xia *et al.* (2022) used scStereo-seq to generate a high-resolution spatial cell atlas and reveal spatial gradients of gene expression. Xu *et al.* (2021) combined scRNA-seq with other omics to study maize ear development more comprehensively. Farmer *et al.* (2021) integrated snRNA-seq with assay for transposase accessible chromatin sequencing (ATAC-seq) in roots to enable simultaneous profiling of gene expression and chromatin accessibility at a single-cell resolution. Overall, integrating single-cell transcriptomics with other spatial, epigenomic, and proteomic information reduces key limitations (such as loss of spatial context and redundancy across gene families), providing a more complete understanding of plant cellular regulation.

8.2.3 Spatial multi-omics

Spatial multi-omics provides a major advance for studying plant GRNs because it connects tissue-level organization with molecular regulation. By preserving the spatial relationship between gene expression, TF activity, and chromatin accessibility, this approach provides a clearer view of how plant cells coordinate development and respond to environmental signals. As technologies and computational methods improve, spatial multi-omics is expected to become a core tool in plant systems biology, enabling deeper insights into the structure and behavior of plant GRNs. Traditional GRN prediction methods based on single-cell transcriptomics often fail to reflect the true chromatin accessibility landscape. In many cases, GRN inference is not performed at true single-cell resolution but rather at the level of cell clusters. Additionally, many approaches rely heavily on TF motif databases, which limits their ability to detect cell-type-specific regulation, especially for co-regulators (CRs), the binding motifs of which are poorly described.

To address these issues, SCRIP (Dong *et al.*, 2022) was introduced as a method to infer transcriptional regulator (TR) activity and reconstruct GRNs using scATAC-seq data, supported by large ChIP-seq collections and motif databases (Dong *et al.*, 2022). However, its performance depends strongly on the availability and quality of scATACseq datasets. With ongoing progress in single-cell multi-omics, newer tools have emerged that integrate paired datasets to improve transcriptional regulator activity prediction. For example, FigR (Kartha *et al.*, 2022), GRaNIE (Kamal *et al.*, 2023), DIRECT-NET (Zhang *et al.*, 2022), and GLUE (Cao and Gao, 2022) use multi-ome data and apply linear or nonlinear regression-based strategies for GRN modeling. In addition, tools such as CellOracle (Kamimoto *et al.*, 2023) allow GRN reconstruction using scATAC-seq-derived information, with options for user-defined network design. Concurrently, MICA (Alanis-Lobato *et al.*, 2024) uses bulk ATAC-seq to identify potential regulator binding sites and uses them to improve the interpretation of single-cell transcriptomic patterns (Badia *et al.*, 2023). However, although these innovations are useful, many still depend on motif-dependent methods, which limit accurate prediction of regulator activity when motifs are unknown or poorly annotated. Therefore, better accuracy and stronger cell type specificity are still required.

The advent of spatial multi-omics tools like spatial ATAC-RNA-seq (Zhang *et al.*, 2023), which can profile gene expression and chromatin accessibility simultaneously, provides a major opportunity to estimate TR activity more accurately and infer GRNs directly in spatial tissue contexts. To test and benchmark these approaches in controlled environments, scMultiSim was designed as an adaptive simulation framework for multimodal single-cell datasets. It works in two phases: first, it generates biologically realistic gene expression levels using kinetic models; and second, it adds technical artifacts like batch effects and library preparation to simulate real-world single-cell RNA and ATAC-seq data. scMultiSim represents biological complexity using three major components: cell identity factors (CIFs), gene identity vectors (GIVs), and region identity vectors (RIVs). These factors include information on cellular identity, GRNs structure, chromatin accessibility, and cell–cell interactions (CCI) (Li *et al.*, 2025). In scMultiSim, gene expression dynamics are controlled by kinetic parameters (kon, koff, and s) that are modulated by CIFs, GIVs, and their respective chromatin accessibility. This design enables the simulation of gene activity patterns under multiple regulatory layers. The tool provides two expression simulation options: a detailed kinetic model that captures transcriptional bursts and RNA velocity, and a beta-Poisson model that improves computational efficiency. It also includes an intrinsic noise (σ_i) parameter to evaluate how biological variability influences downstream analytical. By allowing precise control of both biological and technical parameters, scMultiSim is an effective tool for designing and benchmarking multimodal GRN reconstruction workflows.

To further enhance GRN inference from spatial multi-omics, SCRIPro provides an advanced computational framework for analyzing single-cell and spatial datasets. It reduces the sparse signal by applying density-based clustering that captures both transcriptomic similarity and spatial proximity. It then creates SuperCells, which are groups of transcriptionally similar cells that improve data coverage and signal quality. Marker genes are determined for every SuperCell and are used in the LISA framework (Qin *et al.*, 2020) for *in silico* deletion (ISD) analysis (Zhang *et al.*, 2019).

Beyond development-focused research, multi-omics approaches have also improved our understanding of the regulatory mechanisms governing plant–microbe interactions. By jointly studying genes, proteins, and metabolites involved in symbiotic processes, researchers have shown that secondary metabolites and phytohormones can regulate plant defense and stimulate growth upon microbial association (Jain *et al.*, 2024). Advancements in molecular biology and analytical instrumentation have enabled the integrated study of plant–microbe interactions using genomics, metagenomics, transcriptomics, proteomics, and metabolomics (Diwan *et al.*, 2022). These omics technologies help clarify gene structure and function, microbial community structure, gene expression patterns, protein interaction networks, and metabolic pathway regulation (Zulfiqar *et al.*, 2024). Taken together, these findings enhance our understanding of how multi-omics technologies shed light on the intricate dynamics between plants and their microbial partners.

8.2.4 Mass spectrometry imaging

MSI is an advanced analytical technique that enables the spatial mapping of molecules present inside the plant tissues. This technology helps in understanding the plant regulatory networks by showing how molecular distributions correlate with physiological processes, stress responses, and intercellular signaling (Chen *et al.*, 2024b; Vats *et al.*, 2024). This technique supports the study of plant regulatory networks by connecting spatial molecular patterns with physiological processes, stress responses, and intercellular signaling. MSI works by combining the chemical specificity of mass spectrometry with spatial sampling, which produces two-dimensional maps showing the distribution of compounds across a tissue section. It is particularly valuable for studying plant regulatory networks because it can detect and localize key metabolites, phytohormones, and signaling molecules that control growth, development, and stress responses. This capability helps researchers link spatial changes in molecular abundance with regulatory pathways and networks (Chen *et al.*, 2024a). Furthermore, MSI is highly valuable for investigating stress biology because it can capture tissue-level or cell-level metabolic changes during stress conditions such as pathogen infection or drought. In this way, MSI helps elucidate defense mechanisms and adaptive responses at the tissue or cellular level by showing changes in metabolite and hormone distribution patterns (Bhandari *et al.*, 2015). For instance, spatial mapping of antifungal compounds in wheat seed coats has revealed their contribution to passive defense. Similarly, MSI studies of maize roots exposed to mycotoxins have shown how these toxins disrupt hormone signaling pathways and weaken defense systems (Righetti *et al.*, 2022; Chen *et al.*, 2024b).

8.3 Deciphering Stress Response Networks in Plants Through Spatial Omics Approaches

Unravelling plant stress responses requires detailed knowledge of the molecular networks that control gene activity and signaling under adverse environments. Several advanced genomic and molecular biology approaches have contributed to this goal, including high-throughput expressed sequence tag (EST) analysis, large-scale gene expression profiling, mutagenesis (targeted or random), and functional genomics studies based on loss/gain-of-function experiments and mutant complementation (Roychowdhury and Tah, 2013). These approaches generate huge datasets, creating the need for specialized computational algorithms, computer codes, and data repositories for efficient storage, processing, analysis, and dissemination. The 'omics' technologies have modernized biological research by integrating molecular biology with system-level and computational approaches (Deshmukh *et al.*, 2014; Khandagale *et al.*, 2020; Sharma and Gupta, 2022). Omics includes genomics, transcriptomics, proteomics, metabolomics, epigenomics, bioinformatics, proteogenomics, lipidomics, ionomics, interactomics, and phenomics. Altogether, these methods provide high-resolution datsets describing plant behavior under stress conditions (Muthuramalingam *et al.*, 2022; Raza, 2022). Overall, it enables us to understand plant stress tolerance mechanisms and support the development of strategies to reduce the negative impacts of environmental stresses.

However, a single omics layer alone is often insufficient to fully explain the complexity of plant stress responses. Therefore, researchers increasingly adopt multi-omics approaches, where two or more omics datasets are integrated to generate a more complete biological interpretation. Such integrative techniques, such as panomics, integrate genomics, transcriptomics, proteomics, and metabolomics, for stronger crop improvement strategies (Weckwerth *et al.*, 2020). By linking these biological layers, scientists can obtain a systems-level understanding of stress signaling and identify key regulators that drive stress resilience. The availability of large multi-omics datasets, coupled with high-throughput screening tools and central repositories, has supported the development of data integration and visualization tools (Subramanian *et al.*, 2020; Derbyshire *et al.*, 2022). These platforms enable panomics-based analyses through data integration, functional annotation, pathway mapping, data mining, systems biology modeling, and machine learning. As a result, this approach helps identify genes, proteins, and metabolites involved in stress adaptation, as well as biomarkers and molecular targets appropriate for crop improvement. In abiotic stress research, marker-assisted breeding (MAB) can be strengthened when combined with multi-omics analysis. However, compared with classical MAB alone, omics technologies provide broader and deeper systems-level information (Khan *et al.*, 2023). Omics technologies allow researchers to explore complex interactions among genes, proteins, and metabolites, and they can reveal new pathways and regulatory elements that may be missed by conventional breeding methods (Vennapusa *et al.*, 2023). Additionally, omics platforms enable high-throughput screening and selection of the germplasm based on genomic features, gene expression patterns, proteomic signatures, and metabolite content. This allows plant breeders to select genotypes with stronger stress tolerance traits more efficiently (Parida *et al.*, 2024).

The integrative capability of the omics makes it possible to find regulatory networks, candidate genes, and molecular markers for targeted breeding programs. When combined with bioinformatics tools and machine learning methods, multi-omics datasets can also support predictive modeling and precision breeding strategies (Parray *et al.*, 2019). Multi-omics research has also improved the understanding of stress signaling pathways in plants. For instance, integrated studies in *Arabidopsis* and *Oryza* sp. have identified key transcription factors, protein kinases, and metabolites involved in drought and salinity tolerance, highlighting the importance of both abscisic acid (ABA)-dependent and ABA-independent pathways (Roychowdhury *et al.*, 2023; Jain *et al.*, 2024). Such insights are critical for understanding how plants perceive and respond to environmental stress, ultimately supporting the development of stress-resilient crop varieties.

8.4 Challenges and Future Perspectives

Spatial omics tools have greatly improved our ability to map plant regulatory networks by linking gene expression, protein localization, and metabolic activity with the spatial organization of plant tissues. However, several challenges still limit the full use of these technologies in plant research. One major limitation is the technical difficulty related to spatial resolution and the complex structure of plant tissue. Plant cells possess rigid and heterogeneous cell walls, which make enzymatic digestion and protoplast isolation difficult, time-consuming, and highly dependent on plant species and tissue type. If cell wall disruption is incomplete, cell viability decreases and contamination risks increase, which can affect downstream omics data quality (Chen *et al.*, 2023; Rhaman *et al.*, 2024). In addition, protoplasts are highly sensitive to mechanical damage and environmental stress, requiring fast handling to avoid artificial expression changes during processing (Rhaman *et al.*, 2024). Another important issue is that many spatial transcriptomics platforms, such as 10 x Genomics Visium or MERFISH, were initially designed for mammalian tissues and often require plant-specific optimization. During cryosectioning, ice crystal formation can damage plant cells and impair spatial data quality, especially because plant tissues contain high levels of water (Chen *et al.*, 2023). Moreover, differences in tissue fixation, preparation, and sectioning protocols lead to inconsistent results across studies. This problem becomes more severe in plant species with high levels of secondary metabolites or mucilage, which can interfere with RNA integrity and signal detection.

Despite strong progress, there are still issues associated with obtaining high spatial resolution, data integration from different omics datasets, and interpreting the large and complex datasets generated (Yu *et al.*, 2022; Chen *et al.*, 2023). Therefore, there is a growing need for further refinement of user-friendly software, standardized pipelines, and computational workflows specifically designed for plant spatial and multi-omics studies (Roychowdhury *et al.*, 2023). In addition, the field is currently systematic, with high-resolution spatial gene expression atlases, particularly for non-model crop plants. Although extensive atlases in *A. thaliana* provide strong reference data, it remains difficult to generate similar resources for important crops like maize, rice, or sugarcane. Another major difficulty is multi-omics integration across STs, proteomics, metabolomics, and epigenomics. This is challenging because these datasets often differ in spatial resolution, data scale, signal complexity, and measurement units. At the computational level, spatial omics datasets are large and image-rich, requiring advanced image processing, clustering algorithms, and 3D reconstruction methods to precisely determine cell types and identify regulatory elements. Additionally, functional validation of predicted regulatory interactions remains difficult. In many cases, researchers can detect strong correlations, but proving cause-and-effect relationships is limited by the lack of reliable cell-type-specific genetic manipulation tools in many plant systems. Currently, no universally standardized computational workflows exist for processing and integrating plant spatial omics datasets. High noise levels, limited resolution, and sparse signals make regulatory network inference more challenging. Integrating spatial, single-cell, and multi-omics datasets into complete regulatory network models is still computationally demanding and methodologically challenging (Fang *et al.*, 2023).

Looking forward, many of these limitations can be reduced through plant-specific optimization of spatial omics platforms and improved experimental protocols. Advances in tissue permeabilization, sectioning quality, and hybridization will improve compatibility with diverse plant samples. Combining spatial transcriptomics with single-cell omics methods such as single-cell RNA-seq or ATAC-seq will provide a more accurate and dynamic understanding of cell-type-specific regulatory networks. A key future goal is to develop high-resolution plant cell atlases across developmental stages, tissue types, and stress responses. Such atlases will become valuable reference resources for plant systems biology and crop improvement approaches. Artificial intelligence and machine learning will further support cell-type classification, multi-omics integration, and predictive network modeling, enabling deeper insights into gene regulation at the organ and tissue level.

8.5 Conclusion

The convergence of spatial and multi-omics technologies has transformed plant GRNs by providing deeper insights into how gene regulation controls plant development, physiology, and environmental stresses. These advances have addressed major limitations of bulk transcriptomics, which often hide cell type diversity and remove spatial information. Spatial omics approaches, such as laser capture microdissection, *in situ* hybridization, etc., enable researchers to localize gene expression directly within tissues, supporting the discovery of cell type-specific regulatory networks. These spatial technologies have played a pivotal role in identifying tissue-specific GRNs involved in key developmental processes such as root meristem maintenance, floral organ specification, and vascular differentiation. Moreover, spatial omics is crucial for deciphering how plants perceive and respond to environmental gradients, including light, nutrients, and pathogens, in a tissue-dependent manner. Importantly, the stress-driven reprogramming of GRNs under drought, salinity, and temperature extremes is now being studied at much higher cellular resolution. This improves our ability to detect regulatory hubs and key network components that can be targeted for developing stress-resilient crops. In addition, integrating spatial omics with other omics layers helps connect GRNs with physiological traits and metabolic pathways. These integrated approaches have enabled the identification of master regulators, co-expression networks, and signaling modules involved in complex traits, such as flowering time, hormone signaling, and stress tolerance. In crops like rice, maize, and *Arabidopsis*, integrative GRN studies have revealed important regulatory nodes controlling yield, biomass, and resistance traits. Moreover, network modeling, machine learning, and systems biology tools are increasingly applied to predict and validate GRN interactions, improving the reliability and practical value of regulatory networks for crop improvement and plant biology.

References

Ahmed, M.B. and Kabir, A.H. (2022) Understanding of the various aspects of gene regulatory networks related to crop improvement. *Gene* 833, 146556. DOI: 10.1016/j.gene.2022.146556.

Alanis-Lobato, G., Bartlett, T.E., Huang, Q., Simon, C.S., McCarthy, A. *et al.* (2024) MICA: A multi-omics method to predict gene regulatory networks in early human embryos. *Life Science Alliance* 7(1), e202302415. DOI: 10.26508/lsa.202302415.

Andersson, A., Bergenstråhle, J., Asp, M., Bergenstråhle, L., Jurek, A. *et al.* (2020) Single-cell and spatial transcriptomics enables probabilistic inference of cell type topography. *Communications Biology* 3(1), 565. DOI: 10.1038/s42003-020-01247-y.

Badia-I-Mompel, P., Wessels, L., Müller-Dott, S., Trimbour, R., Ramirez Flores, R.O. *et al.* (2023) Gene regulatory network inference in the era of single-cell multi-omics. *Nature Reviews. Genetics* 24(11), 739–754. DOI: 10.1038/s41576-023-00618-5.

Balaguer, M.A., Fisher, A.P., Clark, N.M., Fernandez-Espinosa, M.G., Möller, B.K. *et al.* (2017) Predicting gene regulatory networks by combining spatial and temporal gene expression data in *Arabidopsis* root stem cells. *Proceedings of the National Academy of Sciences* 114(36), E7632–7640. DOI: 10.1073/pnas.1707566114.

Bhandari, D.R., Wang, Q., Friedt, W., Spengler, B., Gottwald, S. *et al.* (2015) High resolution mass spectrometry imaging of plant tissues: Towards a plant metabolite atlas. *The Analyst* 140(22), 7696–7709. DOI: 10.1039/C5AN01065A.

Birnbaum, K., Shasha, D.E., Wang, J.Y., Jung, J.W., Lambert, G.M. *et al.* (2003) A gene expression map of the *Arabidopsis* root. *Science* 302(5652), 1956–1960. DOI: 10.1126/science.1090022.

Brady, S.M., Orlando, D.A., Lee, J.Y., Wang, J.Y., Koch, J. *et al.* (2007) A high resolution root spatiotemporal map reveals dominant expression patterns. *Science* 318(5852), 801–806. DOI: 10.1126/science.1146265.

Brewer, P.B., Heisler, M.G., Hejátko, J., Friml, J. and Benková, E. (2006) *In situ* hybridization for mRNA detection in *Arabidopsis* tissue sections. *Nature Protocols* 1(3), 1462–1467. DOI: 10.1038/nprot.2006.226.

Cao, Z.J. and Gao, G. (2022) Multi-omics single-cell data integration and regulatory inference with graph-linked embedding. *Nature Biotechnology* 40(10), 1458–1466. DOI: 10.1038/s41587-022-01284-4.

Chen, A., Liao, S., Cheng, M., Ma, K., Wu, L. *et al.* (2022) Spatiotemporal transcriptomic atlas of mouse organogenesis using DNA nanoball-patterned arrays. *Cell* 185(10), 1777–1792. DOI: 10.1016/j.cell.2022.04.003.

Chen, C., Ge, Y. and Lu, L. (2023) Opportunities and challenges in the application of single-cell and spatial transcriptomics in plants. *Frontiers in Plant Science* 14, 1185377. DOI: 10.3389/fpls.2023.1185377.

Chen, D., Yan, W., Fu, L.Y. and Kaufmann, K. (2018) Architecture of gene regulatory networks controlling flower development in *Arabidopsis thaliana*. *Nature Communications* 9(1), 4534. DOI: 10.1038/s41467-018-06772-3.

Chen, L., Zhang, Y., Bu, Y., Zhou, J., Man, Y. *et al.* (2024a) Imaging the spatial distribution of structurally diverse plant hormones. *Journal of Experimental Botany* 75(22), 6980–6997. DOI: 10.1093/jxb/erae384.

Chen, Y.J., Zeng, H.S., Jin, H.L. and Wang, H.B. (2024b) Applications of mass spectrometry imaging in botanical research. *Advanced Biotechnology* 2(1), 6. DOI: 10.1007/s44307-024-00014-y.

Conde, D., Triozzi, P.M., Balmant, K.M., Doty, A.L., Miranda, M. *et al.* (2021) A robust method of nuclei isolation for single-cell RNA sequencing of solid tissues from the plant genus *Populus*. *PLoS ONE* 16(5), e0251149. DOI: 10.1101/2021.02.11.430521.

Derbyshire, M.C., Batley, J. and Edwards, D. (2022) Use of multiple 'omics techniques to accelerate the breeding of abiotic stress tolerant crops. *Current Plant Biology* 32, 100262. DOI: 10.1016/j.cpb.2022.100262.

Deshmukh, R., Sonah, H., Patil, G., Chen, W., Prince, S. *et al.* (2014) Integrating omic approaches for abiotic stress tolerance in soybean. *Frontiers in Plant Science* 5, 244. DOI: 10.3389/fpls.2014.00244.

Diwan, D., Rashid, M.M. and Vaishnav, A. (2022) Current understanding of plant-microbe interaction through the lenses of multi-omics approaches and their benefits in sustainable agriculture. *Microbiological Research* 265, 127180. DOI: 10.1016/j.micres.2022.127180.

Dong, R. and Yuan, G.C. (2021) SpatialDWLS: Accurate deconvolution of spatial transcriptomic data. *Genome Biology* 22(1), 145. DOI: 10.1186/s13059-021-02362-7

Dong, X., Tang, K., Xu, Y., Wei, H., Han, T. *et al.* (2022) Single cell gene regulation reference by large scale data integration. *Nucleic Acids Research* 50, e126.

Doroshkov, A.V., Konstantinov, D.K., Afonnikov, D.A. and Gunbin, K.V. (2019) The evolution of gene regulatory networks controlling *Arabidopsis thaliana* L. trichome development. *BMC Plant Biology* 19, 53.

Duncan, S., Olsson, T.S.G., Hartley, M., Dean, C. and Rosa, S. (2016) A method for detecting single mRNA molecules in *Arabidopsis thaliana*. *Plant Methods* 12(1), 13–x. DOI: 10.1186/s13007-016.

Elosua-Bayes, M., Nieto, P., Mereu, E., Gut, I. and Heyn, H. (2021) SPOTlight: Seeded NMF regression to deconvolute spatial transcriptomics spots with single-cell transcriptomes. *Nucleic Acids Research* 49, e50. DOI: 10.1093/nar/gkab043.

Fang, S., Chen, B., Zhang, Y., Sun, H., Liu, L. *et al.* (2023) Computational approaches and challenges in spatial transcriptomics. *Genomics, Proteomics & Bioinformatics* 21(1), 24–47. DOI: 10.1016/j.gpb.2022.10.001.

Farmer, A., Thibivilliers, S., Ryu, K.H., Schiefelbein, J. and Libault, M. (2021) Single-nucleus RNA and ATAC sequencing reveals the impact of chromatin accessibility on gene expression in *Arabidopsis* roots at the single-cell level. *Molecular Plant* 14(3), 372–383. DOI: 10.1016/j.molp.2021.01.001.

Femino, A.M., Fay, F.S., Fogarty, K. and Singer, R.H. (1998) Visualization of single RNA transcripts *in situ*. *Science (New York, N.Y.)* 280(5363), 585–590. DOI: 10.1126/science.280.5363.585.

Giacomello, S. (2021) A new era for plant science: Spatial single-cell transcriptomics. *Current Opinion in Plant Biology* 60, 102041. DOI: 10.1016/j.pbi.2021.102041.

Graeff, M., Rana, S., Wendrich, J.R., Dorier, J., Eekhout, T. *et al.* (2021) A single-cell morpho-transcriptomic map of *Brassinosteroid* action in the *Arabidopsis* root. *Molecular Plant* 14(12), 1985–1999. DOI: 10.1016/j.molp.2021.07.021.

Giacomello, S. and Lundeberg, J. (2018) Preparation of plant tissue to enable spatial transcriptomics profiling using barcoded microarrays. *Nature Protocols* 13(11), 2425–2446. DOI: 10.1038/s41596-018-0046-1.

Giacomello, S., Salmén, F., Terebieniec, B.K., Vickovic, S., Navarro, J.F. *et al.* (2017) Spatially resolved transcriptome profiling in model plant species. *Nature Plants* 3, 17061. DOI: 10.1038/nplants.2017.61.

Guo, X., Liang, J., Lin, R., Zhang, L., Wu, J. *et al.* (2021) Series-spatial transcriptome profiling of leafy head reveals the key transition leaves for head formation in Chinese cabbage. *Frontiers in Plant Science* 12. DOI: 10.3389/fpls.2021.787826.

Gurazada, S.G.R., Cox, K.L., Czymmek, K.J. and Meyers, B.C. (2021) Space: The final frontier - achieving single-cell, spatially resolved transcriptomics in plants. *Emerging Topics in Life Sciences* 5(2), 179–188. DOI: 10.1042/ETLS20200274.

Gyllborg, D.>, Langseth, C.M., Qian, X., Choi, E., Salas, S.M. *et al.* (2020) Hybridization-based *in situ* sequencing (HybISS) for spatially resolved transcriptomics in human and mouse brain tissue. *Nucleic Acids Research* 48(19), e112–e112. DOI: 10.1093/nar/gkaa792.

Hejátko, J., Blilou, I., Brewer, P.B., Friml, J., Scheres, B. *et al.* (2006) *In situ* hybridization technique for mRNA detection in whole mount *Arabidopsis* samples. *Nature Protocols* 1(4), 1939–1946. DOI: 10.1038/nprot.2006.333.

Ichihashi, Y., Aguilar-Martínez, J.A., Farhi, M., Chitwood, D.H., Kumar, R. *et al.* (2014) Evolutionary developmental transcriptomics reveals a gene network module regulating interspecific diversity in plant leaf shape. *Proceedings of the National Academy of Sciences* 111(25), E2616–E2621. DOI: 10.1073/pnas.1402835111.

Ikeuchi, M., Shibata, M., Rymen, B., Iwase, A., Bågman, A.-M. *et al.* (2018) A gene regulatory network for cellular reprogramming in plant regeneration. *Plant and Cell Physiology* 59(4), 770–782. DOI: 10.1093/pcp/pcy013.

Jain, A., Sarsaiya, S., Singh, R., Gong, Q., Wu, Q. *et al.* (2024) Omics approaches in understanding the benefits of plant-microbe interactions. *Frontiers in Microbiology* 15. DOI: 10.3389/fmicb.2024.1391059.

Jean-Baptiste, K., McFaline-Figueroa, J.L., Alexandre, C.M., Dorrity, M.W., Saunders, L. *et al.* (2019) Dynamics of gene expression in single root cells of *Arabidopsis thaliana*. *The Plant Cell* 31(5), 993–1011. DOI: 10.1105/tpc.18.00785.

Kamal, A., Arnold, C., Claringbould, A., Moussa, R., Servaas, N.H. *et al.* (2023) GRaNIE and GRaNPA: Inference and evaluation of enhancer-mediated gene regulatory networks. *Molecular System Biology* 19, e11627. DOI: 10.15252/msb.202311627.

Kamimoto, K., Stringa, B., Hoffmann, C.M., Jindal, K., Solnica-Krezel, L. *et al.* (2023) Dissecting cell identity via network inference and *in silico* gene perturbation. *Nature* 614(7949), 742–751. DOI: 10.1038/s41586-022-05688-9.

Kartha, V.K., Duarte, F.M., Hu, Y., Ma, S., Chew, J.G. *et al.* (2022) Functional inference of gene regulation using single-cell multi-omics. *Cell Genomics* 2(9), 100166. DOI: 10.1016/j.xgen.2022.100166.

Ke, R., Mignardi, M., Pacureanu, A., Svedlund, J., Botling, J. *et al.* (2013) *In situ* sequencing for RNA analysis in preserved tissue and cells. *Nature Methods* 10(9), 857–860. DOI: 10.1038/nmeth.2563.

Khan, M.K.R., Ditta, A., Wang, B., Fang, L., Anwar, Z. *et al.* (2023) The intervention of multi-omics approaches for developing abiotic stress resistance in cotton crop under climate change. In: Prakash, C.S., Fiaz, S., Nadeem, M.A., Baloch, F.S. and Qayyum, A. (eds) *Sustainable Agriculture in the Era of the OMICs Revolution*. Springer, Cham, Switzerland, pp. 37–82.

Khandagale, K., Krishna, R., Roylawar, P., Ade, A.B., Benke, A. *et al.* (2020) Omics approaches in allium research: Progress and way ahead. *Peer J* 8, e9824. DOI: 10.7717/peerj.9824.

Krouk, G., Mirowski, P., LeCun, Y., Shasha, D.E. and Coruzzi, G.M. (2010) Predictive network modeling of the high-resolution dynamic plant transcriptome in response to nitrate. *Genome Biology* 11(12), R123. DOI: 10.1186/gb-2010-11-12-r123.

Kulkarni, S.R. and Vandepoele, K. (2020) Inference of plant gene regulatory networks using data-driven methods: A practical overview. *Biochimica et Biophysica Acta (BBA) - Gene Regulatory Mechanisms* 1863(6), 194447. DOI: 10.1016/j.bbagrm.2019.194447.

Lee, J.H., Daugharthy, E.R., Scheiman, J., Kalhor, R., Yang, J.L. *et al.* (2014) Highly multiplexed subcellular RNA sequencing *in situ*. *Science (New York, N.Y.)* 343(6177), 1360–1363. DOI: 10.1126/science.1250212.

Li, B., Tang, M., Caseys, C., Nelson, A., Zhou, M. *et al.* (2020) Epistatic transcription factor networks differentially modulate *Arabidopsis* growth and defense. *Genetics* 214(2), 529–541. DOI: 10.1534/genetics.119.302996.

Li, H., Zhang, Z., Squires, M., Chen, X. and Zhang, X. (2025) scMultiSim: Simulation of single-cell multi-omics and spatial data guided by gene regulatory networks and cell–cell interactions. *Nature Methods* 22, 982–993.

Liu, C., Leng, J., Li, Y., Ge, T., Li, J. *et al.* (2022) A spatiotemporal atlas of organogenesis in the development of orchid flowers. *Nucleic Acids Research* 50(17), 9724–9737. DOI: 10.1093/nar/gkac773.

Liu, Z., Zhou, Y., Guo, J., Li, J., Tian, Z. *et al.* (2020) Global dynamic molecular profiling of stomatal lineage cell development by single-cell RNA sequencing. *Molecular Plant* 13(8), 1178–1193. DOI: 10.1016/j.molp.2020.06.010.

Lopez-Anido, C.B., Vatén, A., Smoot, N.K., Sharma, N., Guo, V. *et al.* (2021) Single-cell resolution of lineage trajectories in the *Arabidopsis* stomatal lineage and developing leaf. *Developmental Cell* 56(7), 1043–1055. DOI: 10.1016/j.devcel.2021.03.014.

Merritt, C.R., Ong, G.T., Church, S.E., Barker, K., Danaher, P. *et al.* (2020) Multiplex digital spatial profiling of proteins and RNA in fixed tissue. *Nature Biotechnology* 38(5), 586–599. DOI: 10.1038/s41587-020-0472-9.

Moreno-Villena, J.J., Zhou, H., Gilman, I.S., Tausta, S.L., Cheung, C.Y.M. *et al.* (2022) Spatial resolution of an integrated C4+CAM photosynthetic metabolism. *Science Advances* 8(31), eabn2349. DOI: 10.1126/sciadv.abn2349.

Moses, L. and Pachter, L. (2022) Museum of spatial transcriptomics. *Nature Methods* 19(5), 534–546. DOI: 10.1038/s41592-022-01409-2.

Muthuramalingam, P., Jeyasri, R., Rakkammal, K., Satish, L., Shamili, S. *et al.* (2022) Multi-omics and integrative approach towards understanding salinity tolerance in rice: A review. *Biology* 11(7), 1022. DOI: 10.3390/biology11071022.

Necsulea, A. and Kaessmann, H. (2014) Evolutionary dynamics of coding and non-coding transcriptomes. *Nature Reviews. Genetics* 15(11), 734–748. DOI: 10.1038/nrg3802.

Parida, S.K., Mondal, N., Yadav, R., Vishwakarma, H., Yadav, R. *et al.* (2024) Mining legume germplasm for genetic gains: An Indian perspective. *Frontiers in Genetics* 14, 996828. DOI: 10.3389/fgene.2023.996828.

Parray, J.A., Yaseen Mir, M., Shameem, N., Parray, J.A., Yaseen Mir, M. *et al.* (2019) Advancement in sustainable agriculture: Computational and bioinformatics tools. In: Parray, J.A., Yaseen Mir, M. and Shameem, N. (eds) *Sustainable Agriculture: Biotechniques in Plant Biology*. Springer, Singapore, pp. 465–547.

Qin, Q., Fan, J., Zheng, R., Wan, C., Mei, S. *et al.* (2020) Lisa: Inferring transcriptional regulators through integrative modeling of public chromatin accessibility and ChIP-seq data. *Genome Biology* 21(1), 32. DOI: 10.1186/s13059-020-1934-6.

Raj, A., van den Bogaard, P., Rifkin, S.A., Oudenaarden, A. and Tyagi, S. (2008) Imaging individual mRNA molecules using multiple singly labeled probes. *Nature Methods* 5(10), 877–879. DOI: 10.1038/nmeth.1253.

Raza, A. (2022) Plant biotechnological tools: Solutions for raising climate-resilient crop plants. *Modern Phytomorphol* 15, 132–133.

Reynoso, M.A., Kajala, K., Bajic, M., West, D.A., Pauluzzi, G. *et al.* (2019) Evolutionary flexibility in flooding response circuitry in angiosperms. *Science (New York, N.Y.)* 365(6459), 1291–1295. DOI: 10.1126/science.aax8862.

Rhaman, M.S., Ali, M., Ye, W. and Li, B. (2024) Opportunities and challenges in advancing plant research with single-cell omics. *Genomics, Proteomics & Bioinformatics* 22(2), qzae026. DOI: 10.1093/gpbjnl/qzae026.

Righetti, L., Gottwald, S., Tortorella, S., Spengler, B. and Bhandari, D.R. (2022) Mass spectrometry imaging disclosed spatial distribution of defense-related metabolites in *Triticum* spp. *Metabolites* 12(1), 48. DOI: 10.3390/metabo12010048.

Robertson, S.M. and Wilkins, O. (2023) Spatially resolved gene regulatory networks in Asian rice (*Oryza sativa* cv. Nipponbare) leaves. *Plant Journal* 116(1), 269–281.

Romero, L.C., Aroca, M.Á., Laureano-Marín, A.M., Moreno, I., García, I. *et al.* (2012) Cysteine and cysteine-related signaling pathways in *Arabidopsis thaliana*. *Molecular Plant* 7(2), 264–276. DOI: 10.1093/mp/sst168.

Roychowdhury, R. and Tah, J. (2013) Mutagenesis—a potential approach for crop improvement. In: Hakeem, K.R., Ahmad, P. and Ozturk, M. (eds) *Crop Improvement—New Approaches and Modern Techniques*, 1st ed. Springer, Boston, MA, USA, pp. 149–187.

Roychowdhury, R., Das, S.P., Gupta, A., Parihar, P., Chandrasekhar, K. *et al.* (2023) Multi-omics pipeline and omics-integration approach to decipher plant's abiotic stress tolerance responses. *Genes* 14(6), 1281. DOI: 10.3390/genes14061281.

Sanz-Carbonell, A., Marques, M.C., Martinez, G. and Gomez, G. (2020) Dynamic architecture and regulatory implications of the miRNA network underlying the response to stress in melon. *RNA Biology* 17(2), 292–308. DOI: 10.1080/15476286.2019.1697487.

Sharma, T. and Gupta, A. (2022) Integrated OMICS approaches to ameliorate the abiotic stress in *Brassica napus*. In: Roy, S., Mathur, P., Chakraborty, A.P. and Saha, S.P. (eds) *Plant Stress: Challenges and Management in the New Decade*. Springer-Nature, Basel, Switzerland, pp. 361–373. DOI: 10.1007/978-3-030-95365-2_23.

Shi, H., He, Y., Zhou, Y., Huang, J., Maher, K. *et al.* (2023) Spatial atlas of the mouse central nervous system at molecular resolution. *Nature* 622(7983), 552–561. DOI: 10.1038/s41586-023-06569-5.

Shulse, C.N., Cole, B.J., Ciobanu, D., Malley, R.C.O., Brady, S.M. *et al.* (2019) Profiling of plant cell types resource profiling of plant cell types. *Cell Reports* 27, 2241–2247. DOI: 10.1016/j.celrep.2019.04.054.

Song, Q., Ando, A., Jiang, N., Ikeda, Y. and Chen, Z.J. (2020) Single-cell RNA-seq analysis reveals ploidy-dependent and cell-specific transcriptome changes in *Arabidopsis* female gametophytes. *Genome Biology* 21(1), 178. DOI: 10.1186/s13059-020-02094-0.

Ståhl, P.L., Salmén, F., Vickovic, S., Lundmark, A., Navarro, J.F. *et al.* (2016) Visualization and analysis of gene expression in tissue sections by spatial transcriptomics. *Science (New York, N.Y.)* 353(6294), 78–82. DOI: 10.1126/science.aaf2403.

Subramanian, I., Verma, S., Kumar, S., Jere, A. and Anamika, K. (2020) Multi-omics data integration, interpretation, and its application. *Bioinformatics and Biology Insights* 14, 1177932219899051. DOI: 10.1177/1177932219899051.

Taylor-Teeples, M., Lin, L., Lucas, M.D., Turco, G., Toal, T.W. *et al.* (2015) An *Arabidopsis* gene regulatory network for secondary cell wall synthesis. *Nature* 517, 571–575. DOI: 10.1038/nature14099.

Van Moerkercke, A., Duncan, O., Zander, M., Šimura, J., Broda, M. *et al.* (2019) A MYC2/MYC3/MYC4-dependent transcription factor network regulates water spray-responsive gene expression and Jasmonate levels. *Proceedings of the National Academy of Sciences of the United States of America* 116(46), 23345–23356. DOI: 10.1073/pnas.1911758116.

Vats, M., Cillero-Pastor, B., Cuypers, E. and Heeren, R.M.A. (2024) Mass spectrometry imaging in plants, microbes, and food: A review. *The Analyst* 149(18), 4553–4582. DOI: 10.1039/D4AN00644E.

Vennapusa, A.R., Nimmakayala, P., Zaman-Allah, M.A. and Ratnakumar, P. (2023) Editorial: Physiological, molecular and genetic perspectives of environmental stress response in plants. *Frontiers in Plant Science* 14, 1213762. DOI: 10.3389/fpls.2023.1213762.

Wang, J., Yu, H., Weng, X., Xie, W., Xu, C. *et al.* (2014) An expression quantitative trait loci-guided co-expression analysis for constructing regulatory network using a rice recombinant inbred line population. *Journal of Experimental Botany* 65(4), 1069–1079. DOI: 10.1093/jxb/ert464.

Wang, M., Hu, Q., Lv, T., Wang, Y., Lan, Q. *et al.* (2022) High-resolution 3D spatiotemporal transcriptomic maps of developing drosophila embryos and larvae. *Developmental Cell* 57(10), 1271–1283. DOI: 10.1016/j.devcel.2022.04.006.

Weckwerth, W., Ghatak, A., Bellaire, A., Chaturvedi, P. and Varshney, R.K. (2020) PANOMICS meets germplasm. *Plant Biotechnology Journal* 18(7), 1507–1525. DOI: 10.1111/pbi.13372.

Wendrich, J.R., Yang, B., Vandamme, N.S., Verstaen, K., Smet, W. *et al.* (2020) Vascular transcription factors guide plant epidermal responses to limiting phosphate conditions. *Science* 370, 6518. DOI: 10.1126/science.aay4970.

Wittkopp, P.J. (2007) Variable gene expression in eukaryotes: A network perspective. *The Journal of Experimental Biology* 210(Pt 9), 1567–1575. DOI: 10.1242/jeb.002592.

Xia, K., Sun, H.X., Li, J., Li, J., Zhao, Y. *et al.* (2022) The single-cell stereo-seq reveals region-specific cell subtypes and transcriptome profiling in *Arabidopsis* leaves. *Developmental Cell* 57(10), 1299–1310. DOI: 10.1016/j.devcel.2022.04.011.

Xu, X., Crow, M., Rice, B.R., Li, F., Harris, B. *et al.* (2021) Single-cell RNA sequencing of developing maize ears facilitates functional analysis and trait candidate gene discovery. *Developmental Cell* 56(4), 557–568. DOI: 10.1016/j.devcel.2020.12.015.

Yu, X., Liu, Z. and Sun, X. (2022) Single-cell and spatial multi-omics in the plant sciences: Technical advances, applications, and perspectives. *Plant Communications* 4(3), 100508. DOI: 10.1016/j.xplc.2022.100508.

Zhang, T.-Q., Xu, Z.-G., Shang, G.-D. and Wang, J.-W. (2019) A single-cell RNA sequencing profiles the developmental landscape of *Arabidopsis* root. *Molecular Plant* 12(5), 648–660. DOI: 10.1016/j.molp.2019.04.004.

Zhang, B., Srivastava, A., Mimitou, E., Stuart, T., Raimondi, I. *et al.* (2022) Characterizing cellular heterogeneity in chromatin state with scCUT&Tag-pro. *Nature Biotechnology* 40(8), 1220–1230. DOI: 10.1038/s41587-022-01250-0.

Zhang, D., Deng, Y., Kukanja, P., Agirre, E., Bartosovic, M. *et al.* (2023) Spatial epigenome–transcriptome co-profiling of mammalian tissues. *Nature* 616(7955), 113–122. DOI: 10.1038/s41586-023-05795-1.

Zhang, S., Xu, M., Qiu, Z., Wang, K., Du, Y. *et al.* (2016) Spatiotemporal transcriptome provides insights into early fruit development of tomato (*Solanum lycopersicum*). *Scientific Reports* 6(1), 23173. DOI: 10.1038/srep23173.

Zheng, H., Fu, X., Shao, J., Tang, Y., Yu, M. *et al.* (2023) Transcriptional regulatory network of high-value active ingredients in medicinal plants. *Trends in Plant Science* 28(4), 429–446. DOI: 10.1016/j.tplants.2022.12.007.

Zulfiqar, B., Raza, M.A.S., Saleem, M.F., Ali, B., Aslam, M.U. *et al.* (2024) Abscisic acid improves drought resilience, growth, physio-biochemical and quality attributes in wheat (*Triticum aestivum* L.) at critical growth stages. *Scientific Reports* 14(1), 20411. DOI: 10.1038/s41598-024-71404-4.

9 Spatial and Multi-omics Approaches for Exploring Plant Evolutionary and Developmental Biology

Shilpa Malakar[1], Varucha Misra[2*] and A.K. Mall[2]
[1]Department of Microbiology, Kalinga University, Raipur, India; [2]Division of Crop Improvement, ICAR-Indian Sugarcane Research Institute, Lucknow, India

9.1 Introduction

Plants are the root of life on the planet. Understanding plant evolutionary–developmental biology (evo–devo) – the interplay between evolutionary forces and developmental pathways – is essential for addressing major global challenges, including population growth, biodiversity loss, and climate change (Irish and Benfey, 2004). For example, crop domestication through artificial selection has reshaped plant morpho-anatomy, anatomy, and physiology over the course of years (Begna, 2021). However, modern agriculture now faces unprecedented challenges: by 2050, global food production may need to nearly double to feed an estimated 9.7 billion people, while farming systems must also adapt to climate instability and widespread soil degradation (Berners-Lee *et al.*, 2018). In this context, evo–devo research can accelerate the development of stress-tolerant crops by identifying conserved genetic mechanisms that regulate traits such as drought resistance or disease resistance (Feng *et al.*, 2022). From an ecological perspective, plant evo–devo also explains how lineages adapt to different environments. For example, the repeated evolution of C4 photosynthesis in grasses enabled colonization of dry, high-light environments (Sage *et al.*, 2012). These evolutionary adaptations are highly relevant for predicting how plant species may shift their geographic ranges under future climates. Climate resilience further depends on understanding plant–microbe interactions. Mycorrhizal symbioses, which originated >450 million years ago, improve nutrient acquisition and stress resistance in plants (Cairney, 2000). Translating such mechanisms into agriculture could reduce dependence on synthetic fertilizers and lower the environmental footprint of food production. So, plant evo–devo provides a critical link between fundamental biology and long-term sustainability. Classical plant biology has traditionally relied on reductionist approaches, where complex traits are examined through isolated components such as individual genes or proteins (Brinton and Uauy, 2019). Although this strategy has identified main plant regulators like auxins in root development, it often fails to capture interactions across biological scales (Saini *et al.*, 2013). One major limitation arises from pleiotropy, where a single gene influences multiple traits (Hughes and Leips, 2017). For instance, in flowering plants, the flowering locus T (*FT*) gene controls the flowering time but also

*Corresponding author: misra.varucha@gmail.com

© CAB International 2026. *Plant Spatial Biology* (ed. Chen)
DOI: 10.1079/9781836994572.0009

influences stress response and stomatal development (Almeida de Jesus *et al.*, 2022). Such multi-trait control complicates trait engineering because altering *FT* may cause unintended trade-offs. In addition, classical techniques often lack sufficient spatiotemporal resolution (Moeinizade *et al.*, 2020). Bulk RNA sequencing averages gene expression across many cells, masking cell-type-specific expression patterns that are critical for understanding processes such as meristem function or vascular differentiation (Steen *et al.*, 2020). Finally, reductionist strategies are often inadequate for evolutionary questions because comparing homologous genes alone does not fully capture the roles of non-coding regulatory sequences, chromatin remodeling, and metabolic innovation in diversification (Cooper, 2024). As an example, the evolution of crassulacean acid metabolism (CAM) in *Agave* involved concurrent changes in gene expression, enzyme kinetics, and leaf anatomy, a complexity that is difficult to resolve using single-omics research (Heyduk, 2022).

The emergence of high-throughput omics technologies, including genomics, transcriptomics, proteomics, metabolomics, and epigenomics, has transformed plant biology by enabling systems-based analysis (Misra *et al.*, 2018). Genomics provided the foundation for this transformation through major processes, for example, with projects like the 1001 Genomes Initiative that inventoried *Arabidopsis thaliana* diversity, which catalogued natural genetic variation (Weigel and Mott, 2009). Today, multiple omics approaches provide a more comprehensive link between DNA variation and phenotype by capturing how genetic information interacts with regulatory networks and metabolic pathways (Hanson *et al.*, 2018). A particularly important advancement is the rise of spatial omics, which connects molecular information with tissue context. Spatial transcriptomics platforms (e.g.10 x Visium) gene expression directly within tissue sections, whereas matrix-assisted laser desorption/ionization (MALDI) imaging mass spectrometry reveals metabolite distribution maps (Ståhl *et al.*, 2016). For instance, spatial omics in maize roots treated under salt stress identified localized accumulation of osmoprotectants like proline in the endodermis, highlighting tissue-specific targets for engineering salt tolerance (Zhang *et al.*, 2021a). Large-scale integration platforms, such as the Plant Cell Atlas (PCA), further reflect the shift towards multi-omics frameworks (Fahlgren *et al.*, 2023). PCA integrates transcriptomic, proteomic, and imaging datasets to model plant cell states in 4D and supports the identification of developmental checkpoints (Kim *et al.*, 2018). Such frameworks are increasingly necessary for dissecting complex traits like nitrogen-use efficiency or shade avoidance.

This chapter discusses how spatial and multi-omics technologies are addressing long-standing questions in plant evo–devo by bridging biological scales from molecules to organs and ecosystems. For example, Kranz anatomy in C4 plants enhances carbon fixation in hot and dry environments (Lundgren *et al.*, 2014). Earlier research implicated developmental regulators such as *SCARECROW* in the development of the bundle sheath (BS) tissues; however, the formation of spatial expression gradients and their metabolic coordination remained unclear (Hughes and Langdale, 2020). Recent multi-omics studies in a C4 model grass, *Setaria viridis*, have begun to resolve these mechanisms by integrating spatial multi-omics to understand the evolution of the Kranz anatomy (Lammi and Qu, 2024). The spatial analysis confirmed that the higher expression of GOLDEN2-LIKE (*GLK*) transcription factors in mesophyll cells consitent with their roles in chloroplast biogenesis (Bravo *et al.*, 2009), whereas the bundle sheath cells showed increased expression of the *CARBONIC ANHYDRASE*, supporting CO_2 concentration mechanisms (DiMario *et al.*, 2022). Metabolomic profiling also revealed spatial patterns of C4 acids (such as malate and aspartate) that aligned with the known localization of enzymes (Blätke and Bräutigam, 2019). The epigenomic analysis further identified lineage-specific enhancer elements, supporting the concept of convergent regulatory evolution (Javierre *et al.*, 2016). This systems-level perspective showed how the spatial organization of gene regulation and metabolism co-evolved to optimize photosynthetic efficiency. Integrative multi-omics approaches also help explain domestication-driven trait evolution. For example, multi-omics analysis coupled with physiological phenotyping has linked *PROG1* gene mutations (genomics) with changes in auxin gradients (transcriptomics) and lignin deposition (metabolomics), helping to explain alterations in tiller architecture during crop domestication (Sanches *et al.*, 2024). These tools provide a powerful framework to connect molecular regulation with spatial and temporal data, opening new opportunities to study plant evolution and improve crop performance.

9.2 The Shift to Multi-dimensional Data

Plant biology has evolved from single-omics to multi-dimensional models that better explain the structural and functional complexity of living things. Although genomics or transcriptomics, when used independently, provide useful molecular snapshots, they often fail to explain how DNA variation leads to phenotypic traits or how dynamic interactions among transcripts, proteins, and metabolites shape biological outcomes (Cavill *et al.*, 2016). These gaps across multi-omics layers can be addressed through multi-omics integration, which enables researchers to connect regulatory networks (e.g. transcription factor–metabolite crosstalk) and to interpret the cellular processes in their tissue or environmental conditions (Kang *et al.*, 2022). The functions of biological systems, from cells to whole organisms, rely on complex interactions, including gene-to-gene, transcript-to-transcript, protein-to-protein, metabolite-to-metabolite, and environmental-cue-to-environmental-cue relationships, where changes in one layer affect the others within a finely regulated yet flexible network (Bunnik and Le Roch, 2013). Conventional single-omics strategies, which examine molecular layers in isolation, provide only a partial and often incomplete view of this complexity (Baysoy *et al.*, 2023). For example, while genomics can identify genetic variants associated with traits, it cannot fully describe how these variants interact with epigenetic regulation to control the expression of genes, or how downstream proteins and metabolites ultimately generate observable phenotypes (Eichler, 2019). Multi-omics integration, or the concomitant use of several molecular layers for analysis, has become essential for understanding biological complexity, especially in plants, where developmental plasticity and environmental adaptability add further regulatory depth (Picard *et al.*, 2021). Plants serve as models of biological complexity because their sessile nature requires constant adjustment to fluctuating environments. A single stress, such as a drought, initiates cascaded responses at the molecular level. Alterations to DNA methylation (epigenomics) modulate stress-repressed gene activity, transcription factors (transcriptomics) activate drought-responsive pathways, and enzymes (proteomics) drive metabolic reprogramming to produce osmoprotectants such as proline. Studying these layers separately can obscure the system-wide nature of these stress adaptations (Talarico *et al.*, 2024). For instance, the transcriptional up-regulation of the *OsNAC6* gene in rice is well documented under drought conditions (Lee *et al.*, 2017). However, its exact contribution to stress tolerance depends not only on transcript abundance but also on post-translational regulation and interaction with key metabolites such as abscisic acid (ABA) (Ramazi and Zahiri, 2021). Multi-omics approaches have shown that ABA may not only induce *OsNAC6* expression but also enhance stability or activity of the Ots protein product, while simultaneously activating downstream genes involved in proline biosynthesis (Ma *et al.*, 2020). Such cross-layer dependencies are often missed in single-omics pipelines, highlighting the integrated multi-omics frameworks for deciphering plant stress biology.

9.2.1 The need for multi-omics integration to capture biological complexity

Biological systems are highly dynamic and interconnected, and therefore, multi-omics integration is essential to fully understand complex biological processes. Plant development progresses over time, and molecular events occurring at one stage often shape responses at later stages. For example, vernalization, or the cold-induced flowering in winter wheat involves coordinated regulation across multiple omics layers: it begins with chromatin level control (epigenomics) that silences the floral repressor *VRN2* (Yan *et al.*, 2004) followed by transcriptional activation (transcriptomics) of its promoter *VRN1*, and is further supported by metabolic adjustments (metabolomics) that enable successful flowering (Wang and Deng, 2022). Such temporal dependencies, such as how stress-induced transcriptomic changes reshape the proteome hours later and ultimately alter metabolite pools, can be effectively captured using time series multi-omics approaches. In one heat-stress study in *Brachypodium distachyon*, early induction of heat-shock proteins (proteomics) occurred before the accumulation of protective polyamines (metabolomics), and both responses were influenced by non-coding RNAs (transcriptomics) (Chen and Li, 2017). However, integrating multi-omics datasets is challenging because different technologies generate data with unequal resolution and coverage. For example, transcriptomics often provides high throughput and broader coverage, whereas single-cell proteomics may have lower depth and incomplete detection, making cross-layer alignment difficult (Gao *et al.*, 2021). Additional barriers include appropriate normalization across platforms, high computational requirements, and the difficulty of distinguishing causal relationships from correlations. To address these challenges, computational frameworks such as weighted gene co-expression network analysis (WGCNA) (Langfelder and Horvath, 2008) and multi-omics factor analysis (MOFA) can be applied to identify co-varying genes, proteins, and metabolite modules (Argelaguet *et al.*, 2018). For instance, in a drought study in sorghum, WGCNA reported co-expressed gene clusters (transcriptomics) associated with metabolomic signatures, including pathways related to lignin biosynthesis, and highlighted key hub genes that could be targeted for genetically designing drought-resilient cultivars (Ngara *et al.*, 2021). Advances in multi-omics integration strengthen the foundation of systems biology and support emerging applications in precision agriculture (Fig. 9.1).

Engineered pathways may likely fail because of unforeseen interactions of introduced genes with endogenous regulatory networks. Multi-omics approaches help detect such failures early by revealing off-target metabolic shifts, unintended pathway re-routing, or competition for cellular resources. For example, when transcriptomics and metabolomics were integrated in tomato studies (Jia *et al.*, 2022), researchers found that overexpression of *Delila* (an anthocyanin biosynthesis gene) altered phenylpropanoid fluxes, which decreased lignin production and consequently weakened stem mechanical strength (Sunil and Shetty, 2022). Similarly, multi-omics-guided breeding strategies in rice have optimized nitrogen-use efficiency by connecting root transcriptome signatures with soil metabolomic profiles, enabling more effective nutrient acquisition and utilization (Fiaz *et al.*, 2021).

9.2.2 Principles of spatial omics

Spatial omics is an emerging set of technologies that integrates spatial context with molecular profiling, enabling researchers to map the biomolecules (e.g. RNA, proteins, metabolites) within intact tissues. Unlike bulk omics approaches, where tissues are homogenized and spatial information is lost, spatial omics preserves tissue architecture and reveals how molecular processes vary in localized microenvironments (Dezem *et al.*, 2024). This is particularly important in plant biology because developmental gradients (root tips, meristem) and stress responses (pathogen invasion, drought) are strongly linked by spatial organization. Spatial omics platforms achieve this by combining molecular measurement with location-specific tagging using approaches, such as bar-coded oligonucleotide arrays, *in situ* sequencing, and imaging mass spectrometry, which link molecular signals to precise regions within tissues (Ren *et al.*, 2024). In this way, spatial omics fills the gap between molecular signatures and where they are located, providing new insights into how plants coordinate growth, adaptation, and evolution in 3D tissue space. Spatial transcriptomics combines spatial mapping with high-throughput RNA sequencing, enabling researchers to visualize gene expression landscapes within intact plant tissues (Bressan *et al.*, 2023). One widely used platform is 'Visium' from 10x Genomics, which employs spatially bar-coded oligonucleotide probes on glass slides (Wang *et al.*, 2023). When a tissue section is placed on the slide, mRNA molecules bind to these bar-coded probes, allowing sequencing reads to be assigned back to their original spatial coordinates (Ståhl *et al.*, 2016). This approach

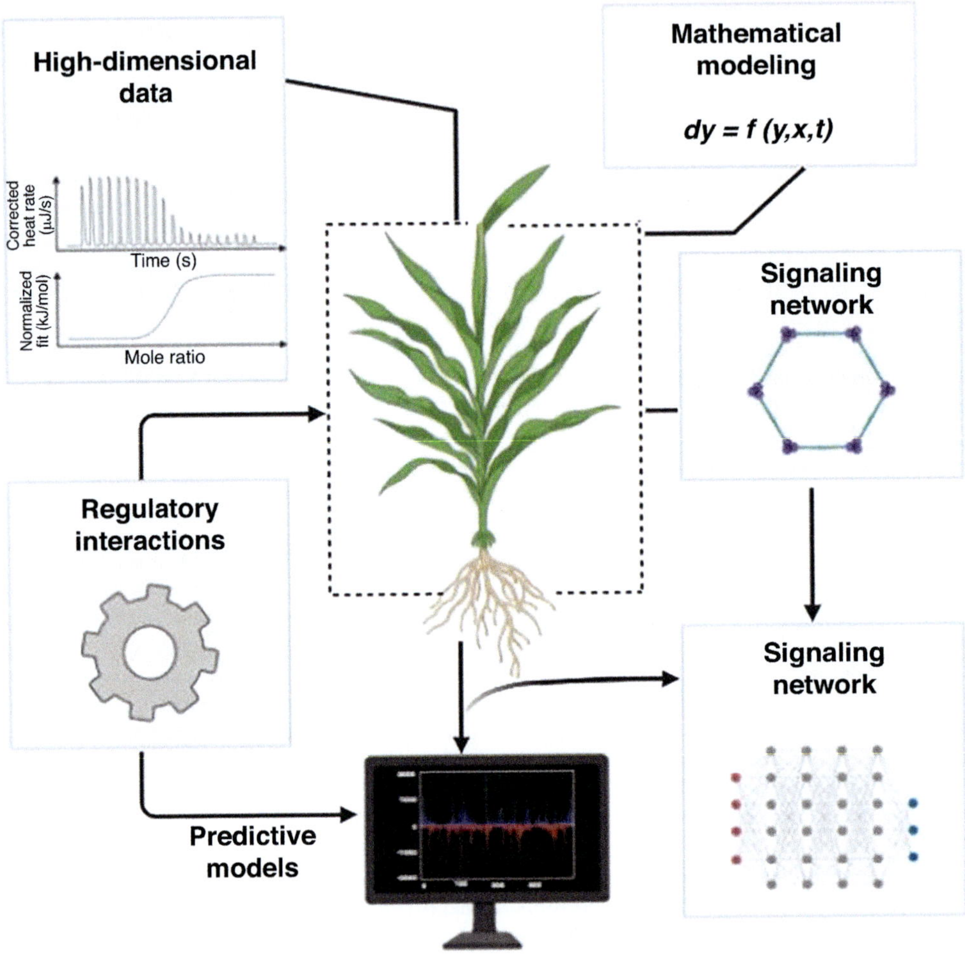

Fig. 9.1. The use of systems biology for studying plants by combining data from different omics, modelling, and network analyses to solve difficult plant-related processes. Genes, metabolic processes, signals from the environment, and the traits of an organism all connect, guided by advanced tools and statistics. The arrows show the pathway from data obtained from experiments to predictive modelling, showing how systems biology allows for a complete understanding of plant growth and development. (Created by Biorender.)

has been particularly valuable for studying plant developmental gradients (Fig. 9.2). For instance, spatial transcriptomics profiling of *A. thaliana* root tips identified distinct expression zones of auxin signaling genes, which corresponded to specific stages of cell differentiation and tissue patterning (Shahan *et al.*, 2022).

The study demonstrated that the spatial distribution and coordinated activity of PIN-FORMED (PIN) auxin transporters generate local auxin maxima, which ultimately promote root elongation and pattern formation (Krecek *et al.*, 2009). Another advanced technique, multiplexed error-robust fluorescence *in situ* hybridization (MERFISH), applies the sequential fluorescent probes to identify hundreds of RNA transcripts simultaneously within intact tissues (Moffitt and Zhuang, 2016). In maize, MERFISH has been applied to map drought-responsive genes throughout leaf vasculature and mesophyll, revealing spatially restricted activation of *NAC* transcription factors in vascular bundles

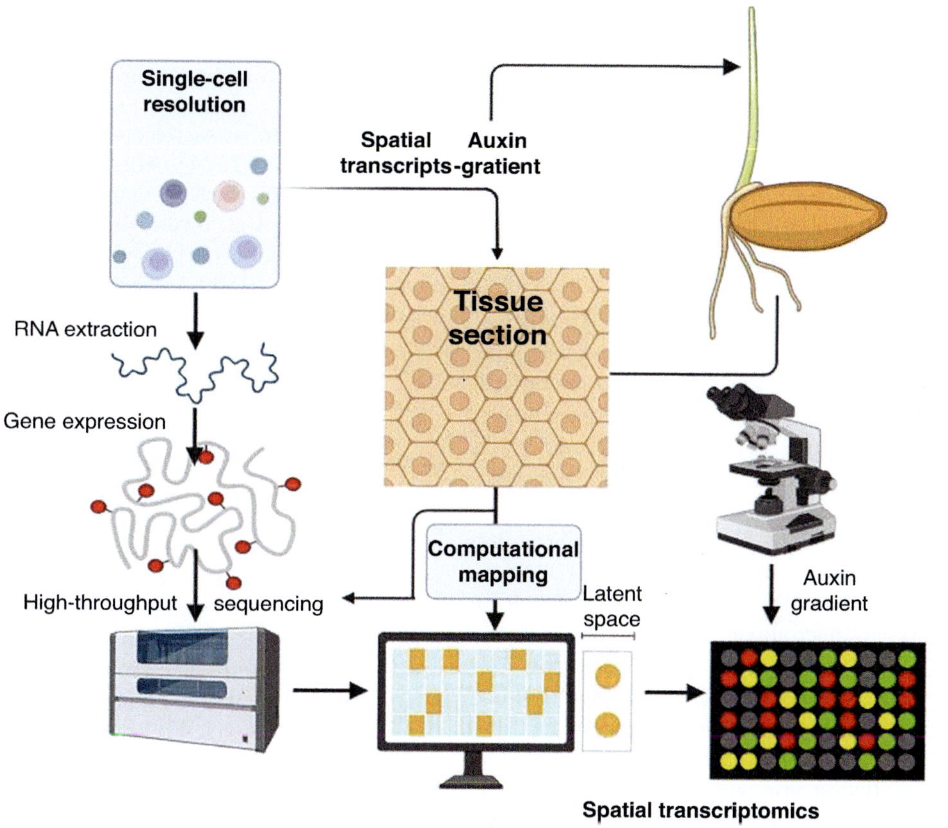

Fig. 9.2. Various spatial omics technologies used for examining leaf and root development at the individual cell level. Key methods, including scRNA-seq, MERFISH, Slide-seq, and imaging mass spectrometry, are highlighted for studying the distribution of auxin and the expression of related genes in plants. They help pinpoint the location and type of cells involved in development, allowing for detailed study. Each step in modern spatial omics is explained with arrows, highlighting the role in exploring plant development and the arrangement of tissues. (Created by Biorender.)

under water-deficit stress conditions (Joshi *et al.*, 2016). Overall, these findings highlighted that spatial transcriptomics enables researchers to examine how individual cell populations respond to stress within their native tissue context.

Imaging-based spatial omics approaches, including matrix-assisted laser desorption/ionization (MALDI) imaging (Chen *et al.*, 2024) and laser micro-dissection sequencing (LMD-seq), offer complementary insight because metabolites and transcripts can be directly mapped onto tissue sections (Kerk *et al.*, 2003). MALDI imaging ionizes molecules from target regions of tissue and produces spatial metabolomic profiles. A pioneering application in plant biology involved mapping alkaloid distributions in *Catharanthus roseus* leaves (Dugé de *et al.*, 2020). MALDI imaging revealed that vindoline and catharanthine were accumulated in distinct cell types, with biosynthesis occurring in epidermal and laticifer cells, respectively (Tang *et al.*, 2022). This spatial separation was essential for understanding the cellular organization of the biosynthesis pathway underlying these anticancer compounds. LMD-seq combines laser micro-dissection with RNA sequencing, enabling transcriptomic profiling of specific cell types or microregions (Kerk *et al.*, 2003). In rice, LMD-seq was used to isolate stomatal guard cells from surrounding epidermal tissue under high CO_2 conditions, revealing cell type expression of *CARBONIC ANHYDRASE1* (CA1) associated with stomatal closure responses (Luo *et al.*, 2012). Such imaging-based techniques are also valuable for studying plant–microbe interactions. As an example, LMD-seq showed that *Piriformospora indica* colonizes the barley root cortical cells and induces localized expression of defense-related genes (Varma *et al.*, 1999).

Despite their utility, these approaches often require specialized instrumentation, and their throughput can be limited compared with newer high-throughput spatial transcriptomic platforms. Beyond conventional plant omics, recent advances in single-cell and near sub-cellular spatial technologies are enabling molecular profiling at much finer sub-cellular levels (Bonev *et al.*, 2024). Methods such as Slide-seqV2 and DBiT-seq utilize DNA-barcoded beads or microfluidic channels to achieve near-cellular mapping of gene expression. For instance, Slide-seqV2 was adapted for profiling the maize shoot apical meristem, resolving transcriptional disparity among stem cell niches and differentiating primordia (Duan *et al.*, 2023). A spatially limited *WUSCHEL*-expressing niche that sustains the meristematic activity has been identified, offering mechanistic insight into the shoot development (Lopes *et al.*, 2021). Another important advance is spatially-resolved chromatin accessibility profiling, including single-cell ATAC sequencing (scATAC-seq). scATAC-seq in *Arabidopsis* showed that root endodermal cells exhibit open chromatin regions near *SCARECROW* loci, consistent with its role in endodermal identity and suberin-related processess (Zhang *et al.*, 2025). These technologies also revealed the investigations of evolutionary conservation. For example, spatial single-cell proteomics in *Selaginella moellendorffii* compared stomatal protein localization with that of flowering plants, revealing the conserved signaling modules despite 400 million years of divergence (Liberali and Schier, 2024). However, applying these methodologies at the scale of whole plant organs remains computationally expensive, and continued advances in data integration algorithms will be required.

9.3 Data Integration Frameworks

Data integration frameworks are computational strategies that integrate heterogeneous multi-omics datasets to extract biological meaning (Abdelaziz *et al.*, 2024). These frameworks address challenges such as data dimensionality, noise, and biological variability by applying statistical models, network-based analyses, and machine learning (ML) to reveal correlations between cross-omics associations and regulatory relationships. Widely-used tools include WGCNA and MOFA, which reduce complexity while preserving biological context (Morabito *et al.*, 2025). Such frameworks support hypothesis generation regarding regulatory mechanisms, evolutionary adaptations, and genotype–phenotype relationships in plants. WGCNA constructs co-expression networks by identifying correlations among gene expression profiles across conditions or samples (Botía *et al.*, 2017). For instance, it has been applied to identify hub genes controlling drought responses in *Arabidopsis*, integrating transcriptomics and metabolomics datasets (Kim *et al.*, 2003). A key strength of WGCNA is its ability to organize genes into modules with shared expression patterns and associate these modules with traits or environmental variables. However, WGCNA primarily captures pair-wise relationships, which may limit its performance when dealing with highly complex multi-dimensional datasets. In contrast, MOFA applies factor analysis to decompose multi-omics datasets into latent factors that represent shared variability across multiple omics layers. This unsupervised integration approach is especially useful for analyzing time-series or spatial omic datasets (Argelaguet *et al.*, 2018). For example, MOFA has helped uncover conserved transcription and metabolism programs associated with leaf development in maize and rice (Chang *et al.*, 2019). Whereas WGCNA is data-driven, MOFA is well-suited for the exploratory discovery of multi-layer structure. Nevertheless, both approaches may face scalability limitations on large datasets and require rigorous statistical validation to prevent over-fitting. Continued improvements in cloud computing infrastructure and algorithm optimization, and scalable statistical pipelines are helping overcome these limitations, making integrative frameworks increasingly indispensable in plant multi-omics research.

9.3.1 Machine learning for cross-omics pattern recognition

ML has significantly improved cross-omics pattern recognition by enabling the automated discovery of non-linear relationships and predictive biomarkers across complex datasets. In supervised ML, labeled multi-omics datasets are used to train models such as random forests and neural networks to predict phenotypes and classify biological states (Picard *et al.*, 2021). In contrast, unsupervised methods such as clustering and dimensionality reduction (e.g. t-SNE, UMAP) uncover latent patterns in unlabeled data, including cell-type-specific regulatory and metabolic programs in plant single-cell datasets (Derkarabetian *et al.*, 2019). Deep learning architectures such as convolutional neural networks (CNNs) are highly effective for analyzing imaging-based spatial omics datasets. These models can extract spatial features from tissue images and help gene expression gradients, for example, within the root meristem (Koo and Ploenzke, 2021). Despite their strong performance, ML approaches often require large and well-annotated datasets, and this approach in plant studies may face a higher risk of overfitting due to strong biological variability across genotypes, environments, and developmental stages. To address these challenges, hybrid methods that incorporate previous biological knowledge into ML models are gaining attention (e.g. graph neural networks can be structured around known or predicted gene regulatory networks) and are the new trending, promising approaches for improving interpretability and biological relevance (Liu *et al.*, 2022). As the plant omics dataset increases, ML-based integration is expected to play a central role in decoding developmental programs and evolutionary adaptation. Connecting genotype to phenotype requires integrating multi-omics data to connect mechanistic regulation with observed traits. For example, the integration of genome-wide association studies (GWAS) with metabolomics can identify genetic variants associated with metabolic pathways, whereas time-series multi-omics can reveal dynamic molecular changes that occur during processes such as domestication and stress responses (Adamski, 2012). These strategies strengthen our understanding of how genetic diversity produces phenotypic diversity in plants.

9.3.2 Synthesizing GWAS and metabolomics to detect adaptive traits

The combination of GWAS and metabolomics has become an effective strategy for identifying the genetic basis of metabolic traits that support plant adaptation and resilience (Smith *et al.*, 2022). GWAS detects the genomic loci related to the phenotypic variation by examining single-nucleotide polymorphisms (SNPs) across diverse populations, whereas metabolomics provides direct measurement of biochemical phenotypes, such as metabolite abundance and pathway activity. When combined, GWAS and metabolomics enable researchers to link SNP variation to metabolite profiles, thereby uncovering gene metabolic relationships and metabolic regulatory networks (Chang *et al.*, 2022). For example, the identification of a *GRAIN* YIELD 1 locus in rice with flavonoid-related metabolic variation provides insight into how this locus may contribute to drought tolerance (Zhang *et al.*, 2021b). By associating SNPs with leaf metabolite profiles under drought stress, scientists identified alleles involved in enhancing antioxidant production, supporting stress protection mechanisms (Kang *et al.*, 2019). On the same note, in the context of tomatoes (Fig. 9.3), GWAS–metabolomics integration identified regulatory variation associated with alkaloid synthesis, including transcription factors that are responsible for pest resistance (Szymański *et al.*, 2020).

This approach also supports crop improvement. In wheat, GWAS-aided metabolomics identified starch biosynthesis genes underlying grain quality, thereby supporting marker-assisted breeding. Key challenges include false positive associations and the need for high-resolution metabolomics platforms. However, progress in genome annotation and pathway databases (e.g. KEGG, PlantCyc) strengthens causal inference and improves biological interpretation (Mao *et al.*, 2005). As metabolomics becomes more

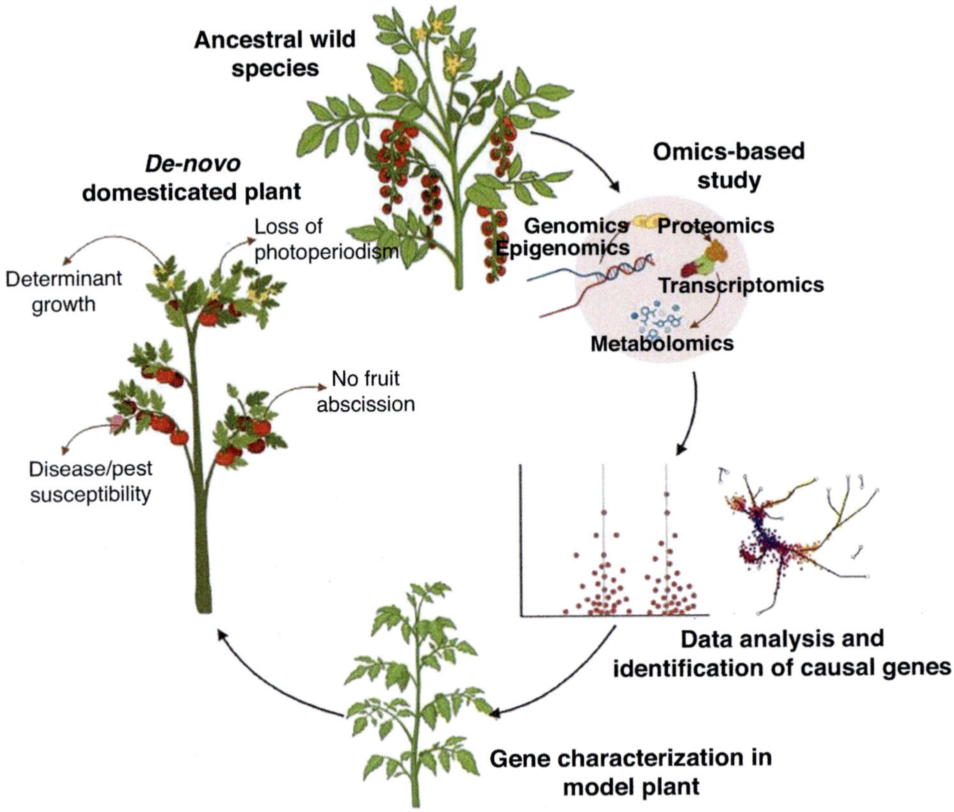

Fig. 9.3. Omics technologies and GWAS are integrated to investigate changes in the traits of tomato plants. Wild ancestral species are used to study the characteristics during domestication, including the loss of sensitivity to daylight and patterns of diseases. Metabolomics and GWAS studies identify genes that link to traits, which are further investigated in model plants. Using this approach leads to the development of domestic tomato varieties that are both healthier and easier to grow. (Created by Biorender.)

affordable, GWAS–metabolomics integration will accelerate the discovery of adaptive traits in non-model plant species at a higher pace.

9.3.3 Time-series multi-omics in studying domestication

Time-series multi-omics breaks down the molecular dynamics of domestication by profiling genomic, transcriptomic, and metabolic changes across development stages. Selective sweeps and regulatory rewiring are part of domestication, and they can be uncovered through the analysis of longitudinal datasets (Jendoubi, 2021). For instance, a study on maize compared gene expression and metabolite profiles of wild teosinte and modern maize at stem elongation, flowering, and grain-filling stages. This integrated analysis identified that selection on *tb1* (teosinte branched1) influenced hormone signaling networks, leading to lowered branching and apical dominance (Xu *et al.*, 2019). In the same way, in domesticated barley, time-series transcriptomics and proteomics revealed that circadian clock genes were co-opted for flowering synchronization under cultivation environments (Haas *et al.*, 2019). Furthermore, metabolomic time course analyses emphasized how artificial selection increased amino acid accumulation in grains, thereby improving nutritional content. In these studies, clustering algorithms (for example, k-means) are used to cluster co-varying genes and metabolites, and enrichment analyses are then performed to connect these modules to domestication-associated traits (Botía *et al.*, 2017). Time-series multi-omics not only reconstructs domestication history but also directs *de novo* domestication of wild species to enhance climate resilience.

9.4 Phylogenomic Studies Using Multi-omics to Deduce Plant Phylogeny

Plant evolutionary relationships have been investigated for decades; however, complex phylogenetic questions remain due to insufficient datasets or conflicting signals from individual molecular markers. Traditional phylogenetic techniques relied heavily on either single-gene sequences (such as ribosomal RNA) or morphological characteristics, which often produced inconsistent results due to convergent evolution (Lee and Palci, 2015), incomplete lineage sorting, or horizontal gene transfer (HGT). Multi-omics provides a comprehensive framework to address these challenges. For instance, whole-genome sequencing has enabled phylogenetic reconstruction at an unprecedented resolution, facilitating the identification of cryptic species and clarifying relationships within controversial clades such as the Brassicaceae family (Xu *et al.*, 2024). By incorporating transcriptomic data, researchers can also identify conserved regulatory networks controlling shared developmental pathways to development (e.g. flower or root formation) that may not be obvious from genomic sequence comparisons alone.

A milestone case is the reconstruction of the angiosperm 'tree of life' by the One Thousand Plant Transcriptomes Initiative (1KP), which integrated genomic and transcriptomic information from more than 1000 species and resolved long-standing uncertainties regarding flower evolution (Guo *et al.*, 2023). This initiative also highlighted the role of whole-genome duplications (WGDs) in diversification, including ancient polyploidy events that shaped the genomes of grasses and crucifers (MacKintosh and Ferrier, 2017). In addition, epigenomic data can strengthen phylogenomic inference by revealing epigenetic modifications (for example, DNA methylation) that occur without changes in the underlying DNA sequence but can regulate gene expression through repression and/or activation (Arora and Tollefsbol, 2021). For example, comparative methylome studies in *Oryzae* (rice) species revealed that epigenetic variation contributes to ecological adaptation and reproductive barriers, providing mechanistic insight into speciation (Gallo-Franco *et al.*, 2022). Although less commonly incorporated, metabolomics adds functional context by relating genetic variation to biochemical innovations such as the evolution of secondary metabolites, including the glucosinolates in the *Brassica* species, which contribute to defense and niche specialization (Liu *et al.*, 2024). Despite these advances, challenges remain in reconciling discordant signals derived from different omics layers. For instance, HGT, which occurs in plants due to interaction with microbes, may bring in foreign genes that distort phylogenetic signals (Aubin *et al.*, 2021). Multi-omics methods help resolve such complexities through cross-validation datasets. The genomic signatures of HGT can be supported by transcriptomic data showing the functional expression of transferred genes. Increasingly, ML algorithms are also being applied to integrate heterogeneous datasets, for example, combining gene trees with species trees to resolve phylogenetic incongruencies (Swenson and El-Mabrouk, 2012).

9.5 Multi-omics Dissection of Symbiosis (e.g. Mycorrhizae, Rhizobia)

Symbiotic associations between plants and microorganisms (e.g. N-fixing rhizobia and arbuscular mycorrhizal fungi (AMF)-acquiring nutrients that enhance nutrient acquisition) are foundational to terrestrial ecosystems and global food production (Xu and Wang, 2023). These partnerships involve complex molecular conversations, in which plants and microorganisms exchange signals to establish mutualistic partnerships. Multi-omics approaches, including metagenomics, transcriptomics, proteomics, and metabolomics, have become essential for dissecting the dynamic relationships among genes, proteins, and metabolites that drive symbiosis (Sanches *et al.*, 2024). For instance, in the rhizobia–legume symbiosis, plant roots secrete flavonoids that attract compatible rhizobia, which in turn synthesize Nod factors that initiate root nodule formation (Porter *et al.*, 2024). Key genes such as *Nodulation Signaling Pathway 1* (*NSP1*) in plants and *nodABC* in rhizobia have been identified using genomics and transcriptomics (Smit *et al.*, 2005). Metabolomic profiling has further shown that root exudates such as strigolactones function as chemo-attractants for AMF, while proteomics has identified effector proteins secreted by colonizing fungi that suppress plant immunity (Salem *et al.*, 2022). A multi-omics analysis of the model legume *Medicago truncatula* and its symbiont *Sinorhizobium meliloti* showed that shifts in the plant transcriptome coincide with metabolic reprogramming in the symbiotic bacterium (Marx *et al.*, 2016). During nodulation, the plant increases expression of genes linked to auxin signaling and cell division, whereas rhizobia enhance the activity of enzymes that convert atmospheric nitrogen into plant-useable forms. Studies of nodule proteomes have also revealed that they often exchange carbon resources with microorganisms in exchange for ammonium (Sin *et al.*, 2024). Similarly, integrated omics studies in mycorrhizal symbiosis indicate that AMF require fatty acids and obtain them through modulation of the host lipid metabolism (Vijayakumar *et al.*, 2016). These findings provide opportunities to enhance the plant use of symbiosis, potentially reducing reliance on synthetic fertilizers.

9.6 Challenges

A major challenge in investigating symbiosis arises from variation in where and when interactions take place. For instance, root nodules and mycorrhizal arbuscules are transient interactions, making it necessary to collect omics samples at multiple time points over short durations (Mony *et al.*, 2024). Using spatial transcriptomics, researchers can generate a detailed map of how symbiosis-related genes are expressed across various regions of the root. Metagenomics also provides a key advantage by enabling the identification of microbial companions, specifically endophytic bacteria that may coexist with AMFs. For instance, fungi residing in tomato roots facilitated the bacterial growth, and these bacteria subsequently stimulated plant growth (Wei *et al.*, 2024). Integrating multi-omics data layers can complicate analysis and limit biological interpretation. The primary challenge lies in synchronizing genomic, transcriptomic, proteomic, and metabolomic datasets due to the data types differing in noise levels, measurement depth, and statistical properties. In addition, transcript abundance does not correlate strongly with protein abundance, which complicates the interpretation of how genetic regulation translates into cellular function. Spatial mapping of molecular activity must also be aligned with changes occurring over time and within environmental contexts, requiring integration with genomic and proteomic responses (Kausar and Falcao, 2019). These discrepancies demand robust normalization and standardization procedures; however, many current tools still struggle to integrate such datasets without introducing bias or misinterpreting biological mechanisms. Another challenge is that biological complexity can obscure key signals due to variability among cell types. In bulk datasets, signals from distinct cell populations may become mixed, whereas single-cell profiling can help resolve cell-type-specific responses more clearly (Trapnell, 2015). Furthermore, multi-omics studies often identify associations rather than causation. For example, if metabolite peaks are observed alongside increased gene expression in drought-stressed roots, the metabolite change may reflect indirect regulation or environmental influence rather than direct genetic control, requiring additional experiments to confirm causality (Kern *et al.*, 2021). Moreover, ML methods can behave as 'black box' models, limiting interpretability and making it difficult to explain how predicted networks lead to phenotypes (Loyola-Gonzalez, 2019). Additionally, the large volume of data generated creates computational challenges, especially in species lacking high-quality reference genomes. Therefore, combining systems biology, bioinformatics, and traditional experimentation remains essential for extracting reliable conclusions about plant evolution and development from large-scale datasets.

9.7 Conclusion

Spatial and multi-omics technologies have substantially advanced plant biology by providing detailed insights into how different molecular layers operate within living systems. These approaches enable the study of development, adaptation, and conservation from single cells up to whole organisms. Spatial omics has revealed how different plant tissues coordinate control gene expression and metabolites in C4 photosynthesis, which explains how Kranz anatomy evolved to enhance carbon fixation under stressful environments. In addition, multi-omics studies have clarified how rhizobia and legumes communicate during symbiosis, guiding strategies to reduce dependence on chemical fertilizers. Multi-omics has also strengthened phylogenetic inference, as demonstrated by the One Thousand Plant Transcriptomes Initiative, which used genomic and epigenomic data to reconstruct angiosperm evolution. While data generation is critical, biological insight depends strongly on effective integration across omics layers. Both WGCNA and MOFA support this integration by helping resolve pleiotropy, detect signatures of HGT, and identify co-regulated molecular modules. ML further strengthens these efforts by enabling the prediction of adaptive traits through approaches such as GWAS metabolomics integration. Nevertheless, major challenges remain, including limited spatiotemporal resolution and the lack of explainability in some algorithms. Addressing these issues will require continued interdisciplinary collaboration among bioinformatics, systems biologists, and experimental researchers to translate multi-omics data into biological knowledge. Future developments, including the integration of single-cell spatial omics, CRISPR-linked omics approaches, and AI-driven phenomics, are expected to deepen our understanding of plant plasticity. Open and collaborative initiatives such as the Plant Cell Atlas represent important progress by working toward 4D cellular mapping and accelerating advances in plant engineering. Ultimately, understanding plant responses to drought, disease, and nutrient limitation will support the development of climate-resistant crops and the protection of biodiversity.

References

Abdelaziz, E.H., Ismail, R., Mabrouk, M.S. and Amin, E. (2024) Multi-omics data integration and analysis pipeline for precision medicine: Systematic review. *Computational Biology and Chemistry* 113, 108254. DOI: 10.1016/j.compbiolchem.2024.108254.

Adamski, J. (2012) Genome-wide association studies with metabolomics. *Genome Medicine* 4(4), 34. DOI: 10.1186/gm333.

Almeida de Jesus, D., Batista, D.M., Monteiro, E.F., Salzman, S., Carvalho, L.M. *et al.* (2022) Structural changes and adaptative evolutionary constraints in FLOWERING LOCUS T and TERMINAL FLOWER1-like genes of FLOWERING plants. *Frontiers in Genetics* 13, 954015. DOI: 10.3389/fgene.2022.954015.

Argelaguet, R., Velten, B., Arnol, D., Dietrich, S., Zenz, T. *et al.* (2018) Multi-omics factor analysis-a framework for unsupervised integration of multi-omics data sets. *Molecular Systems Biology* 14(6), e8124. DOI: 10.15252/msb.20178124.

Arora, I. and Tollefsbol, T.O. (2021) Computational methods and next-generation sequencing approaches to analyze epigenetics data: Profiling of methods and applications. *Methods* 187, 92–103. DOI: 10.1016/j.ymeth.2020.09.008.

Aubin, E., El Baidouri, M. and Panaud, O. (2021) Horizontal gene transfers in plants. *Life* 11(8), 857. DOI: 10.3390/life11080857.

Baysoy, A., Bai, Z., Satija, R. and Fan, R. (2023) The technological landscape and applications of single-cell multi-omics. *Nature Reviews. Molecular Cell Biology* 24, 695–713. DOI: 10.1038/s41580-023-00615-w.

Begna, T. (2021) Effects of crop evolution under domestication and narrowing genetic bases of crop species. *Open Journal of Plant Science* 6, 049–054. DOI: 10.17352/ojps.000032.

Berners-Lee, M., Kennelly, C., Watson, R. and Hewitt, C.N. (2018) Current global food production is sufficient to meet human nutritional needs in 2050 provided there is radical societal adaptation. *Elementa* 6, 1–14. DOI: 10.1525/elementa.310.

Blätke, M.A. and Bräutigam, A. (2019) Evolution of C4 photosynthesis predicted by constraint based modelling. *eLife* 8, e49305. DOI: 10.7554/eLife.4930.

Bonev, B., Castelo-Branco, G., Chen, F., Codeluppi, S., Corces, M.R. *et al.* (2024) Opportunities and challenges of single-cell and spatially resolved genomics methods for neuroscience discovery. *Nature Neuroscience* 27(12), 2292–2309. DOI: 10.1038/s41593-024-01806-0.

Botía, J.A., Vandrovcova, J., Forabosco, P., Guelfi, S., D'Sa, K. *et al.* (2017) An additional k-means clustering step improves the biological features of WGCNA gene co-expression networks. *BMC Systems Biology* 11(1), 47. DOI: 10.1186/s12918-017-0420-6.

Bravo-Garcia, A., Yasumura, Y. and Langdale, J.A. (2009) Specialization of the golden2-like regulatory pathway during land plant evolution. *The New Phytologist* 183(1), 133–141. DOI: 10.1111/j.1469-8137.2009.02829.x.

Bressan, D., Battistoni, G. and Hannon, G.J. (2023) The dawn of spatial omics. *Science* 381(6657), eabq4964. DOI: 10.1126/science.abq4964.

Brinton, J. and Uauy, C. (2019) A reductionist approach to dissecting grain weight and yield in wheat. *Journal of Integrative Plant Biology* 61, 337–358. DOI: 10.1111/jipb.12741.

Bunnik, E.M. and Le Roch, K.G. (2013) An introduction to functional genomics and systems biology. *Advances in Wound Care* 2(9), 490–498. DOI: 10.1089/wound.2012.0379.

Cairney, J.W. (2000) Evolution of mycorrhiza systems. *Die Naturwissenschaften* 87(11), 467–475. DOI: 10.1007/s001140050762.

Cavill, R., Jennen, D., Kleinjans, J. and Briedé, J.J. (2016) Transcriptomic and metabolomic data integration. *Briefings in Bioinformatics* 17(5), 891–901. DOI: 10.1093/bib/bbv090.

Chang, L., Zhou, G., Ou, H. and Xia, J. (2022) MGWAS-explorer: Linking SNPs, genes, metabolites, and diseases for functional insights. *Metabolites* 12(6), 526. DOI: 10.3390/metabo12060526.

Chang, Y.-M., Lin, H.-H., Liu, W.-Y., Yu, C.-P., Chen, H.-J. *et al.* (2019) Comparative transcriptomics method to infer gene coexpression networks and its applications to maize and rice leaf transcriptomes. *Proceedings of the National Academy of Sciences of the United States of America* 116(8), 3091–3099. DOI: 10.1073/pnas.1817621116.

Chen, S. and Li, H. (2017) Heat stress regulates the expression of genes at transcriptional and post-transcriptional levels, revealed by RNA-seq in *Brachypodium distachyon*. *Frontiers in Plant Science* 7, 2067.

Chen, L.C., Lee, C. and Hsu, C.C. (2024) Towards developing a matrix-assisted laser desorption/ionization mass spectrometry imaging (MALDI MSI) compatible tissue expansion protocol. *Analytica Chimica Acta* 1297, 342345. DOI: 10.1016/j.aca.2024.342345.

Cooper, K.L. (2024) The case against simplistic genetic explanations of evolution. *Development (Cambridge, England)* (20), dev203077. DOI: 10.1242/dev.203077.

Derkarabetian, S., Castillo, S., Koo, P.K., Ovchinnikov, S. and Hedin, M. (2019) A demonstration of unsupervised machine learning in species delimitation. *Molecular Phylogenetics and Evolution* 139, 106562. DOI: 10.1016/j.ympev.2019.106562.

Dezem, F.S., Arjumand, W., DuBose, H., Morosini, N.S. and Plummer, J. (2024) Spatially resolved single-cell omics: Methods, challenges, and future perspectives. *Annual Review of Biomedical Data Science* 7(1), 131–153. DOI: 10.1146/annurev-biodatasci-102523-103640.

DiMario, R.J., Giuliani, R., Ubierna, N., Slack, A.D., Cousins, A.B. *et al.* (2022) Lack of leaf carbonic anhydrase activity eliminates the c_4 carbon-concentrating mechanism requiring direct diffusion of CO_2 into bundle sheath cells. *Plant, Cell & Environment* 45(5), 1382–1397. DOI: 10.1111/pce.14291.

Duan, H., Cheng, T. and Cheng, H. (2023) Spatially resolved transcriptomics: Advances and applications. *Blood Science* 5(1), 1–14. DOI: 10.1097/BS9.0000000000000141.

Dugé de Bernonville, T., Maury, S., Delaunay, A., Daviaud, C., Chaparro, C. *et al.* (2020) Developmental methylome of the medicinal plant *Catharanthus roseus* unravels the tissue-specific control of the monoterpene indole alkaloid pathway by DNA methylation. *International Journal of Molecular Sciences* 21(17), 6028. DOI: 10.3390/ijms21176028.

Eichler, E.E. (2019) Genetic variation, comparative genomics, and the diagnosis of disease. *New England Journal of Medicine* 381(1), 64–74. DOI: 10.1056/NEJMra1809315.

Fahlgren, N., Kapoor, M., Yordanova, G., Papatheodorou, I., Waese, J. *et al.* (2023) Toward a data infrastructure for the Plant Cell Atlas. *Plant Physiology* 191(1), 35–46.

Feng, L., Dong, T., Jiang, P., Yang, Z., Dong, A. *et al.* (2022) An eco-evo-devo genetic network model of stress response. *Horticulture Research* 9, uhac135. DOI: 10.1093/hr/uhac135.

Fiaz, S., Wang, X., Khan, S.A., Ahmar, S., Noor, M.A. *et al.* (2021) Novel plant breeding techniques to advance nitrogen use efficiency in rice: A review. *GM Crops & Food* 12(2), 627–646. DOI: 10.1080/21645698.2021.1921545.

Gallo-Franco, J.J., Ghneim-Herrera, T., Tobar-Tosse, F., Romero, M., Chaura, J. *et al.* (2022) Whole-genome DNA methylation patterns of *Oryza sativa* (L.) and *Oryza glumaepatula* (steud) genotypes associated with aluminum response. *Plant Direct* 6(8), e430. DOI: 10.1002/pld3.430.

Gao, M., Ling, M., Tang, X., Wang, S., Xiao, X. *et al.* (2021) Comparison of high-throughput single-cell RNA sequencing data processing pipelines. *Briefings in Bioinformatics* 22(3), bbaa116. DOI: 10.1093/bib/bbaa116.

Guo, C., Luo, Y., Gao, L.-M., Yi, T.-S., Li, H.-T. *et al.* (2023) Phylogenomics and the flowering plant tree of life. *Journal of Integrative Plant Biology* 65(2), 299–323. DOI: 10.1111/jipb.13415.

Haas, M., Schreiber, M. and Mascher, M. (2019) Domestication and crop evolution of wheat and barley: Genes, genomics, and future directions. *Journal of Integrative Plant Biology* 61(3), 204–225. DOI: 10.1111/jipb.12737.

Hanson, C., Cairns, J., Wang, L. and Sinha, S. (2018) Principled multi-omic analysis reveals gene regulatory mechanisms of phenotype variation. *Genome Research* 28, 1207–1216. DOI: 10.1101/gr.227066.117.

Heyduk, K. (2022) Evolution of crassulacean acid metabolism in response to the environment: Past, present, and future. *Plant Physiology* 190, 19–30. DOI: 10.1093/plphys/kiac303.

Hughes, K.A. and Leips, J. (2017) Pleiotropy, constraint, and modularity in the evolution of life histories: Insights from genomic analyses. *Annals of the New York Academy of Sciences* 1389(1), 76–91. DOI: 10.1111/nyas.13256.

Hughes, T.E. and Langdale, J.A. (2020) SCARECROW gene function is required for photosynthetic development in maize. *Plant Direct* 4, e00264. DOI: 10.1002/pld3.264.

Irish, V.F. and Benfey, P.N. (2004) Beyond *Arabidopsis*. Translational biology meets evolutionary developmental biology. *Plant Physiology* 135, 611–614. DOI: 10.1104/pp.104.041632.

Javierre, B.M., Burren, O.S., Wilder, S.P., Kreuzhuber, R., Hill, S.M. *et al.* (2016) Lineage-specific genome architecture links enhancers and non-coding disease variants to target gene promoters. *Cell* 167(5), 1369–1384. DOI: 10.1016/j.cell.2016.09.037.

Jendoubi, T. (2021) Approaches to integrating metabolomics and multi-omics data: A primer. *Metabolites* 11(3), 184. DOI: 10.3390/metabo11030184.

Jia, C., Guo, B., Wang, B., Li, X., Yang, T. *et al.* (2022) Integrated metabolomic and transcriptomic analysis reveals the role of phenylpropanoid biosynthesis pathway in tomato roots during salt stress. *Frontiers in Plant Science* 13, 1023696. DOI: 10.3389/fpls.2022.1023696.

Joshi, R., Wani, S.H., Singh, B., Bohra, A., Dar, Z.A. *et al.* (2016) Transcription factors and plants response to drought stress: Current understanding and future directions. *Frontiers in Plant Science* 7, 1029.

Kang, M., Ko, E. and Mersha, T.B. (2022) A roadmap for multi-omics data integration using deep learning. *Briefings in Bioinformatics* 23(1), bbab454. DOI: 10.1093/bib/bbab454.

Kang, Z., Babar, M.A., Khan, N., Guo, J., Khan, J. *et al.* (2019) Comparative metabolomic profiling in the roots and leaves in contrasting genotypes reveals complex mechanisms involved in post-anthesis drought tolerance in wheat. *PLoS ONE* 14(3), e0213502. DOI: 10.1371/journal.pone.0213502.

Kausar, S. and Falcao, A.O. (2019) A visual approach for analysis and inference of molecular activity spaces. *Journal of Cheminformatics* 11(1), 63. DOI: 10.1186/s13321-019-0386-z.

Kerk, N.M., Ceserani, T., Tausta, S.L., Sussex, I.M. and Nelson, T.M. (2003) Laser capture microdissection of cells from plant tissues. *Plant Physiology* 132(1), 27–35. DOI: 10.1104/pp.102.018127.

Kern, A.F., Yang, G.X., Khosla, N.M., Ang, R.M.L., Snyder, M.P. *et al.* (2021) Divergent patterns of selection on metabolite levels and gene expression. *BMC Ecology and Evolution* 21(1), 185. DOI: 10.1186/s12862-021-01915-5.

Kim, K.-N., Cheong, Y.H., Grant, J.J., Pandey, G.K. and Luan, S. (2003) *CIPK3*, a calcium sensor–associated protein kinase that regulates abscisic acid and cold signal transduction in *Arabidopsis*. *The Plant Cell* 15(2), 411–423. DOI: 10.1105/tpc.006858.

Kim, S., Kang, D., Huo, Z., Park, Y. and Tseng, G.C. (2018) Meta-analytic principal component analysis in integrative omics application. *Bioinformatics* 34(8), 1321–1328. DOI: 10.1093/bioinformatics/btx765.

Koo, P.K. and Ploenzke, M. (2021) Improving representations of genomic sequence motifs in convolutional networks with exponential activations. *Nature Machine Intelligence* 3(3), 258–266. DOI: 10.1038/s42256-020-00291-x.

Krecek, P., Skupa, P., Libus, J., Naramoto, S., Tejos, R. *et al.* (2009) The PIN-FORMED (PIN) protein family of auxin transporters. *Genome Biology* 10(12), 249. DOI: 10.1186/gb-2009-10-12-249.

Lammi, M.J. and Qu, C. (2024) Spatial transcriptomics, proteomics, and epigenomics as tools in tissue engineering and regenerative medicine. *Bioengineering* 11(12), 1235. DOI: 10.3390/bioengineering11121235.

Langfelder, P. and Horvath, S. (2008) WGCNA: An R package for weighted correlation network analysis. *BMC Bioinformatics* 9, 559. DOI: 10.1186/1471-2105-9-559.

Lee, M.S.Y. and Palci, A. (2015) Morphological phylogenetics in the genomic age. *Current Biology* 25(19), R922–R929. DOI: 10.1016/j.cub.2015.07.009.

Lee, D., Chung, P.J., Jeong, J.S., Jang, G., Bang, S.W. *et al.* (2017) The rice *OsNAC6* transcription factor orchestrates multiple molecular mechanisms involving root structural adaptions and nicotianamine biosynthesis for drought tolerance. *Plant Biotechnology Journal* 15(6), 754–764. DOI: 10.1111/pbi.12673.

Liberali, P. and Schier, A.F. (2024) The evolution of developmental biology through conceptual and technological revolutions. *Cell* 187(14), 3461–3495. DOI: 10.1016/j.cell.2024.05.053.

Liu, F., Heiner, M. and Gilbert, D. (2022) Hybrid modelling of biological systems: Current progress and future prospects. *Briefings in Bioinformatics* 23(3), bbac081. DOI: 10.1093/bib/bbac081.

Liu, S., Wu, Z., Chen, X., Chen, Z., Shen, Y. *et al.* (2024) Evolution and comparative transcriptome analysis of glucosinolate pathway genes in *Brassica napus* L. *Frontiers in Plant Science* 15, 1483635. DOI: 10.3389/fpls.2024.1483635.

Lopes, F.L., Galvan-Ampudia, C. and Landrein, B. (2021) WUSCHEL in the shoot apical meristem: Old player, new tricks. *Journal of Experimental Botany* 72, 1527–1535. DOI: 10.1093/jxb/eraa572.

Loyola-Gonzalez, O. (2019) Black-box vs. white-box: Understanding their advantages and weaknesses from a practical point of view. *IEEE Access* 7, 154096–154113. DOI: 10.1109/ACCESS.2019.2949286.

Lundgren, M.R., Osborne, C.P. and Christin, P.A. (2014) Deconstructing kranz anatomy to understand C4 evolution. *Journal of Experimental Botany* 65, 3357–3369. DOI: 10.1093/jxb/eru186.

Luo, L., Li, W., Miura, K., Ashikari, M. and Kyoxuka, J. (2012) Control of tiller growth of rice by Os SPL14 and stringolactones, which work in two independent pathways. *Plant and Cell Physiology* 53(10), 1793–1801.

Ma, A., McDermaid, A., Xu, J., Chang, Y. and Ma, Q. (2020) Integrative methods and practical challenges for single-cell multi-omics. *Trends in Biotechnology* 38, 1007–1022. DOI: 10.1016/j.tibtech.2020.02.013.

MacKintosh, C. and Ferrier, D.E.K. (2017) Recent advances in understanding the roles of whole genome duplications in evolution. *F1000Research* 6, 1623. DOI: 10.12688/f1000research.11792.2.

Mao, X., Cai, T., Olyarchuk, J.G. and Wei, L. (2005) An automated genotyping system for analysis of HIV-1 and other microbial sequences. *Bioinformatics* 21(19), 3797–3800.

Marx, H., Minogue, C.E., Jayaraman, D., Richards, A.L., Kwiecien, N.W. *et al.* (2016) A proteomic atlas of the legume *Medicago truncatula* and its nitrogen-fixing endosymbiont *Sinorhizobium meliloti*. *Nature Biotechnology* 34(11), 1198–1205. DOI: 10.1038/nbt.3681.

Misra, B.B., Langefeld, C.D., Olivier, M. and Cox, L.A. (2018) Integrated omics: Tools, advances and future approaches. *Journal of Molecular Endocrinology* 62(1), R21–R45. DOI: 10.1530/JME-18-0055.

Moeinizade, S., Kusmec, A., Hu, G., Wang, L. and Schnable, P.S. (2020) Multi-trait genomic selection methods for crop improvement. *Genetics* 215(4), 931–945. DOI: 10.1534/genetics.120.303305.

Moffitt, J.R. and Zhuang, X. (2016) RNA imaging with multiplexed error-robust fluorescence *in situ* hybridization (MERFISH). *Methods in Enzymology* 572, 1–49. DOI: 10.1016/bs.mie.2016.03.020.

Mony, C., Vannier, N., Burel, F., Ernoult, A. and Vandenkoornhuyse, P. (2024) The root microlandscape of arbuscular mycorrhizal fungi. *New Phytologist* 244, 394–406. DOI: 10.1111/nph.20048.

Morabito, A., Simone, G.D., Roberta Pastorelli, R., Brunelli, L. and Ferrario, M. (2025) Algorithms and tools for data-driven omics integration to achieve multilayer biological insights: A narrative review. *Journal of Translational Medicine* 23, 425. DOI: 10.1186/s12967-025-06446-x.

Ngara, R., Goche, T., Swanevelder, D.Z.H. and Chivasa, S. (2021) Sorghum's whole-plant transcriptome and proteome responses to drought stress: A review. *Life (Basel, Switzerland)* 11(7), 704. DOI: 10.3390/life11070704.

Picard, M., Scott-Boyer, M.P., Bodein, A., Périn, O. and Droit, A. (2021) Integration strategies of multi-omics data for machine learning analysis. *Computational and Structural Biotechnology Journal* 19, 3735–3746. DOI: 10.1016/j.csbj.2021.06.030.

Porter, S.S., Dupin, S.E., Denison, R.F., Kiers, E.T. and Sachs, J.L. (2024) Host-imposed control mechanisms in legume–rhizobia symbiosis. *Nature Microbiology* 9(8), 1929–1939. DOI: 10.1038/s41564-024-01762-2.

Ramazi, S. and Zahiri, J. (2021) Posttranslational modifications in proteins: Resources, tools and prediction methods. *Database* baab012. DOI: 10.1093/database/baab012.

Ren, J., Luo, S., Shi, H. and Wang, X. (2024) Spatial omics advances for *in situ* RNA biology. *Molecular Cell* 84(19), 3737–3757. DOI: 10.1016/j.molcel.2024.08.002.

Sage, R.F., Sage, T.L. and Kocacinar, F. (2012) Photorespiration and the evolution of C4 photosynthesis. *Annual Review of Plant Biology* 63, 19–47. DOI: 10.1146/annurev-arplant-042811-105511.

Saini, S., Sharma, I., Kaur, N. and Pati, P.K. (2013) Auxin: A master regulator in plant root development. *Plant Cell Reports* 32(6), 741–757. DOI: 10.1007/s00299-013-1430-5.

Salem, M.A., Wang, J.Y. and Al-Babili, S. (2022) Metabolomics of plant root exudates: From sample preparation to data analysis. *Frontiers in Plant Science* 13, 1062982. DOI: 10.3389/fpls.2022.1062982.

Sanches, P.H.G., Melo, N.C., Porcari, A.M. and Carvalho, L.M. (2024) Integrating molecular perspectives: Strategies for comprehensive multi-omics integrative data analysis and machine learning applications in transcriptomics, proteomics, and metabolomics. *Biology* 13, 848. DOI: 10.3390/biology13110848.

Shahan, R., Hsu, C.W., Nolan, T.M., Cole, B.J., Taylor, I.W. *et al.* (2022) A single-cell *Arabidopsis* root atlas reveals developmental trajectories in wild-type and cell identity mutants. *Developmental Cell* 57, 543–560. DOI: 10.1016/j.devcel.2022.01.008.

Sin, W. C., Liu, J., Zhong, J.Y., Lam, H.-M. and Lim, B.L. (2024) Comparative proteomics analysis of root and nodule mitochondria of soybean. *Plant, Cell & Environment* 1–18. DOI: 10.1111/pce.15026.

Smit, P., Raedts, J., Portyanko, V., Debellé, F., Gough, C. *et al.* (2005) NSP1 of the GRAS protein family is essential for rhizobial nod factor-induced transcription. *Science (New York, N.Y.)* 308(5729), 1789–1791. DOI: 10.1126/science.1111025.

Smith, C.J., Sinnott-Armstrong, N., Cichońska, A., Julkunen, H., Fauman, E.B. *et al.* (2022) Integrative analysis of metabolite GWAS illuminates the molecular basis of pleiotropy and genetic correlation. *eLife* 11, e79348. DOI: 10.7554/eLife.79348.

Ståhl, P.L., Salmén, F., Vickovic, S., Lundmark, A., Navarro, J.F. *et al.* (2016) Visualization and analysis of gene expression in tissue sections by spatial transcriptomics. *Science (New York, N.Y.)* 353(6294), 78–82. DOI: 10.1126/science.aaf2403.

Steen, C.B., Liu, C.L., Alizadeh, A.A. and Newman, A.M. (2020) Profiling cell type abundance and expression in bulk tissues with CIBERSORTx. *Molecular Biology (Clifton, N.J.)* 135–157. DOI: 10.1007/978-1-0716-0301-7_7.

Sunil, L. and Shetty, N.P. (2022) Biosynthesis and regulation of anthocyanin pathway genes. *Applied Microbiology and Biotechnology* 106, 1783–1798. DOI: 10.1007/s00253-022-11835-z.

Swenson, K.M. and El-Mabrouk, N. (2012) Gene trees and species trees: Irreconcilable differences. *BMC Bioinformatics* S15. DOI: 10.1186/1471-2105-13-S19-S15.

Szymański, J., Bocobza, S., Panda, S., Sonawane, P., Cárdenas, P.D. *et al.* (2020) Analysis of wild tomato introgression lines elucidates the genetic basis of transcriptome and metabolome variation underlying fruit traits and pathogen response. *Nature Genetics* 52(10), 1111–1121. DOI: 10.1038/s41588-020-0690-6.

Talarico, E., Zambelli, A., Araniti, F., Greco, E., Chiappetta, A. *et al.* (2024) Unravelling the epigenetic code: DNA methylation in plants and its role in stress response. *Epigenomes* 8(3), 30. DOI: 10.3390/epigenomes8030030.

Tang, W., Liu, X., He, Y. and Yang, F. (2022) Enhancement of vindoline and catharanthine accumulation, antioxidant enzymes activities, and gene expression levels in *Catharanthus roseus* leaves by chitooligosaccharides elicitation. *Marine Drugs* 20(3), 188. DOI: 10.3390/md20030188.

Trapnell, C. (2015) Defining cell types and states with single-cell genomics. *Genome Research* 25(10), 1491–1498. DOI: 10.1101/gr.190595.115.

Varma, A., Verma, S., Sudha, A.A. and Sahay, N. (1999) *Piriformospora indica*, a cultivable plant-growth-promoting root endophyte. *Applied and Environmental Microbiology* 65(6), 2741–2744. DOI: 10.1128/AEM.65.6.2741-2744.1999.

Vijayakumar, V., Liebisch, G., Buer, B., Xue, L., Gerlach, N. *et al.* (2016) Integrated multi-omics analysis supports role of lysophosphatidylcholine and related glycerophospholipids in the *Lotus japonicus–Glomus intraradices* mycorrhizal symbiosis. *Plant, Cell & Environment* 39(2), 393–415. DOI: 10.1111/pce.12624.

Wang, Z. and Deng, W. (2022) Dynamic transcription regulation at the single-molecule level. *Developmental Biology* 482, 67–81. DOI: 10.1016/j.ydbio.2021.11.004.

Wang, Y., Liu, B., Zhao, G., Lee, Y., Buzdin, A. *et al.* (2023) Spatial transcriptomics: Technologies, applications and experimental considerations. *Genomics* 115(5), 110671. DOI: 10.1016/j.ygeno.2023.110671.

Wei, Y., Chen, S., Zhou, X., Ding, D., Song, J. *et al.* (2024) Endophytic microorganisms in tomato roots, changes in the structure and function of the community at different growing stages. *Microorganisms* 12(6), 1251. DOI: 10.3390/microorganisms12061251.

Weigel, D. and Mott, R. (2009) The 1001 genomes project for *Arabidopsis thaliana*. *Genome Biology* 10(5), 107. DOI: 10.1186/gb-2009-10-5-107.

Xu, P. and Wang, E. (2023) Diversity and regulation of symbiotic nitrogen fixation in plants. *Current Biology* 33(11), R543–R559. DOI: 10.1016/j.cub.2023.04.053.

Yan, L., Loukoianov, A., Blechl, A., Tranquilli, G., Ramakrishna, W. *et al.* (2004) The wheat *VRN2* gene is a flowering repressor down-regulated by vernalization. *Science* 303(5664), 1640–1644. DOI: 10.1126/science.1094305.

Xu, G., Cao, J., Wang, X., Chen, Q., Jin, W. *et al.* (2019) Evolutionary metabolomics identifies substantial metabolic divergence between maize and its wild ancestor, teosinte. *The Plant Cell* 31(9), 1990–2009. DOI: 10.1105/tpc.19.00111.

Xu, J.-W., Wang, J. and Dong, Y.-W. (2024) Genetic determination of a cryptic species in the *Littoraria* genus with whole-genome molecular resolution. *Ecology and Evolution* 14(12), e70715. DOI: 10.1002/ece3.70715.

Zhang, X., Liu, P., Qing, C., Yang, C., Shen, Y. *et al.* (2021a) Comparative transcriptome analyses of maize seedling root responses to salt stress. *PeerJ* 9, e10765. DOI: 10.7717/peerj.10765.

Zhang, X., Luo, Z., Marand, A.P., Yan, H., Jang, H. *et al.* (2025) A spatially resolved multi-omic single-cell atlas of soybean development. *Cell* 188(2), 550–567. DOI: 10.1016/j.cell.2024.10.050.

Zhang, Y.-M., Yu, H.-X., Ye, W.-W., Shan, J.-X., Dong, N.-Q. *et al.* (2021b) A rice QTL GS3.1 regulates grain size through metabolic-flux distribution between flavonoid and lignin metabolons without affecting stress tolerance. *Communications Biology* 4(1), 1171. DOI: 10.1038/s42003-021-02686-x.

10 Plant–Microbe Interactions Studied Using Spatial and Multiple Omics Tools

Ilham Dehbi[1,2], Khadija Benamar[1], Mouna Janati[2], Hajar Zennouhi[1] and Rachid Lahlali[1]*

[1]Phytopathology Unit, Department of Plant Protection, Ecole National of Agriculture Meknes, Meknes, Morocco; [2]Laboratory of Biotechnology and Bioresources Valorization, Moulay Ismail University of Meknes, Zitoune, Meknes, Morocco

10.1 Introduction

Plants constitute complex biological assemblages known as holobionts, comprising the host plant and its associated microbial communities. Plant–microbe interactions occurring in the rhizosphere, phyllosphere, and endosphere are highly interconnected and dynamically regulated (Hassani *et al.*, 2018). These associations represent an evolutionarily refined system in which plants actively invest in shaping their microbiota, allocating up to 40% of their photosynthetically fixed carbon to root exudates that support and modulate microbial populations (Feierabend and Töpfer, 2025). Increasingly, this plant–microbiome continuum is recognized as a fundamental component of plant biology, exerting a profound influence on plant health, development, and agricultural productivity (Rosier *et al.*, 2018).

Plant–microbe interactions span a wide spectrum, ranging from mutualistic and commensal relationships to antagonistic and pathogenic associations. Numerous beneficial interactions fall under mutualistic partnerships, where resources are exchanged or shared, or where microbial presence alone enhances plant performance (Hassani *et al.*, 2018; Sarsan *et al.*, 2021). Among the most well-characterized examples are symbioses between nitrogen-fixing bacteria, such as *Rhizobium* spp., and leguminous plants, in which atmospheric nitrogen is converted into plant-accessible forms. Similarly, associations with mycorrhizal fungi (MF) expand the functional root system, facilitating the acquisition of essential nutrients, particularly phosphorus and water (Feierabend and Töpfer, 2025). Additional beneficial microbes contribute to plant growth by producing phytohormones, enhancing nutrient cycling, and conferring protection against pathogens through mechanisms such as induced systemic resistance (ISR) (Olowe *et al.*, 2020; Yu *et al.*, 2022). Conversely, pathogenic microorganisms cause bacterial and fungal diseases that result in substantial yield losses in crops worldwide (Kumar *et al.*, 2024).

Understanding plant–microbe interactions is therefore critical for addressing global challenges related to food security and climate change (Muhammad *et al.*, 2024). Harnessing the plant microbiome offers a promising pathway toward sustainable agriculture by reducing reliance on synthetic inputs while

*Corresponding author: rlahlali@enameknes.ac.ma

© CAB International 2026. *Plant Spatial Biology* (ed. Chen)
DOI: 10.1079/9781836994572.0010

maintaining or enhancing crop productivity (Trivedi *et al.*, 2021; Pandey and Saharan, 2025). For example, adding microbes that are good for the plant, such as plant growth-promoting rhizobacteria (PGPRs) and mycorrhizal fungi, can help the crop and the plant's strength to deal not only with issues from bugs but from problems caused by environmental adversities and natural phenomena (Nadeem *et al.*, 2014; Lahlali *et al.*, 2022). Beyond plant health, microbes play central roles in global biogeochemical cycles, including those of carbon (C), nitrogen (N), and phosphorus (P), and are essential for maintaining soil structure, fertility, and long-term ecosystem sustainability (Sahu *et al.*, 2017; Sharma *et al.*, 2025).

Transitioning from descriptive observations to a mechanistic understanding of these complex interactions requires advanced analytical tools. The application of omics technologies – genomics, transcriptomics, proteomics, and metabolomics – enables the simultaneous investigation of thousands of genes, proteins, and metabolites, providing unprecedented insight into biological processes (Mishra *et al.*, 2025; Singh *et al.*, 2025b). Integrating these datasets within a systems biology framework is essential for capturing the multilayered complexity of plant–microbe interactions (Mishra *et al.*, 2019). While traditional bulk omics approaches prioritize throughput, they often obscure spatial context; in contrast, emerging spatial omics technologies preserve the native organization of cells and molecules, allowing interactions to be studied *in situ* and in real time (Barmukh *et al.*, 2025). This represents a major conceptual and technological advance in plant research.

Looking forward, the integration of multi-omics and spatial omics approaches into a unified framework – termed holo-omics – promises to generate a comprehensive, high-resolution map of plant–microbiome interactions (Xu *et al.*, 2021). Such integrative strategies will enable the development of predictive models and targeted microbiome engineering approaches, ultimately enhancing plant performance and resilience. By advancing our understanding and management of plant–microbe holobionts, holo-omics offers a powerful avenue toward sustainable land use, improved agricultural productivity, and long-term food system resilience in the face of global environmental change (Patyal *et al.*, 2025).

10.2 Omics Technologies in Plant–Microbe Interactions

The field of plant–microbe interaction research has been profoundly transformed by the advent of high-throughput omics technologies. These approaches enable a holistic view of biological systems by examining integrated molecular networks rather than isolated components. When applied to the complex plant microbiome, omics tools offer unprecedented opportunities to decipher the intricate molecular dialogues that occur between plants and their associated microbial communities, shifting our understanding from descriptive correlations to mechanistic insights at the molecular level. The integration of genomics, transcriptomics, proteomics, and metabolomics allows researchers to comprehensively map functional potential, gene expression patterns, protein activity, and metabolite dynamics underlying these interactions. Together, these complementary layers provide a systems-level framework for unraveling the regulatory and metabolic processes that govern plant–microbe associations and their impact on plant health, development, and productivity (Table 10.1).

10.2.1 Genomics and metagenomics

Genomics and metagenomics are fundamental approaches for understanding plant–microbe interactions and characterizing the microbial communities associated with plants. This chapter highlights the potential of metagenomics and its methodological applications in contemporary agricultural research. Metagenomics enables the comprehensive exploration of microbial diversity across plant-associated habitats, including the rhizosphere (the soil region surrounding roots) and the phyllosphere (leaf-associated surfaces), allowing researchers to profile microbial communities across multiple ecological niches (Rosselli and Squartini, 2015; Ramadurai *et al.*, 2021). Beyond taxonomic profiling, metagenomic approaches provide functional insights by revealing microbial genes involved in nutrient acquisition, plant growth promotion, and stress tolerance, thereby contributing to plant adaptation under changing environmental conditions

Table 10.1. Omics technologies in plant–microbe interactions.

Omics technology	What it measures	Application in plant–microbe interactions
Genomics and metagenomics	The complete DNA sequence of an organism (genomics) or an entire community (metagenomics)	Identifying the full repertoire of genes for a root-colonizing bacterium, or characterizing the overall microbial diversity and functional genes in a soil sample
Transcriptomics	The set of all RNA transcripts representing active gene expression	Observing the rapid upregulation of plant defense genes after a fungal attack, or identifying the expression of a pathogen's virulence genes during infection
Proteomics	The complete set of proteins expressed by an organism or community	Discovering secreted effector proteins used by a pathogen to suppress host immunity, or identifying plant enzymes involved in root growth promotion
Metabolomics	All small-molecule metabolites present in a biological sample	Detecting the production of a specific phytoalexin (defense compound) by a plant in response to stress, or identifying a microbe's secretion of a growth-promoting phytohormone

(Kaur *et al.*, 2022; Mohan and Sudhakar, 2022). Commonly used techniques, such as 16S rRNA gene amplicon sequencing and whole-genome shotgun sequencing, facilitate the identification and functional annotation of plant-associated microorganisms (Mohan and Sudhakar, 2022; Saxena and Bharadwaj, 2024). Bioinformatics pipelines and tools, including QIIME and Kraken, play a critical role in microbial community profiling, taxonomic assignment, and functional gene prediction, thereby advancing our understanding of plant–microbe interactions (Kaur *et al.*, 2022; Mohan and Sudhakar, 2022). Insights gained from these analyses can be leveraged to enhance plant productivity, improve stress resilience, and promote sustainable agricultural practices by fostering beneficial soil microbial communities (Rosselli and Squartini, 2015; Kaur *et al.*, 2022; Saxena and Bharadwaj, 2024).

Genomics, in contrast, focuses on the complete DNA sequence of a single organism and provides foundational knowledge of the genetic architecture underlying plant traits or microbial functions (Jain *et al.*, 2024). However, in complex and heterogeneous environments such as the rhizosphere, metagenomics offers a more comprehensive perspective by capturing the collective genomic content of entire microbial communities within an environmental sample. This culture-independent approach enables the characterization of microbial functional potential without the need for laboratory cultivation, thereby uncovering previously inaccessible or unculturable taxa (Rajguru *et al.*, 2024). Importantly, metagenomics facilitates the discovery of novel genes and metabolic pathways associated with plant growth promotion, nutrient cycling, and, in some cases, pathogenicity, providing valuable insights for both agricultural innovation and plant disease management (Parnell *et al.*, 2016; Vishwakarma *et al.*, 2020).

10.2.2 Transcriptomics

Plant–microbe interactions are highly dynamic and complex, playing a central role in plant performance and ecosystem functioning. Recent advances in two major research directions – single-cell–level analyses and spatially resolved functional studies – have substantially improved our understanding of how plants and microorganisms interact at fine biological scales. These approaches, which enable precise visualization of gene expression patterns and functional activity within specific tissues and cells, have provided new insights into the spatial and temporal organization of plant–microbe interactions (Serrano *et al.*, 2024). Single-cell analyses have revealed how distinct plant cell types respond to microbial colonization, allowing researchers to dissect cell-specific transcriptional reprogramming associated with microbial recognition and accommodation (Serrano *et al.*, 2024). Complementarily, spatial mapping of gene expression across plant tissues has clarified where and how key regulatory genes are activated during plant–microbe interactions, thereby linking molecular function to anatomical context (Serrano *et al.*, 2024). At the community level, functional profiling of rhizosphere microbiomes provides critical insights into microbial metabolic capabilities and their contributions to plant nutrition and stress resilience (Bharti *et al.*, 2021). Such knowledge is essential for understanding plant-driven biogeochemical cycling and for the rational development of biofertilizers aligned with sustainable agricultural practices (Bharti *et al.*, 2021). Integrating plant–microbe interaction data with multi-omics datasets that capture transcriptomic, proteomic, and metabolomic profiles enables a comprehensive view of the molecular crosstalk underlying these associations. This integrative framework elucidates how plants and microbes coordinate signaling pathways and metabolic processes during symbiosis or stress responses (Jain *et al.*, 2024). Transcriptome-based studies, in particular, have identified key plant genes that are differentially regulated during microbial interactions and environmental challenges, highlighting regulatory networks that mediate either cooperative responses or defense-related trade-offs (Damiani *et al.*, 2012).

10.2.3 Proteomics

Proteomics, particularly mass spectrometry (MS)-based approaches, enables the identification and quantification of proteins that play beneficial or detrimental roles in plant–microbe interactions. This methodology provides critical insights into plant cellular responses to microbial colonization by revealing dynamic changes in protein abundance, localization, and function during both symbiotic and pathogenic interactions.

One of the principal strengths of plant proteomics lies in its ability to characterize proteome-wide changes and post-translational modifications associated with microbial interactions, thereby elucidating regulatory mechanisms underlying plant responses to both beneficial and harmful microorganisms (Balotf *et al.*, 2022). Proteomic studies in crop species, including tomato, have demonstrated distinct protein expression profiles in response to beneficial microbes (e.g. *Bacillus subtilis*) compared with pathogenic organisms (e.g. *Alternaria solani*), identifying specific biomarkers and interaction-specific signaling pathways (Singh *et al.*, 2025a). For instance, proteins involved in defense signaling, such as cyclin b559, are upregulated during pathogen challenge, reflecting activation of immune-related pathways (Singh *et al.*, 2025a). In contrast, proteins associated with photosynthesis, primary metabolism, and growth are more prominently expressed during beneficial interactions, indicating a metabolic shift toward growth promotion and physiological optimization (Singh *et al.*, 2025a).

The integration of proteomics with complementary omics approaches, including genomics and metabolomics, provides a more comprehensive understanding of plant–microbe interactions at the systems level and supports the development of advanced, knowledge-driven agricultural practices. Nevertheless, challenges remain in multi-omics data integration, resolution, and functional interpretation. Addressing these limitations will be essential for fully elucidating the complexity of plant–microbe interactions and for translating omics-based discoveries into sustainable agricultural applications (Jain *et al.*, 2024).

10.2.4 Metabolomics

Metabolomics is a powerful approach for elucidating the biochemical basis of plant–microbe interactions, as it directly captures the dynamic metabolic exchanges that underpin mutualistic and antagonistic relationships. By employing advanced analytical platforms such as MS and nuclear magnetic resonance (NMR) spectroscopy, researchers can profile low-molecular-weight metabolites involved in plant–microbe communication, nutrient exchange, and plant responses to abiotic stress. These technologies provide critical insights into signaling pathways, metabolic coordination, and adaptive responses within plant–microbiome systems. MS enables the sensitive detection, identification, and quantification of a wide range of metabolites, making it particularly effective for characterizing metabolic shifts associated with microbial colonization and plant physiological responses (Gupta *et al.*, 2022). Complementarily, NMR spectroscopy offers robust structural elucidation and metabolic fingerprinting, facilitating the identification of metabolite classes and biosynthetic pathways while providing highly reproducible quantitative data (Fan, 2024). Advances in spatial metabolomics further allow visualization of metabolite distribution within plant tissues, revealing how metabolic processes are spatially organized during plant–microbe interactions and how microbes influence local metabolic landscapes (Fan, 2024).

Metabolomic analyses have been instrumental in uncovering metabolites associated with plant growth promotion and enhanced stress tolerance mediated by beneficial microorganisms (Gupta *et al.*, 2022). In addition, specific metabolic signatures and biomarkers have been identified that correlate with improved resistance to pathogens and environmental stressors, offering valuable indicators of plant health and resilience (Demiwal *et al.*, 2024). Despite its strong potential, metabolomics remains less widely applied than other omics approaches in plant–microbe interaction studies. Expanding its integration with genomics, transcriptomics, and proteomics will be essential to fully exploit its capacity to advance our understanding of crop performance and ecosystem functioning, ultimately supporting sustainable agricultural and environmental management (Gupta *et al.*, 2022).

10.3 Advancements in Spatial Omics

10.3.1 Spatial transcriptomics

Spatial transcriptomics (ST) has emerged as a transformative approach for mapping messenger RNA (mRNA) distributions *in situ*, integrating quantitative gene expression profiling with spatial context. Unlike bulk RNA sequencing, which averages transcriptional signals across heterogeneous cell populations and thereby obscures cell-to-cell variability, ST preserves native tissue architecture and cellular heterogeneity. Although single-cell RNA sequencing (scRNA-seq) offers high resolution for defining cell types, the dissociation of cells disrupts spatial relationships, eliminating critical information regarding tissue organization, intercellular interactions, and positional cues within native microenvironments (You *et al.*, 2024). ST bridges this gap by profiling transcripts directly within intact tissue sections, maintaining gene expression information within its morphological and spatial framework (Wang *et al.*, 2023). By retaining positional information, ST enables the investigation of intercellular signaling, non–cell-autonomous regulation, immune infiltration patterns, and developmental expression gradients that cannot be resolved using dissociated-cell approaches. Current ST technologies broadly fall into two conceptual categories: sequencing-based and imaging-based approaches. Sequencing-based methods employ localized transcript capture on spatially barcoded arrays or surfaces, followed by next-generation sequencing (NGS) to decode transcript identity. These approaches offer transcriptome-wide coverage but often compromise spatial resolution, particularly at the cellular or subcellular level, or sensitivity for low-abundance transcripts. In contrast, imaging-based methods rely on fluorescence *in situ* hybridization (FISH) or direct *in situ* sequencing, using iterative imaging cycles or combinatorial barcoding schemes to localize and quantify transcripts at single-molecule resolution. While imaging-based techniques typically achieve superior spatial precision, they are often limited to smaller gene panels (Wang *et al.*, 2023; Lim *et al.*, 2025).

From a technical perspective, however, ST methodologies can be classified into four major strategies. Microdissection-based approaches, such as laser capture microdissection (LCM), involve the physical isolation of defined tissue regions followed by bulk RNA sequencing. LCM enables region-resolved transcriptomic profiling of tens to hundreds of cells and has been further refined through variants such as LCM-seq and Geo-seq, which combine spatial precision with deep sequencing (Yin *et al.*, 2024; Lim *et al.*, 2025). However, these methods remain limited by low-throughput, labor-intensive workflows, and challenges in achieving fine spatial resolution across large tissue areas.

Another strategy is imaging-based FISH and optical methods, which are centered on single-molecule FISH (smFISH) that uses sets of short, fluorophore-labeled oligonucleotide probes tiled along target RNAs to visualize individual transcripts as diffraction-limited spots. Owing to its high quantitative accuracy, smFISH is widely regarded as a gold standard for RNA quantification (Haimovich and Gerst, 2018). Nonetheless, smFISH is constrained by limited multiplexing capacity due to fluorophore availability, complex probe design requirements, and limited scalability to genome-wide profiling. Furthermore, smFISH is generally incompatible with live-cell imaging.

In plant tissues, smFISH presents additional technical challenges, particularly high levels of autofluorescence – especially in chlorophyll-rich tissues – which elevate background signal and reduce sensitivity (Duncan and Rosa, 2018; Zhao *et al.*, 2023). As a result, its application in plant systems has remained relatively restricted, often limited to species with well-annotated genomes and tissues amenable to optical clearing or signal amplification strategies.

In situ sequencing (ISS) and optical barcoding approaches represent a third major class of ST technologies. These methods use padlock probes, rolling circle amplification (RCA), or combinatorial barcoding schemes to convert transcript identity into localized optical or sequencing-based signals. In many implementations, padlock probes hybridize to target RNAs, undergo ligation and *in situ* amplification, and are subsequently decoded through iterative fluorescence imaging or sequencing cycles (Luce *et al.*, 2025). This category includes techniques such as ISS, HybISS, STARmap, and DART-FISH, as well as commercial platforms like Xenium. These approaches enable higher multiplexing than smFISH while maintaining spatial resolution, but they require complex probe design, multi-step chemistries, and are susceptible to technical limitations such as incomplete ligation, amplification bias, and probe mispriming.

Spatial barcoding/spatial capture/array-based methods form another category. In sequencing-based ST systems, intact tissue sections are mounted onto surfaces containing spatially barcoded oligonucleotides arranged in micropatterned spots. Upon tissue permeabilization, transcripts diffuse locally and hybridize to these barcoded probes. The spatial barcodes are incorporated into the resulting cDNA during reverse transcription and subsequently decoded through high-throughput sequencing, enabling reconstruction of spatially-resolved gene expression maps. While these approaches enable transcriptome-wide profiling across large tissue areas, they typically provide relatively coarse spatial resolution, often on the order of tens of micrometers per capture spot, and reduced sensitivity for low-abundance transcripts compared with imaging-based methods (Wadie *et al.*, 2024).

Despite their promise, ST technologies face several technical and analytical challenges. Variability in transcript capture efficiency, diffusion-related artifacts, probe design and hybridization biases, fluctuations in signal-to-noise ratios, segmentation errors, and difficulties in cross-platform comparison continue to limit reproducibility and scalability. Although community-wide benchmarking initiatives are emerging, standardized evaluation metrics and robust comparative frameworks across platforms remain under development (You *et al.*, 2024).

ST serves as a critical bridge between molecular profiling and histological analysis by capturing both transcriptional heterogeneity and tissue architecture. Its performance depends on balancing key parameters such as multiplexing capacity, spatial resolution, sensitivity, and throughput. Reliable implementation requires careful probe and array design, rigorous platform benchmarking, and appropriate statistical error modeling. Table 10.2 summarizes the major ST and genomics sub-technologies, outlining their underlying principles, key advantages, and practical limitations. Overall, these methods differ primarily in their trade-offs between spatial resolution, molecular coverage, and experimental throughput, rather than in their fundamental objective: preserving positional information while directly quantifying nucleic acids within intact biological structures.

Early imaging-based approaches, particularly smFISH, established a benchmark for precise RNA quantification at subcellular resolution. smFISH employs multiple fluorophore-labeled oligonucleotide probes that bind specifically

Table 10.2. Sub-technologies in spatial transcriptomics/genomics: methods, principles, merits, drawbacks, and updated references.

Domain	Method/sub-technology	Analyte/principle	Key features/strengths	Key limitations/caveats
Spatial transcriptomics/ imaging/ISS	smFISH	*In situ* hybridization of oligo probes (reverse-complementary) labeled with fluorophores	Very precise subcellular localization, direct counting of transcripts	Lower throughput, limited multiplexing, background fluorescence, and probe specificity challenges
	MERFISH (multiplexed error-robust FISH)	Combinatorial barcoding + sequential rounds of smFISH	High plex (hundreds to > 1000), error robustness via coding schemes	Strong fluorescence background, complexity of barcoding and imaging rounds
	seqFISH/seqFISH+	Sequential fluorescence hybridization + imaging rounds	High coding efficiency, scalable multiplexing, and relatively high hybridization efficiency	High cost, long experiment times, and many probes needed
	FISSEQ (*in situ* sequencing)	Sequencing of crosslinked cDNA amplicons *in situ*	Untargeted/transcriptome-scale *in situ* sequencing possibility	Low capture/sensitivity; relatively high noise, complexity
	STARmap	Amplicon readout mapping via padlock probes + *in situ* sequencing/ hybridization	High efficiency, accuracy, and reproducibility for targeted panels	Does not cover the full transcriptome, limited gene coverage
	BaristaSeq	Gap-fill padlock probe + sequencing/ readout	Good accuracy for targeted barcoded sets	Less flexible for broad panels, tissue compatibility limitations
	454 Chemiluminescence cascade	Enzyme cascade + chemiluminescent readout	Potential for throughput/cost advantages	Short readout fragments, limited spatial resolution
	Illumina/Solexa (*in situ*/array hybridization coupling)	Array-based sequencing or *in situ* capture + NGS	High accuracy, high throughput, high sensitivity	Short readouts, complex splicing/ alignment/deconvolution
	SOLiD sequencing (*in situ* adaptation)	Fluorescent-label recognition via ligation chemistry	High accuracy (in principle), high multiplexing potential	Slow kinetics, adaptation to spatial context is challenging
	PacBio long-read (*in situ*/spatial variant)	Single molecule long reads (e.g. via nanopore/real-time)	Ability to capture full isoforms, structural variants, and low bias (e.g. GC)	Higher error rates in single reads, technical complexity, spatial adaptation, and nontrivial
	ONT (Oxford Nanopore, spatial adaptations)	Real-time electrical signal sequencing of RNA/cDNA	Long read length, lower equipment cost, direct RNA possibilities	Challenges in mapping, error correction, adaptation to spatial capture or *in situ* methods
Spatial genomics/DNA	IGS (*in situ* genome sequencing/imaging)	Spatial mapping of paired-end/*in situ* sequencing of genome fragments	Enables imaging + sequencing of DNA *in situ* within intact samples	Hard to cover full genome, sampling limits, low sensitivity
	Slide-DNA-seq (or *in situ* fragmentation + barcoded adapters)	*In situ* fragmentation + spatial barcode adapters + NGS	Spatially resolved DNA sequencing with some resolution	Low efficiency/capture rate relative to bulk, difficulty scaling

to target transcripts, enabling highly accurate single-molecule detection (Raj *et al.*, 2008). However, its broader application is constrained by limited multiplexing capacity, elevated background fluorescence in complex tissues, and labor-intensive probe design. To overcome these limitations, advanced techniques such as multiplexed error-robust FISH (MERFISH) (Bai *et al.*, 2021) and sequential FISH (seqFISH and seqFISH+) (Khan *et al.*, 2025) have been developed. These methods utilize combinatorial barcoding strategies and multiple imaging rounds to expand gene coverage from dozens to thousands of transcripts per cell. While they offer exceptional spatial resolution and coding efficiency, they require prolonged imaging times, sophisticated optical instrumentation, and complex computational decoding pipelines.

In parallel, ISS-based approaches convert RNA molecules into spatially localized DNA amplicons that can be sequenced directly within preserved tissue sections. The pioneering fluorescent *in situ* sequencing (FISSEQ) method demonstrated the feasibility of unbiased transcriptome-wide spatial profiling but suffers from low capture efficiency and substantial technical complexity (Lee *et al.*, 2015). Subsequent refinements, including STARmap (Wang *et al.*, 2018) and BaristaSeq, have improved sensitivity and throughput through optimized padlock probe design, rolling-circle amplification, hydrogel embedding, and compatibility with Illumina sequencing chemistry. Although these methods achieve high spatial precision, they remain constrained by multicycle chemistry, ligation inefficiencies, photobleaching, and heavy computational demands for image segmentation and barcode assignment.

Sequencing-by-synthesis and sequencing-by-ligation platforms – such as 454 pyrosequencing, Illumina/Solexa, and SOLiD – form the technological backbone of most spatial barcoding strategies. Among these, Illumina sequencing has become the dominant readout platform due to its high accuracy and cost efficiency, whereas 454 and SOLiD technologies are now largely obsolete (Chen *et al.*, 2015). The short-read nature of these platforms necessitates computational reconstruction of transcript isoforms and cell-type mixtures. To address these limitations, third-generation long-read sequencing technologies, including Pacific Biosciences (PacBio) (Eid *et al.*, 2009) and Oxford Nanopore Technologies (ONT) (Jain *et al.*, 2016), offer the ability to resolve full-length transcripts and structural variants. While highly complementary to ST, their direct *in situ* implementation remains limited due to technical and chemical constraints.

Finally, spatial genomics approaches extend spatial omics principles from RNA to DNA. Methods such as *in situ* genome sequencing (IGS) (Payne *et al.*, 2021) and Slide-DNA-seq enable the visualization of genomic copy-number variations and chromatin organization within intact tissue sections. These techniques demonstrate the expanding scope of spatial omics beyond transcriptional profiling; however, they currently face challenges related to low capture efficiency and incomplete genome coverage (Yin *et al.*, 2024). Collectively, the ST and genomic sub-technologies summarized here represent complementary solutions to a shared experimental challenge: linking molecular identity with precise spatial organization in complex biological systems.

10.3.2 Imaging mass spectrometry

Metabolites play essential roles in sustaining life, with primary metabolites supporting growth, development, and reproduction, and secondary metabolites contributing to ecological functions such as defense, signaling, and pigmentation (Salam *et al.*, 2023). Gaining insight into complex biological interactions, including metabolite distribution, exchange, and transport across space and time, requires advanced analytical technologies capable of resolving molecular dynamics within heterogeneous biological systems. Modern analytical platforms, such as liquid chromatography–mass spectrometry (LC–MS) and liquid chromatography–nuclear magnetic resonance (LC–NMR), enable the identification and dereplication of metabolites under specific temporal or environmental conditions, facilitating the investigation of biochemical processes as they occur (Wadie *et al.*, 2024). When integrated with genomic and transcriptomic data, these approaches provide powerful opportunities to predict and contextualize the molecular mediators of chemical and biological interactions. Nevertheless, resolving the precise spatial distribution and localization of metabolites within biological samples remains technically challenging.

Imaging mass spectrometry (IMS) addresses this limitation by enabling *in situ* detection and spatial mapping of molecules across a wide range of biological scales, from tissues and organs to single cells, and from macromolecules to small metabolites (Yu *et al.*, 2023). Despite substantial progress in IMS methodologies, several technical challenges persist, particularly with respect to the sensitive and accurate analysis of low-molecular-weight metabolites (Yu *et al.*, 2023). Successful IMS applications depend critically on optimized sample preparation protocols and continuous advancements in instrumentation.

Comprehensive reviews have documented key developments in IMS, including advances in vacuum ionization techniques (Franck *et al.*, 2009), ambient ionization strategies, and applications across diverse disciplines such as histology, pathology, pharmaceutical sciences, plant biology, and microbiology (Gowda and Djukovic, 2014; Mahrous and Farag, 2015).

10.3.3 Single-cell multi-omics

The technique of scRNA-seq was first introduced in 2009 (Tang *et al.*, 2009), marking a paradigm shift by enabling the interrogation of gene expression at single-cell resolution and revealing cellular heterogeneity that bulk RNA sequencing could not resolve. Building on this breakthrough, single-cell multimodal omics and spatially resolved transcriptomics were later recognized as Nature Methods 'Method of the Year' in 2019 and 2020, respectively (Teichmann and Efremova, 2020; Marx, 2021). Together, these advances have given rise to a new generation of technologies, single-cell multi-omics and ST, that allow detailed dissection of tissue architecture, cell identity, and functional interactions within complex biological systems (Baysoy *et al.*, 2023).

Despite its transformative impact, scRNA-seq relies on tissue dissociation, leading to the loss of spatial context and native cellular organization. Because cell fate, function, and communication are tightly regulated by spatial positioning and microenvironmental cues, this limitation has driven the development of spatial multi-omics approaches that integrate molecular profiling with spatial localization (Vandereyken *et al.*, 2023). ST, for example, maps gene expression directly within intact tissues, revealing spatially-defined cell populations and local interaction networks that are essential for understanding tissue-level biology (Williams *et al.*, 2022).

Early spatial RNA detection methods, such as smFISH, established a benchmark for transcript quantification at subcellular resolution. By using multiple fluorescently-labeled probes targeting individual RNA molecules, smFISH enables highly accurate transcript counting within single cells (Raj *et al.*, 2008). However, traditional smFISH approaches suffer from limited multiplexing capacity, high background signal in complex tissues, and demanding probe design requirements.

To overcome these constraints, advanced multiplexed imaging techniques, including MERFISH and seqFISH-based approaches, were developed. These methods use combinatorial barcoding and sequential imaging to dramatically increase the number of detectable transcripts, scaling from tens to thousands of genes per cell while retaining spatial precision (Lubeck *et al.*, 2014; Hodgson, 2016; Eng *et al.*, 2019). Although these technologies substantially improve throughput and cellular coverage, they remain resource-intensive, requiring extensive imaging time, sophisticated optics, complex barcode decoding, and advanced computational pipelines.

In parallel, ISS-based approaches have emerged as alternative strategies for spatial transcript profiling. Early methods, such as FISSEQ, demonstrated the feasibility of untargeted RNA sequencing directly within fixed tissues but were limited by low capture efficiency and technical complexity. Subsequent innovations, including STARmap and BaristaSeq, improved sensitivity and spatial fidelity through optimized probe design, tissue stabilization, and advanced imaging chemistries (Chen *et al.*, 2018). Nevertheless, ISS technologies still face challenges related to signal loss, multi-cycle chemistry, and computational decoding of complex spatial patterns.

Modern spatial transcriptomics platforms predominantly rely on NGS chemistries, with

Illumina sequencing emerging as the dominant readout due to its high accuracy and cost efficiency (Bentley *et al.*, 2008; Yin *et al.*, 2016). While short-read NGS enables transcriptome-wide coverage, it requires computational reconstruction of isoforms and mixed cell populations. Long-read sequencing technologies, such as PacBio and Oxford Nanopore, address this limitation by capturing full-length transcripts and structural variants, thereby complementing spatial datasets; however, their direct *in situ* application remains technically challenging (Eid *et al.*, 2009; Jain *et al.*, 2016).

The scope of spatial omics is now expanding beyond transcriptomics to include spatial genomics. Emerging approaches such as *in situ* genome sequencing and Slide-DNA-seq extend spatial barcoding principles to DNA, enabling visualization of genomic copy-number variation and chromatin organization within intact tissues (Payne *et al.*, 2021). While these methods hold great promise, they are currently limited by capture inefficiency, incomplete genome coverage, and the computational demands of large-scale spatial DNA datasets. In summary, spatial omics technologies aim to integrate molecular identity with spatial context. Imaging-based approaches offer exceptional spatial resolution but limited gene throughput, whereas NGS-based methods provide comprehensive molecular coverage at lower spatial resolution. Long-read and DNA-based techniques add molecular completeness but increase technical complexity. The future of spatial biology lies in integrative frameworks that combine targeted imaging for validation, short-read sequencing for breadth, and long-read technologies for isoform and structural resolution, yielding spatially resolved, molecularly rich representations of biological systems.

10.4 Integration of Multi-omics Approaches

Integrating multiple analytical approaches represents a major advance in our understanding of plant–microbe interactions. While early technologies offered fragmented insights into individual cellular components, their true power emerges when datasets are combined, enabling the reconstruction of causal relationships and functional mechanisms across biological scales (Argelaguet *et al.*, 2020). This integrative strategy moves beyond a reductionist view of plants and their microbiomes as isolated parts and instead adopts a systems-level perspective, in which coordinated interactions give rise to emergent biological functions. The overarching objective is to interpret microbial features, metabolomic signatures, and host genetic architecture as interconnected entities that can be translated into predictive models of biological behavior. These models typically operate across contrasting analytical directions: genotype-to-phenotype frameworks that link genetic variation to functional outcomes, and phenotype-to-genotype approaches that identify genetic determinants underlying observed traits. Together, these complementary strategies enable robust inference of cause-and-effect relationships along complex biological pathways once molecular profiles, phenotypes, and regulatory programs are integrated (Weckwerth *et al.*, 2020).

Through multi-omics integration, it becomes possible not only to identify correlations but also to infer mechanistic drivers of biological processes. For instance, in rice, the integration of transcriptomic and proteomic datasets has enabled the identification of candidate trait-associated genes whose protein products are efficiently translated, providing promising molecular targets for crop improvement (Zhu *et al.*, 2023). Similarly, combining chromatin accessibility profiling using ATAC-seq with metabolite-based genome-wide association studies (mGWAS) has revealed genomic regions that regulate metabolite accumulation, thereby establishing direct links between genetic variation and metabolic phenotypes (Zhao *et al.*, 2021). In other studies, the joint analysis of transcriptomic and MS-based proteomic data uncovered previously unknown biosynthetic pathways in *Podophyllum hexandrum* and *P. peltatum* (Yu *et al.*, 2023), as well as developmental metabolic shifts during late-stage embryogenesis in elm (*Ulmus pumila*) (Qi *et al.*, 2022).

A further conceptual advance in plant–microbe research arises from integrating microbial community data with host gene expression and metabolite profiles, an approach that is increasingly recognized as essential for next-generation breeding strategies aimed at enhancing stress resilience and adaptive capacity (Deek *et al.*, 2024). By coupling spatially resolved metabolomics, transcriptomics, and proteomics, these approaches capture the dynamic organization of plant–microbe interactions within tissues, substantially improving the precision with which functional networks can be resolved (Wang *et al.*, 2021; Zhang *et al.*, 2021). Such integrative analyses have already elucidated regulatory cascades underlying adaptive traits, including salinity tolerance in grapevine (Yang *et al.*, 2021), and have supported predictive trait modeling in major crops to accelerate breeding programs aligned with agricultural demands (Fu *et al.*, 2022).

Given the breadth and complexity of multi-omics datasets, robust analytical pipelines are required to fully exploit their potential. Integrative frameworks that seamlessly connect genomic variation, phenotypic traits, and metabolic and regulatory processes are essential for predicting functional outcomes and identifying core biological pathways and cellular signatures. Such approaches ensure conceptual continuity between systems-level insights into plant–microorganism interactions and their translation into practical applications in crop improvement and ecosystem management (Argelaguet *et al.*, 2020). A representative overview of relevant methodologies, bioinformatics platforms, and specialized databases is provided in Table 10.3.

10.5 Applications in Plant–Microorganism Research

10.5.1 Beneficial plant–microorganism interactions

The application of diverse omics approaches has substantially advanced our understanding of beneficial plant–microbe interactions, underscoring their crucial roles in plant growth, nutrient acquisition, and tolerance to both biotic and abiotic stresses (Sindelar, 2013). In particular, these approaches have revealed how microorganisms enhance plant

Table 10.3. Main computational tools for multi-omics data integration.

Categories	Tools/methods	Purpose
Statistical methods	CCA, PLS, MOFA	Identify correlations or latent factors across omics modalities
Feature-wise methods	Analyte-to-analyte correlations, contrastive log-linear regression, Dirichlet-multinomial regression	Detection of analyte-to-analyte associations
Network analyses	SPIEC-EASI, MiMeNet	Identify microbial or metabolic hubs
Longitudinal methods	Dynamic Bayesian networks, mixed models	Temporal tracking of interactions
Causal mediation analyses	Mendelian randomization, SEM	Distinguish direct and indirect effects
GWAS–omics integration	GWAS coupled with high-throughput phenotyping	Identification of loci controlling agronomic
Bioinformatics platforms	mixOmics, OmicsPLS, SNF, MCIA, GpemDB	Multi-omics integration and visualization
Databases	WheatOmics, MBKbase, ZEAMAP	Exploration of genetic and metabolic networks, functional annotation of host–microbe interactions
AI and deep learning	Various machine learning/deep learning models	Prediction of complex phenotypes, heterosis

performance by mediating key functions such as biological nitrogen fixation, phosphorus solubilization, phytohormone biosynthesis (including auxins, cytokinins, and salicylic acid), and the regulation of metal availability and homeostasis within the rhizosphere (Jain *et al.*, 2024).

Among the most extensively studied beneficial plant–microbe interactions is the symbiosis between legumes and rhizobia. Integrated transcriptomic and proteomic analyses have demonstrated that flavonoids secreted by plant roots activate rhizobial nodulation (nod) genes, thereby initiating nodule formation. In parallel, MSI has enabled the spatial mapping of metabolite distributions within nitrogen-fixing and non–nitrogen-fixing nodules, providing insights into the metabolic compartmentalization underlying symbiotic nitrogen fixation (Liu and Murray, 2016). Proteomic investigations have further shown that inoculation of rice with *Pseudomonas fluorescens* induces the accumulation of antioxidant-related proteins, such as glutathione S-transferases, leading to enhanced plant growth. Similarly, association with *Sinorhizobium meliloti* has been linked to increased abundance of photosynthesis-related proteins in rice. In *Medicago truncatula*, metabolomic analyses revealed a marked accumulation of amino acids (notably glutamine and arginine) and isoflavonoids in response to arbuscular mycorrhizal colonization (Sharma *et al.*, 2020). Moreover, the integration of transcriptomic and proteomic datasets has enabled the dissection of soybean–arbuscular mycorrhizal fungi (AMF) symbiosis, highlighting the contribution of AMF to enhanced disease resistance and host defense responses (Zhang *et al.*, 2020b).

Omics approaches provide powerful frameworks for deciphering how plants detoxify harmful compounds and cope with environmental stress. For instance, proteomic analyses of *Pseudomonas putida* UW4 have revealed molecular mechanisms involved in stress perception, detoxification, and adaptive responses that support plant–microbe interactions (Weckwerth *et al.*, 2020). In poplar trees associated with beneficial fungal endophytes, enhanced accumulation of heat shock proteins (notably Hsp70) has been observed, alongside reduced metal accumulation in plant tissues, collectively contributing to improved stress tolerance and detoxification capacity (Qi *et al.*, 2022).

Plant metabolites also play a decisive role in shaping root-associated microbial communities. In *Arabidopsis thaliana* and *Echinacea purpurea*, specialized metabolites such as triterpenoid glycosides and signaling molecules like glycerol-3-phosphate have been shown to selectively recruit and structure beneficial microbial populations in the rhizosphere (Kumar *et al.*, 2023). Furthermore, high-throughput analyses have demonstrated that *Arabidopsis* can modulate root branching patterns independently of classical auxin signaling, highlighting alternative regulatory pathways controlling root system architecture (Gonin *et al.*, 2023).

Certain microbial consortia further enhance plant performance under suboptimal conditions, including low-light environments, while simultaneously priming host immune responses against pathogens (Hou *et al.*, 2021). In addition, integrated metagenomic and metatranscriptomic analyses have revealed that root- and fruit-associated microbiomes, such as those observed in watermelon, play critical roles in nutrient acquisition, plant growth promotion, and tolerance to abiotic and biotic stresses (Saminathan *et al.*, 2018).

10.5.2 Host–pathogen interactions and disease resistance

Multi-omics approaches have profoundly transformed the study of host–pathogen interactions by uncovering the molecular, genetic, and metabolic mechanisms that underpin plant defense and pathogen virulence. The integration of transcriptomics, proteomics, metabolomics, and spatial analyses now enables the identification of resistance genes, pathogen effectors, and defense-related molecules involved in plant immunity (Scossa *et al.*, 2021; Saini *et al.*, 2022). Pathosystem transcriptomics, in particular, has facilitated the characterization of dynamic gene expression profiles during plant–pathogen interactions. For instance, in the rice–*Magnaporthe oryzae* system, more than 240 fungal transcripts encoding putative effectors, including glycosyl hydrolases, have been identified (Gui *et al.*, 2020). Similarly, transcriptomic analysis of the tomato–*Cladosporium fulvum* interaction revealed key regulatory genes associated with resistance mediated by the *Cf-19* gene (Zhao *et al.*, 2019). In addition, multi-omic analyses have revealed the enrichment of chitinase, polyketide synthase (PKS), and non-ribosomal peptide synthetase (NRPS) genes within the endophytic microbiome of sugar beet in response to fungal infection (Carrión *et al.*, 2019).

Proteomics and metabolomics further enable the identification of defense-related proteins and antimicrobial or immunomodulatory metabolites produced during infection. In barley resistant to *Puccinia hordei*, LC–MS/MS analyses revealed the accumulation of proteins involved in carbohydrate metabolism and pathogen protein degradation. Likewise, in grapevine affected by flavescence dorée, 48 differentially abundant proteins were detected, including the antioxidant enzyme isocitrate dehydrogenase (Sharma *et al.*, 2020). Moreover, certain endophytic bacteria, such as *Klebsiella pneumoniae*, express enzymes including superoxide dismutases, catalases, and peroxidases that contribute to reactive oxygen species (ROS) detoxification, thereby enhancing plant tolerance to oxidative stress (Medina-Cordoba *et al.*, 2021).

Metabolomics also enables the identification of chemical mediators involved in infection. For example, an untargeted metabolomic study in grapevine revealed specific lipid markers associated with *Plasmopara viticola* infection (Negrel *et al.*, 2018). Similarly, in the soybean–*Phytophthora sojae* interaction, metabolomic profiling uncovered 90 metabolites that accumulated differentially between resistant and susceptible lines (Zhu *et al.*, 2018).

Co-transcriptomic approaches allow the simultaneous analysis of plant and pathogen responses. Nobori *et al.* (2020) integrated transcriptomic and proteomic data to demonstrate how plant immune responses shape bacterial behavior. Additionally, MSI has been applied to localize and identify peptides, proteins, and metabolites associated with pathogen infection. In tomato, for instance, MSI revealed the molecular processes underlying the formation of feeding sites by root-knot nematodes (Barbosa *et al.*, 2018).

Multi-omics approaches also facilitate the identification of resistance genes that can be exploited for crop improvement. In rice, genome-wide association studies (GWAS) identified the gene *OsTPS1*, regulated by the demethylase *JMJ705*, which enhances resistance to bacterial leaf blight (Zhan *et al.*, 2023). In cotton, the MYB transcription factor RVE2 is critical for defense against *Verticillium dahliae* (Liu *et al.*, 2023), while in buckwheat (*Fagopyrum tataricum*), cytochrome P450 genes are linked to resistance against *Rhizoctonia solani* (He *et al.*, 2023). Integration of transcriptomic and metabolomic data further showed that the gene *Bph30* confers resistance to the brown planthopper (*Nilaparvata lugens*) through modulation of the shikimate pathway (Shi *et al.*, 2023). Moreover, comparative transcriptomic and proteomic analyses identified over 350 genes secreted by the salivary glands of rice insect pests, highlighting their crucial role in plant–insect interactions (Fan *et al.*, 2025).

10.5.3 Responses to environmental stresses

Interactions between plants and beneficial microbes play a critical role in mitigating the adverse effects of abiotic stresses such as metal toxicity, drought, and salinity. Omics approaches have demonstrated that plant-associated microbes activate a range of physiological and molecular processes, including the production of antioxidants, osmoprotectants, and signaling molecules, thereby enhancing crop resilience to stress (Hussain, 2019). For example, metabolomic studies have shown that certain microbes produce siderophores to improve iron acquisition under stress and synthesize phytoalexins and terpenes that contribute to plant defense (Trivedi *et al.*, 2021). Proteomic analyses have revealed that beneficial bacteria, such as *Pseudomonas fluorescens*, enhance plant growth under drought, heat, or metal stress by modulating proteins involved in photosynthesis, transport, and molecular chaperones such as GroEL and DnaK (Gomes *et al.*, 2014). Similarly, transcriptomic analysis of the halotolerant endophyte *Halomonas sp.* MC1 demonstrated its role in promoting salt tolerance in *Mesembryanthemum crystallinum* by regulating stress-related pathways (Zhang *et al.*, 2020a).

In plants, the integration of multiple omics layers reveals key genes and pathways underlying stress tolerance. In cotton, combined transcriptomic–metabolomic analyses identified genes associated with salt tolerance (Peng *et al.*, 2019). In rice, integrative genomic–transcriptomic studies highlighted modifications in the DROT1 gene, which is essential for drought adaptation (Zhang *et al.*, 2020a). In the Nagina 22 rice variety, transcriptomic–metabolomic integration elucidated the regulation of carbohydrate metabolism and L-phenylalanine biosynthesis in response to water deficit (Dwivedi *et al.*, 2023). In rapeseed, GWAS combined with RNA-seq identified a single-nucleotide polymorphism (SNP) associated with the *ABCG16* gene, linked to drought tolerance (Salami *et al.*, 2024). For cold stress, GWAS revealed the *OsSEH1* gene as a key regulator of phenylpropanoid and flavonoid metabolism (Gu *et al.*, 2023), while proteomic analyses showed that TaPGK contributes to cold tolerance in wheat (Zhang *et al.*, 2023).

Spatial omics approaches complement these studies by visualizing biochemical adjustments *in situ*. MSI of barley roots exposed to salt stress revealed lipid redistribution associated with tolerance (Sarabia *et al.*, 2018). Likewise, nanoparticle-assisted imaging in rice roots mapped critical phytohormones including abscisic acid, auxin, and jasmonate, highlighting their central roles in regulating plant stress responses (Shiono and Taira, 2020).

10.6 Challenges and Future Perspectives

10.6.1 Limitations of current omics technologies

During the past decade, advances in omics technologies have transformed plant biology, providing unprecedented molecular resolution across genomes, transcriptomes, proteomes, and metabolomes. However, each omics layer captures only a portion of the biological question and comes with its own technical and interpretative limitations. Table 10.4 summarizes these complementary perspectives, outlining the principal assays (omics analysis), along with their strengths and weaknesses that define contemporary multi-omics studies.

Genomics provides the most robust molecular signals, with the identification of SNPs, indels, copy number variations (CNVs), and structural variants (SVs) demonstrating high reproducibility across laboratories. Current long-read sequencing technologies resolve repetitive regions and complex haplotypes, clarifying loci that are fragmented or ambiguous in short-read datasets (Li *et al.*, 2020). Nevertheless, sequence information alone is often insufficient to predict phenotype, as post-transcriptional regulation, epigenetic modifications, and environmental influences play critical roles. Moreover, noncoding variants remain difficult to interpret without functional context (Bodea *et al.*, 2018; Abdalla and Abdalla, 2022).

Epigenomics and chromatin-based assays, including DNA methylation mapping, chromatin accessibility profiling, protein–DNA interaction analyses, and three-dimensional genome conformation studies, address this gap by revealing regulatory states and chromatin architecture that influence transcriptional output (Bartha *et al.*, 2018). However, whole-epigenome profiles can obscure cell-type-specific differences and are susceptible to variability in antibody or enzyme efficiency. Single-cell approaches improve resolution but are constrained by data sparsity and cost, highlighting the importance of integrated experimental designs that combine epigenomic and transcriptional readouts (Yan *et al.*, 2022; Kwok *et al.*, 2025).

Transcriptomics provides a dynamic view of gene activity, with several approaches available for its analysis. Bulk RNA-seq remains cost-effective for many applications and is useful for detecting differential gene expression; however, it measures only RNA, not protein, and short-read data can limit isoform resolution (Ren *et al.*, 2023). Single-cell and spatial transcriptomic methods offer higher resolution, capturing gene expression at the level of individual cells and tissues, and are particularly valuable for analyzing heterogeneous tissues and plant–microbe interfaces. These approaches, however, have their own limitations, including dropout events in scRNA-seq/single nucleus (sc)RNA-seq and trade-offs in resolution, sample throughput, and temporal coverage when using spatial methods (Nobori, 2025). Integrating scRNA-seq for

Table 10.4. Advantages and disadvantages of omics technologies.

Omics domain	Representative assays/platforms	Major advantages	Typical disadvantages/ caveats	Practical notes
Genomics	WGS/WES; SNP and CNV arrays; GWAS; pangenomes and graph genomes; long-read (PacBio HiFi, ONT); single-cell DNA-seq; Hi-C	Stable blueprint; resolves SNPs/indels/CNVs/SVs; long reads span repeats, TEs, and complex loci; pangenomes capture intraspecies diversity	Genotype→phenotype gap (regulatory/epigenetic layers); non-coding variant interpretation; complex/large plant genomes (polyploidy, TEs) raise cost and assembly difficulty; organellar–nuclear DNA confounds	Use SNP arrays/targeted panels for large panels; long reads + Hi C for SVs/ haplotypes and polyploid phasing; pangenome graphs improve GWAS/QTL in diverse germplasm
Epigenomics/chromatin	WGBS/ oxBS (DNA methylation); ATAC-seq; ChIP-seq/CUT&RUN/ CUT&Tag; Hi-C/ HiChIP; single-cell/snATAC	Captures regulatory state, chromatin architecture, and stress memory; explains gene–environment effects	Tissue heterogeneity confounds bulk; antibody/ enzymatic biases; high chloroplast DNA may dominate; methylation context (CG/CHG/CHH) interpretation is species- and tissue-specific	Co-profile with RNA-seq or spatial assays for regulatory inference; use nuclei-based preps to reduce plastid contamination; single-cell boosts resolution but increases sparsity/cost
Transcriptomics (bulk)	RNA-seq (poly(A), rRNA-depletion); targeted panels (e.g. NanoString); ribosome profiling	Dynamic readout of gene activity; robust stats/QC; cost-effective across treatments/time-courses	mRNA ≠ protein; batch effects; isoform quant under short reads can be ambiguous; plastid/rRNA carryover; strong diel/ photoperiod effects in plants	Include ERCC spike-ins/UMIs for quantitation; randomize harvesting times or model circadian effects; pair with proteomics for functional read-through
Single-cell/spatial transcriptomics	sc/snRNA-seq (droplet/plate); spatial barcoding (Visium-class); imaging ST (MERFISH/ seqFISH/ISS/ smFISH)	Resolves cell types/states and developmental gradients; preserves tissue niches with ST	Droplet dropout; cell wall requires protoplasting (artifacts) → prefer snRNA-seq for plants; spatial trade-offs among plex, resolution, area; chloroplast transcripts can dominate	Integrate snRNA-seq with ST to deconvolve spots; validate markers with smFISH/ ISS; gentle nuclei isolation improves plant data quality

Continued

Table 10.4. Continued

Omics domain	Representative assays/platforms	Major advantages	Typical disadvantages/ caveats	Practical notes
Proteomics (discovery)	LC–MS/MS (DDA); DIA/ SWATH; top-down MS; phospho-/ubi-enrichment	Closer to phenotype; detects PTMs and proteoforms; DIA improves depth and quantitative completeness	Large dynamic range; missing values (DDA); PTM site localization ambiguity; tough plant matrices (phenolics, polysaccharides)	Prefer DIA for cohorts; enrich phospho/ubi for signaling; add PVPP/TCA-acetone or phenol extraction to remove phenolics; strict FDR and iRT standards
Proteomics (targeted/multiplexed)	PRM/SRM; isobaric tags (TMT/iTRAQ)	Precise, reproducible quant; high multiplexing with TMT for many tissues/time points	Ratio compression (isobaric); instrument time; peptide→protein inference limits; batch effects across TMT sets	Internal standards (heavy peptides) and match-between-runs mitigate missingness; balance TMT batches across genotypes and organs
Metabolomics/lipidomics (untargeted)	LC–MS (HILIC/ RP), GC–MS; IMS (MALDI/DESI/ SIMS) for spatial; NMR	Closest to phenotype; captures stress and pathway end-products; spatial IMS maps specialized metabolites *in situ*	Annotation uncertainty (isomers/adducts); matrix effects/ion suppression; absolute quant is hard; strong tissue and diel effects	Use on-tissue MS/MS + GNPS libraries + orthogonal LC-MS to raise ID level; isotope-labeled standards for key analytes; adopt imzML for FAIR spatial data
Metabolomics (targeted/ quantitative)	LC–MRM/PRM; stable-isotope dilution; GC–MS SIM	High specificity/accuracy; field-deployable panels for hormones, defense metabolites	Limited breadth; need authentic standards; extraction recovery is matrix-dependent	Ideal for validating discovery hits and hormone panels; includes QA pools and matrix-matched calibration
Metagenomics/ metatranscriptomics	Shotgun WGS (rhizosphere/ phyllosphere/ endosphere); 16S/ITS; meta-RNA-seq	Profiles plant-associated microbiomes; links taxa→functions; tracks pathogen/beneficials under treatments	Assembly/annotation bias; compositionality; high host DNA/RNA contamination; strain-level resolution still hard	Host depletion and spike-ins help; pair with host spatial omics (ST, IMS) to map microbe–host interfaces and metabolites in tissues

Multi-omics integration	Joint assays (ATAC+RNA, CITE-seq-like for plants), multi-modal ST (co-detect proteins/metabolites); computational integration	Cross-layer insights (regulation→expression→protein→metabolite); stronger causal hypotheses; improves candidate prioritization	Cost and data scale; batch alignment across organs/stages; differing dynamic ranges/units; complex statistics	Use robust integration frameworks and pre-registered QC; design shared samples/timepoints across layers; orthogonal validation (genetics, reporters, imaging)
Phenomics/imaging (bonus)	Hyperspectral and thermal imaging; 3D/laser scanning; fluorescence reporters; root imaging (rhizotrons/X-ray CT)	High-throughput, non-destructive trait capture; links genotype→field performance; enables time-series under stress	Expensive platforms; environmental confounders; feature extraction and ground truthing needed	Align imaging timepoints with omics sampling; standardized lighting and calibration targets; integrate with weather/soil sensors for covariates

deconvolution with spatially resolved validation methods, such as smFISH or ISS, represents a current best practice for accurately mapping gene expression in context (Yu *et al.*, 2023).

Proteomics provides a closer link to the phenotype by capturing post-translational regulation and the diversity of proteins produced in a biological system (Aebersold and Mann, 2016). Data-independent acquisition (DIA/SWATH) minimizes missing data and is particularly advantageous for studies involving large sample sets (Daley and Smith, 2013; Ludwig *et al.*, 2018), while targeted approaches (PRM/SRM) and multiplexed workflows (TMT/iTRAQ) enable precise quantification of the same proteins across multiple samples (Peterson *et al.*, 2012). Nevertheless, proteomic analyses remain constrained by current technical challenges, including limited dynamic range, uncertainties in peptide-to-protein inference, and ratio compression associated with isobaric labeling, emphasizing the need for careful experimental design and stringent false discovery rate (FDR) control (Savitski *et al.*, 2013).

Metabolomics and lipidomics capture the end-products of biological pathways, placing them closest to the phenotype and reflecting the influence of environmental conditions. Untargeted LC– or gas chromatography (GC)–MS approaches detect a wide range of metabolites but face challenges in small-molecule identification due to isomers, adduct formation, matrix effects, and variable metabolite abundances (Dunn and Hankemeier, 2013). Targeted analyses using stable-isotope dilution provide accurate quantification for selected analytes (Bueschl *et al.*, 2014), while IMS adds spatial resolution, revealing metabolite distributions within tissues. Combining on-tissue MS/MS with curated spectral libraries (e.g. GNPS) and complementary LC–MS analyses enhances confidence in metabolite identification (Alexandrov *et al.*, 2019). Data formats such as imzML and platforms like METASPACE facilitate data sharing and reproducibility (Schramm *et al.*, 2012).

Metagenomics and metatranscriptomics extend this molecular perspective to the plant-associated microbiome, linking microbial community composition with encoded and expressed functions (Graham *et al.*, 2016). These approaches are invaluable for elucidating plant–microbe interactions but require careful handling of compositional biases, assembly errors, and host contamination (Mishra *et al.*, 2022). Integrating host and microbial omics, particularly through spatial co-analyses, helps contextualize microbial signals within plant anatomy and tissue microenvironments (Zhang *et al.*, 2024).

Finally, multi-omics integration leverages complementary data layers to construct causal chains from regulatory states to transcription, to protein/post-translational modifications, to metabolites, and ultimately to phenotype (Kumar *et al.*, 2023). Joint assays such as ATAC-seq combined with RNA-seq, CITE-seq analogs, and spatial multi-omics, alongside mature computational frameworks, now enable aligned profiling across multiple modalities (Kimotho and Maina, 2024). Successful studies emphasize rigorous quality control and cross-layer validation, including smFISH for key transcripts, targeted parallel reaction monitoring (PRM)/tandem mass tag (TMT) for proteins, and isotope-dilution for metabolites, transforming associative maps into mechanistic, testable models (Xu *et al.*, 2021).

10.6.2 Advances in artificial intelligence and machine learning for omics integration

Recent advances in artificial intelligence (AI) and machine learning (ML) have created new opportunities for integrating multi-omics datasets in plant systems biology (Fig. 10.1; Table 10.5) (Cembrowska-Lech *et al.*, 2023). Beyond conventional deep learning architectures such as convolutional and recurrent neural networks, emerging AI paradigms including generative adversarial networks (GANs), reinforcement learning (RL), automated machine learning (AutoML), transfer learning, federated learning, and attention-based frameworks, provide powerful tools for modeling the complexity of plant molecular systems and enhancing predictive accuracy (Libbrecht and Noble, 2015; Huang *et al.*, 2023). These approaches facilitate cross-domain learning, integration of heterogeneous data types, and comprehensive interpretation of multi-layered datasets spanning genomics,

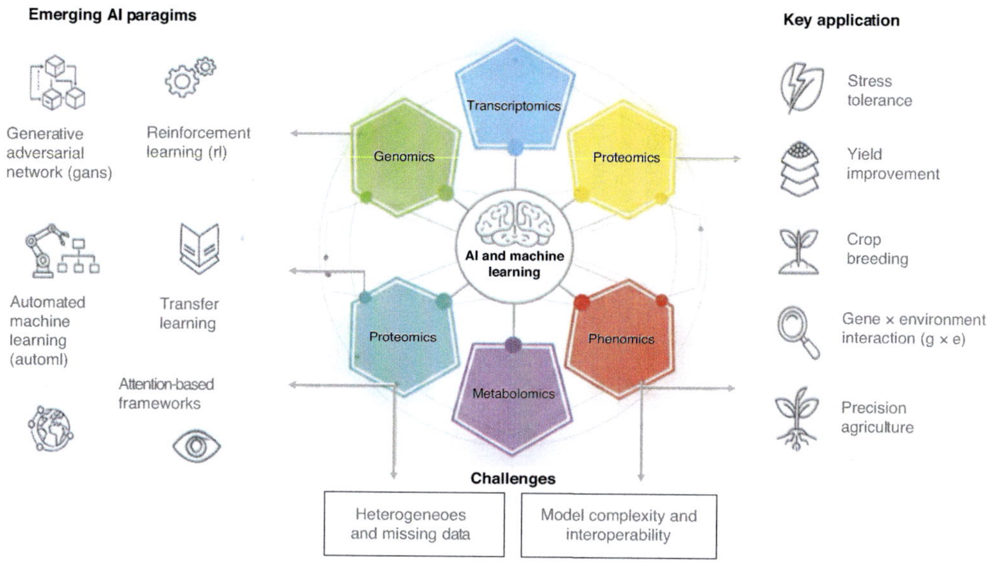

Fig. 10.1. AI and machine learning integration for plant multi-omics.

transcriptomics, proteomics, metabolomics, and phenomics.

The use of AI and ML models is increasingly widespread in plant science, with applications ranging from understanding plant stress responses and predicting crop yield to guiding breeding strategies and deciphering gene–environment interactions (Zhang *et al.*, 2022; Murmu *et al.*, 2024). Tools such as AutoML and transfer learning facilitate model development and enable knowledge transfer from well-studied species to less-characterized plants. Federated learning allows data from multiple laboratories to be integrated and analyzed collaboratively without exposing sensitive information. GANs can generate synthetic but realistic datasets, which are particularly useful for rare plant varieties or when sample sizes are limited (Bai *et al.*, 2024). Despite these advances, a major challenge remains: integrating multi-omics datasets with heterogeneous data types, missing values, or differing sample sizes can complicate analysis, interpretation, and predictive modeling (Flores *et al.*, 2023).

Overall, progress is steadily advancing in linking AI with plant omics, shifting research from descriptive analyses toward predictive modeling of dynamic biological processes. These approaches have the potential to accelerate crop improvement, promote sustainable agriculture, and provide a deeper understanding of plant–environment interactions across multiple organizational levels (Edger, 2022; Mishra *et al.*, 2024). Future efforts will need to balance model complexity with interpretability, ensure that integrated datasets remain biologically meaningful, and maintain transparency to allow reproducibility and validation by the broader research community.

Advances in handling diverse data types, including genomic sequences, gene expression profiles, proteomics, metabolomics, epigenetics, and integrative physiological measurements, have become essential for studying complex plant traits, long-distance signaling, and stress responses. Traditional analytical methods and shallow ML approaches often fall short in this context, as multi-omics datasets are high-dimensional, interdependent, and frequently contain missing values that complicate analysis. Deep learning and generative AI have emerged as transformative solutions, offering the ability to generate representative synthetic data, integrate heterogeneous datasets, augment existing information, and handle multiple data modalities with greater flexibility and accuracy (Libbrecht and Noble, 2015).

Table 10.5. Opportunities, challenges, and future perspectives of using bioinformatics and AI/ML techniques in plant omics.

AI/ML approach	Representative models/ examples	Data types used	Key findings/use cases (plant focus)	Challenges/caveats	Future recommendations
Generative adversarial networks (GANs)	omicsGAN; CycleGAN for cross-modal mapping; cGAN/AC-GAN for synthetic data generation	Multi-omics (transcriptome, proteome, metabolome, methylome); stress–trait datasets	Graph-aware GANs improve phenotype prediction, reconstruct missing omics, and simulate rare stress phenotypes (e.g. drought, salinity)	Mode collapse; training instability; biological plausibility of synthetic omics; limited crop-specific benchmarking	Embed curated gene–metabolite and regulatory graphs; constrain GAN loss by biological priors; test on independent species
Variational autoencoders (VAEs)/multi-modal generative models	scVI/totalVI; MOFA+; Plant-VAE prototypes	Single-cell transcriptome + metabolome; bulk multi-omics under stress	Learn joint latent spaces capturing coordinated metabolic–transcriptional responses; denoise expression data; reveal hidden stress modules	Latent factors are difficult to interpret biologically; batch effects and label drift	Combine VAEs with gene-set priors and graph regularization; report latent-factor stability and pathway enrichment
Graph neural networks (GNNs)	GLUE; DeepMOCCA; GCN/GraphSAGE with gene–protein–metabolite networks	Regulatory and metabolic networks; transcriptomics; proteomics	Encode biological topology for stress-gene prioritization, yield prediction, and regulatory-module discovery	Noisy or incomplete plant interaction graphs; scalability; over-smoothing	Use tissue-specific curated networks; integrate graph uncertainty; contrastive pretraining across species
Transformers and foundation models	DNABERT; Enformer; PlantBERT; multimodal transformers for omics + imaging	Genomic sequences, epigenomic marks, spectral or imaging data	Capture long-range regulatory dependencies; unify gene expression with phenomics and hyperspectral imaging	High pretraining cost; domain shift across crops; data hunger	Develop lightweight adapters and domain-adaptive pretraining; fine-tune on crop-specific tasks with small datasets
Reinforcement learning (RL)	DDPG/actor–critic; bandits for adaptive experimentation	Temporal omics + phenotyping time series	Conceptual frameworks for adaptive irrigation, nutrient, or stress-response optimization guided by omics feedback	Sparse reward signals; lack of plant-specific simulators	Build biologically realistic simulators; apply safe-RL for greenhouse or field automation

Automated machine learning (AutoML)	Auto-sklearn; TPOT; AutoGluon for plant metabolomics and gene-expression pipelines	Transcriptome, proteome metabolome, metagenome	Automates model and feature selection; identifies robust gene signatures for stress or yield	Overfitting and data leakage risk; black-box pipelines; weak temporal modeling	Include time-aware models (transformers/RNNs); enforce nested CV; document provenance via model cards
Deep transfer learning	scArches; cross-species expression transfer; domain-adversarial models	Cross-crop transcriptomes; GWAS; variant function prediction	Transfers knowledge from model plants (*Arabidopsis*, rice) to orphan crops; boosts generalization with limited labeled data	Negative transfer; dataset shift; label mismatch	Domain-adversarial alignment; cross-species calibration; explicit uncertainty quantification
Federated learning (FL)/privacy-preserving AI	Cross-institute FL with secure aggregation; split learning on plant omics consortia	Distributed multi-omics from breeding centers or labs	Enables collaboration without raw-data exchange; supports data sovereignty across regions	Non-IID data; network latency; privacy leakage via gradients	Personalized FL; differential-privacy noise; standardized evaluation across crops

A clear example is omicsGAN, a type of generative adversarial network that integrates multiple omics layers and their interactions to predict phenotypes and identify novel features (Ahmed *et al.*, 2021). Graph-aware deep learning models further enhance multi-omics analysis by imputing missing molecular layers, reducing experimental noise, and controlling model complexity, enabling robust applications across diverse crops and stress conditions (Ballard *et al.*, 2024; Baião *et al.*, 2025). Additionally, techniques such as contrastive learning and cycle-consistency losses are increasingly used to align heterogeneous omics datasets, facilitating accurate generative modeling of plant systems.

In addition to generative AI, other advanced forms of AI are gaining traction in plant science. Although still emerging in plant bioinformatics, reinforcement learning offers the potential for self-optimizing experiments, such as improving nutrient delivery or fine-tuning environmental conditions to precisely study plant phenotypes. Transfer learning enables knowledge gained from well-annotated species, such as *A. thaliana*, to be applied to orphan crops or poorly annotated plants, mitigating challenges associated with limited data (Yang *et al.*, 2023). Federated learning provides a distributed approach, allowing models to learn from datasets located in different laboratories or institutions without centralizing sensitive or proprietary data, such as genomic sequences (Calvino *et al.*, 2024).

Multiple deep learning approaches intersect with graph-based reasoning, transformers, and multimodal fusion to accelerate plant omics analyses. Graph neural networks capture topological relationships among genes, metabolites, and signaling proteins. Multimodal transformers integrate textual annotations, imaging, and multi-omics layers to predict phenotypes and stress responses (Han *et al.*, 2023; Xiao *et al.*, 2023). These methods enhance functional annotation, regulatory network inference, and crop-trait prediction. However, key challenges remain, particularly regarding model interpretability and validation. Black-box models require biologically meaningful modules for feature attribution and pathway-level analysis to ensure transparency and trust, which are critical for practical deployment in decision-support systems.

In summary, the development of sophisticated AI models, capable of generating data, learning from feedback, transferring knowledge across tasks, and integrating distributed datasets, combined with innovations in multi-omics, is transforming our understanding of plant systems at the whole-organism level (Hossen *et al.*, 2025). When paired with explainable tools, rigorous validation, and open-access data, these approaches will accelerate the development of predictive and congruent workflows for crop improvement, enable systems-level models for resilience, and support sustainable agricultural innovations (Abdul Razak *et al.*, 2024).

The rich layers of multi-omics data in plants can be effectively leveraged only with robust computational infrastructure, reliable analytical methods, clear data management protocols, and sustainable workflows (Hasin *et al.*, 2017). Any predictive model intended for crop improvement or agricultural applications must be rigorously validated before deployment. The success of AI in integrating omics data depends on its ability to handle the heterogeneity, large scale, and complex relationships inherent in multi-omics datasets (Leng *et al.*, 2022). For example, integrating gene expression patterns with metabolite profiles to predict stress responses can yield misleading results if baseline references are inaccurate, data normalization is insufficient, or temporal adjustments are poorly applied, potentially generating false associations or misdirecting growth predictions. Advanced computational approaches, including near-certain model fitting, image-based integration, and multi-level forecasting, are required to capture the intricate communication between plant tissues and systems (Hasin *et al.*, 2017).

Another significant challenge lies in the heterogeneity and lack of standardization across omics datasets, which limits data sharing and interoperability among laboratories. Adoption of common formats such as imzML for imaging data, mzML for mass spectrometry, and HDF5/AnnData for single-cell or spatial transcriptomics, alongside adherence to data reporting standards like MIAME and MIBBI, can greatly facilitate integration, transparency, and reproducibility (Wilkinson *et al.*, 2016). Workflow management platforms such as Galaxy, Bioconductor, and Nextflow further enable reproducible analyses, handling fragmented data types and maintaining continuity across multi-omics studies in crops.

A final consideration is that many AI models function as 'black boxes,' posing challenges for interpretability in plant bioinformatics. Effective integration of computational tools with biological insight requires collaboration between plant scientists, data analysts, and model developers to ensure predictions are biologically meaningful and actionable for agriculture. Explainable AI tools, such as LIME (local interpretable model-agnostic explanations) and SHAP (Shapley additive explanations), can be incorporated into multi-omics analyses to identify key genes, metabolites, or spatial–temporal features driving model predictions (Perdih *et al.*, 2021; Karim *et al.*, 2022).

Looking forward, AI-driven plant omics will not only enhance predictive accuracy but also prioritize transparency, ethical standards, and accessibility. These approaches will enable the development of crops that are resilient and sustainable, while ensuring that the benefits of these technologies are equitably shared and implemented globally.

10.7 Conclusion

The integration of spatial omics with existing multi-omics tools represents a major advancement in studying plant–microbe interactions, shifting from broad, averaged datasets to high-resolution, localized insights into the biochemical 'conversations' between host and microbe. Combining genomics, transcriptomics, proteomics, and metabolomics with spatial maps provides detailed mechanistic information on both beneficial and pathogenic partnerships. This integrated approach enables: identification of precise genetic variations within specific plant cells; mapping the production and movement of small molecules, including signaling compounds and secondary metabolites, within defined tissue zones; and tracking the spatial distribution of microbial populations across plant anatomy. Importantly, visualizing these molecules reveals how microbes establish localized interactions with the host, manipulate environmental or cellular conditions, and adapt specifically to niches such as the rhizosphere or infection sites, enhancing our understanding of host support, microbial performance, and system-level integration.

These high-resolution molecular insights have direct applications in precision agriculture. By pinpointing the specific molecules and interactions that promote beneficial plant–microbe partnerships, such as enhanced nutrition or strengthened defense, researchers can identify optimal bio-inoculants tailored to individual farm environments, improving predictability and efficiency. Likewise, detailed *in situ* knowledge of pathogen virulence and host defense mechanisms can inform the development of smarter, environmentally friendly control strategies, reducing reliance on synthetic chemicals, supporting soil and plant health, and minimizing overall crop damage.

For crop improvement, spatial multi-omics provides a rich foundation for more targeted breeding or engineering. Beyond selecting based solely on observable phenotypes, breeders can leverage localized interaction data and cell-type-specific gene maps to enhance beneficial partnerships or suppress pathogenic effects. Future applications include precise modulation of membrane sensors to optimize microbial signaling, or nucleus-level gene editing using tools such as CRISPR-Cas to fine-tune plant defenses. These strategies aim to support the assembly of enhanced synthetic microbial communities (SynComs). Achieving these outcomes will depend on the next generation of technologies, integrating vast spatially resolved molecular datasets with advanced machine learning to generate predictive models for resilient and productive crop systems.

Acknowledgements

This study was financially supported by the Phytopathology Unit, Ecole Nationale d'Agriculture de Meknès, Morocco.

References

Abdalla, M. and Abdalla, M. (2022) A general framework for predicting the transcriptomic consequences of non-coding variation and small molecules. *PLOS Computational Biology* 18(4), e1010028. DOI: 10.1371/journal.pcbi.1010028.

Abdul Razak, S.F., Yogarayan, S., Sayeed, M.S. and Mohd Derafi, M.I.F. (2024) Agriculture 5.0 and explainable AI for smart agriculture: A scoping review. *Emerging Science Journal* 8(2), 744–760. DOI: 10.28991/ESJ-2024-08-02-024.

Aebersold, R. and Mann, M. (2016) Mass-spectrometric exploration of proteome structure and function. *Nature* 537(7620), 347–355. DOI: 10.1038/nature19949.

Ahmed, K.T., Sun, J., Cheng, S., Yong, J. and Zhang, W. (2021) Multi-omics data integration by generative adversarial network. *Bioinformatics (Oxford, England)* 38(1), 179–186. DOI: 10.1093/bioinformatics/btab608.

Alexandrov, T., Ovchinnikova, K., Palmer, A., Kovalev, V., Tarasov, A. *et al.* (2019) METASPACE: A community-populated knowledge base of spatial metabolomes in health and disease. *BioRxiv*. DOI: 10.1101/539478.

Argelaguet, R., Arnol, D., Bredikhin, D., Deloro, Y., Velten, B. *et al.* (2020) MOFA+: A statistical framework for comprehensive integration of multi-modal single-cell data. *Genome Biology* 21(1), 111. DOI: 10.1186/s13059-020-02015-1.

Bai, D.-P., Lin, X.-Y., Hu, Y.-Q., Chen, Z.-Z., Chen, L. *et al.* (2021) Metagenomics approach to identify lignocellulose-degrading enzymes in the gut microbiota of the Chinese bamboo rat cecum. *Electronic Journal of Biotechnology* 50, 29–36. DOI: 10.1016/j.ejbt.2020.12.001.

Bai, W., Li, C., Li, W., Wang, H., Han, X. *et al.* (2024) Machine learning assists prediction of genes responsible for plant specialized metabolite biosynthesis by integrating multi-omics data. *BMC Genomics* 25(1), 418. DOI: 10.1186/s12864-024-10258-6.

Baião, A.R., Cai, Z., Poulos, R.C., Robinson, P.J., Reddel, R.R. *et al.* (2025) A technical review of multi-omics data integration methods: From classical statistical to deep generative approaches. *Briefings in Bioinformatics* 26(4), bbaf355. DOI: 10.1093/bib/bbaf355.

Ballard, J.L., Wang, Z., Li, W., Shen, L. and Long, Q. (2024) Deep learning-based approaches for multi-omics data integration and analysis. *BioData Mining* 17(1), 38. DOI: 10.1186/s13040-024-00391-z.

Balotf, S., Wilson, R., Tegg, R.S., Nichols, D.S. and Wilson, C.R. (2022) Shotgun proteomics as a powerful tool for the study of the proteomes of plants, their pathogens, and plant-pathogen interactions. *Proteomes* 10(1), 5. DOI: 10.3390/proteomes10010005.

Barbosa, E.A., Bonfim, M.F., Jr, Bloch, C., Jr, Engler, G., Rocha, T. *et al.* (2018) Imaging mass spectrometry of endogenous polypeptides and secondary metabolites from galls induced by root-knot nematodes in tomato roots. *Molecular Plant-Microbe Interactions* 31(10), 1048–1059. DOI: 10.1094/MPMI-02-18-0049-R.

Barmukh, R., Garg, V., Liu, H., Chitikineni, A., Xin, L. *et al.* (2025) Spatial omics for accelerating plant research and crop improvement. *Trends in Biotechnology* 43(8), 1904–1920. DOI: 10.1016/j.tibtech.2025.03.007.

Bartha, I., di Iulio, J., Venter, J.C. and Telenti, A. (2018) Human gene essentiality. *Nature Reviews. Genetics* 19(1), 51–62. DOI: 10.1038/nrg.2017.75.

Baysoy, A., Bai, Z., Satija, R. and Fan, R. (2023) The technological landscape and applications of single-cell multi-omics. *Nature Reviews. Molecular Cell Biology* 24(10), 695–713. DOI: 10.1038/s41580-023-00615-w.

Bentley, D.R., Balasubramanian, S., Swerdlow, H.P., Smith, G.P., Milton, J. *et al.* (2008) Accurate whole human genome sequencing using reversible terminator chemistry. *Nature* 456(7218), 53–59. DOI: 10.1038/nature07517.

Bharti, M.K., Siddique, R.A., Ranjan, K., Chandra, D. and Singh, N.P. (2021) Relevance of metatranscriptomics in symbiotic associations between plants and rhizosphere microorganisms. In: *Microbial Metatranscriptomics Belowground*. Springer Singapore, pp. 59–90. DOI: 10.1007/978-981-15-9758-9_3.

Bodea, C.A., Mitchell, A.A., Bloemendal, A., Day-Williams, A.G., Runz, H. *et al.* (2018) PINES: Phenotype-informed tissue weighting improves prediction of pathogenic noncoding variants. *Genome Biology* 19(1), 173. DOI: 10.1186/s13059-018-1546-6.

Bueschl, C., Kluger, B., Lemmens, M., Adam, G., Wiesenberger, G. *et al.* (2014) A novel stable isotope labelling assisted workflow for improved untargeted LC-HRMS based metabolomics research. *Metabolomics* 10(4), 754–769. DOI: 10.1007/s11306-013-0611-0.

Calvino, G., Peconi, C., Strafella, C., Trastulli, G., Megalizzi, D. *et al.* (2024) Federated learning: Breaking down barriers in global genomic research. *Genes* 15(12), 1650. DOI: 10.3390/genes15121650.

Carrión, V.J., Perez-Jaramillo, J., Cordovez, V., Tracanna, V., Hollander, M. *et al.* (2019) Pathogen-induced activation of disease-suppressive functions in the endophytic root microbiome. *Science (New York, N.Y.)* 366(6465), 606–612. DOI: 10.1126/science.aaw9285.

Cembrowska-Lech, D., Krzemińska, A., Miller, T., Nowakowska, A., Adamski, C. *et al.* (2023) An integrated multi-omics and artificial intelligence framework for advance plant phenotyping in horticulture. *Biology* 12(10), 1298. DOI: 10.3390/biology12101298.

Chen, K.H., Boettiger, A.N., Moffitt, J.R., Wang, S. and Zhuang, X. (2015) RNA imaging. Spatially resolved, highly multiplexed RNA profiling in single cells. *Science (New York, N.Y.)* 348(6233), aaa6090. DOI: 10.1126/science.aaa6090.

Chen, T.Y., You, L., Hardillo, J.A.U. and Chien, M.P. (2023) Spatial transcriptomic technologies. *Cells* 12(16), 1–13. DOI: 10.3390/cells12162042.

Chen, X., Sun, Y.-C., Church, G.M., Lee, J.H. and Zador, A.M. (2018) Efficient *in situ* barcode sequencing using padlock probe-based Baristaseq. *Nucleic Acids Research* 46(4), e22–e22. DOI: 10.1093/nar/gkx1206.

Daley, T. and Smith, A.D. (2013) Predicting the molecular complexity of sequencing libraries. *Nature Methods* 10(4), 325–327. DOI: 10.1038/nmeth.2375.

Damiani, I., Baldacci-Cresp, F., Hopkins, J., Andrio, E., Balzergue, S. *et al.* (2012) Plant genes involved in harbouring symbiotic rhizobia or pathogenic nematodes. *The New Phytologist* 194(2), 511–522. DOI: 10.1111/j.1469-8137.2011.04046.x.

Deek, R.A., Ma, S., Lewis, J. and Li, H. (2024) Statistical and computational methods for integrating microbiome, host genomics, and metabolomics data. *eLife* 13, e88956. DOI: 10.7554/eLife.88956.

Demiwal, P., Tayade, S., Yadav, S.R. and Sircar, D. (2024) A metabolomics perspective on root-derived plant immunity and phytohormone interaction. *Physiologia Plantarum* 176(1). DOI: 10.1111/ppl.14150.

Duncan, S. and Rosa, S. (2018) Gaining insight into plant gene transcription using smFISH. *Transcription* 9(3), 166–170. DOI: 10.1080/21541264.2017.1372043.

Dunn, W.B. and Hankemeier, T. (2013) Mass spectrometry and metabolomics: Past, present and future. *Metabolomics* 9(S1), 1–3. DOI: 10.1007/s11306-013-0507-z.

Dwivedi, A.K., Singh, V., Anwar, K., Pareek, A. and Jain, M. (2023) Integrated transcriptome, proteome and metabolome analyses revealed secondary metabolites and auxiliary carbohydrate metabolism augmenting drought tolerance in rice. *Plant Physiology and Biochemistry* 201, 107849. DOI: 10.1016/j.plaphy.2023.107849.

Edger, P.P. (2022) The power of chromosome-scale, haplotype-resolved genomes. *Molecular Plant* 15(3), 393–395. DOI: 10.1016/j.molp.2022.02.010.

Eid, J., Fehr, A., Gray, J., Luong, K., Lyle, J. *et al.* (2009) Real-time DNA sequencing from single polymerase molecules. *Science (New York, N.Y.)* 323(5910), 133–138. DOI: 10.1126/science.1162986.

Eng, C.-H.L., Lawson, M., Zhu, Q., Dries, R., Koulena, N. *et al.* (2019) Transcriptome-scale super-resolved imaging in tissues by RNA seqfish. *Nature* 568(7751), 235–239. DOI: 10.1038/s41586-019-1049-y.

Fan, K.-T. (2024) Decoding nature's conversations: Metabolomics and the intriguing world of plant-microbe interactions. *Preprints*. DOI: 10.22541/au.171934398.85448535/v1.

Fan, B.-L., Chen, L.-H., Chen, L.-L. and Guo, H. (2025) Integrative multi-omics approaches for identifying and characterizing biological elements in crop traits: Current progress and future prospects. *International Journal of Molecular Sciences* 26(4), 1466. DOI: 10.3390/ijms26041466.

Feierabend, M. and Töpfer, N. (2025) *In silico* encounters: Harnessing metabolic modelling to understand plant-microbe interactions. *FEMS Microbiology Reviews* 49, fuaf030. DOI: 10.1093/femsre/fuaf030.

Flores, J.E., Claborne, D.M., Weller, Z.D., Webb-Robertson, B.-J.M., Waters, K.M. *et al.* (2023) Missing data in multi-omics integration: Recent advances through artificial intelligence. *Frontiers in Artificial Intelligence* 6, 1098308. DOI: 10.3389/frai.2023.1098308.

Franck, J., Arafah, K., Elayed, M., Bonnel, D., Vergara, D. *et al.* (2009) MALDI imaging mass spectrometry: State of the art technology in clinical proteomics. *Molecular & Cellular Proteomics* 8(9), 2023–2033. DOI: 10.1074/mcp.R800016-MCP200.

Fu, L.-Y., Zhu, T., Zhou, X., Yu, R., He, Z. *et al.* (2022) ChIP-hub provides an integrative platform for exploring plant regulome. *Nature Communications* 13(1), 3413. DOI: 10.1038/s41467-022-30770-1.

Gomes, D.F., da Silva Batista, J.S., Rolla, A.A.P., da Silva, L.P., Bloch, C. *et al.* (2014) Proteomic analysis of free-living bradyrhizobium diazoefficiens: Highlighting potential determinants of a successful symbiosis. *BMC Genomics* 15(1), 643. DOI: 10.1186/1471-2164-15-643.

Gonin, M., Salas-González, I., Gopaulchan, D., Frene, J.P., Roden, S. *et al.* (2023) Plant microbiota controls an alternative root branching regulatory mechanism in plants. *Proceedings of the National Academy of Sciences* 120(15), 15. DOI: 10.1073/pnas.2301054120.

Gowda, G.A.N. and Djukovic, D. (2014) Overview of mass spectrometry-based metabolomics: Opportunities and challenges. *Methods in Molecular Biology (Clifton, N.J.)* 3–12. DOI: 10.1007/978-1-4939-1258-2_125270919.

Graham, E.B., Crump, A.R., Resch, C.T., Fansler, S., Arntzen, E. *et al.* (2016) Coupling spatiotemporal community assembly processes to changes in microbial metabolism. *Frontiers in Microbiology* 7, 1949. DOI: 10.3389/fmicb.2016.01949.

Haimovich, G. and Gerst, J.E. (2018) Single-molecule fluorescence *in situ* hybridization (smFISH) for RNA detection in adherent animal cells. *Bio-Protocol* 8(21), e3070. DOI: 10.21769/BioProtoc.3070.

Gui, S., Yang, L., Li, J., Luo, J., Xu, X. *et al.* (2020) ZEAMAP, a comprehensive database adapted to the maize multi-omics era. *iScience* 23(6), 101241. DOI: 10.1016/j.isci.2020.101241.

Gupta, S., Schillaci, M. and Roessner, U. (2022) Metabolomics as an emerging tool to study plant-microbe interactions. *Emerging Topics in Life Sciences* 6(2), 175–183. DOI: 10.1042/ETLS20210262.

Gu, S., Zhuang, J., Zhang, Z., Chen, W., Xu, H. *et al.* (2023) Multi-omics approach reveals the contribution of *osseh1* to rice cold tolerance. *Frontiers in Plant Science* 13, 1110724. DOI: 10.3389/fpls.2022.1110724.

Han, X., Wang, B., Situ, C., Qi, Y., Zhu, H. *et al.* (2023) ScapGNN: A graph neural network–based framework for active pathway and gene module inference from single-cell multi-omics data. *PLOS Biology* 21(11), e3002369. DOI: 10.1371/journal.pbio.3002369.

Hasin, Y., Seldin, M. and Lusis, A. (2017) Multi-omics approaches to disease. *Genome Biology* 18(1), 83. DOI: 10.1186/s13059-017-1215-1.

Hassani, M.A., Durán, P. and Hacquard, S. (2018) Microbial interactions within the plant holobiont. *Microbiome* 6(1), 58. DOI: 10.1186/s40168-018-0445-0.

He, Y., Zhang, K., Li, S., Lu, X., Zhao, H. *et al.* (2023) Multiomics analysis reveals the molecular mechanisms underlying virulence in Rhizoctonia and jasmonic acid–mediated resistance in Tartary buckwheat (*Fagopyrum tataricum*). *The Plant Cell* 35(8), 2773–2798. DOI: 10.1093/plcell/koad118.

Hodgson, J. (2016) When biotech goes bad. *Nature Biotechnology* 34(3), 284–291. DOI: 10.1038/nbt.3510.

Hossen, M.I., Awrangjeb, M., Pan, S. and Mamun, A. (2025) Transfer learning in agriculture: A review. *Artificial Intelligence Review* 58(4), 97. DOI: 10.1007/s10462-024-11081-x.

Hou, S., Thiergart, T., Vannier, N., Mesny, F., Ziegler, J. *et al.* (2021) A microbiota-root-shoot circuit favours *Arabidopsis* growth over defence under suboptimal light. *Nature Plants* 7(8), 1078–1092. DOI: 10.1038/s41477-021-00956-4.

Huang, W., Hu, X., Ren, Y., Song, M., Ma, C. *et al.* (2023) IPOP: An integrative plant multi-omics platform for cross-species comparison and evolutionary study. *Molecular Biology and Evolution* 40(12), msad248. DOI: 10.1093/molbev/msad248.

Hussain, S.S. (2019) Microbe-mediated tolerance in plants against biotic and abiotic stresses. In: *Microbial Interventions in Agriculture and Environment*. Springer Singapore, pp. 173–217. DOI: 10.1007/978-981-13-8391-5_7.

Jain, A., Sarsaiya, S., Singh, R., Gong, Q., Wu, Q. *et al.* (2024) Omics approaches in understanding the benefits of plant-microbe interactions. *Frontiers in Microbiology* 15, 1391059. DOI: 10.3389/fmicb.2024.1391059.

Jain, M., Olsen, H.E., Paten, B. and Akeson, M. (2016) The Oxford Nanopore MinION: Delivery of nanopore sequencing to the genomics community. *Genome Biology* 17(1), 239. DOI: 10.1186/s13059-016-1103-0.

Karim, M.R., Islam, T., Shajalal, M., Beyan, O., Lange, C. *et al.* (2022) Explainable AI for bioinformatics: Methods, tools and applications. *Briefings in Bioinformatics* 24(5). DOI: 10.1093/bib/bbad236.

Kaur, M., Kaur, R. and Sodhi, H.S. (2022) Utilization of omics approaches for underpinning plant-microbe interaction. In: *New and Future Developments in Microbial Biotechnology and Bioengineering*. Elsevier, pp. 217–242. DOI: 10.1016/B978-0-323-85577-8.00009-3.

Khan, M., Arslanturk, S. and Draghici, S. (2025) A comprehensive review of spatial transcriptomics data alignment and integration. *Nucleic Acids Research* 53(12), gkaf536. DOI: 10.1093/nar/gkaf536.

Kimotho, R.N. and Maina, S. (2024) Unraveling plant-microbe interactions: Can integrated omics approaches offer concrete answers? *Journal of Experimental Botany* 75(5), 1289–1313. DOI: 10.1093/jxb/erad448.

Kumar, S., Choudhary, M., J.R., K., Vishwakarma, V.K., Kashyap, V.K. *et al.* (2024) A review on the impact of climate change on plant pathogen interactions. *Journal of Advances in Microbiology* 24(8), 11–27. DOI: 10.9734/jamb/2024/v24i8843.

Kumar, U., Raj, S., Sreenikethanam, A., Maddheshiya, R., Kumari, S. *et al.* (2023) Multi-omics approaches in plant–microbe interactions hold enormous promise for sustainable agriculture. *Agronomy* 13(7), 1804. DOI: 10.3390/agronomy13071804.

Kwok, A.W.C., Shim, H. and McCarthy, D.J. (2025) A hierarchical, count-based model highlights challenges in scATAC-seq data analysis and points to opportunities to extract finer-resolution information. *Genome Biology* 26(1), 282. DOI: 10.1186/s13059-025-03735-y.

Lahlali, R., Ezrari, S., Radouane, N., Kenfaoui, J., Esmaeel, Q. *et al.* (2022) Biological control of plant pathogens: A global perspective. *Microorganisms* 10(3), 596. DOI: 10.3390/microorganisms10030596.

Lee, J.H., Daugharthy, E.R., Scheiman, J., Kalhor, R., Ferrante, T.C. *et al.* (2015) Fluorescent *in situ* sequencing (FISSEQ) of RNA for gene expression profiling in intact cells and tissues. *Nature Protocols* 10(3), 442–458. DOI: 10.1038/nprot.2014.191.

Leng, D., Zheng, L., Wen, Y., Zhang, Y., Wu, L. *et al.* (2022) A benchmark study of deep learning-based multi-omics data fusion methods for cancer. *Genome Biology* 23(1), 171. DOI: 10.1186/s13059-022-02739-2.

Li, C., Xiang, X., Huang, Y., Zhou, Y., An, D. *et al.* (2020) Long-read sequencing reveals genomic structural variations that underlie creation of quality protein maize. *Nature Communications* 11(1), 17. DOI: 10.1038/s41467-019-14023-2.

Libbrecht, M.W. and Noble, W.S. (2015) Machine learning applications in genetics and genomics. *Nature Reviews Genetics* 16(6), 321–332. DOI: 10.1038/nrg3920.

Lim, H.J., Wang, Y., Buzdin, A. and Li, X. (2025) A practical guide for choosing an optimal spatial transcriptomics technology from seven major commercially available options. *BMC Genomics* 26(1), 47. DOI: 10.1186/s12864-025-11235-3.

Liu, C.-W. and Murray, J. (2016) The role of flavonoids in nodulation host-range specificity: An update. *Plants* 5(3), 33. DOI: 10.3390/plants5030033.

Liu, F., Cai, S., Ma, Z., Yue, H., Xing, L. *et al.* (2023) RVE2, a new regulatory factor in jasmonic acid pathway, orchestrates resistance to *Verticillium wilt*. *Plant Biotechnology Journal* 21(12), 2507–2524. DOI: 10.1111/pbi.14149.

Lubeck, E., Coskun, A.F., Zhiyentayev, T., Ahmad, M. and Cai, L. (2014) Single-cell *in situ* RNA profiling by sequential hybridization. *Nature Methods* 11(4), 360–361. DOI: 10.1038/nmeth.2892.

Luce, J.J., Reardon-Lochbaum, C.A., Cadinu, P., Xu, R.J., Aceves-Salvador, J. *et al.* (2025) Protocol optimization improves the performance of multiplexed RNA imaging. *Scientific Reports* 15(1), 32030. DOI: 10.1038/s41598-025-17477-1.

Ludwig, C., Gillet, L., Rosenberger, G., Amon, S., Collins, B.C. *et al.* (2018) Data-independent acquisition-based. *Molecular Systems Biology* 14(8). DOI: 10.15252/msb.20178126.

Mahrous, E.A. and Farag, M.A. (2015) Two dimensional NMR spectroscopic approaches for exploring plant metabolome: A review. *Journal of Advanced Research* 6(1), 3–15. DOI: 10.1016/j.jare.2014.10.003.

Marx, V. (2021) Method of the year: Spatially resolved transcriptomics. *Nature Methods* 18(1), 9–14. DOI: 10.1038/s41592-020-01033-y.

Medina-Cordoba, L.K., Chande, A.T., Rishishwar, L., Mayer, L.W., Valderrama-Aguirre, L.C. *et al.* (2021) Genomic characterization and computational phenotyping of nitrogen-fixing bacteria isolated from Colombian sugarcane fields. *Scientific Reports* 11(1), 9187. DOI: 10.1038/s41598-021-88380-8.

Mishra, B., Kumar, N. and Mukhtar, M.S. (2019) Systems biology and machine learning in plant-pathogen interactions. *Molecular Plant-Microbe Interactions* 32(1), 45–55. DOI: 10.1094/MPMI-08-18-0221-FI.

Mishra, M., Choudhury, P. and Pati, B. (2024) IoT enabled plant leaf disease segmentation and multi-classification using mayfly bald eagle optimization-enabled machine learning. *Multimedia Tools and Applications* 83(21), 59747–59781. DOI: 10.1007/s11042-023-17680-0.

Mishra, A.K., Mishra, S., Gupta, S. and Tiwari, S. (2025) Decoding plant metabolomics: Integrative insights into metabolic regulation. *Theoretical and Experimental Plant Physiology* 37(1), 1–26. DOI: 10.1007/s40626-025-00380-5.

Mishra, A.K., Sudalaimuthuasari, N., Hazzouri, K.M., Saeed, E.E., Shah, I. *et al.* (2022) Tapping into plant-microbiome interactions through the lens of multi-omics techniques. *Cells* 11(20), 3254. DOI: 10.3390/cells11203254.

Mohan, S.M. and Sudhakar, P. (2022) Metagenomic approaches for studying plant–microbe interactions. In: *Understanding the Microbiome Interactions in Agriculture and the Environment*. Springer Nature Singapore, pp. 243–254. DOI: 10.1007/978-981-19-3696-8_12.

Muhammad, A., Kong, X., Zheng, S., Bai, N., Li, L. *et al.* (2024) Exploring plant-microbe interactions in adapting to abiotic stress under climate change: A review. *Frontiers in Plant Science* 15. DOI: 10.3389/fpls.2024.1482739.

Murmu, S., Sinha, D., Chaurasia, H., Sharma, S., Das, R. *et al.* (2024) A review of artificial intelligence-assisted omics techniques in plant defense: Current trends and future directions. *Frontiers in Plant Science* 15, 1292054. DOI: 10.3389/fpls.2024.1292054.

Nadeem, S.M., Ahmad, M., Zahir, Z.A., Javaid, A. and Ashraf, M. (2014) The role of mycorrhizae and plant growth promoting rhizobacteria (PGPR) in improving crop productivity under stressful environments. *Biotechnology Advances* 32(2), 429–448. DOI: 10.1016/j.biotechadv.2013.12.005.

Negrel, L., Halter, D., Wiedemann-Merdinoglu, S., Rustenholz, C., Merdinoglu, D. *et al.* (2018) Identification of lipid markers of *Plasmopara viticola* infection in grapevine using a non-targeted metabolomic approach. *Frontiers in Plant Science* 9, 360. DOI: 10.3389/fpls.2018.00360.

Nobori, T. (2025) Exploring the untapped potential of single-cell and spatial omics in plant biology. *The New Phytologist* 247(3), 1098–1116. DOI: 10.1111/nph.70220.

Nobori, T., Wang, Y., Wu, J., Stolze, S.C., Tsuda, Y. *et al.* (2020) Multidimensional gene regulatory landscape of a bacterial pathogen in plants. *Nature Plants* 6(7), 883–896. DOI: 10.1038/s41477-020-0690-7.

Olowe, O.M., Akanmu, A.O. and Asemoloye, M.D. (2020) Exploration of microbial stimulants for induction of systemic resistance in plant disease management. *Annals of Applied Biology* 177(3), 282–293. DOI: 10.1111/aab.12631.

Pandey, K. and Saharan, B.S. (2025) Soil microbiomes: A promising strategy for boosting crop yield and advancing sustainable agriculture. *Discover Agriculture* 3(1), 1–24. DOI: 10.1007/S44279-025-00208-5.

Parnell, J.J., Berka, R., Young, H.A., Sturino, J.M., Kang, Y. *et al.* (2016) From the lab to the farm: An industrial perspective of plant beneficial microorganisms. *Frontiers in Plant Science* 7, AUG2016. DOI: 10.3389/fpls.2016.01110.

Patyal, U., Bala, R., Kaur, M., Faizan, M. and Alam, P. (2025) Phyto-microbiome engineering: Designing plant-microbe interactions for improved crop performance. *The Microbe* 6, 100272. DOI: 10.1016/j.microb.2025.100272.

Payne, A.C., Chiang, Z.D., Reginato, P.L., Mangiameli, S.M., Murray, E.M. *et al.* (2021) *In situ* genome sequencing resolves DNA sequence and structure in intact biological samples. *Science (New York, N.Y.)* 371(6532), eaay3446. DOI: 10.1126/science.aay3446.

Peng, H., Wang, K., Chen, Z., Cao, Y., Gao, Q. *et al.* (2019) MBKbase for rice: An integrated omics knowledgebase for molecular breeding in rice. *Nucleic Acids Research*. DOI: 10.1093/nar/gkz921.

Perdih, T.S., Lavrac, N. and Skrlj, B. (2021) Semantic reasoning from model-agnostic explanations. In: *2021 IEEE 19th World Symposium on Applied Machine Intelligence and Informatics (SAMI)*, Herl'any, Slovakia, pp. 000105–000110. DOI: 10.1109/SAMI50585.2021.9378668.

Peterson, A.C., Russell, J.D., Bailey, D.J., Westphall, M.S. and Coon, J.J. (2012) Parallel reaction monitoring for high resolution and high mass accuracy quantitative, targeted proteomics. *Molecular & Cellular Proteomics* 11(11), 1475–1488. DOI: 10.1074/mcp.O112.020131.

Qi, X., Chen, L., Hu, Z., Shen, W., Xu, H. *et al.* (2022) Cytology, transcriptomics, and mass spectrometry imaging reveal changes in late-maturation elm (*Umus pumila*) seeds. *Journal of Plant Physiology* 271, 153639. DOI: 10.1016/j.jplph.2022.153639.

Raj, A., Bogaard, P., Rifkin, S.A., van Oudenaarden, A. and Tyagi, S. (2008) Imaging individual mRNA molecules using multiple singly labeled probes. *Nature Methods* 5(10), 877–879. DOI: 10.1038/nmeth.1253.

Rajguru, B., Shri, M. and Bhatt, V.D. (2024) Exploring microbial diversity in the rhizosphere: A comprehensive review of metagenomic approaches and their applications. *3 Biotech* (10), 1–18. DOI: 10.1007/S13205-024-04065-9.

Ramadurai, S., Moorthy, A. and Balasundaram, U. (2021) Metagenomic approach in relation to plant–microbe and microbe–microbe interactions. In: *Omics Science for Rhizosphere Biology*. Springer, Singapore, pp. 21–40. DOI: 10.1007/978-981-16-0889-6_2.

Ren, Y., Tseng, E., Smith, T.P.L., Hiendleder, S., Williams, J.L. *et al.* (2023) Long read isoform sequencing reveals hidden transcriptional complexity between cattle subspecies. *BMC Genomics* 24(1), 108. DOI: 10.1186/s12864-023-09212-9.

Rosier, A., Medeiros, F.H.V. and Bais, H.P. (2018) Defining plant growth promoting rhizobacteria molecular and biochemical networks in beneficial plant-microbe interactions. *Plant and Soil* 428(1–2), 35–55. DOI: 10.1007/s11104-018-3679-5.

Rosselli, R. and Squartini, A. (2015) Metagenomics of plant–microbe interactions. In: *Advances in the Understanding of Biological Sciences Using Next Generation Sequencing (NGS) Approaches*. Springer International Publishing, pp. 135–153. DOI: 10.1007/978-3-319-17157-9_9.

Sahu, N., Vasu, D., Sahu, A., Lal, N. and Singh, S.K. (2017) Strength of microbes in nutrient cycling: A key to soil health. *Agriculturally Important Microbes for Sustainable Agriculture* 1, 69–86. DOI: 10.1007/978-981-10-5589-8_4.

Saini, D.K., Chopra, Y., Singh, J., Sandhu, K.S., Kumar, A. *et al.* (2022) Comprehensive evaluation of mapping complex traits in wheat using genome-wide association studies. *Molecular Breeding* 42(1), 1. DOI: 10.1007/s11032-021-01272-7.

Salam, U., Ullah, S., Tang, Z.-H., Elateeq, A.A., Khan, Y. *et al.* (2023) Plant metabolomics: An overview of the role of primary and secondary metabolites against different environmental stress factors. *Life* 13(3), 706. DOI: 10.3390/life13030706.

Salami, M., Heidari, B., Alizadeh, B., Batley, J., Wang, J. *et al.* (2024) Dissection of quantitative trait nucleotides and candidate genes associated with agronomic and yield-related traits under drought stress in rapeseed varieties: Integration of genome-wide association study and transcriptomic analysis. *Frontiers in Plant Science* 15. DOI: 10.3389/fpls.2024.1342359.

Saminathan, T., García, M., Ghimire, B., Lopez, C., Bodunrin, A. *et al.* (2018) Metagenomic and metatranscriptomic analyses of diverse watermelon cultivars reveal the role of fruit associated microbiome in carbohydrate metabolism and ripening of mature fruits. *Frontiers in Plant Science* 9, 4. DOI: 10.3389/fpls.2018.00004.

Sarabia, L.D., Boughton, B.A., Rupasinghe, T., van de Meene, A.M.L., Callahan, D.L. *et al.* (2018) High-mass-resolution MALDI mass spectrometry imaging reveals detailed spatial distribution of metabolites and lipids in roots of barley seedlings in response to salinity stress. *Metabolomics* 14(5), 63. DOI: 10.1007/s11306-018-1359-3.

Sarsan, S., Pandiyan, A., Rodhe, A.V. and Jagavati, S. (2021) Synergistic interactions among microbial communities. *Microbes in Microbial Communities: Ecological and Applied Perspectives* 1–37. DOI: 10.1007/978-981-16-5617-0_1.

Savitski, M.M., Mathieson, T., Zinn, N., Sweetman, G., Doce, C. *et al.* (2013) Measuring and managing ratio compression for accurate iTRAQ/TMT quantification. *Journal of Proteome Research* 12(8), 3586–3598. DOI: 10.1021/pr400098r.

Saxena, A. and Bharadwaj, A. (2024) Role of soil metagenomics in plant–microbe interaction. In: *Plant Endophytes and Secondary Metabolites*. Elsevier, pp. 25–33. DOI: 10.1016/B978-0-443-13365-7.00005-1.

Schramm, T., Hester, Z., Klinkert, I., Both, J.-P., Heeren, R.M.A. *et al.* (2012) ImzML — A common data format for the flexible exchange and processing of mass spectrometry imaging data. *Journal of Proteomics* 75(16), 5106–5110. DOI: 10.1016/j.jprot.2012.07.026.

Scossa, F., Alseekh, S. and Fernie, A.R. (2021) Integrating multi-omics data for crop improvement. *Journal of Plant Physiology* 257, 153352. DOI: 10.1016/j.jplph.2020.153352.

Serrano, K., Tedeschi, F., Andersen, S.U. and Scheller, H.V. (2024) Unraveling plant–microbe symbioses using single-cell and spatial transcriptomics. *Trends in Plant Science* 29(12), 1356–1367. DOI: 10.1016/j.tplants.2024.06.008.

Sharma, M., Sudheer, S., Usmani, Z., Rani, R. and Gupta, P. (2020) Deciphering the omics of plant-microbe interaction: Perspectives and new insights. *Current Genomics* 21(5), 343–362. DOI: 10.2174/1389202921999200515140420.

Shiono, K. and Taira, S. (2020) Imaging of multiple plant hormones in roots of rice (*Oryza sativa*) using nanoparticle-assisted laser desorption/ionization mass spectrometry. *Journal of Agricultural and Food Chemistry* 68(24), 6770–6775. DOI: 10.1021/acs.jafc.0c00749.

Sharma, D., Astapati, A.D., Dey, S., Bhattacharjee, D., Singha, D.M. *et al.* (2025) Microbial interactions in soil ecosystems: Facilitating plant growth, nutrient cycling, and environmental dynamics. In: Babalola, O.O. and Ayangbenro, A.S. (eds) *Microbial Allies*. Springer Nature, Switzerland, pp. 21–48. DOI: 10.1007/978-3-031-90530-8_2.

Shi, S., Zha, W., Yu, X., Wu, Y., Li, S. *et al.* (2023) Integrated transcriptomics and metabolomics analysis provide insight into the resistance response of rice against brown planthopper. *Frontiers in Plant Science* 14. DOI: 10.3389/fpls.2023.1213257.

Sindelar, R.D. (2013) Genomics, other "omic" technologies, personalized medicine, and additional biotechnology-related technique. In: *Pharmaceutical Biotechnology*. Springer New York, pp. 179–221. DOI: 10.1007/978-1-4614-6486-0_8.

Singh, D.P., Maurya, S., Yerasu, S.R., Prabha, R., Renu, S. *et al.* (2025a) Comparative proteomics uncovers distinct biomarkers, protein networks and defense responses in tomato during beneficial and pathogenic microbial interactions. *Plant Biology*. DOI: 10.1101/2025.01.23.634544.

Singh, L., Srivastav, S., Avasthi, A., Rathi, B., Gärtner, W. *et al.* (2025b) Translational omics: Precision and all "Coming of age. In: *Multi-Omics in Biomedical Sciences and Environmental Sustainability*. pp. 1–32. DOI: 10.1007/978-981-96-7067-3_1.

Tang, F., Barbacioru, C., Wang, Y., Nordman, E., Lee, C. *et al.* (2009) MRNA-seq whole-transcriptome analysis of a single cell. *Nature Methods* 6(5), 377–382. DOI: 10.1038/nmeth.1315.

Teichmann, S. and Efremova, M. (2020) Method of the year 2019: Single-cell multimodal omics. *Nature Methods* 17(1), 1–1. DOI: 10.1038/s41592-019-0703-5.

Trivedi, P., Mattupalli, C., Eversole, K. and Leach, J.E. (2021) Enabling sustainable agriculture through understanding and enhancement of microbiomes. *New Phytologist* 230(6), 2129–2147. DOI: 10.1111/nph.17319.

Vandereyken, K., Sifrim, A., Thienpont, B. and Voet, T. (2023) Methods and applications for single-cell and spatial multi-omics. *Nature Reviews Genetics* 24(8), 494–515. DOI: 10.1038/s41576-023-00580-2.

Vishwakarma, K., Kumar, N., Shandilya, C., Mohapatra, S., Bhayana, S. *et al.* (2020) Revisiting plant–microbe interactions and microbial consortia application for enhancing sustainable agriculture: A review. *Frontiersin.Org* 11. DOI: 10.3389/FMICB.2020.560406/FULL.

Wadie, B., Stuart, L., Rath, C.M., Drotleff, B., Mamedov, S. *et al.* (2024) METASPACE-ML: Context-specific metabolite annotation for imaging mass spectrometry using machine learning. *Nature Communications* 15(1), 9110. DOI: 10.1038/s41467-024-52213-9.

Wang, Y., Huan, Q., Li, K. and Qian, W. (2021) Single-cell transcriptome atlas of the leaf and root of rice seedlings. *Journal of Genetics and Genomics* 48(10), 881–898. DOI: 10.1016/j.jgg.2021.06.001.

Wang, X., Allen, W.E., Wright, M.A., Sylwestrak, E.L., Samusik, N. *et al.* (2018) Three-dimensional intact-tissue sequencing of single-cell transcriptional states. *Science* 361(6400), 6400. DOI: 10.1126/science.aat5691.

Wang, Y., Liu, B., Zhao, G., Lee, Y., Buzdin, A. *et al.* (2023) Spatial transcriptomics: Technologies, applications and experimental considerations. *Genomics* 115(5), 110671. DOI: 10.1016/j.ygeno.2023.110671.

Weckwerth, W., Ghatak, A., Bellaire, A., Chaturvedi, P. and Varshney, R.K. (2020) PANOMICS meets germplasm. *Plant Biotechnology Journal* 18(7), 1507–1525. DOI: 10.1111/pbi.13372.

Wilkinson, M.D., Dumontier, M., Aalbersberg, Ij.J., Appleton, G., Axton, M. *et al.* (2016) The FAIR guiding principles for scientific data management and stewardship. *Scientific Data* 3(1), 160018. DOI: 10.1038/sdata.2016.18.

Williams, C.G., Lee, H.J., Asatsuma, T., Vento-Tormo, R. and Haque, A. (2022) An introduction to spatial transcriptomics for biomedical research. *Genome Medicine* 14(1), 68. DOI: 10.1186/s13073-022-01075-1.

Xiao, S., Lin, H., Wang, C., Wang, S. and Rajapakse, J.C. (2023) Graph neural networks with multiple prior knowledge for multi-omics data analysis. *IEEE Journal of Biomedical and Health Informatics* 27(9), 4591–4600. DOI: 10.1109/JBHI.2023.3284794.

Xu, L., Pierroz, G., Wipf, H.M.-L., Gao, C., Taylor, J.W. *et al.* (2021) Holo-omics for deciphering plant-microbiome interactions. *Microbiome* 9(1). DOI: 10.1186/s40168-021-01014-z.

Yan, S., Bhawal, R., Yin, Z., Thannhauser, T.W. and Zhang, S. (2022) Recent advances in proteomics and metabolomics in plants. *Molecular Horticulture* 2(1), 17. DOI: 10.1186/s43897-022-00038-9.

Yang, Y., Saand, M.A., Huang, L., Abdelaal, W.B., Zhang, J. *et al.* (2021) Applications of multi-omics technologies for crop improvement. *Frontiers in Plant Science* 12. DOI: 10.3389/fpls.2021.563953.

Yang, T., Zhou, S., Huang, Z., Xu, A., Ye, J. *et al.* (2023) Urban street tree dataset for image classification and instance segmentation. *Computers and Electronics in Agriculture* 209, 107852. DOI: 10.1016/j.compag.2023.107852.

Yin, S., Liu, J. and Teng, L. (2016) An improved ant colony algorithm used for unscented Kalman filter. *ICIC Express Letters, Part B: Applications* 7(11), 2411–2417. DOI: 10.1038/nature07517.Accurate.

Yin, Z., Huang, W., Li, K., Fernie, A.R. and Yan, S. (2024) Advances in mass spectrometry imaging for plant metabolomics—Expanding the analytical toolbox. *The Plant Journal* 119(5), 2168–2180. DOI: 10.1111/tpj.16924.

You, Y., Fu, Y., Li, L., Zhang, Z., Jia, S. *et al.* (2024) Systematic comparison of sequencing-based spatial transcriptomic methods. *Nature Methods* 21(9), 1743–1754. DOI: 10.1038/s41592-024-02325-3.

Yu, X., Liu, Z. and Sun, X. (2023) Single-cell and spatial multi-omics in the plant sciences: Technical advances, applications, and perspectives. *Plant Communications* 4(3), 100508. DOI: 10.1016/j.xplc.2022.100508.

Yu, Y., Gui, Y., Li, Z., Jiang, C., Guo, J. *et al.* (2022) Induced systemic resistance for improving plant immunity by beneficial microbes. *Plants* 11(3), 386. DOI: 10.3390/plants11030386.

Zhan, C., Lei, L., Guo, H., Zhou, S., Xu, C. *et al.* (2023) Disease resistance conferred by components of essential chrysanthemum oil and the epigenetic regulation of OsTPS1. *Science China Life Sciences* 66(5), 1108–1118. DOI: 10.1007/s11427-022-2241-0.

Zhang, B., Zhang, H. and Xia, Y. (2024) Harnessing spatial transcriptomics for advancing plant regeneration research. *Trends in Plant Science* 29(7), 718–720. DOI: 10.1016/j.tplants.2024.02.004.

Zhang, R., Zhang, C., Yu, C., Dong, J. and Hu, J. (2022) Integration of multi-omics technologies for crop improvement: Status and prospects. *Frontiers in Bioinformatics* 2. DOI: 10.3389/fbinf.2022.1027457.

Zhang, J., Wang, P., Tian, H., Tao, Z. and Guo, T. (2020a) Transcriptome analysis of ice plant growth-promoting endophytic *Bacterium halomonas* sp. Strain MC1 to identify the genes involved in salt tolerance. *Microorganisms* 8(1), 88. DOI: 10.3390/microorganisms8010088.

Zhang, X.-Q., Bai, L., Sun, H.-B., Yang, C. and Cai, B.-Y. (2020b) Transcriptomic and proteomic analysis revealed the effect of *Funneliformis mosseae* in soybean roots differential expression genes and proteins. *Journal of Proteome Research* 19(9), 3631–3643. DOI: 10.1021/acs.jproteome.0c00017.

Zhang, T.-Q., Chen, Y., Liu, Y., Lin, W.-H. and Wang, J.-W. (2021) Single-cell transcriptome atlas and chromatin accessibility landscape reveal differentiation trajectories in the rice root. *Nature Communications* 12(1), 2053. DOI: 10.1038/s41467-021-22352-4.

Zhang, N., Wang, S., Zhao, S., Chen, D., Tian, H. *et al.* (2023) Global crotonylatome and GWAS revealed a *TaSRT1-TaPGK* model regulating wheat cold tolerance through mediating pyruvate. *Science Advances* 9(19), 19. DOI: 10.1126/sciadv.adg1012.

Zhao, H., Li, J., Yang, L., Qin, G., Xia, C. *et al.* (2021) An inferred functional impact map of genetic variants in rice. *Molecular Plant* 14(9), 1584–1599. DOI: 10.1016/j.molp.2021.06.025.

Zhao, L., Fonseca, A., Meschichi, A., Sicard, A. and Rosa, S. (2023) Whole-mount smFISH allows combining RNA and protein quantification at cellular and subcellular resolution. *Nature Plants* 9(7), 1094–1102. DOI: 10.1038/s41477-023-01442-9.

Zhao, T., Liu, W., Zhao, Z., Yang, H., Bao, Y. *et al.* (2019) Transcriptome profiling reveals the response process of tomato carrying Cf-19 and *Cladosporium fulvum* interaction. *BMC Plant Biology* 19(1), 572. DOI: 10.1186/s12870-019-2150-y.

Zhu, L., Zhou, Y., Li, X., Zhao, J., Guo, N. *et al.* (2018) Metabolomics analysis of soybean hypocotyls in response to *Phytophthora sojae* infection. *Frontiers in Plant Science* 9. DOI: 10.3389/fpls.2018.01530.

Zhu, X.-T., Zhou, R., Che, J., Zheng, Y.-Y., Tahir ul Qamar, M. *et al.* (2023) Ribosome profiling reveals the translational landscape and allele-specific translational efficiency in rice. *Plant Communications* 4(2), 100457. DOI: 10.1016/j.xplc.2022.100457.

11 Spatial and Multiple Omics in Horticultural Plants: Advances and Prospects

Gyana Ranjan Mohanty[1], Soumya Ranjan Mohanty[2], Sandip Debnath[1*], Nihar Ranjan Chakraborty[1], Aritabha Kole[1], Soumyajit Paikray[3], Swayanshu Satyaprajna Das[4], Saroj Kumar Mohanty[4] and Prahlad Deb[5]

[1]Department of Genetics and Plant Breeding, Institute of Agriculture, Visva-Bharati University, Sriniketan, West Bengal, India; [2]College of Horticulture, OUAT, Chiplima, Odisha, India; [3]Department of Genetics and Plant Breeding, College of Agriculture, OUAT, Bhubaneswar, Odisha, India; [4]Odisha University of Agriculture and Technology, Bhubaneswar, Odisha, India; [5]Department of Horticulture and Postharvest Technology, Institute of Agriculture, Visva-Bharati University, Sriniketan, West Bengal, India

11.1 Introduction

11.1.1 Background and significance

Horticulture is a rapidly growing branch of agriculture that allows people to grow and preserve a wide variety of plants, including fruits, vegetables, flowers, and herbs. It significantly transformed human society from a nomadic way of life to civilization. The speed of multi-omics and *in silico* research in horticulture plants is being accelerated by recent developments in genome sequencing (Payne *et al.*, 2021). Such breakthroughs link high-throughput molecular technologies with powerful computational techniques to unravel complicated physiological processes, adaptive strategies, and molecular interactions on a plant scale (Chen *et al.*, 2022). One such revolutionary advancement in this field is spatial multi-omics, which retains spatial information inside tissues while overcoming the main drawback of single-cell sequencing.

A whole new understanding of cellular communication, tissue architecture, and functional heterogeneity is provided by spatial multi-omics, which combines data from genomes, transcriptomics, proteomics, metabolomics, and epigenomics with spatial localization (Choe *et al.*, 2020). The use of this technique is widespread, going beyond plant biology to developmental research and biomedicine (Clark *et al.*, 2023), thereby providing insights into intricate biological systems (Nakabayashi and Saito, 2020). The use of spatial multi-omics in conjunction with advanced computational methods for genotype–phenotype mapping and the identification of genetic markers governing the crucial characteristics of horticulture crops has become the norm in present-day research practice (You *et al.*, 2019; Muhammad Aslam *et al.*, 2022; Jin *et al.*, 2024).

*Corresponding author: Sandip.debnath@visva-bharati.ac.in

© CAB International 2026. *Plant Spatial Biology* (ed. Chen)
DOI: 10.1079/9781836994572.0011

11.1.2 Objectives of the chapter

This chapter aims to survey pioneering horticultural technologies involved in the spatial mapping of biological tissues and multi-omics analysis, along with their application to the molecular decoding of organ/tissue systems. The emphasis here is placed on combining spatial scenarios and omics to gain insights into developmental biology, phytopathology, and novel trait formation, thus facilitating the progress of precision breeding and eco-friendly horticultural improvements.

11.2 Spatial Omics Technologies in Horticultural Research

Spatially determined omics is reshaping the plant sciences by providing information on one single-cell gene expression, metabolism, and epigenetic changes in fruits, vegetables, and ornamental plants. The combination of spatial transcriptomics and metabolomics helps to understand the location of active genes that are associated with quality attributes such as taste, color, nutrition, and disease resistance. These techniques allow identification at the cellular level of the coordinators of stress and pathogen responses. Integrating spatial omics into horticultural breeding not only increases phenotyping accuracy but also invigorates the evolutionary process of high-quality and resilient crops that satisfy the requirements of the farmer and the consumer.

11.2.1 Spatial genomics

Spatial genomics maps genomic variations within intact tissues, linking them to local microenvironments to reveal how mutations, copy number changes, and chromatin states drive cell evolution and abiotic stress resistance (Hosea *et al.*, 2024).

11.2.1.1 In situ *genome sequencing (IGS)*

The workflow for IGS involves: DNA fragmentation using *in vitro* transposons → circularization of DNA with hairpin splices → rolling circle amplification by Phi29 polymerase → sequencing both *in situ* and ectopically (Bouwman *et al.*, 2022). The technique preserves tissue structure while enabling unbiased, localized genomic sequencing, which provides high-resolution images of chromosomal structure, permitting genome-wide paired-end mapping directly within tissues.

11.2.1.2 Slide-DNA-seq

The technique of slide-DNA-seq involves: fresh-frozen thin tissue slice transferred to barcode-bead array → photolysis-mediated barcoding → PCR-based amplification → high-throughput sequencing (Bouwman *et al.*, 2022). This method uses spatially barcoded bead arrays to link genomic fragments to specific tissue coordinates. This also allows quantification of clonal diversity, copy number variations (CNVs), and their spatial distributions across tissue. It is powerful for tracking tissue evolution and clonal competition *in situ* in particular (Payne *et al.*, 2021). In comparison:

- IGS is the exact mapping of the local genome and chromosomal architecture in fixed tissues. Slide DNA sequencing offers more thorough tissue-scale genetic variation profiling. (Payne *et al.*, 2021). In contrast to IGS, slide-DNA-seq uses barcoded indexing to identify CNV patterns and genome-wide clonal variation in vast tissue landscapes. The smooth integration of spatial technology into multiscale genomic maps is one of the future goals. Genetic diversity from microscopic to macroscopic scales can be captured by combining IGS structural features with the tissue-level mapping of slide DNA sequencing (Bouwman *et al.*, 2022).
- Linking spatial genomics to spatial transcriptomics and epigenomics will permit *in situ* analysis of genome structure, gene expression, and chromatin states; thus, the concept of molecular anatomy will advance. (Payne *et al.*, 2021; Hosea *et al.*, 2024), enabling the generation of spatially resolved atlases of tissue organization, clonal evolution, and disease mechanisms.

11.2.2 Spatial transcriptomics

Spatial transcriptomics, aided through exhaustive mapping of gene expression in tissues, has broadened the understanding of cellular structure and communication to the closest neighbors (Cao *et al.*, 2024a). The latest innovations feature that an instrument can detect more signals, has better resolution and sensitivity, is more efficient in terms of operation, and can cover a larger tissue area, thus extending the range of the application in the developmental biology sector (Choe *et al.*, 2023; Jin *et al.*, 2024) along with diseases (Cao *et al.*, 2024a).

The use of transcriptomics in space enables the most exact identification of RNA molecules in tissue. This section presents the main technological bases of the field: (i) microdissection; (ii) visualization-based fluorescence *in situ* hybridization (FISH); (iii) *in situ* sequencing; and (iv) spatial barcoding.

11.2.2.1 Microdissection

Microdissection techniques physically isolate molecules from specific tissue regions to enable spatially resolved molecular profiling. Laser-capture microdissection (LCM) commonly extracts small tissue sections comprising tens to hundreds of cells for downstream RNA sequencing (Emmert-Buck *et al.*, 1996). In plants, LCM (Fig. 11.1) is widely used to examine tissue gene expression in combination with spatial data (Choe *et al.*, 2020; Liew *et al.*, 2020). Single-cell resolution may be possible using emerging techniques such as geographical location sequencing (Geo-seq) (Chen *et al.*, 2017) and LCM-seq (Nichterwitz *et al.*, 2018). LCM isolates targeted tissue sections for subsequent gene expression analysis, but it is a labor-intensive process and low throughput limits efficiency.

11.2.2.2 Imaging-based fluorescence in situ *hybridization (FISH)*

FISH is a spatial transcriptomics method that uses fluorescent probes to directly observe RNA molecules in their original tissue environment. According to Zhao *et al.* (2023), single-molecule FISH (smFISH) (Fig. 11.1) achieves single-copy resolution and great sensitivity, allowing precise localization of gene expression and the presence of different cell types while maintaining the tissue's architecture.

These techniques use a combination of short oligonucleotide probes that are tailored to the target RNA sequences (Duncan *et al.*, 2023). Each probe is connected to a fluorophore, allowing for highly multiplexed and quantitative analysis (Duncan *et al.*, 2016). However, tissue autofluorescence, particularly in green plant tissues, and the complexities of probe

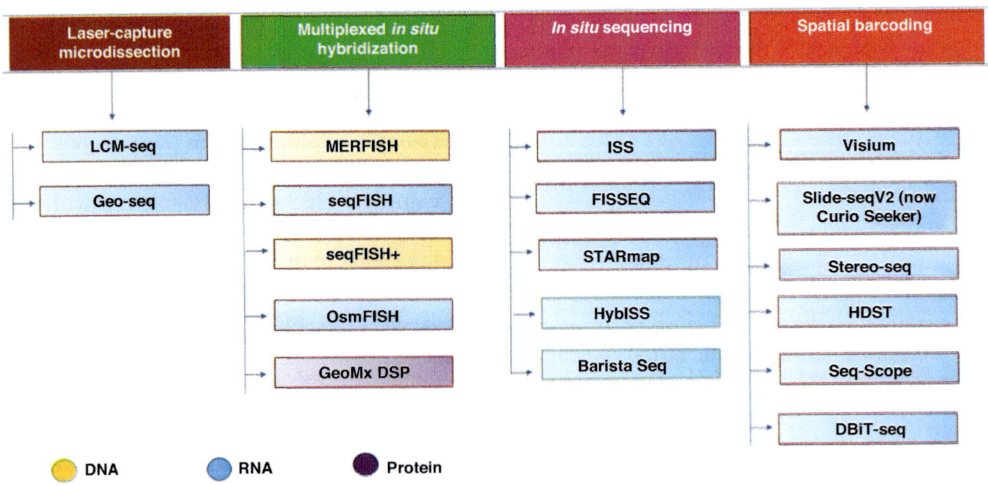

Fig. 11.1. Key spatial transcriptomics technologies in plant research.

design are still challenges for the approach to overcome (Rosa *et al.*, 2016; Solanki *et al.*, 2020). Fluorophore spectral overlap is one of the factors that limit the number of transcripts that can be identified, and it depends on having a well-annotated genome.

FISH is, however, still considered the most reliable method for confirming spatial gene expression because it offers direct visualization, high specificity, and accuracy that are very helpful in the study of gene regulation and cellular interactions in plant tissues.

11.2.2.3 In situ sequencing

In situ sequencing is a spatial transcriptomics technique that uses high-throughput sequencing directly on preserved tissue sections to capture spatial gene expression by chemically decoding RNA molecules in their native microenvironment (Laureyns *et al.*, 2022). This process provides precise localization along with sequence information. For example, several rounds of probe hybridization and sequencing are used in fluorescence *in situ* sequencing (FISSEQ) (Lee *et al.*, 2014) and hybridization-based *in situ* sequencing (HybISS) (Fig. 11.1) (Gyllborg *et al.*, 2020) to visualize and quantify transcripts at subcellular resolution. Single-cell sequencing is free from alterations that occur during dissociation using this method so the molecular and structural integrity is retained.

The advancements in chemistry and spatial resolution have made *in situ* sequencing a pioneering platform for fusing genomics with spatial mapping to uncover plant gene networks and responses with remarkable spatial–genomic precision, despite the possibility of low capture efficiency and challenging probe design in poorly annotated genomes.

11.2.2.4 Spatial barcoding

Spatial barcoding captures spatial gene expression by employing arrays that have unique molecular barcodes (Giacomello *et al.*, 2017; Xia *et al.*, 2022), which, in turn, are generated by the RNA that is transferred from the tissue sections. Spatial barcodes identify transcripts of the exact tissue location (Chen *et al.*, 2023), thus allowing high-throughput sequencing to quickly and extensively chart the spatial distributions transcriptome wide (Hao *et al.*, 2023).

This method is so robust that it almost completely removes the constraints that are typical for single-cell RNA sequencing, including the fragmentation of cells leading to the loss of spatial information and changes in transcriptomics during sample processing. The technologies such as Visium (Polański *et al.*, 2024), slide-seq (Rodriques *et al.*, 2019), and stereo-seq (Chen *et al.*, 2022) are the platforms of spatial barcoding, achieving from micrometer to submicron resolutions.

This resolution, along with the scalability and multiplex capacity for thousands of genes, makes it possible to carry out advanced studies of plant development and stress adaptation. The improved efficiency and resolution are evidence of the successful use of this method as a landmark instrument for plant genomics and precision breeding (Fu *et al.*, 2023).

11.2.3 Spatial proteomics

Spatial proteomics traces protein presence and changes directly in native tissues, hence uncovering new aspects not visible through bulk analyses (Duggan *et al.*, 2021). Since the role of a protein is largely dependent on where it is, localization becomes the key to understanding plant physiology. The main strategies outlined are: (i) mass spectrometry (MS)-based methods such as multiplexed ion beam imaging (MIBI) and imaging mass cytometry (IMC), which provide targeted, high-dimensional protein detection with nanoscale precision (Lundberg and Borner, 2019); (ii) imaging-based multiplexed techniques, e.g. iterative bleaching extends multiplexity (IBEX) (Fig. 11.1) and codetection by indexing (CODEX), (Breckels *et al.*, 2024), that allow the simultaneous visualization of tens to >60 protein markers within the intact tissues by sequential staining and imaging cycles; and (iii) proximity labeling techniques such as TurboID-based enzymatic tagging (Tang *et al.*, 2020; Choi and Rhee, 2022), which, by covalently tagging the proteins that are physically adjacent to the target protein, transform the proteins present in subcellular microenvironments for further MS analysis (Wallner *et al.*,

2024). These techniques illustrate regional heterogeneity and dynamic changes throughout plant growth and stress responses by correctly depicting protein localization and interactions directly in the organism (Wallner *et al.*, 2024).

11.2.4 Spatial metabolomics

Spatial metabolomics rapidly advances the *in situ* mapping and detection of metabolites in tissues, providing crucial insights into metabolism and its location. Mass spectrometry imaging (MSI) methods, which vary in their mass analyzers and ionization sources, are typically used for this type of study (Pareek *et al.*, 2020; Alexandrov, 2023).

Key MSI techniques are demonstrated by: secondary ion mass spectrometry (SIMS) (Righetti *et al.*, 2022), a method that employs an ion beam to achieve submicron resolution and map subcellular metabolites; matrix-assisted laser desorption/ionization (MALDI)–MSI (Sarabia *et al.*, 2018) is a laser ionization of matrix-coated materials that allows high sensitivity for up to 100 kDa and ~5 μm resolution, but with a limited mass range; and DESI–MSI, which allows ambient ionization of frozen materials for a quick, moderate-resolution analysis (Unsihuay *et al.*, 2021). Desorption electrospray ionization (DESI)–MSI is distinguished by its user friendliness, SIMS is a technique that provides the greatest spatial precision, and MALDI–MSI is a method that provides coverage over a broad range. Higher ion yield, improved resolution, and more efficient isomer discrimination are becoming possible, particularly by technical developments in preparation, ion sources, and hybrid systems.

To provide the necessary information for functional research and targeted breeding, researchers have used spatial metabolomics to track the distribution of hexose and sucrose in strawberries at various stages of maturation (Wang *et al.*, 2021a) and to record metabolic changes in both stress and the development of the plant (Enomoto, 2021).

11.3 Single-cell Omics

Single-cell omics enables high-resolution analysis of gene expression, chromatin accessibility, and epigenetic states in individual plant cells (Yu *et al.*, 2024), revealing hidden heterogeneity, rare cell types, and fine-scale dynamics of development and stress responses.

11.3.1 Single-cell transcriptomics

Single-cell transcriptomics is an important method for resolving cellular diversity in horticultural crops because it provides extensive insights into development and stress responses. The approaches for single-cell RNA sequencing (scRNA-seq) in plants are mostly droplet- or well-based. Droplet platforms, particularly 10 × Genomics, are gaining popularity due to their high molecular capture efficiency. Swift *et al.* (2024) found > 50% more unique molecular identifiers (UMIs) per cell in tomato compared to combinatorial indexing, indicating better depth.

Smart-seq-based approaches may yield full-length transcriptome profiling at isoform resolution (Ramskold *et al.*, 2012); nevertheless, the strength of multimodal indexing lies in its ability to handle huge cell counts, which can be accomplished via split and pool barcoding methodologies (Birnbaum, 2018; Nolan and Shahan, 2023).

For horticultural samples, where cell wall debris is the primary reason for the failure of traditional isolation, particle-templated instant partition sequencing, a novel and useful technology, has been applied by vortexing in water–oil emulsions rather than microfluidics (Clark *et al.*, 2023).

These techniques enable the identification of uncommon or transient cell states, such as protophloem differentiation in *Arabidopsis*, from nearly nonexistent cells. It is also feasible to map cell changes during infection and therapy using scRNA-seq (Tenorio Berrío and Dubois, 2024). This method may show immune-activated cells as well as transitional states that are tightly linked to resistance or susceptibility genes (Amini *et al.*, 2023; Rusnak *et al.*, 2024). The primary differences in transcriptomes of evolutionary origin may be identified by cross-species studies, such as those of light-reactive cells in rice (C3) and sorghum (C4) (Swift *et al.*, 2024). This degree of specificity is changing our knowledge of plant cell types and making it possible to improve certain agricultural traits. Figure 11.2

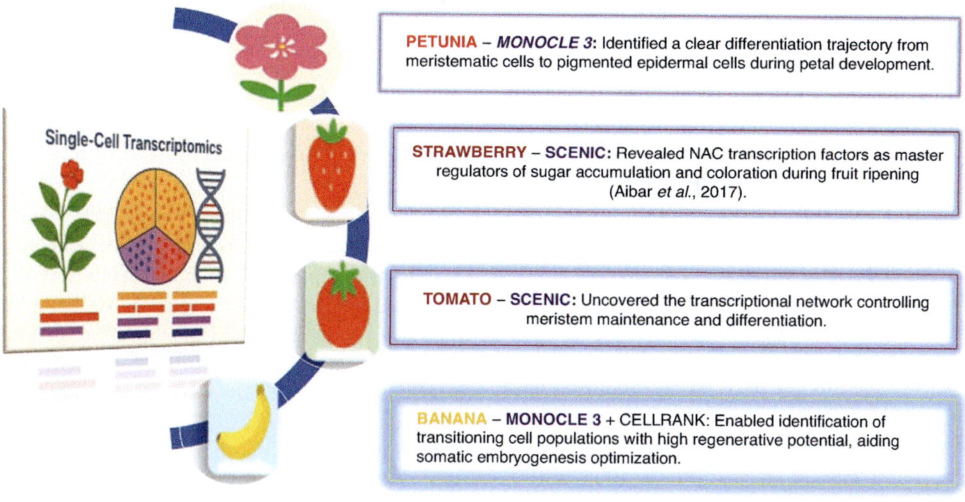

Fig. 11.2. Single-cell transcriptomics in horticultural crops.

shows the technological developments that have been used, and cellular and physiological processes in horticultural crops.

11.3.2 Single-cell proteomics

Single-cell proteomics sheds light on cellular activity in horticultural crops, offering insight into the proteins that are directly functioning in individual plant cells and complementing transcriptome and epigenomic data (Fulcher *et al.*, 2024; Montes *et al.*, 2024). The advances are mostly due to the improved methods of sample preparation, mass spectrometry, and experimental design, which have not only made it feasible but also simpler to quantify the proteins of thousands of cells in hundreds of trials.

The research of Montes *et al.* (2024), which used liquid chromatography (LC)–MS/MS for protein identification after fluorescence-activated sorting of root protoplasts, is a prime example. Consequently, the differences between the cell types of the endodermis and cortex were discovered to comprise more than 3200 proteins. Similarly, Fulcher *et al.* (2024) performed label-free proteomics on *Arabidopsis* leaf mesophyll cells using the nano-POTS sample preparation technique. They were able to identify as many as 3000 proteins under each of the two situations – control and drought stress – under evaluation.

Protoplasting is an essential process in plant single-cell proteomics, but it is not practical for many tissues because it can drastically alter protein profiles. Conversely, protoplast-free approaches such as proximity labeling offer advantages because they apply TurboID under cell-type-specific promoters to tag the proteins of intact cells, hence avoiding the issue (Anderton and Uhrig, 2025). The methodology is useful for determining the cell-type proteome in complex tissues, according to Wallner *et al.* (2024), who used it in guard cells. However, single-nucleus proteomics still encounters hurdles because of the small quantity of nuclear proteins.

11.3.3 Chromatin accessibility

This is a dynamic single-cell method to comprehend gene regulation of horticultural crops. Generally, this process is carried out by single-nucleus assay for transposase-accessible chromatin using sequencing (snATAC-seq), which opens chromatin regions maps where proteins, which are the regulators, bind, thus providing clues of cell-type-specific regulation. The same droplet as well as the well-based snATAC-seq methods have been applied to plant cells with great success (Buenrostro *et al.*, 2015), thus enabling the classification of major cell types and, in combination with scRNA-seq (Dorrity *et al.*, 2020), identification of previously unknown cellular subtypes.

However, approximately one-third of the accessible chromatin regions (ACRs) in maize are cell specific and associated with phenotypic variation and key features (Dorrity *et al.*, 2020; Marand *et al.*, 2021). The single-cell chromatin analysis reveals that the distal ACRs – which are frequently species and cell-type specific – tend to be clustered near the crucial genes, such as those involved in C4 photosynthesis (Mendieta *et al.*, 2024). The snATAC-seq technique reveals chromatin heterogeneity at single-cell resolution, making gene regulation in horticultural crops evident (Buenrostro *et al.*, 2015). This enables precision-driven crop modification.

11.3.4 Single-cell multi-omics

Single-cell multi-omics provides a comprehensive understanding of gene regulation in horticultural crops by integrating several molecular layers inside individual cells (Argelaguet *et al.*, 2021). The most commonly used approach, combining scRNA-seq with scATAC-seq (Vandereyken *et al.*, 2023), correlates chromatin accessibility with gene expression on droplet- or well-based platforms. In *Arabidopsis*, 10 × Genomics single-nucleus multi-omics under drought stress mapped cell-type-specific regulatory networks, whereas pathogen investigations identified transcription factor-chromatin-gene modules active during immunity (Liu *et al.*, 2024; Nobori *et al.*, 2025). Nobori *et al.* (2025) found that around half of immune-responsive genes lacked a link between accessibility and expression, suggesting additional regulation through DNA methylation and histone changes.

11.4 Emerging Technologies with Potential to Advance Plant Science

Although sc/snRNA-seq, snATAC-seq, and spatial transcriptomics are gaining traction in plant biology, numerous advanced modalities and innovations remain underutilized. This section aims to describe these emerging tactics, focusing on methodological aspects over biological applications.

11.4.1 Advanced spatial omics

Advanced spatial omics combines spatial transcriptomics, proteomics, and metabolomics to map the molecules in plant tissues with a preserved spatial context. This allows the identification of specific cell types in gene, protein, and metabolite patterns that characterize developmental and stress-responsive processes (Gandin *et al.*, 2025). These techniques, by fusing imaging with computational reconstruction, uncover the structural and functional organization of plant systems.

11.4.1.1 Spatial transcriptomics

Cellular heterogeneity and spatial organization in intact tissues can be represented by 3D spatial transcriptomics using either direct thick-tissue imaging or serial sectioning with computational reconstruction. According to Sampath Kumar *et al.* (2023), serial sectioning profiles successive thin slices and digitally assembles them to create 3D map sections at 10 μm. Newer methods such as Open-ST provide high-resolution spatial transcriptomics in 3D (Schott *et al.*, 2024). The gene expression of tissues up to 300 μm thick can be mapped using direct imaging techniques like expansion-assisted iterative (EASI)-FISH and cycle-hybridization chain reaction (cycle-HCR) without sectioning (Wang *et al.*, 2021c; Gandin *et al.*, 2025). Deep STAR-map profiles tens to >1000 transcripts on 200 μm thick tissues using *in situ* sequencing (Sui *et al.*, 2024), whereas 3D MERFISH combines confocal microscopy with deep learning to achieve 200 + gene detection in samples with a thickness of up to 200 μm (Fang *et al.*, 2024).

These imaging-based methods produce massive datasets that need significant processing power and have intrinsically limited gene throughput. The field of 3D spatial transcriptomics in plants is still in its infancy; PHYTOMap multiplexes up to 28 transcripts in intact *Arabidopsis* root tips, revealing patterns specific to regions and cell types (Nobori *et al.*, 2023), while HCR only visualizes a few genes in whole-mount tissues (Yu *et al.*, 2024).

11.4.1.2 Spatial proteomics

Through local mapping of protein expression in plant tissues, proteomics localized in the plant system has demonstrated the functional structure of the plant beyond the data of the transcript. The simultaneous detection of dozens of proteins without the issue of spectral overlap is made possible by protein quantification using multiplexed antibody-based techniques such as cycIF or CODEX, i.e. DNA-barcoded antibodies, and iterative fluorescence. MSI methods, for example, MALDI, DESI, and IR–MALDESI, are suitable for spatial profiling of proteins, metabolites, and lipids (Alexandrov, 2023).

Recent advances, such as deep visual proteomics (DVP) (Mund *et al.*, 2022), include the use of laser microdissection and AI-driven tissue image processing. With a cellular resolution of up to 10,000 proteins, this is incredibly sensitive proteome profiling (Mund *et al.*, 2022; Nordmann *et al.*, 2024). By producing root proteome maps of plants at a very high resolution for the physiological, stress-responsive, and developmental stages of proteins, these techniques will increase our knowledge of the multifunctionality of proteins and the organization of tissues.

11.4.1.3 Spatial metabolomics

Spatial metabolomics enables the visualization of metabolite distribution in the plant tissue, thereby unearthing the localized metabolism and nutrient dynamics. MSI-based approaches such as MALDI–MSI, DESI–MSI, and IR–MALDESI are responsible for spatially resolved detection of metabolites, peptides, and lipids (Yin *et al.*, 2024a). These techniques are crucial for deciphering the intricacies of metabolism in diverse tissues.

In addition to making subcellular visualization easier, cutting-edge tools like cryo-nano-SIMS enable incredibly accurate elemental nutrient localization in *Arabidopsis* roots, revealing new processes for nutrient allocation (Kavipriya *et al.*, 2025). Cotton somatic embryogenesis molecular changes have been identified when spatial metabolomics is combined with scRNA-seq and spatial transcriptomics (Ge *et al.*, 2021). Although it is less expensive than other omics, it is constrained by poor annotation confidence and insufficient metabolite coverage, which calls for methodological advancements to obtain precise and thorough plant metabolite profiling (Alexandrov, 2023; Yin *et al.*, 2024b).

11.4.2 Sample multiplexing

Multiplexed single-cell sequencing is a method that allows the parallel analysis of multiple samples and thus greatly improves the number of cells studied without losing the ability to detect the rare ones. Additional levels such as scifi-RNA-seq (where scifi is single-cell combinatorial fluidic indexing) and UDA-seq boost the throughput capacity by pre-labeling cells, and scifi-ATAC-seq allows plant chromatin profiling (Alexandrov, 2023).

In the case of horticultural crops, these methods have been developed into effective instruments that enable researchers to identify region-specific regulatory networks and expedite the breeding process aimed at resilience, flavor, and nutritional quality by tracking changes in gene expression at the cell-specific level caused by stress, treatment, or genetic variation.

11.4.3 Diverse RNA species

Current techniques allow a range of RNA species to be sampled at much higher resolution. While MAS-ISO-seq uses long-read sequencing to generate whole transcriptomes, Smart-seq3 allows isoforms to be followed for the entire length of the RNA molecule and alleles to be detected (Hagemann-Jensen *et al.*, 2020). Furthermore, smart-seq-total and its droplet variant, VASA-seq, provide the noncoding regions of the RNA among newly identified RNAs (Isakova *et al.*, 2021; Salmen *et al.*, 2022). Spatial total RNA sequencing combines the two methods of direct tissue mapping of both coding and noncoding RNAs with polyadenylation and spatial transcriptomics (McKellar, 2023).

Research on *Arabidopsis* root miRNA profiling specific to cell types has shown miRNA specialization, and FlsnRNA-seq has made it possible for single-nucleus full-length analysis of splicing and polyadenylation (Long *et al.*, 2022). In the horticulture sector, the use of these tools can be advantageous in exposing RNA regulatory networks that govern the processes of ripening, flavor, nutrition, color, stress tolerance, and metabolite synthesis to molecular targets for accurate facilitation.

11.4.4 Translational regulation

Translation allows plants to have an exciting developmental and ecological response range. Cell-type and state-specific translational regulation in plants remains mostly unknown. A tool called TRAP-seq enables researchers to characterize ribosome-bound RNA species using ribosomal protein tagging (Mustroph *et al.*, 2009). Selecting for focused translation analysis is made possible by the use of cell-type or condition-specific promoters. The synergistic integration of TRAP-seq with single-cell transcriptomics and epigenomics data can potentially uncover as yet unrecognized regulatory elements. Such profiling allows the discovery of ribosomal heterogeneity, uORF regulation, and tissue-specific translation, which in turn can impact ripening, flavor, stress tolerance, and nutritional quality, leading to trait improvement (Mustroph *et al.*, 2009).

11.4.5 Epigenome analyses

Epigenome analysis is the study of the transmission of molecular modifications such as DNA methylation and changes in histones, which regulate gene expression but do not change the DNA sequence. It reveals the mechanisms by which plants manage growth, respond to stress, and change their appearance through the restructuring of chromatin.

11.4.5.1 DNA methylation

In plants, DNA methylation regulates genome stability, cell specialization, and gene behavior (Luo *et al.*, 2018). Bisulfite sequencing, which converts unmethylated cytosines to uracils while preserving those that are methylated, is the primary method used for single-cell profiling. This enables the development of genome-wide, base-resolution maps (Liu *et al.*, 2023; Tian *et al.*, 2023). The development of enzymatic methyl sequencing, which is less destructive, as well as commercial assays such as the single-cell bisulfite platform by ScaleBio, has widened the range of users. However, it is still very difficult and costly to obtain coverage of the full single-cell methylome because the precision required for the low volume of liquid handling in the library preparation makes it the most challenging step.

Different types of cells are at various stages in their DNA methylome profiles in plants, which is why this has not yet accomplished in single cells. The results of bulk studies show that methylation patterns that are cell-type specific are very distinct. If these first steps are understood, the application of dCas9-based methyltransferase or demethylase editing could be made feasible. Such methods, in the case of horticultural crops, could be used to determine the connection between epigenetic regulation and the phenomena of ripening, color, stress, and metabolic traits, thus allowing the development of a precise and sustainable future for crop improvement.

11.4.5.2 Histone modifications

Histone modifications allow gene regulation through changing chromatin accessibility and transcriptional activity. Correspondingly, in plants, stress involves the increase of activating marks such as H3K4me2/3 and H3K9ac along with the decrease of repressive H3K27me3, which consequently results in stress-responsive transcription (Peterson *et al.*, 2025). CUT&Tag and chromatin integration labeling are examples of exceptionally sensitive low-input methods that, unlike ChIP-seq, allow for single-cell investigations. In animals, single-cell CUT&Tag has been utilized to locate cell-specific histone landscapes (Bartosovic *et al.*, 2021), and, with the introduction of plant-focused variants, the application area is expanding. INTACT-ChIP-seq identified histone states of the meristem connected to blooming (You *et al.*, 2019), and the optimized CUT&Tag procedures achieve almost single-cell precision in plants (Tao *et al.*, 2022).

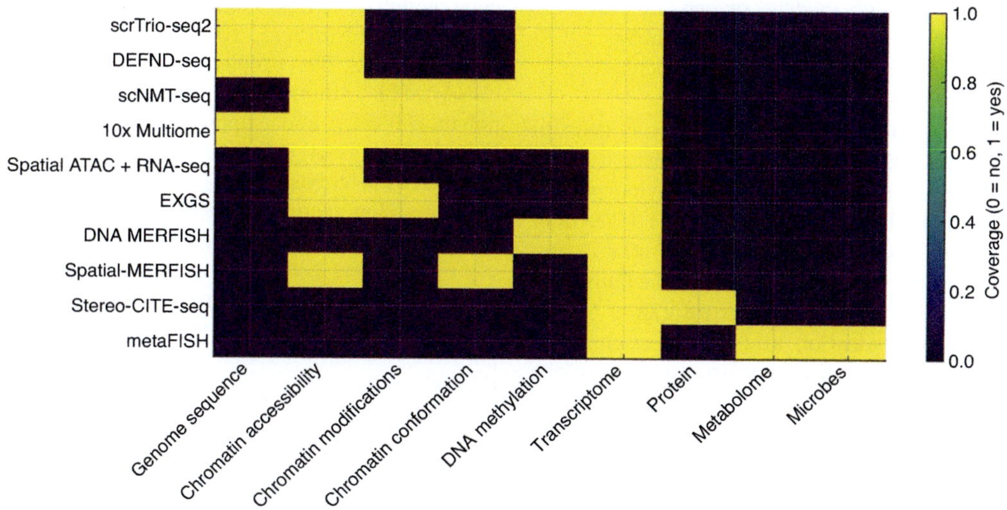

Fig. 11.3. Modality combinations compatible with simultaneous profiling in single-cell and spatial omics methods, highlighting those applied to plant tissues.

Currently, multiplexed versions can simultaneously profile multiple histone marks simultaneously (Meers *et al.*, 2022; Xiong *et al.*, 2024). Notably, CRISPR-based systems such as dCas9 or SunTag, when combined with histone-modifying enzymes, may already convert histone states in plants, allowing for target chromatin mark editing for functional studies of specific cell types (Wang *et al.*, 2023).

Single-cell histone profiling in horticultural crops can detect chromatin states that control tomato ripening, grape and citrus color formation, strawberry scent chemical generation, pepper stress tolerance, and leafy vegetable quality enhancement. The integration from profiling to targeted editing not only provides the possibility for accurate validation and perfecting epigenetic switches, but it also paves the way for the development of new precision breeding technologies (Wang *et al.*, 2023).

Figure 11.3 compares various multi-omics methods across distinct biological layers, highlighting their genome sequencing. Each yellow-marked cell indicates the specific omic layer that a given technique can simultaneously analyze, illustrating the integrative capacity of methods such as scNMT-seq, 10 × Multiome, and Stereo-CITE-seq (Fig. 11.3).

11.5 Multiple Omics Approaches

Multiple omics approaches interlink genomic, transcriptomic, proteomic, metabolomic, and epigenomic data for understanding complex biological systems. They identify molecular networks that connect genotype to phenotype and demonstrate plant responses to developmental and environmental cues.

11.5.1 Genomics and epigenomics in horticulture

Genomics and pangenomics have greatly improved the ability to breed desirable traits in horticulture through the use of genomics data to establish connections between genotypes and phenotypes, which is a fundamental step for the targeted improvement of traits (Shendure *et al.*, 2017). The completion of *de novo* assemblies, the generation of high-quality reference genomes, and the implementation of large-scale whole-genome resequencing (WGRS) for exotic varieties or landraces of different crops, such as rice (Wing *et al.*, 2018), soybean (Zhang *et al.*, 2022), tomato (The Tomato Genome Consortium, 2012), and rapeseed (Chalhoub *et al.*, 2014), are now possible through modern sequencing platforms comprising next-generation sequencing (NGS) and also third-generation systems such as PacBio and Oxford Nanopore.

The interest in pangenomes has been to show the wide-ranging genetic diversity that surpasses single-nucleotide polymorphisms (SNPs), with structural variations (SVs) and presence–absence variations (PAVs) stimulating the changes of complicated traits like flowering, stress adaptation, architecture (Zhou *et al.*, 2009), domestication, and disease resistance.

For instance, there are structural variations that are the root cause for virus resistance in maize (Scmv1 locus) (Liu *et al.*, 2019) and salt tolerance (Luo *et al.*, 2019), genes resistant to blight derived from wild potato (Du *et al.*, 2015), deletion of a sulfotransferase gene in sorghum responsible for resistance to Striga, retrotransposon insertions related to salt sensitivity in soybean, and copy number variation of HvCBF that contributes to the ability of barley to survive frost (Francia *et al.*, 2016). In the context of horticultural crops, which are mostly pangenomic, the graph-based models have been especially instrumental in revealing large SVs (Liu *et al.*, 2019; Liu *et al.*, 2022a; Qin *et al.*, 2023) and gene copy number variations that traditional single-reference genomes fail to detect, thus further facilitating the identification of alleles related to domestication and adaptation (Qin *et al.*, 2023).

This method works by combining *de novo* genome sequencing, WGRS, comparative genomics, pangenome assembly, and graph-based algorithms, with GWAS serving as the link to the trait. Furthermore, transcriptomics, epigenomics, and CRISPR-Cas validation have a significant impact on horticultural genetic diversity as well as the rate of breeding for high-yield, nutrient-dense, and climate-resistant crops (Della Coletta *et al.*, 2021; James *et al.*, 2021).

11.5.2 Transcriptomics for functional insights

Transcriptomics, involving all coding and non-coding RNAs in cells or tissues, is pivotal for horticultural crop breeding by revealing gene activity during development and stress responses (Schiessl *et al.*, 2020). Low-resolution early methods like droplet digital (DD)-PCR, cDNA–AFLP (amplified fragment length polymorphism), and suppression subtractive hybridization (SSH) (Mir *et al.*, 2019) have been replaced by NGS-based RNA-seq, which delivers high-precision, genome-wide profiling beyond microarray limits. This has enabled the discovery of seed size genes in rapeseed (Niu *et al.*, 2020), links between signal peptides, immunity, and development in rice, and ion homeostasis pathways for salt tolerance in Chinese cabbage (Li *et al.*, 2021).

Transcription factors such as MYB, bZIP, and WRKY regulate maize root responses to salt stress (Zhang *et al.*, 2021), while WRKY genes in apple respond to various biotic stresses (Zhang *et al.*, 2025). Transcriptomic analyses also reveal regulatory roles of noncoding RNAs (ncRNAs) – miRNAs, lncRNAs, circRNAs, rRNAs, and siRNAs – in transcriptional and post-transcriptional control. Of these, miRNAs mediate mRNA cleavage, lncRNAs regulate diverse pathways, and circRNAs act as miRNA sponges (Lai *et al.*, 2018). Whole-transcriptome sequencing has elucidated mRNA expression profiles in maize (Liu *et al.*, 2022a), citrus (Fu *et al.*, 2019), and Chinese cabbage (Shi *et al.*, 2022), offering insights into crop-specific regulation. Additionally, isolated *Arabidopsis* female gamete transcriptomes have advanced the understanding of reproduction, polyploidization, and hybridization in the context of horticultural improvement (Song *et al.*, 2020).

11.5.3 Epigenomics

By demonstrating alterations in DNA methylation, histone modifications, and chromatin accessibility, epigenomics presents techniques for crops to modify their genes. A number of methods, including whole-genome bisulfite sequencing (WGBS), methylation-sensitive amplification polymorphism (MSAP), and methylated DNA immunoprecipitation (MeDIP), have been employed to examine DNA methylation, one of the elements that influences how an organism adapts to its surroundings and develops (Sun *et al.*, 2022).

In mulberry, drought resistance was found to be associated with methylation patterns that were both stable and differentially methylated regions (DMRs) correlated with stress genes (Ackah *et al.*, 2022). In alfalfa, the increase of methylation was found to strengthen salt resistance, whereas 5-AzaC treatment weakened it. Re-replication, connected with methylation, is also demonstrated to be a condition that assists the organisms exposed to heavy metal stress (Raza *et al.*, 2022; Sun *et al.*, 2022). In general, it is found to be a mechanism that regulates seed development, organ differentiation, and fruit ripening. The techniques used in profiling the histone marks include ChIP-seq, ChIP-exo, ChIP-mentation, CUT&Tag, and DamID for chromatin state mapping in horticultural plants.

In tea plants, post-harvest desiccation enhanced ABA accumulation through increased histone acetylation and reduced H3K9me2 (Gu *et al.*, 2021), linking stress-induced histone changes to transcriptional reprogramming. Chromatin accessibility profiling using DNase-seq, MNase-seq, FAIRE-seq, and ATAC-seq provide further evidence of genome remodeling. In grapevine, ATAC-seq integrated with Hi-C and ChIP-seq revealed open chromatin regions correlated with gene regulation (Schwope *et al.*, 2021). DNA methylation, histone modification, and chromatin accessibility constitute key epigenetic mechanisms driving stress resilience, fruit ripening, and developmental regulation in horticultural crops.

11.5.4 Proteomics applications

Proteomics enables the comprehensive analysis of protein sequence, structure, function, and abundance to understand stress responses in horticultural crops (Aizat and Hassan, 2018). Sequence proteomics uses high-performance liquid chromatography (HPLC), while structural proteomics applies nuclear magnetic resonance (NMR), electron microscopy, and X-ray crystallography (Opdensteinen *et al.*, 2022). Functional proteomics, through Y1H, Y2H, and protein microarrays, identifies key protein–protein and protein–DNA interactions in stress signaling. Earlier methods using chromatography and ELISA evolved into quantitative gel-based platforms such as SDS–PAGE, 2-D electrophoresis, and 2-D difference gel electrophoreses (2D-DIGE).

Protein abundance differences between treatments may be precisely measured using sophisticated quantitative proteomics techniques such as ICAT, SILAC, and iTRAQ. Protein microarrays and chips provide easy, high-throughput, limited-sample-type profiling, whereas high-throughput NMR can yield structural suggestions for protein functioning. Single-cell proteomics still lags behind transcriptomics in terms of development; however, recent MS innovations have led to an improvement of protein quantification at single-cell resolution (Vistain and Tay, 2021).

These technological advances are indeed necessary for dealing with the stress symptoms in the areas of horticulture because they uncover the protein changes and the regulatory mechanisms that form the basis of tolerance. The study of differential protein expression under abiotic stress has been instrumental in the determination of stress-responsive proteins, the unraveling of the regulatory networks, and also the establishment of breeding programs for resilience, yield, and quality improvements (Ueda and Seki, 2020).

11.5.5 Metabolomics and phenotyping integration

Metabolomics enables inclusive metabolite profiling to reveal mechanisms of crop resilience and quality. Plant metabolites include primary metabolites, lipids, sugars, and amino acids essential for growth, and secondary metabolites such as flavonoids, carotenoids, alkaloids (e.g. atropine), phytic acid, antioxidants, coenzymes, and ROS, which are critical for stress defense and nutritional value. Metabolites are the end products of metabolic pathways that connect phenotypes to genetic, transcriptomic, and proteomic data. With ~ 3,91,000 plant metabolites identified (Nakabayashi and Saito, 2020), advanced analytical tools are essential for targeted and untargeted profiling.

Common techniques in horticultural metabolomics include LC–MS, gas chromatography (GC)–MS, capillary electrophoresis (CE)–MS,

HPLC, NMR, and direct flow injection (DFI) (Raza *et al.*, 2022). Metabolomics, when combined with NGS, connects metabolic networks to stress-induced gene regulation (Abdelrahman *et al.*, 2018) and aids crop improvement under changing climates (Pan *et al.*, 2020; Scossa *et al.*, 2021).

Trichoderma-induced biotic stress in tomatoes increased isoprenoid accumulation via reshaping metabolic networks (Coppola *et al.*, 2019). Metabolomics, when combined with transcriptomics, has elucidated the mechanisms behind stress tolerance and facilitated the evaluation of nutritional quality characteristics essential for horticultural progress (Pott *et al.*, 2021). Metabolomics, which integrates genomes, transcriptomics, and proteomics, provides high-resolution insights into plant stress responses, development, and nutritional composition. It is exposing complex molecular interactions that are critical for precision breeding and crop quality enhancement.

11.6 Integration of Spatial Multi-omics

Single-cell multi-omics reveals regulatory complexity across omics layers (Chen *et al.*, 2019) but lacks spatial scenarios within tissues. Advances in spatial transcriptomics, proteomics, genomics, epigenomics, and metabolomics now enable high-resolution mapping of biological processes (Chen *et al.*, 2022; Bian *et al.*, 2024), though most remain single-omic, and computational integration is insufficient to link cross-modal mechanisms (Lu *et al.*, 2024). Spatial multi-omics addresses this issue by simultaneously assessing gene expression, proteins, chromatin states, variations, and metabolites within a single tissue segment, offering integrated insight into cellular function, microenvironments, and regulatory dynamics under normal and stress circumstances.

11.6.1 Integration of spatial transcriptomics and epigenomics

Spatial multi-omics platforms such as spatial ATAC & RNA-seq and spatial CUT&Tag (Deng *et al.*, 2022a, b) have transformed tissue biology by enabling simultaneous mapping of epigenomic and transcriptomic landscapes at cellular resolution (Vicari *et al.*, 2023). These methods couple chromatin accessibility or histone modification chemistry with mRNA expression using deterministic co-barcoding in microfluidic DBiT-seq systems (Vicari *et al.*, 2023). Spatial ATAC-seq, employing Tn5 transposase to insert barcoded oligonucleotides into accessible chromatin, provides genome-wide and high-resolution maps of regulatory architecture, revealing cell identity, lineage relationships, and state dynamics in development and disease.

In order to preserve the structural integrity of the tissue and disclose spatially distinct chromatin states throughout embryonic organogenesis and brain development, spatial CUT&Tag enables local profiling of histone modifications directly *in situ* inside frozen tissue slices (Takei *et al.*, 2021), a technique that could be applied to spatial epigenomic profiling in horticultural plant systems.

To create geographic atlases of transcriptional and regulatory activity, barcoded DNA and cDNA fragments are sequenced and computationally integrated with tissue imaging. In addition to transcript–epigenome mapping, DNA-seq FISH+ (Fig. 11.3) advances the field of spatial genomics by mapping thousands of genomic loci and chromatin marks in individual cells. This reveals nuclear domain interactions, 3D genome organization, and both conserved and cell-type-specific chromatin features (Vicari *et al.*, 2023).

As a result, these methods provide a detailed picture of how chromatin accessibility, histone modification, nuclear architecture, and transcriptional regulation are coordinated across biological systems, and thus how they govern the physiological processes.

11.6.2 Integration of spatial proteomics and transcriptomics

Integrating spatial proteomics with transcriptomics resolves the spatial gap in single-cell multi-omics, adding related information to molecular profiling. Because RNA and protein levels offer complementary views of gene regulation and cell state (Slyper *et al.*, 2020), their joint mapping enhances functional understanding.

SM-Omics employs DNA-barcoded antibodies and automated liquid handling for high-throughput, *in situ* co-profiling of transcripts and proteins (Vickovic *et al.*, 2022). Similarly, using oligonucleotide barcode counting, NanoString's GeoMxDSP allows high-plex, non-destructive quantification of RNA and proteins from frozen or formalin-fixed paraffin-embedded samples, facilitating slide reuse for further analysis.

Stereo-CITE-seq (Fig. 11.3) integrates transcriptomic and proteomic profiling by combining antibody-derived tags for approximately 200–300 proteins with deterministic tissue barcoding of DNA tags and mRNAs, assigning spatial address codes to each molecule (Hernandez *et al.*, 2022; Liu *et al.*, 2022b). The resultant libraries produce high-resolution *in situ* molecular atlases by enabling the computational reconstruction of spatially resolved gene and protein expression maps. When combined, these technologies provide profound insights into cell states, tumor microenvironments, and disease pathways by combining spatial transcriptomics and proteomics.

11.6.3 Integration of spatial transcriptomics and metabolomics

The integration of metabolomics with spatial transcriptomics provides a unique opportunity to investigate the relationship between metabolite distribution and gene expression. Cell type, microenvironment, signaling, and regulatory pathways all influence this interaction, and spatial information needs to be maintained. On the same tissue segment, the spatial multimodal analysis (SMA) procedure combines MSI with spatially resolved transcriptomics (Vicari *et al.*, 2023), thus identifying the relationships between gene activity and metabolite patterns across areas while preserving the sensitivity and specificity of both methods.

11.6.4 Integration of spatial transcriptomics, genomics, and proteomics

The transcriptome, genome, epigenome, proteome, and metabolome may all be simultaneously profiled because of the integration of spatial multi-omics. This enables scientists to identify molecular pathways and mechanistic connections essential to cellular function. By enabling high-throughput, subcellular resolution detection of DNA, RNA, proteins, and biomolecules inside the same tissue, multi-omics *in situ* pairwise sequencing (MiP-seq) offers a significant development (Wu *et al.*, 2024). Using dual barcoded padlock probes, MiP-seq targets antibody-conjugated nucleic acids for protein and biomolecule detection and directly binds nucleic acids for DNA and RNA analysis.

In contrast to traditional *in situ* sequencing, MiP-seq improves signal fidelity by increasing decoding capacity while reducing sequencing rounds (4N vs 10N) (Tian *et al.*, 2023), saving sequencing time by about 50%, lowering imaging expenses, and minimizing laser damage. MiP-seq offers precise spatial maps of proteins, RNA, and DNA at single-cell resolution due to its sensitivity and adaptability. This provides an effective means of examining intercellular connections, disease processes, and cell states in complex tissues (Muster *et al.*, 2017).

11.7 Artificial Intelligence and Machine Learning

11.7.1 Potential of AI and machine learning in horticulture research

Horticultural crop research has been transformed by artificial intelligence (AI) and machine learning (ML) (Table 11.1). The capacity to combine and analyze intricate multi-omics information is the main innovation here. While ML, a subset of AI, develops algorithms and statistical models that enable machines to learn from data without explicit programming (Collins and Moons, 2019; Wang *et al.*, 2022), AI generally imitates human cognitive functions, such as learning (Nturambirwe and Opara, 2020), reasoning, and the ability to correct mistakes. Unlike traditionally programmed systems, ML uses trends in the input and output data to create prediction algorithms (Wang *et al.*, 2021b).

Supervised learning (working on labeled data) (Jacob and Darney, 2021), unsupervised

Table 11.1. Machine learning and AI techniques in horticultural crop analysis.

ML/AI technique	Techniques/algorithms	Description	Examples in horticulture
Supervised learning	Decision trees, Naive Bayes (NB), support vector machines (SVM) (Lanckriet *et al.*, 2004), LASSO, boosted regression trees, random forests, artificial neural networks (ANNs) (Bavafaye Haghighi *et al.*, 2019), K-Star (Cleary and Trigg, 1995), SVR (support vector regression)	Algorithms are trained on labeled data to predict outcomes or classify inputs based on learned patterns	Plant disease identification, yield prediction, fruit quality and size estimation, phenotype classification
Unsupervised learning	K-Means Clustering, MOFA2 (multi-omics factor analysis) (Argelaguet *et al.*, 2018), moCluster (Meng *et al.*, 2016), iCluster-Bayes (Mo *et al.*, 2018), iCluster, JIVE, joint Bayes factor (JBF), Bayesian consensus clustering (BCC), similarity network fusion (SNF) (Wang *et al.*, 2014)	Algorithms that find hidden patterns or clusters in unlabeled data, including integration of multi-omics datasets	Clustering of plant genotypes or phenotypes, pattern detection in multi-omics data, and optimization of environmental conditions
Reinforcement learning (RL)	RL agents with reward-based optimization (Jaiswal *et al.*, 2020)	Software agents learn to take actions in an environment to maximize long-term cumulative reward	Automated regulation of greenhouse parameters such as lighting, irrigation, and humidity control (You *et al.*, 2019)
Deep learning	Artificial neural networks (ANNs), deep neural networks (DNNs) (Holzinger *et al.*, 2023)	Multi-layer neural networks modeling complex and nonlinear interactions in data	Complex trait prediction, stress and disease identification (Muhammad Aslam *et al.*, 2022), automated phenotyping using image or multi-omics data
Convolutional neural networks (CNNs)	CNN architectures specialized for grid-structured data-like images (Lele, 2019)	A type of deep learning model highly effective in image analysis and feature extraction	Leaf disease detection, plant species identification from leaf images, detecting early fruit defects and pathogens

learning (identifying new patterns in data without labels) (Karimi *et al.*, 2020), semi-supervised learning (just a combination of both), and reinforcement learning (learning the best moves by obtaining the most rewards) (Novakovsky *et al.*, 2023) are some of the various ML classes. The vast volumes of high-dimensional, heterogeneous data that arise from multi-omics and high-throughput phenotyping technologies in horticulture suit these methods well.

The applications of AI and ML in phenotyping, disease detection (Stafford *et al.*, 2020), stress identification (Gill *et al.*, 2022), yield prediction (Kim *et al.*, 2015), genomic selection, multi-omics integration, and environmental monitoring revolutionize horticultural research. Image and sensor data analysis are fully automated by ML; thus, the quantification of traits is accurate and unprejudiced.

One of the crucial technologies that is the basis for early disease diagnosis and stress detection is pattern recognition. Here, visual and multi-omics inputs are used (Ben Ayed and Hanana, 2021). Accelerating breeding by pinpointing the essential markers that are linked with the traits is one of the feats of AI-driven genomic selection (Kar *et al.*, 2021). The combination of AI/ML with multi-omics and real-time sensors has a positive effect on precision agriculture, that allows the most effective water, nutrient, and pest management for improved resilience and sustainability.

11.7.2 Current applications of multi-omics and AI in plant phenotyping

The integration of multi-omics data with AI and ML is increasingly advancing plant phenotyping (Kar *et al.*, 2021), deepening biological insights, and enhancing breeding strategies. AI/ML models coupled with genomics have improved genomic selection and complex trait prediction, as demonstrated by Montesinos-López *et al.* (2019), whose deep learning framework outperformed conventional methods in maize and other crops. High-throughput phenotyping has similarly benefited, with Pinu *et al.* (2019) developing *Deep Root*, a CNN-based system (Table 11.1), that analyzes X-ray computed tomography images to precisely quantify root architecture traits.

AI and ML-driven image analysis have demonstrated great potential in early disease detection, by applying deep learning models to identify plant diseases from leaf images, enabling timely and targeted interventions. Emerging research, such as the integration of genomic, transcriptomic, and epigenomic data using AI and ML in *Arabidopsis thaliana* (Argueso *et al.*, 2019), reveals complex interactions underlying plant stress responses. Furthermore, ML approaches like those used by Sarabia *et al.* (2018) on hyperspectral imaging data facilitate the identification and quantification of various biotic and abiotic stresses, contributing to enhanced understanding and management of plant health.

11.8 Data Integration and Systems Biology

Data integration and systems biology combine multi-omics information to understand molecular interactions within plant systems. They reveal regulatory and metabolic networks that synchronize development and environmental responses.

11.8.1 Computational methods for multi-omics integration

Multi-omics integration has transformed horticultural science by revealing how plants respond to environmental stress and guiding the breeding of resilient varieties (Clyde, 2021). This approach uncovers complex regulatory networks and identifies key genetic and metabolic traits, directly enhancing crop quality and sustainability for modern horticultural practices. The techniques discussed here include matrix factorization, neural networks, Bayesian models, and manifold alignment (Table 11.2), which allow the integration of different omics layers at single-cell and cross-sample levels. The main points of the discussion are captured in Table 11.2 that these integration strategies enhance the understanding, extend the processing capacity, and increase the precision of linking molecular networks.

11.8.2 Case studies: real-world application in horticulture

- LncRNA1840 is essential in tomato fruit ripening, contributing to gene expression and regulatory networks in multiple ways. CRISPR-Cas9-mediated lncRNA1840 mutants caused the delayed tomato fruit ripening. ChIRP-seq suggested that lncRNA1840 had occupancy on 40 genes, but none of them are differentially expressed genes in transcriptomic analysis, which indicated lncRNA1840 might indirectly modulate gene expression. ChIRP-mass spectrometry analysis identified potential protein interactors of lncRNA1840, Pre-mRNA processing splicing factor 8, highlighting its involvement in post-transcriptional regulatory pathways. (Zhu *et al.*, 2024).
- The length of introns flanking grape circRNAs was closely related to exon circularization in grapevine. Although the

Table 11.2. Classification of computational methodologies for multi-omics data integration.

Classification	Methodology type	Algorithm	Examples	Advantages
Integration (same single cells)	Matrix factorization	Bayesian factor analysis, matrix decomposition	MOFA+ (Clyde, 2021), scAI (Macaulay *et al.*, 2015)	Interpretability, scalability
	Neural networks	Autoencoders, variational autoencoders	scMVAE, totalVI (Bian *et al.*, 2018), BABEL (Zhang *et al.*, 2017)	Capture nonlinear relations, handle noise, and impute missing data
	Network-based	Similarity network fusion, weighted nearest neighbors	citeFUSE (Kim *et al.*, 2020), Seurat v4 (Stanojevic *et al.*, 2022)	Robust to noise, leverage neighborhood info
	Bayesian and metric learning	Bayesian mixtures, metric learning	BREM-SC (Hao *et al.*, 2021), SCHEMA (Wang *et al.*, 2020)	Statistical rigor, computational efficiency
Alignment (different single cells)	Bayesian	Latent variable models	Clonealign (Campbell *et al.*, 2019), MUSIC (Duan *et al.*, 2019)	Probabilistic cell assignment
	Manifold alignment	Pseudotime, kernel methods	MATCHER (Welch *et al.*, 2017), MMD-MA (Demetci *et al.*, 2022)	Low-dimensional alignment
	Distance-based matching	Graph optimization, optimal transport	UNION-Com, SCOT, Pamona (Liu *et al.*, 2019)	Preserve dataset structures
	Neural networks	Adversarial autoencoders, generative adversarial networks (GANs)	SCIM (Stark *et al.*, 2020), MULTIGRATE, MAGAN (Amodio and Krishnaswamy, 2018)	Handle nonlinear mappings, adversarial learning
	Others (CCA/NMF)	Canonical correlation, mutual nearest neighbors	Seurat v3 (Tran *et al.*, 2020), LIGER (Dou *et al.*, 2020)	Widely used, effective for known data types

longer introns flanking grape circRNAs appeared to circularize more efficiently, a 20- to 50-nt region seemed large enough to drive grape circRNA biogenesis. In addition, the endogenous introns flanking circularized exon(s) in conjunction with reverse complementary sequences could support the accurate and efficient circularization of various exons in grape, which constitutes a new tool for exploring the functional consequences caused by circRNA expression. It is identified that 475 differentially expressed circRNAs in grape leaves under cold stress (Gao *et al.*, 2019).

- *Fragaria nilgerrensis*, a wild strawberry from Southwest China, exhibits a distinct peach-like aroma. Researchers profiled volatile organic compounds (VOCs) and gene expression across four fruit developmental stages, revealing γ-decalactone, β-decalactone, and γ-undecalactone as key contributors. Their levels were strongly correlated with FngFAD2 enzyme activity. Silencing FngFAD2 via TRV reduced both lactone content and aroma intensity, confirming its pivotal role. Three transcription factors – FngDOF1.2, FngWRKY3, and FngWRI1 – were identified as positive regulators of FngFAD2, validated through dual luciferase assays, Y1H, and subcellular localization. This study establishes the molecular regulatory network of peach-like aroma in *F. nilgerrensis* and highlights its potential for improving the flavor of cultivated strawberries (Cao *et al.*, 2024b).

11.9 Challenges

When advanced technology is used to merge data from several labs, platforms, or experimental batches, it typically introduces batch effects or technical biases that make it difficult to see the true biological differences. To preserve the data purity and enable meaningful comparisons, it is crucial to remove these variables (Luecken *et al.*, 2022). Furthermore, an appropriate algorithm that can analyze partial data without bias must be used since some omics layers or cell populations may be missing or incompletely represented (Swechha, Mendonca *et al.*, 2021).

Furthermore, it is crucial to distinguish the genuine biological variation from the technological noise since minimal correction permits noise to continue, while excessive correction may cause significant signals to be missed (Chung and Kang, 2019). Large-scale, intricate single-cell datasets demand effective and scalable computational techniques (Saelens *et al.*, 2019). In the meantime, the use of these technologies in horticulture raises ethical and legal questions, such as how to protect data privacy and property rights, ensure process openness, and mitigate the detrimental effects of technology on employment (Cobianchi *et al.*, 2022).

The technological and moral conundrums discussed here are the same as those identified as the threats and vulnerabilities in SWOT analysis (Fig. 11.4), and as such, they are important concerns. In order to avoid impeding the potential use of spatial and multi-omics techniques in horticulture research, there is a strong need to reduce the challenges of data integration, the lack of qualified personnel, and ethical concerns.

11.10 Prospects of Spatial and Multi-omics in Next-generation Horticulture

The integration of multi-omics into horticultural plant phenotyping brings a new era of possibilities. This approach allows precision cultivation tailored to the unique genetics and environment of each plant, enhancing productivity, sustainability, and crop quality. Predictive breeding is becoming faster and more efficient by using layered multi-omics data to develop better varieties (Chen *et al.*, 2022). It also improves plant health management through early warnings of diseases and stresses. Additionally, multi-omics reveals complex interactions between plants, soil, and climate, promoting eco-friendly practices, conserving biodiversity, and discovering new genetic resources. As these technologies become more accessible, they are revolutionizing horticultural research globally and providing new frontiers in sustainable crop improvement.

The Indian spatial omics market is at a nascent yet rapidly evolving stage, as shown in Fig. 11.5 driven by advances in genomics, precision medicine, and agricultural biotechnology.

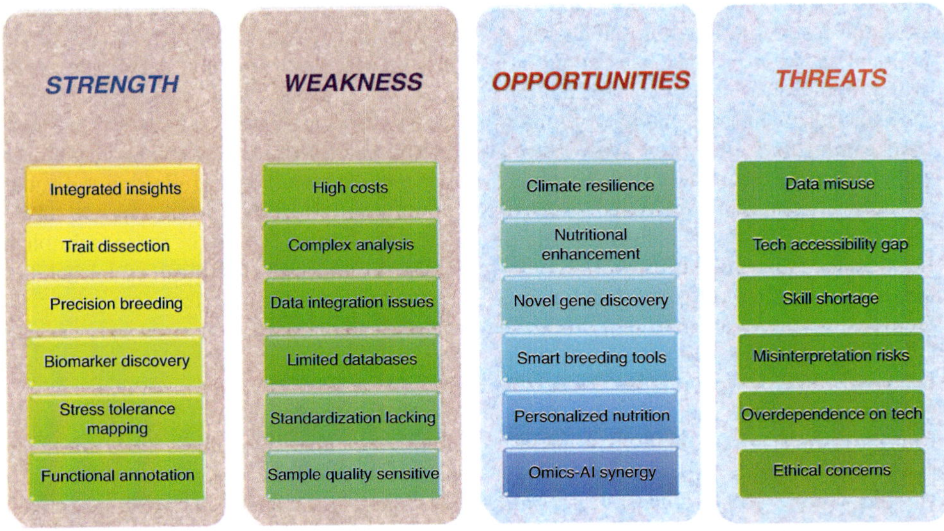

Fig. 11.4. SWOT analysis for spatial omics and multi-omics in horticulture.

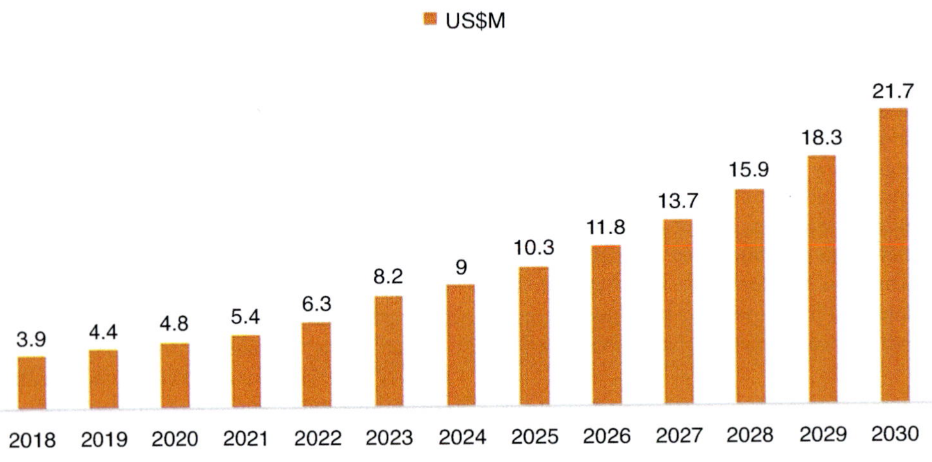

Fig. 11.5. Bar chart of Indian spatial omics market values in 2018–2030 (estimate) (Grand View Research, 2024).

Increasing investments in research infrastructure, government initiatives, and global collaborations are fostering the adoption of spatial and multi-omics technologies (Grand View Research, 2024).

The rapid advancement of spatial omics is enabling a deeper understanding of biological processes within their native spatial perspectives. While these cutting-edge technologies have been widely adopted in animal studies, their application in plant research is relatively recent. Future developments are expected to focus on lowering costs, increasing throughput, and integrating multiple omics layers into unified assays (Yin *et al.*, 2024a).

In the near future, untargeted *in situ* sequencing (ISS) and spatial barcoding methods are projected to offer improved resolution and specificity. Among these, spatial barcoding is likely to become a key tool for large-scale investigations. MSI has already shown great promise in mapping the spatial distribution of metabolites, offering valuable insights into their localization, associated pathways, and biological functions (Barmukh *et al.*, 2025).

Early warning systems supplemented by digital technologies have improved the efficiency and resilience of farms (Yin *et al.*, 2024b). The use of multi-omics methods to combine data from crops, soil, and climate leads to sustainability, biodiversity, and the discovery of new genes (Tao *et al.*, 2022).

ML, when integrated with spatial multi-omics, results in the following advantages: instant identification of pathogens through biosensors and mobile phones (Zheng *et al.*, 2023); rapid breeding and trait prediction (Kou *et al.*, 2020; Mushtaq *et al.*, 2024; Singh *et al.*, 2024); and the improvement of CRISPR/Cas9 precision via off-target performance simulation (Sherkatghanad *et al.*, 2023). Digital twins, big data, and cloud computing technologies illustrate real-time plant–environment interactions (van Voorn *et al.*, 2023). Furthermore, the amalgamation of omics with climate models not only aids the breeding of climate-resilient, location-specific horticulture crops but also helps to achieve sustainability objectives (Wang *et al.*, 2023).

11.11 Conclusion: Toward a Future Omics-driven Horticulture

Horticulture's future is being transformed at pace by the adoption of multi-omics technologies alongside cutting-edge computational tools, which are bringing unprecedented insights into plant biology. When data from genomics, transcriptomics, proteomics, metabolomics, and spatial phenotyping are integrated/correlated, researchers can decipher the complex molecular networks that regulate growth, quality, and stress responses of horticultural crops. These revelations provide the basis for precision breeding and predictive modeling by using AI and machine learning, thereby supporting the rapid creation of varieties that are resilient, high-yielding, and sustainable. Such technological breakthroughs are the prerequisite for solving global issues such as climate change and food insecurity. This vision will need teamwork, open access to data without any discrimination, and the removal of the technological barriers that will make it possible for smarter, more sustainable horticultural systems to develop, and that will be advantageous not only to farmers but also to consumers and ecosystems. This multi-omics-driven approach marks a new era in the sector of green and innovation-led horticulture.

References

Abdelrahman, M., Al-Sadi, A.M., Pour-Aboughadareh, A., Burritt, D.J. and Tran, L.S.P. (2018) Genome editing using CRISPR/Cas9–targeted mutagenesis: An opportunity for yield improvements of crop plants grown under environmental stresses. *Plant Physiology and Biochemistry* 131, 31–36.

Ackah, M., Guo, L., Li, S., Jin, X., Asakiya, C. *et al.* (2022) DNA methylation changes and its associated genes in mulberry (*Morus alba* L.) Yu-711 response to drought stress using MethylRAD sequencing. *Plants* 11(2), 190. DOI: 10.3390/plants11020190.

Alzat, W.M. and Hassan, M. (2018) Proteomics in systems biology. *Advances in Experimental Medicine and Biology* 1102, 31–49. DOI: 10.1007/978-3-319-98758-3_3.

Alexandrov, T. (2023) Spatial metabolomics: From a niche field towards a driver of innovation. *Nature Metabolism* 5(9), 1443–1445. DOI: 10.1038/s42255-023-00881-0.

Amini, S., Doyle, J.J. and Libault, M. (2023) The evolving definition of plant cell type. *Frontiers in Plant Science* 14, 1271070. DOI: 10.3389/fpls.2023.1271070.

Amodio, M. and Krishnaswamy, S. (2018) MAGAN: Aligning biological manifolds. *arXiv Preprint, arXiv:1803.00385.*

Anderton, C.R. and Uhrig, R.G. (2025) The promising role of proteomes and metabolomes in defining the single-cell landscapes of plants. *The New Phytologist* 245(3), 945–948. DOI: 10.1111/nph.20303.

Argelaguet, R., Velten, B., Arnol, D., Dietrich, S., Zenz, T. *et al.* (2018) Multi-omics factor analysis-a framework for unsupervised integration of multi-omics data sets. *Molecular Systems Biology* 14(6), e8124. DOI: 10.15252/msb.20178124.

Argelaguet, R., Cuomo, A.S.E., Stegle, O. and Marioni, J.C. (2021) Computational principles and challenges in single-cell data integration. *Nature Biotechnology* 39(10), 1202–1215. DOI: 10.1038/s41587-021-00895-7.

Argueso, C.T., Assmann, S.M., Birnbaum, K.D., Chen, S., Dinneny, J.R. *et al.* (2019) Directions for research and training in plant omics: Big questions and big data. *Plant Direct* 3(4), e00133.

Barmukh, R., Garg, V., Liu, H., Chitikineni, A., Xin, L. *et al.* (2025) Spatial omics for accelerating plant research and crop improvement. *Trends in Biotechnology* 43(8), 1904–1920. DOI: 10.1016/j.tibtech.2025.03.007.

Bartosovic, M., Kabbe, M. and Castelo-Branco, G. (2021) Single-cell CUT&Tag profiles histone modifications and transcription factors in complex tissues. *Nature Biotechnology* 39(7), 825–835.

Bavafaye Haghighi, E., Knudsen, M., Elmedal Laursen, B. and Besenbacher, S. (2019) Hierarchical classification of cancers of unknown primary using multi-omics data. *Cancer Informatics* 18, 1176935119872163. DOI: 10.1177/1176935119872163.

Ben Ayed, R. and Hanana, M. (2021) Artificial intelligence to improve the food and agriculture sector. *Journal of Food Quality* 2021, 1–7. DOI: 10.1155/2021/5584754.

Bian, S., Hou, Y., Zhou, X., Li, X., Yong, J. *et al.* (2018) Single-cell multiomics sequencing and analyses of human colorectal cancer. *Science (New York, N.Y.)* 362(6418), 1060–1063. DOI: 10.1126/science.aao3791.

Bian, X., Wang, W., Abudurexiti, M., Zhang, X., Ma, W. *et al.* (2024) Integration analysis of single-cell multi-omics reveals prostate cancer heterogeneity. *Advanced Science* 11, 18. DOI: 10.1002/advs.202305724.

Birnbaum, K.D. (2018) Power in numbers: Single-cell RNA-seq strategies to dissect complex tissues. *Annual Review of Genetics* 52, 203–221. DOI: 10.1146/annurev-genet-120417-031247.

Bouwman, B.A.M., Crosetto, N. and Bienko, M. (2022) The era of 3D and spatial genomics. *Trends in Genetics* 38(10), 1062–1075. DOI: 10.1016/j.tig.2022.05.010.

Breckels, L.M., Hutchings, C., Ingole, K.D., Kim, S., Lilley, K.S. *et al.* (2024) Advances in spatial proteomics: Mapping proteome architecture from protein complexes to subcellular localizations. *Cell Chemical Biology* 31(9), 1665–1687. DOI: 10.1016/j.chembiol.2024.08.008.

Buenrostro, J.D., Wu, B., Litzenburger, U.M., Ruff, D., Gonzales, M.L. *et al.* (2015) Single-cell chromatin accessibility reveals principles of regulatory variation. *Nature* 523, 486–490.

Campbell, K.R., Steif, A., Laks, E., Zahn, H., Lai, D. *et al.* (2019) Clonealign: Statistical integration of independent single-cell RNA and DNA sequencing data from human cancers. *Genome Biology* 20(1), 54. DOI: 10.1186/s13059-019-1645-z.

Cao, J., Li, C., Cui, Z., Deng, S., Lei, T. *et al.* (2024a) Spatial transcriptomics: A powerful tool in disease understanding and drug discovery. *Theranostics* 14(7), 2946–2968. DOI: 10.7150/thno.95908.

Cao, Q., Zhu, N., Lan, G., Yuan, B., Wang, J. *et al.* (2024b) Multi-omics analysis of peach-like aroma formation in fruits of wild strawberry (*Fragaria nilgerrensis*). *Horticultural Plant Journal*. DOI: 10.1016/j.hpj.2024.06.009.

Chalhoub, B., Denoeud, F., Liu, S., Parkin, I.A., Tang, H. *et al.* (2014) Plant genetics. Early allopolyploid evolution in the post-neolithic *Brassica napus* oilseed genome. *Science* 345, 950–953. DOI: 10.1126/science.1253435.

Chen, A., Liao, S., Cheng, M., Ma, K. and Wu, L. (2022) Spatiotemporal transcriptomic atlas of mouse organogenesis using DNA nanoball-patterned arrays. *Cell* 185(10), 1777–1792. DOI: 10.1016/j.cell.2022.04.003.

Chen, C., Wang, J., Pan, D. and Wang, X. (2023) Applications of multi-omics analysis in human diseases. *MedComm*. DOI: 10.1002/mco2.315.6.

Chen, J., Suo, S., Tam, P.P., Han, J.-D.J. and Peng, G. (2017) Spatial transcriptomic analysis of cryosectioned tissue samples with Geo-seq. *Nature Protocols* 12(3), 566–580. DOI: 10.1038/nprot.2017.003.

Chen, S., Lake, B.B. and Zhang, K. (2019) High-throughput sequencing of the transcriptome and chromatin accessibility in the same cell. *Nature Biotechnology* 37(12), 1452–1457. DOI: 10.1038/s41587-019-0290-0.

Choe, G., Hoang, N.V. and Lee, J.-Y. (2020) An optimized protocol of laser capture microdissection for tissue-specific RNA profiling in a radish tap root. *STAR Protocols* 1(3), 100110. DOI: 10.1016/j.xpro.2020.100110.

Choe, K., Pak, U., Pang, Y., Hao, W. and Yang, X. (2023) Advances and challenges in spatial transcriptomics for developmental biology. *Biomolecules* 13(1), 156. DOI: 10.3390/biom13010156.

Choi, C.R. and Rhee, H.W. (2022) Proximity labeling: An enzymatic tool for spatial biology. *Trends in Biotechnology* 40(2), 145–148. DOI: 10.1016/j.tibtech.2021.09.008.

Chung, R.H. and Kang, C.Y. (2019) A multi-omics data simulator for complex disease studies and its application to evaluate multi-omics data analysis methods for disease classification. *GigaScience* 8(5), giz045. DOI: 10.1093/gigascience/giz045.

Clark, I.C., Fontanez, K.M., Meltzer, R.H., Xue, Y., Hayford, C. *et al.* (2023) Microfluidics-free single-cell genomics with templated emulsification. *Nature Biotechnology* 41(11), 1557–1566. DOI: 10.1038/s41587-023-01685-z.

Cleary, J.G. and Trigg, L.E. (1995) K*: An instance-based learner using an entropic distance measure. In: *Machine Learning Proceedings*, Elsevier, Amsterdam, The Netherlands, pp. 108–114. DOI: 10.1016/B978-1-55860-377-6.50022-0.

Clyde, D. (2021) SHARE-seq reveals chromatin potential. *Nature Reviews Genetics* 22(1), 2–2. DOI: 10.1038/s41576-020-00308-6.

Cobianchi, L., Verde, J.M., Loftus, T.J., Piccolo, D., Dal Mas, F. *et al.* (2022) Artificial intelligence and surgery: Ethical dilemmas and open issues. *Journal of the American College of Surgeons* 235, 268–275.

Collins, G.S. and Moons, K.G.M. (2019) Reporting of artificial intelligence prediction models. *Lancet (London, England)* (393), 1577–1579.

Coppola, M., Diretto, G., Digilio, M.C., Woo, S.L., Giuliano, G. *et al.* (2019) Transcriptome and metabolome reprogramming in tomato plants by *Trichoderma harzianum* strain T22 primes and enhances defense responses against aphids. *Frontiers in Physiology* 10, 745.

Della Coletta, R., Qiu, Y., Ou, S., Hufford, M.B. and Hirsch, C.N. (2021) How the pan-genome is changing crop genomics and improvement. *Genome Biology* 22, 3. DOI: 10.1186/s13059-020-02224-8.

Demetci, P., Santorella, R., Sandstede, B., Noble, W.S. and Singh, R. (2022) SCOT: Single-cell multi-omics alignment with optimal transport. *Journal of Computational Biology* 29(1), 3–18. DOI: 10.1089/cmb.2021.0446.

Deng, Y., Bartosovic, M., Kukanja, P., Zhang, D., Liu, Y. *et al.* (2022a) Spatial-CUT&Tag: Spatially resolved chromatin modification profiling at the cellular level. *Science (New York, N.Y.)* 375(6581), 681–686. DOI: 10.1126/science.abg7216.

Deng, Y., Bartosovic, M., Ma, S., Zhang, D., Kukanja, P. *et al.* (2022b) Spatial profiling of chromatin accessibility in mouse and human tissues. *Nature* 609(7926), 375–383. DOI: 10.1038/s41586-022-05094-1.

Dorrity, M.W., Alexandre, C.M., Hamm, M.O., Vigil, A.-L.B.G., Fields, S. *et al.* (2020) The regulatory landscape of *Arabidopsis thaliana* roots at single-cell resolution. *Genomics* 3334. DOI: 10.1101/2020.07.17.204792.

Dou, J., Liang, S., Mohanty, V., Cheng, X., Kim, S. *et al.* (2020) Unbiased integration of single cell multi-omics data. *BioInformatics*. DOI: 10.1101/2020.12.11.422014.

Du, J., Verzaux, E., Chaparro-Garcia, A., Bijsterbosch, G., Keizer, L.C.P. *et al.* (2015) Elicitin recognition confers enhanced resistance to phytophthora infestans in potato. *Nature Plants* 1, 15034. DOI: 10.1038/NPLANTS.2015.34.

Duan, B., Zhou, C., Zhu, C., Yu, Y., Li, G. *et al.* (2019) Model-based understanding of single-cell CRISPR screening. *Nature Communications* 10(1), 2233. DOI: 10.1038/s41467-019-10216-x.

Duggan, C., Moratto, E., Savage, Z., Hamilton, E., Adachi, H. *et al.* (2021) Dynamic localization of a helper NLR at the plant-pathogen interface underpins pathogen recognition. *Proceedings of the National Academy of Sciences of the United States of America* 118(34), e2104997118. DOI: 10.1073/pnas.2104997118.

Duncan, S., Johansson, H.E. and Ding, Y. (2023) Reference genes for quantitative *Arabidopsis* single molecule RNA fluorescence *in situ* hybridization. *Journal of Experimental Botany* 74(7), 2405–2415. DOI: 10.1093/jxb/erac521.

Duncan, S., Olsson, T.S.G., Hartley, M., Dean, C. and Rosa, S. (2016) A method for detecting single mRNA molecules in *Arabidopsis thaliana*. *Plant Methods* 12, 13. DOI: 10.1186/s13007-016-0114-x.

Emmert-Buck, M.R., Bonner, R.F., Smith, P.D., Chuaqui, R.F., Zhuang, Z. *et al.* (1996) Laser capture microdissection. *Science* 274(5289), 998–1001. DOI: 10.1126/science.274.5289.998.

Enomoto, H. (2021) Adhesive film applications help to prepare strawberry fruit sections for desorption electrospray ionization-mass spectrometry imaging. *Bioscience, Biotechnology, and Biochemistry* 85(6), 1341–1347. DOI: 10.1093/bbb/zbab033.

Fang, R., Halpern, A., Rahman, M.M., Huang, Z. and Lei, Z. (2024) Three-dimensional single-cell transcriptome imaging of thick tissues. *eLife* 12, RP90029. DOI: 10.7554/eLife.90029.3.

Francia, E., Morcia, C., Pasquariello, M., Mazzamurro, V., Milc, J.A. *et al.* (2016) Copy number variation at the HvCBF4- HvCBF2 genomic segment is a major component of frost resistance in barley. *Plant Molecular Biology* 92, 161–175. DOI: 10.1007/s11103-016-0505-4.

Fu, M., Zou, B., An, K., Yu, Y., Tang, D. *et al.* (2019) Anti-asthmatic activity of alkaloid compounds from Pericarpium Citri Reticulatae (Citrus reticulata'Chachi'). *Food and Function* 10(2), 903–911. DOI: 10.1039/c8fo01753k.

Fu, Y., Xiao, W., Tian, L., Guo, L. and Ma, G. (2023) Spatial transcriptomics uncover sucrose post-phloem transport during maize kernel development. *Nature Communications* 14(1), 7191. DOI: 10.1038/s41467-023-43006-7.

Fulcher, J.M., Dawar, P., Balasubramanian, V.K., Williams, S.M., Nadeau, K.D. *et al.* (2024) Single-cell proteomics of *Arabidopsis* leaf mesophyll identifies drought stress-related proteins. *Plant Biology*. DOI: 10.1101/2024.07.01.601628.

Gandin, V., Kim, J., Yang, L.-Z., Lian, Y., Kawase, T. *et al.* (2025) Deep-tissue transcriptomics and subcellular imaging at high spatial resolution. *Science (New York, N.Y.)* 5, eadq2084.

Gao, Z., Li, J., Luo, M., Li, H., Chen, Q. *et al.* (2019) Characterization and cloning of grape circular RNAs identified the cold resistance–related Vv-circATS1. *Plant Physiology* 180(2), 966–985. DOI: 10.1104/pp.18.01331.

Ge, C., Wang Y., G., Lu, S., Zhao X., Y., Hou B., K. *et al.* (2021) Multi-omics analyses reveal the regulatory network and the function of ZmUGTs in maize defense response. *Frontiers in Plant Science* 12. DOI: 10.3389/fpls.2021.738261.

Giacomello, S., Salmén, F., Terebieniec, B.K., Vickovic, S., Navarro, J.F. *et al.* (2017) Spatially resolved transcriptome profiling in model plant species. *Nature Plants* 3, 17061. DOI: 10.1038/nplants.2017.61.

Gill, T., Gill, S.K., Saini, D.K., Chopra, Y., de Koff, J.P. *et al.* (2022) A comprehensive review of high throughput phenotyping and machine learning for plant stress phenotyping. *Phenomics (Cham, Switzerland)* 2(3), 156–183. DOI: 10.1007/s43657-022-00048-z.

Grand View Research (2024) Spatial omics market size, share & trends analysis report. Available at: https://www.grandviewresearch.com/horizon/outlook/spatial-omics-market-size/global (accessed 12 October 2025).

Gu, D., Yang, J., Wu, S., Liao, Y., Zeng, L. *et al.* (2021) Epigenetic regulation of the phytohormone abscisic acid accumulation under dehydration stress during postharvest processing of tea (*Camellia sinensis*). *Journal of Agricultural and Food Chemistry* 69(3), 1039–1048. DOI: 10.1021/acs.jafc.0c07220.

Gyllborg, D., Langseth, C.M., Qian, X., Choi, E., Salas, S.M. *et al.* (2020) Hybridization-based *in situ* sequencing (HybISS) for spatially resolved transcriptomics in human and mouse brain tissue. *Nucleic Acids Research* 48(19), e112. DOI: 10.1093/nar/gkaa792.

Hagemann-Jensen, M., Ziegenhain, C., Chen, P., Ramsköld, D., Hendriks, G.J. *et al.* (2020) Single-cell RNA counting at allele and isoform resolution using Smart-seq3. *Nature Biotechnology* 38(6), 708–714.

Hao, F., Liu, X., Zhou, B., Tian, Z., Zhou, L. *et al.* (2023) Chromosome-level genomes of three key allium crops and their trait evolution. *Nature Genetics* 55(11), 1976–1986. DOI: 10.1038/s41588-023-01546-0.

Hao, Y., Hao, S., Andersen-Nissen, E., Mauck, W.M., Zheng, S. *et al.* (2021) Integrated analysis of multimodal single-cell data. *Cell* 184(13), 3573–3587. DOI: 10.1016/j.cell.2021.04.048.

Hernandez, S., Lazcano, R., Serrano, A., Powell, S., Kostousov, L. *et al.* (2022) Challenges and opportunities for immunoprofiling using a spatial high-plex technology: The NanoString GeoMx® digital spatial profiler. *Frontiers in Oncology* 12. DOI: 10.3389/fonc.2022.890410.

Holzinger, A., Keiblinger, K., Holub, P., Zatloukal, K. and Müller, H. (2023) AI for life: Trends in artificial intelligence for biotechnology. *New Biotechnology* 74, 16–24. DOI: 10.1016/j.nbt.2023.02.001.

Hosea, R., Hillary, S., Naqvi, S., Wu, S. and Kasim, V. (2024) The two sides of chromosomal instability: Drivers and brakes in cancer. *Signal Transduction and Targeted Therapy* 9(1), 75. DOI: 10.1038/s41392-024-01767-7.

Isakova, A., Neff, N. and Quake, S.R. (2021) Single-cell quantification of a broad RNA spectrum reveals unique noncoding patterns associated with cell types and states. *Proceedings of the National Academy of Sciences* 118(51), e2113568118.

Jacob, I.J. and Darney, P.E. (2021) Design of deep learning algorithm for IoT application by image based recognition. *Journal of ISMAC* 3(3), 276–290. DOI: 10.36548/jismac.2021.3.008.

Jaiswal, A.K., Alkan, N., Elad, Y., Sela, N., Philosoph, A.M. *et al.* (2020) Molecular insights into biochar-mediated plant growth promotion and systemic resistance in tomato against *Fusarium crown* and root rot disease. *Scientific Reports* 10(1), 13934. DOI: 10.1038/s41598-020-70882-6.

James, D., Rennya, P.R., Mallavarapu, M.D., Panigrahi, R.C. and Patel, H.K. (2021) Integrating pan-omics data in a systems approach for crop improvement: Opportunities and challenges. In: Kumar, A., Kumar, R., Shukla, P. and Patel, H.K. (eds) *Omics Technologies for Sustainable Agriculture and Global Food Security (Vol II)*. Springer, Singapore, pp. 215–246. DOI: 10.1007/978-981-16-2956-3_8.

Jin, Y., Zuo, Y., Li, G., Liu, W., Pan, Y. *et al.* (2024) Advances in spatial transcriptomics and its applications in cancer research. *Molecular Cancer* 23(1). DOI: 10.1186/s12943-024-02040-9.

Kar, S., Purbey, V.K., Suradhaniwar, S., Korbu, L.B., Kholová, J. *et al.* (2021) An ensemble machine learning approach for determination of the optimum sampling time for evapotranspiration assessment from high-throughput phenotyping data. *Computers and Electronics in Agriculture* 182, 105992. DOI: 10.1016/j.compag.2021.105992.

Karimi, D., Dou, H., Warfield, S.K. and Gholipour, A. (2020) Deep learning with noisy labels: Exploring techniques and remedies in medical image analysis. *Medical Image Analysis* 65, 101759. DOI: 10.1016/j.media.2020.101759.

Kavipriya, S., Suresh Kumar, K., Saini, C. and RamaKrishna, B.V. (2025) The impact of cloud computing on machine learning applications for environmental startups. *International Journal of Environmental Sciences* 2252–2260.

Kim, H.J., Lin, Y., Geddes, T.A., Yang, J.Y.H. and Yang, P. (2020) CITE-Fuse enables multi-modal analysis of CITE-seq data. *Bioinformatics (Oxford, England)* 36(14), 4137–4143. DOI: 10.1093/bioinformatics/btaa282.

Kim, N., Ha, K.-J., Park, N.-W., Cho, J., Hong, S. *et al.* (2015) A comparison between major artificial intelligence models for crop yield prediction: Case study of the Midwestern United States, 2006–2015. *ISPRS International Journal of Geo-Information* 8(5), 240. DOI: 10.3390/ijgi8050240.

Kou, X., Tong, L., Shen, Y., Zhu, W., Yin, L. *et al.* (2020) Smartphone-assisted robust enzymes@MOFs-based paper biosensor for point-of-care detection. *Biosensors and Bioelectronics* 156, 112095. DOI: 10.1016/j.bios.2020.112095.

Lai, X., Bazin, J., Webb, S., Crespi, M., Zubieta, C. *et al.* (2018) CircRNAs in plants. In: Xiao, J. (ed.) *Circular RNAs. Advances in Experimental Medicine and Biology, Vol. 1087*. Springer, Singapore, p. 370.

Lanckriet, G.R.G., De Bie, T., Cristianini, N., Jordan, M.I. and Noble, W.S. (2004) A statistical framework for genomic data fusion. *Bioinformatics (Oxford, England)* 20(16), 2626–2635. DOI: 10.1093/bioinformatics/bth294.

Laureyns, R., Joossens, J., Herwegh, D., Pevernagie, J., Pavie, B. *et al.* (2022) An *in situ* sequencing approach maps PLASTOCHRON1 at the boundary between indeterminate and determinate cells. *Plant Physiology* 188(2), 782–794. DOI: 10.1093/plphys/kiab533.

Lee, J.H., Daugharthy, E.R., Scheiman, J., Kalhor, R., Yang, J.L. *et al.* (2014) Highly multiplexed subcellular RNA sequencing *in situ*. *Science (New York, N.Y.)* 343(6177), 1360–1363. DOI: 10.1126/science.1250212.

Lele, A. (2019) Artificial intelligence (AI). In: *Disruptive Technologies for the Militaries and Security*. Springer International Publishing, Singapore, pp. 139–154. DOI: 10.1007/978-981-13-3384-2_8.

Li, N., Zhang, Z., Chen, Z., Cao, B. and Xu, K. (2021) Comparative transcriptome analysis of two contrasting Chinese cabbage (*Brassica rapa* L.) genotypes reveals that ion homeostasis is a crucial biological pathway involved in the rapid adaptive response to salt stress. *Frontiers in Plant Science* 12. DOI: 10.3389/fpls.2021.683891.

Liew, L.C., Wang, Y., Peirats-Llobet, M., Berkowitz, O., Whelan, J. *et al.* (2020) Laser-capture microdissection RNA-sequencing for spatial and temporal tissue-specific gene expression analysis in plants. *Journal of Visualized Experiments* 162(162), e61517. DOI: 10.3791/61517.

Liu, J., Huang, Y., Singh, R., Vert, J.P. and Noble, W.S. (2019) Jointly embedding multiple single-cell omics measurements. *Bioinformatics*. DOI: 10.1101/644310.

Liu, J., Majeed, A., Kumar, N., Pajerowska-Mukhtar, K.M. and Mukhtar, M.S. (2024) Single nuclei multiomics reveals the drought-driven gene regulatory atlas in *Arabidopsis*. *Plant Biology*. DOI: 10.1101/2024.01.11.575118.

Liu, Y., DiStasio, M., Su, G., Asashima, H., Enninful, A. *et al.* (2022a) Spatial-CITE-seq: Spatially resolved high-plex protein and whole transcriptome co-mapping. *Research Square* rs.3.rs-1499315. DOI: 10.21203/rs.3.rs-1499315/v1.

Liu, Y., Yang, M., Deng, Y., Su, G., Enninful, A. *et al.* (2022b) High-spatial-resolution multi-omics sequencing via deterministic barcoding in tissue. *Cell* 185(9), 1631–1642. DOI: 10.1016/j.cell.2022.04.003.

Liu, Y., Zhang, X., Yang, L., Zhou, S., Li, Y. *et al.* (2023) Proteomics and transcriptomics explore the effect of mixture of herbal extract on diabetic wound healing process. *Phytomedicine* 116. DOI: 10.1016/j.phymed.2023.154892.

Long, H., Yan, L., Pu, J., Liu, Y., Zhong, X. *et al.* (2022) Multi-omics analysis reveals the effects of microbiota on oral homeostasis. *Frontiers in Immunology* 13, 1005992.

Lu, S., Xu, Y., Song, X., Li, J., Jiang, J. *et al.* (2024) Multi-omics reveal the effects and regulatory mechanism of dietary neutral detergent fiber supplementation on carcass characteristics, amino acid profiles, and meat quality of finishing pigs. *Food Chemistry* 445, 138765.

Luecken, M.D., Büttner, M., Chaichoompu, K., Danese, A., Interlandi, M. *et al.* (2022) Benchmarking atlas-level data integration in single-cell genomics. *Nature Methods* 19(1), 41–50. DOI: 10.1038/s41592-021-01336-8.

Lundberg, E. and Borner, G.H.H. (2019) Spatial proteomics: A powerful discovery tool for cell biology. *Nature Reviews Molecular Cell Biology* 20(5), 285–302. DOI: 10.1038/s41580-018-0094-y.

Luo, C., Rivkin, A., Zhou, J., Sandoval, J.P., Kurihara, L. *et al.* (2018) Robust single-cell DNA methylome profiling with snmC-seq2. *Nature Communications* 9, 3824. DOI: 10.1038/s41467-018-06355-2.

Luo, M.J., Zhang, Y.X., Chen, K., Kong, M.S., Song, W. *et al.* (2019) Mapping of quantitative trait loci for seedling salt tolerance in maize. *Molecular Breeding* 39, 64–76. DOI: 10.1007/s11032-019-0974-7.

Macaulay, I.C., Haerty, W., Kumar, P., Li, Y.I., Hu, T.X. *et al.* (2015) G&T-seq: Parallel sequencing of single-cell genomes and transcriptomes. *Nature Methods* 12, 519–522.

Marand, A.P., Chen, Z., Gallavotti, A. and Schmitz, R.J. (2021) A *cis*-regulatory atlas in maize at single-cell resolution. *Cell* 184, 3041–3055.

McKellar, D.W. (2023) *Expanding the Scope of Single-Cell and Spatial Transcriptomics Technologies*. Cornell University, Cornell, USA.

Meers, M.P., Llagas, G., Janssens, D.H., Codomo, C.A. and Henikoff, S. (2022) Multifactorial profiling of epigenetic landscapes at single-cell resolution using MulTI-Tag. *Nature Biotechnology* 41, 708. DOI: 10.1038/s41587-022-01522-9.

Mendieta, J.P., Tu, X., Jiang, D., Yan, H., Zhang, X. *et al.* (2024) Investigating the *cis*-regulatory basis of C3 and C4 photosynthesis in grasses at single-cell resolution. In: *Proceedings of the National Academy of Sciences of the United States of America* (Vol. 121). DOI: 10.1073/pnas.2402781121.

Meng, C., Zeleznik, O.A., Thallinger, G.G., Kuster, B., Gholami, A.M. *et al.* (2016) Dimension reduction techniques for the integrative analysis of multi-omics data. *Briefings in Bioinformatics* 17(4), 628–641. DOI: 10.1093/bib/bbv108.

Mir, R.R., Reynolds, M., Pinto, F., Khan, M.A. and Bhat, M.A. (2019) High-throughput phenotyping for crop improvement in the genomics era. *Plant Science* 282, 60–72. DOI: 10.1016/j.plantsci.2019.01.007.

Montes, C., Zhang, J., Nolan, T.M. and Walley, J.W. (2024) Single-cell proteomics differentiates *Arabidopsis* root cell types. *New Phytologist* 244(5), 1750–1759. DOI: 10.1111/nph.19923.

Mo, Q., Shen, R., Guo, C., Vannucci, M., Chan, K.S. *et al.* (2018) A fully Bayesian latent variable model for integrative clustering analysis of multi-type omics data. *Biostatistics* 19(1), 71–86.

Montesinos-López, O.A., Martín-Vallejo, J., Crossa, J., Gianola, D., Hernández-Suárez, C.M. *et al.* (2019) A benchmarking between deep learning, support vector machine and Bayesian threshold best linear unbiased prediction for predicting ordinal traits in plant breeding. *G3 Genes Genomes Genetics* 9, 601–618.

Muhammad Aslam, M., Waseem, M., Jakada, B.H., Okal, E.J., Lei, Z. *et al.* (2022) Mechanisms of abscisic acid-mediated drought stress responses in plants. *International Journal of Molecular Sciences* 23(3), 1084. DOI: 10.3390/ijms23031084.

Mund, A., Coscia, F., Kriston, A., Hollandi, R., Kóvacs, F. *et al.* (2022) Deep visual proteomics defines single-cell identity and heterogeneity. *Nature Biotechnology* 40, 1231–1240. DOI: 10.1038/s41587-022-01302-5.

Mushtaq, M.A., Ahmed, H.G.M.-D. and Zeng, Y. (2024) Applications of artificial intelligence in wheat breeding for sustainable food security. *Sustainability* 16(13), 5688. DOI: 10.3390/su16135688.

Muster, B., Rapp, A. and Cardoso, M.C. (2017) Systematic analysis of DNA damage induction and DNA repair pathway activation by continuous wave visible light laser micro-irradiation. *AIMS Genetics* 4(1), 47–68. DOI: 10.3934/genet.2017.1.47.

Mustroph, A., Zanetti, M.E., Jang, C.J., Holtan, H.E., Repetti, P.P. *et al.* (2009) Profiling translatomes of discrete cell populations resolves altered cellular priorities during hypoxia in *Arabidopsis*. *Proceedings of the National Academy of Science USA* 106, 18843–18848.

Nakabayashi, R. and Saito, K. (2020) Higher dimensional metabolomics using stable isotope labeling for identifying the missing specialized metabolism in plants. *Current Opinion in Plant Biology* 55, 84–92. DOI: 10.1016/j.pbi.2020.02.009.

Nichterwitz, S., Benitez, J.A., Hoogstraaten, R., Deng, Q. and Hedlund, E. (2018) LCM-seq: A method for spatial transcriptomic profiling using laser capture microdissection coupled with polya-based RNA sequencing. In: *Methods in Molecular Biology*, Springer New York, New York, NY, pp. 95–110 (Vol. 1649). DOI: 10.1007/978-1-4939-7213-5_6.

Niu, Y., Wu, L., Li, Y., Huang, H., Qian, M. *et al.* (2020) Deciphering the transcriptional regulatory networks that control size, color, and oil content in *Brassica rapa* seeds. *Biotechnology for Biofuels* 13, 90. DOI: 10.1186/s13068-020-01728-6.

Nobori, T., Oliva, M., Lister, R. and Ecker, J.R. (2023) Multiplexed single-cell 3D spatial gene expression analysis in plant tissue using phytomap. *Nature Plants* 9(7), 1026–1033. DOI: 10.1038/s41477-023-01439-4.

Nobori, T., Monell, A., Lee, T.A., Sakata, Y., Shirahama, S. *et al.* (2025) A rare PRIMER cell state in plant immunity. *Nature* 638(8049), 197–205. DOI: 10.1038/s41586-024-08383-z.

Nolan, T.M. and Shahan, R. (2023) Resolving plant development in space and time with single-cell genomics. *Current Opinion in Plant Biology* 76, 102444. DOI: 10.1016/j.pbi.2023.102444.

Nordmann, T.M., Anderton, H., Hasegawa, A., Schweizer, L., Zhang, P. *et al.* (2024) Spatial proteomics identifies JAKi as treatment for a lethal skin disease. *Nature* 635, 1001–1009. DOI: 10.1038/s41586-024-08061-0.

Novakovsky, G., Dexter, N., Libbrecht, M.W., Wasserman, W.W. and Mostafavi, S. (2023) Obtaining genetics insights from deep learning via explainable artificial intelligence. *Nature Reviews Genetics* 24(2), 125–137. DOI: 10.1038/s41576-022-00532-2.

Nturambirwe, J.F.I. and Opara, U.L. (2020) Machine learning applications to non-destructive defect detection in horticultural products. *Biosystems Engineering* 189, 60–83. DOI: 10.1016/j.biosystemseng.2019.11.011.

Opdensteinen, P., Sperl, L.E., Mohamadi, M., Kündgen-Redding, N., Hagn, F. *et al.* (2022) The transient expression of recombinant proteins in plant cell packs facilitates stable isotope labelling for NMR spectroscopy. *Plant Biotechnology Journal* 20, 1928–1939. DOI: 10.1111/pbi.13873.

Pan, Y., Ji, T., Li, Y. and Ma, L. (2020) Omics biomarkers for frailty in older adults. *Clinica Chimica Acta* 510, 363–372.

Pareek, V., Tian, H., Winograd, N. and Benkovic, S.J. (2020) Metabolomics and mass spectrometry imaging reveal channeled *de novo* purine synthesis in cells. *Science* 368(6488), 283–290. DOI: 10.1126/science.aaz6465.

Payne, A.C., Chiang, Z.D., Reginato, P.L., Mangiameli, S.M., Murray, E.M. *et al.* (2021) *In situ* genome sequencing resolves DNA sequence and structure in intact biological samples. *Science* 371, 6532. DOI: 10.1126/science.aay3446.

Peterson, J.R., Harrison, E.L., Miller, C.D., Bennett, S.M. and Yusuff, M. (2025) Multi-omics approach in personalized reconstruction. *International Journal of Reconstructive Surgery*.

Pinu, F.R., Beale, D.J., Paten, A.M., Kouremenos, K., Swarup, S. *et al.* (2019) Systems biology and multi-omics integration: Viewpoints from the metabolomics research community. *Metabolites* 9(4), 76.

Polański, K., Bartolomé-Casado, R., Sarropoulos, I., Xu, C., England, N. *et al.* (2024) Bin2cell reconstructs cells from high resolution Visium HD data. *Bioinformatics (Oxford, England)* 40(9), btae546. DOI: 10.1093/bioinformatics/btae546.

Pott, D.M., Durán-Soria, S., Osorio, S. and Vallarino, J.G. (2021) Combining metabolomic and transcriptomic approaches to assess and improve crop quality traits. *CABI Agriculture and Bioscience* 2(1), 1.

Qin, X., Ning, Y., Zhou, L. and Zhu, Y. (2023) Oral *Submucous fibrosis*: Etiological mechanism, malignant transformation, therapeutic approaches and targets. *International Journal of Molecular Sciences* 24(5), 4992. DOI: 10.3390/ijms24054992.

Ramskold, D., Luo, S., Wang, Y.-C., Li, R., Deng, Q. *et al.* (2012) Full-length mRNA-seq from single-cell levels of RNA and individual circulating tumor cells. *Nature Biotechnology* 30, 777–782.

Raza, A., Tabassum, J., Zahid, Z., Charagh, S., Bashir, S. *et al.* (2022) Advances in “omics” approaches for improving toxic metals/metalloids tolerance in plants. *Frontiers in Plant Science* 12, 794373.

Righetti, L., Gottwald, S., Tortorella, S., Spengler, B. and Bhandari, D.R. (2022) Mass spectrometry imaging disclosed spatial distribution of defense-related metabolites in *Triticum* spp. *Metabolites* 12(1), 48. DOI: 10.3390/metabo12010048.

Rodriques, S.G., Stickels, R.R., Goeva, A., Martin, C.A., Murray, E. *et al.* (2019) Slide-seq: A scalable technology for measuring genome-wide expression at high spatial resolution. *Science (New York, N.Y.)* 363(6434), 1463–1467. DOI: 10.1126/science.aaw1219.

Rosa, S., Duncan, S. and Dean, C. (2016) Mutually exclusive sense-antisense transcription at FLC facilitates environmentally induced gene repression. *Nature Communications* 7, 13031. DOI: 10.1038/ncomms13031.

Rusnak, B., Clark, F.K., Vadde, B.V.L. and Roeder, A.H.K. (2024) What is a plant cell type in the age of single-cell biology? It's complicated. *Annual Review of Cell and Developmental Biology* 40, 321–328.

Saelens, W., Cannoodt, R., Todorov, H. and Saeys, Y. (2019) A comparison of single-cell trajectory inference methods. *Nature Biotechnology* 37(5), 547–554. DOI: 10.1038/s41587-019-0071-9.

Salmen, F., De Jonghe, J., Kaminski, T.S., Alemany, A., Parada, G.E. *et al.* (2022) High-throughput total RNA sequencing in single cells using VASA-seq. *Nature Biotechnology* 40(12), 1780–1793.

Sampath Kumar, A., Tian, L., Bolondi, A., Hernández, A.A., Stickels, R. *et al.* (2023) Spatiotemporal transcriptomic maps of whole mouse embryos at the onset of organogenesis. *Nature Genetics* 55(7), 1176–1185. DOI: 10.1038/s41588-023-01435-6.

Sarabia, L.D., Boughton, B.A., Rupasinghe, T., van de Meene, A.M.L., Callahan, D.L. *et al.* (2018) High-mass-resolution MALDI mass spectrometry imaging reveals detailed spatial distribution of metabolites and lipids in roots of barley seedlings in response to salinity stress. *Metabolomics* 14(5), 63. DOI: 10.1007/s11306-018-1359-3.

Schiessl, S.V., Quezada-Martinez, D., Orantes-Bonilla, M. and Snowdon, R.J. (2020) Transcriptomics reveal high regulatory diversity of drought tolerance strategies in a biennial oil crop. *Plant Science* 297, 110515. DOI: 10.1016/j.plantsci.

Schott, M., León-Periñán, D., Splendiani, E., Strenger, L., Licha, J.R. *et al.* (2024) Open-ST: High-resolution spatial transcriptomics in 3D. *Cell* 187(15), 3953–3972. DOI: 10.1016/j.cell.2024.05.055.

Schwope, R., Magris, G., Miculan, M., Paparelli, E., Celii, M. *et al.* (2021) Open chromatin in grapevine marks candidate CREs and with other chromatin features correlates with gene expression. *The Plant Journal* 107, 1631–1647. DOI: 10.1111/tpj.15404.

Scossa, F., Alseekh, S. and Fernie, A.R. (2021) Integrating multi-omics data for crop improvement. *Journal of Plant Physiology* 257, 153352.

Shendure, J., Balasubramanian, S., Church, G.M., Gilbert, W., Rogers, J. *et al.* (2017) DNA sequencing at 40: Past, present and future. *Nature* 550, 345–353. DOI: 10.1038/nature24286.

Sherkatghanad, Z., Abdar, M., Charlier, J. and Makarenkov, V. (2023) Using traditional machine learning and deep learning methods for on-and off-target prediction in CRISPR/Cas9: A review. *Briefings Bioinform* 24(3), bbad131.

Shi, F., Pang, Z., Liu, C., Zhou, L., Tan, C. *et al.* (2022) Whole-transcriptome analysis and construction of an anther development-related ceRNA network in Chinese cabbage (*Brassica campestris* L. ssp. pekinensis). *Science Reports* 12, 2667. DOI: 10.1038/s41598-022-06556-2.

Singh, N., Ujinwal, M. and Singh, N.K. (2024) Ppomicsdb: A multi-omics database for genetic and molecular breeding applications in pigeonpea. *Legume Science* 6(2), e220.

Slyper, M., Porter, C.B.M., Ashenberg, O., Waldman, J., Drokhlyansky, E. *et al.* (2020) A single-cell and single-nucleus RNA-seq toolbox for fresh and frozen human tumors. *Nature Medicine* 26(5), 792–802. DOI: 10.1038/s41591-020-0844-1.

Solanki, S., Ameen, G., Zhao, J., Flaten, J., Borowicz, P. *et al.* (2020) Visualization of spatial gene expression in plants by modified rnascope fluorescent *in situ* hybridization. *Plant Methods* 16, 71. DOI: 10.1186/s13007-020-00614-4.

Song, Q., Ando, A., Jiang, N., Ikeda, Y. and Chen, Z.J. (2020) Single-cell RNA-seq analysis reveals ploidy-dependent and cell-specific transcriptome changes in *Arabidopsis* female gametophytes. *Genome Biology* 21, 178. DOI: 10.1186/s13059-020-02094-0.

Stafford, I.S., Kellermann, M., Mossotto, E., Beattie, R.M., MacArthur, B.D. *et al.* (2020) A systematic review of the applications of artificial intelligence and machine learning in autoimmune diseases. *NPJ Digital Medicine* 3, 30. DOI: 10.1038/s41746-020-0229-3.

Stanojevic, S., Li, Y., Ristivojevic, A. and Garmire, L.X. (2022) Computational methods for single-cell multi-omics integration and alignment. *Genomics, Proteomics & Bioinformatics* 20, 836–849. DOI: 10.1016/j.gpb.2022.11.013.

Stark, S.G., Ficek, J., Locatello, F., Bonilla, X., Chevrier, S. *et al.* (2020) SCIM: Universal single-cell matching with unpaired feature sets. *Bioinformatics (Oxford, England)* 36(Suppl_2), i919–i927. DOI: 10.1093/bioinformatics/btaa843.

Sui, X., Lo, J.A., Luo, S., He, Y., Tang, Z. *et al.* (2024) Scalable spatial single-cell transcriptomics and translatomics in 3D thick tissue blocks. *Genomics.* DOI: 10.1101/2024.08.05.606553.

Sun, G., Ma, S., Zheng, Z., Wang, X., Chen, S. *et al.* (2022) Multi-omics analysis of expression and prognostic value of NSUN members in prostate cancer. *Frontiers in Oncology* 12, 965571.

Swechha, Mendonca, D., Focsa, O., Díaz-Mejía, J.J. and Cooper, S. (2021) scMARK an 'MNIST' like benchmark to evaluate and optimize models for unifying scRNA data. *bioRxiv* 2021–12.

Swift, J., Luginbuehl, L.H., Hua, L., Schreier, T.B., Donald, R.M. *et al.* (2024) Exaptation of ancestral cell-identity networks enables C4 photosynthesis. *Nature* 636(8041), 143–150. DOI: 10.1038/s41586-024-08204-3.

Takei, Y., Yun, J., Zheng, S., Ollikainen, N., Pierson, N. *et al.* (2021) Integrated spatial genomics reveals global architecture of single nuclei. *Nature* 590(7845), 344–350. DOI: 10.1038/s41586-020-03126-2.

Tang, Y., Huang, A. and Gu, Y. (2020) Global profiling of plant nuclear membrane proteome in *Arabidopsis*. *Nature Plants* 6(7), 838–847. DOI: 10.1038/s41477-020-0700-9.

Tao, H., Xu, S., Tian, Y., Li, Z., Ge, Y. *et al.* (2022) Proximal and remote sensing in plant phenomics: 20 years of progress, challenges, and perspectives. *Plant Communications* 3(6), 100344. DOI: 10.1016/j.xplc.2022.100344.

Tenorio Berrío, R. and Dubois, M. (2024) Single-cell transcriptomics reveals heterogeneity in plant responses to the environment: A focus on biotic and abiotic interactions. *Journal of Experimental Botany* 75(17), 5188–5203. DOI: 10.1093/jxb/erae107.

The Tomato Genome Consortium (2012) The tomato genome sequence provides insights into fleshy fruit evolution. *Nature* 485, 635–641. DOI: 10.1038/nature11119.

Tian, L., Chen, F. and Macosko, E.Z. (2023) The expanding vistas of spatial transcriptomics. *Nature Biotechnology* 41(6), 773–782.

Tran, H.T.N., Ang, K.S., Chevrier, M., Zhang, X., Lee, N.Y.S. *et al.* (2020) A benchmark of batch-effect correction methods for single-cell RNA sequencing data. *Genome Biology* 21(1), 12. DOI: 10.1186/s13059-019-1850-9.

Ueda, M. and Seki, M. (2020) Histone modifications form epigenetic regulatory networks to regulate abiotic stress response. *Plant Physiology* 182, 15–26. DOI: 10.1104/pp.19.00988.

Unsihuay, D., Yin, R., Sanchez, D.M., Yang, M., Li, Y. *et al.* (2021) High-resolution imaging and identification of biomolecules using nano-DESI coupled to ion mobility spectrometry. *Analytica Chimica Acta* 1186, 339085. DOI: 10.1016/j.aca.2021.339085.

van Voorn, G.A.K., Boer, M.P., Truong, S.H., Friedenberg, N.A., Gugushvili, S. *et al.* (2023) A conceptual framework for the dynamic modeling of time-resolved phenotypes for sets of genotype-environment-management combinations: A model library. *Frontiers in Plant Science* 14, 1172359. DOI: 10.3389/fpls.2023.1172359.

Vandereyken, K., Sifrim, A., Thienpont, B. and Voet, T. (2023) Methods and applications for single-cell and spatial multi-omics. *Nature Reviews Genetics* 6, 1–22.

Vicari, M., Mirzazadeh, R., Nilsson, A. and Shariatgorji, R. (2023) Spatial multimodal analysis of transcriptomes and metabolomes in tissues. *Nature Biotechnology*. DOI: 10.1038/s41587-023-01937-y.

Vickovic, S., Lötstedt, B., Klughammer, J., Mages, S., Segerstolpe, Å. *et al.* (2022) SM-omics is an automated platform for high-throughput spatial multi-omics. *Nature Communications* 13(1), 795.

Vistain, L.F. and Tay, S. (2021) Single-cell proteomics. *Trends in Biochemical Sciences* 46(8), 661–672. DOI: 10.1016/j.tibs.2021.01.013.

Wallner, E.-S., Mair, A., Handler, D., McWhite, C., Xu, S.-L. *et al.* (2024) Spatially resolved proteomics of the *Arabidopsis* stomatal lineage identifies polarity complexes for cell divisions and stomatal pores. *Developmental Cell* 59, 1096–1109.

Wang, B., Mezlini, A.M., Demir, F., Fiume, M., Tu, Z. *et al.* (2014) Similarity network fusion for aggregating data types on a genomic scale. *Nature Methods* 11(3), 333–337. DOI: 10.1038/nmeth.2810.

Wang, F., Harindintwali, J.D., Wei, K., Shan, Y., Mi, Z. *et al.* (2023) Climate change: Strategies for mitigation and adaptation. *The Innovation Geoscience* 1, 100015. DOI: 10.59717/j.xinn-geo.2023.100015.

Wang, J., Yang, E., Chaurand, P. and Raghavan, V. (2021a) Visualizing the distribution of strawberry plant metabolites at different maturity stages by MALDI-TOF imaging mass spectrometry. *Food Chemistry* 345, 128838. DOI: 10.1016/j.foodchem.2020.128838.

Wang, P., Fan, E. and Wang, P. (2021b) Comparative analysis of image classification algorithms based on traditional machine learning and deep learning. *Pattern Recognition Letters* 141, 61–67. DOI: 10.1016/j.patrec.2020.07.042.

Wang, X., Sun, Z., Zhang, Y., Xu, Z., Xin, H. *et al.* (2020) BREM-SC: A Bayesian random effects mixture model for joint clustering single cell multi-omics data. *Nucleic Acids Research* 48, 5814–5824.

Wang, X., Xu, R., Tong, X., Zeng, J., Chen, M. *et al.* (2022) Characterization of different meat flavor compounds in Guangdong small-ear spotted and Yorkshire pork using two-dimensional gas chromatography–time-of-flight mass spectrometry and multi-omics. *LWT* 169, 114010. DOI: 10.1016/j.lwt.2022.114010.

Wang, Y., Eddison, M., Fleishman, G., Weigert, M., Xu, S. *et al.* (2021c) EASI-FISH for thick tissue defines lateral hypothalamus spatio-molecular organization. *Cell* 184(26), 6361–6377. DOI: 10.1016/j.cell.2021.11.024.

Welch, J.D., Hartemink, A.J. and Prins, J.F. (2017) MATCHER: Manifold alignment reveals correspondence between single-cell transcriptome and epigenome dynamics. *Genome Biology* 18, 1–19.

Wing, R.A., Purugganan, M.D. and Zhang, Q. (2018) The rice genome revolution: From an ancient grain to green super rice. *Nature Reviews Genetics* 19, 505–517. DOI: 10.1038/s41576-018-0024-z.

Wu, X., Xu, W., Deng, L., Li, Y., Wang, Z. *et al.* (2024) Spatial multi-omics at subcellular resolution via high-throughput in situ pairwise sequencing. *Nature Biomedical Engineering* 8(7), 872–889. DOI: 10.1038/s41551-024-01205-7.

Xia, K., Sun, H.-X., Li, J., Li, J., Zhao, Y. *et al.* (2022) The single-cell stereo-seq reveals region-specific cell subtypes and transcriptome profiling in *Arabidopsis* leaves. *Developmental Cell* 57(10), 1299–1310. DOI: 10.1016/j.devcel.2022.04.011.

Xiong, H., Wang, Q., Li, C.C. and He, A. (2024) Single-cell joint profiling of multiple epigenetic proteins and gene transcription. *Science Advances* 10, eadi3664. DOI: 10.1126/sciadv.adi3664.

Yin, X., Ren, Z., Jia, R., Wang, X., Yu, Q. *et al.* (2024a) Metabolic profiling and spatial metabolite distribution in wild soybean (*G. soja*) and cultivated soybean (*G. max*) seeds. *Food Chemistry* 23, 101717. DOI: 10.1016/j.fochx.2024.101717.

Yin, Z., Huang, W., Li, K., Fernie, A.R. and Yan, S. (2024b) Advances in mass spectrometry imaging for plant metabolomics-expanding the analytical toolbox. *The Plant Journal: For Cell and Molecular Biology* 119, 2168–2180.

You, J., Zhang, Y., Liu, A., Li, D., Wang, X. *et al.* (2019) Transcriptomic and metabolomic profiling of drought-tolerant and susceptible sesame genotypes in response to drought stress. *BMC Plant Biology* 19(1), 267. DOI: 10.1186/s12870-019-1880-1.

Yu, Y., Huss, D., Li, M., Wickramanayake, J.S., Bélanger, S. *et al.* (2024) Fluorescence hybridization chain reaction enables localization of multiple molecular classes combined with plant cell ultrastructure. *Plant Biology*. DOI: 10.1101/2024.01.29.577761.

Zhang, R., Isola, P. and Efros, A.A. (2017) Split-brain autoencoders: Unsupervised learning by cross-channel prediction. In: *2017 IEEE Conference on Computer Vision and Pattern Recognition (CVPR)*, Honolulu, HI, pp. 645–654. DOI: 10.1109/CVPR.2017.76.

Zhang, X., Liu, P., Qing, C., Yang, C., Shen, Y. *et al.* (2021) Comparative transcriptome analyses of maize seedling root responses to salt stress. *PeerJ* 9, e10765. DOI: 10.7717/peerj.10765.

Zhang, R., Zhang, C., Yu, C., Dong, J. and Hu, J. (2022) Integration of multi-omics technologies for crop improvement: Status and prospects. *Frontiers in Bioinformatics* 2, 1027457. DOI: 10.3389/fbinf.2022.1027457.

Zhang, X., Liu, W., Yin, Y., Zheng, J., Li, J. *et al.* (2025) WRKY transcription factors participate in abiotic stress responses mediated by sugar metabolism. *Frontiers in Plant Science* 16, 1646357. DOI: 10.3389/fpls.2025.1646357.

Zhao, L., Fonseca, A., Meschichi, A., Sicard, A. and Rosa, S. (2023) Whole-mount smfish allows combining RNA and protein quantification at cellular and subcellular resolution. *Nature Plants* 9(7), 1094–1102. DOI: 10.1038/s41477-023-01442-9.

Zheng, P., Zhang, N., Ren, D., Yu, C., Zhao, B. *et al.* (2023) Integrated spatial transcriptome and metabolism study reveals metabolic heterogeneity in human injured brain. *Cell Reports Medicine* 4(6), 101057. DOI: 10.1016/j.xcrm.2023.101057.

Zhou, Y., Zhu, J., Li, Z., Yi, C., Liu, J. *et al.* (2009) Deletion in a quantitative trait gene qPE9-1 associated with panicle erectness improves plant architecture during rice domestication. *Genetics* 183, 315–324. DOI: 10.1534/genetics.109.102681.

Zhu, G., Li, R., Zhang, L., Ma, L., Li, J. *et al.* (2024) RNA-protein interactions reveals the pivotal role of lncRNA1840 in tomato fruit maturation. *The Plant Journal* 120(2), 526–539. DOI: 10.1111/tpj.16998.

12 Spatial and Multiple Omics Approaches for Advancing Climate-smart Crop Breeding

Augustine Antwi-Boasiako[1,2*], Daniel Oppong-Sekyere[3], Joseph Adomako[1,4], Augustine Boakye Boateng[1], Ernest Baafi[1*], Kennedy Agyeman[1], Isaac Osei Bonsu[1], Daouda Coulibaly[5] and David Appiah Abban[1]

[1]Council for Scientific and Industrial Research-Crops Research Institute (CSIR-CRI), Kumasi, Ghana; [2]Key Laboratory of Biology and Genetics Improvement of Soybean, Ministry of Agriculture, Zhongshan Biological Breeding Laboratory (ZSBBL), National Innovation Platform for Soybean Breeding and Industry-Education Integration, State Key Laboratory of Crop Genetics & Germplasm Enhancement and Utilization, Key Laboratory of Biology and Genetics Improvement of Soybean, Ministry of Agriculture, Zhongshan Biological Breeding Laboratory (ZSBBL), National Innovation Platform for Soybean Breeding and Industry-Education Integration, State Key Laboratory of Crop Genetics & Germplasm Enhancement and Utilization, Nanjing, China; [3]Department of Ecological Agriculture, Bolgatanga Technical University, Bolgatanga, Ghana; [4]Department of Crop and Soil Science Education, Akenten Appiah-Menka University of Skill Training and Entrepreneurial Development (AAMUSTED), Mampong, Ghana; [5]Laboratory of Agro-physio-genetics and Plant Biotechnology, Department of Agricultural Technical Science (STA), Rural Polytechnic Institute for Training and Applied Research (IPR/IFRA) of Katibougou, Koulikoro, Mali

12.1 Introduction

Agriculture and the climate have a fundamental and delicate relationship. Global food and nutrition security is seriously threatened by climate change due to elevated temperatures, changes in precipitation patterns, increased atmospheric CO_2 concentrations, and an increasing incidence of severe weather events (Tirado *et al.*, 2010; Nunes, 2023). These changes, however, present a great opportunity for agricultural systems to undergo innovation and transformation (Panda, 2018). One of the most pressing issues is ensuring global food security for a population that is predicted to be 10 billion by 2050 (Hall *et al.*, 2017) while preserving the sustainability of natural resources and agricultural systems. To meet the demand, agricultural production must almost be doubled by 2050 (Beltran-Peña *et al.*, 2020; Hemathilake and Gunathilake, 2022).

*Corresponding authors: bbbantwi@yahoo.com and e.baafi@gmail.com

© CAB International 2026. *Plant Spatial Biology* (ed. Chen)
DOI: 10.1079/9781836994572.0012

However, crop yield improvements are not keeping up with this demand (Begna, 2022; Bal and Sharangi, 2025). There is also an increasing concern that the current productivity gaps could be worsened by climate change (Farooq *et al.*, 2022). It has been reported that crop yields are greatly impacted not only by mean climate trends, but also via intricate and regionally specific pathways (Ray *et al.*, 2015; Najafi *et al.*, 2019).

The deterioration of fundamental natural resources makes matters worse. Soil fertility is declining due to nutrient depletion, salinization, erosion, and unsustainable management (Lal, 2009; FAO, 2021). Water resources are also under increasing stress from changing rainfall patterns and rising temperatures (Rockström *et al.*, 2009). Climate change is causing new pests and diseases to emerge, and the incidence and severity of biotic and abiotic factors are increasing, causing unpredictable effects on crop productivity (Deutsch *et al.*, 2018). To evaluate the implications of global warming and to direct crop breeding and management efforts, mechanistic crop growth models that consider climate, soil, management, and genetic factors are crucial (Challinor *et al.*, 2014). In light of this, plant breeding continues to be an important driver of agricultural innovations and evolution.

The history of plant breeding has been influenced by multiple technological revolutions, from early domestication to contemporary hybrid breeding (Gepts, 2004; Moose and Mumm, 2008). This is because plant breeding has advanced through several stages, each distinguished by a greater incorporation of new technologies (Varshney *et al.*, 2021). Modern plant breeding uses molecular markers, genomic selection (GS), and increasingly extensive multi-omics datasets. Breeders are now using sophisticated computational tools such as artificial intelligence (AI), in addition to machine learning (ML), to analyze and comprehend intricate biological data, and facilitate informed decision making (Kumar *et al.*, 2022a). Global plant breeding efforts are currently concentrated on creating climate-resilient crops that can maintain or enhance productivity and quality under adverse environmental circumstances (Bailey-Serres *et al.*, 2019). These crops must endure fluctuating biotic threats and many abiotic challenges, such as heat, salt, drought, and nutritional deficits (Beillouin *et al.*, 2020). AI has great potential because it presents new possibilities for processing high-throughput phenotyping data, determining gene function, and incorporating multi-dimensional environmental (envirotyping) data into breeding pipelines (Montesinos-López *et al.*, 2022).

Our knowledge of plant biology and stress responses is also being completely transformed by the advent of multi-omics (proteomics, phenomics, ionomics, metabolomics, transcriptomics, genomics) and its integration with spatial technologies (Rai *et al.*, 2021). Finding the dynamic spatiotemporal interactions that support stress tolerance and adaptation is made possible by spatial omics techniques, which allow the mapping of molecular profiles across various tissues and cellular contexts. New developments in spatial multi-omics technologies open the door to precise molecular breeding techniques and offer previously unheard-of resolution into plant physiology (Lee *et al.*, 2023). There is a unique chance to hasten the development of climate-smart crops by combining AI-driven analytics with spatial and multiple omics approaches. In the light of the escalating urgency of global warming and food insecurity, such integration may improve the efficiency, validity, and exactness of breeding efforts (Varshney *et al.*, 2018). To fully realize the possibilities of these advances in technology, interdisciplinary partnerships between plant physiologists, breeders, computational biologists, and data scientists will be essential in the future.

Spatial multiple omics integrates several omics layers (genome, transcriptome, proteome, etc.) with spatial information, providing a comprehensive understanding of cellular and molecular organization in tissues. This approach is essential for deciphering the intricate mechanisms of development, disease, and tissue function by analyzing molecular profiles within a specific spatial context. Spatial multi-omics provides details on tissue organization, which the traditional omics approaches lack. Thus, spatial multi-omics incorporates spatial information, allowing researchers to see how cells and molecules are arranged within tissues. Moreover, by analyzing the spatial organization of cells and molecules, researchers can gain insights into mechanisms governing drought resistance and identify potential genes. Hence, it supports targeted gene editing tools, such as clustered regularly interspaced short palindromic repeats (CRISPR)-Cas9, transcription activator-like effector nucleases (TALENs), and zinc finger nucleases (ZFNs), which offer varying levels of precision and efficiency in modifying DNA. Therefore, spatial studies integrating with multi-omics have emerged as a viable methodology for the thorough investigation of cells inside tissues, facilitating the simultaneous examination of many data modalities such as the transcriptome and metabolome, either in parallel or within the same tissue segment. The current status of spatial and multiple omics approaches in crop improvement, their contribution to the development of climate-resilient cultivars, and the potential for incorporating these tactics into next-generation breeding pipelines are all covered in this chapter. This chapter seeks to provide a thorough picture of how state-of-the-art research can help create agricultural systems that are more resilient and sustainable.

12.2 Consequences of Climate Change on Crops

A change in climate significantly affects crop growth, yield, and quality (Kumar, 2016). Increasing temperatures, shifting precipitation patterns, and a heightened occurrence of extreme weather events are modifying the environment for the cultivation of cereals, vegetables, legumes, and fruits, leading to a range of consequences. This is expected to diminish the yield of cereals, legumes, vegetables, and fruits by 5–10% before mid-century (Pereira, 2017). Increased temperatures accelerate phenology, impair pollination, reduce grain filling, and ultimately lower yields (Moretti *et al.*, 2010). Climate change and its variability significantly impact vegetable cultivation and food security by altering competitive dynamics and generally exacerbating pathogen and pest infections (Newton *et al.*, 2011). Climate change has influenced the interactions between crop plants and insect pests. It is affecting the prevalence of diseases and pests, pathogen and host relationship interactions, insect spread and ecology, phenology, migration to new places, and overwintering capabilities, alongside physiological and biochemical alterations (Ayyogari *et al.*, 2014). Thus, variations in precipitation and temperature can increase the pressure of pests and diseases, which lowers vegetable yields and degrades fruit quality.

Temperature fluctuations are anticipated to reduce world rice and soybean yields by 3.2% and 3.1%, respectively, for each 1°C rise in global temperatures (Zhao *et al.*, 2017). Alterations in precipitation and temperature patterns interrupt conventional agricultural calendars, impacting planting and harvesting timelines. This may result in diminished yield and modified cultivation zones. Variations in temperature and precipitation patterns can modify the behavior and distribution of insects, promoting fungal plant diseases and pest migrations to new areas. Up to 40% of crops could be lost due to pests and diseases exacerbated by climate change. Furthermore, reduced soil fertility, organic matter loss, and soil erosion are all caused by altered precipitation patterns and rising temperatures. This may result in reduced agricultural production and yields of crops. Also, drought or waterlogging brought on by altered precipitation patterns and increased evapotranspiration can exert a deleterious impact on agricultural output and plant health. Additionally, increased atmospheric CO_2 levels also have the potential to cause widespread deficiencies by lowering the content of protein, iron (Fe), and zinc (Zn), along with other minerals in cereals, legumes, vegetables, and fruits (Kidane *et al.*, 2025; Wang *et al.*, 2025; Yaseen *et al.*, 2025).

To mitigate these consequences, farmers can adopt adaptive strategies such as cultivating diverse crops to reduce reliance on a single crop and enhance ecosystem resilience. Planting a variety of crops with diverse developmental needs and stress tolerances enables farmers to mitigate risk and minimize susceptibility to the impact of climate change. Using cultivars developed to withstand changing climate conditions, farmers may reduce yield losses and have a guarantee for sustainable crop yield despite climate-related challenges. Furthermore, altering planting schedules to accommodate shifting growing seasons and utilizing appropriate irrigation techniques and water management strategies can minimize these consequences.

For vegetables, heat and water stresses alter secondary metabolite synthesis, diminishing nutritional value and postharvest quality (Mattos *et al.*, 2014; Dumitru *et al.*, 2023). Legumes are highly susceptible to heat, drought, and salinity, which impair nitrogen fixation and exacerbate pest and disease pressures (Ahmed *et al.*, 2010; Vadez *et al.*, 2012). Similarly, fruit crops face yield losses due to disrupted flowering and fruit set, insufficient chilling hours, and irregular maturation cycles (Kuden, 2018; Haokip *et al.*, 2020). Climate change influences the geographic spread, incidence, and severity of plant diseases and pests (Ahmed *et al.*, 2010). Shifting pest ranges and disease outbreaks further affect sustainable crop production. The ramifications of global warming on crop production are extensive and intricate. Comprehending these effects is essential for formulating appropriate adaptation and mitigation strategies for food security and sustainable agriculture against climate change. Promoting climate-resilient agriculture and assisting farmers in adapting to climate change can lead to a more food-secure future.

12.2.1 Vegetables

Climate change and its variability significantly impact vegetable cultivation and food security by altering competitive dynamics and generally exacerbating pathogen and pest infections and resistance (Newton *et al.*, 2011). Vegetables are particularly prone to climate change owing to their high sensitivity to environmental fluctuations and short growth cycles (Ayyogari *et al.*, 2014; Dumitru *et al.*, 2023). They are vital components of human diets, providing essential micronutrients, antioxidants, and dietary fiber (Saeed *et al.*, 2023). However, climate-induced stresses pose significant risks to vegetable production, quality, and supply chains (Porter *et al.*, 2019). Among the primary impacts of climate change on vegetables are yield reductions, shifts in growing seasons, water scarcity, loss of nutritional quality, and rising pest and disease threats. Climate change also affects the nutritional and sensory quality of vegetables. High temperatures can reduce the accumulation of key phytochemicals, including vitamins, carotenoids, and phenolic compounds, thereby diminishing the nutritional value of produce (Mattos *et al.*, 2014). In the postharvest context, fluctuations in temperature and relative humidity accelerate senescence and spoilage, resulting in substantial quality losses (Moretti *et al.*, 2010). Vegetables are predominantly cultivated in open fields, exposing them directly to the effects of rising temperatures, drought, flooding, and salinity (Table 12.1). Heat stress can disrupt flowering, pollination, and fruit set in crops such as tomato, pepper, and broccoli, leading to yield losses (Mattos *et al.*, 2014) (Table 12.1). Increased temperatures also accelerate respiration rates, potentially reducing carbohydrate accumulation and altering flavor profiles (Moretti *et al.*, 2010).

Changing climatic conditions influence pest and disease dynamics in vegetable systems. Increased temperatures and modified precipitation patterns can enhance the distribution and reproduction of insect pests and pathogens, resulting in greater severity and frequency of infestations (Dumitru *et al.*, 2023). This requires the development of crop cultivars with enhanced resilience to new biotic and abiotic stresses. To tackle these issues, climate-smart breeding techniques must incorporate geographical and omics data to drive genetic advancements. High-resolution climate models can identify target environments and inform genotype × environment interactions, guiding breeding decisions (Dumitru *et al.*, 2023). Genomics, transcriptomics, proteomics, and metabolomics tools can uncover molecular mechanisms underlying tolerance to stress, facilitating marker-assisted selection (MAS) and genomic selection (GS) of resilient varieties (Mattos *et al.*, 2014). For example, transcriptomic analyses can reveal heat- and drought-responsive gene networks in crops such as tomato and lettuce, while metabolomic studies can identify stress-induced changes in bioactive compounds (Moretti *et al.*, 2010). Coupled with advanced phenomics platforms, these insights accelerate the creation of vegetable varieties that can flourish in future climate scenarios.

Alterations in the patterns of precipitation and temperature may result in diminished vegetable yields and inferior quality food, as well as decreased crop productivity. Temperature, whether independently or in conjunction with

Table 12.1. Impact of climate change on vegetable, legume, and fruit crops.

Crop type	Major climate change effects	Specific impacts	References
Vegetables	Increased temperature, drought, elevated CO_2, and extreme weather events	Accelerated senescence Reduced nutrient density (e.g. vitamin C, minerals) Decreased postharvest quality and shelf life Disrupted secondary metabolite synthesis (e.g. antioxidants)	Moretti *et al*. (2010); Mattos *et al*. (2014); Dumitru *et al*. (2023)
Legumes	Heat stress, drought, salinity, and changing pest/disease dynamics	Impaired nodulation and nitrogen fixation Reduced seed set and pod filling Greater vulnerability to pests and foliar diseases Yield and quality losses	Ahmed *et al*. (2010); Vadez *et al*. (2012); Rahman *et al*. (2022); Khan *et al*. (2025)
Fruit crops	Higher temperatures, altered rainfall, reduced chilling hours, and extreme events	Disrupted flowering and fruit set Increased fruit cracking and sunburn Irregular ripening and reduced fruit size Lower fruit sugar and nutrient content	Malhotra (2017); Kuden (2018); Haokip *et al*. (2020); Bhattacharjee *et al*. (2022)

other environmental factors, profoundly influences the development and ripening phases in tomatoes (Driedonks *et al.*, 2016). Elevated humidity can lead to considerable reductions in tomato yield due to diminished fruit set and the production of smaller, inferior-quality fruits (Shamshiri *et al.*, 2018). In pepper, elevated temperatures following pollination reduced fruit set, demonstrating the vulnerability of fertilization to thermal stress (Erickson and Markhart, 2002; Abou Hussein, 2012). In cauliflower, higher temperatures may result in inadequate vernalization to delay head induction (Bisbis *et al.*, 2018).

The disruption of typical growth seasons caused by warmer temperatures and shifting precipitation patterns may impact the planting and harvesting schedule of vegetables. Although climate change directly impacts rainfed vegetable farming, it also has an impact on water storage and irrigation water availability (Ayyogari *et al.*, 2014). Additionally, insufficient water supplies resulting from altered precipitation patterns and elevated evapotranspiration can affect the amount and nutritional value of vegetable production. Vegetables growing in environments with higher CO_2 levels may have lower levels of vital vitamins, minerals, and other nutrients. Nonetheless, elevated CO_2 levels can enhance photosynthesis, particularly in C3 plants (Bisbis *et al.*, 2018). Climate change could consequently end in increased yields under this scenario. According to Jin *et al.* (2009), the yield of Chinese cabbage, celery, and lettuce produced nearly doubled when grown at CO_2 values of 800 to 1000 ppm. Bettoni *et al.* (2014) also reported that higher CO_2 enhanced the dry matter content of the shoots and roots of onions grown at 700 ppm. When atmospheric CO_2 doubled, yields in lettuce, tomatoes, and cucumbers increased by up to 44% (Korres *et al.*, 2016).

12.2.2 Legumes

Legumes represent an essential component of global agriculture and nutrition. They provide significant sources of plant-derived fiber, protein, minerals, and vitamins. Legumes are crucial for sustainable agricultural systems by facilitating the biological fixation of nitrogen for soil fertility improvement (Rahman *et al.*, 2022). Legumes are important for global food security, serving as significant sources of plant-based protein that facilitate growth and development in both humans and animals (Dave *et al.*, 2024). Their production, food security, and human health are all at risk because of environmental changes. Climate change, owing to rising temperatures, altered precipitation patterns, and elevated atmospheric carbon dioxide levels, poses a danger to legume production and productivity, as does the resultant increase in pest and pathogen activity, including fungal diseases and insect pests (Dave *et al.*, 2024). Legumes are highly sensitive to climatic variability, which significantly impacts their yield stability, quality, and ecosystem services (Table 12.1). It is predicted that the consequences of environmental changes on legumes will result in a reduction of yields and lower nutritional quality. If adaptation measures are not taken, it is predicted that mean temperatures will rise above the ideal thresholds for growing legumes by 2050, resulting in yield losses of 10–49% for cowpea, lentils, beans, and soybeans (Dave *et al.*, 2024).

Legumes thrive in temperatures between 10 and 36°C, and any higher temperatures can result in significant reductions in grain yield (Farooq *et al.*, 2017). Research indicates that crop yields are greatly impacted by climate variability, frequently via intricate and regionally specific pathways (Ray *et al.*, 2015; Najafi *et al.*, 2019). High temperatures can cause heat stress, leading to reduced plant growth, leaf scorching, and premature senescence. The severity of destruction or harm resulting from heat stress is primarily dependent on the defense mechanism of the crop and the stage of growth at which the high temperature was experienced (Ul Hassan *et al.*, 2022). Grain development in grain legumes is impacted by heat stress because it causes poor pod development and shattered embryos, which in turn reduce grain yield (Dave *et al.*, 2024). Bishop *et al.* (2016) reported a combination of decreased pod and seed set, frequently resulting in yield losses under heat stress. Elevated temperatures can disrupt photosynthesis, impairing the ability of the plant to assimilate carbon dioxide for energy. The degradation of photosynthetic compounds, a marked reduction in net photosynthesis, and an escalation in senescence of leaves are the primary consequences of thermal stress. Genotypes of common bean (*Phaseolus vulgaris* L.) subjected to day temperatures of 32°C during floral development displayed decreased yield mass, seed set, and pod set (Verma *et al.*, 2025).

Heat stress, especially during flowering and pod filling stages, is a major threat to legume productivity. Grain legumes such as chickpea, lentil, and faba bean are particularly vulnerable to terminal heat stress, which shortens the growing period to disrupt seed set and development (Vadez *et al.*, 2012), concurrently affecting grain yield and nutritional quality. Heat and drought stress can reduce seed protein content and alter oil and carbohydrate profiles (Khan *et al.*, 2025). Additionally, legumes grown under elevated CO_2 may exhibit imbalances in C:N ratios, potentially lowering protein concentrations and compromising their role in human diets.

Drought stress, which is increasingly common in dry and semi-arid environments, constrains water availability, reduces stomatal conductance, and impairs photosynthesis, ultimately resulting in significant output reductions (Vadez *et al.*, 2012). The incidence and severity of diseases affecting legumes are also influenced by climate change. Warmer temperatures and altered humidity patterns promote the proliferation of foliar and root pathogens, such as *Ascochyta blight*, *Fusarium wilt*, and *Rhizoctonia* species (Ahmed *et al.*, 2010). New disease outbreaks and shifts in pathogen virulence further complicate disease management under future climate scenarios. Despite these challenges, legumes offer substantial potential for climate change adaptation and mitigation. Legume-based cropping systems—such as intercropping, relay cropping, and rotations improve soil health, enhance water utilization efficiency, and support greenhouse gas mitigation through nitrogen fixation and reduced fertilizer needs (Rahman *et al.*, 2022). These systems support greater agroecological resilience, especially in smallholder and rainfed farming systems. Climate change increases susceptibility to pests and diseases. This is because heat-stressed leguminous crops may become more susceptible to pests and diseases, impacting plant health, which further reduces yield. The disease spectrum of legumes is expected to be impacted by climate change, specifically the dissemination, onset of epidemics, and formation of various pathotypes and disorders that affect agricultural and disease management strategies (Ahmed *et al.*, 2010).

Pathogen transmission and resistance of the host in *Ascochyta*–lentil and *Ascochyta*–chickpea pathosystems can be impacted by varying temperature regimes. Late rains can also result in significant quality losses due to *Ascochyta* blight infection of chickpea pods.

Harnessing the power of spatial data and multiple omics approaches is critical to climate-smart legume speed-breeding. Progress in genomic selection, genome-wide association studies (GWAS), and transcriptome profiling has identified key loci and regulatory networks linked to stress endurance in legumes (Vadez *et al.*, 2012; Rahman *et al.*, 2022). Proteomic and metabolomic analyses further elucidate stress-induced metabolic adjustments and adaptive responses. Combined with phenomics tools and spatially informed breeding strategies, these insights can facilitate the development of legume cultivars that are resilient, high-yielding, and nutritionally robust under a changing climate.

12.2.3 Fruit crops

Fruit crops occupy a vital niche in global agriculture, providing essential nutrients, dietary fiber, and phytochemicals that contribute to human health and well-being. However, they are particularly vulnerable to climate variability due to their perennial growth habit, long reproductive cycles, and sensitivity to environmental cues (Haokip *et al.*, 2020) (Table 12.1). Climate change impacts fruit crops at multiple stages, from flowering, fruit set, maturation, quality, and postharvest. Temperature fluctuations are among the most critical threats to fruit crop production. Elevated temperatures during the flowering stage diminish the viability of pollen and fruit development in fruit crops such as apple, pear, and citrus (Bhattacharjee *et al.*, 2022). For temperate fruits like apple and cherry, insufficient chilling hours caused by warmer winters disrupt bud break and flowering synchrony, leading to irregular yields (Malhotra, 2017). Conversely, sudden frost events following early bud break can cause severe crop losses.

Climate change also significantly impacts fruit production. Rising temperatures alter flowering patterns, reduce fruit set, and increase pollination failures, ultimately leading to lower yields and decreased fruit quality (Karagatiya *et al.*, 2023). Changes in morphological traits like vigor, canopy development, and reproductive traits, such as fruiting capacity and diminished fruit size, alongside quality factors (insufficient color development, reduced juice content, shortened shelf life), and susceptibility to pests, can result in substandard fruit quality and diminished fruit yield (Rajan *et al.*, 2020). An increase in temperature will have a negative impact on pollination, which will inhibit the initiation of flower buds in most plants. Extremely high temperatures will often cause flower abortions in flowers and fruits, which reduce fruit yield (Mehmood *et al.*, 2025). Warmer temperatures can lead to decreased color development and reduced sugar content in fruits like grapes (Karagatiya *et al.*, 2023). Additionally, high temperatures exacerbate disorders such as spongy tissue in mangoes, fruit cracking in litchis, and black spot in custard apples. Furthermore, climate change may alter the suitability of locations for specific fruit crops, potentially shifting production areas.

Climate change may lead to early flowering in mangoes and crop harvest, with potential yield losses. Higher relative humidity can be linked to higher rates of powdery mildew and mango hopper infestation. Elevated CO_2 and high temperatures can reduce the fruit yield of strawberries by 12–35% (Fischer *et al.*, 2022). Rising temperatures can also accelerate grape maturation and pigmentation. Warmer winters may compromise chilling requirements, affecting cherry productivity. In citrus, variations in relative humidity levels may have a negative impact on fruit set, causing an excessive amount of fruit to drop.

Drought stress and altered rainfall patterns also pose significant challenges. Water scarcity during fruit development reduces fruit size, increases fruit drop, and affects sugar accumulation and flavor (Kuden, 2018). In grapes and citrus, drought conditions can lead to increased concentrations of bitter compounds and undesirable flavor profiles (Bhattacharjee *et al.*, 2022). Moreover, water stress weakens plant defenses, making fruit crops more susceptible to pests and diseases.

Elevated CO_2 levels have a mixed impact on fruit crops. While CO_2 enrichment can enhance photosynthesis and biomass accumulation, it may also lead to imbalances in carbon and nitrogen metabolism, affecting fruit nutritional quality (Myers *et al.*, 2014). Additionally, changes in temperature and humidity influence postharvest physiology, accelerating spoilage and reducing shelf life (Moretti *et al.*, 2010). The spatial distribution of fruit production is also shifting due to climate change. Regions previously unsuitable for certain fruit crops may become viable, while traditional growing areas may face new climatic constraints (Haokip *et al.*, 2020). For example, grape cultivation is expanding into higher latitudes to escape excessive heat (Kuden, 2018). To develop climate-resilient fruit crops, breeders are increasingly adopting spatial and multi-omics approaches. Genomic methodologies, including GWAS and genomic selection, are employed to pinpoint alleles associated with resilience to stress, fruit quality, and phenological adaptation (Malhotra, 2017). Transcriptomics and metabolomics elucidate the molecular processes that govern crop reactions towards temperature extremes, water stress, and increased CO_2 levels (Bhattacharjee *et al.*, 2022). Phenomics, combined with remote sensing and spatial data, enables high-throughput phenotyping under variable field conditions, accelerating the selection of superior genotypes (Rahman *et al.*, 2022).

The following adaptive techniques can be employed to minimize these adverse consequences: developing climate-resilient varieties, introducing sustainable production practices, and implementing climate-smart agricultural techniques. Utilizing cultivars specifically bred for resilience against fluctuating climatic conditions, adopting integrated pest and disease management strategies, conservation agriculture, and efficient water utilization, alongside implementing climate-smart agricultural practices that account for anticipated climate change effects on fruit production, can alleviate the detrimental impacts of climate change. Through integration of spatial intelligence with omics technologies, breeders can design and develop cultivars that not only withstand the stresses imposed by climate change but also deliver consistent yield and quality. This comprehensive strategy is vital for guaranteeing the sustainability and economics of fruit cultivation processes against an increasingly unpredictable climate future.

12.3 Conventional Breeding Versus Novel Approaches for Developing Climate-smart Crops

Traditional breeding has historically been fundamental to crop enhancement, utilizing germplasm collection, phenotypic selection, and hybridization to develop cultivars with advantageous characteristics such as drought resistance, insect tolerance, and increased yield (Tester and Langridge, 2010; Chen *et al.*, 2019). Although effective in some circumstances, conventional approaches are frequently time-consuming, labor-intensive, and constrained by the genetic resources available within a breeding pool. The intricate polygenic characteristics of traits that enhance climate adaptation such as tolerance of heat, efficiency in water use, and nutrient-use efficiency, present considerable obstacles to traditional breeding, owing to genotype-by-environmental interactions and low heritability (Varshney *et al.*, 2021).

To address these limitations, novel approaches leveraging molecular biology and other high-throughput technologies have emerged. These include MAS, GS, and, more recently, the multi-omics methodologies that combine genomes, transcriptomics, proteomics, metabolomics, and phenomics data to elucidate the intricate networks of stress adaptation (Varshney *et al.*, 2021). Such integrative strategies enable breeders to identify key genes, regulatory elements, and biomarkers linked to climate resilience traits, thereby accelerating the advancement of climate-resilient cultivars characterized by enhanced precision and efficiency.

12.3.1 Omics for crop improvement

Omics technology has transformed plant breeding by facilitating extensive profiling of molecular constituents across several biological levels (Fig. 12.1). Genomics elucidates DNA sequence variation, enabling the identification of quantitative trait loci (QTLs) and genome-wide association studies (GWAS) to identify loci linked to stress tolerance (Varshney *et al.*, 2021). Transcriptomics elucidates gene expression dynamics in response to environmental challenges, facilitating the identification of genes

Fig. 12.1. Overview of omics strategies for crop improvement.

responsive to stress and regulatory networks. Proteomics and metabolomics further characterize protein abundance and metabolite profiles, elucidating biochemical pathways impacted by abiotic and biotic stresses. Phenomics, powered by high-throughput imaging and sensor technologies, quantifies complex phenotypic traits under field conditions, offering crucial data for trait validation (Araus and Cairns, 2014).

The combined use of various information involving multi-omics has demonstrated immense value in analyzing intricate features and forecasting plant performance amid climate fluctuations. For instance, combined transcriptome and metabolome analyses have identified candidate genes and metabolites linked to tolerance to drought in maize and rice (Gupta *et al.*, 2020). Additionally, advances in bioinformatics and ML enable the synthesis and analysis of complex multi-layered omics data, empowering predictive breeding and precision agriculture (Kumar *et al.*, 2022a).

12.3.2 Harnessing spatial omics approaches for developing climate-smart crops

Spatial omics represents a cutting-edge dimension in crop science, combining traditional omics with spatial information about cellular and tissue-level molecular distributions. This approach allows the mapping of gene expression, protein localization, and metabolite accumulation with high spatial resolution, elucidating how plants orchestrate stress responses across several tissues and developmental phases. In climate-smart breeding, spatial omics enables the identification of tissue-specific and cell-type-specific mechanisms conferring resilience. For example, spatial transcriptomics has revealed differential gene expression patterns in root and leaf tissues under drought stress, uncovering localized adaptation strategies that are invisible in bulk tissue analyses (Wang *et al.*, 2023). Furthermore, spatial metabolomics facilitates the visualization of metabolite gradients critical for osmoprotection and antioxidative responses during heat and salinity stresses. Integrating spatial omics with phenotyping and environmental data allows breeders to link molecular heterogeneity with field-level performance, enhancing selection accuracy for climate resilience traits. Such spatially resolved molecular insights are essential for designing targeted genetic interventions and for developing cultivars that can withstand the spatially variable stresses imposed by climate change.

12.3.3 Recent advancements in spatial multi-omics technologies

Recent technological innovations have propelled spatial multi-omics from proof-of-concept to practical applications in plant science. High-throughput spatial transcriptomics platforms – such as Slide-seq, 10 × Genomics Visium, and MERFISH – facilitate the acquisition of gene expression around nearly single-cell resolution throughout whole tissues (Ståhl *et al.*, 2016). Concurrently, development in imaging mass spectrometry and matrix-assisted laser desorption/ionization (MALDI) has enabled spatially resolved proteomics and metabolomics to map the distribution of thousands of molecules simultaneously (Boughton *et al.*, 2016). Multimodal data integration, which combines spatial transcriptomics with proteomics and metabolomics, reveals a detailed molecular atlas of plant tissues subjected to stress conditions, offering insights into regulatory cascades and metabolic rewiring. Additionally, ML and AI tools have been applied to analyze complex spatial omics datasets, enabling predictive modeling of plant stress responses and guiding the rational design of climate-resilient genotypes (Kumar *et al.*, 2022a).

Emerging single-cell multi-omics techniques further allow simultaneous profiling of the genome, epigenome, transcriptome, and proteome at cellular resolution, to reveal intricate regulatory networks underpinning stress tolerance (Ryu *et al.*, 2019). Together, these advances mark a transformative shift in crop breeding, equipping researchers and breeders with unprecedented resolution and depth of understanding to expedite the advancement of climate-resilient crop improvement. Integration of spatial with multiple omics approaches is transforming climate-smart crop breeding by offering unparalleled insights into the complex molecular and spatial dynamics by which plants react under stress conditions. These tools, combined with AI and high-throughput phenotyping, enable the precise selection of resilient traits across diverse crops. Mitigating the many effects of global warming on vegetables, legumes, and fruit crops necessitates comprehensive breeding techniques based on sophisticated technology. Moving forward, an interdisciplinary approach and continued innovations will be essential to expedite the creation of climate-resilient cultivars to guarantee sustainable agricultural output and global food security amid an increasingly unpredictable climatic future.

12.4 Overview of Omics Strategies for Crop Improvement

Progress in next-generation sequencing (NGS) has enabled the development of several omics, including genomics, transcriptomics, proteomics, metabolomics, and phenomics, which have been thoroughly documented in the domain of plant research. Advanced multi-omics methodologies have significantly improved the understanding of development, senescence, yield, and responses to stressors for several crops. These omics methodologies have been employed in numerous crops such as tomato, maize, millet, cotton, soybean, and rice. Next-generation high-throughput screening methods, commonly referred to as 'omics,' are being developed to expedite the enhancement of plant features (Castillejo *et al.*, 2023), while simultaneously increasing agricultural sustainability (Fig. 12.1). Genomics, transcriptomics, proteomics, metabolomics, and phenomics, among others, are large-scale methodologies that have provided substantial datasets for this purpose, and these multidisciplinary approaches have expedited trait improvement (Fig. 12.1). The combined use of multi-omics data through panomics systems and platforms in biology may boost our comprehension of the complicated regulatory networks governing genetic enhancement and facilitate the prediction of intricate agricultural traits. Simultaneously, novel bioinformatics and ML techniques are facilitating the comprehension of this complex hyper-dimensional data (Libbrecht and Noble, 2015), while elucidating the mechanisms of agronomic traits within the contemporary plant breeding triangle (genomics versus phenomics versus enviromics) (Crossa *et al.*, 2021).

12.5 Harnessing Spatial Omics Approaches for Developing Climate-smart Crops

Spatial omics tools have been instrumental in unraveling the complexity of individual cells and defining cell types, while maintaining their spatial organization. To acquire a thorough comprehension of the molecular chain from genome to phenome, employing multi-omics technologies with spatial resolution is considered essential (Vickovic *et al.*, 2022). Spatial omics has the potential to uncover multi-omics alterations during the life cycle of plants, along with their interactions with the surrounding environment. This method enhances our comprehension of the molecules and cellular interactions in essential biological processes, thus expediting investigations into the fundamental molecular mechanisms. Consequently, it is increasingly certain that spatial omics will have a crucial role in enhancing crop development. A detailed analysis of tissue-specific variations in expression at a given degree of spatial resolution can show how critical cell localization and interplay are directing complex regulatory networks (Neumann *et al.*, 2022). Understanding how specific plant cells respond across many omics layers during their interaction within a range of tissue domains can reveal complex spatial linkages at a level of resolution never before possible.

The process of biologically modifying crops to bring about an improvement in their yield and biotic and abiotic stress resilience is known as crop improvement. Crop improvement started with selection, along with breeding for beneficial traits through genetic inheritance and crossover. The advent of genetic engineering techniques has reduced the spread of undesirable traits and production challenges typically associated with conventional breeding methods, while facilitating the swift and precise integration of favorable characteristics into valuable crops (Zhang *et al.*, 2018). The emergence of DNA sequencing technologies has enabled the mapping of QTLs and the identification of candidate genes through precise mapping (Jaganathan *et al.*, 2020). Identifying candidate genes using genetic mapping may necessitate more than one generation and a long time, and these genes often have limited or partial impacts on complex traits, along with undesirable phytotoxic consequences, affecting their efficiency and commercial potential. The adoption of spatial omics approaches promises to accelerate the development of climate-smart crops. Climate change exacerbates the frequency and intensity of stresses, hence significantly undermining global crop yields, threatening food security, and hindering the attainment of the Sustainable Development Goals. Hence, the development of climate-smart crops seeks to improve yield, nutritional value, and drought tolerance, and this improvement can be more accurate with the use of spatial omics data (Yu *et al.*, 2023; Li *et al.*, 2024). The breakthroughs in single-cell and spatial transcriptomics approaches have significantly improved our understanding of the molecular pathways underlying crop life.

The production of 1 kg of rice consumes around 3000 liters of water, and drought results in a yield reduction of 65–100% contingent upon the development cycle of the crop (Oo *et al.*, 2021). Similarly, sorghum is a resilient crop; however, when dryness occurs at the vegetative stage, it results in yield decreases of 50–60%, and at the reproductive stage, yield loss is 87–100% (Craufurd and Peacock, 1993). Data from field studies conducted from 1980 to 2015 indicated that drought, characterized by an estimated 40% reduction in water availability, diminished the quantity of maize and wheat production significantly by 21% and 39%, respectively (Daryanto *et al.*, 2016). Drought may negatively influence the development and reproduction of maize. High temperatures resulting from drought stress can also impede the initiation of blossoming in male flowering plants and disrupt pollen formation (Kim and Lee, 2023). These challenges would be addressed by utilizing the power of multi-omics.

12.5.1 Genomics approach

In the present accelerated research landscape, genomics is central to crop improvement and is extensively utilized for detecting genetic variability related to phenotypic variations, novel traits, and crucially, for characterizing the pathways associated with stress tolerance in support of sustainable agriculture (Bohra *et al.*, 2020; Mishra *et al.*, 2024). Genomics represents the extensive examination of an organism's whole DNA repertoire, encompassing the sequencing, mapping, and functional assessment of genes and non-coding regions (Huq *et al.*, 2016). Genomics approaches offer comprehensive insights into genes associated with desirable qualities, their responses to changes in the environment, and their interactions with other critical features. To effectively target and optimize genetic options for crop improvement during climate change, it is essential to understand unique genotype-to-phenotype correlations (Bohra *et al.*, 2020; Mishra *et al.*, 2024).

The construction of pangenomes, along with the presence or lack of variation-based GWAS research, has enhanced the capability of genomic and genetic analyses for investigating genomic resources. Furthermore, genomic methods (GWAS, GS, MAS, and gene editing technologies) have been crucial in finding genetic variations associated with significant agronomic features. For instance, GWAS analyses have been conducted to investigate drought-related features in rice, sorghum, and sunflower (Ma *et al.*, 2016; Maina *et al.*, 2022; Wu *et al.*, 2022), and temperate-related traits in wheat and maize (Zhang *et al.*, 2020; Devate *et al.*, 2022), thus helping detect genes or QTLs that may regulate drought or heat stress, supporting the advancement of climate-resilient crops. Conversely, these investigations exhibit problems in delineating numerous alleles and significantly varied population patterns (Zhou and Huang, 2019). MAS is an essential method for creating climate-resilient crops that have enhanced drought tolerance by expediting the identification of plants with specific desirable genes (Azrai *et al.*, 2024). This is achieved with the use of molecular markers to discover advantageous features, such as enhanced roots or osmotic adjustment, in the plant DNA, circumventing the necessity for laborious conventional techniques, thereby enabling breeders to expedite the development of drought-resistant crops. To contribute to food security amidst a changing climate, GWAS has successfully been applied in maize to detect drought-tolerant loci (Tian and Qin, 2023). The gene editing tool CRISPR-Cas9 has also been used in editing tomato genes to offer drought tolerance (Larriba *et al.*, 2024).

By employing advanced genotyping and phenotyping, in conjunction with bioinformatics techniques, researchers can analyze complex quantitative traits into discrete genetic components (Varshney *et al.*, 2019). This facilitates the identification of QTLs and single-nucleotide polymorphisms (SNPs), as well as potential genes that govern substantial drought resistance. Through the utilization of high-throughput genotyping and phenotyping, alongside bioinformatics methodologies, researchers can deconstruct intricate quantitative traits into manageable genetic elements (Varshney *et al.*, 2018). This enables the discovery of QTLs, SNPs, and potential genes that regulate significant drought resistance. The extensive resources and information acquired through structural genetics are ultimately employed by functional genomics. Functional genomics, in conjunction with other sequencing-based omics technologies, elucidates the relationship connecting crop genomes and phenotypic traits under specific physiological conditions and environmental situations (Varshney *et al.*, 2019, 2021).

12.5.2 Transcriptomics approach

Transcriptomics enables the examination of gene expression in response to environmental stresses, including the regulatory networks implicated in plant adaptation to these stresses. RNA sequencing (RNA-seq) effectively identifies genes that respond to stresses. Consequently, this serves as an effective method for capturing dynamic transcriptional alterations in plants under environmental stress, facilitating the identification of crucial gene regulators that are either induced or inhibited in response to particular stimuli (Kumar *et al.*, 2022b). The application of whole transcriptome sequencing has identified genes associated with drought resistance in upland cotton (Zheng *et al.*, 2022) and common bean (Cai *et al.*, 2025). Moreover, transcriptome analysis has identified pathways and essential genes responsive to drought conditions in oat (Xu *et al.*, 2024), soybean (Li *et al.*, 2022), wheat (Niu *et al.*, 2023), and sweet potato (Cheng *et al.*, 2024), which set the pace for breeding for climate-smart crops. The combination of transcriptomic data with bioinformatic analyses enables researchers to elucidate complex controlling relationships and feedback loops, providing a systems-level perspective on plant stress biology. Comprehensive network models are essential for identifying key regulators and hub genes, which serve as potential targets for genetic modification to improve crop resilience to drought. RNA-seq is a crucial instrument for functional genomics, facilitating the identification of potential genes along with their regulatory networks pertinent for molecular breeding or genome editing aimed at enhancing agricultural stress resilience.

12.5.3 Proteomics approach

Proteins directly contribute to the formation of the plant's epigenome, transcriptome, and metabolome by executing structural or regulatory functions that affect plant phenotypes. Proteomics is the thorough examination of the entire set of proteins, known as the proteome, produced or modified in an organism at a specific time under defined conditions. Unlike the relatively stable DNA, the proteome is highly dynamic, reflecting the organism's physiological state, developmental stage, and environmental interactions. Proteins facilitate nearly all biological functions, particularly growth, development, signal transduction, and stress responses. In crop improvement, proteomics is crucial for understanding how genetic information generates complex molecular phenotypes. Proteomics provides essential insights into the regulation of metabolic processes, the activation of defense systems, and the modulation of developmental programs under diverse environmental situations (Liu *et al.*, 2019; Tian *et al.*, 2021; Martínez-Esteso *et al.*, 2024).

Proteomic study provides a comprehensive perspective on plant responses to stressful conditions at the protein level. Investigation into stress-responsiveness within plants is a critical pathway to the identification of genes that confer tolerance to stress and their application in breeding systems. The reactions of plants to water deficiency are more easily characterized using proteomics methods. A more robust defense against oxidative stress and an effective photosynthesis machinery are the factors that promote drought tolerance in plants. For example, proteomic analysis in soybean indicated that the plant may respond to arid stress by effectively implementing a stay-green mechanism by means of regulated gene expression throughout the vegetative stage (Yahoueian *et al.*, 2021). The drought resistance of plants is closely correlated with the morphology and physiological traits of their roots (Zarzyńska *et al.*, 2017), and analysis of these could be enhanced by proteomic analysis. The pivotal functions of phosphorylation pathways in coping with drought or ubiquitination-induced proteostasis at heat stress have been revealed by proteomic studies. This concentrates on post-translational change research and offers new targets for crop improvement interventions that are designed to optimize metabolic homeostasis, developmental regulation, and stress signaling pathways. The detection of differently expressed proteins associated with particular agronomic traits is facilitated by proteomic profiling across various developmental phases, tissues, or environmental treatments. For instance, drought-tolerant genotypes frequently exhibit upregulation of proteins that are involved in osmotic adjustment, antioxidative defense, and cellular repair mechanisms (Wang *et al.*, 2022), thus, establishing a theoretical foundation for the development of crops that are climate-smart.

12.5.4 Phenomics approach

Plant phenomics, enabled by cutting-edge genomics and AI technology, addresses the phenotype–genotype gap in the creation of climate-resilient crops. Phenomics is a powerful tool for analyzing complex features across spatial and temporal dimensions, encompassing a comprehensive examination of the morphological, physiological, and biochemical phenomes within plants. In addition to understanding and improving agricultural resilience to climate shocks, it is crucial to grasp plant–microbiome interactions, plant–environment relationships, and input utilization efficiency to boost crop productivity and ensure global food security. The latest developments in imaging techniques and high-throughput platforms for real-time monitoring of agricultural traits (such as drought and heat stress) in both controlled and open-field settings are outlined in phenotype analysis and AI technologies for understanding crop resilience. Morphological traits (size, shape, and architecture of the plant), physiological traits (nutrient uptake efficiency and stress tolerance of plants), and biochemical traits (nitrogen and chlorophyll content of plants) are identified by employing imaging sensors depending on various electromagnetic responses, which allow various phenotypic features to be extracted using the right sensors (Jiang and Zhu, 2024). For instance, remote sensing has been employed to monitor the reaction of plants to drought conditions (Maes and Steppe, 2019). The amalgamation of plant phenomics alongside cutting-edge technologies, such as the Internet of Things and blockchain, improves existing comprehensive strategies for monitoring and managing plant growth, particularly in regulated environmental conditions. Employing modern phenotyping methods and AI-based analytics allows researchers to accelerate genetic improvements, enhance crop resilience, and promote sustainable agriculture practices to secure future food supply (Miranda *et al.*, 2025). The demand for phenotypic information regarding root architecture and the dynamic characteristics of plant growth and development is significant. Additionally, there is a need for microscale phenotypic data to understand processes within organs, tissues, or cells. Furthermore, there is a call for the development of more economical and mobile phenotyping platforms to facilitate broader application (Yang *et al.*, 2020).

Multiple omics pipelines elucidate molecular processes, identify biomarkers, target genetic engineering, and map regulatory networks, providing precision agriculture solutions to enhance crop tolerance to drought stress and to ensure food security in changing environmental conditions. Various omics techniques have proven effective in improving drought tolerance in plants. Recent breakthroughs in omics technologies have facilitated the generation of various omics data, including transcriptomics, genomics, phenomics, and proteomics, consequently enhancing the understanding of plant reactions to drought stress.

12.6 Recent Advances in Spatial Multi-omics Technologies

Recent advancements in spatial biology, along with multi-omics, have created a novel avenue for exploring intricate biological systems. Conventional omics technologies, such as genomics, proteomics, metabolomics, and transcriptomics, offered comprehensive insights into cellular activities; yet, they have failed to provide a detailed picture of how cells function in a particular part of the organism. Spatial multi-omics has helped bridge this gap by mapping out cellular activities in great detail through the measurement of multiple molecular layers within intact tissues, thereby preserving spatial relationships among cells. Traditional methods like bulk sequencing and single-cell sequencing often lose this spatial context because they require breaking tissues apart. While single-cell methods have revolutionized our understanding of cellular diversity, spatial techniques have opened researchers' eyes to a broader perspective, where intact tissues can be analyzed and the natural architecture of cells well maintained. This has enabled the measurements of genes being expressed, abundance of proteins, or metabolite levels directly within tissue (Park *et al.*, 2022).

Advances in spatial multi-omics technologies have unveiled a wide range of tools/platforms, each designed to study different molecular layers with spatial precision (Fig. 12.2). These emerging tools are making it possible to combine multiple omics layers in a single analysis, offering a deeply integrated and spatial view of biology (Krull *et al.*, 2025). The tools/platforms provide different types of data depending on their resolution (how detailed the image is), throughput (how many features or samples can be analyzed at once), and modality (what type of molecule is being measured). While some offer near single-cell or subcellular detail, others rapidly scan larger tissue regions.

Typically, these instruments are grouped according to the biomolecules they detect (RNA, proteins, or metabolites) and by the spatial

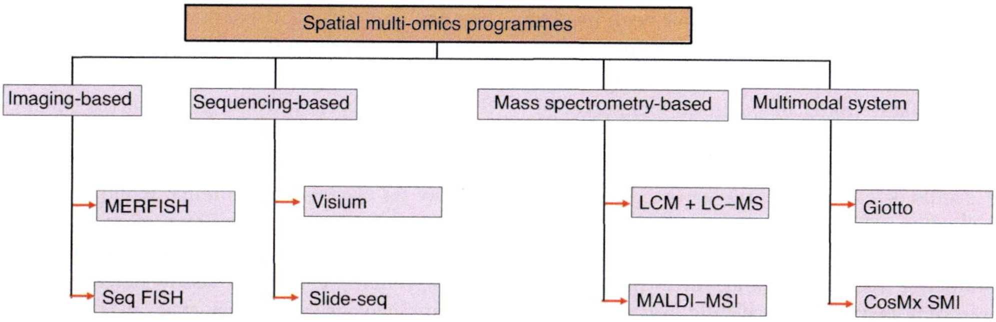

Fig. 12.2. Classification of spatial multi-omics platforms based on fundamental methods. MERFISH, multiplexed error-robust fluorescence *in situ* hybridization; FISH, fluorescence *in situ* hybridization; LC, liquid chromatography; LCM, laser capture microdissection; MALDI, matrix-assisted laser desorption/ionization; MS, mass spectrometry; MSI, mass spectrometry imaging; SMI, spatial molecular imager.

resolution they offer, which can range from whole-tissue overviews to subcellular mapping. Platforms such as spatial transcriptomics provide a detailed understanding of cellular behavior within organized tissues in their native spatial context. There are two main types of spatial transcriptomics tools: image based and sequencing based. Image-based techniques such as the MERFISH and seqFISH are noted for their ability to offer high spatial resolution, often down to the subcellular level, but ideally focusing on pre-selected genes. Sequencing-based techniques, such as 10×Genomics Visium and Slide-seq, however, provide genome-wide transcriptomic coverage but at a lower spatial resolution. Advances in the field have bridged this gap by combining both high-resolution and comprehensive gene coverage techniques. For instance, microfluidic array chip-based approaches have been developed to increase throughput and accuracy in spatial transcriptomic mapping to allow for scalable and spatially resolved RNA-seq directly from tissues without losing architectural information (Chen *et al.*, 2023a; Shi *et al.*, 2023). Additionally, spatial proteomics tools such as MIBI, imaging mass cytometry (IMC), and CODEX provide multiplexed protein imaging with high spatial resolution that facilitates the detection of proteins within tissue samples. Furthermore, spatial proteomics has been combined with mass spectrometry (MS) approaches for deeper protein profiling. The application of laser capture microdissection (LCM), followed by liquid chromatography–mass spectrometry (LC–MS), for example, has enabled the accurate selection and analysis of protein content from specific tissue regions. This integration, according to Fan (2024) and Lukowski *et al.* (2025), has facilitated the study of protein composition and microenvironments. The integration of key innovations and computational tools for spatial multi-omics applications has reformed and advanced research by fast-tracking discoveries in plants, human, and animal sciences and unveiling regulatory landscapes that were previously inaccessible (Krull *et al.*, 2025; Nobori, 2025).

Despite the extensive research and progress on existing spatial platforms, attaining high spatial resolution, deep molecular profiling, and high throughput simultaneously remains a looming challenge. Also, biological noise, tissue heterogeneity, and autofluorescence have made signal detection more difficult. Integrating large, multimodal datasets poses significant queries about data normalization, dimensionality reduction, and interpretability (Velten and Stegle, 2023). Increased/enhanced knowledge in the field, however, has facilitated the development of technology-agnostic computational ecosystems that enable the integration of data from different platforms. These tools align, analyze, and correct technical variations across multiple omics layers to generate a more comprehensive view of tissue organization and function. One such platform is the Giotto suite, which Chen *et al.* (2023b) reported to facilitate the generation of tissue maps by analyzing transcriptomic and protein signals together to reveal coordinated biological patterns. Again, AI and ML have been unified into spatial multi-omics systems to interpret complex datasets. These tools have been used to identify spatial gene expression patterns, classify tissue regions, predict cell–cell interactions, and even simulate biological systems under perturbation. Datasets in spatial multi-omics are usually affected by autofluorescence, background noise, or events such as missing gene counts. The application of deep learning algorithms, for example, has proved useful for denoising data, enhancing resolution, and inferring missing molecular features. Specifically, auto encoders and denoising convolutional neural networks (CNNs) deep learning models have helped reduce noise, improve signal-to-noise ratio, and preserve biological relevance (Shao *et al.*, 2024). AI has been essential in identifying spatially structured gene and protein expression patterns inside tissues, facilitating valuable translational applications like precision pathology, where these spatial patterns can guide diagnostic and therapeutic choices.

The field continues to face challenges in standardization, despite recent advancements. There is still a lack of common data formats, benchmarking protocols, and a curated reference atlas. As the volume and diversity of spatial omics data grow, efforts to establish community standards and open-access databases will be critical to ensure reproducibility and interoperability across studies (Velten and Stegle, 2023). The key to maximizing spatial multi-omics lies in the implementation of computational innovations to examine biological systems with new depth and clarity through the integration of several molecular layers in a spatially resolved way and the use of AI to analyze results. As spatial datasets become increasingly accessible, their integration into precision agriculture is poised to expand. For example, the technology has gained much interest in plant biology, where it aids in investigating tissue-specific responses to stress and pathogen interactions (Nobori, 2025). This progress may soon enable routine spatial profiling of samples, enhancing diagnostic accuracy and precision in management strategies across diverse diseases.

While spatial multi-omics technologies have revolutionized research by enabling high-resolution molecular mapping within tissues, several key challenges hinder their widespread adoption and application due to technical limitations, data integration complexities, analytical hurdles, and practical barriers to implementation. A persistent challenge has been the trade-off between resolution, molecular coverage, and throughput across existing platforms. Some techniques provide single-cell or subcellular spatial resolution, but are limited to profiling a narrow set of genes or proteins, while methods offering genome-scale profiling often sacrifice spatial granularity, making it difficult to capture a comprehensive view of tissue heterogeneity in a single experiment (Park *et al.*, 2022; Velten and Stegle, 2023). Additionally, inconsistencies in sample preparation, detection chemistry, and imaging protocols across platforms complicate cross-study comparisons and multi-omic data integration (Chen *et al.*, 2023a). Although computational tools like the Giotto Suite facilitate data harmonization, merging disparate datasets remains technically demanding and computationally intensive (Fan, 2024). Analytical and computational bottlenecks arising from high volume and complex spatial multi-omics datasets, which combine molecular profiles with precise spatial coordinates, pose significant analytical challenges. Many laboratories still rely on custom bioinformatics pipelines or manual curation, limiting reproducibility and scalability (Lukowski *et al.*, 2025). Again, the field lacks standardized analytical frameworks, requiring interdisciplinary expertise in both biology and computational science to derive meaningful insights (Velten and Stegle, 2023). The high cost associated with instrumentation, reagent consumption, and data storage also restricts the adoption of spatial multi-omics, particularly in resource-limited settings. The significant computational infrastructure required for data processing further exacerbates disparities in access, confining advanced applications primarily to well-funded institutions (Park *et al.*, 2022). The absence of a comprehensive reference atlas for healthy tissues also complicates the interpretation of disease-associated spatial alterations (Chen *et al.*, 2023a; Nobori, 2025).

Despite these challenges, ongoing advancements in instrumentation, computational tools, and protocol standardization are steadily addressing these limitations. As technologies advance and costs decline, spatial multi-omics is poised to transition from a specialized research tool to a more widely accessible resource, unlocking new opportunities in precision agriculture.

A major focus should be on developing integrated platforms capable of analyzing multiple molecular layers, such as RNA, proteins, lipids, and metabolites, within single tissue sections while maintaining spatial resolution. Emerging technologies combining imaging, sequencing, and MS approaches should offer comprehensive characterization of tissue microenvironment, to provide unprecedented insights into cellular interactions and tissue function, to overcome current limitations in molecular coverage and analytical resolution, and to broaden applications from basic research (Fan, 2024; Lukowski *et al.*, 2025). AI will persist in its essential function of evaluating spatial data, with ML tools assisting in identifying patterns within large and complex datasets that humans might overlook. In the future, AI could be employed not only for data analysis but also in experimental design and predicting outcomes based on spatial molecular signatures (Shao *et al.*, 2024). More importantly, future spatial multi-omics technologies should be more affordable and accessible. As current technologies remain costly, slow, and require specialized personnel, greater efforts should be directed towards developing new kits and protocols that are easy to adopt and adapt, helping to integrate spatial omics into routine diagnostics and management strategies (Krull *et al.*, 2025). As spatial multi-omics applications expand into new areas like plant biology and environmental science, they should be employed to elucidate how plant roots respond to drought or how microbial communities interact in the soil, addressing major challenges in food security and sustainability (Nobori, 2025). The creation of a spatial reference atlas for various tissues and diseases will be essential as reference tools for comparing healthy and diseased tissues/organs to guide AI models and foster a better understanding of tissue development and disease progression (Velten and Stegle, 2023). Overall, the future of spatial multi-omics should focus more on increased integration, automation, and widespread application to establish it as a key technology for modern science.

12.7 Conclusion

Omics strategies have transformed crop enhancement by offering robust tools to comprehend and modify plant genomes underlying protein activities. This enables the identification of critical genes associated with beneficial traits, including drought resistance and yield enhancement, hence improving MAS, molecular markers, and genome-editing techniques. Proteomics elucidates post-translational modifications and protein interactions, especially stress responses that link genetic details to phenotypic expression. The integration of omics tools enables a systems biology approach, enhancing precise breeding and the development of climate-resilient, nutrient-rich crops. Advancements in NGS, bioinformatics, and advanced proteomic techniques have accelerated the discovery of novel biomarkers and metabolic pathways, promoting sustainable agriculture. Nevertheless, challenges such as data integration, ethical concerns, and the significant costs associated with omics technology remain. Future research should focus on the integration of multi-omics data, the utilization of ML, and translational genomics to maximize agricultural improvement efforts. By employing genomics and proteomics, scientists can address global food security challenges, mitigate environmental impacts, and meet the demands of a growing population. Ultimately, these new technologies have substantial potential to transform agriculture, ensuring resilient and high-yielding crop varieties for future generations. Continuous funding and a multidisciplinary approach will be essential to fully realize its benefits in crop science.

References

Abou-Hussein, S.D. (2012) Climate change and its impact on the productivity and quality of vegetable crops. *Journal of Applied Sciences Research* 8, 4359–4383.

Ahmed, S., Muhammad, I., Kumar, S., Malhotra, R. and Maalouf, F. (2010) Impact of climate change and variability on diseases of food legumes in the dry areas. In: Solh, M. and Saxena, M.C. (eds)

International conference on Food Security and Climate Change in Dry Areas, Amman, Jordan, pp. 157–165.

Araus, J.L. and Cairns, J.E. (2014) Field high-throughput phenotyping: The new crop breeding frontier. *Trends in Plant Science* 19(1), 52–61. DOI: 10.1016/j.tplants.2013.09.008.

Ayyogari, K., Sidhya, P. and Pandit, M.K. (2014) Impact of climate change on vegetable cultivation - A review. *International Journal of Agriculture, Environment and Biotechnology* 7(1), 145. DOI: 10.5958/j.2230-732X.7.1.020.

Azrai, M., Bahrun, A.H., Efendi, R., Andayani, N.N., Jihad, M. *et al.* (2024) Global drought tolerant maize research and development: Analysis and visualization of cutting-edge scientific technologies. *Journal of Agriculture and Food Research* 18, 101323. DOI: 10.1016/j.jafr.2024.101323.

Bailey-Serres, J., Parker, J.E., Ainsworth, E.A., Oldroyd, G.E.D. and Schroeder, J.I. (2019) Genetic strategies for improving crop yields. *Nature* 575(7781), 109–118. DOI: 10.1038/s41586-019-1679-0.

Bal, S. and Sharangi, A.B. (2025) Speed breeding: A "paradigm shift" in crop improvement. In: *Ecologically Mediated Development: Promoting Biodiversity Conservation and Food Security*. Springer Nature, Singapore, pp. 589–607.

Begna, T. (2022) Speed breeding to accelerate crop improvement. *International Journal of Agricultural Science and Food Technology* 8(2), 178–186. DOI: 10.17352/2455-815X.000161.

Beillouin, D., Schauberger, B., Bastos, A., Ciais, P. and Makowski, D. (2020) Impact of extreme weather conditions on European crop production in 2018. *Philosophical Transactions of the Royal Society B* 375(1810), 20190510. DOI: 10.1098/rstb.2019.0510.

Beltran-Peña, A., Rosa, L. and D'Odorico, P. (2020) Global food self-sufficiency in the 21st century under sustainable intensification of agriculture. *Environmental Research Letters* 15(9), 095004. DOI: 10.1088/1748-9326/ab9388.

Bettoni, M.M., Mogor, Á.F., Pauletti, V. and Goicoechea, N. (2014) Growth and metabolism of onion seedlings as affected by the application of humic substances, mycorrhizal inoculation and elevated Co_2. *Scientia Horticulturae* 180, 227–235. DOI: 10.1016/j.scienta.2014.10.037.

Bhattacharjee, P., Warang, O., Das, S. and Das, S. (2022) Impact of climate change on fruit crops- A review. *Current World Environment* 17(2), 319–330. DOI: 10.12944/CWE.17.2.4.

Bisbis, M.B., Gruda, N. and Blanke, M. (2018) Potential impacts of climate change on vegetable production and product quality – A review. *Journal of Cleaner Production* 170, 1602–1620. DOI: 10.1016/j.jclepro.2017.09.224.

Bishop, J., Potts, S.G. and Jones, H.E. (2016) Susceptibility of faba bean (*Vicia faba* L.) to heat stress during floral development and anthesis. *Journal of Agronomy and Crop Science* 202, 508–517. DOI: 10.1111/jac.12172.

Bohra, A., Chand Jha, U., Godwin, I.D. and Kumar Varshney, R. (2020) Genomic interventions for sustainable agriculture. *Plant Biotechnology Journal* 18(12), 2388–2405. DOI: 10.1111/pbi.13472.

Boughton, B.A., Thinagaran, D., Sarabia, D., Bacic, A. and Roessner, U. (2016) Mass spectrometry imaging for plant biology: A review. *Phytochemistry Reviews* 15(3), 445–488. DOI: 10.1007/s11101-015-9440-2.

Cai, D., Zhang, Y., Ding, D., Yang, S., Zhang, J. *et al.* (2025) Whole-transcriptome RNA sequencing reveals the global molecular responses in response to drought stress in common bean (*Phaseolus vulgaris* L.). *Euphytica: Netherlands Journal of Plant Breeding* 221(6), 79. DOI: 10.1007/s10681-025-03535-2.

Castillejo, M.A., Pascual, J., Jorrín-Novo, J.V. and Balbuena, T.S. (2023) Proteomics research in forest trees: A 2012-2022 update. *Frontiers in Plant Science* 14, 1130665. DOI: 10.3389/fpls.2023.1130665.

Challinor, A.J., Watson, J., Lobell, D.B., Howden, S.M., Smith, D.R. *et al.* (2014) A meta-analysis of crop yield under climate change and adaptation. *Nature Climate Change* 4(4), 287–291. DOI: 10.1038/nclimate2153.

Chen, J.G., Chávez-Fuentes, J.C., O'Brien, M., Xu, J., Ruiz, E. *et al.* (2023a) Giotto suite: A multiscale and technology-agnostic spatial multi-omics analysis ecosystem. *Bioinformatics*. DOI: 10.1101/2023.11.26.568752.

Chen, K., Wang, Y., Zhang, R., Zhang, H. and Gao, C. (2019) CRISPR/Cas genome editing and precision plant breeding in agriculture. *Annual Review of Plant Biology* 70(1), 667–697. DOI: 10.1146/annurev-arplant-050718-100049.

Chen, T.-Y., You, L., Hardillo, J.A.U. and Chien, M.-P. (2023b) Spatial transcriptomic technologies. *Cells* 12, 16. DOI: 10.3390/cells12162042.

Cheng, P., Kong, F., Han, Y., Liu, X. and Xia, J. (2024) Transcriptome analysis reveals genes and pathways associated with drought tolerance of early stages in sweet potato (*Ipomoea batatas* (L.) Lam). *Genes* 15(7), 948. DOI: 10.3390/genes15070948.

Craufurd, P. and Peacock, J. (1993) Effect of heat and drought stress on sorghum bicolor grain yield. *Experimental Agriculture* 29, 77–86. DOI: 10.1017/S0014479700020421.

Crossa, J., Fritsche-Neto, R., Montesinos-Lopez, O.A., Costa-Neto, G., Dreisigacker, S. *et al*. (2021) The modern plant breeding triangle: Optimizing the use of genomics, phenomics, and enviromics data. *Frontiers in Plant Science* 12, 651480. DOI: 10.3389/fpls.2021.651480.

Daryanto, S., Wang, L. and Jacinthe, P.-A. (2016) Global synthesis of drought effects on maize and wheat production. *PloS One* 11(5), e0156362. DOI: 10.1371/journal.pone.0156362.

Dave, K., Kumar, A., Dave, N., Jain, M., Dhanda, P.S. *et al*. (2024) Climate change impacts on legume physiology and ecosystem dynamics: A multifaceted perspective. *Sustainability* 16(14), 6026. DOI: 10.3390/su16146026.

Deutsch, C.A., Tewksbury, J.J., Tigchelaar, M., Battisti, D.S., Merrill, S.C. *et al*. (2018) Increase in crop losses to insect pests in a warming climate. *Science* 361(6405), 916–919. DOI: 10.1126/science.aat3466.

Devate, N.B., Krishna, H., Parmeshwarappa, S.K.V., Manjunath, K.K., Chauhan, D. *et al*. (2022) Genome-wide association mapping for component traits of drought and heat tolerance in wheat. *Frontiers in Plant Science* 13, 943033. DOI: 10.3389/fpls.2022.943033.

Driedonks, N., Rieu, I. and Vriezen, W.H. (2016) Breeding for plant heat tolerance at vegetative and reproductive stages. *Plant Reproduction* 29(1–2), 67–79. DOI: 10.1007/s00497-016-0275-9.

Dumitru, E.A., Berevoianu, R.L., Tudor, V.C., Teodorescu, F.-R., Stoica, D. *et al*. (2023) Climate change impacts on vegetable crops: A systematic review. *Agriculture* 13(10), 1891. DOI: 10.3390/agriculture13101891.

Erickson, A.N. and Markhart, A.H. (2002) Flower developmental stage and organ sensitivity of bell pepper (*Capsicum annuum* L.) to elevated temperature. *Plant, Cell & Environment* 25(1), 123–130. DOI: 10.1046/j.0016-8025.2001.00807.x.

Fan, R. (2024) Integrative spatial protein profiling with multi-omics. *Nature Methods* 21(12), 2223–2225. DOI: 10.1038/s41592-024-02533-x.

FAO (2021) State of the world's land and water resources for food and agriculture (SOLAW) – systems at breaking point. *Food and Agriculture Organization of the United Nations, Rome*. DOI: 10.4060/cb7654en.

Farooq, M., Nadeem, F., Gogoi, N., Ullah, A., Alghamdi, S.S. *et al*. (2017) Heat stress in grain legumes during reproductive and grain-filling phases. *Crop and Pasture Science* 68, 985–1005. DOI: 10.1071/CP17012.

Farooq, M.S., Uzair, M., Raza, A., Habib, M., Xu, Y. *et al*. (2022) Uncovering the research gaps to alleviate the negative impacts of climate change on food security: A review. *Frontiers in Plant Science* 13, 927535. DOI: 10.3389/fpls.2022.927535.

Fischer, G., Melgarejo, L.M. and Balaguera-López, H.E. (2022) Review on the impact of elevated CO_2 concentrations on fruit species in the face of climate change. *Ciencia & Tecnología Agropecuaria* 23(2). DOI: 10.21930/rcta.vol23_num2_art:2475.

Gepts, P. (2004) Crop domestication as a long-term selection experiment. *Plant Breeding Reviews* 24, 1–44. DOI: 10.1002/9780470650288.

Gupta, A., Rico-Medina, A. and Caño-Delgado, A.I. (2020) The physiology of plant responses to drought. *Science* 368(6488), 266–269. DOI: 10.1126/science.aaz7614.

Hall, C., Dawson, T.P., Macdiarmid, J.I., Matthews, R.B. and Smith, P. (2017) The impact of population growth and climate change on food security in Africa: Looking ahead to 2050. *International Journal of Agricultural Sustainability* 15(2), 124–135. DOI: 10.1080/14735903.2017.1293929.

Haokip, S.W., Shankar, K. and Lalrinngheta, J. (2020) Climate change and its impact on fruit crops. *Journal of Pharmacognosy and Phytochemistry* 9(1), 435–438.

Hemathilake, D.M.K.S. and Gunathilake, D.M.C.C. (2022) Agricultural productivity and food supply to meet increased demands. In: *Future Foods*. Academic Press, pp. 539–553.

Huq, Md., Akter, S., Jung, Y.-J., Nou, I.S., Cho, Y.-G. *et al*. (2016) Genome sequencing, a milestone for genomic research and plant breeding. *Plant Breeding and Biotechnology* 4(1), 29–39. DOI: 10.9787/PBB.2016.4.1.29.

Jaganathan, D., Bohra, A., Thudi, M. and Varshney, R.K. (2020) Fine mapping and gene cloning in the post-NGS era: Advances and prospects. *Theoretical and Applied Genetics* 133(5), 1791–1810. DOI: 10.1007/s00122-020-03560-w.

Jiang, N. and Zhu, X.G. (2024) Modern phenomics to empower holistic crop science, agronomy, and breeding research. *Journal of Genetics and Genomics* 51(8), 790–800. DOI: 10.1016/j.jgg.2024.04.016.

Jin, C., Du, S., Wang, Y., Condon, J., Lin, X. *et al.* (2009) Carbon dioxide enrichment by composting in greenhouses and its effect on vegetable production. *Journal of Plant Nutrition and Soil Science* 172(3), 418–424. DOI: 10.1002/jpln.200700220.

Karagatiya, F., Patel, S., Parasana, J., Vasava, H., Chaudhari, T.M. *et al.* (2023) Adapting fruit crops to climate change: Strengthening resilience and implementing adaptation measures in fruit crop. *Journal of Pharmaceutical Innovation* 12, 3159–3164.

Khan, M.N., Wahab, S., Wahid, N., Shah, S.N., Ullah, B. *et al.* (2025) Impact of climate change on yield and quality of legumes. In: *Challenges and Solutions of Climate Impact on Agriculture*. Academic Press, Massachusetts, USA, pp. 85–111. DOI: 10.1016/B978-0-443-23707-2.00004-0.

Kidane, B., Urugo, M.M., Hirpha, H.H., Paulos, T., Hundea, W. *et al.* (2025) Nutritional challenges of staple crops due to increasing atmospheric carbon dioxide levels: Case of Sub-saharan Africa. *Journal of Agriculture and Food Research* 19, 101592. DOI: 10.1016/j.jafr.2024.101592.

Kim, K.H. and Lee, B.M. (2023) Effects of climate change and drought tolerance on maize growth. *Plants* 12(20), 3548. DOI: 10.3390/plants12203548.

Korres, N.E., Norsworthy, J.K., Tehranchian, P., Gitsopoulos, T.K., Loka, D.A. *et al.* (2016) Cultivars to face climate change effects on crops and weeds: A review. *Agronomy for Sustainable Development* 36(1), 12. DOI: 10.1007/s13593-016-0350-5.

Krull, D., Haynes, P., Kesarwani, A., Tessier, J., Chen, B.J. *et al.* (2025) A best practices framework for spatial biology studies in drug discovery and development: Enabling successful cohort studies using digital spatial profiling. *Journal of Histotechnology* 48(1), 7–26. DOI: 10.1080/01478885.2024.2391683.

Kuden, A.B. (2018) Climate change affects fruit crops. In: *XXX International Horticultural Congress IHC2018: International Symposium on Cultivars, Rootstocks and Management Systems of 1281*, Belgium, pp. 437–440. DOI: 10.17660/ActaHortic.2020.1281.57.

Kumar, M. (2016) Impact of climate change on crop yield and role of model for achieving food security. *Environmental Monitoring and Assessment* 188(8), 465. DOI: 10.1007/s10661-016-5472-3.

Kumar, A., Sandhu, K.S. and Joshi, S. (2022a) Applications of artificial intelligence and machine learning for crop improvement: Current trends and future perspectives. *Frontiers in Plant Science* 13, 817103. DOI: 10.3389/fpls.2022.817103.

Kumar, S., Seem, K., Kumar, S. and Mohapatra, T. (2022b) RNA-seq analysis reveals the genes/pathways responsible for genetic plasticity of rice to varying environmental conditions on direct-sowing and transplanting. *Scientific Reports* 12(1), 2241. DOI: 10.1038/s41598-022-06009-w.

Lal, R. (2009) Soil degradation as a reason for inadequate human nutrition. *Food Security* 1(1), 45–57. DOI: 10.1007/s12571-009-0009-z.

Larriba, E., Yaroshko, O. and Pérez-Pérez, J.M. (2024) Recent advances in tomato gene editing. *International Journal of Molecular Sciences* 25(5), 2606. DOI: 10.3390/ijms25052606.

Lee, H., Park, J.W. and Seo, P.J. (2023) Spatial multi-omics in plants: Technologies and applications. *Trends in Plant Science* 28(3), 234–247. DOI: 10.1016/j.tplants.2022.10.006.

Li, M., Li, H., Sun, A., Wang, L., Ren, C. *et al.* (2022) Transcriptome analysis reveals key drought-stress-responsive genes in soybean. *Frontiers in Genetics* 13, 1060529. DOI: 10.3389/fgene.2022.1060529.

Li, X., Li, B., Gu, S., Pang, X., Mason, P. *et al.* (2024) Single-cell and spatial RNA sequencing reveal the spatiotemporal trajectories of fruit senescence. *Nature Communications* 15(1), 3108. DOI: 10.1038/s41467-024-47329-x.

Libbrecht, M.W. and Noble, W.S. (2015) Machine learning applications in genetics and genomics. *Nature Reviews. Genetics* 16(6), 321–332. DOI: 10.1038/nrg3920.

Liu, Y., Lu, S., Liu, K., Wang, S., Huang, L. *et al.* (2019) Proteomics: A powerful tool to study plant responses to biotic stress. *Plant Methods* 15(1), 135. DOI: 10.1186/s13007-019-0515-8.

Lukowski, J.K., Cho, B., Calderon, A.Z., Dianati, B., Stumpo, K. *et al.* (2025) Advances in spatial multi-omics: A review of multi-modal mass spectrometry imaging and laser capture microdissection-LCMS integration. *PROTEOMICS* 25(21–22), 151–159. DOI: 10.1002/pmic.202400378.

Ma, X., Feng, F., Wei, H., Mei, H., Xu, K. *et al.* (2016) Genome-wide association study for plant height and grain yield in rice under contrasting moisture regimes. *Frontiers in Plant Science* 7, 1801. DOI: 10.3389/fpls.2016.01801.

Maes, W.H. and Steppe, K. (2019) Perspectives for remote sensing with unmanned aerial vehicles in precision agriculture. *Trends in Plant Science* 24(2), 152–164. DOI: 10.1016/j.tplants.2018.11.007.

Maina, F., Harou, A., Hamidou, F. and Morris, G.P. (2022) Genome-wide association studies identify putative *Pleiotropic locus* mediating drought tolerance in sorghum. *Plant Direct* 6(6), e413. DOI: 10.1002/pld3.413.

Malhotra, S.K. (2017) Horticultural crops and climate change: A review. *The Indian Journal of Agricultural Sciences* 87(1), 12–22. DOI: 10.56093/ijas.v87i1.67138.

Martínez-Esteso, M.J., Morante-Carriel, J., Samper-Herrero, A., Martínez-Márquez, A., Sellés-Marchart, S. *et al.* (2024) Proteomics: An essential tool to study plant-specialized metabolism. *Biomolecules* 14(12), 1539. DOI: 10.3390/biom14121539.

Mattos, L.M., Moretti, C.L., Jan, S., Sargent, S.A., Lima, C.E.P. *et al.* (2014) Climate changes and potential impacts on quality of fruit and vegetable crops. In: *Emerging Technologies and Management of Crop Stress Tolerance*. Academic Press, Massachusetts, USA, pp. 467–486. DOI: 10.1016/B978-0-12-800876-8.00019-9.

Mehmood, M., Tanveer, N.A., Joyia, F.A., Ullah, I. and Mohamed, H.I. (2025) Effect of high temperature on pollen grains and yield in economically important crops: A review. *Planta* 261(6), 141. DOI: 10.1007/s00425-025-04714-0.

Miranda, M.C. de C., Aono, A.H., Fagundes, T.G., Arduini, G.M. and Pinheiro, J.B. (2025) High-throughput phenotyping and machine learning techniques in soybean breeding: Exploring the potential of aerial imaging and vegetation indices. *Agronomy Journal* 117(1), e70012. DOI: 10.1002/agj2.70012.

Mishra, S., Srivastava, A.K., Khan, A.W., Tran, L.S.P. and Nguyen, H.T. (2024) The era of panomics-driven gene discovery in plants. *Trends in Plant Science* 29(9), 995–1005. DOI: 10.1016/j.tplants.2024.03.007.

Montesinos-López, O.A., Montesinos-López, A., Singh, R.P., Crossa, J. and Rutkoski, J. (2022) Applications of machine learning in plant breeding. *Plant Breeding Reviews* 45, 247–302. DOI: 10.1002/9781119876530.ch6.

Moose, S.P. and Mumm, R.H. (2008) Molecular plant breeding as the foundation for 21st century crop improvement. *Plant Physiology* 147(3), 969–977. DOI: 10.1104/pp.108.118232.

Moretti, C.L., Mattos, L.M., Calbo, A.G. and Sargent, S.A. (2010) Climate changes and potential impacts on postharvest quality of fruit and vegetable crops: A review. *Food Research International* 43(7), 1824–1832. DOI: 10.1016/j.foodres.2009.10.013.

Myers, S.S., Zanobetti, A., Kloog, I., Huybers, P., Leakey, A.D.B. *et al.* (2014) Increasing CO_2 threatens human nutrition. *Nature* 510(7503), 139–142. DOI: 10.1038/nature13179.

Najafi, E., Pal, I. and Khanbilvardi, R. (2019) Climate drives variability and joint variability of global crop yields. *Science of The Total Environment* 662, 361–372. DOI: 10.1016/j.dib.2019.103745.

Neumann, M., Xu, X., Smaczniak, C., Schumacher, J., Yan, W. *et al.* (2022) A 3D gene expression atlas of the floral meristem based on spatial reconstruction of single nucleus RNA sequencing data. *Nature Communications* 13(1), 2838. DOI: 10.1038/s41467-022-30177-y.

Newton, A.C., Johnson, S.N. and Gregory, P.J. (2011) Implications of climate change for diseases, crop yields and food security. *Euphytica* 179(1), 3–18. DOI: 10.1007/s10681-011-0359-4.

Niu, Y., Li, J., Sun, F., Song, T., Han, B. *et al.* (2023) Comparative transcriptome analysis reveals the key genes and pathways involved in drought stress response of two wheat (*Triticum aestivum* L.) varieties. *Genomics* 115(5), 110688. DOI: 10.1016/j.ygeno.2023.110688.

Nobori, T. (2025) Exploring the untapped potential of single-cell and spatial omics in plant biology. *New Phytologist* 247(3), 1098–1116. DOI: 10.1111/nph.70220.

Nunes, L.J. (2023) The rising threat of atmospheric CO_2: A review on the causes, impacts, and mitigation strategies. *Environments* 10(4), 66. DOI: 10.3390/environments10040066.

Oo, K.S., Krishnan, S.G., Vinod, K.K., Dhawan, G., Dwivedi, P. *et al.* (2021) Molecular breeding for improving productivity of *Oryza sativa* L. cv. Pusa 44 under reproductive stage drought stress through introgression of a major QTL, qDTY12.1. *Genes* 12(7), 967. DOI: 10.3390/genes12070967.

Panda, A. (2018) Transformational adaptation of agricultural systems to climate change. *Wiley Interdisciplinary Reviews: Climate Change* 9(4), e520. DOI: 10.1002/wcc.520.

Park, J., Kim, J., Lewy, T., Rice, C.M., Elemento, O. *et al.* (2022) Spatial omics technologies at multimodal and single cell/subcellular level. *Genome Biology* 23(1), 256. DOI: 10.1186/s13059-022-02824-6.

Pereira, L. (2017) Climate change impacts on agriculture across Africa. *Oxford Research Encyclopedia of Environmental Science*. DOI: 10.1093/acrefore/9780199389414.013.292.

Porter, J.R., Challinor, A.J., Henriksen, C.B., Howden, S.M., Martre, P. *et al.* (2019) Invited review: Intergovernmental panel on climate change, agriculture, and food—A case of shifting cultivation and history. *Global Change Biology* 25(8), 2518–2529. DOI: 10.1111/gcb.14700.

Rahman, M.M., Alam, M.S., Islam, M.M., Kamal, M.Z.U., Rahman, G.M. *et al.* (2022) Potential of legume-based cropping systems for climate change adaptation and mitigation. In: *Advances in Legumes for Sustainable Intensification*. Academic Press, Massachusetts, USA, pp. 381–402. DOI: 10.1016/B978-0-323-85797-0.00030-6.

Rai, A., Saito, K. and Yamazaki, M. (2021) Integrated omics analysis of specialized metabolism in medicinal plants. *Plant Journal* 106(2), 429–441. DOI: 10.1111/tpj.15201.

Rajan, R., Feza, M., Pandey, K., Aman, A. and Kumar, V. (2020) *Climate Change and Resilience in Fruit Crops*. Biotec Books, India.

Ray, D.K., Gerber, J.S., MacDonald, G.K. and West, P.C. (2015) Climate variation explains a third of global crop yield variability. *Nature Communications* 6(1), 5989. DOI: 10.1038/ncomms6989.

Rockström, J., Steffen, W., Noone, K., Persson, A., Chapin, F.S. *et al.* (2009) A safe operating space for humanity. *Nature* 461, 472–475. DOI: 10.1038/461472a.

Ryu, K.H., Huang, L., Kang, H.M. and Schiefelbein, J. (2019) Single-cell RNA sequencing resolves molecular relationships among individual plant cells. *Plant Physiology* 179(4), 1444–1456. DOI: 10.1104/pp.18.01482.

Saeed, F., Chaudhry, U.K., Raza, A., Charagh, S., Bakhsh, A. *et al.* (2023) Developing future heat-resilient vegetable crops. *Functional & Integrative Genomics* 23(1), 47. DOI: 10.1007/s10142-023-00967-8.

Shamshiri, R.R., Jones, J.W., Thorp, K.R., Ahmad, D., Man, H.C. *et al.* (2018) Review of optimum temperature, humidity, and vapour pressure deficit for microclimate evaluation and control in greenhouse cultivation of tomato: A review. *International Agrophysics* 32(2), 287–302. DOI: 10.1515/intag-2017-0005.

Shao, Y., Lv, X., Ying, S. and Guo, Q. (2024) Artificial intelligence-driven precision medicine: Multi-omics and spatial multi-omics approaches in diffuse large B-cell lymphoma (DLBCL). *Frontiers in Bioscience-Landmark* 29(12), 404. DOI: 10.31083/j.fbl2912404.

Shi, J., Pan, Y., Liu, X., Cao, W., Mu, Y. *et al.* (2023) Spatial omics sequencing based on microfluidic array chips. *Biosensors* 13(7), 712. DOI: 10.3390/bios13070712.

Ståhl, P.L., Salmén, F., Vickovic, S., Lundmark, A., Navarro, J.F. *et al.* (2016) Visualization and analysis of gene expression in tissue sections by spatial transcriptomics. *Science (New York, N.Y.)* 353(6294), 78–82. DOI: 10.1126/science.aaf2403.

Tester, M. and Langridge, P. (2010) Breeding technologies to increase crop production in a changing world. *Science (New York, N.Y.)* 327(5967), 818–822. DOI: 10.1126/science.1183700.

Tian, T. and Qin, F. (2023) CIMBL55: A repository for maize drought resistance alleles. *Stress Biology* 3(1), 13. DOI: 10.1007/s44154-023-00091-4.

Tian, X., Zhu, L., Yang, N., Song, J., Zhao, H. *et al.* (2021) Proteomics and metabolomics reveal the regulatory pathways of ripening and quality in post-harvest kiwifruits. *Journal of Agricultural and Food Chemistry* 69(2), 824–835. DOI: 10.1021/acs.jafc.0c05492.

Tirado, M.C., Clarke, R., Jaykus, L.A., McQuatters-Gollop, A. and Frank, J.M. (2010) Climate change and food safety: A review. *Food Research International* 43(7), 1745–1765. DOI: 10.1016/j.foodres.2010.07.003.

Ul Hassan, M., Rasool, T., Iqbal, C., Arshad, A., Abrar, M. *et al.* (2022) Linking plants functioning to adaptive responses under heat stress conditions: A mechanistic review. *Journal of Plant Growth Regulation* 41(7), 2596–2613. DOI: 10.1007/s00344-021-10493-1.

Vadez, V., Berger, J.D., Warkentin, T., Asseng, S., Ratnakumar, P. *et al.* (2012) Adaptation of grain legumes to climate change: A review. *Agronomy for Sustainable Development* 32(1), 31–44. DOI: 10.1007/s13593-011-0020-6.

Varshney, R.K., Singh, V.K., Kumar, A., Powell, W. and Sorrells, M.E. (2018) Can genomics deliver climate-change ready crops. *Current Opinion in Plant Biology* 45, 205–211. DOI: 10.1016/j.pbi.2018.03.007.

Varshney, R.K., Pandey, M.K., Bohra, A., Singh, V.K., Thudi, M. *et al.* (2019) Toward the sequence-based breeding in legumes in the post-genome sequencing era. *Theoretical and Applied Genetics* 132(3), 797–816. DOI: 10.1007/s00122-018-3252-x.

Varshney, R.K., Bohra, A., Yu, J., Graner, A., Zhang, Q. *et al.* (2021) Designing future crops: Genomics-assisted breeding comes of age. *Trends in Plant Science* 26(6), 631–649. DOI: 10.1016/j.tplants.2021.03.010.

Velten, B. and Stegle, O. (2023) Principles and challenges of modeling temporal and spatial omics data. *Nature Methods* 20(10), 1462–1474. DOI: 10.1038/s41592-023-01992-y.

Verma, K.K., Song, X., Kumari, A., Jagadesh, M., Singh, S.K. *et al.* (2025) Climate change adaptation: Challenges for agricultural sustainability. *Plant, Cell & Environment* 48(4), 2522–2533. DOI: 10.1111/pce.15078.

Vickovic, S., Lötstedt, B., Klughammer, J., Mages, S., Segerstolpe, Å. *et al.* (2022) SM-omics is an automated platform for high-throughput spatial multi-omics. *Nature Communications* 13(1), 795. DOI: 10.1038/s41467-022-28445-y.

Wang, J., Yang, W., Zhang, S., Hu, H., Yuan, Y. *et al.* (2023) A pangenome analysis pipeline provides insights into functional gene identification in rice. *Genome Biology* 24(1), 19. DOI: 10.1186/s13059-023-02861-9.

Wang, X., Wu, Z., Yan, C., Ma, C. and Dong, S. (2022) Proteomics analysis of soybean seedlings under short-term water deficit. *Phyton* 91(7), 1381–1401. DOI: 10.32604/phyton.2022.020251.

Wang, X., Zhang, S., Li, H., Du Laing, G., Odlare, M. *et al.* (2025) Elevated CO_2 decreases micronutrient Zn but not Fe in vegetables – Evidence from a meta-analysis. *Frontiers in Plant Science* 16, 1509102. DOI: 10.3389/fpls.2025.1509102.

Wu, Y., Shi, H., Yu, H., Ma, Y., Hu, H. *et al.* (2022) Combined GWAS and transcriptome analyses provide new insights into the response mechanisms of sunflower against drought stress. *Frontiers in Plant Science* 13, 847435. DOI: 10.3389/fpls.2022.847435.

Xu, W., Guo, L., Wang, C., Wei, L., Wang, Q. *et al.* (2024) Transcriptome analysis reveals drought-responsive pathways and key genes of two oat (*Avena sativa*) varieties. *Plants* 13(2), 177. DOI: 10.3390/plants13020177.

Yahoueian, S.H., Bihamta, M.R., Babaei, H.R. and Bazargani, M.M. (2021) Proteomic analysis of drought stress response mechanism in soybean (*Glycine max* L.) leaves. *Food Science & Nutrition* 9(4), 2010–2020. DOI: 10.1002/fsn3.2168.

Yang, W., Feng, H., Zhang, X., Zhang, J., Doonan, J.H. *et al.* (2020) Crop phenomics and high-throughput phenotyping: Past decades, current challenges, and future perspectives. *Molecular Plant* 13(2), 187–214. DOI: 10.1016/j.molp.2020.01.008.

Yaseen, A.A., Khaleel, N.T., Al-Azzami, A.A., Aldossary, A.T.Y. and Ameen, R.A. (2025) Climate change and its effect on nutritional value: A review. *IOP Conference Series* 1449(1), 012163. DOI: 10.1088/1755-1315/1449/1/012163.

Yu, X., Liu, Z. and Sun, X. (2023) Single-cell and spatial multi-omics in the plant sciences: Technical advances, applications, and perspectives. *Plant Communications* 4(3), 100508. DOI: 10.1016/j.xplc.2022.100508.

Zarzyńska, K., Boguszewska-mańkowska, D. and Nosalewicz, A. (2017) Differences in size and architecture of the potato cultivars root system and their tolerance to drought stress. *Plant, Soil and Environment* 63(4), 159–164. DOI: 10.17221/4/2017-PSE.

Zhang, H., Zhang, J., Xu, Q., Wang, D., Di, H. *et al.* (2020) Identification of candidate tolerance genes to low-temperature during maize germination by GWAS and RNA-seq approaches. *BMC Plant Biology* 20(1), 333. DOI: 10.1186/s12870-020-02543-9.

Zhang, Y., Massel, K., Godwin, I.D. and Gao, C. (2018) Applications and potential of genome editing in crop improvement. *Genome Biology* 19(1), 210. DOI: 10.1186/s13059-018-1586-y.

Zhao, C., Liu, B., Piao, S., Wang, X., Lobell, D.B. *et al.* (2017) Temperature increase reduces global yields of major crops in four independent estimates. *Proceedings of the National Academy of Sciences* 114(35), 9326–9331. DOI: 10.1073/pnas.1701762114.

Zheng, J., Zhang, Z., Liang, Y., Gong, Z., Zhang, N. *et al.* (2022) Whole transcriptome sequencing reveals drought resistance-related genes in upland cotton. *Genes* 13(7), 1159. DOI: 10.3390/genes13071159.

Zhou, X. and Huang, X. (2019) Genome-wide association studies in rice: How to solve the low power problems? *Molecular Plant* 12(1), 10–12. DOI: 10.1016/j.molp.2018.11.010.

13 Omics Tools in Horticultural Research and Recent Applications

Augustine Antwi-Boasiako[1,2*], Isaac Osei Bonsu[1], Joseph Adomako[1,3*], Harry Okyere[1], Augustine Boakye Boateng[1], Ernest Baafi[1], Yaw Danso[1], Daouda Coulibaly[4] and Joshua Obeng[5,6]

[1]Council for Scientific and Industrial Research-Crops Research Institute (CSIR-CRI), Kumasi, Ghana; [2]Key Laboratory of Biology and Genetics Improvement of Soybean, Ministry of Agriculture, Zhongshan Biological Breeding Laboratory (ZSBBL), National Innovation Platform for Soybean Breeding and Industry-Education Integration, State Key Laboratory of Crop Genetics & Germplasm Enhancement and Utilization, College of Agriculture, Nanjing Agricultural University, Nanjing, China; [3]Department of Crop and Soil Science Education, Akenten Appiah-Menka University of Skill Training and Entrepreneurial Development (AAMUSTED), Mampong, Ghana; [4]Laboratory of Agro-physio-genetics and Plant Biotechnology, Rural Polytechnic Institute for Training and Applied Research (IPR/IFRA) of Katibougou, Koulikoro, Mali; [5]Council for Scientific and Industrial Research – Oil Palm Research Institute, Kade, Ghana; [6]College of Agriculture, Tennessee State University, Nashville, USA

13.1 Introduction

Horticulture encompasses the science and art of raising fruit crops, nuts, vegetables, herbs, spices, flowers, plantation crops, and ornamental or avenue plants. Horticultural crops hold significant importance in the economies of many countries as they enhance the incomes of rural inhabitants. These crops require substantial effort to cultivate, hence offering several employment opportunities for rural populations. Horticultural crops provide nutritional security and profitable land use. Fruits and vegetables possess a high concentration of nutritious and health-promoting components, notably vitamins, minerals, and antioxidants (Rajani and Shourabh Joshi, 2017; Jideani *et al.*, 2021). The positive impacts of the horticultural industry include enhanced nourishment for families and children, augmented income from selling horticultural produce, and elevated status and confidence among farmers, especially women, as well as lowering the risk of developing diseases (Borbaruah and Borbaruah, 2023; Ahmed *et al.*, 2024). The horticultural industry therefore has a significant impact on the economy, human nutrition, gender equity, and employment opportunities.

The horticultural industry has made significant advances in the agricultural sector for

*Corresponding authors: bbbantwi@yahoo.com and joeadomako@gmail.com

© CAB International 2026. *Plant Spatial Biology* (ed. Chen)
DOI: 10.1079/9781836994572.0013

decades in most countries. However, many challenges, such as pests, diseases, drought, unpredictable weather conditions, and postharvest losses, have been identified in their production and marketing. Horticulture and climate change are intrinsically linked, as climate change primarily induces or makes the crops susceptible to biotic and abiotic pressures. Biotic and abiotic stresses negatively impact physiological, anatomical, morphological, and biochemical aspects of horticultural crops. Increased temperatures modify plant morphology, anatomy, and physiology, affecting the emergence of seeds, plant development, flower abscission, pollen viability, fertilization, fruit set, fruit weight, size, and quality. It is therefore imperative to develop a horticultural industry that could resist biotic and abiotic stresses. Though high volumes of horticultural crops are produced periodically, shortages in consumption and nutrition exist due to losses in the production value chain. In addition, nutritional deficiencies persist owing to losses throughout the production value chain because of the perishable nature of most produce (Sambo *et al.*, 2016). Postharvest losses and waste (PHLW) of perishable horticultural products are significant globally, particularly in developing nations, and are projected to reach as high as 60%, contingent upon the commodity, season, and production region (Yahia *et al.*, 2019).

Conventional and biotechnological methods, including genetic mutation and molecular markers, have enhanced desirable traits such as new plant architecture, shelf life, floral structure and quality, yield, and stress tolerance in fruit crops, vegetables, landscape plants, and medicinal and aromatic species. Despite this progress, there is an increased necessity to leverage existing spatial and multi-omics technologies in horticultural crops to identify novel genes and regulators associated with significant traits, elucidate metabolic pathways, as well as influence these pathways through genetic regulatory mechanisms. Commencement of integration of data and insights acquired from diverse omics fields into multi-omics has begun the transformation (Jamil *et al.*, 2020). Thus, spatial and multi-omics technologies are transforming horticulture crops by incorporating genomics, transcriptomics, proteomics, phenomics, and metabolomics to elucidate intricate biological mechanisms and refine breeding methodologies. This comprehensive approach facilitates a profound comprehension of plant responses to many environmental conditions, resulting in the cultivation of crops with enhanced characteristics such as nutritional value, yield, quality, and stress adaptability. This focuses on the development and future direction of horticultural crops in relation to the application of spatial and multiple omics.

13.2 Genomics and its Application in Horticultural Plants

Genomics, which involves the study of an organism's complete genome, is essential for the continued advancement of horticulture, plant development, and breeding. It bridges the knowledge gap in the genetics underpinning beneficial traits, such as yield, nutrition, disease resistance, and tolerance to stressors, thereby facilitating the development of more effective breeding strategies. This entails discovering certain genes, comprehending their activities, and applying this information to enhance crops via molecular breeding and genome editing approaches. Principal uses of genomics in horticultural crops encompass: crop enhancement, genome modification, elucidation of plant biology, optimization of breeding programs, and studies on domestication. Hence, the advent and advancement of contemporary genomic methods, coupled with the incorporation of comparative and functional genomics, have resulted in substantial innovations with numerous applications in breeding horticultural crops (Singh *et al.*, 2021a).

13.2.1 Genome sequencing of horticultural plants

A genome is the entirety of an organism's DNA, containing all the information necessary for its functioning. Genomics is a swiftly evolving field that employs high-throughput genome-scale tools and techniques to explore the composition and functionality of plant genomes, with the goal of elucidating the molecular basis of traits, genetic variations, and relationships. The progress and use of genomic-scale resources and tools have been costly, with their applications largely limited to model crops and a few specific annual crops. Recent advancements in next-generation sequencing (NGS) have notably decreased both the cost and time needed for genome-scale sequencing. To align with the constantly evolving customer preferences and industrial requirements, such initiatives must be enhanced with biotechnological instruments. The emergence of sequencing technologies, such as Oxford Nanopore and PacBio, has significantly augmented the quantity of sequenced genomes of horticultural plant species over the past decade, significantly enhancing the progress in the understanding of genetics and genomics of horticultural crops (Rhoads and Au, 2015; Wang *et al.*, 2021). Genome sequencing has been carried out on tomato, watermelon, grape, pear, orange, garlic, onion, cabbage, pepper, carrot, and coffee, among others. By connecting the genome sequence of various crop varieties and their wild relatives, genetic differences related to beneficial traits like disease resistance, fruit quality, and adaptation to environmental conditions can be identified. The genomes of horticultural plants exhibit significant diversity and complexity, frequently characterized by elevated heterozygosity and high ploidy, attributable to their extensive and intricate evolutionary and domestication histories. For instance, the dimensions of horticultural crops, even those of the same family, exhibit significant variation. Typical examples are garlic (*Allium sativum*), which possesses a diploid genome ($2n = 16$) and has an approximate genome size of more than 30 Gb, whereas onion (*Allium cepa*) exhibits a comparable genome size (Peška *et al.*, 2019).

At the end of 2018, a total of 181 genomes of horticultural species had been sequenced, comprising four beverage plants, 47 fruit crops, 44 medicinal plants, 44 ornamental plants, and 42 vegetable crops (Chen *et al.*, 2019). The application of genomic sequencing to horticultural crops has significantly enhanced understanding of their floral and fruit biology. For example, the high-quality apple genome sequencing revealed that a single allele governs red fruit peel coloration (Zhang *et al.*, 2019), whereas the reference genome of rose has elucidated the routes for floral color and aroma (Raymond *et al.*, 2018). Likewise, available genome sequencing data have facilitated the elucidation of the evolutionary and genetic underpinnings of compounds such as vitamin C in tomatoes (Ruggieri *et al.*, 2016). Hence, the combined use of multi-omics and spatial innovations is currently at the forefront of plant research.

13.2.2 Advances and prospects of genomics in horticultural plants

Genomics is the scientific discipline focused on the study of genomes, facilitating research on evolution, variations, and gene regulation rooted in the plant genome, which is a crucial biological process informed by detailed genomic sequence data of the plant. This section presents advances made in genomics on ornamental plants, fruit crops, and vegetables, and prospects for advancing genetic improvement.

13.2.2.1 Ornamental plants

Ornamental plants represent a significant segment of the horticultural sector, valued in monetary terms in billions of dollars globally. The wide variety of ornamental plants has led to a significant advancement in genomic research. Recent breakthroughs in ornamental plant genomics have led to a significant transformation in whole-genome sequencing technologies, which have generated unprecedented opportunities for breeding programs. The Japanese apricot (*Prunus mume*) was the inaugural ornamental plant to undergo sequencing in 2012. However, whole-genome sequencing of more than 100 ornamental plants has been accomplished during the past decade (Zheng *et al.*, 2021; Li *et al.*, 2022).

Whole-genome sequencing has been employed in ornamental plants, enabling gene annotation, comparative genomic studies, quantitative trait loci (QTL) mapping, genome-wide association studies (GWAS), and transcriptome analysis (Fu *et al.*, 2022). Gene annotation has been utilized for locating possible coding genes, noncoding RNAs, proteins, transfer RNAs, and ribosomal RNAs that regulate ornamental traits including blooming regulation, flower scent and color, dormancy, plant type, cold resistance, and resistance to pests and diseases (Li *et al.*, 2024a, b; Pagariya *et al.*, 2025). Similarly, mitochondrial and chloroplast genomes have enhanced our understanding of ornamental plants in terms of genetic diversity, evolutionary history, and phylogenetic relationships (Shen *et al.*, 2019). Comparative studies in genomics in ornamental plants have enhanced the understanding of plant genome architecture and trait-associated genes through genome mapping and sequencing, thereby revealing the evolutionary relationships among ornamental plant species and validating the functionality of significant genes (Chen *et al.*, 2021). QTL mapping is crucial in identifying the precise genomic areas that influence variation in complex traits such as flower color, which is one of the key desirable traits of ornamental plants. This information is valuable for examining the genetic foundation of a specific trait and for identifying prospective genetic markers for the purpose of breeding (Chao *et al.*, 2021; Pagariya *et al.*, 2025). Genomics applications on ornamental plants have facilitated the application of GWAS. GWAS has proven beneficial in screening and identifying pivotal genes linked to significant economic features in ornamental plants, such as the flower color differences, carotenoid metabolic processes, and blooming time (Fu *et al.*, 2022; Pagariya *et al.*, 2025).

The increasing volume of published genomic data presents significant challenges in the analysis, storage, and dissemination of extensive genome sequencing information. A primary issue is addressing the time and cost challenges encountered to minimize redundant studies, enhance the applicability of data, extract research content, and increase the transparency of scientific inquiry and data sharing across various disciplines. Furthermore, it is essential to improve knowledge in bioinformatics and its application in practice. We anticipate that the examination of more gene regulatory and metabolic networks related to attractive features, together with advancements in biological gene synthesis, will lead to the emergence of ornamental plants tailored to customer preferences in the near future. Ornamental plant research currently has significant challenges in systematic sequencing, thorough investigations of vital genetic resources, and ensuring their protection and utilization. As sequencing technology advances, it is anticipated that whole-genome sequencing of ornamental plants will soon enter a phase of rapid development, significantly benefiting the global horticultural business. Consequently, breeders must fully leverage genomic theories and methodologies to mitigate the constraints of the current variety of innovation and utilization, examine the genetic regulatory mechanisms of significant ornamental traits, develop a robust breeding technology framework, and engage in variety development.

13.2.2.2 Fruit crops

Fruit crops are crucial for attaining dietary and nutritional security. The genetic enhancement of fruit crops through conventional breeding remains sluggish due to the lengthy juvenile phase, polyploidy, and high heterozygosity. Hence, accurate and focused breeding methodologies are crucial for the swift and efficient production of cultivars in fruit crops. Advancements in genomics have facilitated the creation of innovative genomics-based breeding techniques that enable targeted and precise genetic alterations in fruit crops. Genome editing, for instance, holds significant promise for accelerating the development of new varieties and could transform the genetic enhancement of fruit crops (van Nocker and Gardiner, 2014; Chen *et al.*, 2020). More than 50 fruit crops with high-quality genome sequences have been reported (Savadi *et al.*, 2021). An exceptional genome can significantly support fruit crop breeders. This is because homologs of candidate genes associated with fruit crop traits can be identified using spatial and integrated ‘omics’ methodologies. In 2007, the complete genome of the grapevine was sequenced for the first time using Sanger sequencing technology, a process recognized for its accuracy (Jaillon *et al.*, 2007). However, this method is costly and time-consuming. Also, the genome of pineapple was assembled in 2015, utilizing the varieties ‘F153’ and ‘MD2’, along with the *Ananas bracteatus* accession CB5, which is a wild relative of pineapple (Ming *et al.*, 2015). The assembly of the mango cultivar Amrapali was conducted using Illumina reads, which indicated redundancy in the assembly process (Singh *et al.*, 2021b). In 2020, the Alphonso cultivar was assembled utilizing PacBio, Illumina, and Hi-C reads, resulting in an enhanced 392 Mb genome

(Wang *et al.*, 2020). The Alphonso serves as the benchmark genome for mango, as it offers superior contiguity and completeness in comparison to other genomes. The advent of NGS systems such as Illumina, SOLiD Life Technologies, and 454 Roche has significantly reduced costs while enhancing the efficiency and data yield of genome sequencing of fruit crops. Recent breakthroughs in long-read sequencing methods like PacBio and RS/Oxford NanoPore provide comprehensive coverage of broad genomic regions, encompassing numerous repeated sequences, thereby effectively mitigating the problems associated with short-read assemblies. In fruit crops, the presence of heterozygosity complicates the precision of genome assembly, and is further exacerbated by the abundance of repetitive sequences. The use of doubled haploid (DH) lines in genome sequencing efforts has addressed the challenge of heterozygosity, leading to superior genome assemblies (Wu *et al.*, 2014; Linsmith *et al.*, 2019). Thus, these advancements in technology have significantly expedited the genome sequencing process for fruit crops. The continuous growth of genomic data necessitates the development of high-density genetic markers, the innovation of genetic mapping strategies, and the identification of genetic variation within genomes that are associated with traits of economic importance for the improvement of fruit crops. For instance, the genome of the 'Tommy Atkins' mango revealed potential genes linked to fruit quality (Bally *et al.*, 2021), serving as a significant resource for the global community to promote the progress of mango cultivation. Despite this, many fruit crops, especially those originating from tropical regions and possessing considerable economic importance, do not have a fully assembled genome sequence. The lack of genomic sequences in these species has hindered molecular genetic studies and the use of genomics for advancing crop improvement.

The application of marker-assisted breeding (MAB) has been shown to enhance fruit development and its quality traits, including aroma, size, pulp and peel color, texture, and shape. The availability of genome sequences for fruit crops enables the identification of single-nucleotide polymorphism (SNP) variants, QTLs, and genes that clarify the genetic determinants of fruit characteristics. Utilizing spatial and multi-omics methodologies, the detection and quantification of transcripts, encompassing non-coding RNAs, related to sugar metabolism, fruit development and ripening, shelf life, as well as the biotic and abiotic stresses affecting fruit quality, were achieved (Mathiazhagan *et al.*, 2021; Patil *et al.*, 2022; Manzoor *et al.*, 2023). The implementation of genomic-assisted breeding techniques, including GWAS, genomic selection (GS), and genetic modifications through CRISPR/Cas9 and transgenics, has significantly advanced the study of gene function and the development of cultivars with preferred fruit characteristics, effectively addressing the challenges posed by lengthy breeding cycles (Penna and Jain, 2023).

13.2.2.3 Vegetables

Vegetables are essential in the human food chain due to their abundance of nutrients and phytochemicals, also known as nutraceuticals. They are the sources of protein, carbohydrates, dietary fiber, minerals, vitamins, and a range of secondary metabolites, each playing a role in bolstering immunity and supporting both physical and mental well-being. Recent progress in sequencing and computational biology has equipped breeders with a wide array of innovative tools and techniques, allowing for an in-depth analysis of entire genomes, enabling the direct investigation of the genotype–phenotype relationship (Tester and Langridge, 2010). Comprehensive information regarding the catalogue of genome-sequenced vegetable plants and their related counterparts is reported (Hao *et al.*, 2020). The accelerated growth of spatial, genomic, and sequencing tools for vegetables has enhanced biological and genetic research, establishing a basis for the implementation of molecular breeding. These hold potential for enhancing populations and advancing hybrid development to tackle biotic and abiotic stresses, overcome production challenges, and deliver superior breeding materials and strategies in response to the promptly evolving pathogen strains and changing climates. The utilization of genomics and marker-assisted selection for enhancing vegetable crops such as tomato, pepper, eggplant, lettuce, spinach, cucumber, and chicory, with an emphasis on genome sequences, mapped genes, and QTLs, is thoroughly documented (Simko *et al.*, 2021). The advancement and utilization of vegetable genomics has enhanced the effectiveness of investigating novel gene resources for vegetables (Li and Lai, 2024). The discovery of the role of significant genes through the progress of whole-genome sequencing has enabled the CRISPR-Cas9 technique to modify specific target genes in various vegetable crops, including broccoli, tomato, pepper, and onion (Han *et al.*, 2021; Wan *et al.*, 2021; Wang and Zhang, 2022). Moreover, the emergence of spatial and omics technologies has significantly enhanced the efficiency and effectiveness of uncovering information related to pathways linked to drought stress resilience in various vegetable crops (Saeed *et al.*, 2023).

Given the significant advancements in genomic data and tools, it is essential to pursue further collaboration in next-generation breeding technologies. For instance, genome-based breeding by design can lead to the development of environmentally adaptive vegetable cultivars that exhibit high yield and quality. The emergence of spatial and omics technologies has significantly enhanced the efficiency and effectiveness of uncovering information related to pathways linked to abiotic stress resilience in various vegetable crops. In summary, it is essential to integrate spatial, omics, transgenic, and genome-editing technologies with traditional and genomics-assisted breeding methods to develop vegetable crops that can cope with stresses to sustain the increasing global population.

13.3 Transcriptomics and its Application in Horticultural Plants

Transcriptomics is the analysis of gene function after the genomic sequence has been identified. This means examining the full spectrum of RNA transcripts produced by the genome under specific circumstances. By examining these transcripts, scientists can get insight into the patterns of gene expression and the regulatory processes that govern them. Determining how genes respond to various stimuli requires an understanding of these patterns. Additionally, transcriptome research might aid in the identification of possible biomarkers for plant stress, which would support initiatives to lessen those pressures. These developments are essential for improving horticultural crops and expanding the knowledge base of complex molecular characteristics. High-throughput transcriptome research will be made possible by the integration of geographical and transcriptomic data as technology advances.

13.3.1 Advances in transcriptomics application in horticultural plants

Transcriptome studies provide information on the gene expression profiles of plants. This enables the identification of crucial genes associated with traits such as fruit quality and disease resistance. Therefore, by employing particular breeding techniques, transcriptomics can aid in the development of improved cultivars. The loss of spatial information in single-cell sequencing is addressed by the spatial transcriptome (ST) approach. This enables scientists to have a better understanding of the geographic distribution of gene expression in tissues. By enhancing conventional breeding techniques with molecular markers, scientists can expedite the selection process. It is anticipated that this all-encompassing strategy for crop improvement will expand the knowledge of plant biology to enable researchers to develop interventions that boost productivity while lessening their negative effects on the environment. The interoperability of multi-omics and spatial technologies may determine the direction of agriculture in the future. This might inspire fresh concepts that increase profitability and environmental friendliness.

Transcriptomics provides deep insights into metabolic processes, stress responses, gene expression, and development in horticultural plants. Recent developments make it possible to analyze transcriptomes at the single-cell level, clarifying patterns of gene expression unique to individual cells. Tissue development, stress reactions, and cell differentiation in horticultural crops, including citrus, tomato, and grape, have all been successfully studied using this technology. For non-model horticultural plants without reference genomes, transcriptomes can be assembled *de novo* using NGS. This technique has accelerated functional genomics in previously understudied species.

NGS-based transcriptomics is applied to identify genes implicated in responses to biotic and abiotic stress, fruit ripening, blooming, and the manufacture of secondary metabolites. NGS affects the development and enhancement of cultivars, the integration of metabolomics and functional genomics, and the comprehension of plant stress responses. Additionally, new genes have been found, breeding has been supported, pathways have been clarified, and stress responses have been studied due to advances in transcriptomics. Functional validation, pricing, accessibility, and handling large volumes of data are a few issues with NGS. There is now a better understanding of gene expression and plant biology because of the tremendous advancement made in transcriptomics on horticultural plants by NGS technologies. Since sequencing technology, bioinformatics, and integrative omics keep evolving, we will be able to further enhance the sustainability and characteristics of horticulture crops.

13.4 Proteomics and its Application in Horticultural Plants

The broad-scale study of proteins and their functions is known as proteomics, and has emerged as a revolutionary tool in plant science. Proteomics decodes the intricate molecular processes involved in plant growth, development, and stress responses (Satrio *et al.*, 2024). Across the horticultural crops, namely fruit crops, vegetables, ornamental plants, and herbs, proteomics has been employed to enhance the yield and quality of produce by improving disease resistance and adaptability to altering environmental conditions. In the exploitation of genetic potential of horticultural crops, proteomics helps to capture a 'now snapshot' of cellular activity and helps to identify critical proteins involved in various metabolism, defense processes, and signaling transduction pathways (Zivy and de Vienne, 2000). The establishment of herb and vegetable horticulture has greatly advanced from studying postharvest plant physiology, plant-pathogen interaction, and abiotic stresses like drought, salinity, and extreme temperatures (García-Mier *et al.*, 2013). Using proteomics, breeders and researchers are enabled to develop higher quality and more resilient cultivars as a result of studying protein expression patterns under various conditions. Moreover, mass spectrometry (MS), two-dimensional gel electrophoresis, and bioinformatics have all contributed to crop improvement programs (Cramer *et al.*, 2011). As the impact of climate change continues to negatively affect the agricultural industry, proteomics offers promising solutions to ensure food and nutrition security, sustainability, and the economic success of ornamental industries worldwide.

13.4.1 Proteomics and crop yield improvement in horticultural plants

Proteomics is critical in increasing the yield of horticultural crops. Applying condition-based analysis of the entire protein profile of a given plant, crucial proteins linked to growth, stress tolerance, and overall productivity can be determined (Ahmad *et al.*, 2016). This assists in breeding plant varieties that have greater adaptability to environmental challenges like drought, salinity, and diseases. Proteomics also assists in elucidating the response of different plant species to fertilizers, hormonal supplements, and other agricultural practices on a molecular scale (Hu *et al.*, 2015). These optimized farming methods increase photosynthesis and nutrient absorption, as well as reproductive activities. In horticulture, where crop yield and quality are equally important, proteomics helps in developing new varieties with enhanced fruit size, taste, and shelf life (Palma *et al.*, 2011), thereby ensuring sustainable crop production and food security.

13.4.2 Proteomics and stress tolerance in horticultural plants

Proteomics plays a key role in improving stress tolerance in horticultural plants by revealing how plants respond to environmental challenges (Jan *et al.*, 2023). By identifying and studying stress-responsive proteins, proteomics helps researchers learn how plants react to abiotic stresses such as drought, salinity, heat, and cold, along with biotic stresses like pest and disease attacks (Varadharajan *et al.*, 2025). Examining changes in protein expression and function during stress, scientists can identify important proteins and pathways that help plants tolerate stress. This knowledge aids in developing better plant varieties through breeding or genetic engineering (Xu *et al.*, 2020). Proteomic studies have uncovered heat-shock proteins, antioxidant enzymes, and signaling molecules that assist plants in maintaining balance under stress. Overall, proteomics helps in developing strategies to improve stress tolerance in horticultural crops, supporting sustainable production even in tough growing conditions.

13.4.3 Proteomics and disease resistance in horticultural plants

Proteomics is essential for understanding how horticultural plants resist disease because proteins are the main products of plant defense responses. Proteomics helps to identify and describe proteins that recognize pathogens and start the immune responses (Liu *et al.*, 2019). In disease-resistant plants, proteomics identifies pathogenesis-related proteins, chitinases, glucanases, and signal transduction proteins that increase during infection (Lopes *et al.*, 2023). Comparing the proteomes of resistant and susceptible plant varieties shows which proteins are expressed differently. This knowledge helps develop resistant cultivars through marker-assisted breeding or genetic engineering. Quantitative proteomics, using technologies like liquid chromatography (LC)–MS/MS and 2-D electrophoresis, detects stress-responsive proteins and post-translational changes that affect plant immunity (Zhang *et al.*, 2016). Insights from proteomics improve our understanding of systemic acquired resistance (SAR) and induced systemic resistance (ISR) in horticultural crops such as tomatoes and cucumbers. This work ultimately reduces the need for pesticide application and supports sustainable agriculture.

13.4.4 Proteomics and nutritional value in horticultural plants

Proteomics provides important insights into improving the nutritional quality of horticultural plants by identifying key proteins involved in nutrient biosynthesis, transport, and storage (Martínez-Esteso *et al.*, 2024). Proteomics analysis shows how different environmental conditions, developmental stages, and genetic backgrounds influence proteins related to vitamins, antioxidants, minerals, and amino acids. In crops like tomatoes, spinach, and carrots, proteomic tools have revealed pathways responsible for producing vitamin C, carotenoids, flavonoids, and essential amino acids (Meléndez-Martínez *et al.*, 2022). This assists breeding programs aimed at creating nutrient-rich varieties. Comparative proteomics also helps identify biofortification markers, which enable targeted selection or modification of varieties with improved nutrient density (Garg *et al.*, 2018). Furthermore, proteomics also helps in the understanding of how postharvest handling and storage affect nutrient-related proteins. This ensures better quality retention throughout the supply chain. Overall, proteomics plays a vital role in food security and public health by enhancing the nutritional value of horticultural crops.

13.5 Metabolomics and its Application in Horticultural Plants

Metabolomics is a thorough study of metabolites that are involved in various cellular processes inside a biological system. Metabolomics has profiled hundreds to thousands of chemicals (metabolites) in plant tissues using high-throughput analytical methods (mostly MS and nuclear magnetic resonance (NMR)) that reflect environmental factors and genotype (Patel *et al.*, 2021). Metabolomics identifies the biochemical processes that are crucial to gene activity and determines the metabolic expression of proteins (Weckwerth and Morgenthal, 2005). In plant systems, metabolomics is especially crucial since plants generate more metabolites than microorganisms and animals. Plant metabolites are tiny compounds found in plant cells that are essential for defense, growth, development, and environmental adaptation (Salam *et al.*, 2023). In addition to affecting plant survival and reproduction, metabolites also have an impact on agricultural productivity and the equilibrium of ecosystem processes. The entire collection of metabolites produced by plant metabolic pathways is referred to as the metabolome.

Knowledge of plant metabolites helps guide developments in biotechnology and plant breeding by providing important insights into how plants react to environmental changes. Understanding the roles and regulating processes of metabolites helps in the understanding of how plants adjust to their changing natural surroundings (Salam *et al.*, 2023). It may result in the discovery of metabolites connected to particular characteristics or abilities. There are two primary methods: targeted metabolomics, which concentrates on a limited number of metabolites or pathways with greater sensitivity and quantitative precision, and untargeted (global) metabolomics, which aims to identify as many metabolites as possible, including unknowns (Han *et al.*, 2023). In practice, combinations of LC–MS, gas chromatography (GC)–MS, and NMR techniques are often applied to capture primary metabolites (sugars, amino acids, lipids, organic acids) and secondary (specialized) metabolites (e.g. phenylpropanoids, alkaloids, terpenoids). Together, these tools allow researchers to study how plants grow and respond to stresses by mapping changes in metabolic pathways (Han *et al.*, 2023).

13.5.1 Metabolomics studies on biotic stress in horticultural plants

Metabolomic profiling has been widely applied to study plant responses to pathogens and pests in horticultural crops. In tomato, for example, *Ralstonia solanacearum* (bacterial wilt) causes distinct metabolic changes. Thus, resistant tomato cultivars showed a significant decrease in valine and a corresponding increase in leucine relative to susceptible lines, as revealed by ^{1}H-NMR metabolomics (Murti *et al.*, 2021). Likewise, Roumani *et al.* (2022) used untargeted LC–MS metabolomics (integrated with transcriptomics) to study tomato attacked by the leafminer *Tuta absoluta* and found a strong induction of various phenolamides in infested leaves and upregulation of polyamine and phenylpropanoid pathways, identifying new defense-related metabolites (Roumani *et al.*, 2022). Similarly, profiling of volatile metabolites from *Pseudomonas syringae*-infected tomato leaves showed that a resistance-triggering (avirulent) infection led to emission of green-leaf volatile esters ((Z) − 3-hexenol esters) and hydroxylated monoterpenes, whereas a virulent infection produced different monoterpenes and salicylic-acid derivatives (López-Gresa *et al.*, 2017). These examples illustrate how targeted and untargeted metabolomics pinpoint specific biochemical responses (amino acids, polyamines, phenolics, volatiles) underlying tomato immunity.

In grapevine, metabolomic studies of fungal diseases have highlighted similar findings. For powdery mildew infection, tolerant and susceptible grape varieties showed shifts in phenolic metabolism: both accumulated flavonoids, phenolic acids, stilbenes, and terpenoids upon infection. The resistant genotype accumulated excessively high levels of stilbenes, whereas the susceptible type showed higher flavonoid accumulation (Estêvão *et al.*, 2024). Work on grape downy mildew or trunk diseases had likewise noted increases in defense-associated compounds (resveratrol, gallic acid, etc.) in infected tissues. In general, metabolomic profiles can distinguish resistant from susceptible grape lines and suggest involvement of phenylpropanoid-derived phytoalexins in disease resistance.

Citrus biotic stress metabolomics has focused on key diseases. For Huanglongbing (HLB, citrus greening), targeted metabolomics revealed that tolerant citrus cultivars maintain growth-related metabolism rather than mounting a strong defense. Thus, in HLB-infected trees, tolerant varieties upregulated amino acid (aspartate, glutamate), purine metabolism, and phytohormone pathways (auxin, cytokinin) while down-regulated salicylic-acid-related metabolites (Suh *et al.*, 2021). This suggests HLB-tolerant trees prioritize nutrient and hormone homeostasis over classic defense signaling. For citrus canker (*Xanthomonas citri*), increased phenylpropanoids in leaves upon infection were observed among several citrus species, especially fluorescent coumarins like umbelliferone, scoparone, and auraptene (Ference *et al.*, 2020). The least susceptible genotypes (kumquat, calamondin) already had high basal levels of the antibiotic dihydrochalcone phloretin-2′-O-C-glucoside, although its level did not change with infection (Ference *et al.*, 2020). In summary, metabolomics in tomato, grape, and citrus have revealed that biotic stresses induce characteristic changes in primary metabolites (amino acids, sugars) and secondary metabolites (phenolics, phytoalexins, volatiles), varying from species and resistance status (Ference *et al.*, 2020; Murti *et al.*, 2021; Suh *et al.*, 2021).

13.5.2 Metabolomics studies on abiotic stresses in horticultural plants

Under abiotic stresses (drought, salinity, heat), plants adjust metabolism to cope with osmotic and oxidative challenges. In tomato, salt stress causes broad metabolic reprogramming. Findings on 847 profiled metabolites in tomato fruits under moderate and high salinity using LC–MS indicated that salt stress differentially affected 54 metabolites (mainly flavanols and phenolamides) (Sun *et al.*, 2024). Key osmoprotectants – notably dopamine, galactitol, and mannitol – were undetected in control fruits but accumulated under salinity, and pathway analysis showed enrichment in flavonoid, betaine, sulfur, and galactose metabolism (Sun *et al.*, 2024). These changes (including known quality traits like increased lycopene in mild salt conditions) reflect the plant's osmotic adjustment and enhanced antioxidant capacity. Heat stress also drives metabolic shifts in tomato. For example, manipulation of the heat-stress transcription factor HsfB1 showed that HsfB1-knockdown leaves accumulated high levels of osmoprotective sugars (sucrose, glucose) and the polyamine putrescine, whereas HsfB1 overexpression plants accumulated phenylpropanoid/flavonoid compounds (e.g. multiple caffeoylquinic acid isomers) (Paupière *et al.*, 2020). This study indicates that heat tolerance involves both osmolyte accumulation and boosted antioxidant metabolite synthesis, controlled by heat-response regulators (Paupière *et al.*, 2020).

In grapevine, water deficit generally causes increases in osmolytes and phenolics. Drought-stressed grape berries and leaves often accumulate proline and polyphenols (e.g. phenylpropanoids and stilbenes) that help maintain osmotic balance and scavenge reactive oxygen species (as reported in several grape studies). Salinity stress in grapes also alters secondary metabolism. For instance, a proteomic/metabolomic study of cabernet suspension cells under salt stress showed induction of genes and metabolites in the flavonoid and anthocyanin pathways, suggesting that high salt triggers antioxidant flavonoid production. In general, salt-grown grapes may upregulate anthocyanins, depending on genotype and conditions. Although comprehensive metabolomic analyses in citrus under abiotic stress are limited, it is expected that citrus responds similarly by accumulating osmolytes (proline, sugars) and antioxidants often influenced by rootstock. Overall, abiotic stresses in horticultural crops elicit broad shifts in both primary (e.g. amino acids, sugars) and secondary (e.g. flavonoids, carotenoids) metabolites, reflecting stress-protection strategies. These metabolic markers (e.g. proline or specific phenolics) help identify stress-tolerant genotypes and can guide breeding for stress resilience (Paupière *et al.*, 2020; Sun *et al.*, 2024).

13.5.3 Advances and prospects of metabolomics in horticultural plants

Recent technological and computational advances are greatly expanding plant metabolomics. On the instrumentation side, ultra-high-resolution mass spectrometers (FT–ICR, Orbitrap, QTOF) and hybrid platforms now allow identification of thousands of metabolites with high accuracy (Hao *et al.*, 2025). In particular, spatial metabolomics (MS imaging; e.g. MALDI–MSI, DESI–MSI) has emerged as a powerful tool, as it enables *in situ* mapping of metabolites across plant tissues and cells with high spatial resolution (Yin *et al.*, 2024; Hao *et al.*, 2025). These methods can reveal, for example, how defensive or nutritional compounds are localized within fruit skins, roots, or infection sites. Additionally, single-cell metabolomics techniques are beginning to resolve cell-type-specific metabolism in plants (Hao *et al.*, 2025).

Computationally, integration and data analysis have advanced. Large metabolomics datasets are now routinely interpreted using multivariate statistics, and machine learning algorithms can uncover hidden metabolite patterns and select key stress markers from complex data (Hao *et al.*, 2025). Moreover, metabolomics is increasingly combined with other omics (genomics, transcriptomics, proteomics), and integrated network analyses link metabolites to genes and regulatory pathways (Hao *et al.*, 2025). For example, metabolite quantitative trait loci (mQTL) mapping and genome-wide association can connect metabolite changes to genetic loci, accelerating the discovery of stress-tolerance genes. Reviews emphasize that such multi-omics integration, possibly augmented by AI, is the future of plant biology (Hao *et al.*, 2025).

These advances have strong prospects for horticultural research and breeding. Comprehensive metabolome databases and improved annotation tools (spectral libraries and bioinformatics pipelines) make it easier to identify novel compounds. Metabolomics-assisted breeding is now feasible. This is because specific metabolites or metabolic profiles can serve as biomarkers for desirable traits (e.g. flavor, nutrient content, or stress resistance) in tomato, grape, citrus, and other crops (Hao *et al.*, 2025). For instance, as noted by recent reviews, metabolomics-driven breeding can add nutritional or stress-tolerance value to germplasm. Breeders can then select for seedlings with favorable metabolite signatures or use genome editing to tweak metabolic pathways once key genes are identified (Hao *et al.*, 2025). In summary, by coupling cutting-edge MS techniques (including spatial imaging) with advanced data integration, plant metabolomics is poised to greatly enhance our understanding of metabolism under stress.

13.6 Phenomics and its Application in Horticultural Plants

Phenomics represents a comprehensive approach to studying the phenome, which is the complete array of observable characteristics in an organism, shaped by both genetic factors and environmental influences. As a relatively new discipline, phenomics continues to evolve in its definition. However, the central premise remains its focus on the intricate relationships between genotype, phenotype, and the environment (Zavafer *et al.*, 2023). Phenomics is characterized by deep phenotyping that focuses on multifaceted and precise traits and/ or high-throughput (HTP) phenotyping, which emphasizes large-scale multi-environment assessment of traits. This methodology has become essential in modern plant science, acting as a critical link that bridges the gap between genetic information and observable traits. The integration of advanced technological tools and big data analytics has greatly enhanced our ability to decipher these complex interactions (Furbank and Tester, 2011). Amidst contemporary challenges such as climate variability, limited resources, and growing demands for food, phenomics provides powerful tools to analyze plant responses to drive improvement in crop performance. Horticultural plants, which include fruits, vegetables, and ornamental plants, stand out for their remarkable phenotypic diversity and heightened sensitivity to environmental fluctuations. Consequently, the application of phenomics in horticulture is vital for advancing crop resilience, enhancing quality, and increasing yields (Fiorani and Schurr, 2013; Rahaman *et al.*, 2015). This section introduces the fundamental concepts underpinning phenomics, reviews the technologies that enable its application, and examines its specific benefits for horticultural crops. Furthermore, it addresses the current challenges facing its application and discusses emerging directions for future research, positioning phenomics within the broader context of spatial and multi-omics advancements in horticultural science.

13.6.1 Principles and technologies in plant phenomics

Phenomics covers the systematic collection of plant phenotypic trait data, which encompasses morphological, physiological, and biochemical features across temporal and spatial scales (Furbank and Tester, 2011). Unlike traditional phenotyping, which deals with a single or a few phenotypic traits accessed at one time on a single or multiple genotypes in a single environment, phenomics leverages automation, imaging, and computational methods to capture complex trait dynamics across large populations and multiple environments. Key principles include linking genetic information (genotype) with observable characteristics (phenotype) under various environmental conditions. The integration of phenotypic data with genomics and environmental data is essential for unravelling genotype-by-environment interactions and identifying candidate genes for trait improvement (Tardieu *et al.*, 2017).

13.6.1.1 Core technologies in plant phenomics

HIGH-THROUGHPUT PHENOTYPING PLATFORMS (HTP). High-throughput phenotyping platforms represent a major technological leap in phenomics. These automated systems, which include conveyor systems in greenhouse and growth chambers, and field-deployed mobile units, allow for precise and non-destructive evaluation of plant traits at multiple scales (Cobb *et al.*, 2013). The deployment of unmanned aerial vehicles (UAVs) further accelerates the measurement process, facilitating rapid data collection from an extensive population of genotypes under varying field conditions (Zaman-Allah *et al.*, 2015). Notable HTP platforms include controlled-environment systems such as LemnaTec and PhenoRig, as well as field-based systems like FieldScan and PhenoRover. These technologies often employ combinations of imaging and sensor modalities, supporting dynamic and continuous monitoring of plant growth, development, and environmental responses.

IMAGING TECHNOLOGIES. The following imaging technologies are indispensable in phenomics, providing multidimensional data for both qualitative and quantitative analyses (Fiorani and Schurr, 2013):

- RGB imaging: captures plant size, shape, and color, supporting structural and morphological assessments (Minervini *et al.*, 2015).
- Hyperspectral imaging: enables detection of biochemical changes linked to nutrient deficiencies or stress responses, offering deep insights into plant health (Mahlein *et al.*, 2013).
- Thermal imaging: monitors canopy temperature, allowing inference of transpiration rates and water stress.
- Fluorescence imaging: used to assess photosynthetic performance by measuring chlorophyll fluorescence, providing data on plant vitality (Murchie and Lawson, 2013).
- X-ray computed tomography (CT): facilitates internal visualization of plant tissues such as roots, leaves, and fruits, enabling non-invasive anatomical studies (Piovesan *et al.*, 2021; Karahara *et al.*, 2023).
- 3D imaging: delivers spatial models to track plant growth and structure over time.

SENSOR TECHNOLOGIES. Sensor systems enhance phenomics by enabling continuous, *in situ* measurement of root, canopy, soil, and microclimate parameters (White *et al.*, 2012). For instance:

- LiDAR sensors provide three-dimensional reconstructions of plant architecture, crucial for assessing structural dynamics.
- Thermal infrared (TIR) sensors measure leaf temperature, offering insights into plant water status and heat stress (Velazquez-Chavez *et al.*, 2024).

These sensors collectively deepen our understanding of plant–environment interactions and stress responses.

ROBOTICS AND AUTOMATION. Robotics and automation are integral to modern phenomics platforms, enabling automated sampling and data acquisition. These innovations boost throughput, minimize human-induced environmental bias, and reduce labor intensity (Rahaman *et al.*, 2015). Automation ensures consistency and reliability in data collection, a necessity for large-scale phenotyping efforts.

DATA ANALYTICS AND INFORMATICS. Phenomics generates vast and heterogeneous datasets, making sophisticated data infrastructure and analytical frameworks essential. Machine learning and big data platforms play a key role in managing, visualizing, and interpreting complex phenotypic datasets (Singh *et al.*, 2016). Artificial intelligence techniques are employed for pattern recognition, trait prediction, and classification, augmenting the discovery process. Image processing and trait quantification are facilitated by software tools such as PlantCV, ImageJ, and Phenotyper. Moreover, integrating phenotypic data with genomic information enables GWAS and GS (Araus and Cairns, 2014), empowering breeders to make data-driven decisions. Effective management and analysis of phenomics data are thus critical for extracting biologically meaningful insights that guide breeding and crop management.

13.6.2 Application of phenomics for crop improvement in horticultural crops

Phenomics has transformed research approaches in horticultural crop improvement, offering new avenues for enhancing genetic advancement, dissecting complex traits, and increasing resilience to environmental stress. Integrating HTP phenotyping technologies and advanced data analytics, phenomics-driven strategies have enabled precision in breeding and selection, leading to the development of superior horticultural cultivars. This section explores the key applications of phenomics in horticultural crops, highlighting recent advances, supported by extensive research and multiple case studies.

13.6.2.1 Trait discovery

Integrating phenomics with transcriptomics, proteomics, and metabolomics provides powerful insights into the molecular basis of complex traits (Fernie and Schauer, 2009; Singh *et al.*, 2016). This multi-omics approach allows the identification of key candidate genes underlying traits such as disease resistance, fruit quality, and environmental adaptation, thereby guiding targeted interventions in horticultural crop breeding. Automated phenomics analysis has greatly expedited the discovery of novel genetic variants associated with important traits such as yield, quality, and stress tolerance (Yang *et al.*, 2013; Araus and Cairns, 2014). Through the deployment of robotics, machine vision, and image-based phenotyping, breeders can efficiently screen large populations and identify alleles linked to desirable characteristics. This approach has proven especially valuable in horticultural crops, where phenotypic diversity is high and complex traits often have polygenic backgrounds (Fiorani and Schurr, 2013).

13.6.2.2 Yield and productivity

Precise quantification of yield-contributing components such as biomass, fruit set, fruit size, and harvest index is vital for effective selection in horticultural crops (Fahlgren *et al.*, 2015; Rahaman *et al.*, 2015). High-throughput platforms utilizing multispectral imaging, laser scanning, and thermal sensors enable breeders to capture quantitative data on yield parameters with high accuracy and throughput. These tools facilitate rapid assessment across different environmental conditions and growth stages, supporting the identification of genotypes with stable and enhanced productivity (Li *et al.*, 2014).

13.6.2.3 Quality attributes

Modern imaging and sensor technologies are instrumental in the evaluation of quality traits, such as color, shape, texture, nutritional content, and flavor, to align breeding objectives closely with market and consumer expectations (Fiorani and Schurr, 2013; Lu *et al.*, 2017). For example, hyperspectral and fluorescence imaging can assess pigment concentrations and surface characteristics, while electronic noses and tongues provide insights into volatile compounds and taste profiles. All these are critical for the horticultural sector (Vanaraj *et al.*, 2025).

13.6.2.4 Abiotic and biotic stress tolerance

Phenomics enables real-time monitoring of plant responses to abiotic stresses such as drought, salinity, and extreme temperatures, as well as biotic factors including pest and pathogen pressure (Dhondt *et al.*, 2013; Mir *et al.*, 2019). Through time-lapse imaging and environmental sensors, researchers can track physiological changes, identify tolerant genotypes, and accelerate the breeding of robust cultivars. The dynamic nature of environments where horticultural crops grow necessitates ongoing stress monitoring, and phenomics provides the tools needed for rapid response and selection (Großkinsky *et al.*, 2015).

13.6.2.5 Developmental analysis

Time-series imaging coupled with advanced software reveals patterns of growth and development, allowing optimization of crop cycles and cultivation practices (Cobb *et al.*, 2013). In horticultural crops where developmental timing and phenology are tightly linked to market windows and quality, phenomics support the fine-tuning of growth parameters and harvest scheduling (Fahlgren *et al.*, 2015).

13.6.2.6 Integration with genomics

Phenomics data significantly enhance GWAS and GS, strengthening marker-assisted breeding strategies for horticultural improvement (Poland and Rutkoski, 2016; Tardieu *et al.*, 2017). The combination of rich phenotypic datasets with genomic information enables the identification of QTLs and accelerates the selection process, making breeding programs more efficient and data driven.

13.6.3 Specific application of phenomics in horticultural crops

The use of advanced imaging modalities and sensor platforms enables breeders, physiologists, and technologists to monitor plant growth, detect stress symptoms, assess disease resistance, and evaluate post-harvest quality with unprecedented precision and efficiency. Such innovations not only accelerate breeding cycles but also improve crop management, postharvest handling, and quality assurance for a wide array of horticultural species. Key examples of phenomics applications across different horticultural crops, organized by crop category, are presented in Table 13.1. This highlights the extent of phenomics application across vegetables, fruit crops, and ornamental plants, providing a comprehensive view of its impact. Collectively, these applications are pivotal for both upstream breeding and downstream quality management in horticulture, fostering the development of resilient, high-quality crops through precise and efficient selection strategies. The ongoing integration of phenomics with other disciplines such as genomics, bioinformatics, and artificial intelligence is expected to further enhance the precision and scalability of horticultural research and practice. As presented in Table 13.1, the scope of phenomics is rapidly expanding, supporting both fundamental discovery and practical application in diverse horticultural crops.

13.6.4 Challenges and limitations of phenomics applications in horticultural plants

Challenges with the application of phenomics in horticultural plants range from genetic and phenotypic diversity, technical and financial constraints, environmental variability, data management, and translation to breeding, among others. These are elaborated in the subsequent paragraphs. Horticultural plants frequently exhibit varied morphologies, growth patterns, and developmental stages, complicating standardized phenotyping. Also, the extraordinary diversity found within horticultural species presents a formidable challenge to the field of phenomics. Horticultural plants exhibit a wide range of genetic backgrounds and physical traits, making it difficult to develop standardized phenotyping protocols that are universally applicable. This variation complicates efforts to compare traits across species and cultivars, as protocols often require adaptation to account for unique characteristics. As a result, researchers must continually refine their methods to ensure comprehensive and accurate data collection (Tardieu *et al.*, 2017).

Implementing advanced phenomics technologies demands considerable financial investment and technical expertise. High costs are associated with acquiring sophisticated imaging equipment, automation tools, and data analysis software. These barriers are particularly pronounced for programs with limited resources, which may lack the means to access or maintain cutting-edge technologies. Thus, the progress and impact of phenomics in horticultural crops are often constrained by the availability of funding and technical support (Dhondt *et al.*, 2013). Heterogeneous field conditions introduce a significant degree of noise into the phenomics dataset. Environmental factors such as soil type, temperature, humidity, and light can vary greatly across different locations and seasons, affecting plant growth and development. This variability complicates

Table 13.1. Summary of application of phenomics in horticultural crops.

Crop category	Crop	Phenomics application	Reference
Vegetables	Tomato	Hyperspectral imaging and UAV for disease detection	Abdulridha *et al.*, 2020
	Cucumber	High-throughput image-based phenotyping to analyze fruit shape, color, and growth rate	Yang *et al.*, 2020
	Eggplant	Early disease diagnosis and resistance evaluation via hyperspectral imaging	Mahlein *et al.*, 2012
	Pepper	Disease detection and resistance screening using hyperspectral reflectance	Mahlein *et al.*, 2012
	Carrot	Automated image analysis for root shape uniformity and color assessment	Turner *et al.*, 2018
Fruits	Blueberry	Non-destructive assessment of soluble solids by modeling based on hyperspectral imaging	Meng *et al.*, 2025
	Mango	Robustness of near-infrared (NIR) spectroscopy for postharvest quality assessment: firmness, sweetness, and decay resistance	Jha *et al.*, 2014
	Apple	Differentiation of cultivars and detection of disease symptoms using hyperspectral imaging	Mahlein *et al.*, 2012
	Grape	Machine learning and hyperspectral imaging technology for quality prediction	Wei *et al.*, 2022
	Strawberry	Hyperspectral and multispectral imaging for assessment of ripening stage and internal defects	Liu *et al.*, 2014; Shao *et al.*, 2020
Ornamentals	Orchid	Image-based phenotyping for color pattern analysis and flower morphology	Apriyanti *et al.*, 2021
	Rose	Automated image analysis for petal shape uniformity and detection of aesthetic defects	Afonso *et al.*, 2024

the attribution of observed phenotypes to either genetic factors or management practices, as environmental influences may mask or distort genetic signals. Consequently, researchers must employ meticulous experimental design and statistical methods to disentangle these effects and achieve meaningful insights (White *et al.*, 2012).

Phenomics generates a vast amount of data that is often highly heterogeneous, encompassing images, sensor readings, genomic sequences, and environmental measurements. Moreover, different types of phenotypic data (pictures, sensor data, and manual records) need strong tools for storing, processing, and analysis. Effective handling, curating, and integrating these datasets requires robust informatics infrastructure and advanced data management strategies. Without such resources, valuable data may be lost or underutilized, and analysis may be hampered by inconsistencies and gaps. The challenge of data management is central to unlocking the full potential of phenomics in horticultural research (Singh *et al.*, 2016). One of the enduring priorities in plant phenomics is the translation of complex, high-dimensional data into actionable outcomes for breeding programs. Researchers must bridge the gap between discoveries emanating from phenomics and practical field-ready solutions that improve crop performance or resilience. This process involves synthesizing large datasets, identifying key traits, and developing selection strategies that are both efficient and effective. While significant progress has been made, the integration of phenomics into routine breeding practices remains a goal that requires ongoing attention and innovation (Tardieu *et al.*, 2017). To fully harness the potential of phenomics, it is critical to prioritize open data initiatives and standardized protocols that facilitate cross-study comparisons and meta-analyses. International consortia and public–private partnerships can play a transformative role in establishing accessible databases, comprehensive phenotyping networks, and training programs for the next generation of scientists. As phenomics matures, the continuous feedback between technological advancements and practical breeding outcomes will accelerate the development of resilient, high-yielding varieties of horticultural crops tailored to diverse environmental challenges. Ultimately, surmounting the barriers outlined above will not only advance the science of horticulture but also contribute to global food security and sustainability. By embracing innovation and fostering collaboration, the horticultural community practice is poised to unlock new frontiers in plant breeding and agricultural productivity.

13.6.5 Future perspectives and emerging trends

As phenomics continues to evolve, its transformative potential for horticultural science becomes ever more apparent. This is because several key trends are shaping the future landscape of phenomics application in horticultural plants. Among these are multi-omics integration, artificial intelligence, precision agriculture, consumer-focused breeding, and climate resilience, as elaborated below.

The convergence of genomics, transcriptomics, metabolomics, and phenomics is ushering in a new era of crop improvement. By integrating data across these diverse biological layers, researchers are now able to gain a holistic understanding of plant systems, identifying genetic markers and molecular pathways that govern complex traits such as yield, quality, and stress tolerance. Multi-omics approaches facilitate precise trait dissection and accelerate the breeding of superior cultivars, ultimately enabling the development of horticultural plants that are both productive and resilient (Mochida *et al.*, 2019).

Artificial intelligence through advanced computational tools, including but not limited to deep learning and sophisticated modeling techniques, is revolutionizing the way scientists analyze phenotypic data. Artificial intelligence enables high-throughput trait prediction and robust pattern recognition, making it possible to accurately map the relationship between genotype and phenotype in large populations. These approaches not only enhance the efficiency of breeding programs but also empower the discovery of subtle trait associations that might otherwise go unnoticed (Singh *et al.*, 2016).

Precision agriculture, which has led to the rise of automated remote phenotyping supported by Internet of Things (IoT) devices and sensor networks, has paved the way for real-time crop management. In precision agriculture, sensors continuously monitor key plant traits and environmental variables, feeding data into decision-support systems that guide irrigation, fertilization, and pest control strategies. This real-time data-driven approach optimizes resource use, maximizes crop performance, and reduces environmental impact, making horticultural crop production more sustainable and adaptive to changing conditions (Yang *et al.*, 2020).

Consumer-focused breeding is the application of phenomics to personalize crop attributes such as nutritional content, flavor, texture, and shelf life that meet the ever-evolving preferences of consumers, shaping consumer-focused breeding. By linking phenotypic data directly to market demands, breeders can more rapidly develop horticultural varieties that resonate with targeted consumer segments, driving innovation and increasing the competitiveness of horticultural products in global markets (Fiorani and Schurr, 2013).

As global climates shift, the ability to breed crops that thrive under increasingly unpredictable conditions becomes critical. Phenomics-guided trait discovery is instrumental in identifying and incorporating traits associated with heat and drought tolerance, disease resistance, and adaptability to new environments. By equipping breeders with the tools to select for climate-resilient traits, phenomics ensures the long-term sustainability of horticultural systems and food security in the face of environmental change (Tardieu *et al.*, 2017).

These emerging trends collectively illustrate how phenomics is positioned at the forefront of horticultural innovation, driving progress towards more productive, resilient, and consumer-responsive crop systems. Phenomics is emerging as a transformative tool in horticultural science, offering unparalleled resolution in the characterization of complex traits. Through integration with genomics and precision agriculture, it is accelerating breeding cycles, enhancing resource-use efficiency, and supporting adaptive management practices. As technologies become more accessible and scalable, phenomics will increasingly underpin innovations aimed at feeding the world sustainably and nutritiously.

13.7 Challenges in the Application of Spatial and Multiple Omics in Horticultural Plants

The merging of spatial and multiple omics – such as genomics, transcriptomics, proteomics, and metabolomics – in horticultural plant study is an exciting frontier that could greatly enhance our understanding of how plants grow, respond to stress, and can be improved for better traits. However, there are some significant hurdles we need to overcome. First and foremost, the biological and technical complexity of horticultural crops is incredibly diverse, which can complicate research. Additionally, the unique structure of plant tissues presents challenges. Different parts of a plant, such as fruit and seeds, have their microenvironments, making sampling spatial resolution challenging (Yu *et al.*, 2023; Rhaman *et al.*, 2024).

Many horticultural species have limited reference genomes and annotations, lack high-quality reference genomes, or have poorly annotated genomes. This deficiency makes it challenging to integrate data from multiple omics effectively, particularly for lesser-known or wild varieties (Chen *et al.*, 2019; Liu *et al.*, 2021; Ahmed *et al.*, 2022; Cheng *et al.*, 2024). Again, the cost and technical demands of advanced spatial omics techniques, such as spatial transcriptomics and proteomics, can be expensive and require specialized skills. Equipment and methods, such as *in situ* hybridization or MALDI imaging, are often not designed with plant tissues in mind. Furthermore, preserving plant tissue and slicing it for analysis can be challenging due to its lignified or watery nature (Vandereyken *et al.*, 2023; Dezem *et al.*, 2024; Hui *et al.*, 2025).

Furthermore, the datasets generated from multi-omics work can be vast, varied, and noisy, complicating the integration of spatial data with other omics data. Successfully matching spatial transcriptomics with metabolomics, for example, requires advanced computational skills, including machine learning and network analysis. There is also a lack of standardized methods and reference atlas specifically for horticultural plants to guide researchers (Dang *et al.*, 2019; Oliveira, 2019; Cembrowska-Lech *et al.*, 2023), and the spatiotemporal resolution of plant growth and reactions to their environment are dynamic and vary over time and space. Many current methods in spatial omics can only provide static snapshots, which do not fully capture the ongoing biological processes (Velten and Stegle, 2023; Liu *et al.*, 2024; Zhang *et al.*, 2024). Due to the lack of public databases and shared resources, there are limited multi-omics datasets available for horticultural plants. The absence of centralized repositories or spatial atlas complicate the sharing, comparing, and reanalysis of data (Großkinsky *et al.*, 2018; Rout and Peter, 2022; Tao *et al.*, 2022; Cheng *et al.*, 2023; Nobori, 2025). In addition, implementing insights gained from omics into practical breeding strategies is difficult. This is partly due to the complex nature of traits, interactions between genotype and environment, and the disconnect between molecular discoveries and observable plant characteristics. Moreover, developing reliable markers from multi-omics research requires testing across different environments and genetic backgrounds (Cembrowska-Lech *et al.*, 2023).

Lastly, the application of omics in horticulture, particularly in areas such as gene editing (e.g. CRISPR), faces various regulatory challenges in different countries. Addressing these challenges is vital for unlocking the potential of omics in horticulture and making strides in plant improvement and sustainability (Ahmad *et al.*, 2021; Shao *et al.*, 2021; Touzdjian Pinheiro Kohlrausch Távora *et al.*, 2022). Ethical and regulatory issues differ across continents. North America largely accepts genetically modified organisms, while this is a major concern for many European countries (Zhang *et al.*, 2020; Munawar *et al.*, 2024).

References

Abdulridha, J., Ampatzidis, Y., Kakarla, S.C. and Roberts, P. (2020) Detection of target spot and bacterial spot diseases in tomato using UAV-based and benchtop-based hyperspectral imaging techniques. *Precision Agriculture* 21(5), 955–978. DOI: 10.1007/s11119-019-09703-4.

Afonso, M., Paulo, M.-J., Fonteijn, H., van den Helder, M., Zwinkels, H. *et al.* (2024) Automatic trait estimation in floriculture using computer vision and deep learning. *Smart Agricultural Technology* 7, 100383. DOI: 10.1016/j.atech.2023.100383.

Ahmad, A., Ghouri, M.Z., Munawar, N., Ismail, M., Ashraf, S. *et al.* (2021) Regulatory, ethical, and social aspects of CRISPR crops. In: *CRISPR Crops: The Future of Food Security*. Springer, Germany. DOI: 10.1007/978-981-15-7142-8_9.

Ahmad, P., Abdel Latef, A.A.H., Rasool, S., Akram, N.A., Ashraf, M. *et al.* (2016) Role of proteomics in crop stress tolerance. *Frontiers in Plant Science* 7, 1336. DOI: 10.3389/fpls.2016.01336.

Ahmed, M., Babayola, M. and Bake, I.D. (2024) Role of horticultural crops in food and nutritional security: A review. *Jurnal of Nutrition and Food Processing* 7(8), 01–06.

Ahmed, U., Xie, Q., Shi, X. and Zheng, B. (2022) Development of reference genes for horticultural plants. *Critical Reviews in Plant Sciences* 41(3), 190–208. DOI: 10.1080/07352689.2022.2084227.

Apriyanti, D.H., Spreeuwers, L.J., Lucas, P.J. and Veldhuis, R.N. (2021) Automated color detection in orchids using color labels and deep learning. *PloS One* 16(10), e0259036. DOI: 10.1371/journal.pone.0259036.

Araus, J.L. and Cairns, J.E. (2014) Field high-throughput phenotyping: The new crop breeding frontier. *Trends in Plant Science* 19(1), 52–61. DOI: 10.1016/j.tplants.2013.09.008.

Bally, I.S.E., Bombarely, A., Chambers, A.H., Cohen, Y., Dillon, N.L. *et al.* (2021) The 'Tommy Atkins' mango genome reveals candidate genes for fruit quality. *BMC Plant Biology* 21(1), 108. DOI: 10.1186/s12870-021-02858-1.

Borbaruah, R. and Borbaruah, N. (2023) Horticultural economics and market trends. *New Horizons and Advancements in Horticulture* 211.

Cembrowska-Lech, D., Krzemińska, A., Miller, T., Nowakowska, A., Adamski, C. *et al.* (2023) An integrated multi-omics and artificial intelligence framework for advance plant phenotyping in horticulture. *Biology* 12(10), 1298. DOI: 10.3390/biology12101298.

Chao, Y.U., Hui-hua, W., Bourke, P.M., Cheng, B.X., Le, L.U.O. *et al.* (2021) High density genetic map and quantitative trait loci (QTLs) associated with petal number and flower diameter identified in tetraploid rose. *Journal of Integrative Agriculture* 20(5), 1287–1301. DOI: 10.1016/S2095-3119(20)63416-5.

Chen, F., Song, Y., Li, X., Chen, J., Mo, L. *et al.* (2019) Genome sequences of horticultural plants: Past, present, and future. *Horticulture Research* 6(1). DOI: 10.1038/s41438-019-0195-6.

Chen, H., Zeng, X., Yang, J., Cai, X., Shi, Y. *et al.* (2021) Whole-genome resequencing of *Osmanthus fragrans* provides insights into flower color evolution. *Horticulture Research* 8(1). DOI: 10.1038/s41438-021-00531-0.

Chen, Y., Mao, W., Liu, T., Feng, Q., Li, L. *et al.* (2020) Genome editing as a versatile tool to improve horticultural crop qualities. *Horticultural Plant Journal* 6(6), 372–384. DOI: 10.1016/j.hpj.2020.11.004.

Cheng, B., Du, W., Bourke, P.M. and Yu, C. (2024) Population genetics of horticultural crops aided by multi-omics technology and its implications for ornamental plants. *Ornamental Plant Research* 4(1), 0–0. DOI: 10.48130/opr-0024-0020.

Cheng, H., Zhang, H., Song, J., Jiang, J., Chen, S. *et al.* (2023) GERDH: An interactive multi-omics database for cross-species data mining in horticultural crops. *The Plant Journal* 116(4), 1018–1029. DOI: 10.1111/tpj.16350.

Cobb, J.N., DeClerck, G., Greenberg, A., Clark, R. and McCouch, S. (2013) Next-generation phenotyping: Requirements and strategies for enhancing our understanding of genotype–phenotype relationships and its relevance to crop improvement. *Theoretical and Applied Genetics* 126(4), 867–887. DOI: 10.1007/s00122-013-2066-0.

Cramer, G.R., Urano, K., Delrot, S., Pezzotti, M. and Shinozaki, K. (2011) Effects of abiotic stress on plants: A systems biology perspective. *BMC Plant Biology* 11(1), 163. DOI: 10.1186/1471-2229-11-163.

Dang, Y., Lin, Q. and Huang, P. (2019) AIOps: Real-world challenges and research innovations. In: *2019 IEEE/ACM 41st International Conference on Software Engineering*, Montreal, QC, Canada. DOI: 10.1109/ICSE-Companion.2019.00023.

Dezem, F.S., Arjumand, W., DuBose, H., Morosini, N.S. and Plummer, J. (2024) Spatially resolved single-cell omics: Methods, challenges, and future perspectives. *Annual Review of Biomedical Data Science* 7(1), 131–153. DOI: 10.1146/annurev-biodatasci-102523-103640.

Dhondt, S., Wuyts, N. and Inzé, D. (2013) Cell to whole-plant phenotyping: The best is yet to come. *Trends in Plant Science* 18(8), 428–439. DOI: 10.1016/j.tplants.2013.04.008.

Estêvão, C., Rodrigues, L., Rato, A.E., Garcia, R., Cardoso, H. *et al.* (2024) Applicability of metabolomics to improve sustainable grapevine production. *Frontiers in Molecular Biosciences* 11, 1395677. DOI: 10.3389/fmolb.2024.1395677.

Fahlgren, N., Gehan, M.A. and Baxter, I. (2015) Lights, camera, action: High-throughput plant phenotyping is ready for a close-up. *Current Opinion in Plant Biology* 24, 93–99. DOI: 10.1016/j.pbi.2015.02.006.

Ference, C.M., Manthey, J.A., Narciso, J.A., Jones, J.B. and Baldwin, E.A. (2020) Detection of phenylpropanoids in citrus leaves produced in response to *Xanthomonas citri* subsp. *Phytopathology* 110(2), 287–296. DOI: 10.1094/PHYTO-06-19-0219-R.

Fernie, A.R. and Schauer, N. (2009) Metabolomics-assisted breeding: A viable option for crop improvement. *Trends in Genetics* 25(1), 39–48. DOI: 10.1016/j.tig.2008.10.010.

Fiorani, F. and Schurr, U. (2013) Future scenarios for plant phenotyping. *Annual Review of Plant Biology* 64(1), 267–291. DOI: 10.1146/annurev-arplant-050312-120137.

Fu, Y., Jiang, S., Zou, M., Xiao, J., Yang, L. *et al.* (2022) High-quality reference genome sequences of two *Cannaceae* species provide insights into the evolution of *Cannaceae*. *Frontiers in Plant Science* 13, 955904. DOI: 10.3389/fpls.2022.955904.

Furbank, R.T. and Tester, M. (2011) Phenomics–technologies to relieve the phenotyping bottleneck. *Trends in Plant Science* 16(12), 635–644. DOI: 10.1016/j.tplants.2011.09.005.

García-Mier, L., Guevara-González, R., Mondragón-Olguín, V., Del Rocío Verduzco-Cuellar, B. and Torres-Pacheco, I. (2013) Agriculture and bioactives: Achieving both crop yield and phytochemicals. *International Journal of Molecular Sciences* 14(2), 4203–4222. DOI: 10.3390/ijms14024203.

Garg, M., Sharma, N., Sharma, S., Kapoor, P., Kumar, A. *et al.* (2018) Biofortified crops generated by breeding, agronomy, and transgenic approaches are improving lives of millions of people around the world. *Frontiers in Nutrition* 5, 12. DOI: 10.3389/fnut.2018.00012.

Großkinsky, D.K., Svensgaard, J., Christensen, S. and Roitsch, T. (2015) Plant phenomics and the need for physiological phenotyping across scales to narrow the genotype-to-phenotype knowledge gap. *Journal of Experimental Botany* 66(18), 5429–5440. DOI: 10.1093/jxb/erv345.

Großkinsky, D.K., Syaifullah, S.J. and Roitsch, T. (2018) Integration of multi-omics techniques and physiological phenotyping within a holistic phenomics approach to study senescence in model and crop plants. *Journal of Experimental Botany* 69(4), 825–844. DOI: 10.1093/jxb/erx333.

Han, F., Liu, Y., Fang, Z., Yang, L., Zhuang, M. *et al.* (2021) Advances in genetics and molecular breeding of broccoli. *Horticulturae* 7(9), 280. DOI: 10.3390/horticulturae7090280.

Han, W., Ward, J.L., Kong, Y. and Li, X. (2023) Editorial: Targeted and untargeted metabolomics for the evaluation of plant metabolites in response to the environment. *Frontiers in Plant Science* 14, 1167513. DOI: 10.3389/fpls.2023.1167513.

Hao, N., Han, D., Huang, K., Du, Y., Yang, J. *et al.* (2020) Genome-based breeding approaches in major vegetable crops. *Theoretical and Applied Genetics* 133(5), 1739–1752. DOI: 10.1007/s00122-019-03477-z.

Hao, Y., Zhang, Z., Luo, E., Yang, J. and Wang, S. (2025) Plant metabolomics: Applications and challenges in the era of multi-omics big data. *aBIOTECH* 6(1), 116–132. DOI: 10.1007/s42994-024-00194-0.

Hu, J., Rampitsch, C. and Bykova, N.V. (2015) Advances in plant proteomics toward improvement of crop productivity and stress resistancex. *Frontiers in Plant Science* 6, 209. DOI: 10.3389/fpls.2015.00209.

Hui, T., Zhou, J., Yao, M., Xie, Y. and Zeng, H. (2025) Advances in spatial omics technologies. *Small Methods* 9(5), e2401171. DOI: 10.1002/smtd.202401171.

Jaillon, O., Aury, J., Noel, B., Policriti, A., Clepet, C. *et al.* (2007) French-Italian public consortium for grapevine genome characterization the grapevine genome sequence suggests ancestral hexaploidization in major angiosperm phyla. *Nature* 449(7161), 463–467. DOI: 10.1038/nature06148.

Jamil, I.N., Remali, J., Azizan, K.A., Nor Muhammad, N.A., Arita, M. *et al.* (2020) Systematic multi-omics integration (MOI) approach in plant systems biology. *Frontiers in Plant Science* 11, 944. DOI: 10.3389/fpls.2020.00944.

Jan, N., Rather, A.M.-U.-D., John, R., Chaturvedi, P., Ghatak, A. *et al.* (2023) Proteomics for abiotic stresses in legumes: Present status and future directions. *Critical Reviews in Biotechnology* 43(2), 171–190. DOI: 10.1080/07388551.2021.2025033.

Jha, S.N., Narsaiah, K., Jaiswal, P., Bhardwaj, R., Gupta, M. *et al.* (2014) Nondestructive prediction of maturity of mango using near infrared spectroscopy. *Journal of Food Engineering* 124, 152–157. DOI: 10.1016/j.jfoodeng.2013.10.012.

Jideani, A.I.O., Silungwe, H., Takalani, T., Omolola, A.O., Udeh, H.O. *et al.* (2021) Antioxidant-rich natural fruit and vegetable products and human health. *International Journal of Food Properties* 24(1), 41–67. DOI: 10.1080/10942912.2020.1866597.

Karahara, I., Yamauchi, D., Uesugi, K. and Mineyuki, Y. (2023) Three-dimensional visualization of plant tissues and organs by X-ray micro–computed tomography. *Microscopy: The Journal of the Quekett Microscopical Club* 72(4), 310–325. DOI: 10.1093/jmicro/dfad026.

Li, X.-X. and Lai, Y.-S. (2024) The development and application of vegetable genomics increase the efficiency of exploring new gene resources for vegetables. *International Journal of Molecular Sciences* 25(13), 6906. DOI: 10.3390/ijms25136906.

Li, H., Zhai, X., Peng, H., Qing, Y., Deng, Y. *et al.* (2024a) Chromosomal level genome assemblies of two *Malus crabapple* cultivars flame and royalty. *Scientific Data* 11(1), 201. DOI: 10.1038/s41597-024-03049-x.

Li, L., Zhang, Q. and Huang, D. (2014) A review of imaging techniques for plant phenotyping. *Sensors* 14(11), 20078–20111. DOI: 10.3390/s141120078.

Li, M., Wen, Z., Meng, J., Cheng, T., Zhang, Q. *et al.* (2022) The genomics of ornamental plants: Current status and opportunities. *Ornamental Plant Research* 2(1), 1–18. DOI: 10.48130/OPR-2022-0006.

Li, Y., Sun, Z., Jin, Z. and Li, J. (2024b) Genome sequencing provides insights into *Caprifoliaceae* genome evolution and the mechanism underlying second blooming phenomenon in *Heptacodium miconioides*. *Horticultural Plant Journal* 10(3), 897–910. DOI: 10.1016/j.hpj.2022.11.008.

Linsmith, G., Rombauts, S., Montanari, S., Deng, C.H., Celton, J.-M. *et al.* (2019) Pseudo-chromosome–length genome assembly of a double haploid "Bartlett" pear (*Pyrus communis* L.). *GigaScience* 8(12), giz138. DOI: 10.1093/gigascience/giz138.

Liu, C., Liu, W., Lu, X., Ma, F., Chen, W. *et al.* (2014) Application of multispectral imaging to determine quality attributes and ripeness stage in strawberry fruit. *PLoS ONE* 9(2), e87818. DOI: 10.1371/journal.pone.0087818.

Liu, T., Li, M., Liu, Z., Ai, X. and Li, Y. (2021) Reannotation of the cultivated strawberry genome and establishment of a strawberry genome database. *Horticulture Research* 8(1). DOI: 10.1038/s41438-021-00476-4.

Liu, X., Peng, T., Xu, M., Lin, S., Hu, B. *et al.* (2024) Spatial multi-omics: Deciphering technological landscape of integration of multi-omics and its applications. *Journal of Hematology & Oncology* 17(1), 72. DOI: 10.1186/s13045-024-01596-9.

Liu, Y., Lu, S., Liu, K., Wang, S., Huang, L. *et al.* (2019) Proteomics: A powerful tool to study plant responses to biotic stress. *Plant Methods* 15(1), 135. DOI: 10.1186/s13007-019-0515-8.

Lopes, N. dos S., Santos, A.S., de Novais, D.P.S., Pirovani, C.P. and Micheli, F. (2023) Pathogenesis-related protein 10 in resistance to biotic stress: Progress in elucidating functions, regulation and modes of action. *Frontiers in Plant Science* 14, 1193873. DOI: 10.3389/fpls.2023.1193873.

López-Gresa, M.P., Lisón, P., Campos, L., Rodrigo, I., Rambla, J.L. *et al.* (2017) A non-targeted metabolomics approach unravels the VOCs associated with the tomato immune response against *Pseudomonas syringae*. *Frontiers in Plant Science* 8, 1188. DOI: 10.3389/fpls.2017.01188.

Lu, Y., Huang, Y. and Lu, R. (2017) Innovative hyperspectral imaging-based techniques for quality evaluation of fruits and vegetables: A review. *Applied Sciences* 7(2), 189. DOI: 10.3390/app7020189.

Mahlein, A.-K., Oerke, E.-C., Steiner, U. and Dehne, H.-W. (2012) Recent advances in sensing plant diseases for precision crop protection. *European Journal of Plant Pathology* 133(1), 197–209. DOI: 10.1007/s10658-011-9878-z.

Mahlein, A.-K., Rumpf, T., Welke, P., Dehne, H.-W., Plümer, L. *et al.* (2013) Development of spectral indices for detecting and identifying plant diseases. *Remote Sensing of Environment* 128, 21–30. DOI: 10.1016/j.rse.2012.09.019.

Manzoor, I., Samantara, K., Bhat, M.S., Farooq, I., Bhat, K.M. *et al.* (2023) Advances in genomics for diversity studies and trait improvement in temperate fruit and nut crops under changing climatic scenarios. *Frontiers in Plant Science* 13, 1048217. DOI: 10.3389/fpls.2022.1048217.

Martínez-Esteso, M.J., Morante-Carriel, J., Samper-Herrero, A., Martínez-Márquez, A., Sellés-Marchart, S. *et al.* (2024) Proteomics: An essential tool to study plant-specialized metabolism. *Biomolecules* 14(12), 1539. DOI: 10.3390/biom14121539.

Mathiazhagan, M., Chidambara, B., Hunashikatti, L.R. and Ravishankar, K.V. (2021) Genomic approaches for improvement of tropical fruits: Fruit quality, shelf life and nutrient content. *Genes* 12(12), 1881. DOI: 10.3390/genes12121881.

Meléndez-Martínez, A.J., Mandić, A.I., Bantis, F., Böhm, V., Borge, G.I.A. *et al.* (2022) A comprehensive review on carotenoids in foods and feeds: Status quo, applications, patents, and research needs. *Critical Reviews in Food Science and Nutrition* 62(8), 1999–2049. DOI: 10.1080/10408398.2020.1867959.

Meng, L., Chen, G., Liu, D. and Tian, N. (2025) Universal modeling for non-destructive testing of soluble solids content in multi-variety blueberries based on hyperspectral imaging technology. *Applied Sciences* 15(7), 3888. DOI: 10.3390/app15073888.

Minervini, M., Scharr, H. and Tsaftaris, S.A. (2015) Image analysis: The new bottleneck in plant phenotyping. *IEEE Signal Processing Magazine* 32(4), 126–131. DOI: 10.1109/MSP.2015.2405111.

Ming, R., VanBuren, R., Wai, C.M., Tang, H., Schatz, M.C. *et al.* (2015) The pineapple genome and the evolution of CAM photosynthesis. *Nature Genetics* 47(12), 1435–1442. DOI: 10.1038/ng.3435.

Mir, R.R., Reynolds, M., Pinto, F., Khan, M.A. and Bhat, M.A. (2019) High-throughput phenotyping for crop improvement in the genomics era. *Plant Science* 282, 60–72. DOI: 10.1016/j.plantsci.2019.01.007.

Mochida, K., Koda, S., Inoue, K., Hirayama, T., Tanaka, S. *et al.* (2019) Computer vision-based phenotyping for improvement of plant productivity: A machine learning perspective. *GigaScience* 8(1), giy153. DOI: 10.1093/gigascience/giy153.

Munawar, N., Faheem, M., Niamat, A., Munir, A. and Khan, S.H. (2024) Regulatory, ethical, social, and biosafety concerns in genome-edited horticultural crops. In: *CRISPRized Horticulture Crops*. Elsevier, Netherlands, pp. 421–438. DOI: 10.1016/B978-0-443-13229-2.00026-0.

Murchie, E.H. and Lawson, T. (2013) Chlorophyll fluorescence analysis: A guide to good practice and understanding some new applications. *Journal of Experimental Botany* 64(13), 3983–3998. DOI: 10.1093/jxb/ert208.

Murti, R.H., Afifah, E.N. and Nuringtyas, T.R. (2021) Metabolomic response of tomatoes (*Solanum lycopersicum* L.) against bacterial wilt (*Ralstonia solanacearum*) using 1H-NMR spectroscopy. *Plants* 10(6), 1143. DOI: 10.3390/plants10061143.

National Academies of Sciences, Engineering, and Medicine, Engineering, Medicines, Division on Earth, Life Studies and Board on Agriculture and Natural Resources *et al.* (2016) *Genetically Engineered crops: Experiences and Prospects*. National Academies Press, Washington, USA.

Nobori, T. (2025) Exploring the untapped potential of single-cell and spatial omics in plant biology. *New Phytologist* 247(3), 1098–1116. DOI: 10.1111/nph.70220.

Oliveira, A.L. (2019) Biotechnology, big data and artificial intelligence. *Biotechnology Journal* 14(8), e1800613. DOI: 10.1002/biot.201800613.

Pagariya, M.C., Jadhav, P.R., Gawade, N.S., Sarode, D.K., Karkute, S.G. *et al.* (2025) Advancing ornamental plant breeding through genomic technologies: Opportunities, challenges, and future directions. *Functional & Integrative Genomics* 25(1), 1–23. DOI: 10.1007/s10142-025-01640-y.

Palma, J.M., Corpas, F.J. and del Río, L.A. (2011) Proteomics as an approach to the understanding of the molecular physiology of fruit development and ripening. *Journal of Proteomics* 74(8), 1230–1243. DOI: 10.1016/j.jprot.2011.04.010.

Patel, M.K., Pandey, S., Kumar, M., Haque, M.I., Pal, S. *et al.* (2021) Plants metabolome study: Emerging tools and techniques. *Plants* 10(11), 2409. DOI: 10.3390/plants10112409.

Patil, P.G., Singh, N., Parashuram, S., Sowjanya, P.R., Sharma, J. *et al.* (2022) Recent advances in pomegranate genomics: Status and prospects. In: *Omics in Horticultural Crops*. Academic Press, Netherlands. DOI: 10.1016/B978-0-323-89905-5.00023-9.

Paupière, M.J., Tikunov, Y., Schleiff, E., Bovy, A. and Fragkostefanakis, S. (2020) Reprogramming of tomato leaf metabolome by the activity of heat stress transcription factor HsfB1. *Frontiers in Plant Science* 11, 610599. DOI: 10.3389/fpls.2020.610599.

Penna, S. and Jain, S.M. (2023) Fruit crop improvement with genome editing, *in vitro* and transgenic approaches. *Horticulturae* 9(1), 58. DOI: 10.3390/horticulturae9010058.

Peška, V., Mandáková, T., Ihradská, V. and Fajkus, J. (2019) Comparative dissection of three giant genomes: *Allium cepa*, *Allium sativum*, and *Allium ursinum*. *International Journal of Molecular Sciences* 20(3), 733. DOI: 10.3390/ijms20030733.

Piovesan, A., Vancauwenberghe, V., Van De Looverbosch, T., Verboven, P. and Nicolaï, B. (2021) X-ray computed tomography for 3D plant imaging. *Trends in Plant Science* 26(11), 1171–1185. DOI: 10.1016/j.tplants.2021.07.010.

Poland, J. and Rutkoski, J. (2016) Advances and challenges in genomic selection for disease resistance. *Annual Review of Phytopathology* 54(1), 79–98.

Rahaman, M.M., Chen, D., Gillani, Z., Klukas, C. and Chen, M. (2015) Advanced phenotyping and phenotype data analysis for the study of plant growth and development. *Frontiers in Plant Science* 6, 619. DOI: 10.3389/fpls.2015.00619.

Rajaniand Shourabh Joshi, R. (2017) The role of horticultural crops in enhancing nutrient security. *International Journal of Current Microbiology and Applied Sciences* 6(9), 311–316. DOI: 10.20546/ijcmas.2017.609.039.

Raymond, O., Gouzy, J., Just, J., Badouin, H., Verdenaud, M. *et al.* (2018) The *Rosa* genome provides new insights into the domestication of modern roses. *Nature Genetics* 50(6), 772–777. DOI: 10.1038/s41588-018-0110-3.

Rhaman, M.S., Ali, M., Ye, W. and Li, B. (2024) Opportunities and challenges in advancing plant research with single-cell omics. *Genomics, Proteomics & Bioinformatics* 22(2), qzae026. DOI: 10.1093/gpbjnl/qzae026.

Rhoads, A. and Au, K.F. (2015) PacBio sequencing and its applications. *Genomics, Proteomics & Bioinformatics* 13(5), 278–289. DOI: 10.1016/j.gpb.2015.08.002.

Roumani, M., Le Bot, J., Boisbrun, M., Magot, F., Péré, A. *et al.* (2022) Transcriptomics and metabolomics analyses reveal high induction of the phenolamide pathway in tomato plants attacked by the leafminer *Tuta absoluta*. *Metabolites* 12(6), 484. DOI: 10.3390/metabo12060484.

Rout, G.R. and Peter, K. (2022) *Omics in horticultural crops*. Academic Press, Netherlands.

Ruggieri, V., Bostan, H., Barone, A., Frusciante, L. and Chiusano, M.L. (2016) Integrated bioinformatics to decipher the ascorbic acid metabolic network in tomato. *Plant Molecular Biology* 91(4–5), 397–412. DOI: 10.1007/s11103-016-0469-4.

Saeed, F., Chaudhry, U.K., Raza, A., Charagh, S., Bakhsh, A. *et al.* (2023) Developing future heat-resilient vegetable crops. *Functional & Integrative Genomics* 23(1), 47. DOI: 10.1007/s10142-023-00967-8.

Salam, U., Ullah, S., Tang, Z.-H., Elateeq, A.A., Khan, Y. *et al.* (2023) Plant metabolomics: An overview of the role of primary and secondary metabolites against different environmental stress factors. *Life (Basel, Switzerland)* 13(3), 706. DOI: 10.3390/life13030706.

Sambo, B.E., Ahmed, A., Omotugba, J.T., Adedapo, J. and Shekwolo, G. (2016) The challenge of infrastructure development on preservation of perishable horticultural produce for sustained urban food nutrition in Metropolitan Kaduna, Nigeria. *International Journal of Horticulture and Forestry Research* 1(1), 13–26.

Satrio, R.D., Fendiyanto, M.H. and Miftahudin, M. (2024) Tools and techniques used at global scale through genomics, transcriptomics, proteomics, and metabolomics to investigate plant stress responses at the molecular level. In: *Molecular Dynamics of Plant Stress and Its Management*. Springer, Germany. DOI: 10.1007/978-981-97-1699-9_25.

Savadi, S., Mangalassery, S. and Sandesh, M.S. (2021) Advances in genomics and genome editing for breeding next generation of fruit and nut crops. *Genomics* 113(6), 3718–3734. DOI: 10.1016/j.ygeno.2021.09.001.

Shao, Y., Wang, Y., Xuan, G., Gao, Z., Hu, Z. *et al.* (2020) Assessment of strawberry ripeness using hyperspectral imaging. *Analytical Letters* 54(10), 1547–1560.

Shao, Y., Wang, Y., Xuan, G., Gao, Z., Hu, Z. *et al.* (2021) Assessment of strawberry ripeness using hyperspectral imaging. *Analytical Letters* 54(10), 1547–1560. DOI: 10.1080/00032719.2020.1812622.

Shen, Y., Sun, T., Pan, Q., Anupol, N., Chen, H. *et al.* (2019) RrMYB5- and RrMYB10-regulated flavonoid biosynthesis plays a pivotal role in feedback loop responding to wounding and oxidation in *Rosa rugosa*. *Plant Biotechnology Journal* 17(11), 2078–2095. DOI: 10.1111/pbi.13123.

Simko, I., Jia, M., Venkatesh, J., Kang, B.-C., Weng, Y. *et al.* (2021) Genomics and marker-assisted improvement of vegetable crops. *Critical Reviews in Plant Sciences* 40(4), 303–365. DOI: 10.1080/07352689.2021.1941605.

Singh, A., Ganapathysubramanian, B., Singh, A.K. and Sarkar, S. (2016) Machine learning for high-throughput stress phenotyping in plants. *Trends in Plant Science* 21(2), 110–124. DOI: 10.1016/j.tplants.2015.10.015.

Singh, D., Singh, A. and Singh, A. (2021a) Plant breeding: Past, present, and future perspectives. *Plant Breeding and Cultivar Development* 1–24. DOI: 10.1016/B978-0-12-817563-7.00004-0.

Singh, N.K., Mahato, A.K. and Jayaswal, P.K. (2021b) The genome sequence and transcriptome studies in mango (*Mangifera indica* L.). In: *The Mango Genome*. Springer, Germany, pp. 165–186. DOI: 10.1007/978-3-030-47829-2_10.

Suh, J.H., Tang, X., Zhang, Y., Gmitter, F.G., Jr. and Wang, Y. (2021) Metabolomic analysis provides new insight into tolerance of Huanglongbing in citrus. *Frontiers in Plant Science* 12, 710598. DOI: 10.3389/fpls.2021.710598.

Sun, Y., Ma, R., Yang, X. and Zhang, G. (2024) A metabolomic analysis of tomato fruits in response to salt stress. *Horticulturae* 10(12), 1303. DOI: 10.3390/horticulturae10121303.

Tao, H., Xu, S., Tian, Y., Li, Z., Ge, Y. *et al.* (2022) Proximal and remote sensing in plant phenomics: 20 years of progress, challenges, and perspectives. *Plant Communications* 3(6), 100344. DOI: 10.1016/j.xplc.2022.100344.

Tardieu, F., Cabrera-Bosquet, L., Pridmore, T. and Bennett, M. (2017) Plant phenomics, from sensors to knowledge. *Current Biology* 27(15), R770–R783. DOI: 10.1016/j.cub.2017.05.055.

Tester, M. and Langridge, P. (2010) Breeding technologies to increase crop production in a changing world. *Science (New York, N.Y.)* 327(5967), 818–822. DOI: 10.1126/science.1183700.

Touzdjian Pinheiro Kohlrausch Távora, F., de Assis Santos Diniz, F., de Moraes Rêgo-Machado, C., Chagas Freitas, N., Barbosa Monteiro Arraes, F. *et al.* (2022) CRISPR/Cas- and topical RNAi-based technologies for crop management and improvement: Reviewing the risk assessment and challenges towards a more sustainable agriculture. *Frontiers in Bioengineering and Biotechnology* 10, 913728. DOI: 10.3389/fbioe.2022.913728.

Turner, S.D., Ellison, S.L., Senalik, D.A., Simon, P.W., Spalding, E.P. *et al.* (2018) An automated image analysis pipeline enables genetic studies of shoot and root morphology in carrot (*Daucus carota* L.). *Frontiers in Plant Science* 9, 1703. DOI: 10.3389/fpls.2018.01703.

van Nocker, S. and Gardiner, S.E. (2014) Breeding better cultivars, faster: Applications of new technologies for the rapid deployment of superior horticultural tree crops. *Horticulture Research* 1(1). DOI: 10.1038/hortres.2014.22.

Vanaraj, R., B., I.P., Mayakrishnan, G., Kim, I.S., Kim, S.-C. *et al.* (2025) A systematic review of the applications of electronic nose and electronic tongue in food quality assessment and safety. *Chemosensors* 13(5), 161. DOI: 10.3390/chemosensors13050161.

Vandereyken, K., Sifrim, A., Thienpont, B. and Voet, T. (2023) Methods and applications for single-cell and spatial multi-omics. *Nature Reviews. Genetics* 24(8), 494–515. DOI: 10.1038/s41576-023-00580-2.

Varadharajan, V., Rajendran, R., Muthuramalingam, P., Runthala, A., Madhesh, V. *et al.* (2025) Multi-omics approaches against abiotic and biotic stress—A review. *Plants* 14(6), 865. DOI: 10.3390/plants14060865.

Velazquez-Chavez, L., Daccache, A., Mohamed, A. and Centritto, M. (2024) Plant-based and remote sensing for water status monitoring of orchard crops: Systematic review and meta-analysis. *Agricultural Water Management* 303, 109051. DOI: 10.1016/j.agwat.2024.109051.

Velten, B. and Stegle, O. (2023) Principles and challenges of modeling temporal and spatial omics data. *Nature Methods* 20(10), 1462–1474. DOI: 10.1038/s41592-023-01992-y.

Wan, L., Wang, Z., Tang, M., Hong, D., Sun, Y. *et al.* (2021) CRISPR-Cas9 gene editing for fruit and vegetable crops: Strategies and prospects. *Horticulturae* 7(7), 193. DOI: 10.3390/horticulturae7070193.

Wang, Y. and Zhang, Y. (2022) Advances in molecular breeding of vegetable crops. *Horticulturae* 8(9), 821. DOI: 10.3390/horticulturae8090821.

Wang, P., Luo, Y., Huang, J., Gao, S., Zhu, G. *et al.* (2020) The genome evolution and domestication of tropical fruit mango. *Genome Biology* 21(1), 60. DOI: 10.1186/s13059-020-01959-8.

Wang, Y., Zhao, Y., Bollas, A., Wang, Y. and Au, K.F. (2021) Nanopore sequencing technology, bioinformatics and applications. *Nature Biotechnology* 39(11), 1348–1365. DOI: 10.1038/s41587-021-01108-x.

Weckwerth, W. and Morgenthal, K. (2005) Metabolomics: From pattern recognition to biological interpretation. *Drug Discovery Today* 10(22), 1551–1558. DOI: 10.1016/S1359-6446(05)03609-3.

Wei, X., Wu, L., Ge, D., Yao, M. and Bai, Y. (2022) Prediction of the maturity of greenhouse grapes based on imaging technology. *Plant Phenomics* 9753427, 1–14. DOI: 10.34133/2022/9753427.

White, J.W., Andrade-Sanchez, P., Gore, M.A., Bronson, K.F., Coffelt, T.A. *et al.* (2012) Field-based phenomics for plant genetics research. *Field Crops Research* 133, 101–112. DOI: 10.1016/j.fcr.2012.04.003.

Wu, G.A., Prochnik, S., Jenkins, J., Salse, J., Hellsten, U. *et al.* (2014) Sequencing of diverse mandarin, pummelo and orange genomes reveals complex history of admixture during citrus domestication. *Nature Biotechnology* 32(7), 656–662. DOI: 10.1038/nbt.2906.

Xu, F., Chen, S., Yang, X., Zhou, S., Chen, X. *et al.* (2020) Genome-wide association study on seminal and nodal roots of wheat under different growth environments. *Frontiers in Plant Science* 11, 602399. DOI: 10.3389/fpls.2020.602399.

Yahia, E.M., Fonseca, J.M. and Kitinoja, L. (2019) Postharvest losses and waste. In: *Postharvest Technology of Perishable Horticultural Commodities*. Woodhead Publishing, United Kingdom. DOI: 10.1016/B978-0-12-813276-0.00002-X.

Yang, W., Duan, L., Chen, G., Xiong, L. and Liu, Q. (2013) Plant phenomics and high-throughput phenotyping: Accelerating rice functional genomics using multidisciplinary technologies. *Current Opinion in Plant Biology* 16(2), 180–187. DOI: 10.1016/j.pbi.2013.03.005.

Yang, W., Feng, H., Zhang, X., Zhang, J., Doonan, J.H. *et al.* (2020) Crop phenomics and high-throughput phenotyping: Past decades, current challenges, and future perspectives. *Molecular Plant* 13(2), 187–214. DOI: 10.1016/j.molp.2020.01.008.

Yin, Z., Huang, W., Li, K., Fernie, A.R. and Yan, S. (2024) Advances in mass spectrometry imaging for plant metabolomics-expanding the analytical toolbox. *The Plant Journal* 119(5), 2168–2180. DOI: 10.1111/tpj.16924.

Yu, X., Liu, Z. and Sun, X. (2023) Single-cell and spatial multi-omics in the plant sciences: Technical advances, applications, and perspectives. *Plant Communications* 4(3), 100508. DOI: 10.1016/j.xplc.2022.100508.

Zaman-Allah, M., Vergara, O., Araus, J.L., Tarekegne, A., Magorokosho, C. *et al.* (2015) Unmanned aerial platform-based multi-spectral imaging for field phenotyping of maize. *Plant Methods* 11(1), 35. DOI: 10.1186/s13007-015-0078-2.

Zavafer, A., Bates, H., Mancilla, C. and Ralph, P.J. (2023) Phenomics: Conceptualization and importance for plant physiology. *Trends in Plant Science* 28(9), 1004–1013. DOI: 10.1016/j.tplants.2023.03.023.

Zhang, D., Hussain, A., Manghwar, H., Xie, K., Xie, S. *et al.* (2020) Genome editing with the CRISPR-Cas system: An art, ethics and global regulatory perspective. *Plant Biotechnology Journal* 18(8), 1651–1669. DOI: 10.1111/pbi.13383.

Zhang, H., Lu, K.H., Ebbini, M., Huang, P., Lu, H. *et al.* (2024) Mass spectrometry imaging for spatially resolved multi-omics molecular mapping. *Npj Imaging* 2(1), 20. DOI: 10.1038/s44303-024-00025-3.

Zhang, L., Hu, J., Han, X., Li, J., Gao, Y. *et al.* (2019) A high-quality apple genome assembly reveals the association of a retrotransposon and red fruit colour. *Nature Communications* 10(1), 1494. DOI: 10.1038/s41467-019-09518-x.

Zhang, W., Zhang, H., Ning, L., Li, B. and Bao, M. (2016) Quantitative proteomic analysis provides novel insights into cold stress responses in *Petunia* seedlings. *Frontiers in Plant Science* 7, 136. DOI: 10.3389/fpls.2016.00136.

Zheng, T., Li, P., Li, L. and Zhang, Q. (2021) Research advances in and prospects of ornamental plant genomics. *Horticulture Research* 8(1). DOI: 10.1038/s41438-021-00499-x.

Zivy, M. and de Vienne, D. (2000) Proteomics: A link between genomics, genetics and physiology. *Plant Molecular Biology* 44(5), 575–580. DOI: 10.1023/a:1026525406953.

14 Spatial and Multiple Omics Approaches for Pest and Disease Management: Opportunities and Challenges

Augustine Antwi-Boasiako[1,2]*, Priscilla Amponsah[3], Joseph Adomako[1,4]*, Kofi Frimpong-Anin[1], Godfried Ohene-Mensah[1], Augustine Boakye Boateng[1], Richard Dormatey[1], Ernest Baafi[1], Esther Afoley Annang[1] and Daouda Coulibaly[5]

[1]Council for Scientific and Industrial Research-Crops Research Institute (CSIR-CRI), Kumasi, Ghana; [2]Key Laboratory of Biology and Genetics Improvement of Soybean, Ministry of Agriculture, Zhongshan Biological Breeding Laboratory (ZSBBL), National Innovation Platform for Soybean Breeding and Industry-Education Integration, State Key Laboratory of Crop Genetics & Germplasm Enhancement and Utilization, College of Agriculture, Nanjing Agricultural University, Nanjing, China; [3]State Key Laboratory of Agricultural and Forestry Biosecurity, College of Plant Protection, Nanjing Agricultural University, Nanjing, PR China; [4]Department of Crop and Soil Science Education, Akenten Appiah-Menka University of Skill Training and Entrepreneurial Development (AAMUSTED), Mampong, Ghana; [5]Laboratory of Agro-physio-genetics and Plant Biotechnology, Department of Agricultural Technical Science (STA), Rural Polytechnic Institute for Training and Applied Research (IPR/IFRA) of Katibougou, Koulikoro, Mali

14.1 Introduction

Plant pests and diseases are among the most critical challenges to global agricultural productivity, food security, environmental sustainability (Chakraborty and Newton, 2011; Singh *et al.*, 2021), and improved livelihoods. Annually, it is estimated that between 20 and 40% of the world's crop yield is lost due to pests and pathogens, with significant outbreaks resulting in even more extensive damage to staple crops such as wheat, rice, and maize (Savary *et al.*, 2019). These losses translate not only into reduced food availability but also into significant economic hardship for farming communities, especially in low-income regions where agriculture remains the backbone of livelihoods and national economies (Oerke and Dehne, 2004; Savary *et al.*, 2019). Beyond agricultural landscapes, pests and pathogens threaten natural ecosystems by disrupting biodiversity and altering ecosystem

*Corresponding author: bbbantwi@yahoo.com; joeadomako@gmail.com

© CAB International 2026. *Plant Spatial Biology* (ed. Chen)
DOI: 10.1079/9781836994572.0014

functions critical to maintaining environmental balance (Diyaolu and Folarin, 2024). The intensification of agriculture, global trade, and climate change has collectively reshaped pest and pathogen distributions, life cycles, and outbreak dynamics (Baker *et al.*, 2022). Climate change facilitates the expansion of pests into new geographic regions, alters their phenology, and intensifies outbreak frequencies (Harvey *et al.*, 2020; Skendžić *et al.*, 2021). These evolving pressures complicate pest and disease management, demanding adaptive and innovative control strategies that extend beyond traditional approaches (Carroll *et al.*, 2014; Lamichhane *et al.*, 2015).

Historically, pest and disease management has relied heavily on inorganic chemicals, crop rotation, and classical resistance breeding (Döring *et al.*, 2012). While these strategies have provided short-term control benefits, they are increasingly becoming insufficient in efficacy due to rapid pest adaptation and environmental concerns (Abate *et al.*, 2000). The widespread emergence of pesticide resistance, caused by genetic mutations and metabolic adaptations in pest populations, has led to a global pesticide treadmill that increases costs and environmental risk (Belinato and Martins, 2016; Hawkins *et al.*, 2019). Moreover, overreliance on chemical controls can disrupt beneficial insects such as pollinators and natural enemy populations, undermining ecosystem services, which are vital to sustainable agriculture (Kremen and Chaplin-Kramer, 2007; Diyaolu and Folarin, 2024). Resistance breeding, based on phenotypic selection of resistant cultivars, remains a fundamental component of holistic pest and disease control strategy, but is often slow and it is vulnerable to resistance breakdown due to a narrow genetic base and the evolutionary plasticity of pests and pathogens (Murugan *et al.*, 2025). The complexity of host–pest–pathogen interactions, coupled with environmental variability, challenges the durability of resistance conferred by single gene or simple loci (Nelson *et al.*, 2018; Lefebvre *et al.*, 2020). Monoculture practices increase the possibility of widespread outbreaks by creating uniform hosts that enable rapid proliferation of pests and pathogens (Ratnadass *et al.*, 2012). In response to these challenges, recent advances in molecular biology and high-throughput omics technologies provide new avenues for dissecting the intricate biological networks that govern plant defense and pest/pathogen virulence (Makhumbila *et al.*, 2022). Multi-omics integration collectively enables comprehensive profiling of the genetic, transcriptional, protein, and metabolic landscapes involved in plant immunity and pest adaptation (Satrio *et al.*, 2024; Prashar *et al.*, 2025). When integrated with emerging spatial omics approaches that retain molecular data within the spatial context of plant tissues, these technologies offer unprecedented resolution to understand how localized and systemic defense responses are orchestrated (Barmukh *et al.*, 2025).

Spatially resolved multi-omics methodologies enable the determination of essential resistance genes, regulatory systems, and biochemical pathways implicated in an effective plant defense system (Fan *et al.*, 2025). They also enable early detection of infections through molecular markers and spatial signatures, which are critical for timely and targeted interventions (Samarskaya *et al.*, 2025). Furthermore, integration of these molecular data with environmental and climatic variables via machine learning (ML) and artificial intelligence (AI) enhances predictive modeling of pest and disease outbreaks, which supports proactive decision making within integrated pest management frameworks (Ali *et al.*, 2024; Delfani *et al.*, 2024). Complementary advances in synthetic biology and genome editing facilitate exact alteration to engineer plant genomes, leading to the development of crops with enhanced and durable resistance attributes, overcoming many limitations of conventional breeding (Wang *et al.*, 2022). These integrative approaches collectively represent a paradigm shift towards sustainable, precise, and resilient pest and disease management systems essential for tackling the intricate issues presented by a shifting global landscape (Wang *et al.*, 2022). This chapter provides an overview of spatial and multiple omics approaches as applied to pest and disease management. It explores current methodologies, practical applications, and ongoing challenges, while outlining future opportunities for translating multi-omics insights into actionable strategies. By harnessing these cutting-edge technologies, agriculture can advance towards more efficient, environmentally friendly, and resilient crop protection, contributing to global food security and ecosystem health.

14.1.1 Pests, diseases, and climate change

Plant pests and diseases constitute a pervasive and complex threat to global agriculture, ecosystems, and human well-being (Spence *et al.*, 2020). Crop losses attributed to pests and pathogens are estimated to account for 20–40% of total agricultural production worldwide, with annual economic costs exceeding hundreds of billions of US dollars (Oerke, 2006; Sharma *et al.*, 2017; Junaid and Gokce, 2024). These losses directly impact food availability, market stability, and nutritional security, disproportionately affecting vulnerable populations in developing countries where agriculture remains the primary source of livelihood and employment (Chaudhry *et al.*, 2023). Beyond mere yield reduction, pests and diseases also deteriorate crop quality through the production of mycotoxins, reduced storage life, and lowered market value (Magan and Aldred, 2007; Neme and Mohammed, 2017). For example, in wheat plants, *Fusarium* head blight diminishes yield and infects grain with harmful toxins, presenting significant health risks to consumers (Matny, 2015; Torres *et al.*, 2019; Mielniczuk and Skwaryło-Bednarz, 2020). In regions where smallholder farmers operate near subsistence thresholds, these quality and quantity losses result in food shortages, malnutrition, and increased poverty (Morton, 2007).

The ecological ramifications of pest and disease outbreaks extend into natural and semi-natural ecosystems, where invasive insects, weeds, and emergent pathogens disrupt native flora and fauna interactions, undermine biodiversity, and alter ecosystem processes (Ielmini and Sankaran, 2021). The loss of keystone species and disruption of pollinator networks have cascading effects on ecosystem stability, reducing resilience to environmental perturbations such as drought and fire (Latty and Dakos, 2019; Bascompte and Scheffer, 2023). These impacts compromise ecosystem services essential for agriculture, creating a feedback loop that exacerbates vulnerability (Power, 2010). Climate change acts as a force multiplier in these dynamics. Changing temperature and precipitation patterns alter pest life cycles, voltinism (number of generations per year), and overwintering success, resulting in shifts in geographical distributions and seasonal timing (Karuppaiah and Sujayanad, 2012; Sunil *et al.*, 2023). For example, the northern expansion of the corn earworm (*Helicoverpa zea*) into temperate zones has been linked to warmer temperatures, exposing previously unaffected crops to new pest pressures (Sarwar, 2023). Similarly, warming accelerates the development rates of pathogens such as *Phytophthora infestans*, the primary cause of late blight disease increases its severity and epidemic potential (Wu *et al.*, 2022).

14.1.2 Rationale for integrative, spatially informed, and omics-driven approaches in pest and disease management

Traditional control of insects and diseases primarily relies on the use of chemical pesticides, agronomic strategies like crop rotation, and the development of resilient cultivars through classical methods (Haq and Ijaz, 2020). While these strategies have contributed to past successes in controlling key pests and diseases, they face escalating challenges (Agarwal and Verma, 2020). Chemical pesticides, though effective, are increasingly compromised by the evolution of resistance in target organisms (Hawkins *et al.*, 2019). The mechanisms of resistance encompass targeted mutations, metabolic detoxification, and behavioral avoidance, reducing pesticide efficacy, often requiring higher doses or more frequent applications that exacerbate environmental contamination and non-target toxicity (Pandian and Ramesh, 2020; Rigon *et al.*, 2020). For instance, resistance to neonicotinoids in *Myzus persicae* aphids and pyrethroid resistance in *Helicoverpa armigera* exemplify these concerns (Devine and Denholm, 2009). Beyond resistance, pesticide applications can harm beneficial organisms such as pollinators and natural enemies, thereby disrupting biological control services essential for sustainable pest management (Roubos *et al.*, 2014).

Classical resistance breeding, dependent on phenotypic selection and crossing with wild relatives, is time- and labor-intensive, often requiring multiple generations to introgress resistance traits (Acquaah, 2015; Bal *et al.*, 2025). Resistance is frequently conferred by major genes that pests and pathogens can overcome rapidly due to their high genetic variability and short generation times (Sharma *et al.*, 2004; Broekgaarden *et al.*, 2011). The 'boom-and-bust' cycles of resistance deployment highlight the need for more durable and complex resistance mechanisms (Rimbaud *et al.*, 2021). The complexity of plant defense responses, involving multi-layered signaling pathways and interactions among diverse molecular players, has been difficult to unravel using traditional genetics alone (Roychowdhury *et al.*, 2025). This gap has been bridged by the emergence of omics technologies, which collectively provide a panoramic view of biological systems (Ritchie *et al.*, 2015).

Genomics allows for the characterization of the entire genome of plants and pests, revealing resistance gene repertoires and virulence determinants (Lindeberg, 2012; Lefebvre *et al.*, 2020). Transcriptomics identifies temporal and spatial patterns of gene expression, highlighting inducible defense networks activation upon attack (Liu *et al.*, 2020; Puri *et al.*, 2023). Proteomics reveal post-translational modifications and protein–protein interactions critical for defense signaling, while metabolomics elucidates the biochemical landscape, including secondary metabolites with antimicrobial or insecticidal properties (Vo *et al.*, 2021). Spatial omics techniques preserve the native tissue context of molecular information, enabling the visualization of defense and infection dynamics within specific cellular and tissue compartments (Carstens *et al.*, 2024). This spatial resolution is vital because plant defense is not uniform but localized and systemic, involving cell-to-cell communication and coordinated responses across organs (Toyota and Betsuyaku, 2022). Integrating multi-omics data across spatial and temporal dimensions offers a holistic understanding of host–pathogen and host–pest interactions, uncovering novel resistance mechanisms, biomarkers, and targets for genetic improvement or biocontrol (Ontoy and Ham, 2024).

14.1.3 Limitations of conventional control strategies and resistance management

The limitations of conventional pest and disease management methods have become more pronounced as pests and pathogens rapidly evolve in response to human interventions, particularly through the overuse of chemical control methods (Ekström and Ekbom, 2011; Baker *et al.*, 2020). As resistance alleles emerge within pest populations, the effectiveness of many pesticide classes declines, leading to what is commonly referred to as the 'pesticide treadmill' (Walsh *et al.*, 2022). This ongoing cycle forces farmers to increase pesticide application rates and/or frequencies, which escalates both costs and environmental contamination. The consequences of over-reliance on pesticides extend far beyond pest resistance, affecting soil health, contaminating water bodies, and harming non-target organisms (Gill and Garg, 2014). Pollinators, vital for the reproduction of many crops, have been especially impacted, with declining populations constituting an increasing risk to the world's food security and agricultural sustainability (Marshman *et al.*, 2019).

Resistance breeding has long been considered a viable strategy to mitigate pest damage, but its success has been limited (Rubiales *et al.*, 2015). Traditionally, breeders have focused on monogenic traits, hoping that introducing a single resistance gene would protect crops from pests. However, pests have demonstrated a remarkable ability to adapt quickly, rendering these strategies ineffective in the long term (Obermeier *et al.*, 2022). This issue is particularly severe in monoculture farming systems, which provide pests with a uniform selective environment. The push toward breeding for quantitative resistance involving multiple genes and pathways represents a more sustainable approach but remains hindered by the complexity of molecular interactions that govern the plant defense system (Pretty and Bharucha, 2015; Lenné and Wood, 2024). Without a comprehensive understanding of these molecular networks, achieving durable and effective resistance remains a significant challenge for crop protection.

Biological control strategies, though promising in their potential for environmentally sustainable pest management, also face barriers to widespread adoption. This is because biological control agents can be unpredictable in their effectiveness, particularly under varying field conditions (Gurr and You, 2016; Baker *et al.*, 2020). In addition, the integration of biological control into comprehensive pest management systems requires significant technical expertise and knowledge. Moreover, farmers often lack the incentives or resources to invest in biological control methods, making them less appealing compared to traditional chemical-based approaches (Shrestha *et al.*, 2024). However, the emergence of spatial and multi-omics methodologies has created new opportunities to address these challenges (Baysoy *et al.*, 2023; Vandereyken *et al.*, 2023; Liu *et al.*, 2024). By leveraging cutting-edge tools in molecular biology and genomics, it is feasible to identify and edit multiple resistance genes with greater precision, reducing unintended genetic consequences such as linkage drag and pleiotropic effects (Buch *et al.*, 2023). This promotes the advancement of crops with improved resistance and increased resilience to changing pests.

Alongside advances in breeding, early detection technologies, powered by molecular markers and spatial signatures, allow for more precise and localized pest management interventions (Van Leeuwen *et al.*, 2020; John *et al.*, 2023). These innovations enable farmers to detect pest problems early and apply targeted treatments, thus reducing the need for blanket chemical applications (Li and Wang, 2024). Additionally, understanding the tissue-specific dynamics of plant defense systems could significantly enhance the efficacy of pesticide and biocontrol agent applications (Lucas, 2011; Li and Wang, 2024). Thus, by tailoring interventions to the specific needs of different tissues or pest species, farmers can reduce the impact on non-target organisms and mitigate the selection pressure that leads to resistance (Akhter *et al.*, 2024). Moreover, predictive modeling, which integrates multi-omics data with environmental and climatic factors, promises to revolutionize pest management (Amin *et al.*, 2025; Fan *et al.*, 2025). A number of multidisciplinary strategies to analyze plant–pest and plant–pathogen interactions across biological hierarchies are presented in Table 14.1.

14.2 Advances in Spatial Omics Technologies

The ever-evolving nature of pathogen infection within a host or attack by a pest, as well as the specific reactions of various plant cells and tissues to infections and attacks, is complicated (Zhang *et al.*, 2020). Plants have evolved an intricate, strong defense system against pests and diseases (Tang *et al.*, 2023). Numerous facets of plant health are influenced by the intricate, dynamic, and diverse relationship that exists between a plant and its pathogens, such as bacteria, fungi, viruses, and insect pests (Lloyd-Price *et al.*, 2016). Pathogens use a range of manipulative approaches, which might change as an infection progresses. These approaches entail suppressing plant defenses and using chemicals and degradative enzymes to stimulate colonization and the release of nutrients (Zhu *et al.*, 2023). Plant pathogen spread is not uniform, leading to uneven disease expression (Daugherty *et al.*, 2010). The development of spatial omics technology has provided new perspectives on complicated biological systems. Spatial omics technologies have made it possible for scientists to analyze intricate connections between tissue function and structure at the cellular or subcellular level (Park *et al.*, 2022). These technologies provide a thorough understanding of cellular activity within their tissue context by combining spatial data with high-throughput omics techniques, such as transcriptomics, proteomics, genomics, epigenomics, and metabolomics (Vandereyken *et al.*, 2023; Jain and Eadon, 2024). In addition to revealing the relationship between metabolic pathways, signal transduction, and defensive reactions, the results rapidly predict pertinent defensive factors under adversity (Wang *et al.*, 2020). This helps to clarify the molecular mechanisms of plant stress resistance and is advantageous for crop improvement programs, which may result in increased resilience and productivity (Chen *et al.*, 2024). These techniques can help with precision breeding tactics to improve disease resistance by determining the

Table 14.1. Multidisciplinary approaches to study plant–pest and plant–pathogen interactions across biological hierarchies.

Approach (technology)	Description	Applications in pest and disease management	References
Genomics	Sequencing and analysis of plant, pest, and pathogen genomes to identify resistance and virulence genes	Identification of resistant genes, virulence factors, and genome-editing targets	Rato *et al.*, 2021
Transcriptomics	Profiling gene expression changes during plant–pest/pathogen interactions	Discovery of defense genes and signaling networks for marker-assisted breeding	Huang *et al.*, 2023
Proteomics	Analysis of proteins involved in plant defense and pest/pathogen virulence	Identification of defense-related proteins and enzymes for targeted control	Kumari *et al.*, 2024
Metabolomics	Profiling of metabolites and secondary defense compounds	Detection of antimicrobial or signaling compounds for enhancing natural resistance	Gao *et al.*, 2020
Spatial omics	Mapping spatial distributions of genes, proteins, and metabolites within plant tissues	Understanding local defense responses and pathogen colonization patterns	Zhu *et al.*, 2023
Imaging techniques	Microscopy and imaging mass spectrometry	Visualizing infection structures, cell death, and defense compound localization	
Phenotyping	High-throughput assessment of resistance traits and disease symptoms	Screening breeding populations and evaluating pest impacts	Goggin *et al.*, 2015
Genome editing	CRISPR/Cas and related gene editing tools	Engineering crops with durable resistance traits without linkage drag	Ali *et al.*, 2022b
Ecology and epidemiology	Monitoring pest/pathogen population dynamics and disease spread	Predicting outbreaks, risk assessment, and designing integrated pest management (IPM) plans	Ciancio and Mukerji, 2007

genetics underpinnings of gene expression patterns, protein functions, and metabolic alterations. These technologies are evolving quickly, providing new ways to fully comprehend the variety and roles of different cell types within organisms in response to diseases or pests (Hui *et al.*, 2025). An extensive overview of cellular dynamics and heterogeneity is provided by the combination of spatial and molecular data, illuminating basic biological processes, including responses to diseases and pests. Although direct uses in disease and pest management are still in their infancy, these technologies have the potential to create solutions.

14.2.1 Spatial transcriptomics: Mapping gene expression in response to pests and diseases

Spatial transcriptomics examines regulatory networks and gene expression patterns to comprehend how mRNA is arranged spatially within cells or tissues. The extensive spatiotemporal networks mediating the perception and development of defense responses to pests and diseases are still poorly understood (Toruño *et al.*, 2016). Plant responses to pests and diseases can be effectively understood through the use of spatial transcriptomics, which can assist in identifying genes and pathways that regulate particular features to withstand biotic stresses (Varadharajan *et al.*, 2025). In the transcriptome study of root-knot nematode-resistant cells, genes implicated in the regulation of auxin balance, salicylic acid, and jasmonic acid signaling pathways in peach were revealed (Cao *et al.*, 2017). Spatial transcriptomics assists in mapping the whole-transcriptome changes directly on tissue sections, for the location of immune pathways that switch on/off during an attack.

The understanding of plants' response to pests and diseases can be completely transformed by the integration of many mapping methodologies employing multi-omics data (Ontoy and Ham, 2024). High-throughput omics information can be superimposed on genetic maps to help researchers identify certain genes, proteins, or metabolites linked to diseases and pest resistance. Researchers can use these technologies to detect non-coding RNAs, identify alternative splicing processes, and quantify the levels of gene expression. Spatial transcriptomics has been used in numerous studies to examine changes in gene expression, find genes that are differentially expressed, and uncover pathways that contribute to rice protection against diverse infections (Ontoy and Ham, 2024). It is observed that both resistant and susceptible cultivars showed overall transcriptome modifications following rice sheath blight (*Rhizoctonia solani*) inoculation, which corresponded to various biological processes, molecular activities, and cell components (Yang *et al.*, 2022). In order to identify the molecular mechanisms behind pathogen resistance and susceptibility, RNA sequencing (RNA-seq) has been widely used in the study of rice diseases. In addition, soybean leaves infected with Asian soybean rust (*Phakopsora pachyrhizi*), paired with *Visium* and single-nuclei RNA-seq, captured two distinct spatial defense states and their distribution across lesions vs surrounding tissues, providing a map of localized vs systemic responses (Hu *et al.*, 2025). The finding of genes that confer resistance to rice sheath blight, the extensive use of various omics technologies in plant–pathogen interactions, and the evolving method of interaction between the sheath blight pathogen and rice have brought a better understanding of plant responses to the pathogen (Yang *et al.*, 2022).

14.2.2 Spatial metabolomics: Localizing and visualizing defense metabolites

Metabolomics involves the comprehensive analysis of an organism's metabolites before and after specific stimulation and interference circumstances, or during distinct stages of growth (Liu *et al.*, 2021). Determining the dynamic variations in small-molecule metabolites in an organism brought on by genetic, physiological, and pathological alterations was the primary purpose of metabolomics. However, it is currently widely used to ascertain how plants react to pests and diseases, investigate the physiology of the maturation of fruit, and the genetic mechanism of metabolic diversity in various parts and their correlation with complex plant traits (Pott *et al.*, 2020; Zuo *et al.*, 2020). It is also used to establish a basis for the investigation of mature phenotypic mechanisms, to enhance plant quality, and to modify breeding techniques (La *et al.*, 2016). The metabolic activity and metabolite-induced phenotypic modulation of complex, multicellular organisms can be better understood due to spatially precise metabolite localization and control.

Metabolomic profiling of pests and disease responses is a thorough examination of the metabolites found in the plant leaves, roots, and seeds under diverse environmental circumstances (Kumari *et al.*, 2024). Significant alterations in plant primary and secondary metabolites brought on by pathogen infection and pest attack can be determined by metabolomic analysis (Kumari *et al.*, 2024). These metabolites are essential indicators that provide insight into the metabolic processes underlying plant defense systems. Plants have developed a variety of metabolites in response to biotic stressors, which serve as biomarkers to regulate biotic stress tolerance across multiple plant species. To achieve a deeper comprehension of the plant's metabolome during development or in response to various stressors, as well as cell-to-cell signaling responses, biological studies may now begin mapping metabolites back to their location within modified tissue (Myers, Jr *et al.*, 2025). As a defense mechanism against biotic stress, maize has significant levels of benzoxazinoids, which are crucial secondary metabolites. Comparative metabolic profiling of unhealthy and healthy plants makes it possible to identify intricate metabolic networks involved in plant–pathogen interaction.

Spatial metabolomics is a sophisticated method that combines single-cell analysis and spatial imaging methods to provide a novel approach to understanding cellular metabolic activity. Metabolomics completes the spatial omics landscape by clarifying the spatial distribution and dynamics of cellular metabolites by analyzing metabolite distribution and change within cells or tissues. This method makes it possible to examine the metabolic properties of individual cells as well as the spatial interactions that exist within and between cells. Spatial metabolomics uses a mass spectrometer to systematically scan tissue slices in order to offer precise information about metabolites in biological samples. Deeper biological insights on plant responses to pests and diseases can be obtained with mass spectrometry imaging (MSI) effective methodology that includes correlating metabolism traits to gene expression to assist functional genomics techniques and pathway elucidation (Dong *et al.*, 2020). MSI makes it possible to spatially resolve metabolite localization at the tissue, cell, and subcellular levels. Details of the application of MSI are shown in Table 14.2. MSI chemical 'hotspots' at infection fronts, stomata, vascular bundles, or lesion margins that are not apparent from bulk extracts. Metabolite mapping may be used to promote stress resilience without undesirable off-target consequences, such as growth and yield decreases, and to target particular cell types for metabolic adjustment utilizing tailored genetic techniques.

14.2.3 Single-cell omics: Cell-type-specific responses to pest/pathogen attack

Single-cell omics provides the highest precision of information on cellular heterogeneity. This technology is an effective method for characterizing rare cell types and improving the comprehension of various cell states since it allows for the investigation of a large number of cells (Stein *et al.*, 2021). The introduction of single-cell methodologies allows for the identification of heterogeneity and cell-to-cell variations from a new perspective by giving a thorough and impartial investigation of each cell. At the genomic level, single-cell DNA sequencing (scDNA-seq) aids in revealing intercellular diversity and heterogeneity. Single-nucleotide variants, copy number variations, and microsatellite variations can currently be studied using scDNA-seq techniques. Single-cell RNA sequencing (scRNA-seq) is a method for sequencing individual genomes to acquire transcriptomic or genomic data, facilitating the identification of heterogeneity within cell populations (Bawa *et al.*, 2022). The scRNA-seq technique offers a comprehensive insight into transcriptome dynamics across various cell populations by facilitating gene expression analysis at the individual cell level (Bawa *et al.*, 2022). Prior transcriptomics research was typically conducted on entire plants or portions of tissues, averaging the variation of thousands of cells and obscuring the individuality of each cell type (Bawa *et al.*, 2022). The complex nature of whole-transcriptome illustrations from a tissue to a cell has been transformed by the development of scRNA-seq (Shaw *et al.*, 2021). Employing gene expression profiles to identify new cell types has been made possible by transitioning from broad-spectrum profiling of entire tissues to more specialized datasets that contain a single type of tissue (Kim *et al.*, 2021; Lopez-Anido *et al.*, 2021). With scRNA-seq, one can characterize both common and uncommon cell types and cell states, discover new cell types, and understand the spatial and developmental relationships between cell types.

Reliable gene expression information from the extremely small amounts of RNA available in a single cell is necessary for switching from homogeneous cell analysis to single-cell transcriptome analysis (Brennecke *et al.*, 2013). Using RNA-seq for single cells can provide important information on the diversity within a population of cells that appear to be homogeneous. The scRNA-seq attains single-cell specificity for transcriptome sequencing of tissues, and its spatiotemporal resolution not only identifies rare cell types but also enables the understanding of the spatial and temporal interactions between cell types (Chen *et al.*, 2024). This technology could identify cell types, their functions, and unearth underlying mechanisms of gene regulation, offering a richer understanding of developmental processes in a variety of biological

Table 14.2. High-throughput spatial omics technologies used in plant–pest/pathogen research.

Technology/platform	Modality	Resolution/scale	Throughput (examples)	Plant–pest/pathogen use case(s) and notes
10 × Genomics Visium (spatial RNA-seq)	Transcriptome (untargeted)	55 μm spots; tissue-wide	Whole-transcriptome per slide	Spatial co-transcriptomics in infected tissues (e.g. parasitic interfaces, rust-infected leaves) requires plant-optimized tissue preparation
MERFISH (Vizgen MERSCOPE)	Targeted RNA (hundreds to thousands of genes)	Single-cell/subcellular	10^5–10^6 molecules imaged	Applied to pathogen-challenged leaves to map immune cell states; integrates well with single-cell atlases
PHYTOMap (plant-optimized multiplexed FISH)	Targeted RNA (tens to hundreds of genes)	Single-cell, 3D whole-mount	Dozens of targets per run	Plant-specific multiplexed spatial gene expression; adaptable to infection-induced genes in intact tissues
SEER-FISH (sequential error-robust FISH)	Targeted rRNA/taxa ID (microbiome)	Single bacterial cell	10–100 s taxa per field	Maps root-associated bacterial communities at single-cell resolution, informative for pathogen/beneficial gradients.
MALDI–MSI (spatial metabolomics/proteomics)	Metabolites, lipids, and some proteins	10–100 μm pixels (typ.)	10^3–10^5 pixels/tissue	Visualizes defense metabolites and lipids in infected tissues; widely used in plant–microbe interaction chemistry.
DESI–MSI (ambient spatial metabolomics)	Metabolites/lipids	50–200 μm	10^3–10^5 pixels	Suitable for disease studies with minimal preparation (ambient ionization); effective on leaves and soft tissues.
LAESI-MSI/ToF-SIMS	Metabolites, elements	1–50 μm (SIMS), 30–200 μm (LAESI)	10^3–10^5 pixels	Maps micro-scale chemistry at host-pathogen interfaces (e.g. cell walls, lesion margins).
Laser capture microdissection + RNAseq (LCM-seq/Geo-seq)	Transcriptome (region/cell-specific)	Single cells to small regions	10^2–10^3 regions per study	Classic spatial method for infection sites (e.g. powdery mildew, nematode-induced syncytia); strong for focused contrasts.

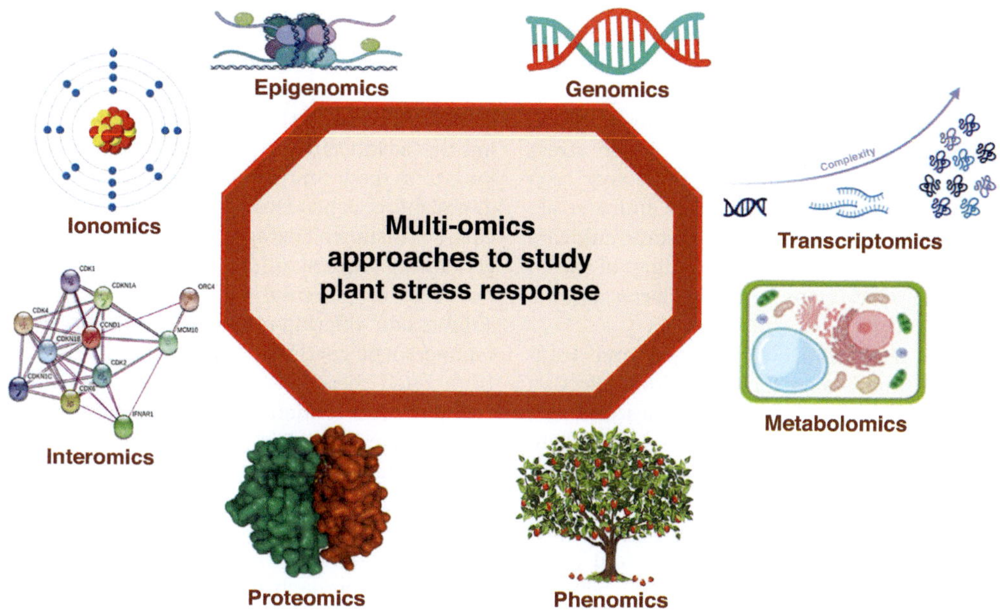

Fig. 14.1. Multi-omics approach to studying plant stress response.

samples (Sun *et al.*, 2023). By comparing the transcriptomes of individual cells, scRNA-seq has been employed to assess transcriptional variations and similarities among a population of cells, when prior publications have shown high levels of heterogeneity in the root (Shaw *et al.*, 2021). Single-cell and whole-tissue responses differ significantly, indicating that the tissue-level response is a composite of fluctuations between cells that are targeted by the pathogen and cells that are not (Thor and Peiter, 2014). Plants have an inherent immune system that includes nucleotide-binding leucine-rich repeat receptors (NLRs) and surface-localized pattern recognition receptors (PRRs) (Ngou *et al.*, 2022). High-throughput spatial omics methodologies employed in plant–pest/pathogen investigations are shown in Table 14.2.

14.3 Multi-omics Integration for Pest and Disease Management

The advent of multi-omics technologies (Fig. 14.1) has expanded knowledge and perspectives on how plants exhibit specific molecular responses to stresses (Shafi *et al.*, 2021; Mourou *et al.*, 2023). This has led to an enhanced comprehension of the complex biological processes participating in plant disease development, the chemical compounds (metabolites) associated with disease progression, the genes, and metabolic pathways affected by pest and pathogen invasions. Advances in multi-omics technologies have paved the way for innovative solutions, including the rise of resilient crops exhibiting improved performance in challenging environments, as well as serving as a vital tool to establish comprehensive pest and disease management strategies in an increasingly demanding environment (Ali *et al.*, 2022a; Tyagi *et al.*, 2024). Spatial and multi-omics technology can be used to discover virulent weapons that pathogens employ, detect certain phenotypic responses that occur during plant–pest or plant–pathogen interaction, and find plant defensive systems. Accurate plant disease detection based on the application of phenomics, metabolomics, proteomics, ionomics, and genomics will have an impact not only on plant pathology but also on agronomy and precision agriculture, which are vital for pest and disease management. A key component of a comprehensive pest and disease management approach is to develop host-resistant cultivars. The multi-omics framework has facilitated the investigation of interactions among various biological molecules and their roles in stress responses, resulting in highly effective crop improvement strategies (Roychowdhury *et al.*, 2023). However, developing resistant cultivars requires a thorough understanding of host–pathogen interactions and the corresponding gene action.

The application of genomic and transcriptomic tools has facilitated the selection and incorporation of desirable genes/markers, contributing to the advancement of resilient plant varieties against pests and diseases. This underscores the significance of multi-omics in pest and disease management, because these technologies can identify resistant genes through examining interactions between plants and pathogens (Bastas and Kumar, 2023), particularly the molecular mechanisms underlying disease development and control. The concurrent modification of three homoeoalleles in hexaploid bread wheat resulted in inheritable resistance to powdery mildew (Wang *et al.*, 2014). In tomato plants, clustered regularly interspaced short palindromic repeats (CRISPR)/Cas9 identified two genes, *SlPDS* and *SlPIF4*, targeted for mutagenesis via an *Agrobacterium tumefaciens*-mediated transformation, which resulted in a high mutation rate (Pan *et al.*, 2016). CRISPR-mediated genome editing was successfully utilized to produce canker-resistant Hamlin sweet orange (Huang *et al.*, 2022). Brassinazole-resistant 1 modulates the expression of the gene *ERF49*, thereby increasing the plant's sensitivity to heat stress (Chen *et al.*, 2022). In *Arabidopsis thaliana*, phenolic quantities increased after exposure to fungi or bacteria, thereby improving immunity against *Pseudomonas syringae* (Soylu, 2006). Further evidence from examining the intersection of stress response pathways using microarray analyses demonstrated that ADR1 stimulation resulted in the increased expression of a multitude of drought-adaptive genes. The findings highlight the complex mechanisms of reactions to stress within plant tissues and the controlling function of DREB proteins in these biological events.

The harmful effects of synthetic pesticides on the environment, human, and animal health have reduced their appeal. Application of metabolomics and metagenomics can assist in identifying beneficial microbial agents, facilitating the advancement of biological controls for long-term pest and disease suppression, thus diminishing dependence on synthetic chemicals. For instance, metagenomics based on next-generation sequencing (NGS) has facilitated the discovery of potential microbes for nematode suppression, offering novel and sustainable pest management methods (Lahlali *et al.*, 2021). Untargeted metabolomics methodologies have effectively differentiated among non-suppressive and suppressive soils for *Rhizoctonia solani*, pinpointing specific metabolites including lipids, sugars, and terpenes that are more prevalent in suppressive soils (Hayden *et al.*, 2019). Metabolomics has been employed in an untargeted approach to uncover bacteria that are challenging to culture in laboratory environments yet generate various bioactive chemicals effective against plant diseases (Topalović *et al.*, 2020; Mokrini *et al.*, 2024). Another crucial area in pest and disease management is early detection and warning strategies. The utilization of omics biomarkers has facilitated early detection, enabling the deployment of management interventions before an outbreak occurs.

14.4 Resistance Breeding Using Synthetic Biology and Genome Editing

In the early days of plant genetic engineering (GE), there was great enthusiasm and a desire to use this innovative technology in agriculture. According to Barton and Brill (1983), GE aims to improve crops by modifying seed proteins to increase nutritional value, fix nitrogen to reduce the need for nitrogen fertilizer, improve photosynthesis to increase productivity, and increase stress tolerance and resistance to pests and diseases to reduce crop losses, production costs, and environmental hazards. Emerging GE technologies over the years include gene editing, in which DNA is altered at specific loci in the genome through gene drives, which increase the frequency of particular genes that affect the reproduction of insects or pathogens, and synthetic biology, in which biological objects, such as organelles or chromosomes, are redesigned purposely to increase the resistance of plants to achieve higher productivity and food security. Plant breeding is being revolutionized by synthetic biology and genome-editing techniques, particularly for increasing crop yields as well as pest and disease resistance. Compared to conventional breeding techniques, these technologies offer faster and more targeted approaches by precisely modifying the plant genome. This approach enables the cultivation of plants that exhibit enhanced disease resistance, thereby necessitating fewer chemical interventions and yielding superior performance (Kumlehn *et al.*, 2018).

With a focus on disease resistance, these technologies can help create crops that provide the world with food, feed, fiber, medicine, aesthetics, ecological controls, and building materials. To satisfy the food and feed requirements of an expanding population, global agricultural production must be enhanced. These demands are in line with the increasing understanding of the impact of agriculture on the environment and the need to change farming practices and consumption habits to ensure sustainable production (Rockström *et al.*, 2009). Furthermore, the scope and intensity of pest and disease outbreaks (Bebber, 2015) will change due to the expected changes in the global climate (IPCC, 2014), which has an impact on yields and yield stability (Nelson *et al.*, 2014). It is postulated that synthetic biology and genome editing are effective techniques for improving crop resistance breeding (Sargent *et al.*, 2022).

Synthetic biology enables the design and construction of novel biological systems, especially those that enhance resistance, while genome editing facilitates precise modifications within plant genomes to augment features, including disease resistance. These technologies provide numerous advantages over traditional breeding methods, including enhanced speed and precision, as well as the capability to transcend species barriers (Pixley *et al.*, 2019). Genome editing could substantially alleviate the limitations of classical resistance breeding. First, genome editing is frequently utilized to directly change a target gene in elite types, thus skipping the mating process. Second, the selection of the target gene is independent of plant populations possessing adequate genetic variation; only the sequencing data for the target gene is required. Third, linkage drag is not an issue because genome editing does not modify regions beyond the target sites. Fourth, resistant breeding utilizing genome editing facilitates swift advancements, as it eliminates the need for either genetic crosses or progeny selection. Currently, the majority of crops resistant to non-viral infections are generated through genome editing that precisely targets modifying genes referred to as 'S-genes', which adversely impact the immune system. Genome editing facilitates the creation of targeted S-gene mutants in various plant species for breeding purposes, circumventing species barriers due to the functional conservation of S-genes across all plant taxa. It is anticipated that additional S-genes will be discovered, thereby augmenting the targets for genome editing. Nonetheless, null mutations in S-genes could sporadically impede normal growth, as S-genes frequently play a role in plant growth and development, and their functionality may be compromised as a consequence. In plant pathology, the disease triangle consists of a virulent pathogen, a susceptible host, and a favorable environment (Pokhrel, 2021). A disease develops only if all three factors are concurrently present. Genome editing might be employed to mitigate diseases by targeting one of the three components and disrupting the relations between plants and pathogens.

14.4.1 CRISPR/Cas-mediated precision breeding for resistance traits

In the context of global warming, the cultivation of staple foods must sustainably address the needs of an increasing population. In 2024, 673 million people, i.e. 8.2%, experienced malnutrition, particularly in Africa and Asia (FAO *et al.*, 2025). Contemporary agricultural practices predominantly depend on growing a limited number of plant species, often in inappropriate regions distant from their areas of domestication (Fernie and Yan, 2019). Conventional plant breeding, based on useful genetic variation, has improved food availability and crop adaptation, while genome engineering offers precision breeding opportunities for domesticating wild species and improving elite germplasm (Zhang *et al.*, 2019). Plant pathogens exert a considerable influence over food security and incur large economic losses (Savary *et al.*, 2019). Pathogens exploit plant immunity by creating effectors and molecular weapons that target host proteins, while plants have a highly developed immune system for protection against infection. Invasion pattern receptors are critical components that recognize signals from microbes on/in the host or that indicate the existence of microbial intruders (Cook *et al.*, 2015). Intracellular invasion pattern receptors predominantly reside within intracellular cells, whereas cell surface receptors, including nucleotide-binding and leucine-rich repeat receptors (NLRs) and receptor-like kinases (RLKs), are situated on cell membranes (Cesari, 2018; Kourelis and Van der Hoorn, 2018). Receptor-like kinases and receptor-like proteins identify pathogen-associated molecular patterns (PAMP) and damage-associated molecular patterns (DAMP), which are generated by host cells under pathogen assault (Boutrot and Zipfel, 2017; Kanyuka and Rudd, 2019).

Intracellular invasion pattern receptors, such as NLRs and RLKs, primarily consist of nucleotide-binding proteins, while cell surface receptors are predominantly receptor-like proteins or kinases. NLRs identify intracellular effectors, whereas RLKs and receptor-like proteins detect external pathogens (Boutrot and Zipfel, 2017; Kanyuka and Rudd, 2019). Microbe-associated molecular patterns (MAMPs) and DAMPs are released by the host cells (Boutrot and Zipfel, 2017; Kanyuka and Rudd, 2019). The swift advancement of genome editing technologies, particularly RNA-guided endonuclease systems, has profoundly influenced the domain of genetics (Zhang *et al.*, 2019). Genome editing has conventionally been employed to produce loss-of-function alleles, exemplified by the creation of a powdery mildew-resistant tomato through the deletion of the gene associated with the powdery mildew-resistant locus O (Nekrasov *et al.*, 2017). However, the resistance to rice blast has been attributed to the functional loss of the transcription factor *OsERF922* in rice (Wang *et al.*, 2016). Nucleotide substitutions can impart characteristics, especially in plant–pathogen relationships, where coevolution promotes mutations in immune receptors for perceptual recovery as well as mutations in pathogen effectors for evasion (Jones *et al.*, 2016). Genome editing techniques, including CRISPR-mediated imitation of natural alleles, are helpful in generating gain-of-function mutants that elucidate pathogen and host component evasion (Bastet *et al.*, 2017). Precision breeding entails specific genome alterations, including modest deletions and nucleotide mutations, employing genetic engineering methods, while ensuring the absence of transgenes in the final plant (Andersen *et al.*, 2015).

14.4.1.1 The CRISPR-Cas9 technology

The CRISPR-SpCas9 system, originating via the adaptive immune mechanism of *Streptococcus pyogenes*, is widely used for genome editing by creating customizable single-guide (sg)RNA (Le Rhun *et al.*, 2019). The 80-bp scaffold of sgRNA enables binding to Cas9, while the spacer sequence provides complex targeting to DNA. The nuclease switches from inactive to DNA-seeking when it binds to SpCas9 (Zhang *et al.*, 2019). The process of cleavage of a spacer sequence by SpCas9 involves the recognition of a protospacer attachment motif (PAM), interaction with neighboring DNA, and the formation of an R-loop structure. This requires perfect base pairing with the target DNA and the spacer sequence, but mismatches can cause unwanted off-target activity (Zhang *et al.*, 2019). The HNH and RuvC nuclease domains of SpCas9 work together to cleave DNA through the slow base pairing that causes conformational changes at the active site. Recent research has shown that the nuclease activity of SpCas9 produces both blunt and offset ends, likely due to the flexibility of RuvC in cutting. Many species now regularly use the CRISPR-SpCas9 system, which is considered the best method for editing plant genomes (Manghwar *et al.*, 2019). Delayed base pairing prompts SpCas9 to transition towards a vigorous conformation, ultimately resulting in DNA cleavage via the synergistic activity of its RuvC and HNH nuclease regions. Recent studies indicate how the nuclease activity of SpCas9 produces equally blunt and staggered ends, presumably owing to the flexibility of the RuvC region. Historically, SpCas9 was believed to produce solely blunt-ended double-stranded DNA breaks (DSBs) roughly 3 bp upstream

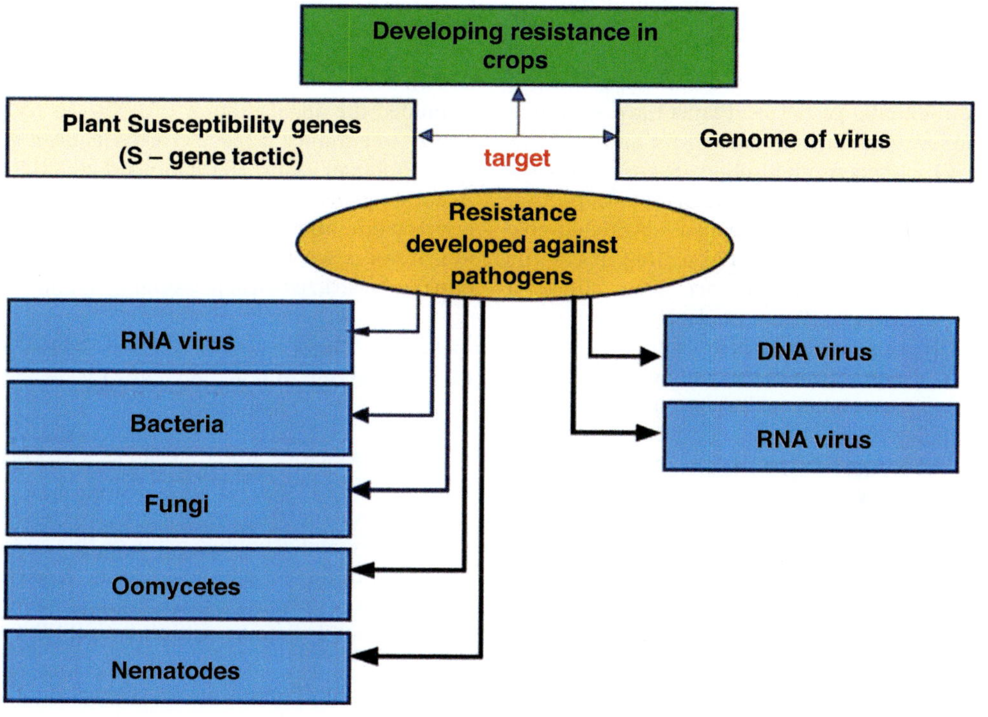

Fig. 14.2. A flowchart to show how resistance develops in plants.

of the PAM (Jiang and Doudna, 2017). Many species now regularly use the CRISPR-SpCas9 system, which is considered the best method for editing plant genomes (Manghwar *et al.*, 2019).

14.4.1.2 A realistic method for generating genome-edited plants with resistance using CRISPR/Cas9

A pressing necessity exists for novel methodologies to increase agricultural productivity due to growing environmental and population pressures and corresponding difficulties in meeting global food security demand (Abdelrahman *et al.*, 2018b). The CRISPR/Cas9 system serves as a potent and versatile tool that employs nucleases with sgRNA to accurately alter targeted genomic areas. The genome of the pathogen or the genes responsible for susceptibility characteristics (S- genes) can be modified to impart disease resistance within plants (Fig. 14.2). Full-length nucleotide sequences for the virus can be obtained from NCBI to produce sgRNA. Conserved viral genome sequences are then selected with a PAM, usually within the range of 2–6 bp of the DNA sequence that follows the DNA sequence targeted by the Cas9 nuclease in the bacterial adaptive CRISPR immune system, and have as few off-targets (Iqbal *et al.*, 2016). To develop resistance to RNA viruses, fungi, bacteria, oomycetes, and nematodes, host susceptibility factors are considered. The evolution of resistance to viruses employs both DNA and RNA targets in the viral genome.

14.4.2 Designing synthetic immune pathways in crop systems

The development of synthetic immune pathways in crops involves adding or modifying existing immune pathways in plants to create new or improved mechanisms for disease resistance (Bigini *et al.*, 2021). Numerous strategies, such as the use of artificial triggers, the modification of immune receptors and signaling pathways, can be used for this purpose (Abdul Malik *et al.*, 2020) with the objective of enhancing plant resilience to diseases and diminishing reliance on traditional pesticides. Plants have an innate immune system that uses PRRs and NLRs to detect pathogens and initiate defense responses (Noman *et al.*, 2019). Nevertheless, natural plant immunity is often not sufficiently durable due to the rapid evolution of pathogens and the limited diversity of tolerance genes among a crop germplasm (Bigini *et al.*, 2021; Pati *et al.*, 2025). This challenge has led to the development of synthetic biology methods that attempt to modify and improve the plant immune system beyond what it can do naturally (Fig. 14.3). Synthetic immune pathways are produced by creating new receptor systems, rewiring signaling cascades, and combining synthetic regulatory circuits to achieve broad-spectrum and long-lasting resistance (Ranf, 2018). Synthetic PRRs that can recognize conserved microbial patterns, chimeric NLRs with better recognition spectra, and synthetic transcription modules that fine-tune the expression of defense genes are all powerful tools for crop protection (Ranf, 2018; Han *et al.*, 2025). These new ideas not only make crops more resistant to new pathogens but also reduce yield losses caused by the constant activation of defense mechanisms. Ultimately, the development of synthetic immune pathways is a new way to improve crops. It is a strategic way to make agricultural systems more resilient amidst climate change and worldwide food insecurity.

14.4.3 Case studies in root and tuber crops: Field-level implementation

Yams (*Dioscorea* spp.), a commercial crop in tropical and subtropical climates, provide tuberous roots, food, and medicinal raw materials (Padhan and Panda, 2020). A number of stress factors, such as diseases and pest infestations, limit yam production. Genetic enhancement can reduce these stress factors, increase productivity, expand markets, and increase economic profits. Modern genetic engineering, including CRISPR/Cas genome editing, has improved yam breeding by studying gene function and addressing inherent plant traits that may hinder yam breeding. CRISPR/Cas technology is reliable, inexpensive, and easy to use (Jinek *et al.*, 2012). Numerous extensive assessments have examined the application of CRISPR/Cas technology to enhance tropical crops (Haque *et al.*, 2018; Tofazzal Islam, 2019; Molla *et al.*, 2020). Accurate gene modification could mitigate the primary limitations of yam production, including disease susceptibility, post-harvest deterioration, and restricted yield potential.

14.4.3.1 Techniques for utilizing CRISPR/Cas technology to enhance yams for disease resistance

Biotic stress factors caused by diseases and pest infestations can reduce yam productivity by up to 25% each year (Anukwuorji *et al.*, 2016). The productivity and economic value of this tuberous crop can therefore be increased by breeding yams that are resistant to commercially important pests and diseases. Utilizing CRISPR/Cas technology to alter the genes of the host plant enables the management of yam diseases.

CRISPR/CAS FOR RESISTANCE TO VIRUSES. The majority of farmers source their planting material from their farms or acquire supplementary material from their nearby farms. Yams are cultivated from their sets. This strategy facilitates the accumulation and dissemination of diseases among infested sets, particularly viruses. Farmers subsequently experience significant yield and quality reductions (Mantell and Haque, 1978; Musa *et al.*, 2023). In addition, yam viruses impede or prevent the international exchange of germplasm. The viral genomes or host-responsive genes may be precisely altered utilizing CRISPR/Cas technology to modulate yam viruses. The replication of plant RNA viruses is aided by eukaryotic translation initiation components like eIF4E, eIF(iso)4E, and eIF4G, which serve as host factors with overlapping activities in plants (Sanfaçon, 2015). Therefore, as shown in other crops such as rice against rice tungro spherical virus and cucumber against cucumber vein yellowing virus, zucchini yellow mosaic virus, and papaya ringspot virus type W, editing the eIF locus in yams may lead to resistance to yam viruses (Chandrasekaran *et al.*, 2016; Macovei *et al.*, 2018).

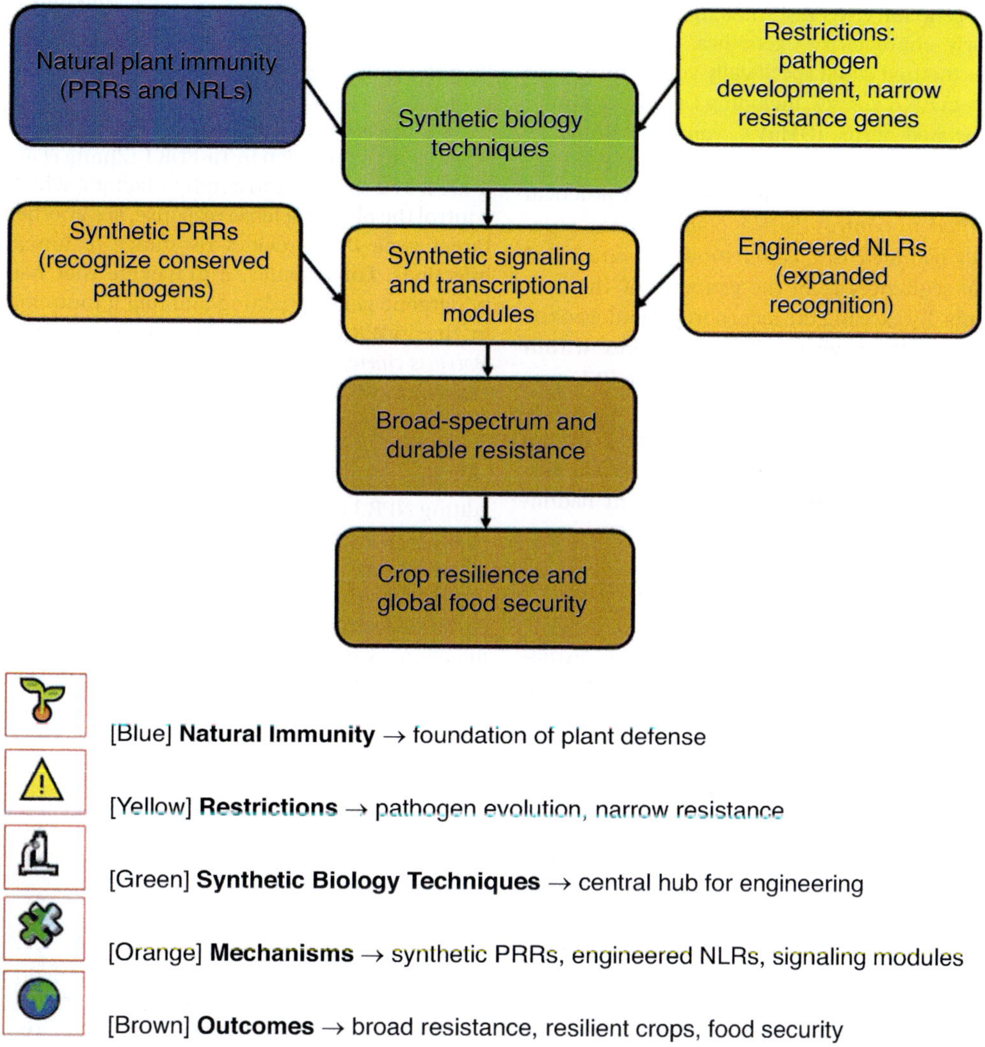

Fig. 14.3. Schematic representation of designing synthetic immune pathways in crop systems. This flowchart shows the basic idea behind the development of synthetic immune responses in plants. Plants depend on natural immunity via pattern recognition receptors (PRRs) and nucleotide-binding leucine-rich repeat receptors (NLRs). However, these defense mechanisms are constrained by the evolution of pathogens and a limited repertoire of resistance genes. Synthetic biology offers innovative solutions to these challenges, including synthetic PRRs, chimeric/modified NLRs, and synthetic signaling/transcriptional modules. These methods work together to strengthen plant defenses, yielding enduring and extensive resistance. The primary objective is to develop plant systems capable of enduring increased biotic and abiotic stressors, hence ensuring food security for everybody.

Likewise, CRISPR/Cas9 is employed to modify two eIF4E isoforms, thereby conferring resistance to the cassava brown streak virus (CBSV) to cassava plants (Gomez *et al.*, 2019). The alterations diminished tuber necrosis, lessened the severity and prevalence of the infection in storage roots, and mitigated CBSV signs in cassava shoots. The effective alteration within the eIF4 gene to confer virus resistance in several crops illustrates the potential to employ this gene for virus resilience in crops that have poorly characterized genomes, such as yams. This method could specifically confer resistance to yam RNA viruses, including *Dioscorea* mosaic-associated virus (DMaV), yam mosaic virus (YMV), yam mild mosaic virus (YMMV), and yam asymptomatic virus 1 (YaV1). A practical method to control dsDNA plant viruses, especially badnaviruses, is to remove the integrated viral sequences from the genomes of the host plants. By excising the incorporated endogenous banana streak virus (BSV) sequences within the host genome, immunity against BSV was conferred in bananas, hence averting the potential stimulation by pathogenic virus particles (Tripathi *et al.*, 2019). This strategy could also be used for the management of yam badnaviruses in West Africa, as their viral sequences are incorporated into the yam genome, turning them into pararetroviruses (Seal *et al.*, 2014).

CRISPR/CAS FOR RESISTANCE TO FUNGAL PATHOGENS. The most commercially relevant field pathogen for *Dioscorea alata*, the world's most widely cultivated yam variety, is a fungus that causes anthracnose disease. Numerous recalcitrant fungal infections belonging to the genus *Colletotrichum* are responsible for the leaf disease anthracnose (Ntui *et al.*, 2021). *Aspergillus niger*, *Aspergillus tamari*, and *Fusarium* spp. cause dry rot of tubers. It is the most common field disease of yams and leads to considerable yield losses worldwide. *Rhizopus* spp., which cause wet rot of tubers, and *Botryodiplodia theobromae* are other fungi that infect yam roots (Syombua *et al.*, 2022). Through the loss of function of the host susceptibility gene (S-gene), genome editing with CRISPR/Cas has shown great potential for the generation of fungal resistance in a variety of crops. For example, many crops, such as wheat and tomato, have developed resistance to fungal pathogens when the proteins of the powdery mildew resistance locus (MLO) were knocked out (Wang *et al.*, 2014; Nekrasov *et al.*, 2017). The MLO locus, which is evolutionarily conserved in monocotyledons and dicotyledons, encodes a plasma membrane protein (Acevedo-Garcia *et al.*, 2014). Targeting ethylene response factor 922 (*ERF922*) and enhanced disease resistance 1 (*EDR1*), which are involved in ethylene signaling and pathogen resistance, has further demonstrated resistance to fungal genes. For example, resistance to rice blast is achieved by altering the *OsERF922* gene (Abdelrahman *et al.*, 2018a). Increased resistance to powdery mildew disease was demonstrated in gene-edited wheat with a mutation in Ta-EDR1 (Zhang *et al.*, 2017). The WRKY transcription factors, which control the plant's defense response, are another target gene that provides resistance to fungal infections. For example, a grapevine that had undergone genome editing and had a mutation in the *VvWRKY52* gene showed resistance to *Botrytis cinerea* (Wang *et al.*, 2018). Since Non-Expressor of Pathogenesis-Related 3 (NPR3) is a negative regulator of the defense pathway, the NPR3 gene could be another target. Cacao showed resistance to *Phytophthora tropicalis* after editing NPR3 (Fister *et al.*, 2018). The orthologs of the S-genes in yams can be targeted to generate resistance to fungal diseases, as these genes are often conserved in nature. To generate novel variations with increased resistance to fungal infections, yam orthologs of known susceptibility genes (*MLO*, *EDR1*, *ERF922*, *NPR3*, and/or *WRKY52*) can be targeted, similar to other crops.

CRISPR/CAS FOR THE CONTROL OF BACTERIAL DISEASES. The two most important bacterial pathogens in yams are *Bacillus pumilus* (Tariq *et al.*, 2024) and *Erwinia carotovora* subsp. carotovora, which causes bacterial wet rot in yam tubers (Tariq *et al.*, 2024). CRISPR/Cas deletion of host susceptibility genes, such as Downy Mildew Resistance 6 (DMR6) and Sugars Will Eventually Be Exported Transporters (SWEET), is shown to confer long-lasting resistance to bacterial infection in plants. The DMR6 is a negative regulator of plant defense responses, and its expression level increases after pathogen infection (Sun *et al.*, 2016). Thus, the resistance of yams to bacterial infection could be generated by altering the expression of *DMR6* gene homologs and/or promoters. Tomatoes and bananas with changes in the *DMR6* ortholog showed increased resistance to bacterial *Xanthomonas infections* (Thomazella *et al.*, 2021; Tripathi *et al.*, 2021). The susceptibility genes of the SWEET family are intracellular and extracellular sucrose transporters that play an important role in the pathophysiology of bacteria. To combat plant diseases, cultivated plants with genetic polymorphisms in the SWEET genes are bred (Gupta, 2020). In a previous study, resistance to the bacterial diseases, *Xanthomonas oryzae* and *pathovar oryzae*, in rice is enabled by precise manipulation of the SWEET susceptibility genes (*OsSWEET14* and *OsSWEET11*) or their promoters (Oliva *et al.*, 2019; Xu *et al.*, 2019). Activation of endogenous genes involved in defense mechanisms, such as bZIP, DELLA, RAV1, and RAV2, is another strategy to improve resistance to bacterial infection (Li *et al.*, 2018; Wei *et al.*, 2018). Resistance to bacterial cassava disease was conferred by these genes. Therefore, CRISPR activation (CRISPRa), which utilizes a modified Cas9 without endonuclease activity (dead Cas proteins; dCas) along with an additional transcriptional activator, could enhance the expression of the desired gene(s) and upregulate the endogenous defense genes of yams.

14.5 Artificial Intelligence in Predictive Modeling and Decision Support in Pest and Disease Management

Pests and pathogens are major biotic stresses in agriculture and food production, accounting for about 20–40% crop losses globally (Junaid and Gokce, 2024). Animal pests and microbial pathogens contribute to about 18% and 16%, respectively, of the losses (Terentev and Dolzhenko, 2023; Akhtar *et al.*, 2024). There is a need to optimize agricultural productivity to meet the growing world population. This is because the projected global population by 2050 is 10 billion, and it is estimated that food production will need to increase by 60% to meet demand (Ma *et al.*, 2023). This means every aspect of agriculture, from plant breeding through agronomic and plant health practices, must meet the current demand to achieve a world of food sufficiency. The quest to improve crop productivity does not just entail breeding high-yielding varieties and climate resilience, but also protecting them from pest and disease attacks to ensure and sustain potential yields.

Plant protection is one of the pivots of agricultural productivity and must be of high precision and reliability, with minimal detrimental footprint in ecosystems. Ever since efforts have been made to cultivate crops, humans have been actively trying to tilt the competition between pests/pathogens and humans in favor of the latter. This has led to the development of an array of strategies, some with detrimental consequences to the target and to humans. Nevertheless, these competing biological agents are key ecosystem linkages that are crucial for the sustenance of biodiversity. For this reason, the idea of complete annihilation of pests and pathogens in any agricultural production system is unwelcome. As such, plant protection has evolved through three stages: natural-farming-based, pesticide-priority-based, and integrated-management-based plant protection (Zheng and Xu, 2023). Natural farming protection systems include intercropping, crop rotation, biological control, cultural practices, bio-extracts, and host resistance, among others. The advent of chemical pesticides drastically changed protection systems because of their superior prowess in exterminating pests and pathogens. The introduction of chemical insecticides and fungicides was an instant success, leading to an unprecedented leap in crop production output due to a huge reduction in crop losses. The impact on human health, environment, other non-target organisms, and resistance build-up fueled the need to re-align plant protection strategies. This led to integrated management, where available management methods are strategically combined, taking other factors besides the target pest or pathogen into consideration. For example, climatic conditions such as temperature, humidity, and rainfall can be used to predict pest or disease incidence, thereby ensuring judicious use of chemical insecticides. Similarly, sensors and cameras can remotely inform about the state of pest or disease incidence following the release of biological control agents or practices such as intercropping or performance of a resistance variety, and when needed, samples can be taken to the laboratory for further analysis to confirm the pathogen expressing disease symptoms. All these steps are usually not seamlessly linked, and only a few are integrated due to challenges associated with data gathering and analysis. Despite this, several specialties from molecular genetics, remote sensing, mathematics, computing, and others are being used to generate data to enhance the effectiveness of pest and disease management. These are advancing pest and disease management pathways to help enhance agricultural productivity and food security.

Pest and disease management strategies can leverage the advancements in AI, spatial, and multi-omics techniques. AI is a suite of techniques used to simulate human intelligence processes by machines. These processes include learning (the acquisition of information and rules for using the information), reasoning (using rules to reach approximate or definite conclusions), and self-correction (Farooq *et al.*, 2024). AI is becoming an indispensable tool in pest and disease management. Several AI techniques, particularly ML, deep learning (DL), and AI-based data analytics platforms, are revolutionizing stress detection, prediction, and management (Varghese *et al.*, 2025). A comprehensive list of the ML techniques in biotic stress analysis has been reported (Mahenge *et al.*, 2023). In this context, specific algorithms such as AI-based predictive analytics, ML (algorithms that are known for performing the assigned task via the learning of the provided data set), DL (highly speculated ML algorithm that trains the machine to learn via multiple processing layers) are collectively termed AI, except in a few situations where specifics are referenced.

14.5.1 AI-driven simulations of plant defense responses

Plants are sessile organisms and must develop warding-off or resilience mechanisms to withstand pests (arthropods and other herbivores), pathogens (fungi, bacteria, viruses, viroids, phytoplasma, and nematodes), and environmental stressors (drought, extreme temperatures, salinity, and nutrient deficiencies) (Khater *et al.*, 2017; Cembrowska-Lech *et al.*, 2023). Among life forms, plants exhibit the greatest suite of adaptations (Agrawal, 2011). Plants' defense mechanisms against pests and diseases are evolutionary adaptive processes, involving complex activities spanning genes, proteins, and metabolites (Cembrowska-Lech *et al.*, 2023). Standalone investigations in these specialized areas offer limited insights into a holistic understanding of plant defense mechanisms. Current trends are forging to integrate these specialized areas to better understand the plant defense system from a diverse perspective to enhance pest and pathogen management tools. The incorporation of genomics, proteomics, and metabolomics provides a more comprehensive perspective of plant defense processes and interactions with bio-mediated stresses. This also means analyzing large and complex datasets to generate valid and useful outputs. This is where AI is an indispensable tool, considering the rapid advancements in recent years and the unlimited projected breakthroughs in the coming years.

Metabolomics involves chemical trails of specific cellular activity and can be used to track responses to stress. Thus, the influence of gene expression, protein production, and metabolic pathways in response to stress is revealed through metabolomics. There is evidence of plants producing specific metabolites in response to biotic stresses such as pathogen attack (Cembrowska-Lech *et al.*, 2023; Zhang *et al.*, 2024). Changes in concentrations of a group of metabolites are linked to pathogen intrusion, and this may occur as early as the first day of inoculation. Identification of these metabolites complements data from other technical methods, such as proteomics and genomics, in facilitating early disease detection (Castro-Moretti *et al.*, 2020). These require high-throughput data analysis, which can reliably be conducted using AI analytical procedures. Plant defense research is now employing AI-driven simulations to analyze spatial and multi-omics datasets and model plant responses to biotic stresses. These practices help understand the complex interplay of genes, proteins, and metabolites involved in pest and disease stress responses. This enhances the identification of key defense mechanisms and potential biomarkers. Ultimately, crops that are resilient to biotic stresses are developed to improve agricultural productivity (Cembrowska-Lech *et al.*, 2023; Farooq *et al.*, 2024).

Due to their versatility, AI methods are also used to fast-track the process of developing new plant varieties with improved pest and disease tolerance by analyzing vast germplasm resources to identify beneficial genes and subsequently for gene pyramiding. Specific genes involved in plant defense pathways can be identified, allowing targeting of stress tolerance triggers

(Cembrowska-Lech *et al.*, 2023). Resistance to a specific disease, for instance, can be predicted and validated experimentally by training ML algorithms on genomic, transcriptomic, proteomic, and metabolomic datasets (Zhou *et al.*, 2021; Cembrowska-Lech *et al.*, 2023), supporting efficient breeding decisions and workflows where stress-controlled gene combinations predicted stress resilience. This involved the integration of spatial and multi-omics data, environmental simulation, and ML for the prediction and enhancement of stress tolerance in these plants. The inherent high scalability creates an avenue to integrate abiotic and biotic stresses driven by climate change (Kaushik, 2025).

14.5.2 AI modeling of plant immune dynamics and pest/pathogen behavior

Plants must cope with more than 10,000 insect species, 1500 plant viruses, 1500 plant diseases, and other biotic stressors such as rodents and weeds (Zheng and Xu, 2023). Plants, through evolutionary drive for adaptation and survival, have developed mechanisms to detect and defend themselves against pathogens, pests, and environmental stressors. The effectiveness of the responses determines the level of resilience against a suite of stresses. Immune responses come in the form of physical barriers, molecular and cellular pathways (Agrawal, 2011; Waterman *et al.*, 2019). A simple constitutive defense strategy involves physical barriers such as trichomes, thorns, spines, waxy cuticle, and thick cell walls to trigger and elaborate induced strategies, where chemicals are released upon attack. In the case of pests, the compounds released are usually secondary metabolites (phenolics, alkaloids, terpenoids, and glucosinolates) that poison or deter attackers. Pathogens are also inhibited, thereby halting inbound diseases through antimicrobial activity. A more complex strategy is cellular-level defense, where cytoplasmic, cellular, or histological structural and chemical defenses are triggered (Rithesh *et al.*, 2023). Chemical defense pathways are intricately complex, and understanding them for their exploitation in plant science requires the incorporation of genomics, transcriptomics, proteomics, and metabolomics data.

Proteases are important in plant immune mechanisms as a function of plant and pathogen interactions. While plants release enzymes to defend against foreign biotic agents such as pathogens through proteolysis, the pathogen also counters this with an antagonist enzyme to neutralize plant enzymes (Paiva-Silva *et al.*, 2025). Insights into protease regulatory channels offer opportunities to leverage for developing plants resilient to biotic stresses. Small secreted peptides (SSPs) that are involved in protein inhibition pathways can now be predicted through simulations (Homma *et al.*, 2023). For example, an AI-based protein structure prediction tool, AlphaFold-Multimeter (AFM), is used to study pathogens and their strategies for suppressing tomato immunity, thus SSPs of the pathogens (fungi, bacteria, and oomycetes) on the one hand and the host's defense-mediated hydrolases on the other hand. At the plant–pathogen interface, it has been demonstrated that pathogen-secreted inhibitors targeted a tomato-secreted immune protease P69B (Homma *et al.*, 2023). In addition, four novel P69B inhibitors were identified, thus demonstrating the cascading power of using AI (Homma *et al.*, 2023). Generally, integration of AI tools into proteomic workflows is an emerging area, and although few inroads have been made, there are great prospects. Proteomics in plant defense covers amino acid sequence, structure of proteins, functions of different proteins at the cell level, and quantifying protein expression levels in response to stresses (Murmu *et al.*, 2024). Potential progress prospects in leveraging proteomics for plant defense applications include early disease detection using AI–proteomics fusion, predictive modeling for disease dynamics, precision crop breeding, and integration of remote sensing data (Murmu *et al.*, 2024).

Evolution of pest traits to plant defenses has been the basis for resistance breakdown in crop varieties that once showed a high level of resistance. Traditional plant immune response studies tend to focus on single-trait-driven defense instead of synergistic or additive effects. Multiple traits may interact synergistically, as shown in cases where toxin combinations provide more effective deterrence than individual compounds (Agrawal, 2011). However, studying multi-trait defenses can be very challenging using traditional techniques. Simulations of insect or pathogen behavior can track how specific plant defense scenarios influence the evolution or selection of pest/pathogen populations over time (Agrawal, 2011; Hardy, 2022). Similarly, AI simulation models that include coevolutionary dynamics, where both plant traits and pest responses evolve, can help predict the long-term effectiveness and stability of defense strategies.

14.5.3 Climate-change-informed pest and disease forecasting using AI

Pest and disease management is historically reactive rather than predictive, partly because dependable instruments for pre-empting occurrences are unavailable. Practitioners had to depend on phenological markers of the plant or conduct physical inspections for damage on plants and symptoms, in addition to assessing pest and vector numbers, to inform their judgments. The basics of quantitative pest forecast systems were established in the 1940s and have undergone progressive improvement through computerization (Singh and Shailbala, 2012). AI-based models for biotic stresses gained momentum in the late 2010s as the influence of climate change took center stage (Kamilaris and Prenafeta-Boldú, 2018). Thus, the adverse impact of climate change on agriculture and food security has been a dominant topic in recent decades (Yousaf *et al.*, 2023; Delfani *et al.*, 2024). It is projected that it could reduce food production due to adverse weather conditions, such as prolonged droughts, flooding, and high temperatures. Besides the direct effect on the crop, it will compound the already precarious situation of pest and disease outbreaks.

The survival, behavior, distribution, and resistance of insect pests and plant pathogens are influenced by climatic factors, including temperature, humidity, and rainfall. Climate change is reshuffling pest distribution as some pests and diseases spread to previously unsuitable locations due to rising temperatures (Saleem *et al.*, 2025). Also, an estimated 10–25% crop loss is projected for every degree rise in temperature (Shrestha, 2019). There is also a disruption of pests and their natural enemy balance due to phenological oscillations (Delfani *et al.*, 2024; Saleem *et al.*, 2025). The traditional timelines for estimating pest and disease incidence based on normal climate patterns in a geographical area would not be reliable, due to continual changes in crop phenology caused by a variable climate. Climatic factors have become less predictable, thereby complicating forecasting approaches using traditional modalities. This means there must be a paradigm shift in managing climate change, pests, and plant pathogens in the agricultural space (Garrett *et al.*, 2022; Delfani *et al.*, 2024). Historical data simulations and models with AI technologies are taking prominence in pest and disease forecasting for monitoring, decision making, and intervention. AI-driven decision support system platforms recommend specific interventions based on multiple factors, such as current pest pressure, crop stage, environmental conditions, and known resistance profiles. Thus, the systems help to optimize timing and dosage of interventions, which leads to efficacy and sustainability improvement.

14.5.3.1 Insect pest forecasting

The rapidly improving AI techniques facilitate the prediction of insect pest behavior under different climate conditions because ML and DL algorithms seamlessly integrate climate data with pest behavior and population dynamics. AI gleans multiple spatial datasets to forecast climate, which is then related to pest dynamics. Algorithms help determine intricate patterns to explain pest behavior (Bukhari *et al.*, 2025). AI-driven climate–pest models provide an important pathway that helps to establish essential instruments that assist in mitigating climate change effects on agriculture through advanced pest management approaches (Bukhari *et al.*, 2025). Increasing temperatures are widening the habitable range of some insect pests, and unrecognized pests in certain regions are now becoming dominant. In the temperate regions, low temperatures limit certain pests but conditions are expected to become conducive (Skendžić *et al.*, 2021). Fall armyworm (FAW) (*Spodoptera frugiperda*), for instance, thrives in warmer climates and cannot overwinter in colder climates, although it can migrate to relatively warmer regions during warm seasons. However, milder winters can expand its range, threatening agricultural lands and food production over a wider area. The influence of climatic factors has been modeled with AI algorithms on the dynamics of FAW in the early years of its introduction to Africa (Early *et al.*, 2018). Similarly, early warning projections to new areas, including Asia, Europe, and Australia, offered opportunities to prepare in advance (Early *et al.*, 2018). Thus, AI models are platforms for the prediction of possible geographical range expansion and risk from major insect pests, providing a climate-smart early warning model for the effective management of such pests.

The African armyworm (*Spodoptera exigua*) is another devastating pest for which the solitary and gregarious phases are driven by climatic factors, particularly temperature, precipitation, and humidity. Early detection is very critical in halting the enormous destruction by the gregarious phase. The mitigating strategy has been the reliance on field scouting, pheromone trapping, and spatial imaging (Sokame *et al.*, 2025). A key limitation over the years has been the ability to reliably predict outbreaks of the pest for swift intervention, primarily due to the intricacy of the influence of ecological, climatic, and human factors. Based on AI-driven data simulation, a reliable predictive model that integrates biological, ecological, and integrated pest management (IPM) strategies has been developed for effective management of the African armyworm (Sokame *et al.*, 2025).

Natural enemies are an important component of IPM, and some AI models consider the efficiency of the natural enemies under agricultural systems and associated practices. In most cases, alien pests move ahead without their natural enemies, and it is also important to be able to predict the geographical range of their host. AI-driven digital twin models are used to create a virtual agroecosystem to simulate how climate change influences the interaction between insect pests and their natural enemies. In these models, ecological shifts, weather patterns, and behavior of pests and natural enemies are integrated to produce responses under diverse climate settings, such as how temperature increases affect pest interactions with their natural enemies (Wang *et al.*, 2024). A field simulation of the effect of increased temperature on the predatory rate of the ladybird beetle (*Harmonia axyridis*) on different wheat aphid species showed aphid densities increased, but ladybird beetle densities did not during an elevated temperature period. The predatory rate of the ladybird beetle was reduced by 25% at elevated temperature (Wang *et al.*, 2024). In a choice experiment in the laboratory, the consumption of all the aphid species except one increased with elevated temperature (Wang *et al.*, 2024). In addition to this classical work, there have been several studies on aphid–ladybird interactions in diverse agrosystems that indicate rising temperature, and thus climate change, influences their interactions. Most of these studies, however, employ mathematical models and statistical approaches to analyze the interactions and have limitations in not accounting for the intricate multiple interactions. Therefore, limiting the scope of interaction and impact of these factors that pertain in the real world (Agboka *et al.*, 2024). These limitations could be addressed by the application of AI models such as ML, DL, and reinforcement learning (RL) to holistically account for pest population dynamics as influenced by environmental factors and resource allocations (Agboka *et al.*, 2024). In demonstrating this, RL was used to simulate reward, climatic, and crop availability matrices to create a real-world condition scenario that provided a good output. Based on spatial and environmental data, the shift of FAW as driven by seasonal factors has been developed with species distribution models, thereby predicting and tracking the local and regional migratory movement (Niassy *et al.*, 2021; Qi *et al.*, 2024), and the cross-continent pest migration trajectory (Wang *et al.*, 2023; Fu *et al.*, 2025). AI-powered applications for the notorious FAW help in the detection, severity, and spread dynamics of the pest (Shaurub, 2024).

14.5.3.2 *Diseases*

Plant diseases may be infectious or non-infectious, depending on the causative agent. The infectious diseases are caused by pathogens (viruses, bacteria, fungi, nematodes, and parasitic plants) and are transmissible. Non-infectious diseases, in contrast, are induced by abiotic factors (nutrient deficiency, extreme temperatures, inadequate or excessive water, among others). An accurate prognosis directly determines the mode and success of the control strategy. Conventional methods for diagnosing plant pathogen-mediated diseases include visual observation, microscopy, mycological analysis, and molecular diagnostics (Kaur and Sharma, 2021; Khakimov *et al.*, 2022). The non-infectious diseases are also detected by visual inspection. Visual observation is usually the most widespread method and also the first diagnostic and predictive action. Visual distinction of disease can be subtle and must be validated by more robust methods, such as microscopy and molecular analysis. The visual observation method is time intensive, and specialized expertise and equipment are required. These make them less practical in many situations, such as in remote areas and large fields.

Phenotypic expressions are outputs from genetics, transcription, metabolites, and proteins. Phenotypic response to insect pest stress can be monitored with tools such as drones, thermometer sensors, and RGB cameras. Traits associated with these stresses can be used to predict associated traits through hypersensitive platforms (Cembrowska-Lech *et al.*, 2023). The fast-paced advancement in technology has integrated various techniques for detection, forecasting, and management solutions for diseases, irrespective of the causative agent. Sensors, cameras, and satellite data are used to analyze real-time data for decision making. These models factor in temperature (the key driver of climate change), humidity, and rainfall, which are influenced consequently. AI algorithms techniques such as ML, DL, convolutional neural networks (CNNs), random forests (RFs), and support vector machines (SVMs) have been used to analyze images and to integrate environmental data for predicting disease outbreaks. Early disease detection, even when there are no visible signs, has been made with high precision (Dhiman *et al.*, 2024). There are several disease detection models based on image detection AI algorithms, but they are scant on forecasting of disease outbreaks (Skelsey, 2021; Kaur *et al.*, 2025). A few early referenced plant disease forecasts based on ML techniques are: wheat tan spot (De Wolf and Franel, 1997), wheat stripe rust (Gang *et al.*, 2006), rice blast (Kaundal *et al.*, 2006), *Fusarium* head blight (Landschoot *et al.*, 2013), wheat yellow rust (Xu *et al.*, 2018), and powdery mildew (Hamer *et al.*, 2020). A weather-mediated forecast model suitable for climate change-focused prediction is simulated for late blight outbreak in potatoes (Skelsey, 2021). Success from this is expected to spur the development of climate change-oriented AI models for other crops.

Unlike insect pests, for which there have been many open-source datasets to simulate climatic factors-driven pest population dynamics, the same cannot be said for disease. The overemphasis on image-processing pathways relative to weather-based analysis leaves a gap in climate-based disease prediction models (Skelsey, 2021; Kaur *et al.*, 2025). Available literature suggests there is much discussion on the early detection of plant diseases using enhanced approaches such as gas chromatography–mass spectrometry (GC–MS), high-performance liquid chromatography–mass spectrometry (HPLC–MS), hyperspectral remote sensing, and spectroscopy, but their practical applications are limited. This is primarily due to difficulty in extracting the appropriate data for disease identification, and a streamlined interdisciplinary approach is required in order to harness the power offered by these modern technologies (Terentev and Dolzhenko, 2023).

14.5.4 Integrating spatial, multi-omics, and environmental data through AI platforms for pest management

The development of technologies for spatial, multi-omics, and environmental knowledge domains is fast paced. This has diversified their applications in an integrated manner for benefits that each could not have achieved without AI applications. Spatial data refer to geo-referenced information about any object, and in agriculture, such data traditionally inform habitats, distribution, and spread patterns of plants, animals, as well as the stresses they encounter. The data are generated through geographic information systems (GIS), remote sensing, and drones, which can provide real-time insights for immediate action. Optical remote sensing methods (RGB imaging, thermography, fluorescence imaging, multi- and hyperspectral imaging) are currently being used to monitor plant health by detecting damage levels, densities, and spread of pests and diseases in plants (Cen *et al.*, 2022; Terentev and Dolzhenko, 2023).

This has the advantage of being fast and able to cover a large landscape, but it has the limitation of accurately detecting some diseases at early stages. These modern techniques are enhanced by the integration of metabolomics for the early detection of biotic stresses (Terentev and Dolzhenko, 2023). There is a classic example of AI-driven early disease detection in the integration of metabolomics with ML as a management tool. Traditional management methods rely on symptom manifestation (yellowing leaves, stunted growth, bitter/misshapen fruit), which makes containment of the disease difficult due to late detection. Already, the use of UAVs with hyperspectral remote sensing and handheld spectrophotometers with embedded AI is used to accurately detect symptomatic Huanglongbing (HLB) (Pourreza *et al.*, 2015; Deng *et al.*, 2020). Integration of AI-powered metabolomics and spectral analysis can therefore offer the capacity to detect HLB at all stages (asymptomatic and symptomatic). There have been suggestions that AI trained in metabolomics and spectral data can be integrated into portable sensing devices such as handheld spectrophotometers and drone-based imaging systems (multispectral/hyperspectral) (Deng *et al.*, 2020).

Remote sensing coupled with imaging has been used to monitor localized pests and diseases of crops (Iost Filho *et al.*, 2022), as well as regionally wide, such as for the desert locust (Islam *et al.*, 2023). AI-based spatial systems for pest management are used for early detection, insect monitoring, and population forecasting. AI-enhanced automated systems based on current technological advancements enhance response time and efficiency, while reducing pesticide use, labor intensity, and errors (Kapetas *et al.*, 2025). They highlighted the accuracy levels of AI-driven imaging models, which generally show a mean average precision (mAP) of 66–99%, depending on the AI technique used. These studies covered a wide range of insect pests, including those notoriously difficult to count in the field (whiteflies and aphids). Drones supplied with appropriate imaging systems and sensors can capture high-resolution and multi-dimensional images of a field trial. This helps spatial variability assessments in plant traits, which is particularly useful in large fields. Mobile phones can be used to capture images of traps or plants to count insects for decision-making purposes. For instance, an AI-driven system in a smartphone is used to detect, monitor, and forecast black aphid populations in greenhouse cucumber (Kapetas *et al.*, 2025). Multi-omics data provides genomic traits of interest and merges with imaging and sensory systems through AI platforms, hastening resistance breeding programs. The integration of multi-omics approaches with these advanced AI modeling techniques provides a better view of the plant system, thereby giving insights into the complex interactions and regulatory mechanisms that underlie observable plant traits (Cembrowska-Lech *et al.*, 2023). Merged spatial and environmental long-term climate data are used to model pest and disease trends and responses, while real-time weather inputs are used to determine real-time status for decision making or the effectiveness of an intervention. Moreover, it is used to forecast biotic stress models, thereby helping anticipate outbreaks before they occur. The key advantage of AI in monitoring is the ability to detect subtle changes in satellite imagery or sensor data for biotic stress.

An emerging field with great potential for the AI moderation of agriculture is swarm intelligence. This system imitates the collective behavior of social organisms such as ants and bees. In the context of plant protection, swarm intelligence can be used to coordinate multiple autonomous agents (e.g. drones, sensors) to survey fields, monitor plant health, and efficiently apply pesticides without centralized control (Sharma *et al.*, 2021). It is also capable of being used in phenotyping plant varieties to track resistant genes. Integration of spatial, molecular, and ML diagnosis tools for diseases and pest management is also receiving attention (Kreuze *et al.*, 2022). For instance, the PestDisPlace platform combines spatial diagnosis of disease symptoms and molecular diagnosis of root pathogens and their vectors on root, tuber, and banana crops. This is used to project disease identification, spread, and early warning to potential areas of new invasion (Cuéllar *et al.*, 2018). Cutting-edge AI systems seamlessly integrate spatial, omics, and environmental data into simplified and user-friendly interfaces. These interfaces allow predictive simulations to run and test different strategies before validating real-world solutions (Zhou *et al.*, 2021). This integration has been used to facilitate early detection of resistance by identifying emerging resistance mutations or shifts in gene expression profiles. Recognition assessment after pest or disease treatment with pesticides usually signals the performance of the pesticide against the pest or pathogen.

Phenotyping is very important in selecting a plant from a germplasm. However, the use of traditional phenotyping is unable to segregate some morphologically identical plants. Marker-assisted and AI algorithms are now used to address such challenges, where the latter is used in predicting quantitative characteristics, genotype classification, and genomic selection (Alipanahi *et al.*, 2015; Yang *et al.*, 2025). AI simulations are used in predicting optimal crosses by analyzing complex genetic, phenotypic, and environmental datasets to identify high-performing parental combinations (Sadeghi-Tehran *et al.*, 2019; Khaki *et al.*, 2020). This facilitates the breeding of resistant crop varieties as part of the IPM drive. Holistic IPM implementation in farms still faces some hurdles, although the ideas have been in the agricultural space for decades. This is primarily due to inherent agricultural landscape factors ranging from variable weather, diverse insect behaviors, and crop variability to soil heterogeneity. These collectively affect the right pest identification, the right method of selection, right control timing, and the right action taken (4Rs) in IPM (Yang *et al.*, 2025). Acoustic methods for the identification of pests are one of the underexplored areas in pest management. However, this has been successfully used to detect insect pest species, estimate population density, and track distributions. Acoustic devices have been used in storage products and wood pest detection (Mankin *et al.*, 2011). The potential for wide-scale improvement in the technology and broader usage has been adopted (Mankin *et al.*, 2011; Jiang *et al.*, 2024). With the current limitations of other methods for detecting subterranean pests, reliance on AI platforms can help improve pest detection and prediction when integrated with other imaging and sensing technologies.

Biotic stresses in agricultural investigations are usually conducted in a laboratory or controlled environments where overall landscape interactions are restricted. This is basically due to the cost, time, and difficulty in measuring and analyzing a wide range of parameters. Although this provides snippets of real-world situations, broader landscape experiments or data integration offer a better view. Current trends in AI-assisted experimental design, simulation, and modeling offer the leverage in carrying out such investigations or integrating data from different domains. Crop models and simulation platforms have been popular over the years, but these did not incorporate biotic stresses until recently (Triki *et al.*, 2023). Some existing crop growth platforms include the environmental policy impact model (EPIC) (Williams *et al.*, 1989), the decision support system for agrotechnology transfer (DSSAT) (Jones *et al.*, 2003), and the agricultural production systems simulator (APSIM) (Keating *et al.*, 2003). Recent AI models are coupling crop, pest, and disease models for enhanced application in the field, although there are teething problems with the generality of such models (Triki *et al.*, 2023). Examples are AgGym, which incorporates crop and biotic stress conditions to simulate yield using an AI framework (Khosravi *et al.*, 2024); a tri-trophic simulation of crop (oil palm), pest, and natural enemy interaction (Qi *et al.*, 2009); and crop (robusta coffee) and pest (coffee berry borer) (Triki *et al.*, 2023).

14.6 Challenges and Future Opportunities

14.6.1 Translating omics insights into practical pest and disease control strategies

Despite significant advancements in spatial and multi-omics technologies, a major challenge lies in translating these insights into actionable pest and disease management interventions. The data generated from genomics, transcriptomics, proteomics, and metabolomics are often voluminous, multifaceted, and context dependent, making it challenging to derive direct applications for field-based management (Dhiman *et al.*, 2024). One primary barrier is the disconnect between laboratory-derived molecular profiles and phenotypic outcomes under diverse agro-ecological conditions (Sanooja *et al.*, 2025). For example, genes or pathways identified as critical under controlled conditions may not exhibit the same expression patterns or practical impact in field environments subject to variable stressors.

Furthermore, the bioinformatics tools and computational infrastructure required to integrate multi-omics data remain limited in low-resource research settings. Many developing countries that are heavily impacted by pest and disease outbreaks lack the technical capacity to analyze and interpret such data effectively (Sanooja *et al.*, 2025). Bridging this capacity gap requires both investment in infrastructure and interdisciplinary training programs that equip plant scientists with the necessary computational and translational biology skills. There is also a need for robust validation channels to assess the real-world effectiveness of omics-derived targets. Marker-assisted selection (MAS), genome-wide association studies (GWAS), and CRISPR/Cas technologies have shown promise, but the deployment of resulting products must be followed by rigorous field trials to re-evaluate their impact on crop yield and pest/disease suppression (Demirjian *et al.*, 2023). Another critical element is stakeholder involvement. Translational success hinges on early engagement with farmers, breeders, agribusinesses, and policy makers to co-develop solutions that are socially acceptable, economically viable, and ecologically sustainable (Radio *et al.*, 2024). Without participatory frameworks, even the most advanced novelties risk limited adoption or outright rejection.

14.6.2 Bridging gaps between laboratory discoveries and field applications

The gap between laboratory innovations and real-world applications remains a longstanding issue in agricultural research. While spatial and multi-omics platforms have greatly expanded our understanding of plant responses to biotic stresses, the path from discovery to deployment is disjointed. Factors contributing to this detachment include the use of model plants that may not accurately reflect responses in economically important crops, lack of interdisciplinary collaboration, and regulatory bottlenecks (Yaschenko *et al.*, 2025). Most multi-omics studies are conducted in highly controlled environments, often without dynamic field variables such as unstable weather conditions, mixed pest/pathogen pressures, and soil microbiome interactions. Consequently, the phenotypic relevance of molecular markers identified under laboratory conditions can be reduced significantly under field conditions (Mochida *et al.*, 2019). Addressing this requires the design of omics experiments that integrate spatial and temporal variability, including multi-location and multi-season field studies supported by precision agriculture technologies like drones, sensors, and satellite imaging.

Bridging the research–application rift also requires stronger linkages between academic institutions, private sector breeders, and extension systems. Public–private partnerships (PPPs) have demonstrated effective acceleration of the commercialization of spatial and multi-omics tools (Thakor and Gautam, 2025). However, these partnerships must be established based on transparent data-sharing policies, unbiased benefit-sharing arrangements, and mutual trust. Another promising avenue is the development of decision-support tools that incorporate multi-omics data with geospatial, climatic, and crop modeling information. Such integration has been successfully used in the development of the pest risk information service (PRISE) by CABI (Day *et al.*, 2024). These tools inform real-time decisions about pest and disease risks, the selection of resistant varieties, and the optimal timing for control interventions (Mahlein, 2016). However, their success depends on user-friendly interfaces and integration with mobile platforms accessible to smallholder farmers.

14.6.3 Ethical, regulatory, and environmental considerations in deploying new technologies

As spatial and multi-omics technologies merge with genome editing and AI, they raise important ethical, regulatory, and environmental concerns. However, the fast pace of technological development often outpaces the evolution of governance frameworks needed to ensure responsible innovation (Molinari *et al.*, 2024). From an ethical perspective, concerns include the patenting and ownership of genetic data, informed consent in the collection and use of plant materials, especially native materials, and equitable access to the derived benefits. Smallholder farmers, especially in the less developed regions, are often excluded from patent-protected biotechnologies that emerge from publicly funded research (Macnaghten and Habets, 2020). To address this, open-source platforms and inclusive innovation models should be encouraged and promoted.

On the regulatory front, the release of genome-edited crops, though potentially free of foreign DNA, remains contentious in many jurisdictions. Some countries recognize these as genetically modified organisms (GMOs), subjecting them to the same arduous and costly approval processes, while others have chosen differentiated regulatory approaches (Mishra, 2024). Harmonizing global biosafety guidelines is crucial to facilitate trade, promote innovation, and prevent regulatory deadlocks. Moreover, environmental considerations are equally dominant. While omics-informed breeding can lessen pesticide use and improve sustainability, incidental ecological impacts must be carefully monitored. For instance, the deployment of broad-spectrum resistance traits may exert selection pressure on pests, leading to the advent of new virulent strains (Li *et al.*, 2020). Sustaining surveillance, ecological risk assessments, and adaptive management frameworks are vital.

14.7 Conclusion

The handling of plant pests and diseases constitutes a multifaceted, evolving task. Extreme weather events weaken their innate defenses, increasing plant susceptibility to pest and pathogen invasion. The convergence of these stresses leads to more frequent and severe outbreaks, challenging traditional management paradigms. The predictive algorithms furnish growers with real-time insights, enabling them to arrive at informed and flexible choices that enhance long-term sustainability. Comprehensive strategies for a more sophisticated and sustainable pest control model correspond with international objectives for food security, environmental sustainability, and biodiversity preservation. Moreover, it provides more effective tools to manage pest resistance, reduce pesticide reliance, and protect ecosystems, ultimately fostering a more resilient agricultural landscape. Developments in spatial and single-cell technologies highlight the biological reactions surrounding pathogen-targeted cells. Researchers can examine cellular changes through indirect transcriptomic techniques, but metabolic profiling offers a more direct description of a cell's functional output. The use of AI in pest and disease management is charting a new course in managing these biotic stresses as it enhances precision, reduces labor, and provides a faster means of reducing crop losses from pests and diseases. The opportunities of AI-driven systems in plant protection are endless, but to the farmer, what is important is accessibility and cost. These issues can be mitigated with smartphone-based AI applications that may notify farmers and accelerate disease detection, potentially averting or constraining pest and disease epidemics. Despite the lack of access to advanced equipment for many farmers in underdeveloped nations, enhanced internet penetration, smartphone usage, and offline models present novel potential for in-field crop disease diagnosis. Technologies such as spatial transcriptomics, AI-enabled phenotyping, and advanced bioinformatics are generally inaccessible in many developing regions due to high costs, lack of standardized data and interoperability, insufficient bandwidth, and lack of technical expertise. Universal equity in innovation demands deliberate investments in infrastructure, open data initiatives, and international research collaborations to gain the full benefits of spatial and multi-omics in managing pests and diseases.

References

Abate, T., Van Huis, A. and Ampofo, J.K. (2000) Pest management strategies in traditional agriculture: An African perspective. *Annual Review of Entomology* 45, 631–659. DOI: 10.1146/annurev.ento.45.1.631.

Abdelrahman, M., Al-Sadi, A.M., Pour-Aboughadareh, A., Burritt, D.J. and Tran, L.-S.P. (2018a) Genome editing using CRISPR/Cas9–targeted mutagenesis: An opportunity for yield improvements of crop plants grown under environmental stresses. *Plant Physiology and Biochemistry* 131, 31–36. DOI: 10.1016/j.plaphy.2018.03.012.

Abdelrahman, M., Jogaiah, S., Burritt, D.J. and Tran, L.S.P. (2018b) Legume genetic resources and transcriptome dynamics under abiotic stress conditions. *Plant, Cell & Environment* 41, 1972–1983. DOI: 10.1111/pce.13123.

Abdul Malik, N.A., Kumar, I.S. and Nadarajah, K. (2020) Elicitor and receptor molecules: Orchestrators of plant defense and immunity. *International Journal of Molecular Sciences* 21(3), 963. DOI: 10.3390/ijms21030963.

Acevedo-Garcia, J., Kusch, S. and Panstruga, R. (2014) Magical mystery tour: MLO proteins in plant immunity and beyond. *The New Phytologist* 204(2), 273–281. DOI: 10.1111/nph.12889.

Acquaah, G. (2015) Conventional plant breeding principles and techniques. In: *Advances in Plant Breeding Strategies: Breeding, Biotechnology and Molecular Tools*. Springer, Switzerland. DOI: 10.1007/978-3-319-22521-0_5.

Agarwal, M. and Verma, A. (2020) Modern technologies for pest control: A review. In: *Heavy Metals-Their Environmental Impacts and Mitigation*. IntechOpen, London, UK. DOI: 10.5772/intechopen.93556.

Agboka, K.M., Peter, E., Bwambale, E. and Sokame, B.M. (2024) Biological reinforcement learning simulation for natural enemy -host behavior: Exploring deep learning algorithms for population dynamics. *MethodsX* 13, 102845. DOI: 10.1016/j.mex.2024.102845.

Agrawal, A.A. (2011) Current trends in the evolutionary ecology of plant defence. *Functional Ecology* 25, 420–432. DOI: 10.1111/j.1365-2435.2010.01796.x.

Akhtar, H., Usman, M., Binyamin, R., Hameed, A., Arshad, S.F. *et al.* (2024) Traditional strategies and cutting-edge technologies used for plant disease management: A comprehensive overview. *Agronomy* 14, 2175. DOI: 10.3390/agronomy14092175.

Akhter, S., Naik, V.K., Naladi, B.J., Rathore, A., Yadav, P. *et al.* (2024) The ecological impact of pesticides on non-target organisms in agricultural ecosystems. *Advances in Bioresearch* 15, 322–334. DOI: 10.15515/abr.0976-4585.15.4.322334.

Ali, A., Altaf, M.T., Nadeem, M.A., Karaköy, T., Shah, A.N. *et al.* (2022a) Recent advancement in OMICS approaches to enhance abiotic stress tolerance in legumes. *Frontiers in Plant Science* 13, 952759. DOI: 10.3389/fpls.2022.952759.

Ali, F., Rehman, A., Hameed, A., Sarfraz, S., Rajput, N.A. *et al.* (2024) Climate change impact on plant pathogen emergence: Artificial intelligence (AI) approach. In: *Plant Quarantine Challenges under Climate Change Anxiety*. Springer Cham, Switzerland. DOI: 10.1007/978-3-031-56011-8_9.

Ali, Q., Yu, C., Hussain, A., Ali, M., Ahmar, S. *et al.* (2022b) Genome engineering technology for durable disease resistance: Recent progress and future outlooks for sustainable agriculture. *Frontiers in Plant Science* 13, 860281. DOI: 10.3389/fpls.2022.860281.

Alipanahi, B., Delong, A., Weirauch, M.T. and Frey, B.J. (2015) Predicting the sequence specificities of DNA-and RNA-binding proteins by deep learning. *Nature Biotechnology* 33, 831–838. DOI: 10.1038/nbt.3300.

Amin, A., Zaman, W. and Park, S. (2025) Harnessing multi-omics and predictive modeling for climate-resilient crop breeding: From genomes to fields. *Genes* 16, 809. DOI: 10.3390/genes16070809.

Andersen, M.M., Landes, X., Xiang, W., Anyshchenko, A., Falhof, J. *et al.* (2015) Feasibility of new breeding techniques for organic farming. *Trends in Plant Science* 20(7), 426–434. DOI: 10.1016/j.tplants.2015.04.011.

Anukwuorji, C., Obianuju, C.M., Ezebo, R. and Anuagasi, C. (2016) Antimicrobial effects of four plant extracts against post harvest spoilage fungi of yam (Dioscorea rotundata poir). *International Journal of Plant & Soil Science* 12(3), 1–10. DOI: 10.9734/IJPSS/2016/27718.

Baker, B.P., Green, T.A. and Loker, A.J. (2020) Biological control and integrated pest management in organic and conventional systems. *Biological Control* 140, 104095. DOI: 10.1016/j.biocontrol.2019.104095.

Baker, R.E., Mahmud, A.S., Miller, I.F., Rajeev, M., Rasambainarivo, F. *et al.* (2022) Infectious disease in an era of global change. *Nature Reviews. Microbiology* 20(4), 193–205. DOI: 10.1038/s41579-021-00639-z.

Bal, S., Chattopadhyay, A. and Hazra, P. (2025) Utilization of crop wild relatives in vegetable breeding programs could enhance crop adaptation to challenging environments. In: *Ecologically Mediated Development: Promoting Biodiversity Conservation and Food Security*. Springer. DOI: 10.1007/978-981-96-2413-3_4.

Barmukh, R., Garg, V., Liu, H., Chitikineni, A., Xin, L. *et al.* (2025) Spatial omics for accelerating plant research and crop improvement. *Trends in Biotechnology*. DOI: 10.1016/j.tibtech.2025.03.007.

Barton, K.A. and Brill, W.J. (1983) Prospects in plant genetic engineering. *Science (New York, N.Y.)* 219, 671–676. DOI: 10.1126/science.6297007.

Bascompte, J. and Scheffer, M. (2023) The resilience of plant–pollinator networks. *Annual Review of Entomology* 68, 363–380. DOI: 10.1146/annurev-ento-120120-102424.

Bastas, K.K. and Kumar, A. (2023) Omics technologies in the plant–microbe interactions. In: *Microbial Biocontrol: Molecular Perspective in Plant Disease Management*. Springer, Singapore. DOI: 10.1007/978-981-99-3947-3_13.

Bastet, A., Robaglia, C. and Gallois, J.-L. (2017) eIF4E resistance: Natural variation should guide gene editing. *Trends in Plant Science* 22(5), 411–419. DOI: 10.1016/j.tplants.2017.01.008.

Bawa, G., Liu, Z., Yu, X., Qin, A. and Sun, X. (2022) Single-cell RNA sequencing for plant research: Insights and possible benefits. *International Journal of Molecular Sciences* 23(9), 4497. DOI: 10.3390/ijms23094497.

Baysoy, A., Bai, Z., Satija, R. and Fan, R. (2023) The technological landscape and applications of single-cell multi-omics. *Nature Reviews. Molecular Cell Biology* 24(10), 695–713. DOI: 10.1038/s41580-023-00615-w.

Bebber, D.P. (2015) Range-expanding pests and pathogens in a warming world. *Annual Review of Phytopathology* 53, 335–356. DOI: 10.1146/annurev-phyto-080614-120207.

Belinato, T.A. and Martins, A.J. (2016) Insecticide resistance and fitness cost. In: *Insecticides Resistance*. Intechopen Rijeka, Croatia. DOI: 10.5772/61826.

Bigini, V., Camerlengo, F., Botticella, E., Sestili, F. and Savatin, D.V. (2021) Biotechnological resources to increase disease-resistance by improving plant immunity: A sustainable approach to save cereal crop production. *Plants* 10, 1146. DOI: 10.3390/plants10061146.

Boutrot, F. and Zipfel, C. (2017) Function, discovery, and exploitation of plant pattern recognition receptors for broad-spectrum disease resistance. *Annual Review of Phytopathology* 55, 257–286. DOI: 10.1146/annurev-phyto-080614-120106.

Brennecke, P., Anders, S., Kim, J.K., Kołodziejczyk, A.A., Zhang, X. *et al.* (2013) Accounting for technical noise in single-cell RNA-seq experiments. *Nature Methods* 10(11), 1093–1095. DOI: 10.1038/nmeth.2645.

Broekgaarden, C., Snoeren, T.A.L., Dicke, M. and Vosman, B. (2011) Exploiting natural variation to identify insect-resistance genes. *Plant Biotechnology Journal* 9(8), 819–825. DOI: 10.1111/j.1467-7652.2011.00635.x.

Buch, K., Kaushik, A., Mishra, U., Beese, S., Samanta, S. *et al.* (2023) Unravelling the complexity of plant breeding through modern genetic techniques and tools: A review. *International Journal of Plant & Soil Science* 35(21), 97–105. DOI: 10.9734/ijpss/2023/v35i213950.

Bukhari, S.A.S., Mushtaq, A.R., Butt, M.H. and Shahid, M. (2025) AI-driven strategies for predicting and managing insect pest dynamics under climate change. *Trends in Animal and Plant Sciences* 5, 46–53. DOI: 10.62324/TAPS/2025.063.

Cao, K., Li, H., Wang, Q., Zhao, P., Zhu, G. *et al.* (2017) Comparative transcriptome analysis of genes involved in the response of resistant and susceptible peach cultivars to nematode infection. *Scientia Horticulturae* 215, 20–27. DOI: 10.1016/j.scienta.2016.11.054.

Carroll, S.P., Jørgensen, P.S., Kinnison, M.T., Bergstrom, C.T., Denison, R.F. *et al.* (2014) Applying evolutionary biology to address global challenges. *Science (New York, N.Y)* 346, 1245993. DOI: 10.1126/science.1245993.

Carstens, J.L., Krishnan, S.N., Rao, A., Sorace, A.G., Seeley, E.H. *et al.* (2024) Spatial multiplexing and omics. *Nature Reviews Methods Primers* 4, 54. DOI: 10.1038/s43586-024-00330-6.

Castro-Moretti, F.R., Gentzel, I.N., Mackey, D. and Alonso, A.P. (2020) Metabolomics as an emerging tool for the study of plant–pathogen interactions. *Metabolites* 10, 52. DOI: 10.3390/metabo10020052.

Cembrowska-Lech, D., Krzemińska, A., Miller, T., Nowakowska, A., Adamski, C. *et al*. (2023) An integrated multi-omics and artificial intelligence framework for advance plant phenotyping in horticulture. *Biology* 12, 1298. DOI: 10.3390/biology12101298.

Cen, Y., Huang, Y., Hu, S., Zhang, L. and Zhang, J. (2022) Early detection of bacterial wilt in tomato with portable hyperspectral spectrometer. *Remote Sensing* 14, 2882. DOI: 10.3390/rs14122882.

Cesari, S. (2018) Multiple strategies for pathogen perception by plant immune receptors. *New Phytologist* 219, 17–24. DOI: 10.1111/nph.14877.

Chakraborty, S. and Newton, A.C. (2011) Climate change, plant diseases and food security: An overview. *Plant Pathology* 60, 2–14. DOI: 10.1111/j.1365-3059.2010.02411.x.

Chandrasekaran, J., Brumin, M., Wolf, D., Leibman, D., Klap, C. *et al*. (2016) Development of broad virus resistance in non-transgenic cucumber using CRISPR/Cas9 technology. *Molecular Plant Pathology* 17, 1140–1153. DOI: 10.1111/mpp.12375.

Chaudhry, U.K., Junaid, M.D., Gökçe, Z.N.Ö. and Gökçe, A.F. (2023) Impact of biotic and abiotic stresses on onion production: potential mitigation approaches in modern era. In: *Smart Plant Breeding for Vegetable Crops in Post-Genomics Era*. Springer nature, Singapore. DOI: 10.1007/978-981-19-5367-5_7.

Chen, X., Xue, H., Zhu, L., Wang, H., Long, H. *et al*. (2022) ERF49 mediates brassinosteroid regulation of heat stress tolerance in *Arabidopsis thaliana*. *BMC Biology* 20(1), 254. DOI: 10.1186/s12915-022-01455-4.

Chen, X., Ru, Y., Takahashi, H., Nakazono, M., Shabala, S. *et al*. (2024) Single-cell transcriptomic analysis of pea shoot development and cell-type-specific responses to boron deficiency. *The Plant Journal* 117(1), 302–322. DOI: 10.1111/tpj.16487.

Ciancio, A. and Mukerji, K.G. (2007) *General Concepts in Integrated Pest and Disease Management*. Springer, The Netherlands. DOI: 10.1007/978-1-4020-6061-8.

Cook, D.E., Mesarich, C.H. and Thomma, B.P.H.J. (2015) Understanding plant immunity as a surveillance system to detect invasion. *Annual Review of Phytopathology* 53(1), 541–563. DOI: 10.1146/annurev-phyto-080614-120114.

Cuéllar, W.J., Mwanzia, L., Lourido, D., Garcia, C., Cruz, P. *et al*. (2018) PestDisPlace; monitoring the distribution of pests and diseases. Version 1.0.0.

Daugherty, M.P., Lopes, J. and Almeida, R.P. (2010) Vector within-host feeding preference mediates transmission of a heterogeneously distributed pathogen. *Ecological Entomology* 35, 300–300. DOI. 10.1111/j.1365-2311.2010.01189.x.

De Wolf, E. and Franel, L. (1997) Neural networks that distinguish infection periods of wheat tan spot in an outdoor environment. *Phytopathology* 87, 83–87. DOI: 10.1094/PHYTO.1997.87.1.83.

Day, C., Murphy, S.T., Styles, J., Taylor, B., Beale, T. *et al*. (2024) Forecasting the population development of within-season insect crop pests in sub-Saharan Africa: The pest risk information service. *Journal of Integrated Pest Management* 15(1), 7. DOI: 10.1093/jipm/pmad026.

Dolfani, P., Thuraga, V., Banerjee, B. and Chawade, A. (2024) Integrative approaches in modern agriculture: IoT, ML and AI for disease forecasting amidst climate change. *Precision Agriculture* 25, 2589–2613. DOI: 10.1007/s11119-024-10164.

Demirjian, C., Vailleau, F., Berthomé, R. and Roux, F. (2023) Genome-wide association studies in plant pathosystems: Success or failure. *Trends in Plant Science* 28, 471–485. DOI: 10.3390/RS12172678.

Deng, X., Zhu, Z., Yang, J., Zheng, Z., Huang, Z. *et al*. (2020) Detection of citrus huanglongbing based on multi-input neural network model of UAV hyperspectral remote sensing. *Remote Sensing* 12, 2678. DOI: 10.3390/RS12172678.

Devine, G.J. and Denholm, I. (2009) Insecticide and acaricide resistance. *Encyclopedia of Insects* 505–511. DOI: 10.1016/B978-0-12-374144-8.00144-2.

Dhiman, V.K., Singh, D., Dhiman, V.K. and Pandey, H. (2024) Multi-omics: An advanced bioinformatics approach for crop improvement in agriculture. *Bioinformatics for Plant Research and Crop Breeding* 75–98. DOI: 10.1002/9781394209965.

Diyaolu, C.O. and Folarin, I.O. (2024) The role of biodiversity in agricultural resilience: Protecting ecosystem services for sustainable food production. *International Journal of Research Publication and Reviews* 5(10), 1560–1573. DOI: 10.55248/gengpi.5.1024.2741.

Dong, Y., Sonawane, P., Cohen, H., Polturak, G., Feldberg, L. *et al*. (2020) High mass resolution, spatial metabolite mapping enhances the current plant gene and pathway discovery toolbox. *New Phytologist* 228(6), 1986–2002. DOI: 10.1111/nph.16809.

Döring, T.F., Pautasso, M., Wolfe, M.S. and Finckh, M.R. (2012) Pest and disease management in organic farming: Implications and inspirations for plant breeding. *Organic Crop Breeding* 39–59. DOI: 10.1002/9781119945932.

Early, R., González-Moreno, P., Murphy, S.T. and Day, R. (2018) Forecasting the global extent of invasion of the cereal pest *Spodoptera frugiperda*, the fall armyworm. *NeoBiota* 40, 25–50. DOI: 10.3897/neobiota.40.28165.

Ekström, G. and Ekbom, B. (2011) Pest control in agro-ecosystems: an ecological approach. *Critical Reviews in Plant Sciences* 30(1–2), 74–94. DOI: 10.1080/07352689.2011.554354.

Fan, B.-L., Chen, L.-H., Chen, L.-L. and Guo, H. (2025) Integrative multi-omics approaches for identifying and characterizing biological elements in crop traits: Current progress and future prospects. *International Journal of Molecular Sciences* 26, 1466. DOI: 10.3390/ijms26041466.

FAO, IFAD, UNICEF, WFP, and WHO (2025) In: *The State of Food Security and Nutrition in the World 2025*. Food and Agriculture Organization of the United Nations, Rome.

Farooq, M.A., Gao, S., Hassan, M.A., Huang, Z., Rasheed, A. *et al.* (2024) Artificial intelligence in plant breeding. *Trends in Genetics* 40, 891–908. DOI: 10.1016/j.tig.2024.07.001.

Fernie, A.R. and Yan, J. (2019) De novo domestication: An alternative route toward new crops for the future. *Molecular Plant* 12(5), 615–631. DOI: 10.1016/j.molp.2019.03.016.

Fister, A.S., Landherr, L., Maximova, S.N. and Guiltinan, M.J. (2018) Transient expression of CRISPR/Cas9 machinery targeting TcNPR3 enhances defense response in *Theobroma cacao*. *Frontiers in Plant Science* 9, 268. DOI: 10.3389/fpls.2018.00268.

Fu, P., Li, X., Sun, Z., Chen, Y., Lu, Z. *et al.* (2025) Population genetics and trajectory simulation reveals the invasion process of the fall armyworm (*Spodoptera frugiperda*) in the Eastern Hemisphere. *Evolutionary Applications* 18(2), e70086. DOI: 10.1111/eva.70086.

Gang, C., Haiguang, W. and Zhanhong, M. (2006) Forecasting wheat stripe rust by discrimination analysis. *Plant Protection* 32, 24–27.

Gao, H., Zhou, Q., Yang, L., Zhang, K., Ma, Y. *et al.* (2020) Metabolomics analysis identifies metabolites associated with systemic acquired resistance in *Arabidopsis*. *PeerJ* 8, e10047. DOI: 10.7717/peerj.10047.

Garrett, K.A., Bebber, D.P., Etherton, B.A., Gold, K.M., Plex Sulá, A.I. *et al.* (2022) Climate change effects on pathogen emergence: Artificial intelligence to translate big data for mitigation. *Annual Review of Phytopathology* 60, 357–378. DOI: 10.1146/annurev-phyto-021021-042636.

Gill, H.K. and Garg, H. (2014) Pesticide: Environmental impacts and management strategies. *Pesticides-Toxic Aspects* 8, 10–5772. DOI: 10.5772/57399.

Goggin, F.L., Lorence, A. and Topp, C.N. (2015) Applying high-throughput phenotyping to plant–insect interactions: Picturing more resistant crops. *Current Opinion in Insect Science* 9, 69–76. DOI: 10.1016/j.cois.2015.03.002.

Gomez, M.A., Lin, Z.D., Moll, T., Chauhan, R.D., Hayden, L. *et al.* (2019) Simultaneous CRISPR/Cas9-mediated editing of cassava eIF 4E isoforms nCBP-1 and nCBP-2 reduces cassava brown streak disease symptom severity and incidence. *Plant Biotechnology Journal* 17, 421–434. DOI: 10.1111/pbi.12987.

Gupta, P.K. (2020) SWEET genes for disease resistance in plants. *Trends in Genetics* 36(12), 901–904. DOI: 10.1016/j.tig.2020.08.007.

Gurr, G.M. and You, M. (2016) Conservation biological control of pests in the molecular era: New opportunities to address old constraints. *Frontiers in Plant Science* 6, 1255. DOI: 10.3389/fpls.2015.01255.

Hamer, W.B., Birr, T., Verreet, J.-A., Duttmann, R. and Klink, H. (2020) Spatio-temporal prediction of the epidemic spread of dangerous pathogens using machine learning methods. *ISPRS International Journal of Geo-Information* 9, 44. DOI: 10.3390/ijgi9010044.

Han, X., Li, S., Zeng, Q., Sun, P., Wu, D. *et al.* (2025) Genetic engineering, including genome editing, for enhancing broad-spectrum disease resistance in crops. *Plant Communications* 6. DOI: 10.1016/j.xplc.2024.101195.

Haq, I.U. and Ijaz, S. (2020) *Plant Disease Management Strategies for Sustainable Agriculture through Traditional and Modern Approaches*. Springer Nature, Singapore.

Haque, E., Taniguchi, H., Hassan, M.M., Bhowmik, P., Karim, M.R. *et al.* (2018) Application of CRISPR/Cas9 genome editing technology for the improvement of crops cultivated in tropical climates: Recent progress, prospects, and challenges. *Frontiers in Plant Science* 9, 617. DOI: 10.3389/fpls.2018.00617.

Hardy, N.B. (2022) Delaying quantitative resistance to pesticides and antibiotics. *Evolutionary Applications* 15, 2067–2077. DOI: 10.1111/eva.13497.

Harvey, J.A., Heinen, R., Gols, R. and Thakur, M.P. (2020) Climate change-mediated temperature extremes and insects: From outbreaks to breakdowns. *Global Change Biology* 26(12), 6685–6701. DOI: 10.1111/gcb.15377.

Hawkins, N.J., Bass, C., Dixon, A. and Neve, P. (2019) The evolutionary origins of pesticide resistance. *Biological Reviews of the Cambridge Philosophical Society* 94(1), 135–155. DOI: 10.1111/brv.12440.

Hayden, H.L., Rochfort, S.J., Ezernieks, V., Savin, K.W. and Mele, P.M. (2019) Metabolomics approaches for the discrimination of disease suppressive soils for *Rhizoctonia solani* AG8 in cereal crops using 1H NMR and LC-MS. *The Science of the Total Environment* 651(Pt 1), 1627–1638. DOI: 10.1016/j.scitotenv.2018.09.249.

Homma, F., Huang, J. and van der Hoorn, R.A.L. (2023) AlphaFold-multimer predicts cross-kingdom interactions at the plant-pathogen interface. *Nature Communications* 14(1), 6040. DOI: 10.1038/s41467-023-41721-9.

Hu, Y., Schaefer, R., Rendleman, M., Slattery, A., Cramer, A. *et al.* (2025) Spatial and single-cell transcriptomics capture two distinct cell states in soybean defense response to *Phakopsora pachyrhizi* infection. *Frontiers in Plant Science* 16, 1637176. DOI: 10.3389/fpls.2025.1637176.

Huang, R., Li, H., Gao, C., Yu, W. and Zhang, S. (2023) Advances in omics research on peanut response to biotic stresses. *Frontiers in Plant Science* 14, 1101994. DOI: 10.3389/fpls.2023.1101994.

Huang, X., Wang, Y. and Wang, N. (2022) Highly efficient generation of canker-resistant sweet orange enabled by an improved CRISPR/Cas9 system. *Frontiers in Plant Science* 12, 769907. DOI: 10.3389/fpls.2021.769907.

Hui, T., Zhou, J., Yao, M., Xie, Y. and Zeng, H. (2025) Advances in spatial omics technologies. *Small Methods* 9, 2401171. DOI: 10.1002/smtd.202401171.

Ielmini, M.R. and Sankaran, K. (2021) Invasive alien species: A prodigious global threat in the anthropocene. In: *Invasive Alien Species: Observations and Issues from around the World*, Vol. 1. Wiley-Blackwell, Hoboken, New Jersey, USA, p. 79. DOI: 10.1002/9781119607045.

Iost Filho, F.H., De Bastos Pazini, J., De Medeiros, A.D., Rosalen, D.L. and Yamamoto, P.T. (2022) Assessment of injury by four major pests in soybean plants using hyperspectral proximal imaging. *Agronomy* 12, 1516. DOI: 10.3390/agronomy12071516.

IPCC (2014) Synthesis Report. Contribution of Working Groups I, II and III to the Fifth Assessment Report of the Intergovernmental Panel on Climate Change. [Core Writing Team, RK Pachauri and LA Meyer] [Core Writing Team, RK Pachauri and LA Meyer]. Geneva, Switzerland: IPCC.

Iqbal, Z., Sattar, M.N. and Shafiq, M. (2016) CRISPR/Cas9: A tool to circumscribe cotton leaf curl disease. *Frontiers in Plant Science* 7, 475. DOI: 10.3389/fpls.2016.00475.

Islam, M.M., Adil, M.A.A., Talukder, M.A., Ahamed, M.K.U., Uddin, M.A. *et al.* (2023) DeepCrop: Deep learning-based crop disease prediction with web application. *Journal of Agriculture and Food Research* 14, 100764. DOI: 10.1016/j.jafr.2023.100764.

Jain, S. and Eadon, M.T. (2024) Spatial transcriptomics in health and disease. *Nature Reviews Nephrology* 20, 659–671. DOI: 10.1038/s41581-024-00841-1.

Jiang, F. and Doudna, J.A. (2017) CRISPR–Cas9 structures and mechanisms. *Annual Review of Biophysics* 46, 505–529. DOI: 10.1146/annurev-biophys-062215-010822.

Jiang, L., Wen, G., Lu, J., Yang, H., Jin, Y. *et al.* (2024) Machine learning in soil nutrient dynamics of alpine grasslands. *Science of The Total Environment* 946, 174295. DOI: 10.1016/j.scitotenv.2024.174295.

Jinek, M., Chylinski, K., Fonfara, I., Hauer, M., Doudna, J.A. *et al.* (2012) A programmable dual-RNA-guided DNA endonuclease in adaptive bacterial immunity. *Science (New York, N.Y.)* 337(6096), 816–821. DOI: 10.1126/science.1225829.

John, M.A., Bankole, I., Ajayi-Moses, O., Ijila, T., Jeje, T. *et al.* (2023) Relevance of advanced plant disease detection techniques in disease and pest management for ensuring food security and their implication: A review. *American Journal of Plant Sciences* 14(11), 1260–1295. DOI: 10.4236/ajps.2023.1411086.

Jones, J.D.G., Vance, R.E. and Dangl, J.L. (2016) Intracellular innate immune surveillance devices in plants and animals. *Science (New York, N.Y.)* 354(6316), aaf6395. DOI: 10.1126/science.aaf6395.

Jones, J.W., Hoogenboom, G., Porter, C.H., Boote, K.J., Batchelor, W.D. *et al.* (2003) The DSSAT cropping system model. *European Journal of Agronomy* 18, 235–265. DOI: 10.1016/S1161-0301(02)00107-7.

Junaid, M. and Gokce, A. (2024) Global agricultural losses and their causes. *Bulletin of Biological and Allied Sciences Research* 66–66. DOI: 10.54112/bbasr.v2024i1.66.

Kamilaris, A. and Prenafeta-Boldú, F.X. (2018) Deep learning in agriculture: A survey. *Computers and Electronics in Agriculture* 147, 70–90. DOI: 10.1016/j.compag.2018.02.016.

Kanyuka, K. and Rudd, J.J. (2019) Cell surface immune receptors: The guardians of the plant's extracellular spaces. *Current Opinion in Plant Biology* 50, 1–8. DOI: 10.1016/j.pbi.2019.02.005.

Kapetas, D., Christakakis, P., Faliagka, S., Katsoulas, N. and Pechlivani, E.M. (2025) AI-driven insect detection, real-time monitoring, and population forecasting in greenhouses. *AgriEngineering* 7, 29. DOI: 10.3390/agriengineering7020029.

Karuppaiah, V. and Sujayanad, G. (2012) Impact of climate change on population dynamics of insect pests. *World Journal of Agricultural Sciences* 8, 240–246.

Kaundal, R., Kapoor, A.S. and Raghava, G.P. (2006) Machine learning techniques in disease forecasting: A case study on rice blast prediction. *BMC Bioinformatics* 7, 485. DOI: 10.1186/1471-2105-7-485.

Kaur, L. and Sharma, S.G. (2021) Identification of plant diseases and distinct approaches for their management. *Bulletin of the National Research Centre* 45, 169. DOI: 10.1186/s42269-021-00627-6.

Kaur, A., Randhawa, G.S., Farooque, A.A., Singh, R., Ali, M. *et al.* (2025) Systematic review of AI and machine learning approaches for predicting crop diseases in the context of climate change and the food security. In: *Climate Change, Food Security, and Land Management: Strategies for a Sustainable Future*. pp. 1–20. DOI: 10.1007/978-3-031-71164-0_35-1.

Kaushik, P. (2025) AI-driven breeding enhances stress tolerance in high-elevation extremophytes: A proof-of-concept study with cross-component validation. *Biorxiv*. DOI: 10.1101/2025.02.21.639605.

Keating, B.A., Carberry, P.S., Hammer, G.L., Probert, M.E., Robertson, M.J. *et al.* (2003) An overview of APSIM, a model designed for farming systems simulation. *European Journal of Agronomy* 18(3–4), 267–288. DOI: 10.1016/S1161-0301(02)00108-9.

Khaki, S., Khalilzadeh, Z. and Wang, L. (2020) Predicting yield performance of parents in plant breeding: A neural collaborative filtering approach. *PLoS ONE* 15, e0233382. DOI: 10.1371/journal.pone.0233382.

Khakimov, A., Salakhutdinov, I., Omolikov, A. and Utaganov, S. (2022) Traditional and current-prospective methods of agricultural plant diseases detection: A review. In: *IOP Conference Series: Earth and Environmental Science*, IOP Publishing, p. 012002 (Vol. 951). DOI: 10.1088/1755-1315/951/1/012002.

Khater, M., De La Escosura-Muñiz, A. and Merkoçi, A. (2017) Biosensors for plant pathogen detection. *Biosensors & Bioelectronics* 93, 72–86. DOI: 10.1016/j.bios.2016.09.091.

Khosravi, M., Carroll, M., Tan, K.L., Van Der Laan, L., Raigne, J. *et al.* (2024) AgGym: An agricultural biotic stress simulation environment for ultra-precision management planning arXiv preprint arXiv:2409.00735. DOI: 10.48550/arXiv.2409.00735.

Kim, J.-Y., Symeonidi, E., Pang, T.Y., Denyer, T., Weidauer, D. *et al.* (2021) Distinct identities of leaf phloem cells revealed by single cell transcriptomics. *The Plant Cell* 33(3), 511–530. DOI: 10.1093/plcell/koaa060.

Kourelis, J. and Van der Hoorn, R.A.L. (2018) Defended to the nines: 25 years of resistance gene cloning identifies nine mechanisms for R protein function. *The Plant Cell* 30(2), 285–299. DOI: 10.1105/tpc.17.00579.

Kremen, C. and Chaplin-Kramer, R. (2007) As providers of ecosystem services: Crop pollination and pest control. In: *Insect conservation biology: Proceedings of the royal entomological society's 23rd symposium*, CABI Publishing Wallingford, UK, pp. 349–382. DOI: 10.1079/9781845932541.0349.

Kreuze, J., Adewopo, J., Selvaraj, M., Mwanzia, L., Kumar, P.L. *et al.* (2022) Innovative digital technologies to monitor and control pest and disease threats in root, tuber, and banana (RT&b) cropping systems: Progress and prospects. In: *Root, Tuber and Banana Food System Innovations: Value Creation for Inclusive Outcomes*. pp. 261–288. DOI: 10.1007/978-3-030-92022-7_9.

Kumari, M., Yagnik, K.N., Gupta, V., Singh, I.K., Gupta, R. *et al.* (2024) Metabolomics-driven investigation of plant defense response against pest and pathogen attack. *Physiologia Plantarum* 176(2), e14270. DOI: 10.1111/ppl.14270.

Kumlehn, J., Pietralla, J., Hensel, G., Pacher, M. and Puchta, H. (2018) The CRISPR/Cas revolution continues: From efficient gene editing for crop breeding to plant synthetic biology. *Journal of Integrative Plant Biology* 60(12), 1127–1153. DOI: 10.1111/jipb.12734.

La, G., Hao, X., Li, X., Ou, M. and Yang, T. (2016) Application of metabolomics in plant research. *Botanical Research* 05(1), 26–33. DOI: 10.12677/BR.2016.51005.

Lahlali, R., Ibrahim, D.S.S., Belabess, Z., Kadir Roni, M.Z., Radouane, N. *et al.* (2021) High-throughput molecular technologies for unraveling the mystery of soil microbial community: Challenges and future prospects. *Heliyon* 7(10), e08142. DOI: 10.1016/j.heliyon.2021.e08142.

Lamichhane, J.R., Barzman, M., Booij, K., Boonekamp, P., Desneux, N. *et al.* (2015) Robust cropping systems to tackle pests under climate change. *A Review. Agronomy for Sustainable Development* 35, 443–459. DOI: 10.1007/s13593-014-0275-9.

Landschoot, S., Waegeman, W., Audenaert, K., Van Damme, P., Vandepitte, J. *et al.* (2013) A field-specific web tool for the prediction of *Fusarium* head blight and deoxynivalenol content in Belgium. 2013, *Computers and Electronics in Agriculture* 93, 140–148. DOI: 10.1016/j.compag.2013.02.011.

Latty, T. and Dakos, V. (2019) The risk of threshold responses, tipping points, and cascading failures in pollination systems. *Biodiversity and Conservation* 28, 3389–3406. DOI: 10.1007/s10531-019-01844-2.

Le Rhun, A., Escalera-Maurer, A., Bratovič, M. and Charpentier, E. (2019) CRISPR-Cas in streptococcus pyogenes. *RNA Biology* 16(4), 380–389. DOI: 10.1080/15476286.2019.1582974.

Lefebvre, V., Boissot, N. and Gallois, J.-L. (2020) Host plant resistance to pests and pathogens, the genetic leverage in integrated pest and disease management. In: *Integrated Pest and Disease Management in Greenhouse Crops*. Springer, Switzerland.

Lenné, J. and Wood, D. (2024) Crop diversity in agroecosystems for pest management and food production. *Plants* 13(8), 1164. DOI: 10.3390/plants13081164.

Li, C. and Wang, M. (2024) Pest and disease management in agricultural production with artificial intelligence: Innovative applications and development trends. *Advances in Resources Research* 4, 381–401. DOI: 10.50908/arr.5.2_971.

Li, W., Deng, Y., Ning, Y., He, Z. and Wang, G.-L. (2020) Exploiting broad-spectrum disease resistance in crops: From molecular dissection to breeding. *Annual Review of Plant Biology* 71, 575–603. DOI: 10.1146/annurev-arplant-010720-022215.

Li, X., Liu, W., Li, B., Liu, G., Wei, Y. *et al.* (2018) Identification and functional analysis of cassava DELLA proteins in plant disease resistance against cassava bacterial blight. *Plant Physiology and Biochemistry* 124, 70–76. DOI: 10.1016/j.plaphy.2017.12.022.

Lindeberg, M. (2012) Genome-enabled perspectives on the composition, evolution, and expression of virulence determinants in bacterial plant pathogens. *Annual Review of Phytopathology* 50(1), 111–132. DOI: 10.1146/annurev-phyto-081211-173022.

Liu, R., Bao, Z.X., Zhao, P.J. and Li, G.H. (2021) Advances in the study of metabolomics and metabolites in some species interactions. *Molecules* 26(11), 3311. DOI: 10.3390/molecules26113311.

Liu, X., Peng, T., Xu, M., Lin, S., Hu, B. *et al.* (2024) Spatial multi-omics: Deciphering technological landscape of integration of multi-omics and its applications. *Journal of Hematology & Oncology* 17(1), 72. DOI: 10.1186/s13045-024-01596-9.

Liu, Y., Xin, J., Liu, L., Song, A., Guan, Z. *et al.* (2020) A temporal gene expression map of *Chrysanthemum* leaves infected with *Alternaria alternata* reveals different stages of defense mechanisms. *Horticulture Research* 7(1). DOI: 10.1038/s41438-020-0245-0.

Lloyd-Price, J., Abu-Ali, G. and Huttenhower, C. (2016) The healthy human microbiome. *Genome Medicine* 8(1), 51. DOI: 10.1186/s13073-016-0307-y.

Lopez-Anido, C.B., Vatén, A., Smoot, N.K., Sharma, N., Guo, V. *et al.* (2021) Single-cell resolution of lineage trajectories in the *Arabidopsis* stomatal lineage and developing leaf. *Developmental Cell* 56(7), 1043–1055. DOI: 10.1016/j.devcel.2021.03.014.

Lucas, J. (2011) Advances in plant disease and pest management. *The Journal of Agricultural Science* 149, 91–114. DOI: 10.1017/S0021859610000997.

Ma, W., Hong, S., Reed, W.R., Duan, J. and Luu, P. (2023) Yield effects of agricultural cooperative membership in developing countries: A meta-analysis. *Annals of Public and Cooperative Economics* 94(3), 761–780. DOI: 10.1111/apce.12411.

Macnaghten, P. and Habets, M.G.J.L. (2020) Breaking the impasse: Towards a forward-looking governance framework for gene editing with plants. *Plants, People, Planet* 2(4), 353–365. DOI: 10.1002/ppp3.10107.

Macovei, A., Sevilla, N.R., Cantos, C., Jonson, G.B., Slamet-Loedin, I. *et al.* (2018) Novel alleles of rice eIF4g generated by CRISPR/Cas9-targeted mutagenesis confer resistance to Rice tungro spherical virus. *Plant Biotechnology Journal* 16(11), 1918–1927. DOI: 10.1111/pbi.12927.

Magan, N. and Aldred, D. (2007) Post-harvest control strategies: Minimizing mycotoxins in the food chain. *International Journal of Food Microbiology* 119(1–2), 131–139. DOI: 10.1016/j.ijfoodmicro.2007.07.034.

Mahenge, M.P.J., Hussein, M., Camilius, A.S., Richard, R.M., Beatrice, M. *et al.* (2023) Artificial intelligence and deep learning based technologies for emerging disease recognition and pest prediction in beans (*Phaseolus vulgaris* L.): A systematic review. *African Journal of Agricultural Research* 19(3), 260–271. DOI: 10.5897/AJAR2022.16226.

Mahlein, A.-K. (2016) Plant disease detection by imaging sensors–parallels and specific demands for precision agriculture and plant phenotyping. *Plant Disease* 100, 241–251. DOI: 10.1094/PDIS-03-15-0340-FE.

Makhumbila, P., Rauwane, M., Muedi, H. and Figlan, S. (2022) Metabolome profiling: A breeding prediction tool for legume performance under biotic stress conditions. *Plants* 11, 1756. DOI: 10.3390/plants11131756.

Manghwar, H., Lindsey, K., Zhang, X. and Jin, S. (2019) CRISPR/Cas system: Recent advances and future prospects for genome editing. *Trends in Plant Science* 24(12), 1102–1125. DOI: 10.1016/j.tplants.2019.09.006.

Mankin, R.W., Hagstrum, D.W., Smith, M.T., Roda, A.L. and Kairo, M.T.K. (2011) Perspective and promise: A century of insect acoustic detection and monitoring. *American Entomologist* 57(1), 30–44. DOI: 10.1093/ae/57.1.30.

Mantell, S.H. and Haque, S.Q. (1978) Incidence of internal brown spot disease in white Lisbon yams (*Dioscorea alata*) during storage. *Experimental Agriculture* 14(2), 167–172. DOI: 10.1017/S0014479700008589.

Marshman, J., Blay-Palmer, A. and Landman, K. (2019) Anthropocene crisis: Climate change, pollinators, and food security. *Environments* 6, 22. DOI: 10.3390/environments6020022.

Matny, O.N. (2015) Fusarium head blight and crown rot on wheat & barley: Losses and health risks. *Advances in Plants & Agriculture Research* 2(1), 00039. DOI: 10.15406/apar.2015.02.00039.

Mielniczuk, E. and Skwaryło-Bednarz, B. (2020) Fusarium head blight, mycotoxins and strategies for their reduction. *Agronomy* 10, 509. DOI: 10.3390/agronomy10040509.

Mishra, M. (2024) Harmonizing regulatory policies for genome-edited crops. Asian biotechnology and development review. *British Medical Journal* 26, 2–3. DOI: 10.18356/9789210059091c021.

Mochida, K., Koda, S., Inoue, K., Hirayama, T., Tanaka, S. *et al.* (2019) Computer vision-based phenotyping for improvement of plant productivity: A machine learning perspective. *GigaScience* 8(1), giy153. DOI: 10.1093/gigascience/giy153.

Mokrini, F., Laasli, S.-E., Ezrari, S., Belabess, Z. and Lahlali, R. (2024) Plant-parasitic nematodes and microbe interactions: A biological control perspective. In: *Sustainable Management of Nematodes in Agriculture, Vol. 2: Role of Microbes-Assisted Strategies*. Springer, Switzerland.

Molinari, M.D.C., Pagliarini, R.F., Florentino, L.H., Lima, R.N., Arraes, F.B.M. *et al.* (2024) Navigating the path from lab to market: Regulatory challenges and opportunities for genome editing technologies for agriculture. *Plant Genome Editing Technologies* 25–63. DOI: 10.1007/978-981-99-9338-3_2.

Molla, K.A., Karmakar, S. and Islam, M.T. (2020) Wide horizons of CRISPR-Cas-derived technologies for basic biology, agriculture, and medicine. In: *CRISPR-Cas Methods*. Springer, Switzerland.

Morton, J.F. (2007) The impact of climate change on smallholder and subsistence agriculture. *Proceedings of the National Academy of Sciences* 104. DOI: 10.1073/pnas.07018551.

Mourou, M., Raimondo, M., Lops, F. and Carlucci, A. (2023) Brassicaceae fungal diseases: Molecular detection and host–plant interaction. *Plants512* 1033. DOI: 10.3390/plants12051033.

Murmu, S., Sinha, D., Chaurasia, H., Sharma, S., Das, R. *et al.* (2024) A review of artificial intelligence-assisted omics techniques in plant defense: Current trends and future directions. *Frontiers in Plant Science* 15, 1292054. DOI: 10.3389/fpls.2024.1292054.

Murugan, S.S., Mohanapriya, R. and Prasanna, G.S. (2025) Host plant resistance and breeding. In: *Plant Health and Disease Management: A Modern Approach*. Textify, India, p. 136.

Musa, M., Shuaib, H. and Ogidan, M. (2023) Food security and produce storage, the nexus: Attaining sustainable yam production in Nigeria. *SSRN 4495555*. DOI: 10.2139/ssrn.4495555.

Myers, Jr, R.J., Tretter, Z.M., Daffron, A.G., Fritschi, E.X., Santos, W.T. *et al.* (2025) Spatially resolved plant metabolomics. *Metabolites* 15(8), 539. DOI: 10.3390/metabo15080539.

Nekrasov, V., Wang, C., Win, J., Lanz, C., Weigel, D. *et al.* (2017) Rapid generation of a transgene-free powdery mildew resistant tomato by genome deletion. *Scientific Reports* 7(1), 482. DOI: 10.1038/s41598-017-00578-x.

Nelson, G.C., Valin, H., Sands, R.D., Havlík, P., Ahammad, H. *et al.* (2014) Climate change effects on agriculture: Economic responses to biophysical shocks. *Proceedings of the National Academy of Sciences* 111(9), 3274–3279. DOI: 10.1073/pnas.1222465110.

Nelson, R., Wiesner-Hanks, T., Wisser, R. and Balint-Kurti, P. (2018) Navigating complexity to breed disease-resistant crops. *Nature Reviews Genetics* 19(1), 21–33. DOI: 10.1038/nrg.2017.82.

Neme, K. and Mohammed, A. (2017) Mycotoxin occurrence in grains and the role of postharvest management as a mitigation strategies. A review. *Food Control* 78, 412–425. DOI: 10.1016/j.foodcont.2017.03.012.

Ngou, B.P.M., Jones, J.D. and Ding, P. (2022) Plant immune networks. *Trends in Plant Science* 27, 255–273. DOI: 10.1016/j.tplants.2021.08.012.

Niassy, S., Agbodzavu, M.K., Kimathi, E., Mutune, B., Abdel-Rahman, E.F.M. *et al.* (2021) Bioecology of fall armyworm *Spodoptera frugiperda* (JE smith), its management and potential patterns of seasonal spread in Africa. *PLoS ONE* 16(6), e0249042. DOI: 10.1371/journal.pone.0249042.

Noman, A., Aqeel, M. and Lou, Y. (2019) PRRs and NB-LRRs: From signal perception to activation of plant innate immunity. *International Journal of Molecular Sciences* 20(8), 1882. DOI: 10.3390/ijms20081882.

Ntui, V.O., Uyoh, E.A., Ita, E.E., Markson, A.-A.A., Tripathi, J.N. *et al.* (2021) Strategies to combat the problem of yam anthracnose disease: Status and prospects. *Molecular Plant Pathology* 22(10), 1302–1314. DOI: 10.1111/mpp.13107.

Obermeier, C., Mason, A.S., Meiners, T., Petschenka, G., Rostás, M. *et al.* (2022) Perspectives for integrated insect pest protection in oilseed rape breeding. *Theoretical and Applied Genetics* 135(11), 3917–3946. DOI: 10.1007/s00122-022-04074-3.

Oerke, E.-C. (2006) Crop losses to pests. *The Journal of Agricultural Science* 144, 31–43. DOI: 10.1017/S0021859605005708.

Oerke, E.-C. and Dehne, H.-W. (2004) Safeguarding production—losses in major crops and the role of crop protection. *Crop Protection* 23(4), 275–285. DOI: 10.1016/j.cropro.2003.10.001.

Oliva, R., Ji, C., Atienza-Grande, G., Huguet-Tapia, J.C., Perez-Quintero, A. *et al.* (2019) Broad-spectrum resistance to bacterial blight in rice using genome editing. *Nature Biotechnology* 37(11), 1344–1350. DOI: 10.1038/s41587-019-0267-z.

Ontoy, J.C. and Ham, J.H. (2024) Mapping and omics integration: Towards precise rice disease resistance breeding. *Plants* 13(9), 1205. DOI: 10.3390/plants13091205.

Padhan, B. and Panda, D. (2020) Potential of neglected and underutilized yams (*Dioscorea* spp.) for improving nutritional security and health benefits. *Frontiers in Pharmacology* 11, 496. DOI: 10.3389/fphar.2020.00496.

Paiva-Silva, C., Proença Pereira, J., Marcolino, F., Figueiredo, A. and Santos, R.B. (2025) Silenced cutters: Mechanisms and effects of protease inhibition in plant-pathogen interactions. *Journal of Experimental Botany* eraf156. DOI: 10.1093/jxb/eraf156.

Pan, C., Ye, L., Qin, L., Liu, X., He, Y. *et al.* (2016) CRISPR/Cas9-mediated efficient and heritable targeted mutagenesis in tomato plants in the first and later generations. *Scientific Reports* 6(1), 24765. DOI: 10.1038/srep24765.

Pandian, S. and Ramesh, M. (2020) Development of pesticide resistance in pests: A key challenge to the crop protection and environmental safety. In: *Pesticides in Crop Production: Physiological and Biochemical Action*, pp. 1–13. DOI: 10.1002/9781119432241.

Park, J., Kim, J., Lewy, T., Rice, C.M., Elemento, O. *et al.* (2022) Spatial omics technologies at multimodal and single cell/subcellular level. *Genome Biology* 23(1), 256. DOI: 10.1186/s13059-022-02824-6.

Pati, R., Sandhu, S., Kawadiwale, A.K. and Kaur, G. (2025) Unveiling the underlying complexities in breeding for disease resistance in crop plants: Review. *Frontiers in Plant Science* 16, 1559751. DOI: 10.3389/fpls.2025.1559751.

Pixley, K.V., Falck-Zepeda, J.B., Giller, K.E., Glenna, L.L., Gould, F. *et al.* (2019) Genome editing, gene drives, and synthetic biology: Will they contribute to disease-resistant crops, and who will benefit? *Annual Review of Phytopathology* 57(1), 165–188. DOI: 10.1146/annurev-phyto-080417-045954.

Pokhrel, A. (2021) Role of individual components of disease triangle in disease development: A review. *Journal of Plant Pathology & Microbiology* 12, 573. DOI: 10.35248/2157-7471.21.12.573.

Pott, D.M., de Abreu E Lima, F., Soria, C., Willmitzer, L., Fernie, A.R. *et al.* (2020) Metabolic reconfiguration of strawberry physiology in response to postharvest practices. *Food Chemistry* 321, 126747. DOI: 10.1016/j.foodchem.2020.126747.

Pourreza, A., Lee, W.S., Etxeberria, E. and Banerjee, A. (2015) An evaluation of a vision-based sensor performance in huanglongbing disease identification. *Biosystems Engineering* 130, 13–22. DOI: 10.1016/j.biosystemseng.2014.11.013.

Power, A.G. (2010) Ecosystem services and agriculture: Tradeoffs and synergies. *Philosophical Transactions of the Royal Society B: Biological Sciences* 365, 2959–2971. DOI: 10.1098/rstb.2010.0143.

Prashar, D.B., Sharma, M., Sharma, D., Mir, Z.A., Sood, H. *et al.* (2025) Exploring the complexity of plant stress response: Role of metabolomics and signalling pathways in metabolic diversification. *Journal of Plant Biochemistry and Biotechnology* 34(4), 851–864. DOI: 10.1007/s13562-025-00986-5.

Pretty, J. and Bharucha, Z.P. (2015) Integrated pest management for sustainable intensification of agriculture in Asia and Africa. *Insects* 6, 152–182. DOI: 10.3390/insects6010152.

Puri, H., Grover, S., Pingault, L., Sattler, S.E. and Louis, J. (2023) Temporal transcriptomic profiling elucidates sorghum defense mechanisms against sugarcane aphids. *BMC Genomics* 24, 441. DOI: 10.1186/s12864-023-09529-5.

Qi, R., Cournède, P.-H., Lecoustre, R. and Reffye, P. (2009) Tri-trophic ecosystem oil palm-pests-auxiliaries: I. modeling and simulation. In: *Third International Symposium on Plant Growth Modeling, Simulation, Visualization and Applications*, IEEE, pp. 87–94. DOI: 10.1109/PMA.2009.49.

Qi, T., Dong, Y., Li, X., Zhao, M. and Huang, W. (2024) Fall armyworm habitat analysis in Africa with multi-source earth observation data. *Computers and Electronics in Agriculture* 225, 109283. DOI: 10.1016/j.compag.2024.109283.

Radio, S., Di Marsico, M., Bersani, C., Malinverni, R., Casacuberta, J. *et al.* (2024) Development of a roadmap for action on the application of omics and associated bioinformatics approaches in risk assessment. *EFSA Supporting Publications* 21(10), 9086E. DOI: 10.2903/sp.efsa.2024.EN-9086.

Ranf, S. (2018) Pattern recognition receptors—versatile genetic tools for engineering broad-spectrum disease resistance in crops. *Agronomy* 8, 134. DOI: 10.3390/agronomy8080134.

Ratnadass, A., Fernandes, P., Avelino, J. and Habib, R. (2012) Plant species diversity for sustainable management of crop pests and diseases in agroecosystems: A review. *Agronomy for Sustainable Development* 32, 273–303. DOI: 10.1007/s13593-011-0022-4.

Rato, C., Carvalho, M.F., Azevedo, C. and Oblessuc, P.R. (2021) Genome editing for resistance against plant pests and pathogens. *Transgenic Research* 30(4), 427–459. DOI: 10.1007/s11248-021-00262-x.

Rigon, C.A.G., Gaines, T.A., Küpper, A. and Dayan, F.E. (2020) Metabolism-based herbicide resistance, the major threat among the non-target site resistance mechanisms. *Outlooks on Pest Management* 31(4), 162–168. DOI: 10.1564/v31_aug_04.

Rimbaud, L., Fabre, F., Papaïx, J., Moury, B., Lannou, C. *et al.* (2021) Models of plant resistance deployment. *Annual Review of Phytopathology* 59, 125–152. DOI: 10.1146/annurev-phyto-020620-122134.

Ritchie, M.D., Holzinger, E.R., Li, R., Pendergrass, S.A. and Kim, D. (2015) Methods of integrating data to uncover genotype–phenotype interactions. *Nature Reviews Genetics* 16, 85–97. DOI: 10.1038/nrg3868.

Rithesh, L., Chandran, D.R. and John, A. (2023) *Plant's Defence Mechanisms against Pathogens-Structural and Biochemical Defence*. Integrated Publishers, India.

Rockström, J., Steffen, W., Noone, K., Persson, Å., Chapin, F.S., III *et al.* (2009) A safe operating space for humanity. *Nature* 461(7263), 472–475. DOI: 10.1038/461472a.

Roubos, C.R., Rodriguez-Saona, C. and Isaacs, R. (2014) Mitigating the effects of insecticides on arthropod biological control at field and landscape scales. *Biological Control* 75, 28–38. DOI: 10.1016/j.biocontrol.2014.01.006.

Roychowdhury, R., Das, S.P., Gupta, A., Parihar, P., Chandrasekhar, K. *et al.* (2023) Multi-omics pipeline and omics-integration approach to decipher plant's abiotic stress tolerance responses. *Genes* 14(6), 1281. DOI: 10.3390/genes14061281.

Roychowdhury, R., Hada, A., Biswas, S., Mishra, S., Prusty, M.R. *et al.* (2025) Jasmonic acid (JA) in plant immune response: Unravelling complex molecular mechanisms and networking of defence signalling against pathogens. *Journal of Plant Growth Regulation* 44(1), 89–114. DOI: 10.1007/s00344-024-11264-4.

Rubiales, D., Fondevilla, S., Chen, W., Gentzbittel, L., Higgins, T.J. *et al.* (2015) Achievements and challenges in legume breeding for pest and disease resistance. *Critical Reviews in Plant Sciences* 34, 195–236. DOI: 10.1080/07352689.2014.898445.

Sadeghi-Tehran, P., Virlet, N., Ampe, E.M., Reyns, P. and Hawkesford, M.J. (2019) DeepCount: In-field automatic quantification of wheat spikes using simple linear iterative clustering and deep convolutional neural networks. *Frontiers in Plant Science* 10, 1176. DOI: 10.3389/fpls.2019.01176.

Saleem, A., Anwar, S., Nawaz, T., Fahad, S., Saud, S. *et al.* (2025) Securing a sustainable future: The climate change threat to agriculture, food security, and sustainable development goals. *Journal of Umm Al-Qura University for Applied Sciences* 11(3), 595–611. DOI: 10.1007/s43994-024-00177-3.

Samarskaya, V., Spechenkova, N., Kalinina, N.O., Love, A.J. and Taliansky, M. (2025) The emerging role of omics-based approaches in plant virology. *Viruses* 17, 986. DOI: 10.3390/v17070986.

Sanfaçon, H. (2015) Plant translation factors and virus resistance. *Viruses* 7(7), 3392–3419. DOI: 10.3390/v7072778.

Sanooja, M., Viji, M. and Dennis Thomas, T. (2025) Applications, challenges, and future perspectives of omics in agriculture. In: *Omics and Genome Editing: Revolution in Crop Improvement for Sustainable Agriculture*. Springer. DOI: 10.1007/978-3-031-81639-0_1.

Sargent, D., Conaty, W.C., Tissue, D.T. and Sharwood, R.E. (2022) Synthetic biology and opportunities within agricultural crops. *Journal of Sustainable Agriculture and Environment* 1, 89–107. DOI: 10.1002/sae2.12014.

Sarwar, M. (2023) Startup of climate-smart integrated pest management against corn earworm *Helicoverpa zea* (Boddie) in maize (*Zea mays* L.) for altering insect risk. *Global Research in Environment and Sustainability* 1, 01–19.

Satrio, R.D., Fendiyanto, M.H. and Miftahudin, M. (2024) Tools and techniques used at global scale through genomics, transcriptomics, proteomics, and metabolomics to investigate plant stress responses at the molecular level. In: *Molecular Dynamics of Plant Stress and Its Management*. Springer. DOI: 10.1007/978-981-97-1699-9_25.

Savary, S., Willocquet, L., Pethybridge, S.J., Esker, P., McRoberts, N. *et al*. (2019) The global burden of pathogens and pests on major food crops. *Nature Ecology & Evolution* 3(3), 430–439. DOI: 10.1038/s41559-018-0793-y.

Seal, S., Turaki, A., Muller, E., Kumar, P.L., Kenyon, L. *et al*. (2014) The prevalence of badnaviruses in West African yams (*Dioscorea cayenensis-rotundata*) and evidence of endogenous pararetrovirus sequences in their genomes. *Virus Research* 186, 144–154. DOI: 10.1016/j.virusres.2014.01.007.

Shafi, A., Zahoor, I. and Habib, H. (2021) Omics technologies to unravel plant-microbe interactions. In: *Plant-Microbe Dynamics*. CRC Press, Forida, USA. DOI: 10.1201/9781003106784-14.

Sharma, A., Jain, A., Gupta, P. and Chowdary, V. (2021) Machine learning applications for precision agriculture: A comprehensive review. *IEEE Access* 9, 4843–4873. DOI: 10.1109/ACCESS.2020.3048415.

Sharma, H.C., Sharma, K.K. and Crouch, J.H. (2004) Genetic transformation of crops for insect resistance: Potential and limitations. *Critical Reviews in Plant Sciences* 23(1), 47–72. DOI: 10.1080/07352680490273400.

Sharma, S., Kooner, R. and Arora, R. (2017) Insect pests and crop losses. In: *Breeding Insect Resistant Crops for Sustainable Agriculture*. Springer, Singapore. DOI: 10.1007/978-981-10-6056-4_2.

Shaurub, E.-S.H. (2024) Remote sensing and geographic information system applications as early-warning tools in monitoring fall armyworm, *Spodoptera frugiperda*: A review. *International Journal of Tropical Insect Science* 44(5), 2241–2258. DOI: 10.1007/s42690-024-01327-0.

Shaw, R., Tian, X. and Xu, J. (2021) Single-cell transcriptome analysis in plants: Advances and challenges. *Molecular Plant* 14, 115–126. DOI: 10.1016/j.molp.2020.10.012.

Shrestha, S. (2019) Effects of climate change in agricultural insect pest. *Acta Scientific Agriculture* 3(12), 74–80. DOI: 10.31080/ASAG.2019.03.0727.

Shrestha, S., Amgain, L.P., Pandey, P., Bhandari, T. and Khatiwada, S. (2024) Adoption status of integrated pest management (IPM) practices among vegetable growers of Lamjung district of Nepal. *Heliyon* 10. DOI: 10.1016/j.heliyon.2024.e37999.

Singh, V.K. and Shailbala, P.V. (2012) Forecasting models: An effective tool for potato late blight management. In: *Eco-Friendly Innovative Approaches in Plant Disease Management*. International Book Distributors and Publisher, New Delhi, pp. 102–112. DOI: 10.13140/RG.2.1.1856.9369.

Singh, V.K., Singh, R., Kumar, A. and Bhadouria, R. (2021) Current status of plant diseases and food security. In: *Food Security and Plant Disease Management*. Elsevier, The Netherlands. DOI: 10.1016/B978-0-12-821843-3.00019-2.

Skelsey, P. (2021) Forecasting risk of crop disease with anomaly detection algorithms. *Phytopathology®* 111, 321–332. DOI: 10.1094/PHYTO-05-20-0185-R.

Skendžić, S., Zovko, M., Živković, I.P., Lešić, V. and Lemić, D. (2021) The impact of climate change on agricultural insect pests. *Insects* 12(5), 440. DOI: 10.3390/insects12050440.

Sokame, B.M., Kipkorir, B., Agboka, K.M., Niassy, S., Belayneh, Y. *et al*. (2025) A system dynamics model for predicting African armyworm occurrence and population dynamics. *Agriculture, Ecosystems & Environment* 380, 109378. DOI: 10.1016/j.agee.2024.109378.

Soylu, S. (2006) Accumulation of cell-wall bound phenolic compounds and phytoalexin in *Arabidopsis thaliana* leaves following inoculation with pathovars of *Pseudomonas syringae*. *Plant Science* 170, 942–952. DOI: 10.1016/j.plantsci.2005.12.017.

Spence, N., Hill, L. and Morris, J. (2020) How the global threat of pests and diseases impacts plants, people, and the planet. *Plants, People, Planet* 2, 5–13. DOI: 10.1002/ppp3.10088.

Stein, C.M., Weiskirchen, R., Damm, F. and Strzelecka, P.M. (2021) Single-cell omics: Overview, analysis, and application in biomedical science. *Journal of Cellular Biochemistry* 122(11), 1571–1578. DOI: 10.1002/jcb.30134.

Sun, C., Shao, Y. and Iqbal, J. (2023) Insect insights at the single-cell level: Technologies and applications. *Cells* 13(1), 91. DOI: 10.3390/cells13010091.

Sun, K., Wolters, A.-M.A., Vossen, J.H., Rouwet, M.E., Loonen, A.E.H.M. *et al.* (2016) Silencing of six susceptibility genes results in potato late blight resistance. *Transgenic Research* 25(5), 731–742. DOI: 10.1007/s11248-016-9964-2.

Sunil, V., Majeed, W., Chowdhury, S., Riaz, A., Shakoori, F.R. *et al.* (2023) Insect population dynamics and climate change. In: *Climate Change and Insect Biodiversity*. CRC Press, Forida, USA. DOI: 10.1201/9781003382089-8.

Syombua, E.D., Tripathi, J.N., Obiero, G.O., Nguu, E.K., Yang, B. *et al.* (2022) Potential applications of the CRISPR/Cas technology for genetic improvement of yam (*Dioscorea* spp.). *Food and Energy Security* 11(1), e330. DOI: 10.1002/fes3.330.

Tang, B., Feng, L., Hulin, M.T., Ding, P. and Ma, W. (2023) Cell-type-specific responses to fungal infection in plants revealed by single-cell transcriptomics. *Cell Host & Microbe* 31, 1732–1747. . DOI: 10.1016/j.chom.2023.08.019.

Tariq, H., Xiao, C., Wang, L., Ge, H., Wang, G. *et al.* (2024) Current status of yam diseases and advances of their control strategies. *Agronomy* 14, 1575. DOI: 10.3390/agronomy14071575.

Terentev, A. and Dolzhenko, V. (2023) Can metabolomic approaches become a tool for improving early plant disease detection and diagnosis with modern remote sensing methods? A review. *Sensors* 23, 5366. . DOI: org/10.3390/s23125366.

Thakor, J. and Gautam, A. (2025) Optimisation of technology transfer driven by multi-omics based research. *International Journal of Technology Transfer and Commercialisation* 21, 278–291. DOI: 10.1504/IJTTC.2025.146549.

Thomazella, D.P.D.T., Seong, K., Mackelprang, R., Dahlbeck, D., Geng, Y. *et al.* (2021) Loss of function of a DMR6 ortholog in tomato confers broad-spectrum disease resistance. *Proceedings of the National Academy of Sciences*118, e2026152118. DOI: 10.1073/pnas.2026152118.

Thor, K. and Peiter, E. (2014) Cytosolic calcium signals elicited by the pathogen-associated molecular pattern flg22 in stomatal guard cells are of an oscillatory nature. *The New Phytologist* 204(4), 873–881. DOI: 10.1111/nph.13064.

Tofazzal Islam, T.I. (2019) CRISPR-Cas technology in modifying food crops. *CABI Reviews* 1–16. DOI: 10.1079/PAVSNNR201914050.

Topalović, O., Hussain, M. and Heuer, H. (2020) Plants and associated soil microbiota cooperatively suppress plant-parasitic nematodes. *Frontiers in Microbiology* 11, 313. DOI: 10.3389/fmicb.2020.00313.

Torres, A.M., Palacios, S.A., Yerkovich, N., Palazzini, J.M., Battilani, P. *et al.* (2019) Fusarium head blight and mycotoxins in wheat: Prevention and control strategies across the food chain. *World Mycotoxin Journal* 12(4), 333–355. DOI: 10.3920/WMJ2019.2438.

Toruño, T.Y., Stergiopoulos, I. and Coaker, G. (2016) Plant-pathogen effectors: Cellular probes interfering with plant defenses in spatial and temporal manners. *Annual Review of Phytopathology* 54, 419–441. DOI: 10.1146/annurev-phyto-080615-100204.

Toyota, M. and Betsuyaku, S. (2022) *In vivo* imaging enables understanding of seamless plant defense responses to wounding and pathogen attack. *Plant and Cell Physiology* 63(10), 1391–1404. DOI: 10.1093/pcp/pcac135.

Triki, H.E., Ribeyre, F., Pinard, F. and Jaeger, M. (2023) Coupling plant growth models and pest and disease models: An interaction structure proposal, MIMIC. *Plant Phenomics* 5, 0077. DOI: 10.34133/plantphenomics.0077.

Tripathi, J.N., Ntui, V.O., Shah, T. and Tripathi, L. (2021) CRISPR/Cas9-mediated editing of *DMR6* orthologue in banana (*Musa* spp.) confers enhanced resistance to bacterial disease. *Plant Biotechnology Journal* 19(7), 1291–1293. DOI: 10.1111/pbi.13614.

Tripathi, L., Ntui, V.O. and Tripathi, J.N. (2019) Application of genetic modification and genome editing for developing climate-smart banana. *Food and Energy Security* 8(4), e00168. DOI: 10.1002/fes3.168.

Tyagi, A., Mir, Z.A., Almalki, M.A., Deshmukh, R. and Ali, S. (2024) Genomics-assisted breeding: A powerful breeding approach for improving plant growth and stress resilience. *Agronomy* 14, 1128. DOI: 10.3390/agronomy14061128.

Van Leeuwen, T., Dermauw, W., Mavridis, K. and Vontas, J. (2020) Significance and interpretation of molecular diagnostics for insecticide resistance management of agricultural pests. *Current Opinion in Insect Science* 39, 69–76. DOI: 10.1016/j.cois.2020.03.006.

Vandereyken, K., Sifrim, A., Thienpont, B. and Voet, T. (2023) Methods and applications for single-cell and spatial multi-omics. *Nature Reviews. Genetics* 24(8), 494–515. DOI: 10.1038/s41576-023-00580-2.

Varadharajan, V., Rajendran, R., Muthuramalingam, P., Runthala, A., Madhesh, V. *et al.* (2025) Multi-omics approaches against abiotic and biotic stress-a review. *Plants (Basel, Switzerland)* 14(6), 865. DOI: 10.3390/plants14060865.

Varghese, R., Shringi, H., Efferth, T. and Ramamoorthy, S. (2025) Artificial intelligence driven approaches in phytochemical research: Trends and prospects. *Phytochemistry Reviews* 24(5), 3649–3664. DOI: 10.1007/s11101-025-10096-8.

Vo, K.T.X., Rahman, M.M., Rahman, M.M., Trinh, K.T.T., Kim, S.T. *et al.* (2021) Proteomics and metabolomics studies on the biotic stress responses of rice: An update. *Rice (New York, N.Y.)* 14(1), 30. DOI: 10.1186/s12284-021-00461-4.

Walsh, T., Heckel, D.G., Wu, Y., Downes, S., Gordon, K. *et al.* (2022) Determinants of insecticide resistance evolution: Comparative analysis among Heliothines. *Annual Review of Entomology* 67, 387–406. DOI: 10.1146/annurev-ento-080421-071655.

Wang, F., Wang, C., Liu, P., Lei, C., Hao, W. *et al.* (2016) Enhanced rice blast resistance by CRISPR/Cas9-targeted mutagenesis of the ERF transcription factor gene *OsERF922*. *PLoS ONE* 11, e0154027. DOI: 10.1371/journal.pone.0154027.

Wang, J., Huang, Y., Huang, L., Dong, Y., Huang, W. *et al.* (2023) Migration risk of fall armyworm (*Spodoptera frugiperda*) from North Africa to Southern Europe. *Frontiers in Plant Science* 14, 1141470. DOI: 10.3389/fpls.2023.1141470.

Wang, X., Tu, M., Wang, D., Liu, J., Li, Y. *et al.* (2018) CRISPR/Cas9-mediated efficient targeted mutagenesis in grape in the first generation. *Plant Biotechnology Journal* 16, 844–855. DOI: 10.1111/pbi.12832.

Wang, X., Li, N., Li, W., Gao, X., Cha, M. *et al.* (2020) Advances in transcriptomics in the response to stress in plants. *Global Medical Genetics* 7, 030–034. DOI: 10.1055/s-0040-1714414.

Wang, Y., Cheng, X., Shan, Q., Zhang, Y., Liu, J. *et al.* (2014) Simultaneous editing of three homoeoalleles in hexaploid bread wheat confers heritable resistance to powdery mildew. *Nature Biotechnology* 32(9), 947–951. DOI: 10.1038/nbt.2969.

Wang, Y., Zafar, N., Ali, Q., Manghwar, H., Wang, G. *et al.* (2022) CRISPR/Cas genome editing technologies for plant improvement against biotic and abiotic stresses: Advances, limitations, and future perspectives. *Cells* 11(23), 3928. DOI: 10.3390/cells11233928.

Wang, Y., Jaworski, C.C., Zi, H., Chen, J., Desneux, N. *et al.* (2024) Increased ladybird predation and metabolism do not counterbalance increased field aphid population growth under experimental warming. *Functional Ecology* 38(5), 1134–1145. DOI: 10.1111/1365-2435.14534.

Waterman, J.M., Cazzonelli, C.I., Hartley, S.E. and Johnson, S.N. (2019) Simulated herbivory: The key to disentangling plant defence responses. *Trends in Ecology & Evolution* 34(5), 447–458. DOI: 10.1016/j.tree.2019.01.008.

Wei, Y., Chang, Y., Zeng, H., Liu, G., He, C. *et al.* (2018) RAV transcription factors are essential for disease resistance against cassava bacterial blight via activation of melatonin biosynthesis genes. *Journal of Pineal Research* 64(1), e12454. DOI: 10.1111/jpi.12454.

Williams, J., Jones, C., Kiniry, J. and Spanel, D.A. (1989) The EPIC crop growth model. *Transactions of the ASAE* 32(2), 0497–0511. DOI: 10.13031/2013.31032.

Wu, E.-J., Wang, Y.-P., Yang, L.-N., Zhao, M.-Z. and Zhan, J. (2022) Elevating air temperature may enhance future epidemic risk of the plant pathogen *Phytophthora infestans*. *Journal of Fungi* 8, 808. DOI: 10.3390/jof8080808.

Xu, W., Wang, Q. and Chen, R. (2018) Spatio-temporal prediction of crop disease severity for agricultural emergency management based on recurrent neural networks. *GeoInformatica* 22, 363–381. DOI: 10.1007/s10707-017-031.

Xu, Z., Xu, X., Gong, Q., Li, Z., Li, Y. *et al.* (2019) Engineering broad-spectrum bacterial blight resistance by simultaneously disrupting variable TALE-binding elements of multiple susceptibility genes in rice. *Molecular Plant* 12, 1434–1446. DOI: 10.1016/j.molp.2019.08.006.

Yang, H., Jin, Y., Jiang, L., Lu, J. and Wen, G. (2025) AI roles in 4R crop pest management—A review. *Agronomy* 15, 1629. DOI: 10.3390/agronomy15071629.

Yang, X., Gu, X., Ding, J., Yao, L., Gao, X. *et al.* (2022) Gene expression analysis of resistant and susceptible rice cultivars to sheath blight after inoculation with *Rhizoctonia solani*. *BMC Genomics* 23(1), 278. DOI: 10.1186/s12864-022-08524-6.

Yaschenko, A.E., Alonso, J.M. and Stepanova, A.N. (2025) *Arabidopsis* as a model for translational research. *The Plant Cell* 37(5), koae065. DOI: 10.1093/plcell/koae065.

Yousaf, A., Kayvanfar, V., Mazzoni, A. and Elomri, A. (2023) Artificial intelligence-based decision support systems in smart agriculture: Bibliometric analysis for operational insights and future directions. *Frontiers in Sustainable Food Systems* 6, 1053921. DOI: 10.3389/fsufs.2022.1053921.

Zhang, J., Coaker, G., Zhou, J.-M. and Dong, X. (2020) Plant immune mechanisms: From reductionistic to holistic points of view. *Molecular Plant* 13(10), 1358–1378. DOI: 10.1016/j.molp.2020.09.007.

Zhang, M., Han, Y., Li, D., Xu, S. and Huang, Y. (2024) Simultaneous modification of three homoeologs of Ta EDR 1 by genome editing enhances powdery mildew resistance in wheat. *The Plant Journal* 10(6), 1257–1273. DOI: 10.1016/j.hpj.2023.03.015.

Zhang, Y., Bai, Y., Wu, G., Zou, S., Chen, Y. *et al.* (2017) Simultaneous modification of three homoeologs of Ta EDR 1 by genome editing enhances powdery mildew resistance in wheat. *The Plant Journal* 91, 714–724. DOI: 10.1111/tpj.13599.

Zhang, Y., Malzahn, A.A., Sretenovic, S. and Qi, Y. (2019) The emerging and uncultivated potential of CRISPR technology in plant science. *Nature Plants* 5(8), 778–794. DOI: 10.1038/s41477-019-0461-5.

Zheng, J. and Xu, Y. (2023) A review: Development of plant protection methods and advances in pesticide application technology in agro-forestry production. *Agriculture* 13(11), 2165. DOI: 10.3390/agriculture13112165.

Zhou, Y., Fan, F., Chaisiri, C., Zhu, Y.-T., Zhao, Y. *et al.* (2021) Sensitivity of *Venturia carpophila* from China to five fungicides and characterization of carbendazim-resistant isolates. *Plant Disease* 105(12), 3990–3997. DOI: 10.1094/PDIS-04-21-0694-RE.

Zhu, J., Moreno-Pérez, A. and Coaker, G. (2023) Understanding plant pathogen interactions using spatial and single-cell technologies. *Communications Biology* 6(1), 814. DOI: 10.1038/s42003-023-05156-8.

Zuo, J., Grierson, D., Courtney, L.T., Wang, Y., Gao, L. *et al.* (2020) Relationships between genome methylation, levels of non-coding RNAs, mRNAs and metabolites in ripening tomato fruit. *The Plant Journal* 103(3), 980–994. DOI: 10.1111/tpj.14778.

15 Single-cell and Spatial Multi-omics to Dissect Regulatory Roles of Plant Functional RNAs under Stress Conditions

Gunna Suresh Babu Suruthi[1], Thiruvengadam Thirumurugan[2], Balasubramanian Balamuralikrishnan[3], Arumugam Maruthupandian[4], Mariadhas Valan Arasu[5], Asirvatham Alwin Robert[6]*, Jeyabal Philomenathan Antony Prabhu[1] and Arumugam Vijaya Anand[1]*

[1]Department of Human Genetics and Molecular Biology, Bharathiar University, Coimbatore, India; [2]Department of Genetics & Plant Breeding, Anbil Dharmalingam Agricultural College & Research Institute, Tamil Nadu Agricultural University, Trichy, India; [3]Department of Food Science and Biotechnology, College of Life Science, Sejong University, Seoul, South Korea; [4]Department of Botany, School of Life Sciences, Periyar University, Salem, India; [5]Department of Botany and Microbiology, College of Science, King Saud University, Riyadh, Saudi Arabia; [6]Department of Endocrinology and Diabetes, Prince Sultan Military Medical City, Ministry of Defense, Riyadh, Kingdom of Saudi Arabia

15.1 Introduction

The study of plant functional RNAs, particularly non-coding RNAs (ncRNAs), has acquired considerable attention recently due to their crucial regulatory roles in various biological processes, including stress responses. Among these, long non-coding RNAs (lncRNAs) and small regulatory RNAs (sRNAs), such as microRNAs (miRNAs) and small interfering RNAs (siRNAs), have appeared to be key in being instrumental in modulating gene expression and facilitating adaptive responses to environmental challenges. For instance, lncRNAs have been involved in regulating developmental processes and stress responses in plants, acting as mediators that influence the expression of developmental regulators during critical phases such as male germline development (Lohani *et al.*, 2022). Similarly, sRNAs are known to control various cellular functions, including defense mechanisms against pathogens and abiotic stressors, by guiding post-transcriptional gene silencing (Dai *et al.*, 2018).

The significance of studying plant strain responses cannot be overstated, as plants are constantly exposed to myriad environmental stressors, like drought, salinity, and pathogen invasion. These stressors can particularly affect plant growth and development, reducing agricultural productivity. Understanding the molecular mechanisms behind plant stress responses is crucial for developing strategies to

*Corresponding authors: aalwinrobert@gmail.com and avahgmb@buc.edu.in

© CAB International 2026. *Plant Spatial Biology* (ed. Chen)
DOI: 10.1079/9781836994572.0015

enhance stress tolerance in plants. Studies indicate that miRNAs contribute to the modulation of plant responses to biotic and abiotic stressors by influencing gene expression at the post-transcriptional stage (Anjali *et al.*, 2017). For example, specific miRNAs have been recognized that target genes involved in stress signaling pathways, thereby modulating the plant's adaptive responses (Sun *et al.*, 2014). Furthermore, the discovery of circular RNAs (circRNAs) and their regulatory interactions with miRNAs adds another layer of complexity to the regulatory networks governing stress responses in plants (Darbani *et al.*, 2016).

The emergence of single-cell and spatial multi-omics technologies has transformed plant biology, allowing researchers to analyze complex regulatory networks with unparalleled precision. These advanced techniques allow for the investigation of several molecular layers, such as transcriptomics, proteomics, and metabolomics, within individual cells or specific tissue regions simultaneously. A comprehensive approach offers a significant understanding of the spatial patterns of gene expression and the functional roles of non-coding RNAs in various cellular situations (Bhogireddy *et al.*, 2021). For instance, the application of single-cell RNA sequencing (scRNA-seq) has identified the heterogeneity of gene expression profiles of various cell types in response to stress, focusing on the significance of spatial context in plant stress responses (Yang *et al.*, 2018). Moreover, incorporating multi-omics information can provide an interpretation of the regulatory interactions between different classes of RNAs, proteins, and metabolites, thereby enhancing our understanding of the complex signaling networks that govern plant responses to environmental challenges (Borges and Martienssen, 2015).

In addition to their roles in stress reactions, functional RNAs are also critical in various developmental processes in plants. Research has demonstrated that miRNAs and lncRNAs regulate flowering time, root development, and hormone signaling pathways (Waseem *et al.*, 2020). The complex interplay between these regulatory RNAs and their target genes underscores the importance of post-transcriptional regulation in shaping plant phenotypes and adapting to changing environments. Furthermore, the identification of transposon-derived small RNAs emphasizes the evolutionary significance of these regulatory elements in plant genomes, suggesting that they can fine-tune gene expression in response to specific developmental signals or stress (Cho, 2018).

As the field of plant functional RNA research continues to evolve, the integration of computational tools and bioinformatics resources will be crucial for the identification and characterization of novel regulatory RNAs. Tools such as psRNATarget facilitate the prediction of miRNA targets, enabling researchers to uncover the intricate regulatory networks that control gene expression in plants (Dai *et al.*, 2018). Additionally, machine-learning approaches are being developed to enhance the prediction of miRNA loci and their functional roles, further advancing our understanding of RNA-mediated control in plants (Bhogireddy *et al.*, 2021). The combination of experimental and computational methodologies will undoubtedly lead to the discovery of new functional RNAs and their role in plant biology.

15.2 Plant Functional RNAs: Categories and Regulatory Roles

The study of functional RNAs in plants has gained a lot of attention due to their diverse roles in regulating gene expression and responding to environmental stresses. Among the various categories of functional RNAs, miRNAs, siRNAs, lncRNAs, and circRNAs are particularly noteworthy (Fig. 15.1). Each of these RNA types plays a specific role in plant biology, controlling processes such as development, stress responses, and pathogen interactions.

For example, miRNAs are small, typically 21–24 nucleotide long, ncRNAs that control gene expression by associating with complementary sequences in target mRNAs, most significantly for their degradation or translational repression. In plants, miRNAs are mainly encoded in intergenic regions and are transcribed by RNA polymerase II, similar to protein-coding genes (Chien *et al.*, 2015). This transcriptional regulation is essential for the proper expression of miRNAs, which regulate several biological functions, including developmental regulation and stress responses (Wang

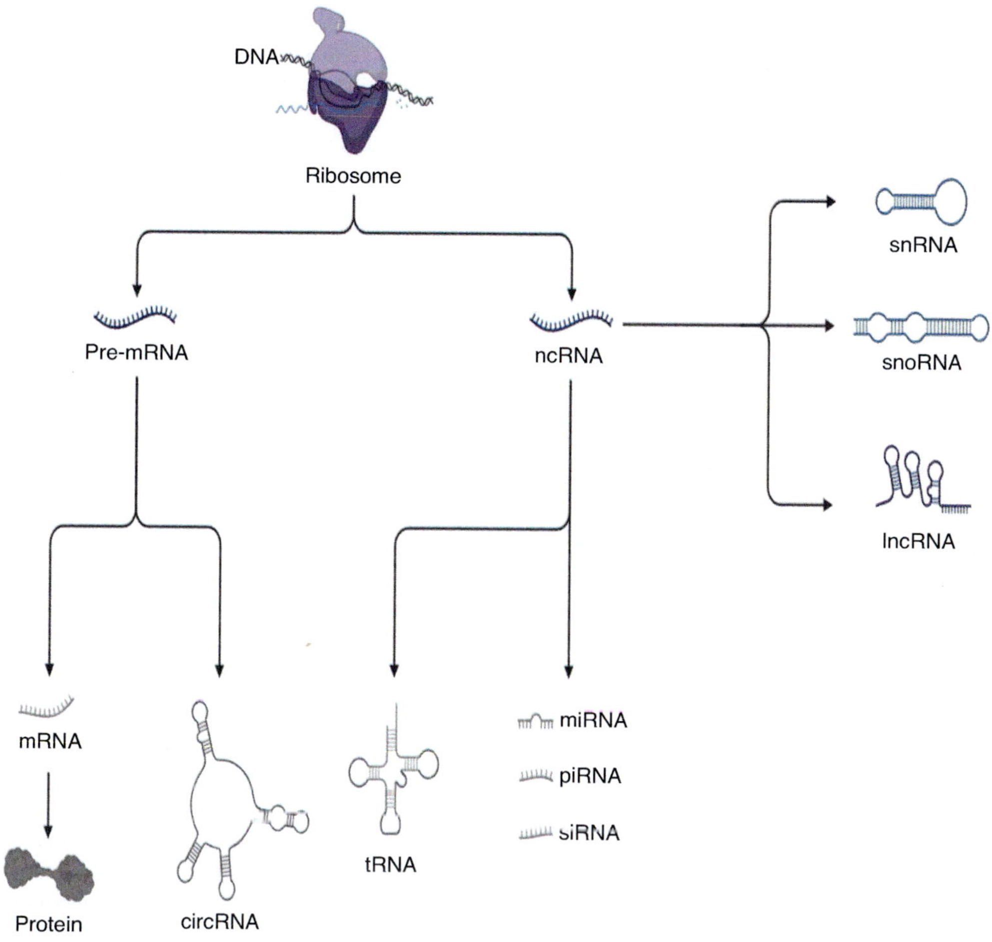

Fig. 15.1. Types of functional RNA. circRNA, circular RNA; lncRNA, long non-coding RNA; mRNA, messenger RNA; miRNA, microRNA; ncRNA, non-coding RNA; piRNA, Piwi-interacting RNA; siRNA, small interacting RNA; snRNA, small nuclear RNA; snoRNA, small nucleolar RNA; tRNA, transfer RNA.

et al., 2019). For instance, miR396 has been recognized as a necessary regulator in response to abiotic stresses by targeting genes associated with stress resistance, thereby influencing plant growth and differentiation (Asefpour Vakilian, 2020).

The mechanisms of RNA-based regulation in plants extend beyond miRNAs to include siRNAs, which originate from double-stranded RNA and play a necessary part in RNA silencing pathways. siRNAs can be categorized into various subclasses, such as heterochromatic siRNAs and trans-acting siRNAs, each with distinct biosynthetic pathways and targets (Willmann *et al.*, 2011). These sRNAs are integral to the plant's defense mechanisms against viruses and transposable elements, as they lead the RNA-induced silencing complex to degrade viral RNA or silence transposons (Deng *et al.*, 2022). The interplay between miRNAs and siRNAs highlights the complexity of RNA-based regulatory networks in plants, where both types of sRNAs can work together to fine-tune the expression of genes related to environmental cues.

In addition, lncRNAs contribute to the regulatory landscape in plants. These RNAs, which exceed 200 nucleotides in length, do not code for proteins but are involved in various regulatory functions, including chromatin remodeling and transcriptional regulation. Recent research has revealed that lncRNAs can interact with miRNAs and other RNA molecules, forming intricate networks that modulate the expression of genes (Schröder and Jullien, 2019). Their role in reactions to stress is particularly noteworthy, as lncRNAs can influence the expression of genes that are related to stress, thereby improving the plant's resilience to adverse conditions.

With a circular structure, circRNAs represent another emerging class of functional RNAs in plants. These RNA molecules are formed by back-splicing events and are involved in regulating gene expression through their interactions with miRNAs and proteins. For example, circRNAs can serve as miRNA sponges, sequestering miRNAs and preventing them from binding to their target mRNAs, thus modulating the availability of miRNAs in the cell (Meng *et al.*, 2012). This regulatory mechanism is particularly relevant in the context of stress responses, where the balance of miRNA activity can determine the plant's ability to cope with environmental challenges.

The role of functional RNAs in stress responses is a critical area of research, as plants are constantly exposed to various environmental stresses. For instance, miRNAs have been shown to play pivotal roles in mediating plant responses to stresses like drought, salinity, and pathogen attacks. For example, specific miRNAs such as miR167 and miR393 have been associated with drought stress regulation in multiple plant species, indicating their conserved function across different taxa (Zhang *et al.*, 2013). Additionally, the identification of stress-responsive miRNAs through high-throughput sequencing has provided knowledge about the complex regulatory networks that govern plant stress responses (Barrera-Figueroa *et al.*, 2012).

Moreover, the interaction between sRNAs and pathogens underscores the importance of RNA-based regulation in plant immunity; sRNAs can target and silence viral RNAs, thereby enhancing the plant's defense mechanisms against viral infections (Deng *et al.*, 2022). This RNA-mediated defense strategy is crucial for maintaining plant health and productivity, especially in the face of increasing biotic stressors in agricultural systems (Table 15.1).

15.3 Understanding Stress Responses in Plants

Comprehending stress responses in plants is crucial for enhancing agricultural productivity and resilience against environmental challenges.

Table 15.1. Key functional RNAs and their roles in plant stress responses.

RNA type	Key characteristics	Regulatory roles	Stress response examples	References
miRNAs	Small (21–24 nucleotides), derived from pri-miRNAs	Post-transcriptional gene silencing by mRNA degradation	The specific miRNA miR167 targets auxin response factors during drought stress	Wang *et al.*, 2019
siRNAs	Derived from double-stranded RNA	RNA silencing, including viral defense and transposon suppression	Silencing viral RNA during pathogen attacks	Deng *et al.*, 2022
lncRNAs	>200 nucleotides, non-coding	Chromatin remodeling, transcriptional and post-transcriptional regulation	Regulating drought and salinity stress-responsive genes	Lu *et al.*, 2016a
circRNAs	Circular structure formed by back-splicing	Acting as miRNA sponges, modulating miRNA availability	Regulating miRNA activity during abiotic stress	Meng *et al.*, 2012

Plants encounter different abiotic and biotic stresses that significantly influence their growth and development. Drought, salinity, severe temperatures, and UV radiation are examples of abiotic pressures, whereas infections, insects, and other creatures that might negatively impact plant health are examples of biotic stresses (Dey and Raichaudhuri, 2022). The responses to these strains are complex and involve intricate signaling pathways, gene expression changes, and regulatory mechanisms, including those mediated by RNA molecules.

Abiotic stresses are particularly detrimental as they can lead to reduced crop yields and compromised plant health. For instance, drought stress triggers a cascade of physiological responses, including the closure of stomata to conserve water, altered root architecture to enhance the uptake of water, and the accumulation of osmoprotectants to maintain cellular integrity (Zhu, 2016). Similarly, salinity stress can disrupt ion homeostasis, leading to the accumulation of toxic ions and osmotic stress, which further exacerbates the plant's capacity to absorb water (Dey and Raichaudhuri, 2022). The interplay between these stresses necessitates a multifaceted approach to understanding plant responses, which often involves the regulation of stress-related genes and metabolic reactions.

Recent investigations have highlighted the importance of lncRNA in mediating plant responses to both stresses. For example, lncRNAs have been recognized as the main factors in regulating gene expression during osmotic and salt stress in *Medicago truncatula*, showing their importance in stress adaptation mechanisms (Wang *et al*., 2015). Additionally, lncRNAs have been implicated in drought stress responses in cotton, where they interact with various metabolic pathways and transcription factors to modulate stress toleration (Lu *et al*., 2016a). These results highlight the role of lncRNAs in coordinating intricate regulatory networks that allow plants to adapt to environmental stressors. The general mechanisms of stress response involve a series of signaling pathways that integrate environmental cues and activate defense mechanisms. One such pathway is the abscisic acid (ABA) signaling pathway, which can mediate the responses to drought and salt stress. ABA accumulation leads to the activation of stress-responsive genes and the modulation of stomatal behavior, thereby enhancing water conservation (Yang *et al*., 2022). Furthermore, the involvement of other plant hormones, such as jasmonic acid and salicylic acid in biotic stress responses, illustrates the interconnectedness of hormonal signaling in plant defense strategies (Yoon *et al*., 2020). The crosstalk between these hormonal pathways is essential for fine-tuning plant responses to simultaneous abiotic and biotic stresses.

Transcription factors (TFs) are key to the regulation of stress-related gene expression. Studies have identified various TF families, such as NAC, MYB, and WRKY, that are activated under stress situations and play critical roles in modulating plant stress tolerance. For instance, NAC transcription factors have been shown to regulate multiple abiotic stress responses in conifers, indicating their evolutionary significance in stress adaptation (Han *et al*., 2023). The functional characterization of these TFs provides insights into their potential applications in developing stress-resistant crops via genetic engineering and breeding programs.

In addition to transcriptional regulation, post-transcriptional mechanisms, including RNA silencing pathways, are vital for stress responses in plants. RNA-dependent RNA polymerases and sRNAs are integral components of these pathways, mediating the regulation of gene expression in response to stress (Kallemi *et al*., 2024). For example, the accumulation of specific sRNAs has been associated with increased stress tolerance, as they can target and silence negative regulators of stress-related genes (Matsui *et al*., 2017). This regulatory layer adds complexity to the plant's ability to respond to environmental challenges, highlighting the significance of RNA-mediated mechanisms in stress adaptation.

Moreover, the endoplasmic reticulum (ER) can integrate signals and maintain cellular homeostasis. During stress situations, the unfolded protein response is triggered to alleviate ER stress, ensuring proper protein folding and function (Park and Park, 2019). This response is crucial for maintaining cellular integrity and function during periods of environmental strain, further emphasizing the interconnectedness of various cellular pathways in plant stress reactions.

The functional positions of RNA chaperones in stress tolerance have also been elucidated, particularly in the context of cold stress. RNA chaperones, such as cold-shock proteins (CSPs), help stabilize RNA structures under low-temperature conditions, facilitating efficient translation and gene expression (Li *et al.*, 2022b). This function is essential for maintaining cellular processes during stress, showcasing the importance of RNA processing and modification in plant resilience.

15.4 Single-cell Omics in Plant Research

Single-cell omics have emerged as a transformative approach in plant research, providing an unprecedented understanding of the cellular heterogeneity and functional dynamics of plant tissues. The scRNA-seq technique has revolutionized our understanding of gene expression at the cellular level. This method allows for the profiling of individual cells, revealing the transcriptional landscape that is often obscured in bulk RNA-seq analyses. For instance, Ryu *et al.* (2019) highlighted the application of scRNA-seq in *Arabidopsis*, where they resolved molecular association among individual plant cells, highlighting the ability of the technique to uncover cellular diversity within tissues. Furthermore, the use of single-nucleus RNA sequencing (snRNA-seq) is effective in studying plant cells, particularly when protoplast isolation is challenging, as illustrated by Xu *et al.* (2021). These advancements underscore the versatility of single-cell techniques in capturing the difficulty of plant cell populations.

In addition to scRNA-seq, single-cell ATAC-seq (assay for transposase-accessible chromatin employing sequencing) has gained traction as a method for assessing chromatin accessibility at the single-cell level. This method complements scRNA-seq by providing insight into the regulatory landscape of gene expression. Researchers can learn in-depth details about the regulatory processes controlling gene expression in plant cells by combining data from scRNA-seq and ATAC-seq. For example, Yang *et al.* (2023) highlight the potential of using multi-omics approaches to generate insights into gene regulation and cellular states, thereby enhancing our understanding of plant biology.

The advantages of single-cell approaches in plant investigation are manifold. One significant benefit is the capability to dissect the cellular heterogeneity that exists within plant tissues. Traditional bulk analyses often conceal the variations in gene expression among individual cells, leading to the loss of critical information. Single-cell techniques allow for the recognition of distinct cell types and states, as demonstrated in studies of *Arabidopsis* roots and floral meristems (Neumann *et al.*, 2022). Moreover, single-cell transcriptomics can elucidate the dynamics of gene expression in response to environmental conditions, providing insights into plant adaptation and resilience (Iqbal *et al.*, 2020). This level of detail is important for comprehending complex biological processes like development, stress responses, and metabolic regulation.

Insights into functional RNA regulation at the single-cell level are particularly valuable for elucidating the mechanisms of gene expression control. For instance, studies have shown that scRNA-seq can reveal the transcriptional responses of particular cells to abiotic stress, highlighting the heterogeneity in gene expression patterns among different cell types (Jean-Baptiste *et al.*, 2019). This is further supported by the findings of Jean-Baptiste *et al.* (2019), who demonstrated that single-cell transcriptomics can uncover subtle differences in gene expression among diverse cell types in response to stress due to heat. Such insights are essential for understanding how plants integrate environmental signals and regulate gene expression accordingly.

Moreover, integrating single-cell omics with computational approaches has opened new avenues for data analysis and interpretation. Techniques such as graph-linked embedding allow for the incorporation of multi-omics information, facilitating the inference of regulatory networks and cellular interactions (Cao and Gao, 2022). This is particularly relevant in plant research, where understanding cell–cell communication and the regulatory landscape is critical for elucidating developmental processes and responses to surrounding challenges. The ability to reconstruct cell lineage and infer cell–cell communications established on single-cell data further enhances our knowledge about plant science at a systems level (Jin *et al.*, 2023).

The benefits of single-cell omics extend beyond basic research and hold substantial prospects for functional applications in agriculture and biotechnology. For instance, in the discovery of cell-type-specific biosynthetic pathways in medicinal plants, as demonstrated by Li *et al.* (2022a), single-cell multi-omics can facilitate the recognition of genes involved in the production of valuable secondary metabolites. This has implications for the development of crops with enriched nutritional or pharmaceutical value, showing the potential of single-cell techniques in handling challenges such as food safety and health.

Furthermore, the understanding acquired from single-cell omics can inform breeding programs intended to enhance plant resilience to stressors like drought or disease. By comprehending the molecular reactions in response to stress at the single-cell level, investigators can identify main regulatory genes and pathways that can be targeted for genetic improvement (Liu *et al.*, 2023). This approach aligns with the growing emphasis on precision agriculture, where data-driven strategies are employed to enhance crop performance and sustainability. Despite considerable advancements in single-cell omics, challenges remain in the field. Issues that are related to cell isolation, data quality, and computational analysis continue to pose hurdles for researchers (Mo *et al.*, 2022). However, ongoing developments in technology and methodology are expected to address these challenges, paving the way for broader applications of single-cell strategies in plant investigation.

15.5 Spatial Multi-omics in Plants

Spatial multi-omics in plants represents a cutting-edge field that integrates various omics technologies to elucidate the spatial organization and functional dynamics of plant tissues. Spatial transcriptomics is one of the emerging transformative technologies that allow investigators to map gene expression profiles directly onto tissue sections, preserving spatial context while providing insights into cellular heterogeneity. For instance, Neumann *et al.* (2022) utilized novoSpaRc to integrate snRNA-seq data with a three-dimensional reconstruction of the *Arabidopsis* floral meristem, revealing the expression patterns of key genes in a spatially resolved manner. This approach underscores the potential of spatial transcriptomics to increase our knowledge of developmental processes in plants. Similarly, Xia *et al.* (2022) demonstrated the utility of single-cell Stereo-seq for characterizing cell type-specific spatial transcriptomes in *Arabidopsis* leaves, providing insights into the differences between vascular and mesophyll cells. These investigations emphasize the ability of spatial transcriptomics to dissect complex tissue architectures and cellular interactions.

In addition to transcriptomics, spatial epigenomics plays an important role in recognizing the regulatory mechanisms governing gene expression in plants. Recent advancements in spatial epigenomics have enabled investigators to profile chromatin accessibility and modifications in a spatial context. For example, the incorporation of spatial epigenomic information with transcriptomic profiles can reveal how epigenetic landscapes influence gene expression patterns across various cells and developmental stages (Chen *et al.*, 2023). This integration is essential for explaining the regulatory networks that dictate plant responses to environmental challenges such as stress. The work of Yang *et al.* (2023) on scCross illustrates the potential of deep generative models to unify single-cell multi-omics data, including epigenomic and transcriptomic data, thereby enhancing our understanding of cellular identity and function.

Proteomics, the investigation of the whole group of proteins expressed in a cell or tissue, is another critical component of spatial multi-omics. Integrating spatial proteomics with transcriptomic and epigenomic data can give a complete view of cellular function and regulation. For instance, totalVI provides a scalable framework for joint analysis of transcriptome and protein measurements at the single-cell level, facilitating the investigation of relationships between RNA and protein expression (Gayoso *et al.*, 2021). This multi-omics approach is particularly valuable in plants, where protein localization and abundance can significantly influence physiological responses and developmental outcomes.

Combining spatial multi-omics with single-cell data is pivotal for advancing our knowledge about plant biology. By integrating spatial transcriptomics with scRNA-seq, researchers can not only identify distinct cell types but also uncover subtle variations in gene expression that may be masked in bulk analyses. Stuart *et al.* (2019) emphasized the importance of integrating single-cell transcriptomics with spatial datasets to capture nuanced changes in cell states and gene expression parameters. This integration is further enhanced by computational tools such as GLUER, which facilitate the analysis of matched datasets from different omics layers, enabling a more holistic understanding of cellular dynamics (Peng *et al.*, 2021).

Moreover, the spatial distribution of functional RNAs under stress conditions is a critical area of investigation in plant biology. The ability to visualize and quantify RNA localization in response to environmental stressors can provide an understanding of the mechanisms by which plants adapt to adverse conditions. For example, the work of Nobori *et al.* (2023) on spatial transcriptomics during pathogen attack highlights how spatial mapping can uncover dynamic modifications in gene expression that are crucial for plant defense responses. This approach allows researchers to recognize specific genes and pathways that are activated in particular cell types or tissue regions, thereby elucidating the spatial aspects of stress responses.

15.6 Technological Advances and Challenges

The occurrence of single-cell and spatial multi-omics methods has revolutionized plant biology, enabling investigators to dissect the intricate molecular networks that govern plant growth, stress responses, and interactions with the environment. These methodologies allow for the detailed analysis of various omic layers such as genomics, proteomics, transcriptomics, and metabolomics at unprecedented resolution. However, despite their potential, the application of these technologies in plant systems is fraught with challenges that necessitate innovative solutions and following paths for research.

Single-cell multi-omics methods have emerged as powerful tools for understanding plant biology at a granular level. These approaches facilitate the simultaneous profiling of multiple omic layers from individual cells, thereby revealing heterogeneity within plant tissues that bulk analyses might obscure. For instance, Hu *et al.* (2018) focus on the potential of single-cell multi-omics to enhance our knowledge of cellular responses to environmental stimuli, particularly in abiotic stress tolerance. This capability is essential for crops that must adapt to fluctuating environmental situations, as it allows researchers to identify specific cellular pathways and regulatory networks involved in stress responses (Zhou *et al.*, 2022).

Despite the promise of these methods, several challenges impede their widespread application in plant systems. One significant hurdle is the complexity of plant genomes, which are often large, polyploid, and poorly annotated. This complicates the interpretation of omics data and necessitates the development of robust bioinformatics tools for data integration and investigation (Jamil *et al.*, 2020). Additionally, the high costs associated with high-throughput sequencing and the need for specialized equipment can limit access to these technologies, particularly in resource-limited settings. Moreover, the inherent biological variability among individual plants and their responses to environmental factors can introduce significant noise into the data, complicating the identification of consistent patterns and biomarkers (Dikobe *et al.*, 2023).

Recent innovations in multi-omics methodologies are addressing some of these challenges. For example, advancements in computational tools and machine learning algorithms are enhancing the integration of multi-omics data, allowing for more accurate modeling of complex biological systems (Cembrowska-Lech *et al.*, 2023). These tools can help researchers uncover hidden patterns and interactions within large datasets, ultimately leading to a greater understanding of plant biology (Murmu *et al.*, 2024). Furthermore, the development of platforms like IPOP, which facilitates cross-species comparisons and evolutionary studies, exemplifies how technological advancements can streamline multi-omics research (Huang *et al.*, 2023). Such platforms can aid in the identification of

Table 15.2. Key technologies in single-cell and spatial multi-omics for plant research.

Technology	Description	Applications in plant research	Key insights	References
scRNA-seq	Single-cell RNA sequencing	Profiling cell-specific expression of genes in roots, leaves, and floral tissues	Revealed heterogeneity in stress response pathways among plant cell types	Ryu *et al.*, 2019
snRNA-seq	Single-nucleus RNA sequencing	Overcomes challenges with protoplast isolation for woody plants	Identified gene expression in hard-to-isolate plant cell types	Xu *et al.*, 2021
Spatial transcriptomics	Spatial mapping of gene expression using tissue sections	Understanding tissue-specific stress responses	Spatial dynamics of gene expression during pathogen attacks	Nobori *et al.*, 2023
Single-cell ATAC-seq	Chromatin attainability profiling at single-cell resolution	Identifying regulatory elements and transcription factor binding sites	Enhanced understanding of epigenomic regulation during stress responses	Yang *et al.*, 2023
Integrated multi-omics	A combination of transcriptomics, proteomics, and epigenomics	Comprehensive analysis of stress response networks	Provided insights into the cross-talk between RNAs and proteins	Gayoso *et al.*, 2021

conserved regulatory mechanisms across different plant species, providing valuable insights for crop improvement efforts.

Looking ahead, the future of single-cell and spatial multi-omics in plant research is promising, with several potential directions for further exploration. One avenue is the continued integration of artificial intelligence (AI) and machine learning with multi-omics approaches, which can enhance data analysis and interpretation (Cembrowska-Lech *et al.*, 2023). By leveraging AI, researchers can develop predictive models that not only elucidate the underlying biological processes but also facilitate the recognition of novel targets for genetic engineering and breeding programs (Murmu *et al.*, 2024). Additionally, incorporating multi-omics information into precision agriculture practices can give farmers actionable knowledge to optimize crop management techniques in the face of climate change (Roychowdhury *et al.*, 2023).

Moreover, as the field progresses, there is a growing emphasis on the ethical implications of omics research, particularly concerning genetic modification and biodiversity conservation (Table 15.2). Addressing these ethical considerations will be crucial as researchers strive to balance innovation with sustainability in agricultural practices (Anjum *et al.*, 2023). Furthermore, collaborative efforts among investigators, agronomists, and policy makers will be important to ensure that the benefits of these technologies are accessible to all stakeholders in the agricultural sector (Choi, 2019).

15.7 Dissecting Functional RNA Regulatory Networks

Dissecting functional RNA regulatory networks is a complex endeavor that involves understanding the intricate interactions between various RNA molecules, proteins, and DNA. These interactions are crucial in regulating gene expression and cellular responses, especially under stress conditions.

RNA–RNA interactions are fundamental to the regulation of gene expression and cellular function. Much research has emphasized the importance of long noncoding RNAs (lncRNAs) in mediating these interactions. For instance, Engreitz *et al.* (2014) demonstrated that lncRNAs can cooperate with chromatin regulatory complexes and RNA processing proteins, implying a dual role in controlling pre-mRNA processing and chromatin regulation. Furthermore, the study by Lu *et al.* (2016b) emphasized that RNA-binding proteins (RBPs) and miRNAs interfere with target RNAs in a structure-dependent manner, revealing the complexity of transcriptome organization and the evolutionary conservation of these interactions. The ability to predict RNA–RNA interactions has been enhanced by tools such as RIsearch2, which allows for large-scale predictions and can infer functional relationships among ncRNAs (Alkan *et al.*, 2017).

The recognition of RNA–protein interactions is equally critical for understanding regulatory networks. RNA-binding proteins are involved in post-transcriptional regulation by interacting with various RNA molecules. For example, the study by Fernandez *et al.* (2011) categorized RNA recognition patterns into groove binding and beta-sheet binding, highlighting the diversity of interactions between RBPs and RNA. Additionally, Iioka *et al.* (2011) developed methods to efficiently detect RNA–protein interactions using tethered RNAs, which can provide insights into the dynamics of these interactions within cellular contexts. The complexity of RNA–protein interactions is further illustrated by Busby *et al.* (2020), who emphasized the regulatory roles of 3′ untranslated regions of mRNAs as hubs for RBPs.

RNA–DNA interactions can also regulate gene expression. These interactions are often mediated by RBPs that can bind to both RNA and DNA, influencing transcriptional regulation. The study by Lin and Miles (2019) discussed the expanding roles of RNA in cellular processes beyond protein coding, emphasizing its involvement in chromatin organization and transcriptional regulation. This interplay between RNA and DNA is essential for maintaining cellular homeostasis, particularly during stress conditions where gene expression must be rapidly adjusted.

The incorporation of multi-omics data is essential for mapping these complex regulatory networks. Multi-omics strategies allow for the simultaneous analysis of various biological layers. This integrative analysis can give a more detailed understanding of cellular responses to stress. For instance, the work by Hawe *et al.* (2019) highlighted the potential of multi-omics information to infer heterogeneous networks, which can reveal the intricate relationships between different molecular entities. Furthermore, the investigation by Henao *et al.* emphasized the importance of uncovering multi-level networks to understand the interplay of their elements, particularly in the context of missing data (Henao *et al.*, 2023).

Stress conditions often lead to the activation of specific RNA regulatory networks that can modulate cellular responses. For example, the investigation by Jacq *et al.* (2021) highlighted that the lncRNA Neat1 is involved in the nuclear retention of RNAs through direct RNA–RNA interactions, contributing to the formation of paraspeckles, which are involved in stress responses. Similarly, the research by Tauber *et al.* (2020) indicated that RNA condensation, influenced by RNA–RNA interaction strengths, can lead to specific compartmentalization of RNAs, which may be relevant during stress responses. These examples illustrate how stress-specific RNA regulatory networks can be identified and characterized through the integration of various experimental approaches.

Moreover, the role of RNA–RNA interactions in stress responses is further supported by the findings of Wenzel *et al.* (2012), who demonstrated that the prediction of miRNA targets can reveal large-scale regulatory networks that connect mRNAs and ncRNAs. This interconnectedness is important for comprehending how cells adapt to stress by modulating gene expression through RNA-mediated mechanisms.

15.8 Integrative Approaches for Functional RNA Analysis

Integrative approaches for functional RNA analysis are becoming increasingly vital in the various aspects of molecular biology, particularly as researchers seek to understand the complex interactions and functions of RNA at both single-cell and spatial levels. The appearance of high-throughput methods has enabled the collection of vast amounts of data, necessitating the development of advanced computational tools and pipelines for data integration. This synthesis of methodologies allows for a multi-scale analysis that can provide a comprehensive understanding of RNA functionality and regulation.

One of the key advancements in RNA analysis is the incorporation of single-cell transcriptomics with spatial multi-omics techniques. For instance, the development of tools like Scanorama has facilitated the efficient integration of heterogeneous single-cell transcriptomes, allowing researchers to assemble a complete picture of biological functions at a single-cell resolution (Hie *et al.*, 2019). This is particularly important in studies of cellular heterogeneity, where understanding the unique transcriptomic profiles of individual cells can reveal insights into cellular differentiation and function. Furthermore, the incorporation of spatial transcriptomics with scRNA-seq provides an influential framework for mapping gene expression in the context of tissue architecture, thereby enhancing our understanding of spatial gene regulation (Hao *et al.*, 2021).

The computational landscape for RNA analysis has also evolved significantly, with various pipelines and tools being developed to streamline the analysis process. For example, the VIPER workflow offers a comprehensive solution for RNA-seq analysis, enabling users to visualize and interpret RNA-seq data efficiently (Cornwell *et al.*, 2018). Additionally, tools like Bento provide scalable analysis of multiplexed spatial transcriptomics data, utilizing geospatial tools to enhance the interpretation of molecular imaging data (Mah *et al.*, 2024). These computational advancements are crucial for managing the complexity of multi-omics data and facilitating integrative analyses.

Moreover, the integration of different high-throughput technologies, such as RNA-seq and ATAC-seq, has proven to be instrumental in elucidating the regulatory networks governing gene expression. The association of these methods allows researchers to identify transcription factor-targeted genes and their corresponding transcripts, providing insights into the regulatory mechanisms at play (Hong *et al.*, 2020). The need for robust computational tools to analyze the vast datasets generated by these high-throughput technologies cannot be overstated. Tools like RNASeqR automate the RNA-seq analysis workflow, making it available for investigators at different levels of computational expertise (Chao *et al.*, 2019). This democratization of data analysis is essential for fostering innovation and collaboration in the field, as it enables a broader range of investigators to contribute to the understanding of RNA biology. Additionally, systems like IRIS-EDA provide integrated platforms for gene expression data analysis, further simplifying the analytical process and enhancing the interpretability of results (Monier *et al.*, 2019).

In the realm of RNA structure analysis, advancements in computational modeling have significantly improved our ability to predict and visualize RNA structures. Tools such as SimRNAweb allow for RNA folding simulations and 3D structure modeling, which are critical for understanding the functional implications of RNA conformations (Moafinejad *et al.*, 2024). Integrating structural data with transcriptomic information can provide a more holistic approach to RNA functionality, especially in the context of RNA–protein interactions and regulatory mechanisms (Li *et al.*, 2020). Furthermore, the benefits of machine learning approaches in predicting RNA regulatory targets have opened new avenues for comprehending the complex interactions between RNA molecules and their regulatory networks (Tjaden, 2023).

As the field continues to evolve, the importance of multi-scale analysis becomes increasingly apparent. By integrating information from various sources and scales, investigators can get a more detailed understanding of RNA biology. For instance, the integration of single-cell transcriptomic information across various conditions and methods has been shown to enhance the identification of shared cellular populations, facilitating comparative analyses that can reveal insights into cellular changes for various stimuli (Butler *et al.*, 2018). This integrative strategy is especially valuable in the context of disease research, where recognizing the heterogeneity of cellular responses can inform therapeutic strategies.

The challenges associated with analyzing ncRNAs also highlight the need for advanced computational methods. The rapid increase in sequencing data for ncRNAs has created significant analytical challenges, including the identification of ncRNA targets and the statistical modeling of their expression (He *et al.*, 2018). Tools developed for this purpose are crucial for unraveling the complex roles that ncRNAs play in gene regulation and cellular functions. Moreover, the detection of RNA modifications using nanopore sequencing technologies has emerged as a critical area of research, as these modifications can significantly influence RNA function and stability (Furlan *et al.*, 2021).

15.9 Challenges and Future Perspectives

The field of single-cell and spatial multi-omics has witnessed rapid advancements, yet it faces significant challenges and limitations that hinder its full potential. One of the primary limitations is the scalability and reproducibility of current techniques. Many single-cell sequencing methods, while powerful, often suffer from low throughput and high costs, which restrict their application in large-scale studies. For instance, the integration of multiple omics layers from single cells can lead to complexities in data handling and analysis, as highlighted by Hu *et al.* (2018). They emphasize that overcoming these limitations is crucial for developing more comprehensive omics measurements on single cells. Furthermore, the issue of reproducibility is exacerbated by the variability in experimental protocols and data analysis methods, which can lead to inconsistent results across studies (Baysoy *et al.*, 2023).

In addition to scalability and reproducibility, data interpretation presents a formidable challenge in single-cell multi-omics. The integration of diverse data types, including transcriptomics, proteomics, and epigenomics, requires advanced computational methods to accurately interpret the relationships between different molecular layers. Efremova and Teichmann (2020) discuss how combined multi-modal omics measurements can reveal associations between genotype and gene expression, yet they also note that the complexity of these datasets can obscure meaningful biological insights. The challenge is further compounded by the presence of noise and batch effects in the data, which can distort the true biological signals (Lancaster *et al.*, 2020). As a result, researchers must employ advanced statistical models and machine learning techniques to disentangle these complexities and derive biologically relevant conclusions (Wang *et al.*, 2023).

Looking toward the future, emerging trends in single-cell and spatial multi-omics are poised to address some of these challenges. One promising direction is the development of novel computational frameworks that facilitate the integration of multi-omics data. For example, the scCross model allows for the seamless integration of multiple omic layers, showcasing its potential for comprehensive data analysis (Yang *et al.*, 2023). This modular approach not only enhances the depth of analysis but also improves the interpretability of complex datasets. Moreover, the advent of probabilistic models, such as totalVI, provides a scalable solution for analyzing paired transcriptome and protein measurements, thereby addressing some of the limitations associated with traditional methods (Gayoso *et al.*, 2021). These advancements indicate a shift towards more robust and interpretable analytical frameworks that can handle the intricacies of multi-omics data.

Another exciting avenue for future research lies in the potential for uncovering novel functional RNAs in plants through single-cell multi-omics methods. Plant development and stress responses can be regulated by previously unidentified regulatory elements and non-coding RNAs (ncRNAs), which can be found by combining transcriptome and epigenomic data. As highlighted by Bravo González-Blas *et al.* (2020), by separating cell-type-specific signals from bulk RNA-seq data, single-cell data can bring insight into the functional roles of these molecules. This capability is relevant in the context of plant biology, especially where comprehending the regulatory networks governing gene expression is essential for improving crop resilience and yield.

Interdisciplinary research and collaborations are also expected to flourish in the realm of single-cell and spatial multi-omics. The complexity of biological systems necessitates a combined approach that brings together experts from various areas, such as computational biology, genomics, and plant sciences. By leveraging diverse expertise, researchers can develop innovative methodologies and analytical tools that enhance the understanding of cellular dynamics and interactions. The integration of high-throughput imaging techniques with multi-omics data, as suggested by Chessel and Carazo Salas (2019), represents a promising frontier that could yield transformative insights into cellular structure and function.

15.10 Conclusion

This chapter highlights the transformative impact of single-cell and spatial multi-omics technologies on understanding the regulatory roles of plant functional RNAs, particularly under stress conditions. The integration of these advanced approaches has provided an unprecedented understanding of the molecular complexity and spatial dynamics of RNA-mediated regulation. Functional RNAs, including miRNAs, lncRNAs, and circRNAs, can modulate gene expression, facilitate plant adaptation to environmental challenges, and govern developmental processes. Despite significant advancements, challenges, including data integration, scalability, and the complexity of plant genomes, persist. Emerging computational tools and interdisciplinary collaborations offer promising solutions to address these issues. The ongoing refinement of single-cell and spatial omics methodologies is expected to deepen our knowledge of the intricate networks governing plant biology. Ultimately, these insights will contribute to developing innovative strategies for enhancing crop resilience and productivity in the face of environmental strains.

Acknowledgements

All the authors are thankful to their respective universities and institutions for their valuable support.

References

Alkan, F., Wenzel, A., Palasca, O., Kerpedjiev, P., Rudebeck, A.F. *et al.* (2017) RIsearch2: Suffix array-based large-scale prediction of RNA-RNA interactions and siRNA off-targets. *Nucleic Acids Research* 45(8), e60–e60. DOI: 10.1093/nar/gkw1325.

Anjali, N., Nadiya, F., Thomas, J. and Sabu, K.K. (2017) Discovery of microRNAs in cardamom (*Elettaria cardamomum* Maton) under drought stress. *Dataset Papers In Science* 2017(1), 9507485. DOI: 10.1155/2017/9507485.

Anjum, N., Masood, A., Thangavel, P. and Khan, N.A. (eds) (2023) *Making Plant Life Easier and Productive Under Salinity-Updates and Prospects: Updates and Prospects*. BoD–Books on Demand.

Asefpour Vakilian, K. (2020) Machine learning improves our knowledge about miRNA functions towards plant abiotic stresses. *Scientific Reports* 10(1), 3041. DOI: 10.1038/s41598-020-59981-6.

Barrera-Figueroa, B.E., Gao, L., Wu, Z., Zhou, X., Zhu, J. *et al.* (2012) High throughput sequencing reveals novel and abiotic stress-regulated microRNAs in the inflorescences of rice. *BMC Plant Biology* 12, 1–11. DOI: 10.1186/1471-2229-12-132.

Baysoy, A., Bai, Z., Satija, R. and Fan, R. (2023) The technological landscape and applications of single-cell multi-omics. *Nature Reviews. Molecular Cell Biology* 24(10), 695–713. DOI: 10.1038/s41580-023-00615-w.

Bhogireddy, S., Mangrauthia, S.K., Kumar, R., Pandey, A.K., Singh, S. *et al.* (2021) Regulatory non-coding RNAs: A new frontier in regulation of plant biology. *Functional & Integrative Genomics* 21(3–4), 313–330. DOI: 10.1007/s10142-021-00787-8.

Borges, F. and Martienssen, R.A. (2015) The expanding world of small RNAs in plants. *Nature Reviews. Molecular Cell Biology* 16(12), 727–741. DOI: 10.1038/nrm4085.

Bravo González-Blas, C., Quan, X.-J., Duran-Romaña, R., Taskiran, I.I., Koldere, D. *et al.* (2020) Identification of genomic enhancers through spatial integration of single-cell transcriptomics and epigenomics. *Molecular Systems Biology* 16(5), e9438. DOI: 10.15252/msb.20209438.

Busby, K.N., Fulzele, A., Zhang, D., Bennett, E.J. and Devaraj, N.K. (2020) Enzymatic RNA biotinylation for affinity purification and identification of RNA-protein interactions. *ACS Chemical Biology* 15(8), 2247–2258. DOI: 10.1021/acschembio.0c00445.

Butler, A., Hoffman, P., Smibert, P., Papalexi, E. and Satija, R. (2018) Integrating single-cell transcriptomic data across different conditions, technologies, and species. *Nature Biotechnology* 36(5), 411–420. DOI: 10.1038/nbt.4096.

Cao, Z.J. and Gao, G. (2022) Multi-omics single-cell data integration and regulatory inference with graph-linked embedding. *Nature Biotechnology* 40(10), 1458–1466. DOI: 10.1038/s41587-022-01284-4.

Cembrowska-Lech, D., Krzemińska, A., Miller, T., Nowakowska, A., Adamski, C. *et al.* (2023) An integrated multi-omics and artificial intelligence framework for advance plant phenotyping in horticulture. *Biology* 12(10), 1298. DOI: 10.3390/biology12101298.

Chao, K.-H., Hsiao, Y.-W., Lee, Y.-F., Lee, C.-Y., Lai, L.-C. *et al.* (2019) RNASeqR: An R package for automated two-group RNA-seq analysis workflow. *IEEE/ACM Transactions on Computational Biology and Bioinformatics* 18(5), 2023–2031. DOI: 10.1109/TCBB.2019.2956708.

Chen, C., Ge, Y. and Lu, L. (2023) Opportunities and challenges in the application of single-cell and spatial transcriptomics in plants. *Frontiers in Plant Science* 14, 1185377. DOI: 10.3389/fpls.2023.1185377.

Chessel, A. and Carazo Salas, R.E. (2019) From observing to predicting single-cell structure and function with high-throughput/high-content microscopy. *Essays in Biochemistry* 63(2), 197–208. DOI: 10.1042/EBC20180044.

Chien, C.H., Chiang-Hsieh, Y.F., Chen, Y.A., Chow, C.N., Wu, N.Y. *et al.* (2015) AtmiRNET: A web-based resource for reconstructing regulatory networks of *Arabidopsis* microRNAs. *Database* bav042. DOI: 10.1093/database/bav042.

Cho, J. (2018) Transposon-derived non-coding RNAs and their function in plants. *Frontiers in Plant Science* 9, 600. DOI: 10.3389/fpls.2018.00600.

Choi, H.K. (2019) Translational genomics and multi-omics integrated approaches as a useful strategy for crop breeding. *Genes & Genomics* 41(2), 133–146. DOI: 10.1007/s13258-018-0751-8.

Cornwell, M., Vangala, M., Taing, L., Herbert, Z., Köster, J. *et al.* (2018) VIPER: Visualization pipeline for RNA-seq, a Snakemake workflow for efficient and complete RNA-seq analysis. *BMC Bioinformatics* 19(1), 135. DOI: 10.1186/s12859-018-2139-9.

Dai, X., Zhuang, Z. and Zhao, P.X. (2018) psRNATarget: A plant small RNA target analysis server (2017 release). *Nucleic Acids Research* 46(W1), W49–W54. DOI: 10.1093/nar/gky316.

Darbani, B., Noeparvar, S. and Borg, S. (2016) Identification of circular RNAs from the parental genes involved in multiple aspects of cellular metabolism in barley. *Frontiers in Plant Science* 7, 776. DOI: 10.3389/fpls.2016.00776.

Deng, Z., Ma, L., Zhang, P. and Zhu, H. (2022) Small RNAs participate in plant-virus interaction and their application in plant viral defense. *International Journal of Molecular Sciences* 23(2), 696. DOI: 10.3390/ijms23020696.

Dey, S. and Raichaudhuri, A. (2022) Abiotic stress in plants. In: *Advances in Plant Defense Mechanisms*. IntechOpen.

Dikobe, T., Masenya, K. and Manganyi, M.C. (2023) Molecular technologies ending with 'omics': The driving force toward sustainable plant production and protection. *F1000Research* 12, 480. DOI: 10.12688/f1000research.131413.1.

Efremova, M. and Teichmann, S.A. (2020) Computational methods for single-cell omics across modalities. *Nature Methods* 17(1), 14–17.

Engreitz, J.M., Sirokman, K., McDonel, P., Shishkin, A.A., Surka, C. *et al.* (2014) RNA-RNA interactions enable specific targeting of noncoding RNAs to nascent Pre-mRNAs and chromatin sites. *Cell* 159(1), 188–199. DOI: 10.1016/j.cell.2014.08.018.

Fernandez, M., Kumagai, Y., Standley, D.M., Sarai, A., Mizuguchi, K. *et al.* (2011) Prediction of dinucleotide-specific RNA-binding sites in proteins. In: *Bmc Bioinformatics*, Vol. 12. BioMed Central, pp. 1–10. DOI: 10.1186/1471-2105-12-S13-S5.

Furlan, M., Delgado-Tejedor, A., Mulroney, L., Pelizzola, M., Novoa, E.M. *et al.* (2021) Computational methods for RNA modification detection from nanopore direct RNA sequencing data. *RNA Biology* 18(sup1), 31–40.

Gayoso, A., Steier, Z., Lopez, R., Regier, J., Nazor, K.L. *et al.* (2021) Joint probabilistic modeling of single-cell multi-omic data with totalVI. *Nature Methods* 18(3), 272–282.

Han, F., Wang, P., Chen, X., Zhao, H., Zhu, Q. *et al*. (2023) An ethylene-induced NAC transcription factor acts as a multiple abiotic stress responsor in conifer. *Horticulture Research* 10(8), uhad130. DOI: 10.1093/hr/uhad130.

Hao, Y., Hao, S., Andersen-Nissen, E., Mauck, W.M., III, Zheng, S. *et al*. (2021) Integrated analysis of multimodal single-cell data. *Cell* 184(13), 3573–3587. DOI: 10.1016/j.cell.2021.04.048.

Hawe, J.S., Theis, F.J. and Heinig, M. (2019) Inferring interaction networks from multi-omics data. *Frontiers in Genetics* 10, 535. DOI: 10.3389/fgene.2019.00535.

He, Q., Liu, Y. and Sun, W. (2018) Statistical analysis of non-coding RNA data. *Cancer Letters* 417, 161–167. DOI: 10.1016/j.canlet.2017.12.029.

Henao, J.D., Lauber, M., Azevedo, M., Grekova, A., Theis, F. *et al*. (2023) Multi-omics regulatory network inference in the presence of missing data. *Briefings in Bioinformatics* 24(5), bbad309. DOI: 10.1093/bib/bbad309.

Hie, B., Bryson, B. and Berger, B. (2019) Efficient integration of heterogeneous single-cell transcriptomes using *Scanorama*. *Nature Biotechnology* 37(6), 685–691. DOI: 10.1038/s41587-019-0113-3.

Hong, M., Tao, S., Zhang, L., Diao, L.-T., Huang, X. *et al*. (2020) RNA sequencing: New technologies and applications in cancer research. *Journal of Hematology & Oncology* 13(1), 166. DOI: 10.1186/s13045-020-01005-x.

Hu, Y., An, Q., Sheu, K., Trejo, B., Fan, S. *et al*. (2018) Single cell multi-omics technology: Methodology and application. *Frontiers in Cell and Developmental Biology* 6, 28. DOI: 10.3389/fcell.2018.00028.

Huang, W., Hu, X., Ren, Y., Song, M., Ma, C. *et al*. (2023) IPOP: An integrative plant multi-omics platform for cross-species comparison and evolutionary study. *Molecular Biology and Evolution* 40(12), msad248. DOI: 10.1093/molbev/msad248.

Iioka, H., Loiselle, D., Haystead, T.A. and Macara, I.G. (2011) Efficient detection of RNA-protein interactions using tethered RNAs. *Nucleic Acids Research* 39(8), e53–e53. DOI: 10.1093/nar/gkq1316.

Iqbal, M.M., Hurgobin, B., Holme, A.L., Appels, R. and Kaur, P. (2020) Status and potential of single-cell transcriptomics for understanding plant development and functional biology. *Cytometry. Part A* 97(10), 997–1006. DOI: 10.1002/cyto.a.24196.

Jacq, A., Becquet, D., Guillen, S., Boyer, B., Bello-Goutierrez, M.M. *et al*. (2021) Direct RNA–RNA interaction between Neat1 and RNA targets, as a mechanism for RNAs paraspeckle retention. *RNA Biology* 18(11), 2016–2027. DOI: 10.1080/15476286.2021.1889253.

Jamil, I.N., Remali, J., Azizan, K.A., Nor Muhammad, N.A., Arita, M. *et al*. (2020) Systematic multi-omics integration (MOI) approach in plant systems biology. *Frontiers in Plant Science* 11, 944. DOI: 10.3389/fpls.2020.00944.

Jean-Baptiste, K., McFaline-Figueroa, J.L., Alexandre, C.M., Dorrity, M.W., Saunders, L. *et al*. (2019) Dynamics of gene expression in single root cells of *Arabidopsis thaliana*. *The Plant Cell* 31(5), 993–1011. DOI: 10.1105/tpc.18.00785.

Jin, J., Yu, S., Lu, P. and Cao, P. (2023) Deciphering plant cell-cell communications using single-cell omics data. *Computational and Structural Biotechnology Journal* 21, 3690–3695. DOI: 10.1016/j.csbj.2023.06.016.

Kallemi, P., Verret, F., Andronis, C., Ioannidis, N., Glampedakis, N. *et al*. (2024) Stress-related transcriptomic changes associated with GFP transgene expression and active transgene silencing in plants. *Scientific Reports* 14(1), 13314.

Lancaster, S.M., Sanghi, A., Wu, S. and Snyder, M.P. (2020) A customizable analysis flow in integrative multi-omics. *Biomolecules* 10(12), 1606.

Li, B., Cao, Y., Westhof, E. and Miao, Z. (2020) Advances in RNA 3D structure modeling using experimental data. *Frontiers in Genetics* 11, 574485. DOI: 10.3389/fgene.2020.574485.

Li, C., Wood, J.C., Vu, A.H., Hamilton, J.P., Rodriguez Lopez, C.E. *et al*. (2022a) Single-cell multi-omics enabled discovery of alkaloid biosynthetic pathway genes in the medical plant *Catharanthus roseus*. *Plant Biology*. DOI: 10.1101/2022.07.04.498697.

Li, S., Cheng, Z., Dong, S., Li, Z., Zou, L. *et al*. (2022b) Global identification of full-length cassava lncRNAs unveils the role of cold-responsive intergenic lncRNA 1 in cold stress response. *Plant, Cell & Environment* 45(2), 412–426. DOI: 10.1111/pce.14236.

Lin, C. and Miles, W.O. (2019) Beyond CLIP: Advances and opportunities to measure RBP-RNA and RNA-RNA interactions. *Nucleic Acids Research* 47(11), 5490–5501. DOI: 10.1093/nar/gkz295.

Liu, X., Zhu, K. and Xiao, J. (2023) Recent advances in understanding of the epigenetic regulation of plant regeneration. *Abiotech* 4(1), 31–46. DOI: 10.1007/s42994-022-00093-2.

Lohani, N., Golicz, A.A., Allu, A.D., Bhalla, P.L. and Singh, M.B. (2022) Genome-wide analysis of lncRNAs points to their roles in the modulation of developmental regulator expression during plant male germline development. *BioRxiv* 2022–08. DOI: 10.1101/2022.08.03.502631.

Lu, X., Chen, X., Mu, M., Wang, J., Wang, X. *et al.* (2016a) Genome-wide analysis of long noncoding RNAs and their responses to drought stress in cotton (*Gossypium hirsutum* L.). *PLoS ONE* 11(6), e0156723. DOI: 10.1371/journal.pone.0156723.

Lu, Z., Zhang, Q.C., Lee, B., Flynn, R.A., Smith, M.A. *et al.* (2016b) RNA duplex map in living cells reveals higher-order transcriptome structure. *Cell* 165(5), 1267–1279. DOI: 10.1016/j.cell.2016.04.028.

Mah, C.K., Ahmed, N., Lopez, N.A., Lam, D.C., Pong, A. *et al.* (2024) Bento: A toolkit for subcellular analysis of spatial transcriptomics data. *Genome Biology* 25(1), 82. DOI: 10.1186/s13059-024-03217-7.

Matsui, A., Iida, K., Tanaka, M., Yamaguchi, K., Mizuhashi, K. *et al.* (2017) Novel stress-inducible antisense RNAs of protein-coding loci are synthesized by RNA-dependent RNA polymerase. *Plant Physiology* 175(1), 457–472. DOI: 10.1104/pp.17.00787.

Meng, Y., Shao, C., Wang, H. and Jin, Y. (2012) Target mimics: An embedded layer of microRNA-involved gene regulatory networks in plants. *BMC Genomics* 13, 1–12. DOI: 10.1186/1471-2164-13-197.

Monier, B., McDermaid, A., Wang, C., Zhao, J., Miller, A. *et al.* (2019) IRIS-EDA: An integrated RNA-Seq interpretation system for gene expression data analysis. *PLoS Computational Biology* 15(2), e1006792. DOI: 10.1371/journal.pcbi.1006792.

Mo, Y., Wu, Y., Yang, X., Liu, F. and Liao, Y. (2022) Review the state-of-the-art technologies of semantic segmentation based on deep learning. *Neurocomputing* 493, 626–646. DOI: 10.1016/j.neucom.2022.01.005.

Moafinejad, S.N., de Aquino, B.R.H., Boniecki, M.J., Pandaranadar Jeyeram, I.P.N., Nikolaev, G. *et al.* (2024) SimRNAweb v2. 0: A web server for RNA folding simulations and 3D structure modeling, with optional restraints and enhanced analysis of folding trajectories. *Nucleic Acids Research* 52(W1), W368–W373. DOI: 10.1093/nar/gkae356.

Murmu, S., Sinha, D., Chaurasia, H., Sharma, S., Das, R. *et al.* (2024) A review of artificial intelligence-assisted omics techniques in plant defense: Current trends and future directions. *Frontiers in Plant Science* 15, 1292054. DOI: 10.3389/fpls.2024.1292054.

Neumann, M., Xu, X., Smaczniak, C., Schumacher, J., Yan, W. *et al.* (2022) A 3D gene expression atlas of the floral meristem based on spatial reconstruction of single nucleus RNA sequencing data. *Nature Communications* 13(1), 2838. DOI: 10.1038/s41467-022-30177-y.

Nobori, T., Monell, A., Lee, T.A., Zhou, J., Nery, J. *et al.* (2023) Time-resolved single-cell and spatial gene regulatory atlas of plants under pathogen attack. *BioRxiv* 2023–04. DOI: 10.1101/2023.04.10.536170.

Park, C.J. and Park, J.M. (2019) Endoplasmic reticulum plays a critical role in integrating signals generated by both biotic and abiotic stress in plants. *Frontiers in Plant Science* 10, 399. DOI: 10.3389/fpls.2019.00399.

Peng, T., Chen, G.M. and Tan, K. (2021) GLUER: Integrative analysis of single-cell omics and imaging data by deep neural network. *BioRxiv* 2021–01. DOI: 10.1101/2021.01.25.427845.

Roychowdhury, R., Das, S.P., Gupta, A., Parihar, P., Chandrasekhar, K. *et al.* (2023) Multi-omics pipeline and omics-integration approach to decipher plant's abiotic stress tolerance responses. *Genes* 14(6), 1281. DOI: 10.3390/genes14061281.

Ryu, K.H., Huang, L., Kang, H.M. and Schiefelbein, J. (2019) Single-cell RNA sequencing resolves molecular relationships among individual plant cells. *Plant Physiology* 179(4), 1444–1456. DOI: 10.1104/pp.18.01482.

Schröder, J.A. and Jullien, P.E. (2019) The diversity of plant small RNAs silencing mechanisms. *Chimia* 73(5), 362–367. DOI: 10.2533/chimia.2019.362.

Stuart, T., Butler, A., Hoffman, P., Hafemeister, C., Papalexi, E. *et al.* (2019) Comprehensive integration of single-cell data. *Cell* 177(7), 1888–1902. DOI: 10.1016/j.cell.2019.05.031.

Sun, F., Guo, G., Du, J., Guo, W., Peng, H. *et al.* (2014) Whole-genome discovery of miRNAs and their targets in wheat (*Triticum aestivum* L.). *BMC Plant Biology* 14(1), 142. DOI: 10.1186/1471-2229-14-142.

Tauber, D., Tauber, G., Khong, A., Van Treeck, B., Pelletier, J. *et al.* (2020) Modulation of RNA condensation by the DEAD-box protein eIF4A. *Cell* 180(3), 411–426. DOI: 10.1016/j.cell.2019.12.031.

Tjaden, B. (2023) TargetRNA3: Predicting prokaryotic RNA regulatory targets with machine learning. *Genome Biology* 24(1), 276. DOI: 10.1186/s13059-023-03117-2.

Wang, J., Mei, J. and Ren, G. (2019) Plant microRNAs: Biogenesis, homeostasis, and degradation. *Frontiers in Plant Science* 10, 360. DOI: 10.3389/fpls.2019.00360.

Wang, T.Z., Liu, M., Zhao, M.G., Chen, R. and Zhang, W.H. (2015) Identification and characterization of long non-coding RNAs involved in osmotic and salt stress in *Medicago truncatula* using genome-wide high-throughput sequencing. *BMC Plant Biology* 15(1), 1–13. DOI: 10.1186/s12870-015-0530-5.

Wang, X., Hu, Z., Yu, T., Wang, Y., Wang, R. *et al.* (2023) Con-AAE: Contrastive cycle adversarial autoencoders for single-cell multi-omics alignment and integration. *Bioinformatics* 39(4), btad162. DOI: 10.1093/bioinformatics/btad162.

Waseem, M., Liu, Y. and Xia, R. (2020) Long non-coding RNAs, the dark matter: An emerging regulatory component in plants. *International Journal of Molecular Sciences* 22(1), 86. DOI: 10.3390/ijms22010086.

Wenzel, A., Akbasli, E. and Gorodkin, J. (2012) RIsearch: Fast RNA-RNA interaction search using a simplified nearest-neighbor energy model. *Bioinformatics (Oxford, England)* 28(21), 2738–2746. DOI: 10.1093/bioinformatics/bts519.

Willmann, M.R., Endres, M.W., Cook, R.T. and Gregory, B.D. (2011) The functions of RNA-dependent RNA polymerases in *Arabidopsis*. *The Arabidopsis Book* 9, e0146. DOI: 10.1199/tab.0146.

Xia, K., Sun, H.-X., Li, J., Li, J., Zhao, Y. *et al.* (2022) The single-cell stereo-seq reveals region-specific cell subtypes and transcriptome profiling in *Arabidopsis* leaves. *Developmental Cell* 57(10), 1299–1310. DOI: 10.1016/j.devcel.2022.04.011.

Xu, M., Du, Q., Tian, C., Wang, Y. and Jiao, Y. (2021) Stochastic gene expression drives mesophyll protoplast regeneration. *Science Advances* 7(33), eabg8466. DOI: 10.1126/sciadv.abg8466.

Yang, X., Mann, K.K., Wu, H. and Ding, J. (2023) scCross: Bridging modalities in single–cell multi–omics–seamless integration, cross–modal synthesis, and *in–silico* exploration. *Bioinformatics*. DOI: 10.1101/2023.11.22.568376.

Yang, X., Yang, M., Deng, H. and Ding, Y. (2018) New era of studying RNA secondary structure and its influence on gene regulation in plants. *Frontiers in Plant Science* 9, 671. DOI: 10.3389/fpls.2018.00671.

Yang, X., Liu, C., Niu, X., Wang, L., Li, L. *et al.* (2022) Research on lncRNA related to drought resistance of Shanlan upland rice. *BMC Genomics* 23(1), 336. DOI: 10.1186/s12864-022-08546-0.

Yoon, Y., Seo, D.H., Shin, H., Kim, H.J., Kim, C.M. *et al.* (2020) The role of stress-responsive transcription factors in modulating abiotic stress tolerance in plants. *Agronomy* 10(6), 788. DOI: 10.3390/agronomy10060788.

Zhang, S., Yue, Y., Sheng, L., Wu, Y., Fan, G. *et al.* (2013) PASmiR: A literature-curated database for miRNA molecular regulation in plant response to abiotic stress. *BMC Plant Biology* 13, 1–8. DOI: 10.1186/1471-2229-13-33.

Zhou, R., Jiang, F., Niu, L., Song, X., Yu, L. *et al.* (2022) Increase crop resilience to heat stress using omic strategies. *Frontiers in Plant Science* 13, 891861. DOI: 10.3389/fpls.2022.891861.

Zhu, J.K. (2016) Abiotic stress signaling and responses in plants. *Cell* 167(2), 313–324. DOI: 10.1016/j.cell.2016.08.029.

16 Structural and Functional Analysis of Plant Non-coding RNAs: Highlighting the Advances in RNA Structures

Kalibulla Syed Ibrahim[1*], Vijayan Praveena[2], Mohanraj Lavanya[2], Raju Prabakaran[2], Thiruvengadam Thirumurugan[3], Sathyalingam Sathya Trisha[4], Sahul Ameed Nihaal Fathima[4] and Arumugam Vijaya Anand[4]

[1]Department of Biotechnology, PSG College of Arts & Science, Coimbatore, India; [2]PG & Research Department of Botany, PSG College of Arts & Science, Coimbatore, India; [3]Department of Genetics & Plant Breeding, Anbil Dharmalingam Agricultural College and Research Institute, Trichirappalli, India; [4]Department of Human Genetics and Molecular Biology, Bharathiar University, Coimbatore, India

16.1 Introduction

Plants are vital to maintaining life and environmental balance on Earth. Their growth and development are largely shaped by external environmental conditions. Numerous genes have been identified that govern both developmental processes and the response to abiotic stress in plants. As well as protein-coding genes, growing research emphasizes the important regulatory roles of non-coding RNAs (ncRNAs) in plant systems. RNAs are involved in almost every aspect of genome organization, cellular structure, and gene regulation. They interact with other RNA molecules, DNA, and proteins to carry out their functions. These interactions frequently involve repetitive elements, such as small interspersed nuclear elements located in the 3′ untranslated regions of genes (Zucchelli *et al.*, 2015). Such interactions contribute to chromatin remodeling and transcriptional regulation (explored in later sections), as well as splicing, protein synthesis and localization, and various aspects of RNA processing, modification, transport, and stability (Gong and Maquat, 2011).

16.1.1 What are ncRNAs?

Non-protein-coding RNAs (ncRNAs) are RNA molecules that have various cellular and ribosomal roles, such as regulating plant growth, development, and responses to stress. Unlike conventional RNAs, ncRNAs differ in their origin and mechanisms of action on target genes. They are essential in several aspects of plant development, including leaf formation, flowering, and meristem organization. Additionally, they are involved in numerous biological processes such

*Corresponding author: syedibrahim.k@gmail.com

© CAB International 2026. *Plant Spatial Biology* (ed. Chen)
DOI: 10.1079/9781836994572.0016

as transcriptional and post-transcriptional regulation, translation, epigenetic modifications, stress adaptation, gene expression control, and maintaining genome stability (Yadav *et al.*, 2024). The first ncRNA, identified as alanine tRNA, was described by Holley *et al.* (1965) in baker's yeast. Later, research in *Caenorhabditis elegans* by Lee *et al.* (1993) expanded our understanding of the diverse roles and types of ncRNAs across different organisms, including animals and plants (Holley *et al.*, 1965; Lee *et al.*, 1993; Yu *et al.*, 2019). Nevertheless, our understanding of the functions of ncRNAs in plants is still limited compared to the knowledge available for animals.

16.2 Classification and Types of ncRNAs

In plants, ncRNAs are categorized into distinct classes based on their origin, pathways of biogenesis, and modes of action. Broadly, they are divided into two major groups: structural ncRNAs, also referred to as regular, housekeeping, or constitutive ncRNAs, and regulatory ncRNAs (Fig. 16.1). Specific traits, including size, biogenesis, polarity, and functional responsibilities, are used to sub-categorize each of these categories (Khan *et al.*, 2023). Structural ncRNAs are primarily involved in essential cellular

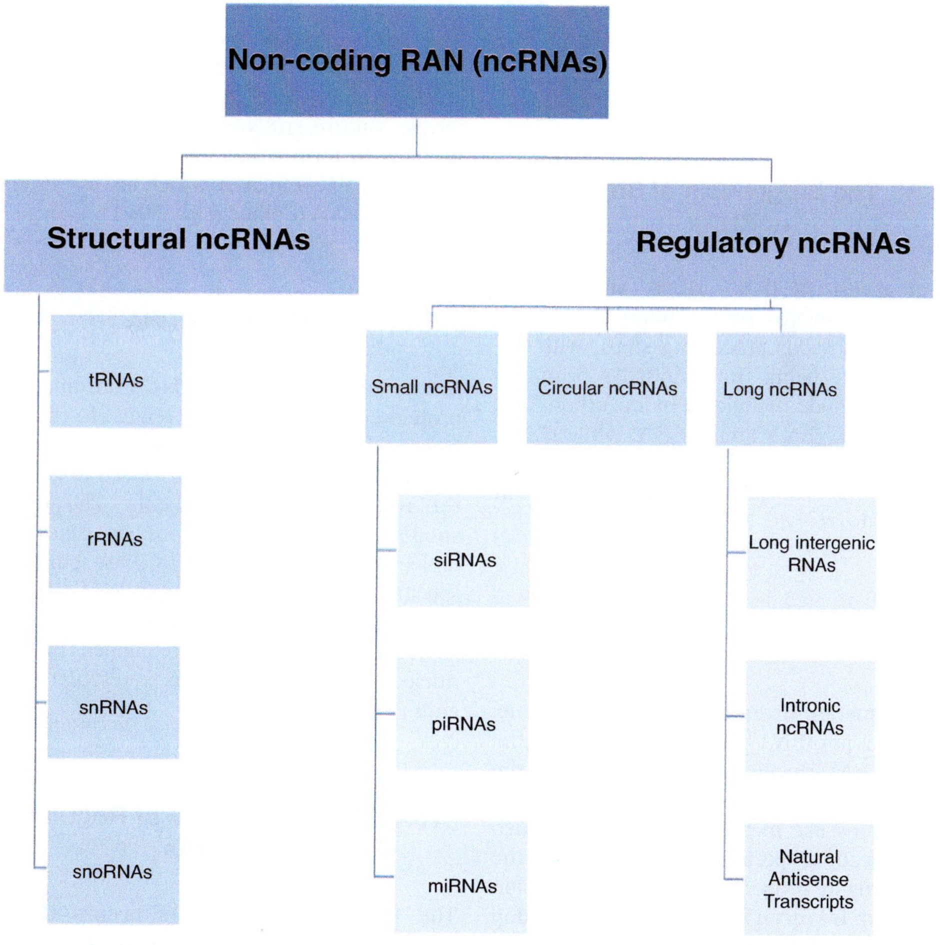

Fig. 16.1. Broad classification of non-coding RNAs.

and ribosomal activities and are continuously expressed. This group encompasses transfer RNAs (tRNAs), ribosomal RNAs (rRNAs), small nuclear RNAs (snRNAs), and small nucleolar RNAs (snoRNAs). Regulatory ncRNAs comprise microRNAs (miRNAs), small interfering RNAs (siRNAs), piwi-interacting RNAs (piRNAs), and long non-coding RNAs (lncRNAs) (Morris and Mattick, 2014) (Fig. 16.1). Of these, lncRNAs are a subclass of ncRNAs longer than 200 nucleotides that lack an open reading frame. Present across a wide range of organisms, they are expressed in a temporally and spatially specific manner. Although derived from DNA, these regulatory ncRNAs do not encode proteins (Brosnan and Voinnet, 2009). Nevertheless, they play vital roles in plant growth, development, and responses to abiotic stress at both transcriptional and post-transcriptional levels (D'Ario *et al.*, 2017).

16.3 The Biogenesis of Structural ncRNAs

The biogenesis of tRNA, rRNA, snRNA, and snoRNA all involve transcription from DNA followed by various processing steps, with the key difference being that snoRNAs primarily function to guide modifications on other RNA molecules like rRNA and snRNA, while tRNA, rRNA, and snRNA are directly involved in protein synthesis or splicing mechanisms within the cell (Alberts *et al.*, 2002).

16.3.1 tRNA

RNA polymerase III transcribes the tRNA gene, creating a pre-tRNA molecule. To create the mature tRNA form, enzymes precisely cleave the pre-tRNA at both the 5′ and 3′ ends. Chemical modifications like methylation and pseudouridylation occur at specific nucleotides within the tRNA to ensure proper function. The anticodon loop, which is crucial for recognizing the codon on mRNA, is formed through the structural folding of the tRNA molecule (Phizicky and Hopper, 2023).

16.3.2 rRNA

The precursor of the large rRNA is transcribed by RNA polymerase I as a single continuous molecule. In the process, snoRNAs guide specific modifications like methylation and pseudouridylation on the pre-rRNA to ensure correct ribosome assembly. The pre-rRNA is cleaved into mature rRNA subunits (18S, 5.8S, 25S/28S, depending on the organism) by endonucleases. Ribosomal proteins bind to the mature rRNA subunits to form the functional ribosome (Böğürcü-Seidel *et al.*, 2023).

16.3.3 snRNA

The snRNA genes are transcribed by RNA polymerase II in the nucleus, and some of these snRNAs are spliced to produce their mature forms. Mature snRNAs combine with specific proteins to form small nuclear ribonucleoprotein (snRNP) complexes, which are essential for splicing mRNA (Collins *et al.*, 2011).

16.3.4 snoRNA

Most snoRNAs are transcribed from introns of protein-coding genes, often related to ribosome biogenesis. The snoRNA precursor is spliced out of the intron and further processed to generate the mature snoRNA molecule. The mature snoRNA guides modifications on other RNA molecules (mainly rRNA) by base pairing to specific sequences on the target RNA (Wajahat *et al.*, 2021). The enhanced stability of snoRNAs is because they have been capped by small nucleolar ribonucleoproteins (snoRNPs) at both ends (Wilusz *et al.*, 2008).

16.4 The Biogenesis of Regulatory ncRNAs

The formation of ncRNAs occurs through two primary pathways, canonical and non-canonical (Fig. 16.2). Each of these pathways follows unique mechanisms and is essential for

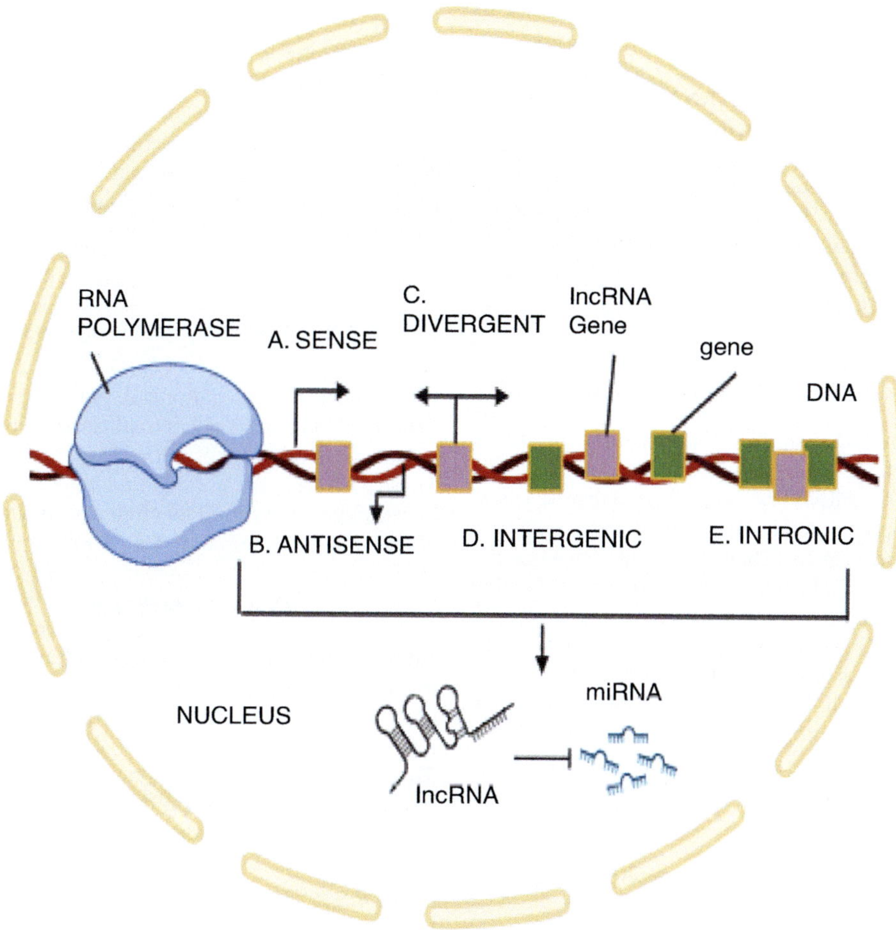

Fig. 16.2. Biogenesis of non-coding RNAs, including (A) sense; (B) antisense; (C) divergent; (D) intergenic; (E) intronic.

regulating gene expression in plants (Asghar *et al.*, 2024).

16.4.1 Canonical biogenesis

The canonical pathway refers to the traditional and well-characterized processes through which various types of ncRNAs, particularly miRNAs and siRNAs, are synthesized. This pathway primarily involves the transcription of long primary RNA transcripts, which are then processed into functional ncRNAs (Hassan and Khan, 2024).

16.4.1.1 MicroRNA (miRNA)

The biogenesis of miRNAs starts with the transcription of primary miRNA transcripts (pri-miRNAs) by RNA polymerase II. These transcripts feature characteristic hairpin-loop structures essential for their processing. A protein complex comprising DICER-LIKE 1 (DCL1), Hyponastic Leaves 1 (HYL1), and SERRATE (SE) cleaves the pri-miRNAs into precursor miRNAs (pre-miRNAs). This step is critical for the production of mature miRNA duplexes. The pre-miRNAs are subsequently exported from the nucleus to the cytoplasm, where they undergo further processing by either DCL1 or DCL4 to form mature miRNA–miRNA* duplexes. The guide strand (mature miRNA) is then selectively incorporated into the RNA-induced silencing complex (RISC), allowing the complex to recognize and bind to complementary target mRNA transcripts. Gene regulation is achieved mainly through mRNA cleavage or translational inhibition. The passenger strand (miRNA*), which is not integrated into RISC, is usually degraded (Kanwal *et al.*, 2023).

16.4.1.2 Small interfering RNA (siRNA)

Like miRNAs, small siRNAs are also produced from long RNA precursors. However, unlike miRNAs, siRNAs can arise from a diverse range of sources, such as transposable elements and viral RNAs. Their biogenesis involves the generation of double-stranded RNA (dsRNA) intermediates, which are then processed by DCL proteins into siRNA duplexes, typically 21 or 24 nucleotides in length. These siRNA duplexes are subsequently loaded into the RISC, where they guide the complex to complementary target RNAs to induce gene silencing (Sanan-Mishra *et al.*, 2021).

16.4.1.3 Natural antisense RNAs (natsiRNAs)

Forming another class of ncRNAs, natsiRNAs are derived from overlapping or complementary genomic loci. They are generated through the action of polymerase II or IV, forming dsRNA intermediates that are processed into functional natsiRNAs. These ncRNAs regulate gene expression by binding to complementary mRNA sequences, leading to their degradation or suppression of translation (Singh *et al.*, 2022).

16.4.1.4 Trans-acting siRNAs (tasiRNAs)

A similar process utilizing dsRNA intermediates produces tasiRNAs, which are formed from certain tasiRNA genes. They are processed by DCL proteins and function in a manner analogous to miRNAs, targeting mRNAs for silencing (Voloudakis *et al.*, 2021).

16.4.1.5 Heterochromatic siRNAs (hc-siRNAs)

The hc-siRNAs originate from the heterochromatic regions of the genome and play a key role in chromatin remodeling. Their biogenesis is initiated by RNA polymerase IV and RNA-dependent RNA polymerase 2, resulting in the production of 24-nucleotide-long hc-siRNAs. (Ali and Tang, 2024).

16.4.2 Non-canonical biogenesis

In contrast to the canonical pathway, non-canonical biogenesis refers to alternative mechanisms through which ncRNAs are generated. These pathways are less well understood but are increasingly recognized for their significance in plant biology (Bhogireddy *et al.*, 2021).

16.4.2.1 Mirtrons

One of the notable non-canonical pathways involves the generation of mirtrons, which are pre-miRNA-like structures formed through splicing rather than the classical DCL1 processing. This splicing mechanism allows for the direct production of mature miRNAs from intronic regions of mRNAs (Panda and Prajapati, 2024).

16.4.2.2 Dicer-independent pathways

Some ncRNAs are produced through dicer-independent mechanisms, which do not rely on the DCL proteins for processing. These pathways can involve the direct back-splicing of precursor mRNAs to form circular RNAs (circRNAs), which are characterized by their covalently closed structures. circRNAs can function as miRNA sponges by binding to miRNAs and preventing them from interacting with their target mRNAs, thus influencing the regulation of gene expression (Holvoet and Holvoet, 2021).

16.4.2.3 Alternative processing of pri-miRNAs

Non-canonical pathways may also include the alternative processing of pri-miRNAs, in which the cleavage proceeds in the opposite direction from the loop toward the base of the hairpin structure. This can result in the generation of multiple miRNA duplexes rather than a single one (Panda and Prajapati, 2024).

16.4.2.4 Diverse ncRNA classes

Non-canonical RNA biogenesis involves several types of ncRNAs, including lncRNAs and circRNAs. Of these, lncRNAs are categorized according to their genomic origin, such as intergenic, intronic, or antisense regions, and they perform a wide range of functions in regulating gene expression at both the transcriptional and post-transcriptional stages (Lakhotia *et al.*, 2020). Additionally, lncRNAs can be grouped based on their location within the cell, including cytoplasmic, nuclear, and mitochondrial types (Alessio *et al.*, 2020).

16.5 Functions of ncRNAs

The ncRNAs play vital roles in regulating numerous biological processes in plants. Unlike protein-coding genes, ncRNAs do not produce proteins; instead, they function at the molecular level to control gene expression and cellular activities. They influence the expression of genes that govern plant growth, development, and responses to environmental stresses such as drought, salinity, heat, and cold. Through the regulation of stress-responsive genes, they contribute to plant adaptation mechanisms (Statello *et al.*, 2021). Additionally, ncRNAs interact not only with one another but also with other RNA types, including mRNAs, to establish intricate regulatory networks. These networks are crucial for fine-tuning gene expression and supporting proper development across various plant tissues such as roots, leaves, and seeds, particularly in the regulation of growth and nutrient dynamics. Furthermore, certain ncRNAs participate in epigenetic regulation, affecting gene activity without modifying the DNA sequence, and thereby contributing to sustained gene expression control (Liang *et al.*, 2020).

Some of the ncRNAs, such as rRNA and tRNA, play essential roles in the process of protein synthesis (Lodish *et al.*, 2005; Bhagavan and Ha, 2015). The most abundant RNAs in the cell, rRNAs, combine with ribosomal proteins to form ribosomes, which are responsible for translating genetic information into proteins using amino acids. In addition, tRNAs contribute to this process by interpreting the mRNA sequence and delivering the corresponding amino acids for polypeptide chain formation. Catalytic RNAs, known as ribozymes, are involved in generating mature tRNAs; for example, ribonuclease P is a ribozyme that cleaves pre-tRNA to produce its mature form. Mature tRNAs can also be processed into tRNA-derived fragments, which are evolutionarily conserved and can regulate gene expression either through the RNA interference (RNAi) pathway or by inhibiting protein synthesis (Keam and Hutvagner, 2015). RNA splicing and processing depend on snRNAs, which are mostly found in the nucleus. Usually acting as ribonucleoprotein particles, these snRNAs can also combine with other proteins, such as the spliceosome, to create complexes. The snoRNAs, meanwhile, are involved in rRNA processing and are localized in the nucleolus. The following sections explore specific regulatory roles of ncRNAs in plants.

16.5.1 miRNAs

The miRNA subgroup constitutes a significant class of regulatory ncRNAs, generally 20–24 nucleotides long, that play crucial roles in post-transcriptional regulation of gene expression. They are involved in a wide range of biological functions, including development, stress responses, and hormone signaling (Li *et al.*, 2024). miRNAs regulate gene activity by targeting transcription factors and various genes, contributing to the formation of intricate gene regulatory networks that control plant developmental processes (Das *et al.*, 2015). These small RNAs are particularly important in managing developmental transitions, such as the shift from vegetative to reproductive growth. For example, the miR156 family influences flowering time by targeting SPL (SQUAMOSA PROMOTER BINDING PROTEIN-LIKE) genes, which play a key role in floral development (Tripathi *et al.*, 2024). Different miRNA-target modules function at various stages of plant development, acting either as enhancers or repressors. In *Arabidopsis*, the miR172-AP2 module promotes transitions from germination to vegetative and then to reproductive stages, while the miR156-SPL module suppresses these transitions (Das *et al.*, 2015).

In addition, miRNAs help plants respond to environmental challenges such as drought, salinity, and pathogen invasion. For instance, miR393 modulates auxin signaling by targeting auxin receptors, promoting root growth during drought stress (Liang *et al.*, 2024). Likewise, miR162 plays a role in regulating genes involved in defense against pathogens (Tiwari, 2024). By regulating the expression of hormone-responsive genes, these short RNAs also play a significant role in hormonal signaling, affecting how plants react to hormones such as auxins, gibberellins, and abscisic acid. This affects both growth and stress adaptation (Singh *et al.*, 2021). Research further indicates that miRNAs can coordinate the regulation of multiple biological functions synergistically (Alptekin *et al.*, 2017).

16.5.2 lncRNAs

The lncRNA group are a diverse category of ncRNAs longer than 200 nucleotides, recognized as important regulators of gene expression in plants. Their functions span transcriptional control, epigenetic modifications, and the regulation of RNA splicing and stability (Bhattacharyya *et al.*, 2021). The functional mechanisms of lncRNAs can be grouped into five main categories based on their molecular interactions and outcomes: interactions with proteins, DNA, and RNA, peptide-coding potential, and catalytic activity of the lncRNA itself (Wang and Chang, 2011). Indeed, lncRNAs can associate with chromatin-modifying proteins to influence the transcription of neighboring genes. Certain lncRNAs, for instance, can attract histone methyltransferases to certain genomic loci, changing chromatin states and thus altering gene expression (Gill *et al.*, 2024). In maize, lncRNAs have been linked to drought stress responses, with 1535 lncRNAs identified at different developmental stages during stress exposure (Pang *et al.*, 2019). These molecules contribute to epigenetic gene regulation by guiding the formation of repressive chromatin configurations, thus playing a role in both developmental control and stress adaptation (Nickerson and Momen-Heravi, 2024). Additionally, lncRNAs can affect the splicing and stability of messenger RNAs by interacting with splicing factors, thereby influencing alternative splicing events and generating diverse protein isoforms (Palos *et al.*, 2024).

16.5.3 siRNAs

A class of regulatory ncRNAs, siRNAs typically range from 21 to 24 nucleotides in length, and play a key role in gene silencing and plant defense mechanisms. Their primary function is in post-transcriptional gene silencing through the RNA interference (RNAi) pathway, where they direct the degradation of specific messenger RNAs to regulate gene expression. Maintaining genomic stability and managing transposable element activity depend on this process (Ding, 2023). siRNAs are crucial for plant defense because they help plants fight off viral infections. They restrict viral replication and dissemination inside plant tissues by targeting viral genomes for destruction. They are made from viral RNA. Additionally, siRNAs are instrumental in silencing transposons, preventing their movement within the genome and thus safeguarding genomic integrity. This regulatory role is crucial for maintaining the stability of the plant genome and reducing the risk of mutation (Hung and Slotkin, 2021).

16.5.4 circRNAs

The circRNAs represent a newly identified class of ncRNAs, distinguished by their covalently closed-loop structures. They are produced through a process called back-splicing of precursor mRNAs and play specialized regulatory roles, including acting as molecular sponges, modulating gene expression, and participating in stress responses. One of their key functions involves sequestering miRNAs, thereby preventing these miRNAs from interacting with their target mRNAs. By modulating miRNA availability, circRNAs can indirectly influence gene expression. Additionally, circRNAs can bind to RNA-binding proteins, affecting their function and stability, which in turn can regulate downstream gene expression and various cellular activities (Ma *et al.*, 2024). According to recent research, circRNAs also influence how plants react to environmental stressors. They appear to be involved in controlling the expression of stress-responsive genes, supporting the plant's capacity to adapt to environmental challenges (Pradhan *et al.*, 2024).

16.6 Methods for Analyzing the Structural Configuration of ncRNAs

Recent advancements in the analysis of long noncoding RNAs (lncRNAs) have significantly enhanced our understanding of their structural configurations and functional roles. Given the complexity and variability of ncRNA structures, researchers have developed a variety of experimental and computational methods to elucidate their secondary and tertiary structures. The approaches, highlighting their methodologies, advantages, and limitations, are as follows.

16.6.1 Experimental methods

16.6.1.1 Enzymatic and chemical probing

Enzymatic and chemical probing techniques are foundational methods for studying RNA structures (Gilmer *et al.*, 2021). By using particular chemicals or enzymes to selectively break RNA at particular locations, these methods enable researchers to deduce structural details from the resultant pieces. Ribonucleases (RNases) are used in the enzymatic footprinting process to cleave RNA molecules (Ziehler and Engelke, 2000). Different RNases have varying specificities for single-stranded or double-stranded regions, enabling the mapping of structural features. The cleaved products are then analyzed using denaturing polyacrylamide gel electrophoresis, providing insights into the conformation of the RNA. An advancement of enzymatic footprinting, parallel analysis of RNA structure combines enzymatic digestion with next-generation sequencing. This technique allows for a transcriptome-wide analysis of RNA structures, providing a comprehensive view of RNA conformations across different conditions (Piao *et al.*, 2017).

16.6.1.2 Nuclear magnetic resonance (NMR) spectroscopy

NMR spectroscopy is a powerful tool for determining the three-dimensional structures of RNA molecules in solution. This technique provides atomic-level resolution and can capture dynamic conformational changes. However, NMR is typically limited to smaller RNA molecules due to the complexity of larger structures, which can lead to overlapping signals and difficulties in interpretation (Zhao *et al.*, 2022).

16.6.1.3 X-ray crystallography

While primarily used for proteins, X-ray crystallography can also be applied to RNA structures. This technique necessitates the crystallization of RNA, a process often complicated by the natural flexibility and instability of the molecule. However, when achieved, X-ray crystallography offers detailed, high-resolution structural information (Jackson *et al.*, 2023).

16.6.1.4 Cryo-electron microscopy (cryo-EM)

Cryo-EM has become an important tool for analyzing large RNA–protein assemblies and lncRNAs. It enables the observation of RNA structures in conditions close to their natural state, eliminating the requirement for crystallization. Cryo-EM can capture multiple conformations of RNA, providing insights into their dynamic nature (McRae *et al.*, 2024).

16.6.2 Computational methods

16.6.2.1 RNA structure prediction algorithms

With the exponential growth of RNA sequence data, computational methods for predicting RNA structures have become increasingly important. These algorithms utilize known RNA structures to model the folding of new sequences (Seetin and Mathews, 2012). Secondary structure prediction tools like RNAfold (Gruber *et al.*, 2008) and Mfold (Zuker, 2003) predict secondary structures based on thermodynamic stability. These predictions can guide experimental validation and provide a starting point for understanding ncRNA functions. Tertiary structure prediction with more advanced algorithms, such as those predicted by a transformer network, aims to predict the three-dimensional conformations of RNA. These methods can incorporate experimental data as constraints to improve accuracy (Wang *et al.*, 2023).

16.6.3 Hybrid approaches

Hybrid approaches combine experimental and computational methods and have proven effective in overcoming the limitations of each approach. For instance, experimental data from enzymatic probing can be used to refine computational models, leading to more accurate predictions of ncRNA structures. This integrative strategy enhances the reliability of structural insights and facilitates the exploration of lncRNA functions (Ponce-Salvatierra *et al.*, 2019).

16.7 Challenges and Future Directions

Despite the progress made in ncRNA structural analysis, several challenges remain. The inherent flexibility and dynamic nature of ncRNAs complicate the determination of stable structures. Additionally, the lack of conservation in ncRNA sequences across species makes it difficult to predict their functions based solely on sequence information. Future research focuses on developing more sophisticated computational models that can account for the dynamic behavior of ncRNAs. Integrating machine learning techniques with structural data may enhance predictive capabilities and facilitate the identification of novel ncRNA functions. Furthermore, new developments in CRISPR/Cas9 technologies have intriguing prospects for ncRNA functional research (Ansori *et al.*, 2023). By introducing targeted modifications in ncRNA gene loci, researchers can explore the impact of specific structural domains on ncRNA functions, leading to a deeper understanding of their roles in cellular processes.

16.8 Conclusion

Plants, being stationary organisms, cannot escape unfavorable conditions and must instead rely on avoidance and tolerance strategies to endure constantly changing environments. In recent years, research on ncRNAs has surged, establishing them as crucial bioactive molecules that contribute to both genomic and phenotypic diversity. Since their discovery, ncRNAs have been classified based on their origin, mode of biogenesis, and functional mechanisms. The production of ncRNAs, through both canonical and non-canonical pathways, is a dynamic and complex process that plays a critical role in gene regulation in plants. Gaining a deeper understanding of these biogenetic routes is vital for unraveling the intricacies of plant biology and for identifying potential applications in crop improvement and stress resilience. As scientific exploration advances, the discovery and functional analysis of diverse ncRNA types are expected to continue, shedding light on their pivotal roles in plant growth and adaptation. Numerous biological processes, including development, stress responses, and hormone control, are influenced by regulatory ncRNAs, which are a crucial component of the complex gene regulatory networks in plants. The ongoing exploration of ncRNA functions underscores their importance in enabling plants to adapt and thrive in fluctuating environments. Furthermore, a comprehensive understanding of regulatory ncRNAs not only enriches our knowledge of plant systems but also offers promising pathways for enhancing crop traits and stress tolerance through biotechnology. In parallel, the study of ncRNA structure is emerging rapidly, fueled by advances in both experimental and computational techniques. Integrating these approaches enables scientists to gain meaningful insights into the three-dimensional configurations of ncRNAs and their corresponding functional roles. As our comprehension of ncRNA biology deepens, these structural and functional investigations will be key to revealing the molecular mechanisms behind gene regulation. Ultimately, this integration will support the discovery of novel ncRNA functions and contribute to the development of innovative strategies for crop improvement and potential therapeutic applications.

Acknowledgements

All the authors are thankful to their respective universities and institutions for their valuable support.

References

Alberts, B., Johnson, A., Lewis, J., Martin, R., Keith, R. *et al.* (2002) *Molecular Biology of the Cell*, 4th edn. Garland Science, New York.

Alessio, E., Bonadio, R.S., Buson, L., Chemello, F. and Cagnin, S. (2020) A single cell but many different transcripts: A journey into the world of long non-coding RNAs. *International Journal of Molecular Sciences* 21(1), 302. DOI: 10.3390/ijms21010302.

Ali, S. and Tang, Y. (2024) Noncoding RNA-mediated regulation of DNA methylation: Insights into plant epigenetic mechanisms. *Journal of Plant Growth Regulation* 44(2), 373–388. DOI: 10.1007/s00344-024-11462-0.

Alptekin, B., Langridge, P. and Budak, H. (2017) Abiotic stress miRNomes in the triticeae. *Functional & Integrative Genomics* 17(2–3), 145–170. DOI: 10.1007/s10142-016-0525-9.

Ansori, A.N., Antonius, Y., Susilo, R.J., Hayaza, S., Kharisma, V.D. *et al.* (2023) Application of CRISPR-Cas9 genome editing technology in various fields: A review. *Narra J* 3(2), e184. DOI: 10.52225/narra.v3i2.184.

Asghar, A., Ullah, Z., Ali, T., Iqbal, J., Abbasi, B.A. *et al.* (2024) Introduction to nc-RNA and their crucial role in biological regulation under environmental stress. In: *ncRNAs: Mediated Regulation: The Unarguable Biological Function to Environmental Constrains*. Springer Nature Switzerland, Cham, pp. 1–46. DOI: 10.1007/978-3-031-69354-0_1.

Bhagavan, N.V. and Ha, C.E. (2015) *Essentials of medical biochemistry. With Clinical Cases*, 2nd edn. Academic Press, London, UK.

Bhattacharyya, N., Pandey, V., Bhattacharyya, M. and Dey, A. (2021) Regulatory role of long non coding RNAs (lncRNAs) in neurological disorders: From novel biomarkers to promising therapeutic strategies. *Asian Journal of Pharmaceutical Sciences* 16(5), 533–550. DOI: 10.1016/j.ajps.2021.02.006.

Bhogireddy, S., Mangrauthia, S.K., Kumar, R., Pandey, A.K., Singh, S. *et al.* (2021) Regulatory non-coding RNAs: A new frontier in regulation of plant biology. *Functional & Integrative Genomics* 21(3–4), 313–330. DOI: 10.1007/s10142-021-00787-8.

Böğürcü-Seidel, N., Ritschel, N., Acker, T. and Németh, A. (2023) Beyond ribosome biogenesis: Noncoding nucleolar RNAs in physiology and tumor biology. *Nucleus (Austin, Tex.)* (1), 2274655. DOI: 10.1080/19491034.2023.2274655.

Brosnan, C.A. and Voinnet, O. (2009) The long and the short of noncoding RNAs. *Current Opinion in Cell Biology* 21(3), 416–425. DOI: 10.1016/j.ceb.2009.04.001.

Collins, L.J., Schönfeld, B. and Chen, X.S. (2011) The epigenetics of non-coding RNA. *Handbook of Epigenetics* 49–61. DOI: 10.1016/B978-0-12-375709-8.00004-6.

D'Ario, M., Griffiths-Jones, S. and Kim, M. (2017) Small RNAs: Big impact on plant development. *Trends in Plant Science* 22(12), 1056–1068. DOI: 10.1016/j.tplants.2017.09.009.

Das, S.S., Karmakar, P., Nandi, A.K. and Sanan-Mishra, N. (2015) Small RNA mediated regulation of seed germination. *Frontiers in Plant Science* 6, 828. DOI: 10.3389/fpls.2015.00828.

Ding, S.W. (2023) Transgene silencing, RNA interference, and the antiviral defense mechanism directed by small interfering RNAs. *Phytopathology®* 113(4), 616–625. DOI: 10.1094/PHYTO-10-22-0358-IA.

Gill, S.S., Khan, N.A., Agarwala, N., Singh, K., Sunkar, R. *et al.* (2024) ncRNAs in plant development and stress responses. *Plant Physiology and Biochemistry* 214, 108950. DOI: 10.1016/j.plaphy.2024.108950.

Gilmer, O., Quignon, E., Jousset, A.C., Paillart, J.C., Marquet, R. *et al.* (2021) Chemical and enzymatic probing of viral RNAs: From infancy to maturity and beyond. *Viruses* 13(10), 1894. DOI: 10.3390/v13101894.

Gong, C. and Maquat, L.E. (2011) lncRNAs transactivate STAU1-mediated mRNA decay by duplexing with 3' UTRs via alu elements. *Nature* 470(7333), 284–288. DOI: 10.1038/nature09701.

Gruber, A.R., Lorenz, R., Bernhart, S.H., Neuböck, R. and Hofacker, I.L. (2008) The vienna RNA websuite. *Nucleic Acids Research* 36(Web Server issue), W70–W74. DOI: 10.1093/nar/gkn188.

Hassan, Z. and Khan, G. (2024) Small interfering RNA. In: *Molecular Techniques for Studying Viruses: Practical Notes*. Springer Nature, Singapore, pp. 51–57. DOI: 10.1007/978-981-99-8097-0_8.

Holley, R.W., Apgar, J., Everett, G.A., Madison, J.T., Marquisee, M. *et al.* (1965) Structure of a ribonucleic acid. *Science* 147(3664), 1462–1465. DOI: 10.1126/science.147.3664.1462.

Holvoet, P. and Holvoet, P. (2021) Biogenesis and modes of action of miRs and circular and long non-coding RNAs. In: *Non-Coding RNAs at the Cross-Road of Cardiometabolic Diseases and Cancer*. Springer International Publishing, Cham, pp. 1–19. DOI: 10.1007/978-3-030-68844-8_1.

Hung, Y.H. and Slotkin, R.K. (2021) The initiation of RNA interference (RNAi) in plants. *Current Opinion in Plant Biology* 61, 102014. DOI: 10.1016/j.pbi.2021.102014.

Jackson, R.W., Smathers, C.M. and Robart, A.R. (2023) General strategies for RNA X-ray crystallography. *Molecules* 28(5), 2111. DOI: 10.3390/molecules28052111.

Kanwal, N., Al Samarrai, O.R., Al-Zaidi, H.M.H., Mirzaei, A.R. and Heidari, M.J. (2023) Comprehensive analysis of microRNA (miRNA) in cancer cells. *Cellular, Molecular and Biomedical Reports* 3(2), 89–97. DOI: 10.55705/cmbr.2022.364591.1070.

Keam, S.P. and Hutvagner, G. (2015) TRNA-derived fragments (tRFs): Emerging new roles for an ancient RNA in the regulation of gene expression. *Life* 5(4), 1638–1651. DOI: 10.3390/life5041638.

Khan, I., Khan, S., Akhoundian, M., Shah, D., Shah, S.S. *et al.* (2023) Biogenesis of non-coding RNAs (ncRNAs) and their biological role in rice (*Oryza sativa* L.). *Plant Molecular Biology Reporter* 41(3), 333–344. DOI: 10.1007/s11105-023-01371-z.

Lakhotia, S.C., Mallick, B. and Roy, J. (2020) Non-coding RNAs: Ever-expanding diversity of types and functions. *RNA-Based Regulation in Human Health and Disease* 5–57. DOI: 10.1016/B978-0-12-817193-6.00002-9.

Lee, R.C., Feinbaum, R.L. and Ambros, V. (1993) The c. elegans heterochronic gene lin-4 encodes small RNAs with antisense complementarity to lin-14. *Cell* 75(5), 843–854. DOI: 10.1016/0092-8674(93)90529-y.

Liang, Y., Yang, X., Wang, C. and Wang, Y. (2024) MiRNAs: Primary modulators of plant drought tolerance. *Journal of Plant Physiology* 301, 154313. DOI: 10.1016/j.jplph.2024.154313.

Liang, Z., Riaz, A., Chachar, S., Ding, Y., Du, H. *et al.* (2020) Epigenetic modifications of mRNA and DNA in plants. *Molecular Plant* 13(1), 14–30. DOI: 10.1016/j.molp.2019.12.007.

Li, Z., Yang, J., Cai, X., Zeng, X., Zou, J.J. *et al.* (2024) Withdrawn: A review on the function of miRNAs in plant responses to stresses. *Plant Stress* 100347. DOI: 10.1016/j.stress.2024.100347.

Lodish, H., Berk, A., Matsudaira, P., Kaiser, C.A., Krieger, M. *et al.* (2005) *Molecular Cell Biology*, 5th edn. W.H. Freeman and Co, New York.

Ma, Y., Wang, T., Zhang, X., Wang, P. and Long, F. (2024) The role of circular RNAs in regulating resistance to cancer immunotherapy: Mechanisms and implications. *Cell Death & Disease* 15(5), 312. DOI: 10.1038/s41419-024-06698-3.

McRae, E.K.S., Wan, C.J.K., Kristoffersen, E.L., Hansen, K., Gianni, E. *et al.* (2024) Cryo-EM structure and functional landscape of an RNA polymerase ribozyme. *Proceedings of the National Academy of Sciences* 121(3), e2313332121. DOI: 10.1073/pnas.2313332121.

Morris, K.V. and Mattick, J.S. (2014) The rise of regulatory RNA. *Nature Reviews Genetics* 15(6), 423–437. DOI: 10.1038/nrg3722.

Nickerson, J.A. and Momen-Heravi, F. (2024) Long non-coding RNAs: Roles in cellular stress responses and epigenetic mechanisms regulating chromatin. *Nucleus* 15(1), 2350180. DOI: 10.1080/19491034.2024.2350180.

Palos, K., Nelson Dittrich, A.C., Lyons, E.H., Gregory, B.D. and Nelson, A.D.L. (2024) Comparative analyses suggest a link between mRNA splicing, stability, and RNA covalent modifications in flowering plants. *BMC Plant Biology* 24(1), 768. DOI: 10.1186/s12870-024-05486-7.

Panda, M. and Prajapati, V.K. (2024) Paradigms in miRNA biogenesis pathways. In: *MicroRNA in Human Infectious Diseases*. Academic Press, London, United Kingdom, pp. 3–21. DOI: 10.1016/B978-0-323-99661-7.00003-5.

Pang, J., Zhang, X., Ma, X. and Zhao, J. (2019) Spatio-temporal transcriptional dynamics of maize long non-coding RNAs responsive to drought stress. *Genes* 10(2), 138. DOI: 10.3390/genes10020138.

Phizicky, E.M. and Hopper, A.K. (2023) The life and times of a tRNA. *Rna* 29(7), 898–957. DOI: 10.1261/rna.079620.123.

Piao, M., Sun, L. and Zhang, Q.C. (2017) RNA regulations and functions decoded by transcriptome-wide RNA structure probing. *Genomics, Proteomics & Bioinformatics* 15(5), 267–278. DOI: 10.1016/j.gpb.2017.05.002.

Ponce-Salvatierra, A.Astha0.0Merdas, K., Nithin, C., Ghosh, P. *et al*. (2019) Computational modeling of RNA 3D structure based on experimental data. *Bioscience Reports* 39(2), BSR20180430. DOI: 10.1042/BSR20180430.

Pradhan, U.K., Behera, P., Das, R., Naha, S., Gupta, A. *et al*. (2024) AScirRNA: A novel computational approach to discover abiotic stress-responsive circular RNAs in plant genome. *Computational Biology and Chemistry* 113, 108205. DOI: 10.1016/j.compbiolchem.2024.108205.

Sanan-Mishra, N., Abdul Kader Jailani, A., Mandal, B. and Mukherjee, S.K. (2021) Secondary siRNAs in plants: Biosynthesis, various functions, and applications in virology. *Frontiers in Plant Science* 12, 610283. DOI: 10.3389/fpls.2021.610283.

Seetin, M.G. and Mathews, D.H. (2012) RNA structure prediction: An overview of methods. *Bacterial Regulatory RNA: Methods and Protocols* 99–122. DOI: 10.1007/978-1-61779-949-5_8.

Singh, P., Dutta, P. and Chakrabarty, D. (2021) MiRNAs play critical roles in response to abiotic stress by modulating cross-talk of phytohormone signaling. *Plant Cell Reports* 40(9), 1617–1630. DOI: 10.1007/s00299-021-02736-y.

Singh, R.K., Prasad, A., Maurya, J. and Prasad, M. (2022) Regulation of small RNA-mediated high temperature stress responses in crop plants. *Plant Cell Reports* 41(3), 765–773. DOI: 10.1007/s00299-021-02745-x.

Statello, L., Guo, C.J., Chen, L.L. and Huarte, M. (2021) Gene regulation by long non-coding RNAs and its biological functions. *Nature Reviews Molecular Cell Biology* 22(2), 96–118. DOI: 10.1038/s41580-020-00315-9.

Tiwari, M. (2024) The role of microRNA in stress signaling and adaptive response in plants. In: *Stress Biology in Photosynthetic Organisms: Molecular Insights and Cellular Responses*. Springer Nature, Singapore, pp. 87–106. DOI: 10.1007/978-981-97-1883-2_5.

Tripathi, R.K., Bekele, W.A., Tinker, N.A. and Singh, J. (2024) Differential expression and global analysis of miR156/SQUAMOSA promoter binding-like proteins (SPL) module in oat. *Scientific Reports* 14(1), 9928. DOI: 10.1038/s41598-024-60739-7.

Voloudakis, A.E., Mukherjee, S.K. and Roy, A. (2021) Novel technologies for transgenic management for plant virus resistance. In: *Genome Engineering for Crop Improvement*. Springer International Publishing, Cham, pp. 163–191. DOI: 10.1007/978-3-030-63372-1_7.

Wajahat, M., Bracken, C.P. and Orang, A. (2021) Emerging functions for snoRNAs and snoRNA-derived fragments. *International Journal of Molecular Sciences* 22(19), 10193. DOI: 10.3390/ijms221910193.

Wang, K.C. and Chang, H.Y. (2011) Molecular mechanisms of long noncoding RNAs. *Molecular Cell* 43(6), 904–914. DOI: 10.1016/j.molcel.2011.08.018.

Wang, W., Feng, C., Han, R., Wang, Z., Ye, L. *et al*. (2023) trRosettaRNA: Automated prediction of RNA 3D structure with transformer network. *Nature Communications* 14(1), 7266. DOI: 10.1038/s41467-023-42528-4.

Wilusz, J.E., Freier, S.M. and Spector, D.L. (2008) 3' end processing of a long nuclear-retained noncoding RNA yields a tRNA-like cytoplasmic RNA. *Cell* 135(5), 919–932. DOI: 10.1016/j.cell.2008.10.012.

Yadav, A., Mathan, J., Dubey, A.K. and Singh, A. (2024) The emerging role of non-coding RNAs (ncRNAs) in plant growth, development, and stress response signaling. *Non-Coding RNA* 10(1), 13. DOI: 10.3390/ncrna10010013.

Yu, Y., Zhang, Y., Chen, X. and Chen, Y. (2019) Plant noncoding RNAs: Hidden players in development and stress responses. *Annual Review of Cell and Developmental Biology* 35(1), 407–431. DOI: 10.1146/annurev-cellbio-100818-125218.

Zhao, S., Li, X., Wen, Z., Zou, M., Yu, G. *et al*. (2022) Dynamics of base pairs with low stability in RNA by solid-state nuclear magnetic resonance exchange spectroscopy. *Iscience* 25(11), 105322. DOI: 10.1016/j.isci.2022.105322.

Ziehler, W.A. and Engelke, D.R. (2000) Probing RNA structure with chemical reagents and enzymes. *Current Protocols in Nucleic Acid Chemistry* 00(1), 6–1. DOI: 10.1002/0471142700.nc0601s00.

Zucchelli, S., Cotella, D., Takahashi, H., Carrieri, C., Cimatti, L. *et al*. (2015) SINEUPs: A new class of natural and synthetic antisense long non-coding RNAs that activate translation. *RNA Biology* 12(8), 771–779. DOI: 10.1080/15476286.2015.1060395.

Zuker, M. (2003) Mfold web server for nucleic acid folding and hybridization prediction. *Nucleic Acids Research* 31(13), 3406–3415. DOI: 10.1093/nar/gkg595.

17 Spatial and Multiple Omics in Medicinal Plants: Advances and Prospects

Rahul Gogoi[1], Bhoirob Gogoi[2], Ishani Borthakur[1], Khomdram Niren Singh[2], Sudin Kumar Pandey[2] and Madhumita Barooah[1,2]*

[1]Department of Agricultural Biotechnology, Assam Agricultural University, Jorhat, India; [2]DBT-NECAB, Assam Agricultural University, Jorhat, India

17.1 Introduction

Medicinal plants represent a concentrated source of chemical diversity, which has offered a vast array of highly specialized metabolites. These metabolites, which are also known as natural products, serve as the basic raw materials in various industries such as the pharmaceutical, cosmetic, and food sectors, etc. Major compounds such as the anti-cancerous alkaloids (e.g. vinblastine), anti-malarial terpenoids (e.g. artemisinin), and other health promotion flavonoids form the basis of a global market. But one problem remains: they are often limited in supply owing to certain problems regarding their natural production (El-Saadony *et al.*, 2025). They are generally produced in very small amounts, are highly sensitive to environmental and developmental changes, and are often localized in tissue and/or cells of the plant, which are referred to as 'difficult-to-harvest'. The overall high demand for these natural products in the world, in combination with the problem of over-harvesting and global warming, has increased the need for strong, stable, and sustainable biomanufacturing platforms. Therefore, the in-depth knowledge of their biosynthetic and regulatory processes has become the ultimate goal of modern biotechnology (Benny *et al.*, 2022).

Specialized metabolite production is biologically challenging owing to the complex structure of such metabolites. The pathways are not linear or simple sequences but complex and branched networks that contain dozens of enzyme-catalyzed steps. These steps are regulated by complex networks of transcription factors and may be partitioned into compartments within a particular cellular organelle or different cell types. In addition, the majority of high-value medicinal plants lack the standard genetic resources (including reference genomes and mutant libraries) that exist in the case of commonly used lab species (Singh *et al.*, 2022). Initial attempts at explaining these pathways utilized reductionist methods, which used individual omics layers. Genomic and transcriptomic analyses revealed information regarding the molecular possibilities of biosynthesis, and as a final signal of accumulated products, metabolomic analyses were provided. Though fundamental, these methods were unable to close the vital gap between genetic data and consequent chemical expression. They were also unable to fully explain the genotype-

*Corresponding author: madhumita.barooah@aau.ac.in

© CAB International 2026. *Plant Spatial Biology* (ed. Chen)
DOI: 10.1079/9781836994572.0017

to-phenotype relationship, especially in the framework of biological dynamics and localization (Aarika *et al.*, 2025).

The shortcomings of single-omics analyses led to the emergence of multi-omics. Multi-omics refers to systematic and integrative analysis of different forms of biological data (including genomics, transcriptomics, proteomics, and metabolomics) within the same biological system. This combined system is necessary since the multi-tiered control of specialized metabolite accumulation occurs on a variety of levels (Wang *et al.*, 2023). For example, it is the stability and catalytic activity of resultant enzymes (proteomics), transcription of the pathway genes, flux of the precursor molecule (metabolomics), and the total epigenetic control (epigenomics) that regulate the availability of a given metabolite. The integration of multi-omics with the use of some sophisticated statistical methods such as the weighted gene co-expression network analysis (WGCNA) and network-based modeling can also allow a researcher to observe groups of co-expressed genes, proteins, and metabolites, and, as a result, enable the discovery of complete functional modules. These modules act together to produce the desired compound. This shifts the emphasis from single gene discovery to mapping the entire molecular network, which in turn provides practical implications for metabolic engineering efforts (Sowers *et al.*, 2022).

Another facet that has been very critical and has been neglected in previous research is the spatial arrangement of these metabolic factories. Multiple specialized metabolites are produced in a particular cell type and transported and stored in a different cell type, a phenomenon known as metabolic cross-talk. This area of study has been transformed by the introduction of spatial omics technologies, including mass spectrometry imaging (MSI) to analyze metabolites and single-cell RNA sequencing (scRNA-seq) paired with spatial transcriptomics. These methods have been able to resolve the aforementioned challenges (Trinchese *et al.*, 2024). Such state-of-the-art technologies provide cellular and subcellular resolution, which enables the researcher to locate precisely each biosynthetic step, where metabolites are stored, and which cell types interact to form more complex pathways. For example, spatial omics has become essential in the dissection of biosynthesis in glandular trichomes and the mapping of the partitioned processes of alkaloids in leaf tissues (Taylor *et al.*, 2021).

This chapter provides an overview of how the concept of spatial and multi-omics approaches is currently being exploited to unravel the complex secrets of specialized metabolism in medicinal plants. Recent innovations in bringing these layers of multiple data are discussed, the groundbreaking case studies are presented, current issues are observed, and future perspectives are predicted that can involve the optimization of metabolite production, conservation, and, finally, human health.

17.2 Overview of Omics in Medicinal Plants

Recent advances in high-throughput technologies have changed the way medicinal plants are understood, as they allow a shift from a single-gene/single-metabolite analysis to extensive systems-based analyses. Omics includes genomics, transcriptomics, proteomics, metabolomics, and epigenomics (Fig. 17.1). The technologies have increasing potential as powerful tools for understanding and dissecting the complex molecular machinery operating in biosynthetic pathways, gene regulation, and accumulation of specialized metabolites in medicinal plants (Dai and Shen, 2022). These methods have further advanced the identification of new biosynthetic pathways and provided information regarding the modifications and ecological roles of plant secondary metabolism. This section summarizes the omics platforms and their applications to the research on medicinal plants, highlighting the unique contribution each level of biological information provides to the overall understanding of these important plants.

17.2.1 Genomics: Decoding the blueprint of medicinal plants

Genomics is the study of an organism's complete genetic information. This serves as the foundation for all subsequent omics investigations. The sequencing of plant genomes was pioneered

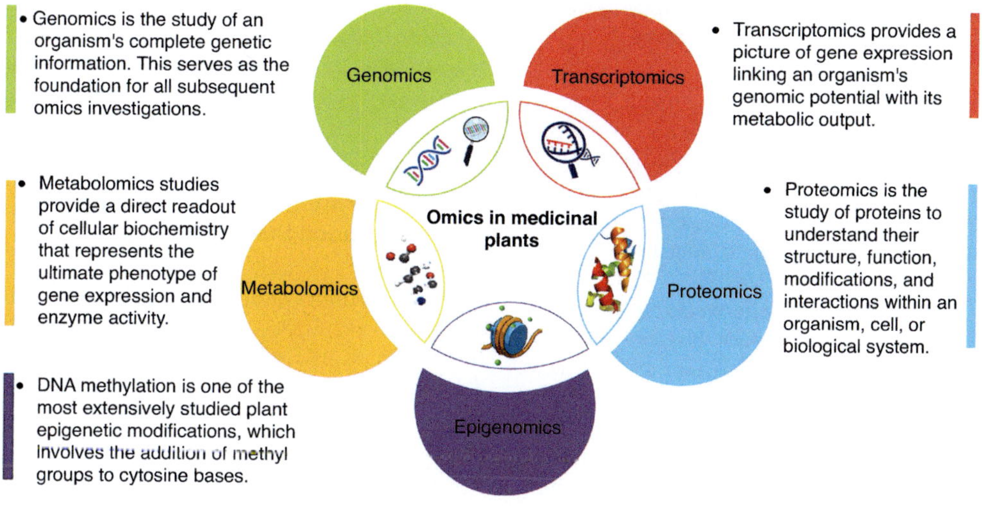

Fig. 17.1. A multi-omics perspective for understanding medicinal plants.

by Sanger sequencing, but it was found to be labor intensive, which led to the rise of rapid next-generation sequencing (NGS) and, now, new technologies such as PacBio and Oxford Nanopore have come into play (Michael and VanBuren, 2020). These technological advances have made it feasible to sequence the genomes of medicinal plants, providing reference frameworks for functional genomics studies.

Several model medicinal plants have had their genomes sequenced and have yielded valuable information. The genome of *Catharanthus roseus* (Madagascar periwinkle), a source of the anticancer alkaloids vinblastine and vincristine, was sequenced to facilitate the elucidation of the complex monoterpenoid indole alkaloid (MIA) biosynthetic pathway (Sharma *et al.*, 2022). Similarly, the genome of *Artemisia annua*, which is the source of the antimalarial drug artemisinin, revealed gene family expansions related to terpenoid biosynthesis and provided genomic resources that helped to engineer enhanced artemisinin production (Shen *et al.*, 2018). The genome of *Cannabis sativa* has been sequenced by multiple groups, which has deciphered the cannabinoid biosynthesis pathway and also the differences in the genetic basis of hemp and drug-type cultivars (Grassa *et al.*, 2021).

Non-model medicinal plants have also benefited from genomic sequencing efforts, but these plants often present greater challenges in sequencing owing to their large genome sizes, high heterozygosity, and polyploidy (Singh *et al.*, 2022). The genome of *Panax ginseng*, which is renowned for its ginsenoside content, was assembled despite its large size (approximately 3.5 Gb). This revealed an extensive expansion of genes that are involved in triterpene saponin biosynthesis (Xu *et al.*, 2017). *Glycyrrhiza uralensis* (licorice) is a traditional medicinal plant that contains bioactive triterpene saponins and flavonoids. It has also been sequenced, which has enabled comparative genomic studies that bring light to the evolution of specialized metabolism in legumes (Mochida *et al.*, 2017).

The understanding of the evolution and diversification of specific specialized biosynthesis pathways has become easier due to the utilization of comparative genomics tools. Gene duplications, neofunctionalizations, and horizontal gene transfer events that shape specialized metabolism in plants have been identified. This has been conducted through genome comparisons between related species (Ji *et al.*, 2024). In addition, genome mining approaches that enable the systematic searching of sequenced genomes for biosynthetic gene clusters (BGCs) and for biosynthetic enzyme signature sequences have also proven to be successful tools for the discovery of new metabolic pathways (Nützmann *et al.*, 2016).

17.2.2 Transcriptomics: Mapping gene expression landscapes

Transcriptomic investigations provide a picture of gene expression linking an organism's genetic potential with its metabolic output. RNA sequencing (RNA-seq) is the new gold standard. It is capable of highly sensitive detection of novel transcripts as compared to microarrays (Lowe *et al.*, 2017).

In the context of the research on medicinal plants, transcriptomics has proved to be instrumental in deciphering various pathways, particularly for the identification of genes that encode for biosynthetic enzymes, transporters, and regulatory transcription factors. For example, RNA-seq studies in *C. roseus* have been able to identify various housekeeping genes involved in MIA biosynthesis, including those encoding for cytochrome P450 enzymes, methyltransferases, and acyltransferases (Liu *et al.*, 2021). Comparative transcriptomics across different tissues has also revealed that MIA biosynthesis is highly compartmentalized, with distinct steps that occur in specific cell types, thereby necessitating intercellular transport of pathway intermediates (Yuan *et al.*, 2021).

Stress-responsive transcriptomics has revealed how medicinal plants are able to modulate specialized metabolism under stress conditions. The elicitation of methyl jasmonate, salicylic acid, or fungal elicitors has aided in the identification of stress-inducible metabolic genes (Goossens *et al.*, 2016). In *Salvia miltiorrhiza*, treated with methyl jasmonate, transcriptomic analysis revealed coordinate upregulation of tanshinone biosynthetic genes and key regulatory transcription factors (Gao *et al.*, 2014).

Temporal transcriptomics has helped in understanding developmental regulation of specialized metabolism in plants. In *A. annua*, developmental profiling revealed that artemisinin biosynthetic genes were predominantly expressed in the glandular trichomes of the young leaves, which declined with maturity (Graham *et al.*, 2010). These insights proved to be of great importance for the optimization of harvest timings. Single-cell RNA sequencing (scRNA-seq) has enabled gene expression profiling at the single-cell level, which reveals cellular heterogeneity (Islam *et al.*, 2024). Despite technical challenges from cell walls and vacuoles, recent advances have made scRNA-seq increasingly feasible for precisely mapping biosynthetic pathways in medicinal plants.

17.2.3 Proteomics: Deciphering the functional machinery

Proteomics studies directly measure the protein abundance and post-translational modifications, that is, the actual catalytic molecules. Advances in mass spectrometry, particularly liquid chromatography tandem mass spectrometry (LC–MS/MS), enable the identification and quantification of various proteins (Karpievitch *et al.*, 2010).

Proteomics profiles the enzyme abundance in an organism and also identifies the bottlenecks in the pathways. Quantitative proteomics of *Papaver somniferum* latex has revealed differential enzyme abundance in the benzylisoquinoline alkaloid pathway, which suggests the occurrence of post-transcriptional regulation (Beaudoin and Facchini, 2014). Subcellular proteomics helps explain the pathway compartmentalization in various organelles such as chloroplasts, mitochondria, and endoplasmic reticulum (Ferro *et al.*, 2010).

Post-translational modifications (PTMs) include phosphorylation, acetylation, ubiquitination, and glycosylation. They regulate enzyme activity, stability, and localization. Phosphoproteomics studies have revealed that specialized metabolism enzymes undergo complex regulatory phosphorylation that modulates pathway flux (Leutert *et al.*, 2021). Phosphoproteomic studies in *Taxus* spp. cultures identified phosphorylation sites on taxol biosynthetic enzymes that suggest a kinase-mediated regulation (Shao *et al.*, 2021).

Interactomics is another branch of proteomics that studies the protein–protein interactions that are essential for understanding metabolic channeling and enzyme complexes. Co-immunoprecipitation coupled with mass spectrometry (Co-IP–MS) and yeast two-hybrid (Y2H) assays have also been instrumental in mapping protein–protein interaction networks, thereby revealing that biosynthetic enzymes frequently assemble into multi-enzyme complexes that enhance catalytic efficiency (Lau and Sattely, 2015). In *C. roseus*, for example, enzymes of the MIA pathway physically interact. This, in turn, supports the concept of metabolic channelling (Li *et al.*, 2023a).

17.2.4 Metabolomics: Profiling the chemical phenotype

Metabolomics studies provide direct information on cellular biochemistry that represents the ultimate phenotype of gene expression and enzyme activity (Alseekh *et al.*, 2021). In medicinal plants, specialized metabolites such as alkaloids, terpenoids, phenolics, and polyketides are the bioactive (therapeutic) compounds.

When the metabolomics is untargeted, it detects and measures as many metabolites as possible. This is conducted in an unbiased manner, which provides a comprehensive metabolome profile using LC–MS, gas chromatography (GC)–MS, and ultra-high-performance LC–MS (UHPLC–MS) with the help of multivariate statistics (Dias *et al.*, 2012). The applications of metabolomics include metabolite fingerprinting, herbal product quality control, and novel bioactive compound discovery (Lee *et al.*, 2017).

Targeted metabolomics is another technique that quantitatively measures already predefined metabolite sets from specific pathways or chemical classes. This method offers higher sensitivity and accuracy for pathway elucidation and metabolic engineering. Nuclear magnetic resonance (NMR) spectroscopy offers complementary information to MS. NMR is a non-destructive, highly reproducible method, and also provides structural information valuable for *de novo* structure elucidation (Roberts *et al.*, 2012). The applications of this technique include herbal medicine quality assessment, adulteration detection, and metabolic profiling (Pauli *et al.*, 2014).

Fluxomics is another type of technique that traces atom flow through metabolic pathways using stable isotope labeling (^{13}C, ^{15}N, ^{2}H). This provides dynamic pathway activity and turnover information that complements static abundance measurements (Srivastava *et al.*, 2016). In medicinal plants, fluxomics identifies biosynthetic routes and also their contributions in rate-limiting steps. These are essential for the metabolic engineering of the medicinal plant (Hashiguchi *et al.*, 2017).

17.2.5 Epigenomics: Unveiling regulatory layers

DNA methylation is one of the most extensively studied plant epigenetic modifications, which involves the addition of methyl groups to cytosine bases. Plant DNA methylation occurs in CG, CHG, and CHH contexts (Where 'H' is A, T, or C), which is maintained by distinct machinery. Studies have shown that specialized metabolism genes can exhibit methylation patterns that are not only specific to the tissue but also the developmental stage of the plant, which correlate with their expression (Kumar and Mohapatra, 2021).

In *Salvia miltiorrhiza*, genome-wide methylation analysis had revealed that the tanshinone biosynthetic genes undergo a dynamic demethylation during the hairy root development, which coincides with their increased expression and accumulation (Li *et al.*, 2023b). In *Camptotheca acuminata*, DNA methylation can regulate alkaloid biosynthetic enzyme gene expression along with the activation of the demethylation-promoting pathway (Salim *et al.*, 2018).

Histone modifications such as acetylation, methylation, phosphorylation, and ubiquitination represent an important step in epigenetic regulation. They modulate chromatin architecture and transcriptional dynamics of the organism. Chromatin immunoprecipitation sequencing (ChIP-seq) is a technique that enables genome-wide profiling of these histone marks and thereby reveals chromatin states that are linked to the activation or repression of certain biosynthetic genes (Sena *et al.*, 2024).

In *C. roseus*, ChIP-seq of H3K9ac (transcriptional activation marker) revealed MIA pathway genes that are enriched for this mark where alkaloid biosynthesis is active (Li *et al.*, 2023a). Histone deacetylase inhibitor treatment increased the MIA gene expression, which demonstrated that the histone acetylation directly regulated the specialized metabolism.

Non-coding RNAs (ncRNAs) include microRNAs (miRNAs), long non-coding RNAs (lncRNAs), and small interfering RNAs (siRNAs). These ncRNAs are able to regulate gene expression through various mechanisms such as mRNA degradation, translational repression, and chromatin modifications (Ratti *et al.*, 2020). sRNA sequences are also able to identify miRNAs, which in turn are able to target biosynthetic genes or regulatory transcription factors in the specialized metabolism of the plant system.

In *A. annua*, miRNA profiling has found that several miRNAs are differentially expressed in young and mature leaves. These tissues vary in the amount of artemisinin present (Guo *et al.*, 2023). Some of these miRNAs have been studied, and it has been found that they target genes that encode for enzymes or transcription factors present in the artemisinin biosynthesis pathway. This suggests that post-transcriptional regulation by miRNAs has a part to play in the developmental control of artemisinin production. Similarly, in the *Salvia miltiorrhiza* plant, identification of miRNAs that targeted phenolic acid biosynthetic genes has provided evidence for miRNA-mediated fine-tuning of the biosynthesis of rosmarinic acid and salvianolic acid (Li *et al.*, 2024a).

The integration of epigenomic data with other omics layers is essential to understand the complex regulatory networks that govern specialized metabolisms. For example, combining DNA methylation profiling with transcriptomics can aid in identifying genes for which expression is under epigenetic control, whereas integrating histone modification maps with transcriptomics and metabolomics can help reveal how the state of the chromatin influences metabolic output (Sajedi *et al.*, 2023).

Overall, the combination of genomics, transcriptomics, proteomics, metabolomics, and epigenomics provides a complete picture of the molecular networks involved in specialized metabolisms in medicinal plants. These multi-omics approaches lead to a synergistic acceleration of pathway discovery, which thereby facilitates metabolic engineering and facilitates the refinement of phytochemical production methodologies for the development of therapeutics. The combined applications of these technologies are summarized in Table 17.1.

17.3 Spatial Omics: Localizing Metabolite Biosynthesis

Medicinal plant research is essentially a study of specialized metabolism that produces the pharmacologically active compounds (such as alkaloids, terpenoids, and phenolics) on which their medicinal effects are built. The synthesis and concentration of such metabolites are carefully coordinated in particular tissues, cell types, or even subcellular compartments. This area has been transformed by the new field of spatial omics, which can afford the required site-specificity and no longer relies on scale analyses of bulk organs, creating molecular maps in high resolution and at the cellular level. Such a transformative approach aids in fully grasping the complicated logistics of chemical factories in plants, providing unique insights into regulatory mechanisms to enable the bioengineering of medicinal plants.

17.3.1 Spatial transcriptomics: Gene expression where it counts

Spatial transcriptomics (ST) is one of the most popular spatial omics technologies, allowing the expression of the entire transcriptome to be visualized and measured while preserving the original spatial context of the tissue section. The approaches that have been used on plants, including slide-based ST (using barcoded oligo (dT) arrays on glass slides) and *in situ* sequencing (ISS) or multiplexed error-robust fluorescence *in situ* hybridization (MERFISH), have been modified to overcome the issues of rigid cell walls and large vacuoles in plants (Li *et al.*, 2023a). Recent work has been able to optimize ST protocols in many different plant tissues, including complex structures such as developing seeds and root tips, and has shown its high-throughput capacity to map cell- and tissue-specific patterns of gene expression to metabolic pathways.

Spatial transcriptomics can be used effectively in medicinal plant research to establish the location of biosynthesis. As an illustration, it can identify the specific group of cells, such as a root epidermis or a glandular trichome or between a vascular bundle, that in large part expresses a whole collection of genes encoding

Table 17.1. Summary of major omics approaches, key techniques/tools, and their primary applications in medicinal plants.

Omics approach	Key techniques/ tools	Primary application in medicinal plants	Representative references
Genomics	Sanger sequencing; next-generation sequencing (NGS); long-read sequencing (PacBio, Oxford Nanopore); comparative genomics; genome mining	Genome assembly of model and non-model species; elucidation of biosynthetic gene clusters (BGCs); identification of gene family expansion and evolution of specialized metabolism	Michael and VanBuren, 2020; Sharma *et al.*, 2022
Transcriptomics	RNA-seq; comparative tissue transcriptomics; stress-responsive transcriptomics; temporal expression profiling; single-cell RNA-seq (scRNA-seq)	Identification of biosynthetic enzymes, transporters, and transcription factors; discovery of stress- and development-induced genes; cell-type-specific expression mapping	Liu *et al.*, 2021; Yuan *et al.*, 2021
Proteomics	2-DE; iTRAQ; LC–MS/MS; phosphoproteomics; subcellular proteomics; Co-IP–MS; yeast two-hybrid (Y2H)	Quantitative and subcellular protein profiling; identification of catalytic enzymes and post-translational modifications; protein–protein interaction mapping; validation of transcriptional data	Leutert *et al.*, 2021; Li *et al.*, 2023a
Metabolomics	LC–MS; GC–MS; UHPLC–MS; NMR; Stable isotope labeling (^{13}C, ^{15}N, ^{2}H) for fluxomics	Untargeted/targeted metabolite profiling; quality control of herbal products; structural elucidation of bioactives; pathway flux and metabolic engineering	Lee *et al.*, 2017; Alseekh *et al.*, 2021
Epigenomics	Whole-genome bisulfite sequencing; ChIP-seq; sRNA-seq; miRNA/lncRNA profiling	Analysis of DNA methylation and histone modifications; ncRNA-mediated regulation of biosynthetic genes; integration with transcriptomics for multi-layer regulation	Kumar and Mohapatra, 2021; Li *et al.*, 2023a

a major portion of an enigmatic metabolic pathway response, such as the synthesis of terpenoid indole alkaloids (TIAs). With high-resolution identification of these expression hot-spots in the plant, scientists are now able to make a distinct linkage between the anatomy and the biochemical role of the plant. This correlation could only be previously observed with the help of slow microdissection methodology. The disclosure acts as a critical, foundational step for advancing metabolic engineering by enabling the identification of novel regulatory mechanisms (Das *et al.*, 2021).

17.3.2 Imaging mass spectrometry (IMS): Localization of the metabolite

In contrast to spatial transcriptomics, IMS, specifically matrix-assisted laser desorption/ionization (MALDI)–MSI, offers the direct, label-free, spatial visualization of metabolites, lipids, and even small peptides in a tissue section. The principle is to scan a beam of laser wavelength through a pre-prepared tissue section in a systematic way, desorb and ionize the endogenous molecules on each point (pixel), producing a mass spectrum, which is identified with the spatial coordinates. The results of this process are high-resolution ion images that present an iconic representation of the specific distribution patterns and relative abundance of hundreds or thousands of compounds at once (Zhang *et al.*, 2024a).

IMS has been essential in the localization of the end-products and intermediates of specialized metabolic processes. As an example, the MALDI–MS imaging of the specific spatial localization of the TIA intermediates in various cell layers of the stem and the leaf in *C. roseus* has been used to identify the routes of transport and sites of accumulation of vinblastine and vincristine. Additionally, in other herbal plants such as *Dendrobium nobile*, IMS has succeeded in mapping the spatial distributions of many bioactive alkaloids and sesquiterpenes in the stem. This has contributed to elucidating their biosynthetic pathways and tissue-specific accumulation patterns. Localization of terpenoids and flavonoids in the glandular trichomes proved that IMS could be used to validate their presence as external storage and secretion sites in defense mechanisms (Li *et al.*, 2024b). Recent developments, such as ambient ionization methods, including direct analysis in real time (DART)–MSI, make it possible to visually inspect macro-scale quantities of alkaloids in whole-plant aerial biomass, as has been conducted in *Ephedra*. These methods can also help determine which parts of the plant are medicinally enriched.

17.3.3 Combining spatial omics and microscopy and microdissection

Spatial omics, combined with the implementation of traditional high-resolution techniques, provides well-rounded insights into the plant. In combination with sophisticated microscopy (e.g. fluorescence, confocal), such techniques can be used to provide molecular maps (transcriptome or metabolome) on top of cell structure to provide contextual information and to enable researchers to compare the molecular results with known anatomical features (Lee *et al.*, 2025). Moreover, laser-capture microdissection (LCM) also plays a crucial role because it provides a method of isolating individual cells or small, histologically identified tissue areas with microscopic guidance. High-quality, targeted omics data can be obtained on extremely rare populations of cells using LCM-based RNA-sequestration or MS analysis that is otherwise not detectable using standard spatial platforms (Rao *et al.*, 2022).

Spatial omics is realized in the true value of multimodal integration. Using a combination of ST data (which reveal the location of gene expression) and IMS data (which show the location of metabolite storage), researchers can obtain a holistic spatiotemporal map of a pathway. This can be anywhere from the first gene transcription event in the synthesis cell to the last metabolite deposition in the storage cell (Wang *et al.*, 2025). It is a systems-level method, sometimes with scRNA-seq to achieve finer cell-type resolution. This enables the determination of the transport mechanisms and cell-to-cell communication networks. These networks then regulate the complex logistics of specialized metabolite production in highly structured plant organs such as roots and vascular tissues (Bian *et al.*, 2024).

The field is progressing through the continued development of computational tools for data integration and visualization. Such tools can overcome issues of data heterogeneity and the scarcity of genomic resources of many non-model medicinal species. This concerted action makes spatial omics a paradigm-shifting driver. It pushes plant biotechnology towards precision breeding and synthetic biology by deciphering the precise molecular blueprints for creating a better way of producing and conserving metabolites.

17.4 Multi-omics Integration in Medicinal Plants

The phytochemicals of medicinal plants are specialized metabolites that are produced by way of complex and multi-leveled regulatory processes within the genome, the transcriptome, the proteome, and the metabolome. Single omics approaches, while vital, are often limited as they only provide a part of the whole picture and are often quite incapable of distinguishing correlation from causation in explaining flux in pathways in a dynamic sense (Liu *et al.*, 2023). The use of multi-omics is a paradigm shift where the field is no longer one of analyses derived from a reductionist approach, but is now of a holistic system-biological perspective of the production of the metabolites. This approach involves the systematic acquisition and computational analysis of data derived from two or more omics approaches, which allow for the elucidation of complete regulatory and metabolic systems, thus making them comprehensible (Luo *et al.*, 2024). This hybrid method has been invaluable in the past 10 years in solving some of the most important problems in medicinal plant research, such as the discovery of new biosynthetic genes, characterization of environmental stress responses, and engineering more desirable chemotypes.

17.4.1 Combining genomics, transcriptomics, and metabolomics

The most widely used and effective combination of genomics, transcriptomics, and metabolomics (G–T–M) has been applied to dissect specific metabolic pathways. The genome contains the blueprint, a complete inventory of genes that encode enzymes, whereas the transcriptome (determined by RNA sequencing or RNA-seq) provides the expression profile of these genes under particular conditions or in specific tissues, which are active components in the pathway. The final functional result of the metabolome, analyzed by MS and NMR, is the identity and the level of the final and intermediate metabolites (Du *et al.*, 2024). The alignment of these three layers allows the researcher to close the gap between a chemical phenotype and a genetic sequence. To give an example, multi-omics analyses have been used in *Salvia miltiorrhiza* by taking RNA-seq datasets of different developmental stages and LC–MS metabolomic data of tanshinone metabolism to distinguish key transcription factors and previously unknown cytochrome P450 enzymes in tanshinone biosynthesis (Yu *et al.*, 2025). The comparison of the level of expression of thousands of genes with the accumulation profiles of dozens of target metabolites has allowed investigators to generate hypotheses that can be tested with great statistical certainty.

One of the major uses of G–T–M integration is building gene–metabolite correlation networks. A variety of algorithms, such as WGCNA, are often used to cluster genes (or transcripts) sharing a similar pattern of expression and to link them highly with the abundance of particular metabolites. These co-expressed sets of genes, frequently encompassing whole biosynthetic pathways or control sub-groups, and highly connected hub genes of these sub-groups are the best candidates to encode rate-limiting proteins or master regulatory transcription factors (Li *et al.*, 2022). An example of this is in the work of the *C. roseus* in which G–T–M analysis has helped to simplify the pathway leading to the production of the potent anticancer alkaloids vinblastine and vincristine, and also elucidate how each pathway component is compartmentalized and regulated in diverse cell types (Salim *et al.*, 2023). This extensive analysis of gene–metabolite association has significantly increased the rate of identification of new steps in the pathway, which is essential since most biosynthetic pathways in non-model medicinal plants are not fully known or even known at all.

17.4.2 Contributions in proteomics and epigenomics

Although G–T–M creates a robust correlational relationship, proteomics and epigenomics layers are important for functional and regulatory validation. Proteomics as a standard activity is conducted using MS/MS to measure the actual products of the protein, which are the enzymes required to catalyze the biosynthetic reactions. Protein abundance and activity are not always tightly linked to the abundance of transcripts because of translational regulation and post-translational modifications (Jiang *et al.*, 2024). When proteomics data are combined with transcriptomics and metabolomics, investigators can verify that transcribed genes were, in fact, translated into functional enzymes in the necessary amounts and locations in sub-cells. As an example, the high level of transcript of an enzyme could be measured, but if the corresponding protein is low, the flux will be minimal. Proteomics has an opportunity to detect bottlenecks, post-translational modifications (e.g. phosphorylation, glycosylation) that enable or disable enzymes, and the exact subcellular location of metabolic machinery. It can also provide a much-needed functional validation of the output of predicted pathways (Liu *et al.*, 2025).

The key upstream regulatory setting of specialized metabolism is given by epigenomics, which addresses the changes in DNA methylation, histone modification, and the non-coding RNA action. These changes do not affect the actual DNA sequence but have a significant impact on gene accessibility and expression. Combining epigenomics information can give us an understanding of the activation or inhibition of whole metabolic pathways by internal and external signals, including stress or developmental signals (Kaszycki and Kim, 2025).

17.4.3 Systems biology approaches

An issue with multi-omics is that the data produced by the diverse experimental platforms are very large volume, highly heterogeneous, and in different formats. Systems biology provides the mathematical tools and statistical power needed to overcome these challenges to transform the large volumes of data into predictive models (Pinu *et al.*, 2019).

Computational strategies are used to normalize, integrate, and model the interrelationships between hundreds of thousands of data points. Network-based integration is still the anchoring point, and all five levels of omics (genomics, transcriptomics, proteomics, metabolomics, epigenomics) have been extended using WGCNA (Watson *et al.*, 2022). In addition to the simpler correlation, more advanced mathematical models (Bayesian networks and machine-learning algorithms, e.g. random forests and support vector machines) are being used to deduce directional or causal interactions between regulatory factors (transcription factors, non-coding RNAs), proteins, and metabolites (Moeckel *et al.*, 2024). Furthermore, pathway rebuilding and regulatory circuit analysis use integrated omics data to build regulatory and metabolic topologies. Mapping known genes, proteins, and metabolites on known biochemical repositories (e.g. KEGG, MetaCyc) can help the researcher visualize the entire carbon and energy flux. This method not only confirms the existing pathways but also shows network hubs that act as points of convergence of several regulatory signals (Sanches *et al.*, 2024). The ultimate goal is to produce *in silico* models that can be used to predict the effect of genetic perturbations (e.g. knock-out or overexpression) or of environmental changes on the final production of a desired metabolite. The future of metabolic engineering and synthetic biology is based on predictive models powered by solid multi-omics data. Therefore, it goes beyond empirical search-and-trial towards the rational design to develop improved production of medicinal products (Zhang *et al.*, 2024b). Overall, multi-omics is not just a means of listing molecules but is the reverse-engineering of a plant's metabolic rationale factory to industrial and therapeutic advantage, which is mediated by the concepts of systems biology.

17.5 Advances in Bioinformatics and Data Integration

The past decade has seen a phenomenal growth in the number of multi-omics datasets (such as genomics, transcriptomics, proteomics, and metabolomics datasets) of medicinal plants, dramatically transforming the field. This has created an immediate need for the development of advanced bioinformatic methods and integrative data-analytic systems. Individual omics modality is too simple to represent the complete range of specialized metabolic complexity, so computational approaches to data analysis have become the key tools for converting unstructured, high-dimensional data into integrated biological information to explain gene–metabolite interactions and regulatory circuitry (Rai *et al.*, 2017). The field has achieved advancement in the form of the iterative optimization of computational pipelines and the development of domain-specific databases, along with the integration of the latest artificial intelligence (AI) and machine learning (ML) methods.

17.5.1 Pipelines of omics integration

The biggest challenge in multi-omics analyses is the heterogeneity of the data; each layer requires specific procedures of preparation, quantification, and normalization. Bioinformaticians, in turn, have developed powerful computation pipelines to unify these contrasting datasets. These pipelines tend to take a stepped-up plan. The first level uses univariate and multivariate statistical methods, including principal component analysis (PCA), partial least-squares discriminant analysis (PLS-DA), and orthogonal partial least-squares discriminant analysis (OPLS-DA) to identify variables (genes, metabolites, proteins) exhibiting the greatest changes in all experimental conditions (stress, developmental stages, etc.). The next, more advanced level will involve the various methods of integration, which entail the integration of the biological context, as well as creating inter-modal connectivity. In the past 10 years, the usefulness of network-based methods has taken center stage, with WGCNA being one of the key tools (Mohr *et al.*, 2024). WGCNA has found wide application in the study of medicinal plants, to cluster transcripts into modules through co-expression and to correlate these modules with metabolite levels to reveal co-regulated pathways, and also to identify hub components (often transcription factors or key biosynthetic enzymes), which are highly connected within the relevant modules (Eshaghi and Rashidi-Monfared, 2025). Also, Bayesian networks and graph–theoretical models have been widely used in analysis to determine causal relationships, as opposed to correlations, between layers. These models allow the determination of whether the change in a transcription factor (transcriptomics) directly triggers an enzyme (proteomics), and subsequently a metabolite (metabolomics). For these advanced computational pipelines to function, noise filtration is required and the isolation of the most biologically relevant targets of metabolic engineering endeavors (Zenere *et al.*, 2022).

17.5.2 Spatial omics and metabolomics visualization tools

Spatial omics has led to these developments: particularly spatial transcriptomics and imaging mass spectrometry. They have enabled the bioinformatics community to produce data that incorporates both spatial coordinates and molecular profiles. The existing omics tools of analysis cannot be used with such data because they presuppose spatial homogeneity. New software packages and modules that are specific to spatial data visualization and analysis have therefore been developed (Lee *et al.*, 2025). These functions include image registration (aligning the omics data with histological images), pixel-level segmentation, and spot-level clustering that combines both molecular and spatial characteristics. In the case of IMS data, the visualization tools are created to generate high-resolution heatmaps on top of the tissue morphology images and overlay metabolite distributions. By using such tools to mathematically resolve overlapping chemical signals, the researcher can precisely identify compounds and visualize their exact distribution within specific microscopic structures of the plant tissue (e.g. a glandular trichome or a vascular bundle) with high spatial resolution (Porta Siegel *et al.*, 2018).

17.5.3 Plant resources/databases: Medicinal

Adequate and carefully maintained data repositories are an essential requirement of multi-omics analyses. During the past 10 years, the number of publicly available and specialized databases explicitly concerning medicinal plant omics information has grown significantly, making them beyond the capability of general biological repositories. Among those, KNApSAcK (chemical constituents and taxonomic relations), METLIN, a metabolite identification catalog, and MPOD (Medicinal Plant Omics Database) may be considered as significant (Zhu *et al.*, 2025).

MPOD is a platform that is both genome and transcriptome wide, and that connects diverse medicinal plant species to their respective metabolic pathways with special focus on the alkaloids, terpenoids, and flavonoids. New genomes and transcriptomes are constantly added to the database; it is thus no longer restricted by the historical shortage of genomic data of non-model medicinal species (Wu *et al.*, 2025).

The site also includes a collection of analytical applications that allow end-users to make online searches, conduct BLAST alignments, and view the gene expression profile using heatmaps. The characteristics facilitate fair access to sophisticated omics data and comparative analysis of a variety of species. As a result, MPOD and similar databases offer crucial annotations of the references that enable the functional interpretation of recently discovered genes and metabolites by mapping them to any known pathway and biological function (He *et al.*, 2022).

17.5.4 Artificial intelligence and machine learning in omics data mining

The use of AI and ML is one of the most radical changes in multi-omics data mining in the past 5 years. Traditional statistical models are unable to represent the non-linearity and high dimensionality of integrated omics data. AI/ML models, on the contrary, are capable of identifying complex patterns and making predictive outputs from complex and multilayered inputs (Rozera *et al.*, 2025).

ML algorithms like random forests and support vector machines are used to build models to predict traits of yield or quality of metabolites using upstream omics data, such as genomic markers and transcript expression levels. As an example, a random forest model fitted on gene expression and metabolomic profiles can predict accurately the concentration of a target compound, which can guide predictive breeding strategies. One sub-area of ML is deep learning, which is finding growing application in the extraction of features and recognition of patterns, especially with complex data, e.g. MS images or uncropped sequencing reads (Yoosefzadeh-Najafabadi *et al.*, 2022).

In addition to predictive modeling, unsupervised methods are used to discover new chemotypes and also to discover functionally related clusters of genes and metabolites, which might have gone unnoticed in traditional analyses. AI can be used not only to analyze existing knowledge but can also be actively involved in the detection of new gene–metabolite networks and prioritization of targets in synthetic biology and metabolic engineering methods by integrating multi-omics data with sophisticated computational methods. This sets the direction of the future for medicinal plant biotechnology (Sharma *et al.*, 2024).

17.6 Challenges and Limitations

Although spatial and multi-omics technologies have brought revolutionary progress in the field of medicine, their application to the study of medicinal plants is challenged by significant technical, biological, and computational barriers. These obstacles will have to be overcome so that the potential of these platforms in biotechnology and metabolic engineering can be fully realized. Figure 17.2 illustrates the major challenges associated with these technologies. The drawbacks can be caused by the inherent biological complexity of non-model plants, the difficulty of obtaining high-quality data with multi-layers of omics, and the computational overhead of performing meaningful integration.

17.6.1 Problems of data heterogeneity and standardization

The most important problem with multi-omics is the variability of data produced on different platforms (genomics, transcriptomics, proteomics, metabolomics, and epigenomics). All omics modalities involve their own sample preparation regimens, methods of detection, and units of quantification, and have their own biases, thus making inter-omics comparison difficult (Aswini *et al.*, 2025).

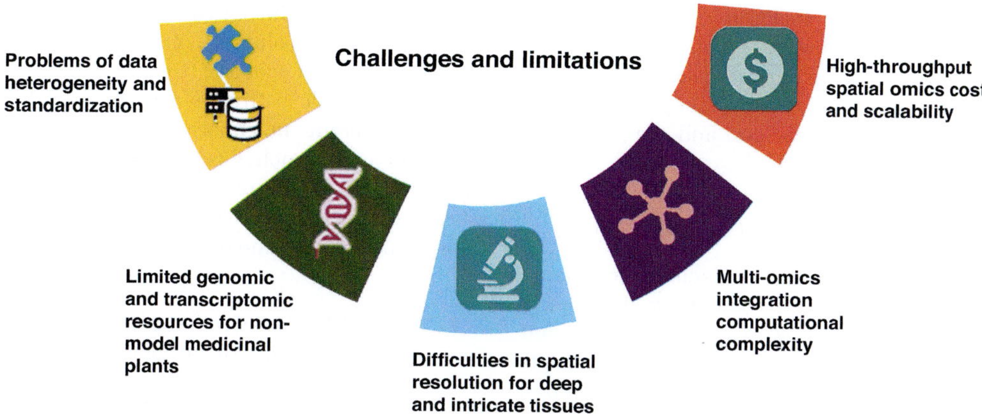

Fig. 17.2. Obstacles in advancing spatial and multi-omics approaches for specialized metabolite discovery.

17.6.1.1 Non-standardized samples are processed by the researcher

Differences in tissue collection procedures, extraction procedures (e.g. protein vs metabolite extraction), and storage conditions introduce systematic noise, making data consolidation hectic. As an example, metabolomics data are very vulnerable to quenching procedures and solvents used to extract, and thus, have a lack of reproducibility when used in different laboratories.

17.6.1.2 Dimensionality and scale disagreements

Multi-omics datasets have extremely dissimilar sizes and numbers of features. A transcriptomics dataset can include a set of about 30,000 gene features, whereas an untargeted metabolomics dataset can include between 5000 and 50,000 mass-to-charge (m/z) features, only a small fraction of which are chemically annotated (Radke *et al.*, 2025). It requires very large-scale normalization and scaling steps to project these heterogeneous and high-dimensional sets of features into a coherent, large-scale network model. In doing so, it can lead to obscuring small but biologically important signals.

17.6.1.3 FAIR and data reporting

The novelty of standardized data reporting and metadata in research groups hinders the process of conducting large-scale meta-analysis. Compliance with FAIR principles (findability, accessibility, interoperability, reusability) can still be considered a major challenge, especially concerning proprietary or large-scale spatial omics data, which restricts the capacity of the rest of the scientific community to reproduce and confirm the findings (Caliskan *et al.*, 2023).

17.6.2 Limited genomic and transcriptomic resources for non-model medicinal plants

Most of the medicinal flora in the world exists as non-model species of plants that are not amenable to the same genetic and functional annotation resources as other staple crops such as *Arabidopsis thaliana* or *Oryza sativa*. This scarcity is a fundamental limitation to integrative multi-omics studies.

17.6.2.1 Challenges in de novo *genome assembly*

De novo genome assembly on medicinal plant genomes is often large, highly heterozygous, polyploid, and highly repetitive, making high-quality, chromosome-scale assembly a technically challenging and expensive process. Poorly constructed genomes thus inherently constrain the accuracy of transcript mapping, gene-prediction pipelines, and regulatory element identification (Rai *et al.*, 2021). Though the barrier to using long-read sequencing technology (e.g. PacBio, Oxford Nanopore) has been mitigated, the financial cost and computational load still pose an issue.

17.6.2.2 Poor functional annotation

Even when a genome is sequenced, functional annotation remains a bottleneck. For newly discovered genes, homology-based annotation often fails, meaning thousands of transcripts detected via RNA-seq remain uncharacterized or are characterized as hypothetical proteins. This void severely restricts the ability to assign an enzyme function to a co-expressed gene and therefore prevents the completion of many biosynthetic pathway maps.

17.6.2.3 Absence of mutants and genetic tools

Non-model medicinal plants typically lack established genetic transformation protocols, T-DNA insertion libraries, or targeted gene-editing resources (CRISPR/Cas9). This makes the crucial step of functional validation, proving that a candidate gene identified via multi-omics actually encodes the hypothesized enzyme, extremely difficult. This often requires laborious transient expression in a heterologous system (McLaughlin *et al.*, 2025).

17.6.3 Difficulties in spatial resolution for deep and intricate tissues

Although spatial omics techniques, such as IMS and ST, are highly capable of localizing compounds at subcellular scales, practical constraints regarding spatial resolution and tissue accessibility persist, especially when analyzing complex tissues that are likely to have specialized metabolites.

17.6.3.1 Cellular heterogeneity and resolution

The true single-cell resolution (5–10 µm) of all omics is not naturally easy to achieve. Several spatial transcriptomics platforms still use larger-sized spots, which include more than one cell (20–100 µm resolution), but IMS is usually limited by the laser spot size (Diop *et al.*, 2025). Since biosynthetic pathways are often divided between different, neighboring cell types, e.g. epidermal and mesophyll cells involved in alkaloid synthesis, partial resolution may result in averaging effects, which suppresses important cell-specific patterns and metabolic cross-talk.

17.6.3.2 Mobility of metabolites during preparation of samples

One of the main presumptions in spatial metabolomics is that the observed metabolite profile has a true *in vivo* localization. However, the sample preparation procedures, including flash-freezing, sectioning, and matrix application, are associated with a chance of metabolite delocalization or redistribution. This is more prominent for small, highly soluble compounds (Andersen *et al.*, 2023). This possible motion may confuse the inferences of the locations of storage and biosynthesis.

17.6.4 Multi-omics integration computational complexity

Multi-omics data are both large and complicated, and specialized computing software is required, but the lack of such software and its accessibility are major obstacles to most plant science laboratories.

17.6.4.1 Absence of integrated integrity prototypes

Although there are now plenty of bioinformatics tools to perform individual omics analyses (e.g. DESeq2 to analyze RNA-seq, XCMS to analyze metabolomics), there is a continuing lack of easy-to-use, unified software platforms that can integrate and simultaneously accept data on all five omics layers (G–T–P–M–E) and achieve statistical rigor (Rosati *et al.*, 2024). To develop custom scripting and a variety of software packages, researchers must use specialized computational skills (e.g. Python, R, ML frameworks) that plant biology might be unlikely to have on hand.

17.6.4.2 *Statistical and causal inference*

Most of the integration strategies, including WGCNA, use correlation, which is unable to distinguish causation and simple association (i.e. that there is a correlation between gene A and metabolite X does not mean that gene A codes the enzyme that catalyzes X). The step towards causal inference with modern methods, such as Bayesian networks or dynamical systems modeling, is computationally intensive, requires large amounts of time-series data, and is susceptible to noise and model assumptions; hence, it is difficult to biologically interpret this information robustly (Yao and Liu, 2025).

17.6.4.3 *Requirement of specialist skills*

Proper multi-omics analysis is challenging because it requires minimal expertise in plant biology, some background in analytical chemistry (in its interpretation of metabolomes and proteomes), and detailed statistics/computer science (Gutierrez Reyes *et al.*, 2024). The lack of scholars who are trained at this interdisciplinary nexus has issues in implementing state-of-the-art integrative approaches. This makes it more challenging and restricts the rigor of data analysis.

17.6.5 High-throughput spatial omics cost and scalability

Lastly, the monetary cost of a spatial and multi-omics study, especially large-scale and high-throughput studies, is a significant limitation to several research programs and institutions.

17.6.5.1 *Capital equipment and maintenance*

ST systems like high-resolution IMS systems (e.g. MALDI-TOF/FT–ICR) and commercial systems are high capital investments that need highly trained technical teams and expensive consumables. As a result, their accessibility is mostly limited to well-funded core facilities. This, in turn increases inequality in the research environment (Kleino *et al.*, 2022).

17.6.5.2 *Per-sample cost*

Deep multi-omics analysis is considerably more expensive than single-omics sequencing. The computation of integrated G–T–P–M–E data of a huge number of biological replicates, which may be required to provide strong statistical models, can soon become cost-prohibitive, forcing scientists to operate on small samples that reduce statistical power (Mansoor *et al.*, 2024).

17.6.5.3 *Translational bottlenecks*

Multi-omics can easily be used to identify targets (genes, transcription factors). But for commercial application (e.g. engineered crops, microbial systems), the cost and difficulty of high-throughput validation and commercialization often hamper this translational approach. This is particularly found to be the case for non-model plants with poorly understood regulatory interactions (Ijoma *et al.*, 2025).

17.7 Future Prospects

Multi-omics is the most popular technique for explaining specialized metabolic pathways of medicinal plants over the past 10 years. The future direction of this field will be characterized by a merger of sophisticated equipment, computational intelligence, and high-throughput biotechnological engineering. The resulting surge of advancement will facilitate an understanding of the spatial and temporal dynamics of metabolic networks, which will enable the sustainable, efficient, and dependable generation of natural products to a greater extent.

17.7.1 Ultra high-resolution space and time omics

One of the future directions that is critically needed is to achieve actual single-cell and subcellular resolution of multi-omics, overcoming the averaging artefacts of many existing spatial platforms. Advances in MSI, particularly systems that combine ion mobility separation with cryogenic techniques, will improve metabolite mapping spatial resolution to below 10 μm and will permit the localization of compounds in individual cellular compartments in a highly controlled manner. This will be complemented by time-series multi-omics, which can track dynamic changes in the genome, transcriptome, and metabolome in response to small-scale changes in the environment (e.g. photoperiodic signal, pathogen challenge, drought stress) (Lee *et al.*, 2023). By constructing 4D data, which encodes spatial positioning and dynamics, researchers can develop dynamic regulatory models that can precisely predict the metabolic flux in real time and switch to dynamic pathway diagrams for interactive and predictive biological simulations, alongside new 3D reconstruction and visualization approaches that combine sequential sectioning with new forms of computational imaging. This will afford the first comprehensive mapping of specialized metabolite biosynthesis in whole structures of the complex plant architecture (Chu *et al.*, 2024).

17.7.2 Intelligent data integration and forecast data modeling

It is the massive scale, fast acquisition velocity, and high dimensionality of multi-omics data that require a paradigm shift to an AI and deep learning approach to ensure strong integration. Traditional methods of computation are often faced with issues related to causality and proper annotation. Future tools will use emerging methods of ML, including Bayesian networks and graph neural networks (GNNs), to go beyond correlative analysis and discover genuine causal regulation relationships between transcription factors, structural genes, and metabolite accumulation (Liu *et al.*, 2024). AI will significantly reduce the bottleneck of metabolite annotation and be based on deep learning to cross-match unannotated mass-spectral signals with large databases and provide real-time proposals of chemical identities and structural modifications depending on built-in gene-expression signatures. This development will accelerate the process of discovering new compounds and branches of pathways that have long since been hidden and are currently limited by computational resources (Beck *et al.*, 2024). The final aim is to build a digital twin of key medicinal species of plants: a holistic, *in silico* architecture that can recapitulate the full metabolic complexity of an entire metabolite-based system, thus allowing researchers to investigate genetic and environmental perturbations virtually before conducting intensive *in vivo* experiments.

17.7.3 Pangenomics, meta-omics and environmental control

It is expected that future studies will expand the bounds of omics to include the whole biological picture of medicinal plants, such as intraspecies variation and the associated microbiome. The systematic study of all genes of a species, or pangenomics, in other words, the study of dispensable genes that are found only in particular strains, is likely to be frequently combined with multi-omics datasets to understand how the genetic variation of different accessions (chemotype) results in different specialized metabolites (Yang *et al.*, 2023). Such integration is critical for the identification and selection of ideal genetic reservoirs to undergo breeding and cultivation programs. Additionally, it is estimated that meta-omics will play a key role in the determination of the impact of the plant microbiome. The plant microbiome, which is a combination of bacteria, fungi, and archaea existing in the rhizosphere and endosphere, affects the host metabolism. Metabolomics integration with metatranscriptomics can map out more complex metabolic cross-talk between the plant and its symbiotic associates and show how the microbes can provide precursors or induce defense status that can increase the accumulation of high-value compounds (Mametja *et al.*, 2025). These findings will supply new possibilities for bio-stimulation and optimization of cultivation.

17.7.4 Accelerating translational biomanufacturing

The key translational potential is the application of multi-omics data of high-resolution to advance the next generation of synthetic biology and metabolic engineering. Spatial omics cell-type-specific and time-regulated pathway data will be used to map out a blueprint of highly effective biosynthetic pathways in a strong host, e.g. yeast, bacterium, or engineered tobacco plants. Future work is likely to focus on pathway modularization and optimization, with cycles of design–build–test–learn (DBTL) processes being directly motivated by omics data to achieve the highest production yield (Zhang *et al.*, 2025). The accuracy of engineering this method will open the way for complicated natural substances, currently hard to synthesize chemically or to harvest on a sustainable basis. They can then be produced in fermentation systems with unparalleled purity and quantity, hence ensuring a sustainable provision of vital plant-derived drugs. The integration of automated data production, automated interpretation of data with AI, and targeted biomining will change the medicinal plant supply chain into a predictable industrial bioprocess (Lawson *et al.*, 2021).

17.8 Conclusion

The use of integrated spatial and multi-omics approaches to medicinal plant research represents a watershed moment in the field, bringing natural product research from traditional phytochemical screening to precision systems biology. Such applications are necessary in the clarification of the complex, specialized metabolic pathways that inform the production of high-value natural products. Combining genomic and transcriptomic models with proteomic and metabolomic data can help scientists to define the molecular components that control biosynthesis, trace the location of biosynthetic pathways in particular tissues of plants, and describe the complex regulation networks that can govern metabolite levels. These integrative, data-focused approaches are crucial to breaking down barriers that have long been present in natural product chemistry. These processes aid in accelerating the retrieval of novel bioactive compounds.

While the transformative impact of these technologies is enormous, the field is still faced with the serious problems associated with, for example, the heterogeneity and the dimensionality of the data, the lack of good quality genomic resources for non-model medicinal plants, and the still existing technological problems. These are the factors that are considered necessary to achieve true single-cell resolution across all layers of omics data. However, the trajectory of future research in this area is clear and promising. The development of ultra-high spatial resolution technologies and the increasing application of AI and deep learning will be a major component of solving those problems, so that we can create dynamic, 4D predictive models or digital twins of plant metabolism.

Eventually, the combination of multi-omics with synthetic biology and metabolic engineering will aid in changing how medicines are produced from plants. The detailed information gained from omics data will act as a clear blueprint, allowing scientists to design and fine tune biosynthetic pathways in engineered microbes or plants. This shift will allow the production of important natural compounds to be moved away from unpredictable agricultural systems towards more sustainable, reliable, and scalable industrial processes. As a result, the pharmacological potential of medicinal plants can be consistently exploited for public health benefit.

References

Aarika, K., Rajyalakshmi, R., Nalla, L.V. and Gajula, S.N.R. (2025) From complexity to clarity: Expanding metabolome coverage with innovative analytical strategies. *Journal of Separation Science* 48(2), e70099. DOI: 10.1002/jssc.70099.

Alseekh, S., Aharoni, A., Brotman, Y., Contrepois, K., D'Auria, J. *et al.* (2021) Mass spectrometry-based metabolomics: A guide for annotation, quantification and best reporting practices. *Nature Methods* 18(7), 747–756. DOI: 10.1038/s41592-021-01197-1.

Andersen, M.K., Giampà, M., Midtbust, E., Høiem, T.S., Krossa, S. *et al.* (2023) Sample preparation for metabolite detection in mass spectrometry imaging. In: Cole, M.L. and Clench, R.M. (eds) *Imaging Mass Spectrometry*. Springer US, New York, NY, pp. 135–146. DOI: 10.1007/978-1-0716-3319-9_12.

Aswini, R., Saranya, B., Gayathri, K. and Karthikeyan, E. (2025) Revolutionizing infectious disease surveillance: Multi-omics technologies and AI-driven integration. *Decoding Infection and Transmission* 3, 100061. DOI: 10.1016/j.dcit.2025.100061.

Beaudoin, G.A.W. and Facchini, P.J. (2014) Benzylisoquinoline alkaloid biosynthesis in opium poppy. *Planta* 240(1), 19–32. DOI: 10.1007/s00425-014-2056-8.

Beck, A.G., Muhoberac, M., Randolph, C.E., Beveridge, C.H., Wijewardhane, P.R. *et al.* (2024) Recent developments in machine learning for mass spectrometry. *ACS Measurement Science Au* 4(3), 233–246. DOI: 10.1021/acsmeasuresciau.3c00060.

Benny, A., Pandi, K. and Upadhyay, R. (2022) Techniques, challenges and future prospects for cell-based meat. *Food Science and Biotechnology* 31(10), 1225–1242. DOI: 10.1007/s10068-022-01136-6.

Bian, J., Zhuang, Z., Ji, X., Tang, R., Li, J. *et al.* (2024) Research progress of single-cell transcriptome sequencing technology in plants. *Agronomy* 14(11), 2530. DOI: 10.3390/agronomy14112530.

Caliskan, A., Dangwal, S. and Dandekar, T. (2023) Metadata integrity in bioinformatics: Bridging the gap between data and knowledge. *Computational and Structural Biotechnology Journal* 21, 4895–4913. DOI: 10.1016/j.csbj.2023.10.006.

Chu, L.-X., Wang, W.-J., Gu, X.-P., Wu, P., Gao, C. *et al.* (2024) Spatiotemporal multi-omics: Exploring molecular landscapes in aging and regenerative medicine. *Military Medical Research* 11(1), 31. DOI: 10.1186/s40779-024-00537-4.

Dai, X. and Shen, L. (2022) Advances and trends in omics technology development. *Frontiers in Medicine* 9. DOI: 10.3389/fmed.2022.911861.

Das, S., Vera, M., Gandin, V., Singer, R.H. and Tutucci, E. (2021) Intracellular mRNA transport and localized translation. *Nature Reviews Molecular Cell Biology* 22(7), 483–504. DOI: 10.1038/s41580-021-00356-8.

Dias, D.A., Urban, S. and Roessner, U. (2012) A historical overview of natural products in drug discovery. *Metabolites* 2(2), 303–336. DOI: 10.3390/metabo2020303.

Diop, M., Davidson, B.R., Fragiadakis, G.K., Sirota, M., Gaudillière, B. *et al.* (2025) Single-cell omics technologies - fundamentals on how to create single-cell looking glasses for reproductive health. *American Journal of Obstetrics and Gynecology* 232(4S), S1–S20. DOI: 10.1016/j.ajog.2024.08.041.

Du, P., Fan, R., Zhang, N., Wu, C. and Zhang, Y. (2024) Advances in integrated multi-omics analysis for drug-target identification. *Biomolecules* 14(6), 692. DOI: 10.3390/biom14060692.

El-Saadony, M.T., Saad, A.M., Mohammed, D.M., Korma, S.A., Alshahrani, M.Y. *et al.* (2025) Medicinal plants: Bioactive compounds, biological activities, combating multidrug-resistant microorganisms, and human health benefits - A comprehensive review. *Frontiers in Immunology* 16, 1491777. DOI: 10.3389/fimmu.2025.1491777.

Eshaghi, M. and Rashidi-Monfared, S. (2025) Network-based identification of hub transcription factors associated with benzylisoquinoline alkaloid biosynthesis in *Papaver somniferum*. *Biochemistry and Biophysics Reports* 43, 102147. DOI: 10.1016/j.bbrep.2025.102147.

Ferro, M., Brugière, S., Salvi, D., Seigneurin-Berny, D., Court, M. *et al.* (2010) AT_CHLORO, a comprehensive chloroplast proteome database with subplastidial localization and curated information on envelope proteins. *Molecular & Cellular Proteomics* 9(6), 1063–1084. DOI: 10.1074/mcp.M900325-MCP200.

Gao, W., Sun, H.-X., Xiao, H., Cui, G., Hillwig, M.L. *et al.* (2014) Combining metabolomics and transcriptomics to characterize tanshinone biosynthesis in *Salvia miltiorrhiza*. *BMC Genomics* 15(1), 73. DOI: 10.1186/1471-2164-15-73.

Goossens, J., Mertens, J. and Goossens, A. (2016) Role and functioning of bHLH transcription factors in jasmonate signalling. *Journal of Experimental Botany* erw440, erw440. DOI: 10.1093/jxb/erw440.

Graham, I.A., Besser, K., Blumer, S., Branigan, C.A., Czechowski, T. *et al.* (2010) The genetic map of *Artemisia annua* L. identifies loci affecting yield of the antimalarial drug artemisinin. *Science (New York, N.Y.)* 327(5963), 328–331. DOI: 10.1126/science.1182612.

Grassa, C.J., Weiblen, G.D., Wenger, J.P., Dabney, C., Poplawski, S.G. *et al.* (2021) A new cannabis genome assembly associates elevated cannabidiol (CBD) with hemp introgressed into Marijuana. *The New Phytologist* 230(4), 1665–1679. DOI: 10.1111/nph.17243.

Guo, Z., Hao, K., Lv, Z., Yu, L., Bu, Q. *et al.* (2023) Profiling of phytohormone-specific microRNA and characterization of the miR160-ARF1 module involved in glandular trichome development and artemisinin biosynthesis in *Artemisia annua*. *Plant Biotechnology Journal* 21(3), 591–605. DOI: 10.1111/pbi.13974.

Gutierrez Reyes, C.D., Alejo-Jacuinde, G., Perez Sanchez, B., Chavez Reyes, J., Onigbinde, S. *et al.* (2024) Multi omics applications in biological systems. *Current Issues in Molecular Biology* 46(6), 5777–5793. DOI: 10.3390/cimb46060345.

Hashiguchi, A., Tian, J. and Komatsu, S. (2017) Proteomic contributions to medicinal plant research: From plant metabolism to pharmacological action. *Proteomes* 5(4), 35. DOI: 10.3390/proteomes5040035.

He, S., Yang, L., Ye, S., Lin, Y., Li, X. *et al.* (2022) MPOD: Applications of integrated multi-omics database for medicinal plants. *Plant Biotechnology Journal* 20(5), 797–799. DOI: 10.1111/pbi.13769.

Ijoma, G.N., Mannie, T.V., Lokwe, G.N., Ogola, H.J.O., Kalu, C.M. *et al.* (2025) The pivotal role of omics technologies in the production of third and fourth generation biodiesel. *Biofuels* 1–26. DOI: 10.1080/17597269.2025.2564504.

Islam, M.T., Liu, Y., Hassan, M.M., Abraham, P.E., Merlet, J. *et al.* (2024) Advances in the application of single-cell transcriptomics in plant systems and synthetic biology. *Biodesign Research* 6, 0029. DOI: 10.34133/bdr.0029.

Ji, W., Osbourn, A. and Liu, Z. (2024) Understanding metabolic diversification in plants: Branchpoints in the evolution of specialized metabolism. *Philosophical Transactions of the Royal Society B* 379(1914), 1914. DOI: 10.1098/rstb.2023.0359.

Jiang, Y., Rex, D.A.B., Schuster, D., Neely, B.A., Rosano, G.L. *et al.* (2024) Comprehensive overview of bottom-up proteomics using mass spectrometry. *ACS Measurement Science Au* 4(4), 338–417. DOI: 10.1021/acsmeasuresciau.3c00068.

Karpievitch, Y.V., Polpitiya, A.D., Anderson, G.A., Smith, R.D. and Dabney, A.R. (2010) Liquid chromatography mass spectrometry-based proteomics: Biological and technological aspects. *The Annals of Applied Statistics* 4(4), 1797–1823. DOI: 10.1214/10-AOAS341.

Kaszycki, J. and Kim, M. (2025) Epigenetic regulation of transcription factors involved in NLRP3 inflammasome and NF-kb signaling pathways. *Frontiers in Immunology* 16, 1529756. DOI: 10.3389/fimmu.2025.1529756.

Kleino, I., Frolovaitė, P., Suomi, T. and Elo, L.L. (2022) Computational solutions for spatial transcriptomics. *Computational and Structural Biotechnology Journal* 20, 4870–4884. DOI: 10.1016/j.csbj.2022.08.043.

Kumar, S. and Mohapatra, T. (2021) Dynamics of DNA methylation and its functions in plant growth and development. *Frontiers in Plant Science* 12, 596236. DOI: 10.3389/fpls.2021.596236.

Lau, W. and Sattely, E.S. (2015) Six enzymes from mayapple that complete the biosynthetic pathway to the etoposide aglycone. *Science (New York, N.Y.)* 349(6253), 1224–1228. DOI: 10.1126/science.aac7202.

Lawson, C.E., Martí, J.M., Radivojevic, T., Jonnalagadda, S.V.R., Gentz, R. *et al.* (2021) Machine learning for metabolic engineering: A review. *Metabolic Engineering* 63, 34–60. DOI: 10.1016/j.ymben.2020.10.005.

Lee, K.M., Jeon, J.Y., Lee, B.J., Lee, H. and Choi, H.K. (2017) Application of metabolomics to quality control of natural product derived medicines. *Biomolecules & Therapeutics* 25(6), 559–568. DOI: 10.4062/biomolther.2016.249.

Lee, S., Vu, H.M., Lee, J.H., Lim, H. and Kim, M.S. (2023) Advances in mass spectrometry-based single cell analysis. *Biology* 12(3), 395. DOI: 10.3390/biology12030395.

Lee, Y., Lee, M., Shin, Y., Kim, K. and Kim, T. (2025) Spatial omics in clinical research: A comprehensive review of technologies and guidelines for applications. *International Journal of Molecular Sciences* 26(9), 3949. DOI: 10.3390/ijms26093949.

Leutert, M., Entwisle, S.W. and Villén, J. (2021) Decoding post-translational modification crosstalk with proteomics. *Molecular & Cellular Proteomics* 20, 100129. DOI: 10.1016/j.mcpro.2021.100129.

Li, C., Wood, J.C., Vu, A.H., Hamilton, J.P., Rodriguez Lopez, C.E. *et al.* (2023a) Single-cell multi-omics in the medicinal plant *Catharanthus roseus*. *Nature Chemical Biology* 19(8), 1031–1041. DOI: 10.1038/s41589-023-01327-0.

Li, H., Guo, Z., Xu, M., Zhao, J. and Xu, D. (2024a) Molecular mechanism of miRNA mediated biosynthesis of secondary metabolites in medicinal plants. *Plant Physiology and Biochemistry* 208, 108524. DOI: 10.1016/j.plaphy.2024.108524.

Li, J., Zhao, R., Yu, K. and Li, X. (2022) Directly self-starting higher-order implicit integration algorithms with flexible dissipation control for structural dynamics. *Computer Methods in Applied Mechanics and Engineering* 389, 114274. DOI: 10.1016/j.cma.2021.114274.

Li, J., Li, C., Deng, Y., Wei, H. and Lu, S. (2023b) Characteristics of *Salvia miltiorrhiza* methylome and the regulatory mechanism of DNA methylation in tanshinone biosynthesis. *Horticulture Research* 10(7). DOI: 10.1093/hr/uhad114.

Li, J., Gao, C., He, Z., Huang, Y., Tan, D. *et al.* (2024b) The chemical structure and pharmacological activity of sesquiterpenoids in *Dendrobium* sw. *Molecules* 29(24), 5851. DOI: 10.3390/molecules29245851.

Liu, J., Han, L., Li, G., Zhang, A., Liu, X. *et al.* (2023) Transcriptome and metabolome profiling of the medicinal plant *Veratrum mengtzeanum* reveal key components of the alkaloid biosynthesis. *Frontiers in Genetics* 14. DOI: 10.3389/fgene.2023.1023433.

Liu, J., Yang, L., Liu, D., Wu, Q., Yu, Y. *et al.* (2025) The role of multi-omics in biomarker discovery, diagnosis, prognosis, and therapeutic monitoring of tissue repair and regeneration processes. *Journal of Orthopaedic Translation* 54, 131–151. DOI: 10.1016/j.jot.2025.07.006.

Liu, X., Shi, J., Jiao, Y., An, J., Tian, J. *et al.* (2024) Integrated multi-omics with machine learning to uncover the intricacies of kidney disease. *Briefings in Bioinformatics* 25(5). DOI: 10.1093/bib/bbae364.

Liu, Y., Patra, B., Singh, S.K., Paul, P., Zhou, Y. *et al.* (2021) Terpenoid indole alkaloid biosynthesis in *Catharanthus roseus*: Effects and prospects of environmental factors in metabolic engineering. *Biotechnology Letters* 43(11), 2085–2103. DOI: 10.1007/s10529-021-03179-x.

Lowe, R., Shirley, N., Bleackley, M., Dolan, S. and Shafee, T. (2017) Transcriptomics technologies. *PLOS Computational Biology* 13(5), e1005457. DOI: 10.1371/journal.pcbi.1005457.

Luo, Y., Zhao, C. and Chen, F. (2024) Multiomics research: Principles and challenges in integrated analysis. *Biodesign Research* 6, 0059. DOI: 10.34133/bdr.0059.

Mametja, N.M., Ramadwa, T.E., Managa, M. and Masebe, T.M. (2025) Recent advances and developments in bacterial endophyte identification and application: A 20-year landscape review. *Plants (Basel, Switzerland)* 14(16), 2506. DOI: 10.3390/plants14162506.

Mansoor, S., Hamid, S., Tuan, T.T., Park, J.E. and Chung, Y.S. (2024) Advance computational tools for multiomics data learning. *Biotechnology Advances* 77, 108447. DOI: 10.1016/j.biotechadv.2024.108447.

McLaughlin, J.E., Kue Foka, I.C., Lawton, M.A. and Di, R. (2025) CRISPR activation: Identifying and using novel genes for plant disease resistance breeding. *Frontiers in Genome Editing* 7, 1596600. DOI: 10.3389/fgeed.2025.1596600.

Michael, T.P. and VanBuren, R. (2020) Building near-complete plant genomes. *Current Opinion in Plant Biology* 54, 26–33. DOI: 10.1016/j.pbi.2019.12.009.

Mochida, K., Sakurai, T., Seki, H., Yoshida, T., Takahagi, K. *et al.* (2017) Draft genome assembly and annotation of *Glycyrrhiza uralensis*, a medicinal legume. *The Plant Journal* 89(2), 181–194. DOI: 10.1111/tpj.13385.

Moeckel, C., Mouratidis, I., Chantzi, N., Uzun, Y. and Georgakopoulos-Soares, I. (2024) Advances in computational and experimental approaches for deciphering transcriptional regulatory networks. *BioEssays* 46(7). DOI: 10.1002/bies.202300210.

Mohr, A.E., Ortega-Santos, C.P., Whisner, C.M., Klein-Seetharaman, J. and Jasbi, P. (2024) Navigating challenges and opportunities in multi-omics integration for personalized healthcare. *Biomedicines* 12(7), 1496. DOI: 10.3390/biomedicines12071496.

Nützmann, H.-W., Huang, A. and Osbourn, A. (2016) Plant metabolic clusters - from genetics to genomics. *The New Phytologist* 211(3), 771–789. DOI: 10.1111/nph.13981.

Pauli, G.F., Chen, S.-N., Simmler, C., Lankin, D.C., Gödecke, T. *et al.* (2014) Importance of purity evaluation and the potential of quantitative ^{1} H NMR as a purity assay. *Journal of Medicinal Chemistry* 57(22), 9220–9231. DOI: 10.1021/jm500734a.

Pinu, F.R., Beale, D.J., Paten, A.M., Kouremenos, K., Swarup, S. *et al.* (2019) Systems biology and multiomics integration: Viewpoints from the metabolomics research community. *Metabolites* 9(4), 76. DOI: 10.3390/metabo9040076.

Porta Siegel, T., Hamm, G., Bunch, J., Cappell, J., Fletcher, J.S. *et al.* (2018) Mass spectrometry imaging and integration with other imaging modalities for greater molecular understanding of biological tissues. *Molecular Imaging and Biology* 20(6), 888–901. DOI: 10.1007/s11307-018-1267-y.

Radke, M.J., Cresswell, S.L. and Leusch, F.D.L. (2025) Combining non-targeted high resolution mass spectrometry with effect-directed analysis to identify contaminants of emerging concern in the field of ecotoxicology: A systematic quantitative literature review. *The Science of the Total Environment* 972, 179122. DOI: 10.1016/j.scitotenv.2025.179122.

Rai, A., Saito, K. and Yamazaki, M. (2017) Integrated omics analysis of specialized metabolism in medicinal plants. *The Plant Journal* 90(4), 764–787. DOI: 10.1111/tpj.13485.

Rai, A., Hirakawa, H., Nakabayashi, R., Kikuchi, S., Hayashi, K. *et al.* (2021) Chromosome-level genome assembly of *Ophiorrhiza pumila* reveals the evolution of camptothecin biosynthesis. *Nature Communications* 12(1), 405. DOI: 10.1038/s41467-020-20508-2.

Rao, B.H., Souček, P. and Hlaváč, V. (2022) Laser capture microdissection: A gear for pancreatic cancer research. *International Journal of Molecular Sciences* 23(23), 14566. DOI: 10.3390/ijms232314566.

Ratti, M., Lampis, A., Ghidini, M., Salati, M., Mirchev, M.B. *et al.* (2020) MicroRNAs (miRNAs) and long non-coding RNAs (lncRNAs) as new tools for cancer therapy: First steps from bench to bedside. *Targeted Oncology* 15(3), 261–278. DOI: 10.1007/s11523-020-00717-x.

Roberts, L.D., Souza, A.L., Gerszten, R.E. and Clish, C.B. (2012) Targeted metabolomics. *Current Protocols in Molecular Biology Chapter* 30(1). DOI: 10.1002/0471142727.mb3002s98.

Rosati, D., Palmieri, M., Brunelli, G., Morrione, A., Iannelli, F. *et al.* (2024) Differential gene expression analysis pipelines and bioinformatic tools for the identification of specific biomarkers: A review. *Computational and Structural Biotechnology Journal* 23, 1154–1168. DOI: 10.1016/j.csbj.2024.02.018.

Rozera, T., Pasolli, E., Segata, N. and Ianiro, G. (2025) Machine learning and artificial intelligence in the multi-omics approach to gut microbiota. *Gastroenterology* 169(3), 487–501. DOI: 10.1053/j.gastro.2025.02.035.

Sajedi, S., Ebrahimi, G., Roudi, R., Mehta, I., Heshmat, A. *et al.* (2023) Integrating DNA methylation and gene expression data in a single gene network using the inetgrate package. *Scientific Reports* 13(1), 21721. DOI: 10.1038/s41598-023-48237-8.

Salim, V., Jones, A.D. and DellaPenna, D. (2018) *Camptotheca acuminata* 10-hydroxycamptothecin O-methyltransferase: An alkaloid biosynthetic enzyme co-opted from flavonoid metabolism. *The Plant Journal* 95(1), 112–125. DOI: 10.1111/tpj.13936.

Salim, V., Jarecki, S.A., Vick, M. and Miller, R. (2023) Advances in metabolic engineering of plant monoterpene indole alkaloids. *Biology* 12(8), 1056. DOI: 10.3390/biology12081056.

Sanches, P.H.G., de Melo, N.C., Porcari, A.M. and de Carvalho, L.M. (2024) Integrating molecular perspectives: Strategies for comprehensive multi-omics integrative data analysis and machine learning applications in transcriptomics, proteomics, and metabolomics. *Biology* 13(11), 848. DOI: 10.3390/biology13110848.

Sena, S., Prakash, A., Van Staden, J. and Kumar, V. (2024) Epigenetic control of plant regeneration: Unraveling the role of histone methylation. *Current Plant Biology* 40, 100408. DOI: 10.1016/j.cpb.2024.100408.

Shao, F., Wilson, I.W. and Qiu, D. (2021) The research progress of taxol in Taxus. *Current Pharmaceutical Biotechnology* 22(3), 360–366. DOI: 10.2174/1389201021666200621163333.

Sharma, A., Tiwari, P., Arora, R. and Sankaranarayanan, A. (2022) Madagascar periwinkle alkaloids: Biosynthesis, ethnobotanical attributes, and pharmacological functions. *South African Journal of Botany* 151, 108–115. DOI: 10.1016/j.sajb.2022.09.039.

Sharma, A., Lysenko, A., Jia, S., Boroevich, K.A. and Tsunoda, T. (2024) Advances in AI and machine learning for predictive medicine. *Journal of Human Genetics* 69(10), 487–497. DOI: 10.1038/s10038-024-01231-y.

Shen, Q., Zhang, L., Liao, Z., Wang, S., Yan, T. *et al.* (2018) The genome of *Artemisia annua* provides insight into the evolution of Asteraceae family and artemisinin biosynthesis. *Molecular Plant* 11(6), 776–788. DOI: 10.1016/j.molp.2018.03.015.

Singh, K.S., van der Hooft, J.J.J., van Wees, S.C.M. and Medema, M.H. (2022) Integrative omics approaches for biosynthetic pathway discovery in plants. *Natural Product Reports* 39(9), 1876–1896. DOI: 10.1039/d2np00032f.

Sowers, M.L., Tang, H., Singh, V.K., Khan, A., Mishra, A. *et al.* (2022) Multi-omics analysis reveals metabolic and epigenetic changes associated with macrophage polarization. *The Journal of Biological Chemistry* 298(10), 102418. DOI: 10.1016/j.jbc.2022.102418.

Srivastava, A., Kowalski, G.M., Callahan, D.L., Meikle, P.J. and Creek, D.J. (2016) Strategies for extending metabolomics studies with stable isotope labelling and fluxomics. *Metabolites* 6(4), 32. DOI: 10.3390/metabo6040032.

Taylor, M.J., Lukowski, J.K. and Anderton, C.R. (2021) Spatially resolved mass spectrometry at the single cell: Recent innovations in proteomics and metabolomics. *Journal of the American Society for Mass Spectrometry* 32(4), 872–894. DOI: 10.1021/jasms.0c00439.

Trinchese, G., Cimmino, F., Catapano, A., Cavaliere, G. and Mollica, M.P. (2024) Mitochondria: The gatekeepers between metabolism and immunity. *Frontiers in Immunology* 15, 1334006. DOI: 10.3389/fimmu.2024.1334006.

Wang, R., Hastings, W.J., Saliba, J.G., Bao, D., Huang, Y. *et al.* (2025) Applications of nanotechnology for spatial omics: Biological structures and functions at nanoscale resolution. *ACS Nano* 19(1), 73–100. DOI: 10.1021/acsnano.4c11505.

Wang, X., Fan, D., Yang, Y., Gimple, R.C. and Zhou, S. (2023) Integrative multi-omics approaches to explore immune cell functions: Challenges and opportunities. *iScience* 26(4), 106359. DOI: 10.1016/j.isci.2023.106359.

Watson, E.R., Taherian Fard, A. and Mar, J.C. (2022) Computational methods for single-cell imaging and omics data integration. *Frontiers in Molecular Biosciences* 8, 768106. DOI: 10.3389/fmolb.2021.768106.

Wu, Z., Hu, Y., Hao, R., Li, R., Lu, X. *et al.* (2025) Research progress of genomics applications in secondary metabolites of medicinal plants: A case study in safflower. *International Journal of Molecular Sciences* 26(8), 3867. DOI: 10.3390/ijms26083867.

Xu, J., Chu, Y., Liao, B., Xiao, S., Yin, Q. *et al.* (2017) *Panax ginseng* genome examination for ginsenoside biosynthesis. *GigaScience* 6(11), 11. DOI: 10.1093/gigascience/gix093.

Yang, L., Yang, Y., Huang, L., Cui, X. and Liu, Y. (2023) From single- to multi-omics: Future research trends in medicinal plants. *Briefings in Bioinformatics* 24(1). DOI: 10.1093/bib/bbac485.

Yao, M. and Liu, Z. (2025) An introduction to causal inference methods with multi-omics data. *Current Protocols* 5(6), e70168. DOI: 10.1002/cpz1.70168.

Yoosefzadeh-Najafabadi, M., Eskandari, M., Torabi, S., Torkamaneh, D., Tulpan, D. *et al.* (2022) Machine-learning-based genome-wide association studies for uncovering QTL underlying soybean yield and its components. *International Journal of Molecular Sciences* 23(10), 5538. DOI: 10.3390/ijms23105538.

Yu, H., Liao, J., Jiang, Y., Zhong, M., Tao, S. *et al.* (2025) Ecotype-specific phenolic acid accumulation and root softness in *Salvia miltiorrhiza* are driven by environmental and genetic factors. *Plant Biotechnology Journal* 23(6), 2224–2241. DOI: 10.1111/pbi.70048.

Yuan, L., Pan, K., Li, Y., Yi, B. and Gao, B. (2021) Comparative transcriptome analysis of *Alpinia oxyphylla* Miq. reveals tissue-specific expression of flavonoid biosynthesis genes. *BMC Genomic Data* 22(1), 19. DOI: 10.1186/s12863-021-00973-4.

Zenere, A., Rundquist, O., Gustafsson, M. and Altafini, C. (2022) Multi-omics protein-coding units as massively parallel Bayesian networks: Empirical validation of causality structure. *iScience* 25(4), 104048. DOI: 10.1016/j.isci.2022.104048.

Zhang, H., Lu, K.H., Ebbini, M., Huang, P., Lu, H. *et al.* (2024a) Mass spectrometry imaging for spatially resolved multi-omics molecular mapping. *Npj Imaging* 2(1), 20. DOI: 10.1038/s44303-024-00025-3.

Zhang, J., Che, Y., Liu, R., Wang, Z. and Liu, W. (2025) Deep learning–driven multi-omics analysis: Enhancing cancer diagnostics and therapeutics. *Briefings in Bioinformatics* 26(4). DOI: 10.1093/bib/bbaf440.

Zhang, N., Li, X., Zhou, Q., Zhang, Y., Lv, B. *et al.* (2024b) Self-controlled *in silico* gene knockdown strategies to enhance the sustainable production of heterologous terpenoid by *Saccharomyces cerevisiae*. *Metabolic Engineering* 83, 172–182. DOI: 10.1016/j.ymben.2024.04.005.

Zhu, T., Li, T., Lü, P. and Li, C. (2025) Single-cell omics in plant biology: Mechanistic insights and applications for crop improvement. *Advanced Biotechnology* 3(3), 20. DOI: 10.1007/s44307-025-00074-8.

Index

Printed and bound by CPI Group (UK) Ltd, Croydon, CR0 4YY

24/08/2026

14947852-0003